D1125999

# HANDBOOK OF ECONOMETRICS
## VOLUME III

# HANDBOOKS
# IN
# ECONOMICS
## 2

*Series Editors*

## KENNETH J. ARROW
## MICHAEL D. INTRILIGATOR

**NORTH-HOLLAND**
AMSTERDAM · NEW YORK · OXFORD · TOKYO

# HANDBOOK OF ECONOMETRICS

## VOLUME III

*Edited by*

**ZVI GRILICHES**

*Harvard University*

and

**MICHAEL D. INTRILIGATOR**

*University of California, Los Angeles*

**1986**
**NORTH-HOLLAND**
AMSTERDAM · NEW YORK · OXFORD · TOKYO

ISBN for this volume 0 444 86185 8
ISBN for this set 0 444 86188 2

HB 139
.H36
1983
V.3

*Publishers*:
ELSEVIER SCIENCE PUBLISHERS B.V.
P.O. Box 1991
1000BZ Amsterdam
The Netherlands

*Sole distributors for the U.S.A. and Canada*:
ELSEVIER SCIENCE PUBLISHING COMPANY, INC.
52 Vanderbilt Avenue
New York, N.Y. 10017
U.S.A

**Library of Congress Cataloging in Publication Data**
(Revised for volume 3)
Main entry under title:

Handbook of econometrics.

(Handbooks in economics ; bk. 2)
Includes bibliographies and indexes.
I. Econometrics--Addresses, essays, lectures.
I. Griliches, Zvi, 1930-          II. Intriligator,
Michael D.
HB139.H36   1983      330′.028         83-2396
ISBN 0-444-86188-2 (set)
ISBN 0-444-86185-8 (v. 1)
ISBN 0-444-86186 6 (v. 2)
ISBN 0-444-86187 4 (v. 3)

PRINTED IN THE NETHERLANDS

# INTRODUCTION TO THE SERIES

The aim of the *Handbooks in Economics* series is to produce Handbooks for various branches of economics, each of which is a definitive source, reference, and teaching supplement for use by professional researchers and advanced graduate students. Each Handbook provides self-contained surveys of the current state of a branch of economics in the form of chapters prepared by leading specialists on various aspects of this branch of economics. These surveys summarize not only received results but also newer developments, from recent journal articles and discussion papers. Some original material is also included, but the main goal is to provide comprehensive and accessible surveys. The Handbooks are intended to provide not only useful reference volumes for professional collections but also possible supplementary readings for advanced courses for graduate students in economics.

# CONTENTS OF THE HANDBOOK

# VOLUME II

## Part 4 – TESTING

## Part 5 – TIME SERIES TOPICS

# PREFACE TO THE HANDBOOK

## Purpose

The *Handbook of Econometrics* aims to serve as a source, reference, and teaching supplement for the field of econometrics, the branch of economics concerned with the empirical estimation of economic relationships. Econometrics is conceived broadly to include not only econometric models and estimation theory but also econometric data analysis, econometric applications in various substantive fields, and the uses of estimated econometric models. Our purpose has been to provide reasonably comprehensive and up-to-date surveys of recent developments and the state of various aspects of econometrics as of the early 1980s, written at a level intended for professional use by economists, econometricians, and statisticians and for use in advanced graduate econometrics courses.

Econometrics is the application of mathematics and statistical methods to the analysis of economic data. Mathematical models help us to structure our perceptions about the forces generating the data we want to analyze, while statistical methods help us to summarize the data, estimate the parameters of our models, and interpret the strength of the evidence for the various hypotheses that we wish to examine. The evidence provided by the data affects our ideas about the appropriateness of the original model and may result in significant revisions of such models. There is, thus, a continuous interplay in econometrics between mathematical-theoretical modeling of economic behavior, data collection, data summarizing, model fitting, and model evaluation. Theory suggests data to be sought and examined; data availability suggests new theoretical questions and stimulates the development of new statistical methods. The examination of theories in light of data leads to their revision. The examination of data in the light of theory leads often to new interpretations and sometimes to questions about its quality or relevance and to attempts to collect new and different data.

In this volume we review only a subset of what might be called "econometrics". The mathematical-theoretical tools required for model building are discussed primarily in the *Handbook of Mathematical Economics*. Issues of sampling theory, survey design, data collection and editing, and computer programming, all important aspects of the daily life of a practicing econometrician, had, by and large, to be left out of the scope of this *Handbook*. We concentrate, instead, on statistical problems and economic interpretation issues associated with the modeling and estimation of economic behavioral relationships from already assembled and often badly collected data. If economists had access to good experimental data, or were able to design and to perform the relevant economic experiments, the topics to be covered in such a *Handbook* would be quite different. The fact that the generation and collection of economic data is mostly outside the hands of the econometrician is the cause of many of the inferential problems which are discussed in this *Handbook*.


## Organization

The organization of the *Handbook* follows in relatively systematic fashion the way an econometric study would proceed, starting from basic mathematical and statistical methods and econometric models, proceeding to estimation and computation, through testing, and ultimately to applications and uses. The *Handbook* also includes a fairly detailed development of time series topics and many other special topics. In particular:

*Part 1* summarizes some basic tools used repeatedly in econometrics, including linear algebra, matrix methods, and statistical theory.

*Part 2* deals with econometric models, their relationship to economic models, their identification, and the question of model choice and specification analysis.

*Part 3* takes up more advanced topics in estimation and computation theory such as non-linear regression methods, biased estimation, and computational algorithms in econometrics. This part also includes a series of chapters on simultaneous equations models, their specification and estimation, distribution theory for such models, and their Bayesian analysis.

*Part 4* considers testing of econometric estimators, including Wald, likelihood ratio, and Lagrange multiplier tests; multiple hypothesis testing; distribution theory for econometric estimators and associated test statistics; and Monte Carlo experimentation in econometrics.

*Part 5* treats various topics in time series analysis, including time series and spectral methods in econometrics, dynamic specification, inference and casuality in economic time series models, continuous time stochastic models, random and changing coefficient models, and the analysis of panel data.

*Parts 6 and 7* present discussions of various special topics in econometrics, including latent variable, limited dependent variable, and discrete choice models; functional forms in econometric model building; economic data issues including longitudinal data issues; and disequilibrium, self selection, and switching models.

Finally, *Part 8* covers selected applications and uses of econometrics. Because of the extremely wide range of applications of econometrics, we could select only a few of the more prominent applications. (Other applications will be treated in later volumes in the "Handbooks in Economics" Series.) Applications discussed here include demand analysis, production and cost analysis, and labor economics. This part includes also chapters on evaluating the predictive accuracy of models, econometric approaches to stabilization policy, the formulation and estimation of models with actors having rational expectations, and the use of econometric models for economic policy formation.

## A brief history of econometrics

A brief review of the history of econometrics will put this *Handbook* in perspective. The historical evolution of econometrics was driven both by the increased availability of data and by the desire of generations of scientists to analyze such data in a rigorous and coherent fashion. There are many historical precursors to that which became "econometrics" in this century. Attempts to interpret economic data "scientifically" go back at least as far as Sir William Petty's "political arithmetic" in the seventeenth century and Engel's studies of household expenditures in the nineteenth. The results of the latter became known as Engel's Law, stating that the proportion of total expenditures devoted to food falls as income rises. This "Law" has been tested extensively for many countries over various time periods, as discussed in Houthakker's (1957) centenary article.

The development of statistical theory has played a critical role in the history of econometrics since econometric techniques are, to a large extent, based on multivariable statistics. Modern statistical theory starts with the work of Legendre and Gauss on least squares, motivated by the attempt to remove errors of observation in astronomy and geodesy. The next great impulse from biology, in particular from evolutionary theory with, among others, Galton's work on regression (a term he invented). Later developments in mathematical statistics included Yule's work on multiple regression, Karl Pearson's formulation of the notions of probable error and of testing hypotheses, the more rigourous small-sample theory of Student and R. A. Fisher, R. A. Fisher's work on the foundations of statistical inference, and the Neyman–Pearson theory of hypothesis testing. All of these developments in mathematical statistics had a significant influence on the development of econometrics.

In the first half of the twentieth century the increased availability of price and quantity data and the interest in price indexes aided by the development of family expenditure surveys generated interest both in theoretical modeling of demand structures and their empirical estimation. Particularly noteworthy were the demand studies of Moore (1914, 1917), Marschak (1931), and Schultz (1928, 1938) and studies of family expenditure by Allen and Bowley (1935). This period also witnessed the initial formulation of the identification problem in econometrics in E. Working (1927); studies of production functions by Cobb and Douglas (1928) [see also Douglas (1948)], and Marschak and Andrews (1944); studies of price determination in agricultural markets by H. Working (1922), Wright (1925), Hanau (1928), Bean (1928), and Waugh (1929), among others; and the statistical modeling of business cycles by Slutsky (1927) and Frisch (1933). Macroeconometric modeling also began in the 1930s by Tinbergen (1935, 1939) and was given additional impetus by the development of National Income Accounts in the United States and other countries and by Keynes' theoretical work.

The growth of data availability and the development of economic and statistical theory generated a demand for more extensive, more rigorous and higher quality data analysis efforts, stimulating significant research into the methodology of economic data analysis. Of great importance in this respect was the founding of the Econometric Society in 1930 and the publication, starting in 1933, of its journal *Econometrica*. Ragnar Frisch played a key role as the first editor of this Journal.

There was a great flourishing of econometric theory and applications in the period after World War II, particularly due to the work of the Cowles Commission at the University of Chicago. The development of the simultaneous equations model in Haavelmo (1943, 1944, 1947), Koopmans (1950), Hood and Koopmans (1953), Theil (1954) and Basmann (1957) provided econometricians with tools designed specifically for them, rather than for by biologists and psychologists. The estimation of simultaneous equations and macroeconometric models in Klein (1950) and Klein and Goldberger (1955) started economic forecasting on a new path. This period also witnessed the important demand studies by Stone (1954a, 1954b) for the United Kingdom and Wold and Jureen (1953) for Sweden and the influential studies by Friedman (1957) of the consumption function and by Theil (1958) of economic forecasts and policy. [For collections of historically important papers in econometrics see Zellner (1968), Hooper and Nerlove (1970), and Dowling and Glahe (1970).]

The more recent period of the 1960s and 1970s has witnessed many important developments in econometric theory and applications. Econometric theory has been refined and extended in many ways. Of particular note is the Bayesian approach to econometrics and the study of special features of econometric models, such as limited dependent variables, latent variables, and non-linear models. Great progress was also made in the statistical analysis of time series. In addition, the development of electronic computers, the great increase in computing power, and the development of sophisticated econometric software packages made it possible to pursue much more ambitious data analysis strategies. These developments expanded the range of applications of econometric methods greatly beyond the earlier applications to household expenditure, demand functions, production and cost functions, and macroeconometric models. Econometrics is now used in virtually every field of economics, including public finance, monetary economics, labor economics, international economics, economic history, health economics, studies of fertility, and studies of criminal behavior, just to mention a few. In all of these fields the greater use of econometric techniques, based in part on increased data availability and more powerful estimation techniques, has led to greater precision in the specification, estimation, and testing of economic data-based models.

Most of the important developments in econometric methods during the 1960s and 1970s are discussed in this *Handbook*. The significant topics under development in this period and the chapters treating them include:

(1) *Bayesian econometrics*, using Bayesian methods in the specification and estimation of econometric models. These topics are discussed in Chapter 2 by Zellner and Chapter 9 by Drèze and Richard.

(2) *Time series methods*, including specialized techniques and problems arising in the analysis of economic time series, such as spectral methods, dynamic specification, and causality. These techniques and problems are discussed in Part 5 on "Time Series Topics," including Chapter 17 by Granger and Watson; Chapter 18 by Hendry, Pagan and Sargan; Chapter 19 by Geweke; Chapter 20 by Bergstrom; and Chapter 21 by Chow. Related issues are discussed in Chapter 33 by Fair.

(3) *Discrete choice models*, in which there is a discrete choice of alternatives available, e.g. buy/don't buy decisions, yes/no responses, or alternative possibilities for urban transportation. Such models are discussed specifically in Chapter 27 by Dhrymes and Chapter 24 by McFadden and are also treated in Chapter 22 by Chamberlain, Chapter 28 by Maddala, and Chapter 29 by Heckman and Singer.

(4) *Latent variables models*, in which certain unmeasurable variables systematically influence measured phenomena, such as ability influencing earnings. This topic is treated in Chapter 23 by Aigner, Hsiao, Kapteyn, and Wansbeek and reappears in various guises in Chapter 22 by Chamberlain and Chapter 32 by Heckman and MaCurdy, among others.

(5) *Specification analysis*, involving problems of model choice and their specification and identification. These issues are treated in Chapter 3 by Intriligator, Chapter 4 by Hsiao, Chapter 5 by Leamer, Chapter 26 by Lau, and Chapter 28 by Maddala. This topic, of course, pervades many other chapters in this *Handbook* and overlaps with chapters which deal with testing and distribution theory.

(6) *Non-linear models and methods*, in which models that are intrinsically nonlinear are specified and estimated. Such models are discussed in Chapter 6 by Amemiya and Chapter 12 by Quandt and surface also in many of the other chapters of this *Handbook*.

(7) *Data analysis issues*, involving various problems with data and how they can be treated. These issues are treated in Chapter 10 by Judge and Bock, in Chapter 11 by Krasker, Kuh, and Welsch, Chapter 22 by Chamberlain, Chapter 25 by Griliches, and Chapter 29 by Heckman and Singer, among others.

(8) *Testing and small sample theory*, including various test procedures and Monte Carlo experimentation. These topics are treated in Part 4 on "Testing," including Chapter 13 by Engle, Chapter 14 by Savin, Chapter 15 by Rothenberg, and Chapter 16 by Hendry. Related issues are discussed in Chapter 8 by Phillips.

(9) *Rational expectations* models which treat economic agents as forming expectations in an optimal fashion, given the information available to them,

impose cross-equation constraints on parameters and lead to new problems of identification and estimation. This topic is discussed in Chapter 34 by Taylor.

ZVI GRILICHES
*Harvard University*

MICHAEL D. INTRILIGATOR
*University of California, Los Angeles*

## References

Allen, R. G. D. and A. L. Bowley (1935) *Family Expenditure*. London: P. S. King

Basmann, R. L. (1957) "A Generalized Classical Method of Linear Estimation of Coefficients in a Structural Equation", *Econometrica*, 25, 77–83.

Bean, L. H. (1928) "Some Interrelationships between the Supply, Price, and Consumption of Cotton", *USDA*, mimeographed.

Cobb, C. W. and P. H. Douglas (1928) "A Theory of Production", *American Economic Review*, 18 (supplement), 139–165.

Douglas, P. H. (1948) "Are There Laws of Production?", *American Economic Review*, 38, 1–41.

Dowling, J. M. and F. R. Glahe (eds.) (1970) *Readings in Econometric Theory*. Boulder: Colorado Associated University Press.

Friedman, M. (1957) *A Theory of the Consumption Function*, National Bureau of Economic Research. Princeton: Princeton University Press.

Frisch, R. (1933) "Propagation Problems and Impulse Problems in Dynamic Economics", in: *Economic Essays in Honor of Gustav Cassel*. London: George Allen & Unwin, pp. 171–205.

Haavelmo, T. (1943) "The Statistical Implications of a System of Simultaneous Equations", *Econometrica*, 11, 1–12.

Haavelmo, T. (1944) "The Probability Approach in Econometrics", *Econometrica*, 12 (supplement), 1–115.

Haavelmo, T. (1947) "Methods of Measuring the Marginal Propensity to Consume", *Journal of the American Statistical Association*, 42, 105–122 (reprinted in Hood and Koopmans (eds) (1953)).

Hanau, A. (1928) "Die Prognose der Schweinepreise", *Vierteljahrshefte zur Konjunkturforschung*, Sonderheft 7, Berlin.

Hood, W. C. and T. C. Koopmans (eds.) (1953) *Studies in Econometric Method*, Cowles Commission Monograph No. 14, New York: John Wiley & Sons.

Hooper, J. W. and M. Nerlove (eds.) (1970) *Selected Readings in Econometrics from Econometrica*. Cambridge: MIT Press.

Houthakker, H. S. (1957) "An International Comparison of Household Expenditure Patterns, Commemorating the Centenary of Engel's Law", *Econometrica*, 25, 532–551.

Klein, L. R. (1950) *Economic Fluctuations in the United States, 1921–1941*, Cowles Commission Monograph No. 11. New York: John Wiley & Sons.

Klein, L. R. and A. S. Goldberger (1955) *An Econometric Model of the United States, 1929–1952*. Amsterdam: North-Holland Publishing Co.

Koopmans, T. C. (ed.) (1950) *Statistical Inference in Dynamic Economic Models*, Cowles Commission Monograph No. 10. New York: John Wiley & Sons.

Marschak, J. (1931) *Elastizität der Nachfrage*. Tübingen: J. C. B. Mohr.

Marschak, J. and W. H. Andrews (1944) "Random Simultaneous Equations and the Theory of Production", *Econometrica* 12, 143–205.

Moore, H. L. (1914) *Economic Cycles: Their Law and Cause*. New York: The Macmillan Company.

Moore, H. L. (1917) *Forecasting the Yield and Price of Cotton*. New York: The Macmillan Company.

Schultz, H. (1928) *Statistical Laws of Demand and Supply*. Chicago: University of Chicago Press.

Schultz, H. (1938) *The Theory and Measurement of Demand*, Chicago: University of Chicago Press.

Slutsky, E. (1927) "The Summation of Random Causes as the Source of Cyclic Processes" (Russian with English summary), in: *Problems of Economic Conditions*, vol. 3. Moscow: Rev. English edn., 1937; *Econometrica*, 5, 105–146.

Stone, R. (1954a) "Linear Expenditure Systems and Demand Analysis: An Application to the Pattern of British Demand", *Economic Journal*, 64, 511–527.

Stone, R. (1954b) *The Measurement of Consumers' Expenditure and Behavior in the United Kingdom, 1920–1938*. New York: Cambridge University Press.

Theil, H. (1954) "Estimation of Parameters of Econometric Models", *Bulletin of the International Statistics Institute*, 34, 122–128.

Theil, H. (1958) *Economic Forecasts and Policy*. Amsterdam: North-Holland Publishing Co. (Second Edition, 1961).

Tinbergen, J. (1935) "Quantitative Fragen der Konjunkturpolitik", *Weltwirtschaftliches Archiv*, 42, 316–399.

Tinbergen, J. (1939) *Statistical Testing of Business Cycle Theories*. Vol. 1: *A Method and its Application to Investment Activity*; Vol. 2: *Business Cycles in the United States of America, 1919–1932*. Geneva: League of Nations.

Waugh, F. V. (1929) *Quality as a Determinant of Vegetable Prices*. New York: Columbia University Press.

Wold, H. and L. Jureen (1953) *Demand Analysis*. New York: John Wiley & Sons.

Working, E. J. (1927) "What do Statistical 'Demand Curves' Show?", *Quarterly Journal of Economics*, 41, 212–235.

Working, H. (1922) "Factors Determining the Price of Potatoes in St. Paul and Minneapolis", University of Minnesota Agricultural Experiment Station Technical Bulletin 10.

Wright, S. (1925) *Corn and Hog Correlations*, Washington, USDA, Bul. 1300.

Zellner, A., Ed. (1968) *Readings in Economic Statistics and Econometrics*. Boston: Little, Brown.

# CONTENTS OF VOLUME III

# PART 7

# SPECIAL TOPICS IN ECONOMETRICS: 2

*Chapter 25*

# ECONOMIC DATA ISSUES

ZVI GRILICHES*

*Harvard University*

## Contents

*I am indebted to the National Science Foundation (SOC78-04279 and PRA81-08635) for their support of my work on this range of topics, to John Bound, Bronwyn Hall, J. A. Hausman, and Ariel Pakes for research collaboration and many discussions, and to O. Ashenfelter, E. Berndt, F. M. Fisher, R. M. Hauser, M. Intriligator, S. Kuznets, J. Medoff, and R. Vernon for comments on an earlier draft.

*Handbook of Econometrics, Volume III, Edited by Z. Griliches and M.D. Intriligator*
© *Elsevier Science Publishers BV, 1986*

## 1.   Introduction: Data and econometricians – the uneasy alliance

> Then the officers of the children of Israel came and cried
>     unto Pharaoh, saying, Wherefore dealest thou thus with thy servants?
> There is no straw given unto thy servants, and they say
>     to us, Make brick: and behold thy servants are beaten; but the fault
>     is in thine own people.
> But he said, Ye are idle, ye are idle: Therefore ye say,
>     Let us go and do sacrifice to the Lord.
> Go therefore now, and work; for there shall no straw be
>     given you, yet shall ye deliver the tale of bricks.
>
> Exodus 5, 15–18

Econometricians have an ambivalent attitude towards economic data. At one level, the "data" are the world that we want to explain, the basic facts that economists purport to elucidate. At the other level, they are the source of all our trouble. Their imperfection makes our job difficult and often impossible. Many a question remains unresolved because of "multicollinearity" or other sins of the data. We tend to forget that these imperfections are what gives us our legitimacy in the first place. If the data were perfect, collected from well designed randomized experiments, there would be hardly room for a separate field of econometrics. Given that it is the "badness" of the data that provides us with our living, perhaps it is not all that surprising that we have shown little interest in improving it, in getting involved in the grubby task of designing and collecting original data sets of our own. Most of our work is on "found" data, data that have been collected by somebody else, often for quite different purposes.

Economic data collection started primarily as a byproduct of other governmental activities: tax and customs collections. Early on, interest was expressed in prices and levels of production of major commodities. Besides tax records, population counts, and price surveys, the earliest large scale data collection efforts were various Censuses, family expenditure surveys, and farm cost and production surveys. By the middle 1940s the overall economic data pattern was set: governments were collecting various quantity and price series on a continuous basis, with the primary purpose of producing aggregate level indicators such as price indexes and national income accounts series, supplemented by periodic surveys of population numbers and production and expenditure patterns to be used primarily in updating the various aggregate series. Little microdata was published or accessible, except in some specific sub-areas, such as agricultural economics.

A pattern was also set in the way the data were collected and by whom they were analyzed.[1] With a few notable exceptions, such as France and Norway, and

---

[1] See Kuznets (1971) and Morgenstern (1950) for earlier expressions of similar opinions. Morgenstern's Cassandra like voice is still very much worth listening to on this range of topics.

until quite recently, econometricians were not to be found inside the various statistical agencies, and especially not in the sections that were responsible for data collection. Thus, there grew up a separation of roles and responsibility. "They" collect the data and "they" are responsible for all of their imperfections. "We" try to do the best with what we get, to find the grain of relevant information in all the chaff. Because of this, we lead a somewhat remote existence from the underlying facts we are trying to explain. We did not observe them directly; we did not design the measurement instruments; and, often we know little about what is really going on (e.g. when we estimate a production function for the cement industry from Census data without ever having been inside a cement plant). In this we differ quite a bit from other sciences (including observational ones rather than experimental) such as archeology, astrophysics, biology, or even psychology where the "facts" tend to be recorded by the professionals themselves, or by others who have been trained by and are supervised by those who will be doing the final data analysis. Economic data tend to be collected (or often more correctly "reported") by firms and persons who are not professional observers and who do not have any stake in the correctness and precision of the observations they report. While economists have increased their use of surveys in recent years and even designed and commissioned a few special purpose ones of their own, in general, the data collection and thus the responsibility for the quality of the collected material is still largely delegated to census bureaus, survey research centers, and similar institutions, and is divorced from the direct supervision and responsibility of the analyzing team.

It is only relatively recently, with the initiation of the negative income tax experiments and various longitudinal surveys intended to follow up the effects of different governmental programs, that econometric professionals had actually become involved in the primary data collection process. Once attempted, the job turned out to be much more difficult than was thought originally, and taught us some humility.[2] Even with relatively large budgets, it was not easy to figure out how to ask the right question and to collect relevant answers. In part this is because the world is much more complicated than even some of our more elaborate models allow for, and partly also because economists tend to formulate their theories in non-testable terms, using variables for which it is hard to find empirical counterparts. For example, even with a large budget, it is difficult to think of the right series of questions, answers to which would yield an unequivocal number of *the* level for "human capital" or "permanent income" of an individual. Thinking about such "alibi-removing" questions should make us a bit more humble, restrain our continuing attacks on the various official data producing agencies, and push us towards formulating theories with more regard to what is observable and what kind of data may be available.

[2] See Hausman and Wise (1985).

Even allowing for such reservations there has been much progress over the years as a result of the enormous increase in the quantity of data available to us, in our ability to manipulate them, and in our understanding of their limitations. Especially noteworthy have been the development of various longitudinal micro-data sets (such as the Michigan PSID tapes, and Ohio State NLS surveys, the Wisconsin high school class follow-up study, and others),[3] the computerization of the more standard data bases and their easier accessibility at the micro, individual response level (I have in mind here such developments as the Public Use Samples from the U.S. Population Census and the Current Population Surveys).[4] Unfortunately, much more progress has been made with labor force and income type data, where the samples are large, than in the availability of firm and other market transaction data. While significant progress has been made in the collection of financial data and security prices, as exemplified in the development of the CRISP and Compustat data bases which have had a tremendous impact on the field of finance, we are still in our infancy as far as our ability to interrogate and get reasonable answers about other aspects of firm behavior is concerned. Most of the available microdata at the firm level are based on legally required responses to questions from various regulatory agencies who do not have our interests exactly in mind.

We do have, however, now a number of extensive longitudinal microdata sets which have opened a host of new possibilities for analysis and also raised a whole range of new issues and concerns. After a decade or more of studies that try to use such data, the results have been somewhat disappointing. We, as econometricians, have learned a great deal from these efforts and developed whole new subfields of expertise, such as sample selection bias and panel data analysis. We know much more about these kinds of data and their limitations but it is not clear that we know much more or more precisely about the roots and modes of economic behavior that underlie them.

The encounters between econometricians and data are frustrating and ultimately unsatisfactory both because econometricians want too much from the data and hence tend to be disappointed by the answers, and because the data are incomplete and imperfect. In part it is our fault, the appetite grows with eating. As we get larger samples, we keep adding variables and expanding our models, until on the margin, we come back to the same insignificance levels.

There are at least three interrelated and overlapping causes of our difficulties: (1) the theory (model) is incomplete or incorrect; (2) the units are wrong, either at too high a level of aggregation or with no way of allowing for the heterogeneity of responses; and, (3) the data are inaccurate on their own terms, incorrect relative

---

[3] See Borus (1982) for a recent survey of longitudinal data sets.

[4] This survey is, perforce, centered on U.S. data and experience, which is what I am most familiar with. The overall developments, however, have followed similar patterns in most other countries.

to what they purport to measure. The average applied study has to struggle with all three possibilities.

At the macro level and even in the usual industry level study, it is common to assume away the underlying heterogeneity of the individual actors and analyze the data within the framework of the "representative" firm or "average" individual, ignoring the aggregation difficulties associated with such concepts. In analyzing microdata, it is much more difficult to evade this issue and hence much attention is paid to various individual "effects" and "heterogeneity" issues. This is wherein the promise of longitudinal data lies – their ability to control and allów for additive individual effects. On the other hand, as is the case in most other aspects of economics, there is no such thing as a free lunch: going down to the individual level exacerbates both some of the left out variables problems and the importance of errors in measurement. Variables such as age, land quality, or the occupational structure of an enterprise, are much less variable in the aggregate. Ignoring them at the micro level can be quite costly, however. Similarly, measurement errors which tend to cancel out when averaged over thousands or even millions of respondents, loom much larger when the individual is the unit of analysis.

It is possible, of course, to take an alternative view: that there are no data problems only model problems in econometrics. For any set of data there is the "right" model. Much of econometrics is devoted to procedures which try to assess whether a particular model is "right" in this sense and to criteria for deciding when a particular model fits and is "correct enough" (see Chapter 5, Hendry, 1983 and the literature cited there). Theorists and model builders often proceed, however, on the assumption that ideal data will be available and define variables which are unlikely to be observable, at least not in their pure form. Nor do they specify in adequate detail the connection between the actual numbers and their theoretical counterparts. Hence, when a contradiction arises it is then possible to argue "so much worse for the facts." In practice one cannot expect theories to be specified to the last detail nor the data to be perfect or of the same quality in different contexts. Thus any serious data analysis has to consider at least two data generation components: the economic behavior model describing the stimulus-response behavior of the economic actors and the measurement model, describing how and when this behavior was recorded and summarized. While it is usual to focus our attention on the former, a complete analysis must consider them both.

In this chapter, I discuss a number of issues which arise in the encounter between the econometrician and economic data. Since they permeate much of econometrics, there is quite a bit of overlap with some of the other chapters in the Handbook. The emphasis here, however, is more on the problems that are posed by the various aspects of economic data than on the specific technological solutions to them.

After a brief review of the major classes of economic data and the problems that are associated with using and interpreting them, I shall focus on issues that are associated with using erroneous or partially missing data, discuss several empirical examples, and close with a few final remarks.

## 2.   Economic data: An overview

> Data: fr. Latin, plural of datum – given.
> Observation: fr. Latin observare – to guard, watch.

It is possible to classify economic data along several different dimensions: (a) Substantive: Prices, Quantities, Commodity Statistics, Population Statistics, Banking Statistics, etc.; (b) Objective versus Subjective: Prices versus expectations about them, actual wages versus self reported opinions about well being; (c) Type and periodicity: Time series versus cross-sections; monthly, quarterly, or annual; (d) Level of aggregation: Individuals, families, or firms (micro), and districts, states, industries, sectors, or whole countries (macro); (e) Level of fabrication: primary, secondary, or tertiary; (f) Quality: Extent, reliability and validity.

As noted earlier, the bulk of economic data is collected and produced by various governmental bodies, often as a by-product of their other activities. Roughly speaking, there are two major types of economic data: aggregate time series on prices and quantities at the commodity, industry, or country level, and periodic surveys with much more individual detail. In recent years, as various data bases became computerized, economic analysts have gained access to the underlying microdata, especially where the governmental reports are based on periodic survey results. This has led to a great flowering of econometric work on various microdata sets including longitudinal panels.

The level of aggregation dimension and the micro–macro dichotomy are not exactly the same. In fact, much of the "micro" data is already aggregated. The typical U.S. firm is often an amalgam of several enterprises and some of the larger ones may exceed in size some of the smaller countries or states. Similarly, consumer surveys often report family expenditure or income data which have been aggregated over a number of individual family members. Annual income and total consumption numbers are also the result of aggregation over more detailed time periods, such as months or weeks, and over a more detailed commodity and sources of income classification. The issues that arise from the mismatch between the level of aggregation at which the theoretical model is defined and expected to be valid and the level of aggregation of the available data have not really received the attention they deserve (see Chapters 20 and 30 for more discussion and some specific examples).

The level of fabrication dimension refers to the "closeness" of the data to the actual phenomenon being measured. Even though they may be subject to various biases and errors, one may still think of reports of hours worked during last week by a particular individual in a survey or the closing price of a specific common stock on the New York Stock Exchange on December 31 as primary observations. These are the basic units of information about the behavior of economic actors and the information available to them (though individuals are also affected by the macro information that they receive). They are the units in which most of our microtheories are denominated. Most of our data are not of this sort, however. They have usually already undergone several levels of processing or fabrication. For example, the official estimate of total corn production in the State of Iowa in a particular year is not the result of direct measurement but the outcome of a rather complicated process of blending sample information on physical yields, reports on grain shipments to and from elevators, benchmark census data from previous years, and a variety of informal Bayes-like smoothing procedures to yield the final official "estimate" for the state as a whole. The final results, in this case, are probably quite satisfactory for the uses they are put to, but the procedure for creating them is rarely described in full detail and is unlikely to be replicable. This is even more true at the aggregated level of national income accounts and other similar data bases, where the link between the original primary observations and the final aggregate numbers is quite tenuous and often mysterious.

I do not want to imply that the aggregate numbers are in some sense worse than the primary ones. Often they are better. Errors may be reduced by aggregation and the informal and formal smoothing procedures may be based on correct prior information and result in a more reliable final result. What needs to be remembered is that the final published results can be affected by the properties of the data generating mechanism, by the procedures used to collect and process the data. For example, some of the time series properties of the major published economic series may be the consequence of the smoothing techniques used in their construction rather than a reflection of the underlying economic reality. (This was brought forceably home to me many years ago while collecting unpublished data on the diffusion of hybrid corn at the USDA when I came across a circular instructing the state agricultural statisticians: "When in doubt – use a growth curve.") Some series may fluctuate because of fluctuations in the data generating institutions themselves. For example, the total number of patents granted by the U.S. Patent Office in a particular year depends rather strongly on the total number of patent examiners available to do the job. For budgetary and other reasons, their number has gone through several cycles, inducing concomitant cycles in the actual number of patents granted. This last example brings up the point that while particular numbers may be indeed correct as far as they go, they do not really mean what we thought they did.

Such considerations lead one to consider the rather amorphous notion of data "quality." Ultimately, quality cannot be defined independently of the intended use of the particular data set. In practice, however, data are used for multiple purposes and thus it makes some sense to indicate some general notions of data quality. Earlier I listed extent, reliability, and validity as the three major dimensions along which one may judge the quality of different data sets. Extent is a synonym for richness: How many variables are present, what interesting questions had been asked, how many years and how many firms or individuals were covered? Reliability is actually a technical term in psychometrics, reflecting the notion of replicability and measuring the relative amount of random measurement error in the data by the correlation coefficient between replicated or related measurement of the same phenomenon. Note that a measurement may be highly reliable in the sense that it is a very good measure of whatever it measures, but still be the wrong measure for our particular purposes.

This brings us to the notion of validity which can be subdivided in turn into representativeness and relevance. I shall come back to the issue of how representative is a body of data when we discuss issues of missing and incomplete data. It will suffice to note here that it contains the technical notion of coverage: Did all units in the relevant universe have the same (or alternatively, different but known and adjusted for) probability of being selected into the sample that underlies this particular data set? Coverage and relevance are related concepts which shade over into issues that arise from the use of "proxy" variables in econometrics. The validity and relevance questions relate less to the issue of whether a particular measure is a good (unbiased) estimate of the associated population parameter and more to whether it actually corresponds to the conceptual variable of interest. Thus one may have a good measure of current prices which are still a rather poor indicator of the currently expected future price and relatively extensive and well measured IQ test scores which may still be a poor measure of the kind of "ability" that is rewarded in the labor market.

## 3. Data and their discontents

> My father would never eat "cutlets" (minced meat patties) in the old country. He would not eat them in restaurants because he didn't know what they were made of and he wouldn't eat them at home because he did.
>
> AN OLD FAMILY STORY

I will be able to touch on only a few of the many serious practical and conceptual problems that arise when one tries to use the various economic data sets. Many of these issues have been discussed at length in the national income and growth measurement literature but are not usually brought up in standard econometrics courses or included in their curriculum. Among the many official and semi-official data base reviews one should mention especially the Creamer GNP Improvement report (U.S. Department of Commerce, 1979), the Rees committee report on productivity measurement (National Academy of Sciences, 1979), the Stigler committee (National Bureau of Economic Research, 1961) and the Ruggles (Council on Wage and Price Stability, 1977) reports on price statistics, the Gordon (President's Committee to Appraise Employment Statistics, 1962), and the Levitan (National Committee on Employment and Unemployment Statistics, 1979) committee reports on the measurement of employment and unemployment, and the many continuous and illuminating discussions reported in the proceedings volumes of the Conference on Research in Income and Wealth, especially in volumes 19, 20, 22, 25, 34, 38, 45, 47, and 48 (National Bureau of Economic Research, 1957...1983). All these references deal almost exclusively with U.S. data, where the debates and reviews have been more extensive and public, but are also relevant for similar data elsewhere.

At the national income accounts level there are serious definitional problems about the borders of economic activity (e.g. home production and the investment value of children) and the distinction between final and intermediate consumption activity (e.g. what fraction of education and health expenditures can be thought of as final rather than intermediate "goods" or "bads"). There are also difficult measurement problems associated with the existence of the underground economy and poor coverage of some of the major service sectors. The major serious problem from the econometric point of view probably occurs in the measurement of "real" output, GNP or industry output in "constant prices," and the associated growth measures. Since most of the output measures are derived by dividing ("deflating") current value totals by some price index, the quality of these measures is intimately connected to the quality of the available price data. Because of this, it is impossible to treat errors of measurement at the aggregate level as being independent across price and "quantity" measures.

The available price data, even when they are a good indicator of what they purport to measure, may still be inadequate for the task of deflation. For productivity comparisons and for production function estimation the observed prices are supposed to reflect the relevant marginal costs and revenues in a, at least temporary, competitive equilibrium. But this is unlikely to be the case in sectors where output or prices are controlled, regulated, subsidized, and sold under various multi-part tariffs. Because the price data are usually based on the pricing of a few selected items in particular markets, they may not correspond well to the average realized price for the industry as a whole during a particular time period, both because "easily priced" items may not be representative of the average price movements in the industry as a whole and because many transactions are made with a lag, based on long term contracts. There are also problems associated with getting accurate transactions prices (Kruskal and Telser, 1960 and

Stigler and Kindahl, 1970), but the major difficulty arises from getting comparable prices over time, from the continued change in the available set of commodities, the "quality change" problem.

"Quality change" is actually a special version of the more general comparability problem, the possibility that similarly named items are not really similar, either across time or individuals. In many cases the source of similarly sounding items is quite different: Employment data may be collected from plants (establishments), companies, or households. In each case the answer to the same question may have a different meaning. Unemployment data may be reported by a teenager directly or by his mother, whose views about it may both differ and be wrong. The wording of the question defining unemployment may have changed over time and so should also the interpretation of the reported statistic. The context in which a question is asked, its position within a series of questions on a survey, and the willingness to answer some of the questions may all be changing over time making it difficult to maintain the assumption that the reported numbers in fact relate to the same underlying phenomenon over time or across individuals and cultures.

The common notion of quality change relates to the fact that many commodities are changing over time and that often it is impossible to construct appropriate pricing comparisons because the same varieties are not available at different times and in different places. Conceptually one might be able to get around this problem by assuming that the many different varieties of a commodity differ only along a smaller number of relevant dimensions (characteristics, specifications), estimate the price-characteristics relationship econometrically and use the resulting estimates to impute a price to the missing model or variety in the relevant comparison period. This approach, pioneered by Waugh (1928) and Court (1936) and revived by Griliches (1961) has become known as the "hedonic" approach to price measurement. The data requirements for the application of this type of an approach are quite severe and there are very few official price indexes which incorporate it into their construction procedures. Actually, it has been used much more widely in labor economics and in the analyses of real estate values than in the construction of price deflator indexes. See Griliches (1971), Gordon (1983), Rosen (1974) and Triplett (1975) for expositions, discussions, and examples of this approach to price measurement.

While the emergence of this approach has sensitized both the producers and the consumers of price data to this problem and contributed to significant improvements in data collection and processing procedures over time, it is fair to note that much still remains to be done. In the U.S. GNP deflation procedures, the price of computers has been kept constant since the early 1960s, for lack of an agreement of what to do about it, resulting in a significant underestimate in the growth of real GNP during the last two decades. Similarly, for lack of a more appropriate price index, aircraft purchases had been deflated by an equally weighted index of gasoline engine, metal door, and telephone equipment prices until the early 1970s, at which point a switch was made to a price index based on data from the CAB on purchase prices for "identical" models, missing thereby the major gains that occurred from the introduction of the jet engine, and the various improvements in operating efficiency over time.[5] One could go on adding to this gallery of horror stories but the main point to be made here is not that a particular price index is biased in one or another direction. Rather, the point is that one cannot take a particular published price index series and interpret it as measuring adequately the underlying notion of a price change for a well specified, unchanging, commodity or service being transacted under identical conditions and terms in different time periods. The particular time series may indeed be quite a good measure of it, or at least better than the available alternatives, but each case requires a serious examination whether the actual procedures used to generate the series do lead to a variable that is close enough to the concept envisioned by the model to be estimated or by the theory under test. If not, one needs to append to the model an equation connecting the available measured variable to the desired but not actually observed correct version of this variable.

The issues discussed above affect also the construction and use of various "capital" measures in production function studies and productivity growth analyses. Besides the usual aggregation issues connected with the "existence" of an unambiguous capital concept (see Diewert, 1980 and Fisher, 1969 on this) the available measures suffer from potential quality change problems, since they are usually based on some cumulated function of past investment expenditures deflated by some combination of available price indexes. In addition, they are also based on rather arbitrary assumptions about the pattern of survival of machines over time and the time pattern of deterioration in the flow of their services. The available information on the reasonableness of such assumptions is very sparse, ancient, and flimsy. In some contexts it is possible to estimate the appropriate pattern from the data rather than impose them a priori. I shall present an example of this type of approach below.

Similar issues arise also in the measurement of labor inputs and associated variables at both the macro and micro levels. At the macro level the questions revolve about the appropriate weighting to be given to different types of labor: young – old, male – female, black – white, educated vs. uneducated, and so forth. The direct answer here as elsewhere is that they should be weighted by their appropriate marginal prices but whether the observed prices actually reflect correctly the underlying differences in their respective marginal productivities is one of the more hotly debated topics in labor economics. (See Griliches, 1970 on the education distinction and Medoff and Abraham, 1980 on the age distinction.)

---

[5] For a recent review and reconstruction of the price indexes for durable producer goods see Gordon's (1985) forthcoming monograph.

Connected to this is also the difficulty of getting relevant labor prices. Most of the usual data sources report or are based on data on average annual, weekly, or hourly earnings which do not represent adequately either the marginal cost of a particular labor hour to the employer or the marginal return to a worker from the additional hour of work. Both are affected by the existence of overtime premia, fringe benefits, training costs, and transportation costs. Only recently has an employment cost index been developed in the United States. (See Triplett, 1983 on this range of issues.) From an individual worker's point of view the existence of non-proportional tax schedules introduces another source of discrepancy between the observed wage rates and the unobserved marginal after tax net returns from working (see Hausman, 1982, for a more detailed discussion).

While the conceptual discrepancy between the desired concepts and the available measures dominates at the macro level the more mundane topics of errors of measurement and missing and incomplete data come to the fore at the micro, individual survey level. This topic is the subject of the next section.

## 4.   Random measurement errors and the classic EVM

> To disavow an error is to invent retroactively.
> Goethe

While many of the macro series may be also subject to errors, the errors in them rarely fit into the framework of the classical errors-in-variables model (EVM) as it has been developed in econometrics (see Chapter 23 for a detailed exposition). They are more likely to be systematic and correlated over time.[6] Micro data are subject to at least three types of discrepancies, "errors," and fit this framework much better:

(a) Transcription, transmission, or recording error, where a correct response is recorded incorrectly either because of clerical error (number transposition, skipping a line or a column) or because the observer misunderstood or misheard the original response.

(b) Response or sampling error, where the correct underlying value could be ascertained by a more extensive sampling, but the actual observed value is not equal to the desired underlying population parameter. For example, an IQ test is based on a sample of responses to a selected number of questions. In principle, the mean of a large number of tests over a wide range of questions would

---

[6] For an "error analysis" of national income account data based on the discrepancies between preliminary and "final" estimates see Cole (1969), Young (1974), and Haitovsky (1972). For an earlier more detailed evaluation based on subjective estimates of the differential quality of the various "ingredients" (series) of such accounts see Kuznets (1954, chapter 12).

converge to some mean level of "ability" associated with the range of subjects being tested. Similarly, the simple permanent income hypothesis would assert that reported income in any particular year is a random draw from a potential population of such incomes whose mean is "permanent income." This is the case where the observed variable is a direct but fallible indicator of the underlying relevant "unobservable," "latent factor" or variable (see Chapter 23 and Griliches, 1974, for more discussion of such concepts).

(c) When one is lacking a direct measure of the desired concept and a "proxy" variable is used instead. For example, consider a model which requires a measure of permanent income and a sample which has no income measures at all but does have data on the estimated market value of the family residence. This housing value may be related to the underlying permanent income concept, but not clearly so. First, it may not be in the same units, second it may be affected by other variables also, such as house prices and family size, and third there may be "random" discrepancies related to unmeasured locational factors and events that occurred at purchase time. While these kinds of "indicator" variables do not fit strictly into the classical EVM framework, their variances, for example, need not exceed the variance of the true "unobservable," they can be fitted into this framework and treated with the same methods.

There are two classes of cases which do not really fit this framework: Occasionally one encounters large transcription and recording errors. Also, sometimes the data may be contaminated by a small number of cases arising from a very different behavioral model and/or stochastic process. Sometimes, these can be caught and dealt with by relatively simple data editing procedures. If this kind of problem is suspected, it is best to turn to the use of some version of the "robust estimation" methods discussed in Chapter 11. Here we will be dealing with the more common general errors-in-measurement problem, one that is likely to affect a large fraction of our observations.

The other case that does not fit our framework is where the true concept, the unobservable is distributed randomly relative to the measure we have. For example, it is clear that the "number of years of school completed" ($S$) is an erroneous measure of true "education" ($E$), but it is more likely that the discrepancy between the two concepts is independent of $S$ rather than $E$. I.e. the "error" of ignoring differences in the quality of schooling may be independent of the measured years of schooling but is clearly a component of the true measure of $E$. The problem here is a left-out relevant variable (quality) and not measurement error in the variable as is (years of school). Similarly, if we use the forecast of some model, based on past data, to predict the expectations of economic actors, we clearly commit an error, but this error is independent of the forecast level (if this forecast is optimal and the actors have had access to the same information). This type of "error" does not induce a bias in the estimated coefficients and can be incorporated into the standard disturbance framework (see Berkson, 1950).

The standard EVM assumes the existence of a true relationship

$$y = \alpha + \beta z + e, \tag{4.1}$$

the absence of direct observations on $z$, and the availability of a fallible measure of it

$$x = z + \varepsilon, \tag{4.2}$$

where $\varepsilon$ is a purely random i.i.d. measurement error, with $E\varepsilon = 0$, and no correlation with either $z$ or $y$. This is quite a restrictive set of assumptions, especially the assumption of the errors not being correlated with anything else in the model including their own past values. But it turns out to be very useful in many contexts and not too far off for a variety of micro data sets. I will discuss the evidence for the existence of such errors further on, when we turn to consider briefly various proposed solutions to the estimation problem in such models, but the required assumptions are not more difficult than those made in the standard linear regression model which requires that the "disturbance" $e$, the model discrepancy, be uncorrelated with all the included explanatory variables.

It may be worthwhile, at this point, to summarize the main conclusions from the EVM for the standard OLS estimates in contexts where one has ignored the presence of such errors. Estimating

$$y = a + bx + u, \tag{4.3}$$

where the true model is the one given above yields $-\beta\lambda$ as the asymptotic bias of the OLS $\hat{b}$, where $\lambda = \sigma_\varepsilon^2/\sigma_x^2$ is a measure of the relative amount of measurement error in the observed $x$ series. The basic conclusion is that the OLS slope estimate is biased towards zero, while the constant term is biased away from zero. Since, in this model one can treat $y$ and $x$ symmetrically, it can be shown (Schultz, 1938, Frisch, 1934, Klepper and Leamer, 1983) that in the "other regression," the regression of $x$ on $y$, the slope coefficient is also biased towards zero, implying a "bracketing" theorem

$$\text{plim } b_{yx} < \beta < 1/\text{plim } b_{xy}. \tag{4.4}$$

These results generalize also to the multivariate case. In the case of two independent variables ($x_1$ and $x_2$), where only one ($x_1$) is subject to error, the coefficient of the other variable (the one not subject to errors of measurement) is also biased (unless the two variables are uncorrelated). That is, if the true model is

$$y = \alpha + \beta_1 z_1 + \beta_2 x_2 + e, \tag{4.5}$$
$$x_1 = z_1 + \varepsilon,$$

then

$$\text{plim}\left(b_{yx_1 \cdot x_2} - \beta_1\right) = -\beta_1 \lambda / (1 - \rho^2), \tag{4.6}$$

where $\rho$ is the correlation between the two observed variables $x_1$ and $x_2$, and if we scale the variables so that $\sigma_{x_1}^2 = \sigma_{x_2}^2 = 1$, then

$$\text{plim}\left(b_{yx_2 \cdot x_1} - \beta_2\right) = \rho \beta_1 \lambda / (1 - \rho^2) \tag{4.7}$$
$$= -\rho [\text{bias}\, \beta_1].$$

That is, the bias in the coefficient of the erroneous variable is "transmitted" to the other coefficients, with an opposite sign (provided, as is often the case, that $\rho > 0$), (see Griliches and Ringstad, 1971, Appendix C, and Fisher, 1980 for the derivation of this and related formulae).

If more than one independent variable is subject to error, the formulae become more complicated, but the basic pattern persists. If both $z_1$ and $z_2$ are unobserved and $x_1 = z_1 + \varepsilon_1$, $x_2 = z_2 + \varepsilon_2$, where the $\varepsilon$'s are independent (of each other) errors of measurement, and we have normalized the variables so that $\sigma_{x_1}^2 = \sigma_{x_2}^2 = 1$, then

$$\text{plim}\left(b_{y1 \cdot 2} - \beta_1\right) = -\beta_1 \lambda_1 / (1 - \rho^2) + \beta_2 \lambda_2 \rho / (1 - \rho^2) \tag{4.8}$$
$$= -\frac{\beta_1 \lambda_1}{1 - \rho^2} \left\{ 1 - \frac{\beta_2 \lambda_2}{\beta_1 \lambda_1} \rho \right\},$$

with a similar symmetric formula for plim $b_{y2 \cdot 1}$. Thus, in the multivariate case, the bias is increased by the factor $1/(1 - \rho^2)$, the reduction in the independent variance of the true signal due to its intercorrelation with the other variable(s), and attenuated by the fact that the particular variable compensates somewhat for the downward bias in the other coefficients caused by the errors in the other variables. Overall, there is still a bias towards zero. For example, in this case the sum of the estimated coefficients is always biased towards zero:

$$\text{plim}\left[(b_{y1 \cdot 2} + b_{y2 \cdot 1}) - (\beta_1 + \beta_2)\right] = -[\beta_1 \lambda_1 + \beta_2 \lambda_2]/(1 + \rho). \tag{4.9}$$

It is a declining function of $\rho$, for $\rho > 0$, which is reasonable it we remember that $\rho$ is defined as the intercorrelation between the observed $x$'s. The higher it is, the smaller must be the role of independent measurement errors in these variables.

The impact of errors in variables on the estimated coefficients can be magnified by some transformations. For example, consider a quadratic equation in the unobserved true $z$:

$$y = \alpha + \beta z + \gamma z^2 + e, \tag{4.10}$$

with the observed

$$x = z + \varepsilon,$$

substituted instead. If both $z$ and $\varepsilon$ are normally distributed, it can be shown (Griliches and Ringstad, 1970) that

$$\operatorname{plim} \hat{b} = \beta(1 - \lambda), \tag{4.11}$$

while

$$\operatorname{plim} \hat{c} = \gamma(1 - \lambda)^2,$$

where $\hat{b}$ and $\hat{c}$ are the estimated OLS coefficients in the $y = a + bx + cx^2 + u$ equation. That is, higher order terms of the equation are even more affected by errors in measurement than lower order ones.

The impact of errors in the levels of the variables may be reduced by aggregation and aggravated by differencing. For example, in the simple model $y = \alpha + \beta z + e$, $x = z + \varepsilon$, the asymptotic bias in the OLS $b_{yx}$ is equal to $-\beta\lambda$, while the bias of the first differenced estimator $[y_t - y_{t-1} = b(x_t - x_{t-1}) + v_t]$ is equal to $-\beta\lambda/(1-\rho)$ where $\rho$ now stands for the first order serial correlation of the $x$'s, and can be much higher than in levels (for $\rho > 0$ and not too small). Similarly, computing "within" estimates in panel data, or differencing across brothers or twins in micro data, can result in the elimination of much of the relevant variance in the observed $x$'s, and a great magnification of the noise to signal ratio in such variables. (See Griliches, 1979, for additional exposition and examples.)

In some cases, errors in different variables cannot be assumed to be independent of each other. To the extent that the form of the dependence is known, one can derive similar formulae for these more complicated cases. The simplest and commonest example occurs when a variable is divided by another erroneous variable. For example, "wage rates" are often computed as the ratio of payroll to total man hours. To the extent that hours are measured with a multiplicative error, so will be also the resulting wage rates (but with opposite sign). In such contexts, the biases of (say) the estimated wage coefficient in a log-linear labor demand function will be towards $-1$ rather than zero.

The story is similar, though the algebra gets a bit more complicated, if the $z$'s are categorical or zero–one variables. In this case the errors arise from misclassification and the variance of the erroneously observed $x$ need not be higher than the variance of the true $z$. Bias formulae for such cases are presented in Aigner (1973) and Freeman (1984).

How does one deal with errors of measurement? As is well known, the standard EVM is not identified without the introduction of additional information, either in the form of additional data (replication and/or instrumental variables) or additional assumptions.

Procedures for estimation with known $\lambda$'s are outlined in Chapter 23. Occasionally we have access to "replicated" data, when the same question is asked on different occasions or from different observers, allowing us to estimate the variance of the "true" variable from the covariance between the different measures of the same concept. This type of an approach has been used in economics by Bowles (1972) and Borus and Nestel (1973) in adjusting estimates of parental background by comparing the reports of different family members about the same concept, and by Freeman (1984) on a union membership variable, based on a comparison of worker and employer reports. Combined with a modelling approach it has been pursued vigorously and successfully in sociology in the works of Bielby, Hauser, and Featherman (1977), Massagli and Hauser (1983), and Mare and Mason (1980). While there are difficulties with assuming a similar error variance on different occasions or for different observers, such assumptions can be relaxed within the framework of a larger model. This is indeed the most promising approach, one that brings in additional independent evidence about the actual magnitude of such errors.

Almost all other approaches can be thought of as finding a reasonable set of instrumental variables for the problem, variables that are likely to be correlated with the true underlying $z$, but not with either the measurement error $\varepsilon$ or the equation error (disturbance) $e$. One of the earlier and simpler applications of this approach was made by Griliches and Mason (1972) in estimating an earnings function and worrying about errors in their ability measure (AFQT test scores). In a "true" equation of the form

$$y = \alpha + \beta s + \gamma a + \delta x + e, \qquad (4.12)$$

where $y = $ log wages, $s = $ schooling, $a = $ ability, and $x = $ other variables, they substituted an observed test score $t$ for the unobserved ability variable and assumed that it was measured with random error: $t = a + \varepsilon$. They used then a set of background variables (parental status, regions of origin) as instrumental variables, the crucial assumption being that these background variables did not belong in this equation on their own accord. Chamberlain and Griliches (1975 and 1977) used "purged" information from the siblings of the respondents as instruments to identify their models (see also Chamberlain, 1971).

Various "grouping" methods of estimation, which use city averages (Friedman, 1957), industry averages (Pakes, 1983), or size class averages (Griliches and Ringstad, 1971), to "cancel out" the errors, can be all interpreted as using the classification framework as a set of instrumental dummy variables which are assumed to be correlated with differences in the underlying true values and uncorrelated with the random measurement errors or the transitory fluctuations.[7]

[7]Grouping methods that do not use an "outside" grouping criterion but are based on grouping on $x$ alone (or using its ranks as instruments) are not in general consistent and need not reduce the EV induced bias. (See Pakes, 1982).

The more complete MIMIC type models (Multiple indicators–multiple causes model, see Hauser and Goldberger, 1971) are basically full information versions of the instrumental variables approaches, with an attempt to gain efficiency by specifying the complete system in greater detail and estimating jointly. In the Griliches–Mason example, such a model would consist of the following set of equations:

$$a = x\delta_1 + g,$$
$$t = a + \varepsilon, \qquad\qquad\qquad (4.13)$$
$$s = x\delta_2 + \gamma_1 a + v,$$
$$y = \beta s + \gamma_2 a + e,$$

where $a$ is an unobserved "ability" factor, and the "unique" disturbances $g$, $e$, $v$, and $\varepsilon$ are assumed all to be mutually uncorrelated. With enough distinct $x$'s and $\delta_1 \neq \delta_2$, this model is estimable either by instrumental variable methods or maximum likelihood methods. The maximum likelihood versions are equivalent to estimating the associated reduced form system:

$$t = x\delta_1 + g + \varepsilon,$$
$$s = x(\delta_2 + \gamma_1\delta_1) + \gamma_1 g + v, \qquad\qquad (4.14)$$
$$y = x[\delta_2 + (\gamma_1\beta + \gamma_2)\delta_1] + (\gamma_1\beta + \gamma_2)g + \beta v + e,$$

imposing the non-linear parameter restrictions across the equations and retrieving additional information about them from the variance–covariance matrix of the residuals, given the no-correlation assumption about the $\varepsilon$'s, $g$'s, $v$'s, and $e$'s. It is possible, for example, to retrieve an estimate of $\beta + \gamma_2/\gamma_1$ from the variance–covariance matrix and pool it with the estimates derived from the reduced form slope coefficients. In larger, more over-identified models, there are more binding restrictions connecting the variance–covariance matrix of the residuals with the slope parameter estimates. Chamberlain and Griliches (1975) used an expanded version of this type of model with sibling data, assuming that the unobserved ability variable has a variance-components structure. Aasness (1983) uses a similar framework and consumer expenditures survey data to estimate Engel functions and the unobserved distribution of total consumption.

All of these models rely on two key assumptions: (1) The original model $y = \alpha + \beta z + e$ is correct for all dimensions of the data. I.e. the $\beta$ parameter is stable and (2) The unobserved errors are uncorrelated in some well specified known dimension. In cross-sectional data it is common to assume that the $z$'s (the "true" values) and the $\varepsilon$'s (the measurement errors) are based on mutually independent draws from a particular population. It is not possible to maintain this assumption when one moves to time series data or to panel data (which are a cross-section of time series), at least as far as the $z$'s are concerned. Identification must hinge then on known differences in the covariance generating functions of the $z$'s and the $\varepsilon$'s. The simplest case is when the $\varepsilon$'s can be taken as white (i.e. uncorrelated over time) while the $z$'s are not. Then lagged $x$'s can be used as valid instruments to identify $\beta$. For example, the "contrast" estimator suggested by Karni and Weisman (1974) which combines the differentially biased level (plim $b = \beta - \beta\lambda$) and first difference estimators [plim $b_\Delta = \beta - \beta\lambda/(1-\rho)$] to derive consistent estimators for $\beta$ and $\lambda$, can be shown, for stationary $x$ and $y$, to be equivalent (asymptotically) to the use of lagged $x$'s as instruments.

While it may be difficult to maintain the hypothesis that errors of measurement are entirely white, there are many different interesting cases which still allow the identification of $\beta$. Such is the case if the errors can be thought of as a combination of a "permanent" error or misperception of or by individuals and a random independent over time error component. The first part can be encompassed in the usual "correlated" or "fixed" effects framework with the "within" measurement errors being white after all. Identification can be had then from contrasting the consequences of differencing over differing lengths of time. Different ways of differencing all sweep out the individual effects (real or errors) and leave us with the following kinds of bias formulae:

$$\text{plim } b_{1\Delta} \simeq \beta\left(1 - 2\sigma_v^2/s_{1\Delta}^2\right), \tag{4.15}$$
$$\text{plim } b_{2\Delta} \simeq \beta\left(1 - 2\sigma_v^2/s_{2\Delta}^2\right),$$

where $\sigma_v^2$ is the variance of the independent over time component of the $\varepsilon$'s, $1\Delta$ denotes the transformation $x_2 - x_1$ while $2\Delta$ indicates differences taken two periods apart: $x_3 - x_1$ and so forth, and the $s^2$'s are the respective variances of such differences in $x$. (4.15) can be solved to yield:

$$\hat{\beta} = \frac{\omega_{2\Delta} - \omega_{1\Delta}}{s_{2\Delta}^2 - s_{1\Delta}^2} \qquad \text{and} \qquad \hat{\sigma}_v^2 = \frac{(\hat{\beta} - b_{2\Delta})s_{2\Delta}^2}{2\hat{\beta}}, \tag{4.16}$$

where $\omega_{j\Delta}$ is the covariance of $j$ period differences in $y$ and $x$. This in turn, can be shown to be equivalent to using past and future $x$'s as instruments for the first differences.[8]

More generally, if one were willing to assume that the true $z$'s are non-stationary, which is not unreasonable for many evolving economic series, but the measurement errors, the $\varepsilon$'s, are stationary, then it is possible to use panel data to identify the parameters of interest even when the measurement errors are corre-

---

[8] See Griliches and Hausman (1984) for details, generalizations, and an empirical example.

lated over time.[9] Consider, for example, the simplest case of $T = 2$. The probability limit of the variance – covariance matrix between $y$ and $x$ is given by:

$$
\begin{array}{c|cc}
 & x_1 & x_2 \\
\hline
y_1 & \beta s_1^2 & \beta s_{12} \\
y_2 & \beta s_{21} & \beta s_2^2 \\
x_1 & s_1^2 + \sigma^2 & s_{12} + \rho\sigma^2 \\
x_1 & s_{21} + \rho\sigma^2 & s_2^2 + \sigma^2
\end{array}
\tag{4.17}
$$

where now $s_{\text{th}}$ stands for the variances and covariances of the true $z$'s, $\sigma^2$ is the variance of the $\varepsilon$'s, and $\rho$ is their first order correlation coefficient. It is obvious that if the $z$'s are non-stationary then $(\text{cov}\, y_1 x_1 - \text{cov}\, y_2 x_2)/(\text{var}\, x_1 - \text{var}\, x_2)$ and $(\text{cov}\, y_1 x_2 - \text{cov}\, y_2 x_1)/(\text{cov}\, x_1 x_2 - \text{cov}\, x_2 x_1)$ yield consistent estimates of $\beta$. In longer panels this approach can be extended to accommodate additional error correlations and the superimposition of "correlated effects" by using its first differences analogue.

Even if the $z$'s were stationary, it is always possible to handle the correlated errors case provided the correlation is *known*. This rarely is the case, but occasionally a problem can be put into this framework. For example, capital measures are often subject to measurement error but these errors cannot be taken as uncorrelated over time, since they are cumulated over time by the construction of such measures. But if one were willing to assume that the errors occur randomly in the measurement of investment and *they* are uncorrelated over time, *and* the weighting scheme (the depreciation rate) used in the construction of the capital stock measure is known, then the correlation between the errors in the stock levels is also known.

For example, if one is interested in estimating the rate of return to some capital concept, where the true equation is

$$
\pi_t = a + rK_t^* + e_t,
\tag{4.18}
$$

$\pi$ is a measure of profits and $K^*$ is defined as a geometrically weighted average of past true investments $I_t^*$:

$$
K_t^* = I_t^* + \lambda K_{t-1}^* = I_t^* + \lambda I_{t-1}^* + \lambda^2 I_{t-2}^* + \cdots,
\tag{4.19}
$$

but we do not observe $I_t^*$ or $K_t^*$ only

$$
I_t = I_t^* + \varepsilon_t,
\tag{4.20}
$$

[9] I am indebted to A. Pakes for this point.

where $\varepsilon_t$ is an i.i.d. error of measurement and the observed $K_t = \Sigma \lambda^i I_{t-1}$ is constructed from the erroneous $I$ series, then if $\lambda$ is taken as known, which is implicit in most studies that use such capital measures, instead of running versions of (4.18) involving $K_t$ and dealing with correlated measurement errors we can estimate

$$\pi_t - \lambda \pi_{t-1} = a(1-\lambda) + rI_t + u_t - \lambda u_{t-1} - r\varepsilon_t, \tag{4.21}$$

which is now in standard EVM form, and use lagged values of $I$ as instruments. Hausman and Watson (1983) use a similar approach to estimate the seasonality in the unemployment series by taking advantage of the known correlation in the measurement errors introduced by the particular structure of the sample design in their data.

One needs to reiterate, that in these kinds of models (as is also true for the rest of econometrics) the consistency of the final estimates depends *both* on the correctness of the assumed economic model and the correctness of the assumptions about the error structure.[10] We tend to focus here on the latter, but the former is probably more important. For example, in Friedman's (1957) classical permanent income consumption function model, the estimated elasticity of consumption with respect to income is a direct estimate of one minus the error ratio (the ratio of the variance of transitory income to the variance of measured income). But this conclusion is conditional on having assumed that the true elasticity of consumption with respect to permanent income is unity. If that is wrong, the first conclusion does not follow. Similarly in the profit–capital stock example above, we can do something because we have assumed that the true depreciation is both known and geometric. All our conclusions about the amount of error in the investment series are conditional on the correctness of these assumptions.

## 5. Missing observations and incomplete data

> This could but have happened once,
> And we missed it, lost it forever.
> Browning

Relative to our desires data can be and usually are incomplete in many different ways. Statisticians tend to distinguish between three types of "missingness": undercoverage, unit non-response, and item non-response (NAS, 1983). Undercoverage relates to sample design and the possibility that a certain fraction of the

---

[10] The usual assumption of normality of such measurement and response errors may not be tenable in many actual situations. See Ferber (1966) and Hamilton (1981) for empirical evidence on this point.

relevant population was excluded from the sample by design or accident. Unit non-response relates to the refusal of a unit or individual to respond to a questionnaire or interview or the inability of the interviewers to find it. Item non-response is the term associated with the more standard notion of missing data: questions unanswered, items not filled in, in a context of a larger survey or data collection effort. This term is usually applied to the situation where the responses are missing for only some fraction of the sample. If an item is missing entirely, then we are in the more familiar omitted variables case to which I shall return in the next section.

In this section I will concentrate on the case of partially missing data for some of the variables of interest. This problem has a long history in statistics and somewhat more limited history in econometrics. In statistics, most of the discussion has dealt with the *randomly missing*, or in newer terminology, *ignorable case* (see Rubin, 1976, and Little, 1982) where, roughly speaking, the desired parameters can be estimated consistently from the complete data subsets and "missing data" methods focus on using the rest of the available data to improve the efficiency of such estimates.

The major problem in econometrics is not just missing data but the possibility (or more accurately, probability) that they are missing for a variety of self-selection reasons. Such "behavioral missing" implies not only a loss of efficiency but also the possibility of serious bias in the estimated coefficients of models that do not take this into account. The recent revival of interest in econometrics in limited dependent variables models, sample-selection, and sample self-selection problems has provided both the theory and computational techniques for attacking this problem. Since this range of topics is taken up in Chapter 28, I will only allude to some of these issues as we go along. It is worth noting, however, that this area has been pioneered by econometricians (especially Amemiya and Heckman) with statisticians only recently beginning to follow in their footsteps (e.g. Little, 1983).

The main emphasis here will be on the no-self-selection ignorable case. It is of some interest, because these kinds of methods are widely used, and because it deals with the question of how one combines scraps of evidence and what one can learn from them. Consider a simple example where the true equation of interest is

$$y = \beta x + \gamma z + e, \tag{5.1}$$

where $e$ is a random term satisfying the usual OLS assumptions and the constant has been suppressed for notational ease. $\beta$ and $\gamma$ could be vectors and $x$ and $z$ could be matrices, but I will think of them at first as scalars and vectors respectively. For some fraction $\lambda[n_2/(n_1 + n_2)]$ of our sample we are missing observations (responses) on $x$. Let us rearrange the data and call the complete data sample $A$ and the incomplete sample $B$. Assume that it is possible to describe the data generating mechanism by the following model

$$d = 1 \quad \text{if} \quad g(x, z, m; \theta) + \varepsilon \geq 0,$$
$$d = 0 \quad \text{if} \quad g(x, z, m; \theta) + \varepsilon < 0, \tag{5.2}$$

where $d = 1$ implies that the observation is in set $A$, it is complete; $d = 0$ implies that $x$ is missing, $m$ is another variable(s) determining the response or sampling mechanism, $\theta$ is a set of parameters, and $\varepsilon$ is a random variable, distributed independently of $x$, $z$, and $m$. The incomplete data problem is *ignorable* if (1) $\varepsilon$ (and $m$) are distributed independently of $e$ and (2) there is no connection or restrictions between the parameters $\theta$ and $\beta$ and $\gamma$. If these conditions hold then one can estimate $\beta$ and $\gamma$ from the complete data subset $A$ and ignore $B$. Even if $\theta$ and $\beta$ and $\gamma$ are connected, if $\varepsilon$ and $e$ are independent, $\beta$ and $\gamma$ can be estimated consistently in $A$ but now some information is lost by ignoring the data generating process. (See Rubin, 1976 and Little, 1982 for more rigorous versions of such statements.)

Note that this notion of ignorability of the data generating mechanism is more general than the simpler notion of *randomly missing x's*. It does not require that the missing $x$'s be similar to the observed ones. Given the assumptions of the model (a constant $\beta$ irrespective of the level of $x$), the $x$'s can be missing "non-randomly," as long as the conditional expectation of $y$ given $x$ does not depend on which $x$'s are missing. For example, there is nothing especially wrong if all "high" $x$'s are missing, provided $e$ and $x$ are independent over the whole range of the data.

Even though with these assumptions $\beta$ and $\gamma$ can be estimated consistently in the $A$ subsample there is still some more information about them in sample $B$. The following questions arise then: (1) How much additional information is there in sample $B$ and about which parameters? (2) How should the missing values of $x$ be estimated (if at all)? What other information can be used to improve these estimates?[11]

Options include using only $z$, using $z$ and $y$, or using $z$ and $m$, where $m$ is an additional variable, related to $x$ but not appearing itself in the $y$ equation.

To discuss this, it is helpful to specify an "auxiliary" equation for $x$:

$$x = \delta z + \phi m + v, \tag{5.3}$$

where $E(v) = 0$ and $E(ve) = 0$. Note that as far as this equation is concerned, the missing data problem is one of missing the dependent variable for sub-sample $B$. If the probability of being present in the sample were related to the size of $v$, we

---

[11] This section borrows heavily from Griliches, Hall and Hausman (1978).

would be in the non-ignorable case as far as the estimation of $\delta$ and $\phi$ are concerned. Assume this is not the case and let us consider at first only the simplest case of $\phi = 0$, with no additional $m$ variables present.

One way of rewriting the model is then

$$y_a = \beta x_a + \gamma z_a + e_a,$$
$$x_a = \delta z_a + v_a,$$                                                                    (5.4)
$$y_b = (\beta + \gamma \delta) z_b + e_b + \beta v_b,$$

How one estimates $\beta$, $\gamma$, and $\delta$ depends on what one is willing to assume about the world that generated such data. There are two kinds of assumptions possible: The first is a "regression" approach, which assumes that the parameters which are constant across different subsamples are the slope coefficients $\beta$, $\gamma$, and $\delta$ but does not impose the restriction that $\sigma_v^2$ and $\sigma_e^2$ are the same across all the various subsamples. There can be heteroscedasticity across samples as long as it is independent from the parameters of interest. The second approach, the maximum likelihood approach, would assume that conditional on $z$, $y$ and $x$ are distributed normally and the missing data are a random sample from such a distribution. This implies that $\sigma_{e_a}^2 = \sigma_{e_b}^2$ and $\sigma_{v_a}^2 = \sigma_{v_b}^2$.

The first approach starts by recognizing that under the general assumptions of the model Sample $A$ yields consistent estimates of $\beta$, $\gamma$, and $\delta$ with variance covariance matrix $\Sigma_a$. Then a "first order" procedure, i.e., one that estimates missing $x$'s by $z$ alone and does not iterate, is equivalent to the following: Estimate $\hat{\beta}_a$, $\hat{\gamma}_a$, $\hat{\delta}_a$ from sample $A$, rewrite the $y$ equation as

$$\begin{pmatrix} y_a - \hat{\beta}_a x_a \\ y_b - \hat{\beta}_a \hat{\delta}_a z_b \end{pmatrix} = \gamma z + \begin{pmatrix} e_a \\ e_b + \beta v \end{pmatrix} + \varepsilon,$$                              (5.5)

where $\varepsilon$ involves terms which are due to the discrepancy between the estimated $\hat{\beta}$ and $\hat{\delta}$ and their true population values. Then just estimate $\gamma$ from this "completed" sample by OLS.

It is clear that this procedure results in no gain in the efficiency of $\beta$, since $\hat{\beta}_a$ is based solely on sample $A$. It is also clear that the resulting estimate of $\gamma$ could be improved somewhat using GLS instead of OLS.[12]

How much of a gain is there in estimating $\gamma$ this way? Let the size of sample $A$ be $N_1$ and of $B$ be $N_2$. The maximum (unattainable) gain in efficiency would be proportional to $(N_1 + N_2)/N_1$ (when $\sigma_v^2 = 0$). Ignoring the contribution of $\varepsilon$'s, which is unimportant in large samples, the asymptotic variance of $\gamma$ from the

[12] See Gourieroux and Monfort (1981).

sample as a whole would be

$$\text{Var}(\hat{\gamma}_{a+b}) \simeq \left[ N_1\sigma^2 + N_2(\sigma^2 + \beta_1^2\sigma_v^2) \right] / (N_1 + N_2)^2\sigma_z^2,$$

and

(5.6)

$$\text{Eff}(\hat{\gamma}_{a+b}) = \frac{\text{var}\,\hat{\gamma}_{a+b}}{\text{Var}(\hat{\gamma}_a)} \simeq (1-\lambda)\left(1 + \lambda\frac{\beta^2\sigma_v^2}{\sigma^2}\right),$$

where $\sigma^2 = \sigma_e^2$; and $\lambda = N_2/(N_1 + N_2)$. Hence efficiency will be improved as long as $\beta^2\sigma_v^2/\sigma^2 < 1/(1-\lambda)$, i.e. the unpredictable part of $x$ (unpredictable from $z$) is not too important relative to $\sigma^2$, the overall noise level in the $y$ equation.[13]

Let us look at a few illustrative calculations. In the work to be discussed below, $y$ will be the logarithm of the wage rate, $x$ is IQ, and $z$ is schooling. IQ scores are missing for about one-third of the sample, hence $\lambda = \frac{1}{3}$. But the "importance" of IQ in explaining wage rates is relatively small. Its independent contribution $(\beta^2\sigma_v^2)$ is small relative to the large unexplained variance in $y$. Typical numbers are $\beta = 0.005$, $\sigma_v = 12$, and $\sigma = 0.4$, implying

$$\text{Eff}(\hat{\gamma}_{a+b}) = 2/3\left[1 + \frac{1}{3}\frac{0.0036}{0.16}\right] = 0.672,$$

which is about equal to the $\frac{2}{3}$'s one would have gotten ignoring the terms in the brackets. Is this a big gain in efficiency? First, the efficiency (squared) metric may be wrong. A more relevant question is by how much can the standard error of $\gamma$ be reduced by incorporating sample $B$ into the analysis. By about 18 percent $(\sqrt{0.672} = 0.82)$ for these numbers. Is this much? That depends how large the standard error of $\gamma$ was to start out with. In Griliches, Hall and Hausman (1978) a sample consisting of about 1,500 individuals with complete information yielded an estimate of $\gamma_a = 0.0641$ with a standard error of 0.0052. Processing another 700 plus observations could reduce this standard error to 0.0043, an impressive but rather pointless exercise, since nothing of substance depends on knowing $\gamma$ within 0.001.

If IQ (or some other missing variable) were more important, the gain would be even smaller. For example, if the independent contribution of $x$ to $y$ were on the order of $\sigma^2$, then with one-third missing, $\text{Eff}(\gamma_{a+b}) \simeq \frac{8}{9}$, and the standard deviation of $\gamma$ would be reduced by only 5.7 percent. There would be no gain at all, if the missing variable was one and a half times as important as the disturbance [or more generally if $\beta^2\sigma_v^2/\sigma^2 > 1/(1-\lambda)$].

---

[13] Thus, remark 2 of Gourieroux and Monfort (1981, p. 583) is in error. The first-order method is not always more efficient. But an "appropriately weighted first-order method," GLS, will be more efficient. See Nijman and Palm (1985).

The efficiency of such estimates can be improved a bit more by allowing for the implied heteroscedasticity in these estimates and by iterating further across the samples. This is seen most clearly by noting that sample B yields an estimate of $\hat{\pi} = \beta + \gamma\delta$ with an estimated standard error $\sigma_{\pi}$. This information can be blended optimally with the sample $A$ estimates of $\beta$, $\gamma$, $\delta$, and $\Sigma_a$ using non-linear techniques and maximum likelihood is one way of doing this.

If additional variables which could be used to predict $x$ but which do not appear on their own accord in the $y$ equation were available, then there is also a possibility to improve the efficiency of the estimated $\beta$ and not just of $\gamma$. Again, unless these variables are very good predictors of $x$ and unless the amount of complete data available is relatively small, the gains in efficiency from such methods are unlikely to be impressive. (See Griliches, Hall and Hausman, 1978, and Haitovsky, 1968, for some illustrative calculations.)

The maximum likelihood approaches differ from the "first-order" ones by using also the dependent variable $y$ to "predict" the missing $x$'s, and by imposing restrictions on equality of the relevant variances across the samples. The latter assumption is not usually made or required by the first order methods, but follows from the underlying likelihood assumption that conditional on $z$, $x$ and $y$ are jointly normally (or some other known distributions) distributed, and that the missing values are missing at random. In the simple case where only one variable is missing (or several variables are missing at exactly the same places), the joint likelihood connecting $y$ and $x$ to $z$, which is based on the two equations

$$y = \beta x + \gamma z + e,$$
$$x = \delta z + v, \tag{5.7}$$

with $Ee = \sigma^2$, $Ev^2 = \eta^2$, $Eev = 0$ can be rewritten in terms of the marginal distribution function of $y$ given $z$, and the conditional distribution function of $x$ given $y$ and $z$, with corresponding equations:

$$y = cz + u,$$
$$x = dy + fz + w, \tag{5.8}$$

and $Eu^2 = g^2$, $Ew^2 = h^2$, $Ewu = 0$. Given the normality assumption, this is just another way of rewriting the same model, with the new parameters related to the old ones by

$$c = \gamma + \beta\delta, \qquad g^2 = \beta\eta^2 + \sigma^2,$$
$$d = \beta\eta^2/(\beta^2\eta^2 + \sigma^2), \qquad f = \delta - cd, \qquad h^2 = \eta^2\sigma^2/g^2. \tag{5.9}$$

In this simple case the likelihood factors and one can estimate $c$ and $g^2$ from the

Table 1
Earnings equations for NLS sisters: Various missing data estimators.[a]

| Estimation method | Y dependent | | T dependent | | |
|---|---|---|---|---|---|
| | S | T | S | $\sigma^2$ | $\eta^2$ |
| OLS on complete data sample N = 366 | 0.0434 (0.0109) | 0.00433 (0.00148) | 3.211 (0.398) | 0.1217 | 152.58 |
| *Total Sample*: N = 520 OLS with predicted IQ in missing portion* | 0.0423 (0.00916) | 0.00433 (0.00148) | | 0.1186 | |
| GLS with predicted IQ* | 0.0432 (0.00915) | 0.00433 (0.00148) | | | |
| Maximum Likelihood | 0.0427 (0.00912) | 0.00421 (0.00144) | 3.205 (0.346) | 0.1177 | 152.48 |

$Y$ = log of wage rate, $S$ = years of schooling completed, $T$ = IQ type test score.

*The standard errors are computed using the Gourieroux–Monfort (1982) formulae. All variables have been conditioned on age, region, race, and year dummy variables. The conditional moment matrices are:

| | Complete data ( N = 366) | | | Incomplete (154) | | |
|---|---|---|---|---|---|---|
| LW | 0.13488 | | | 0.12388 | | |
| IQ | 1.2936 | 187.71 | | — | — | |
| SC | 0.19749 | 11.0703 | 3.4476 | 0.23472 | — | 4.3408 |

[a] Data Source: The National Longitudinal Survey of Young Women (see Center for Human Resource Research, 1979).

complete sample; $d$, $f$, and $h^2$ from the incomplete sample and solve back uniquely for the original parameters $\beta$, $\gamma$, $\delta$, $\sigma^2$, and $\eta^2$. In this way all of the information available in the data is used and computation is simple, since the two regressions ($y$ on $z$ in the whole sample and $x$ on $y$ and $z$ in the complete data portion) can be computed separately. Note, that while $x$ is implicitly "estimated" for the missing portion, no actual "predicted" value of $x$ are either computed or used in this framework.[14]

Table 1 illustrates the results of such computations when estimating a wage equation for a sample of young women from the National Longitudinal Survey, 30 percent of which were missing IQ data. The first row of the table gives

[14] Marini et al. (1980) describe such computations in the context of more than one set of variables missing in a nested pattern.

estimates computed solely from the complete data subsample. The second one uses the schooling variable to estimate the missing IQ values in the incomplete portion of the data and then re-computes the OLS estimates. The third row uses GLS, reweighting the incomplete portion of the data to allow for the increased imprecision due to the estimation of the missing IQ values. The last row reports the maximum likelihood estimates. All the estimates are very close to each other. Pooling the samples and "estimating" the missing IQ values increases the efficiency of the estimated schooling coefficient by 29 percent. Going to maximum likelihood adds another percentage point. While these gains are impressive, substantively not much more is learned from expanding the sample except that no special sample selectivity problem is caused by ignoring the missing data subset. The $\chi_2^2$ test for pooling yields the insignificant value of 0.8. That the samples are roughly similar, also can be seen from computing the biased schooling coefficient (ignoring IQ) in both matrices: it is equal to 0.057 (0.010) in the complete data subset and 0.054 in the incomplete one.

The maximum likelihood computations get more complicated when the likelihood does not factor as neatly as it does in the simple "nested" missing case. This happens in at least two important common cases: (1) If the model is overidentified then there are binding constraints between the $L(y|z, \theta_1)$ and $L(x|y, z, \theta_2)$ pieces of the overall likelihood function. For example, if we have an extra exogenous variable which can help predict $x$ but does not appear on its own in the "structural" $y$ equation, then there is a constraining relationship between the $\theta_1$ and $\theta_2$ parameters and maximum likelihood estimation will require iterating between the two. This is also the case for multi-equation systems where, say, $x$ is itself structurally endogenous because it is measured with error. (2) If the pattern of "missingness" is not nested, if observations on some variables are missing in a number of different patterns which cannot be arranged in a set of nested blocks, then one cannot factor the likelihood function conveniently and one must approach the problem of estimating it directly.

There are two related computational approaches to this problem: The first is the EM algorithm (Dempster et al., 1977). This is a general approach to maximum likelihood estimation where the problem is divided into an iterative two-step procedure. In the $E$-step (estimation), the missing values are estimated on the basis of the current parameter values of the model (in this case starting with all the available variances and covariances) and an $M$-step (maximization) in which maximum likelihood estimates of the model parameters are computed using the "completed" data set from the previous step. The new parameters are then used to solve again for the missing values which are then used in turn to reestimate the model, and this process is continued until convergence is achieved. While this procedure is easy to program, its convergence can be slow, and there are no easily available standard error estimates for the final results (though Beale and Little, 1975, indicate how they might be derived).

An alternative approach, which may be more attractive to model oriented econometricians and sociologists, given the assumption of ignorability of the process by which the data are missing, is to focus directly on pooling the available information from different portions of the sample which under the assumptions of the model are independent of each other. That is, the data are summarized by their relevant variance–covariance matrices (and means, if they are constrained by the model) and the model is expressed in terms of constraints on the elements of such matrices. What is done next is to "fit" the model to the observed matrices. This approach is based on the idea that for multivariate normally distributed random variables the observed moment matrix is a sufficient statistic. Many models can be written in the form $\Sigma(\theta)$, where $\Sigma$ is the true population covariance matrix associated with the assumed multivariate normal distribution and $\theta$ is a vector of parameters of interest. Denote the observed covariance matrix as $S$. Maximizing the likelihood function of the data with respect to the model parameters comes down to maximizing

$$\ln L(\Sigma|S,\theta) = k - \frac{n}{2}\left\{\ln|\Sigma(\theta)| + \operatorname{tr}\Sigma(\theta)^{-1}S\right\}, \tag{5.10}$$

with respect to $\theta$. If $\theta$ is exactly identified, the estimates are unique and can be solved directly from the definition of $\Sigma$ and the assumption that $S$ is a consistent estimator of it. If $\theta$ is over-identified, then the maximum likelihood procedure "fits" the model $\Sigma(\theta)$ to the data $S$ as best as possible. If the observed variables are multivariate normal this estimator is the Full Information Maximum Likelihood estimator for this model. Even if the data are not multivariate normal but follow some other distribution with $E(S|\theta) = \Sigma(\theta)$, this is a pseudo- or quasi-maximum likelihood estimator yielding a consistent $\hat{\theta}$.[15] The correctness of the computed standard errors will depend, however, on the validity of the normality assumption. Robust standard errors for this model can be computed using the approach of White.

There is no conceptual difficulty in generalizing this to a multiple sample situation where the resulting $\Sigma_j(\theta_j)$ may depend on somewhat different parameters. As long as these matrices can be taken as arising independently, their respective contributions to the likelihood function can be added up, and as long as the $\theta_j$'s have parameters in common, there is a return from estimating them jointly. This can be done either utilizing the multiple samples feature of LISREL-V (see Allison, 1981, and Joreskog and Sorbom, 1981) or by extending the MOMENTS program (Hall, 1979) to the connected-multiple matrices case. The estimation procedure combines these different matrices and their associated pieces of the likelihood function, and then iterates across them until a maximum is found. (See Bound, Griliches and Hall, 1984, for more exposition and examples.)

---

[15]See Van Praag (1983).

I will outline this type of approach in a somewhat more complex, multi-equation context: the estimation of earnings functions from sibling data while allowing for an unobserved ability measure and errors of measurement in the variable of interest – schooling. (See Griliches, 1974 and 1979 for an exposition of such models.) The simplest version of such a model can be written as follows:

$$t = a + e_1 = (f + g) + e_1,$$
$$s = \delta a + h + e_2 = \delta(f + g) + (w + v) + e_2, \tag{5.11}$$
$$y = \beta a + \lambda(s - e_2) + e_3 = \pi(f + g) + \gamma(w + v) + e_3,$$

where $t$ is a reported IQ-type test score, $s$ is the recorded years of school completed, and $y = \ln$ wage rate, is the logarithm of the wage rate on the current or last job, $a = (f + g)$ is an unobserved "ability" factor with $f$ being its "family" component. $h = (w + v)$ is the individual opportunity factor (above and beyond $a$ and hence assumed to be orthogonal to it), with $w$, "wealth," as its family component. The $e$'s are all random, uncorrelated and untransmitted measurement errors. That is

$$Eee' = \begin{pmatrix} \sigma_1^2 & 0 & 0 \\ 0 & \sigma_2^2 & 0 \\ 0 & 0 & \sigma_3^2 \end{pmatrix},$$

and $\pi = \beta + \gamma\delta$. In addition, it is convenient to define

$$\begin{array}{ll} \operatorname{Var} a = a^2, & \operatorname{Var} h = h^2, \\ \tau = \operatorname{Var} f / a^2, & \rho = \operatorname{Var} w / h^2, \end{array} \tag{5.12}$$

where $\tau$ and $\rho$ are the ratios of the variance of the family components to total variance in the $a$ and $h$ factors respectively.

Given these assumptions, the expected values of the variance–covariance matrix of all the observed variables across both members of a sib-pair is given by

| | $t_1$ | $s_1$ | $y_1$ | $t_2$ | $s_2$ | $y_2$ |
|---|---|---|---|---|---|---|
| $t_1$ | $a^2 + \sigma_1^2$ | $\delta a^2$ | $\pi a^2$ | $\tau a^2$ | $\tau \delta a^2$ | $\tau \pi a^2$ |
| $s_1$ | | $\delta^2 a^2 + h^2 + \sigma_2^2$ | $\delta \pi a^2 + \gamma h^2$ | $\tau \delta^2 a^2 + \rho h^2$ | $\tau \delta \pi a^2 + \rho \gamma h^2$ |
| $y_1$ | | | $\pi^2 a^2 + \gamma^2 h^2 + \sigma_3^2$ | | $\tau \pi^2 a^2 + \rho \gamma^2 h^2$ |

$$\tag{5.13}$$

where only the 12 distinct terms of the overall $6 \times 6$ matrix are shown, since the others are derivable by symmetry and by the assumption that all the relevant variances (conditional on a set of exogenous variables) are the same across sibs. With 10 unknown parameters this model would be under-identified without sibling data. This type of model was estimated by Bound, Griliches and Hall (1984) using sibling data from the National Longitudinal Surveys of Young Men and Young Women.[16] They had to face, however, a very serious missing data problem since much of the data, especially test scores, were missing for one or both of the siblings. Data were complete for only 164 brothers pairs and 151 sister pairs but additional information subject to various patterns of "missingness" was available for 315 more male and 306 female siblings pairs and 2852 and 3398 unrelated male and female respondents respectively. Their final estimates were based on pooling the information from 15 different matrices for each sex and were used to test the hypothesis that the unobserved factors are the same for both males and females in the sense that their loading (coefficients) are similar in the male and female versions of the model and that the implied correlation between the male and female family components of these factors was close to unity. The latter test utilized the cross-sex cross-sib covariances arising from the brother-sister pairs ($N = 774$) in these panels.

Such pooling of data reduced the estimated standard errors of the major coefficients of interest by about 20 to 40 percent without changing the results significantly from those found solely in their "complete data" subsample. Their major substantive conclusion was that taking out the mean differences in wages between young males and females, one could not detect significant differences in the impact of the unobservables or in their patterns between the male and female portions of their samples. As far as the IQ-Schooling part of the model is concerned, families and the market appeared to be treating brothers and sisters identically.

A class of similar problems occurs in the time series context: missing data at some regular time intervals, the "construction" of quarterly data from annual data and data on related time series, and other "interpolation" type issues. Most of these can be tackled using adaptations of the methods described above, except for the fact that there is usually more information available on the missing values and it makes sense to adapt these methods to the structure of the specific problem. A major reference in this area is Chow and Lin (1971). More recent references are Harvey and Pierse (1982) and Palm and Nijman (1984).

## 6. Missing variables and incomplete models

> "Ask not what you can do to the data but rather what the data can do for you."

Every econometric study is incomplete. The stated model usually lists only the "major" variables of interest and even then it is unlikely to have good measures for all of the variables on the already foreshortened list. There are several ways in

---

[16] The cited paper uses a more detailed 4 equation model based on an additional "early" wage rate.

which econometricians have tried to cope with these facts of life: (1) Assume that the left-out components are random, minor, and independent of all the included exogenous variables. This throws the problem into the "disturbance" and leaves it there, except for possible considerations of heteroscedasticity, variance-components, and similar adjustments, which impinge only on the efficiency of the usual estimates and not on their consistency. In many contexts it is difficult, however, to maintain the fiction that the left-out-variables are unrelated to the included ones. One is pushed than into either, (2), a specification sensitivity analysis where the direction and magnitude of possible biases are explored using prior information, scraps of evidence, and the standard left-out-variable bias formulae (Griliches 1957 and Chapter 5) or (3) one tries to transform the data so as to minimize the impact of such biases.

In this section, I will concentrate on this third way of coping which has used the increasingly available panel data sets to try to get around some of these problems. Consider, then, the standard panel data set-up:

$$y_{it} = \alpha + \beta(i, t)x_{it} + \gamma(i, t)z_{it} + e_{it}, \tag{6.1}$$

where $y_{it}$ and $x_{it}$ are the observed dependent and "independent" variables respectively, $\beta$ is the set of parameters of interest, $z_{it}$ represents various possible misspecifications of the model in the form of left out variables, and $e_{it}$ are the usual random shocks assumed to be well behaved and independently distributed (at this level of generality almost all possible deviations from this can be accommodated by redefining the $z$'s). Two basic assumptions are made very early on in this type of model. The first one, that the relationship is linear, is already implicit in the way I have written (6.1). The second one is that the major parameters of interest, the $\beta$'s, are both stable over time and constant across individuals. I.e.,

$$\beta(i, t) = \beta. \tag{6.2}$$

Both of these assumptions are in principle testable, but are rarely questioned in practice. Unless there is some kind of stability in $\beta$, unless there is some interest in its central moments, it is not clear why one would engage in estimation at all. Since the longitudinal dimension of such data is usually quite short (2–10 years), it makes little sense to allow $\beta$ to change over time, unless one has a reasonably clear idea and a parsimonious parameterization of how such changes happen. (The fact that the $\beta$'s are just coefficients of a first order linear approximation to a more complicated functional relationship and hence *should* change as the level of $x$'s changes can be allowed for by expanding the list of $x$'s to contain higher order terms.)

The assumption that $\beta_i = \beta$, that all individuals respond alike (up to the additive terms, the $z_i$, which can differ across individuals), is one of the more bothersome ones. If longer time series were available, it would be possible to estimate separate $\beta_i$'s for each individual or firm. But that is not the world we find ourselves in at the moment. Right now there are basically three outs from this corner: (1) Assume that all differences in the $\beta_i$'s are random and uncorrelated with everything else. Then we are in the random coefficients world (Chapter 21) and except for issues of heteroscedasticity the problem goes away; (2) Specify a model for the differences in $\beta_i$, making them depend on additional observed variables, either own individual ones or higher-order macro ones (cf. Mundlak 1980). This results in defining a number of additional "interaction" variables with the $x$ set. Unless there is strong prior information on how they differ, this introduces an additional dimension to the "specification search" (in Leamer's terminology) and is not very promising; (3) Ignore it, which is what I shall proceed to do for the moment, focusing instead on the heterogeneity which is implicit in the potential existence of the $z_i$'s, the ignored or unavailable variables in the model.

Even if (6.1) is simplified to

$$y_{it} = \alpha + \beta x_{it} + \gamma_t z_{it} + e_{it} \tag{6.3}$$

$\beta$ is not identified from the data in the absence of direct observations on $z$. Somehow, assumptions have to be made about the source of the $z$'s and their distributional properties, before it is possible to derive consistent estimators of $\beta$. There are (at least) three categories of assumptions that can be made about such $z$'s which lead to different estimation approaches in this context: (a) The $z$'s are random and independent of $x$'s. This is the easy but not too likely case. The $z$'s can be collapsed then into the $e_i$'s with only the heteroscedasticity issue remaining for the "random effects" model to solve. (b) The $z$'s are correlated with the $x$'s but are constant over time and have also constant effects on the $y$'s. I.e.,

$$\gamma(t) z_{it} = z_i, \tag{6.4}$$

where we have normalized $\gamma = 1$. This is the standard "fixed" or "correlated" effects model (see Maddala 1971, and Mundlak 1978) which has been extensively analyzed in the recent literature. This is the case for which the panel structure of the data provides a perfect solution. Letting each individual have its own mean level and expressing all the data as deviations from own means eliminates the $z$'s and leads to the use of "within" estimators.

$$y_{it} - \bar{y}_{i.} = \beta(x_{it} - \bar{x}_{i.}) + e_{it} - \bar{e}_{i.}, \tag{6.5}$$

where $\bar{y}_{i.} = (1/T)\sum_{t=1}^{T} y_{it}$, etc., and yields consistent estimates of $\beta$.

I have only two cautionary comments on this topic: As is true in many other contexts, and as was noted earlier, solving one problem may aggravate another. If there are two reasons for the $z_{it}$, e.g. both "fixed" effects and errors in variables, then

$$z_{it} = \alpha_i - \beta \varepsilon_{it}, \tag{6.6}$$

where $\alpha_i$ is the fixed individual effect and $\varepsilon_{it}$ is the random uncorrelated over time error of measurement in $x_{it}$. In this type of model $\alpha_i$ causes an upward bias in the estimated $\beta$ from pooled samples while $\varepsilon_{it}$ results in a negative one. Going "within" not only eliminates $\alpha_i$ but also increases the second type of bias through the reduction of the signal to noise ratio. This is seen easiest in the simplest panel model where $T = 2$ and within is equivalent to first differencing. Undifferenced, an OLS estimate of $\beta$ would yield

$$\text{plim}(\hat{\beta}_T - \beta) = b_{\alpha_i x} - \beta \lambda_T, \tag{6.7}$$

where $b_{\alpha_i x}$ is the auxiliary regression coefficient in the projection of the $\alpha_i$'s on the $x$'s, while $\lambda_T = \sigma_\varepsilon^2 / \sigma_x^2$ is the error variance ratio in $x$. Going "within", on the other hand, would eliminate the first term and leave us with

$$\text{plim}(\hat{\beta}_w - \beta) = -\beta \lambda_w = -\beta \lambda_T / (1 - \rho), \tag{6.8}$$

where $\rho$ is the first order serial correlation coefficient of the $x$'s. A plausible example might have $\beta = 1$, $\beta_{\alpha_i x} = 0.2$, $\lambda_T = 0.1$, and $\hat{\beta}_T = 1 + 0.2 - 0.1 = 1.1$. Now, as might not be unreasonable, if $\rho = 0.67$, then $\lambda_w = 0.3$ and $\hat{\beta}_w = 0.7$, which is more biased than was the case with the original $\hat{\beta}_T$.

This is not an idle comment. Much of the recent work on production function estimation using panel data (e.g. see Griliches–Mairesse, 1984) starts out worrying about fixed effects and simultaneity bias, goes within, and winds up with rather unsatisfactory results (implausible low coefficients). Similarly, the rather dramatic reductions in the schooling coefficient in earnings equations achieved by analyzing "within" family data for MZ twins is also quite likely the result of originally rather minor errors of measurement in the schooling variable (see Griliches, 1979 for more detail).

The other comment has to do with the unavailability of the "within" solution if the equation is intrinsically non-linear since, for example, the mean of $e^x + \varepsilon$ is not equal to $e^{\bar{x}} + \bar{\varepsilon}$. This creates problems for models in which the dependent variables are outcomes of various non-linear probability processes. In special cases, it is possible to get around this problem by conditioning arguments. Chamberlain (1980) discusses the logit case while Hausman, Hall and Griliches (1984) show how conditioning on the sum of outcomes over the period as a whole converts a Poisson problem into a conditional multinominal logit problem and allows an equivalent "within" unit analysis.

(c) Non-constant effects. The general case here is one of a left out variable(s) and nothing much can be done about it unless more explicit assumptions are made about how the unseen variables behave and/or what their effects are. Solutions are available for special cases, cases that make restrictive enough assumptions on the $\gamma(t)z_{it}$ terms and their correlations with the included $x$ variables (see Hausman and Taylor, 1981).

For example, it is not too difficult to work out the relevant algebra for

$$\gamma(t)z_{it} = \gamma_t \cdot z_i, \tag{6.9}$$

or

$$\gamma(t)z_{it} = -\beta\varepsilon_{it}, \tag{6.10}$$

where $\varepsilon_{it}$ is an i.i.d. measurement error in $x$. The first version, eq. (6.9), is one of a "fixed" common effect with a changing influence over time. Such models have been considered by Stewart (1983) in the estimation of earnings function, by Pakes and Griliches (1984) for the estimation of geometric lag structures in panel data where the unseen truncation remainders decay exponentially over time, and by Anderson and Hsiao (1982) in the context of the estimation of dynamic equations with unobserved initial conditions. The second model, eq. (6.10), is the pure EVM in the panel data context and was discussed in Section IV. It is estimable by using lagged $x$'s as instruments, provided the "true" $x$'s are correlated over time, or by grouping methods if independent (of the errors) information is available which allows one to group the data into groups which differ in the underlying "true" $x$'s (Pakes, 1983). Identification may become problematic when the EVM is superimposed on the standard fixed effects model. Estimation is still possible, in principle, by first differencing to get rid of the $\alpha_i$'s, the fixed effects, and then using past and future $x$'s as instruments. (See Griliches and Hausman, 1984.)

Some of these issues can be illustrated by considering the problem of trying to estimate the form of a lag structure from a relatively short panel.[17] Let us define a flexible distributed lag equation

$$y_{it} = \alpha_i + \beta_0 x_{it} + \beta_1 x_{it-1} + \beta_2 x_{it-2} + \cdots + \varepsilon_{it},$$

$$y_{it} = \alpha_i + \overset{?}{\sum_{\tau=0}} \beta_\tau x_{it-\tau} + \varepsilon_{it}, \tag{6.11}$$

where the constancy of the $\beta$'s is imposed across individuals and across time. The empirical problem is how does one estimate, say, 9 $\beta$'s if one only has four to five

---

[17] The following discussion borrows heavily from Pakes and Griliches (1984).

years history on the $y$'s and $x$'s. In general this is impossible. If the length of the lag structure exceeds the available data, then the data cannot be informative about the unseen tail of the lag distribution without the imposition of stronger a priori restrictions. There are at least two ways of doing this: (a) We can assume something strong about the $\beta$'s. For example, that they decline geometrically after a few free terms, that $\beta_{\tau+1} = \lambda\beta_\tau$. This leads us back to the geometric lag case which we know more or less how to handle.[18] (b) We can assume something about the unseen $x$'s, that they were constant in the past (in which case we are back to the fixed effects with a changing coefficient case), or that they follow some simple low order autoregressive process (in which case their influence on the included $x$'s dies out after a few terms).

Before proceeding along these lines, it is useful to recall the notion of the $\Pi$-matrix, introduced in Chapter 22, which summarizes all the (linear) information contained in the standard time series – cross section panel model. This approach, due to Chamberlain (1982), starts with the set of unconstrained multivariate regressions, relating each year's $y_{it}$ to all of the available $x$'s, past, present, and future. Consider, for example, the case where data on $y$ are available for only three years $(T = 3)$ and on $x$'s for four. Then the $\Pi$ matrix consists of the coefficients in the following set of regressions:

$$y_{1i} = \pi_{13}x_{3i} + \pi_{12}x_{2i} + \pi_{11}x_{1i} + \pi_{10}x_{0i} + v_{1i},$$
$$y_{2i} = \pi_{23}x_{3i} + \pi_{22}x_{2i} + \pi_{21}x_{1i} + \pi_{20}x_{0i} + v_{2i}, \qquad (6.12)$$
$$y_{3i} = \pi_{33}x_{3i} + \pi_{32}x_{3i} + \pi_{31}x_{1i} + \pi_{30}x_{0i} + v_{3i},$$

where we have ignored constants to simplify matters. Now all that we know from our sample about the relationship of the $y$'s to the $x$'s is summarized in these $\pi$'s (or equivalently in the overall correlation matrix between all the $y$'s and the $x$'s), and any model that we shall want to fit will impose a set of constraints on it.[19]

A series of increasingly complex possible worlds can be written as:

a.  $y_{it} = \beta_0 x_{it} + \beta_1 x_{it-1} + e_{it},$

b.  $y_{it} = \beta_0 x_{it} + \beta_1 x_{it-1} + \alpha_i + e_{it},$

c.  $y_{it} = \beta_0 x_{it} + \beta_1(x_{it-1} + \lambda x_{it-2} + \lambda^2 x_{it-3} + \cdots) + e_{it},$

d.  $y_{it} = \beta_0 x_{it} + \beta_1(x_{it-1} + \lambda x_{it-2} + \lambda^2 x_{it-3} + \cdots) + \alpha_i + e_{it},$ $\qquad (6.13)$

e.  $y_{it} = \beta_0 x_{it} + \beta_1 x_{it-1} + \beta_2 x_{it-2} + \beta_3 x_{it-3} + \beta_4 x_{it-4} \cdots + e_{it},$
    $x_{it} = \rho x_{it-1} + \varepsilon_{it},$

f.  $y_{it} = \beta_0 x_{it} + \beta_1 x_{it-1} + \beta_2 x_{it-2} + \beta_3 x_{it-3} + \beta_4 x_{it-4} \cdots + \alpha_i + e_{it},$
    $x_{it} = k\alpha_i + \rho x_{it-1} + \varepsilon_{it},$

---

[18] See Anderson and Hsiao (1982) and Bhargava and Sargan (1983).

[19] There may be, of course, additional useful information in the separate correlation matrices between all of the $y$'s and all the $x$'s respectively.

going from the simple one lag, no fixed effects case (a) to the arbitrary lag structure with the one factor correlated effects structure (f). For each of these cases we can derive the expected value of $\Pi$. It is obvious that (a) implies

$$
\Pi(a) = \begin{pmatrix} 0 & 0 & \beta_0 & \beta_1 \\ 0 & \beta_0 & \beta_1 & 0 \\ \beta_0 & \beta_1 & 0 & 0 \end{pmatrix}.
$$

For the $b$ case, fixed effects with no lags, we need to define the wide sense least squares projection ($E^*$) of the unseen effects ($\alpha_i$) on all the available $x$'s

$$
E^*(\alpha_i | x_{0i} \cdots x_{3i}) = \delta_3 x_{3i} + \delta_2 x_{2i} + \delta_1 x_{1i} + \delta_0 x_{0i}. \tag{6.14}
$$

Then

$$
\Pi(b) = \begin{pmatrix} \delta_3 & \delta_2 & \delta_1 + \beta_0 & \delta_0 + \beta_1 \\ \delta_3 & \delta_2 + \beta_0 & \delta_1 + \beta_1 & \delta_0 \\ \delta_3 + \beta_0 & \delta_2 + \beta_1 & \delta_1 & \delta_0 \end{pmatrix}.
$$

To write down the $\Pi$ matrix for $c$, the geometric lag case, we rewrite (6.11) as

$$
\begin{aligned}
y_{1i} &= \beta_0 x_{1i} + \beta_1 x_{0i} + z_i + e_{1i}, \\
y_{2i} &= \beta_0 x_{2i} + \beta_1 x_{1i} + \beta_1 \lambda x_{0i} + \lambda z_i + e_{2i}, \\
y_{3i} &= \beta_0 x_{3i} + \beta_1 x_{2i} + \beta_1 \lambda x_{1i} + \beta_2 \lambda^2 x_{0i} + \lambda^2 z_i + e_{3i},
\end{aligned} \tag{6.15}
$$

and (6.14) as

$$
E^*(z_i | x) = m'x \tag{6.16}
$$

which gives us the $\Pi$ matrix corresponding to the geometric tail case

$$
\Pi(c) = \begin{pmatrix} m_3 & m_2 & m_1 + \beta_0 & m_0 + \beta_1 \\ \lambda m_3 & \lambda m_2 + \beta_0 & \lambda m_1 + \beta_1 & \lambda(m_0 + \beta_1) \\ \lambda^2 m_3 + \beta_0 & \lambda^2 m_2 + \beta_1 & \lambda^2 m_1 + \lambda \beta_1 & \lambda^2(m_0 + \beta_1) \end{pmatrix}.
$$

This imposes a set of non-linear constraints on the $\Pi$ matrix, but is estimable with standard non-linear multivariate regression software (in SAS or TSP). In this case we have seven unknown parameters to estimate (4 $m$'s, 2 $\beta$'s, and $\lambda$) from the 12 unconstrained $\Pi$ coefficients.[20]

Adding fixed effects on top of this, as in $d$, adds another four coefficients to be estimated and strains identification to its limit. This may be feasible with larger $T$ but the data are unlikely to distinguish well between fixed effects and slowly changing initial effects, especially in short panels.

Perhaps a more interesting version is represented by (6.13e), where we are unwilling to assume an explicit form for the lag distribution since that happens to be exactly the question we wish to investigate, but are willing instead to assume something restrictive about the behavior of the $x$'s in the unseen past; specifically that they follow an autoregressive process of low order. In the example sketched out, we never see $x_{-1}$, $x_{-2}$ and $x_{-3}$, and hence cannot identify $\beta_4$ (or even $\beta_3$) but may be able to learn something about $\beta_0$, $\beta_1$, and $\beta_2$. If the $x$'s follow a first order autoregressive process, then it can be shown (see Pakes and Griliches, 1984) that in the projection of $x_{-\tau}$ on all the observed $x$'s

$$E^*(x_{-\tau}|x_3, x_2, x_1, x_0) = g'x = 0 \cdot x_{3i} + 0 \cdot x_{2i} + 0 \cdot x_{1i} + g_\tau \cdot x_{0i}, \tag{6.17}$$

only the last coefficient is non-zero, since the partial correlation of $x_{-\tau}$ with all the subsequent $x$'s is zero, given its correlation with $x_0$. If the $x$'s had followed a higher order autoregression, say third order, then the last three coefficients would be non-zero. In the first order case the $\Pi$ matrix is

$$\Pi(e) = \begin{pmatrix} 0 & 0 & \beta_0 & \beta_1 + \beta_2 g_1 + \beta_3 g_2 + \beta_4 g_3 \\ 0 & \beta_0 & \beta_1 & \beta_2 + \beta_3 g_1 + \beta_4 g_2 \\ \beta_0 & \beta_1 & \beta_2 & \beta_3 + \beta_4 g_1 \end{pmatrix},$$

where now only $\beta_0$, $\beta_1$ and $\beta_2$ are identified from the data. Estimation proceeds by leaving the last column of $\Pi$ free and constraining the rest of it to yield the parameters of interest.[21] If we had assumed that the $x$'s are $AR(2)$, we would be able to identify only the first two $\beta$'s, and would have to leave the last two columns of $\Pi$ free.

---

[20]An alternative approach would take advantage of the geometric nature of the lag structure, and use lagged values of the dependent variable to solve out the unobserved $z_i$'s. Using the lagged dependent variables formulation would introduce both an errors-in-variables problem (since $y_{t-1}$ proxies for $z$ subject to the $e_{t-1}$ error) and a potential simultaneity problem due to their correlation with the $\alpha_i$'s (even if the $\alpha$'s are not correlated with the $x$'s). Instruments are available, however, in the form of past $y$'s and future $x$'s and such a system is estimable along the lines outlined by Bhargava and Sargan (1983).

[21]This is not fully efficient. If we really believe that the $x$'s follow a low order Markov process with stable coeffiecients over time (which is not necessary for the above), then the equations for $x$ can be appended to this model and the $g$'s would be estimated jointly, constraining this column of $\Pi$ also.

The last case to be considered, represents a mixture of fixed effects and truncated lag distributions. The algebra is somewhat tedious (see Pakes and Griliches, 1984) and leads basically to a mixture of the (c) and (e) case, where the fixed effects have changing coefficients over time, since their relationship to the correlated truncation remainder is changing over time:

$$
\Pi(f) =
\begin{bmatrix}
\delta_3 & \delta_2 & \delta_1 + \beta_0 & \Pi_{10} \\
m_2\delta_3 & m_2\delta_2 + \beta_0 & m_2\delta_1 + \beta_1 & \Pi_{20} \\
m_3\delta_3 + \beta_0 & m_3\delta_2 + \beta_1 & m_3\delta_1 + \beta_2 & \Pi_{30}
\end{bmatrix},
$$

where I have normalized $m_1 = 1$. The first three $\beta$'s should be identified in this model but in practice it may be rather hard to distinguish between all these parameters, unless $T$ is significantly larger than 3, the underlying samples are large, and the $x$'s are not too collinear.

Following Chamberlain, the basic procedure in this type of model is first to estimate the unconstrained version of the $\Pi$ matrix, derive its correct variance–covariance matrix allowing for the heteroscedasticity introduced by our having thrust those parts of the $\alpha_i$ and $z_i$ which are uncorrelated with the $x$'s into the random term (using the formulae in Chamberlain 1982, or White 1980), and then impose and test the constraints implied by the specific version deemed relevant.

Note that it is quite likely (in the context of larger $T$) that the test will reject all the constraints at conventional significance levels. This indicates that the underlying hypothesis of stability over time of the relevant coefficient may not really hold. Nevertheless, one may still use this framework to compare among several more constrained versions of the model to see whether the data indicate, for example, that "if you believe in a distributed lag model with fixed coefficients, then two terms are better than one."

Some of these ideas are illustrated in the following empirical example which considers the ubiquitous question of "capital." What is the appropriate way to define it and measure it? This is, of course, an old and much discussed question to which the theoretical answer is that in general it cannot be done in a satisfactory fashion (Fisher, 1969) and that in practice it depends very much on the purpose at hand (Griliches, 1963). There is no intention of reopening the whole debate here (see the various papers collected in Usher 1980 for a review of the recent state of this topic); the focus is rather on the much narrower question of what is the appropriate functional form for the depreciation or deterioration function used in the construction of conventional capital stock measures. Almost all of the data used empirically are constructed on the basis of conventional "length of life" assumptions developed for accounting and tax purposes and based on very little direct evidence on the pattern of capital services over time. These accounting estimates are then taken to imply rather sharp declines in the service flows of capital over time using either the straight line or double declining balance depreciation formulae. Whatever independent evidence there is on this topic comes largely from used assets markets and is heavily contaminated by the effects of obsolescence due to technical improvements in newer assets.

Pakes and Griliches (1984) present some direct empirical evidence on this question. In particular they asked: What is the time pattern of the contribution of past investments to current profitability? What is the shape of the "deterioration of services with age function" (rather than the "decline in present value" patterns)? All versions of capital stock measures can be thought of as weighted sums of past investments:

$$K_t = \sum w_\tau I_{t-\tau},$$ 
(6.18)

with $w_\tau$ differing according to the depreciation schemes used. Since investments are made to yield profits and assuming that ex ante the expected rate of return comes close to being equalized across different investments and firms, one would expect that

$$\Pi_t = \rho K_t + e_t = \rho \left( \sum w_\tau I_{t-\tau} \right) + e_t,$$ 
(6.19)

where $e_t$ is the ex post discrepancy between expected and actual profits assumed to be uncorrelated with the ex ante optimally chosen $I$'s. Given a series on $\Pi_t$ and $I_t$, in principle one could estimate all the $w$ parameters except for the problem that one rarely has a long enough series to estimate them individually, especially in the presence of rather high multi-collinearity in the $I$'s. Pakes and Griliches used panel data on U.S. firms to get around this problem, which greatly increases the available degrees of freedom. But even then, the available panel data are rather short in the time dimension (at least relative to the expected length of life of manufacturing capital) and hence some of the methods described above have to be used.

They used data on the gross profits of 258 U.S. manufacturing firms for the nine years 1964–72 and their gross investment (deflated) for 11, years 1961–71. Profits were deflated by an overall index of the average gross rate of return (1972 = 100) taken from Feldstein and Summers (1977) and all the observations were weighted inversely to the sum of investment over the whole 1961–71 period to adjust roughly for the great heteroscedasticity in this sample. Model (6.13f) of the previous section was used. That is, they tried to estimate as many unconstrained $w$ terms as possible asking whether these coefficients in fact decline as rapidly as is assumed by the standard depreciation formulae. To identify the model, it was assumed that in the unobserved past the $I$'s followed an autoregressive process. Preliminary calculations indicated that it was adequate to assume a third order autoregression for $I$. Since they had also an accounting measure of capital stock as of the beginning of 1961, it could be used as an additional indicator of the unseen past $I$'s. The possibility that more profitable firms may also invest more was allowed for by including individual firm effects in the model and allowing them to be correlated with the $I$'s and the initial $K$ level. The resulting set of multivariate regressions with non-linear constraints on coefficients and a free covariance matrix was estimated using the LISREL–V program of Joreskog and Sorbom (1981).

Before their results are examined a major reservation should be noted about this model and the approach used. It assumes a fixed and common lag structure (deterioration function) across both different time periods and different firms which is far from being realistic. This does not differ, however, from the common use of accounting or constructed capital measures to compute and compare "rates of return" across projects, firms, or industries. The way "capital" measures are commonly used in industrial organization, production function, finance, and other studies implicitly assumes that there is a stable relationship between earnings (gross or net) and past investments; that firms or industries differ only by a factor of proportionality in the yield on these investments, with the time shape of these yields being the same across firms and implicit in the assumed depreciation formula. The intent of the Pakes–Griliches study was to question only the basic shape of this formula rather than try to unravel the whole tangle at once.

Their main results are presented in Table 2 and can be summarized quickly. There is no evidence that the contribution of past investments to current profits declines rapidly as is implied by the usual straight line or declining balance depreciation formula. If anything, they *rise* during the first three years! Introducing the 1961 stock as an additional indicator improves the estimates of the later $w$'s and indicates no noticeable decline in the contribution of past investments during their first seven years. Compared against a single traditional stock measure (column 3), this model does a significantly better job of explaining the variance of profits across firms and time. But it does not come close to doing as well as the estimates that correspond to the free $\Pi$ matrix, implying that such lag structures may not be stable across time and/or firms. Nevertheless, it is clear that the usual depreciation schemes which assume that the contribution of past investments declines rapidly and immediately with age are quite wrong. If anything, there may be an "appreciation" in the early years as investments are completed, shaken down, and adjusted to.[22]

---

[22] For a methodologically related study see Hall, Griliches and Hausman (1983) which tried to figure out whether there is a significant "tail" to the patents as a function of past R&D expenditures lag structure.

Table 2

The relationship of profits to past investment expenditures for U.S. manufacturing firms: Parameter estimates allowing for heterogeneity.*

| Parameter (standard error) | Without $k^g_{-2}$ | With $k^g_{-2}$ | Comparison model (system 10) | 3 years investment $+ k^n_{i,t-4}$ | 3 years investment $+ k^g_{i,t-4}$ |
|---|---|---|---|---|---|
| | (1) | (2) | (3) | (4) | (5) |
| $w_1$ | 0.067 | 0.068 | | 0.073 | 0.057 |
| | (0.028) | (0.027) | | (0.022) | (0.021) |
| $w_2$ | 0.115 | 0.112 | | 0.104 | 0.077 |
| | (0.033) | (0.032) | | (0.022) | (0.022) |
| $w_3$ | 0.224 | 0.222 | | 0.141 | 0.120 |
| | (0.041) | (0.040) | | (0.024) | (0.024) |
| $w_4$ | 0.172 | 0.208 | | | |
| | (0.046) | (0.046) | | | |
| $w_5$ | 0.072 | 0.198 | | | |
| | (0.049) | (0.050) | | | |
| $w_6$ | 0.096 | 0.277 | | | |
| | (0.062) | (0.057) | | | |
| $w_7$ | $-0.122$ | 0.202 | | | |
| | (0.094) | (0.076) | | | |
| $w_8$ | $-0.259$ | 0.087 | | | |
| | (0.133) | (0.103) | | | |
| Coefficient of: | | | | | |
| $k^n_{i,t}$ | | | 0.095 | | |
| | | | (0.012) | | |
| $k^n_{i,t-4}$ | | | | 0.103 | |
| | | | | (0.011) | |
| $k^g_{i,t-4}$ | | | | | 0.045 |
| | | | | | (0.006) |
| (Trace $\hat{\Omega}$)/253.6[a] | 1.18 | | 2.04 | 1.35 | 1.37 |

[a] $\hat{\Omega}$ = Estimated covariance matrix of the disturbances from the system of profit eqs. (across years). For the free $\Pi$ matrix: trace $\hat{\Omega} = 253.6$

* The dependent variable is gross operating income deflated by the implicit GNP deflator and an index of the overall rate of return in manufacturing (1972 = 1.0). The $w_\tau$ refer to the coefficients of gross investment expenditures in period $t - \tau$ deflated by the implicit GNP producer durable investment deflator. $k^n_{it}$ and $k^g_{it}$ are deflated Compustat measures of net and gross capital at the beginning of the year. $k^g_{-2}$ refers to undeflated gross capital in 1961 as reported by Compustat. All variables are divided by the square root of the firm's mean investment expenditures over the 1961–71 period. Dummy variables for the nine time periods are included in all equations. $N = 258$ and $T = 9$.

The overall fit, measured by $1 - (\text{trace } \hat{\Omega}/1208.4)$, $1208.4 = \sum_1^T s^2_{yt}$, where $s^2_{yt}$ is the sample variance in $y_t$, is 0.72 for the model in Column 2 as against 0.79 for the free $\Pi$ matrix.

*From*: Pakes and Griliches (1984).

## 7. Final remarks

> The dogs bark but the caravan keeps moving.
>
> A Russian proverb

Over 30 years ago Morgenstern (1950) asked whether economic data were accurate enough for the purposes that economists and econometricians were using them for. He raised serious doubts about the quality of many economic series and implicitly about the basis for the whole econometrics enterprise. Years have passed and there has been very little coherent response to his criticisms.

There are basically four responses to his criticism and each has some merit: (1) The data are not that bad. (2) The data are lousy but it does not matter. (3) The data are bad but we have learned how to live with them and adjust for their foibles. (4) That is all there is. It is the only game in town and we have to make the best of it.

There clearly has been great progress both in the quality and quantity of the available economic data. In the U.S. much of the agricultural statistical data collection has shifted from judgment surveys to probability based survey sampling. The commodity converge in the various official price indexes has been greatly expanded and much more attention is being paid to quality change and other comparability issues. Decades of criticisms and scrutiny of official statistics have borne some fruit. Also, some of the aggregate statistics have now much more extensive micro-data underpinnings. It is now routine, in the U.S., to collect large periodic labor force activity and related topics surveys and release the basic micro-data for detailed analysis with relatively short lags. But both the improvements in and the expansion of our data bases have not really disposed of the questions raised by Morgenstern. As new data appear, as new data collection methods are developed, the question of accuracy persists. While quality of some of the "central" data has improved, it is easy to replicate some of Morgenstern's horror stories even today. For example, in 1982 the U.S. trade deficit with Canada was either $12.8 or $7.9 billion depending on whether this number came from U.S. or Canadian publications. It is also clear that the national income statistics for some of the LDC's are more political than economic documents (Vernon, 1983).[23]

Morgenstern did not distinguish adequately between levels and rates of change. Many large discrepancies represent definitional differences and studies that are mostly interested in the movements in such series may be able to evade much of this problem. The tradition in econometrics of allowing for "constants" in most relationships and not over-interpreting them, allows implicitly for permanent

---

[23] See also Prakash (1974) for a collection of confidence shattering comparisons of measures of industrial growth and trade for various developing countries based on different sources.

"errors" in the levels of the various series. It is also the case that in much of economic analysis one is after relatively crude first order effects and these may be rather insensitive even to significant inaccuracies in the data. While this may be an adequate response with respect to much of the standard especially macro-economic analysis, it seems inadequate when we contemplate some of the more recent elaborate non-linear multi-equational models being estimated at the frontier of the subject. They are much more likely to be sensitive to errors and inconsistencies in the data.

In the recent decade there has been a revival of interest in "error" models in econometrics, though the progress in sociology on this topic seems more impressive. Recent studies using micro-data from labor force surveys, negative-tax experiments and similar data sources exhibit much more sensitivity to measurement error and sample selectivity problems. Even in the macro area there has been some progress (see de Leeuw and McKelvey, 1983) and the "rational expectations" wave has made researchers more aware of the discrepancy between observed data and the underlying forces that are presumably affecting behavior. All of this has yet to make a major dent on econometric textbooks and econometric teaching but there are signs that change is coming.[24] It is more visible in the areas of discrete variable analysis and sample selectivity issues, (e.g. note the publication of the Maddala (1983) and Manski–McFadden (1981) monographs) than in the errors of measurement area per se, but the increased attention that is devoted to data provenance in these contexts is likely to spill over into a more general data "aware" attitude.

One of the reasons why Morgenstern's accusations were brushed off was that they came from "outside" and did not seem sensitive to the real difficulties of data collection and data generation. In most contexts the data are imperfect not by design but because that is all there is. Empirical economists have over generations adopted the attitude that having bad data is better than having no data at all, that their task is to learn as much as is possible about how the world works from the unquestionably lousy data at hand. While it is useful to alert users to their various imperfections and pitfalls, the available economic statistics are our main window on economic behavior. In spite of the scratches and the persistent fogging, we cannot stop peering through it and trying to understand

---

[24] Theil (1978) devotes five pages out of 425 to this range of problems. Chow (1983) devotes only six pages out of 400 to this topic directly, but does return to it implicitly in the discussion of rational expectations models. Dhrymes (1974) does not mention it explicitly at all, though some of it is implicit in his discussion of factor analysis. Dhrymes (1978) does devote about 25 pages out of 500 to this topic. Maddala (1977) and Malinvaud (1980) devote separate chapters to the EVM, though in both cases these chapters represent a detour from the rest of the book. The most extensive textbook treatment of the EVM and related topics appears in a chapter by Judge et al. (1980). The only book that has some explicit discussion of economic data is Intriligator (1978). Except for the sample selection literature there is rarely any discussion of the processes that generate economic data and the resultant implications for econometric practice.

what is happening to us and to our environment, nor should we. The problematic quality of economic data presents a continuing challenge to econometricians. It should not cause us to despair, but we should not forget it either.

In this somewhat disjointed survey, I discussed first some of the long standing problems that arise in the encounter between the practicing econometrician and the data available to him. I then turned to the consideration of three data related topics in econometrics: errors of measurement, missing observations and incomplete data sets, and missing variables. The last topic overlapped somewhat with the chapter on panel analysis (Chapter 22), since the availability of longitudinal microdata has helped by providing us with one way of controlling for missing but relatively constant information on individuals and firms. It is difficult, however, to shake off the impression that here also, the progress of econometric theory and computing ability is outracing the increased availability of data and our understanding and ability to model economic behavior in increasing detail. While we tend to look at the newly available data as adding degrees of freedom grist to our computer mills, the increased detail often raises more questions than it answers. Particularly striking is the great variety of responses and differences in behavior across firms and individuals. Specifying additional distributions of unseen parameters rarely adds substance to the analysis. What is needed is a better understanding of the behavior of individuals, better theories and more and different variables. Unfortunately, standard economic theory deals with "representative" individuals and "big" questions and does not provide much help in explaining the production or hiring behavior of a particular plant at a particular time, at least not with the help of the available variables. Given that our theories, while couched in micro-language, are not truly micro-oriented, perhaps we should not be asking such questions. Then what are we doing with microdata? We should be using the newly available data sets to help us find out what is actually going on in the economy and in the sectors that we are analyzing without trying to force our puny models on them.[25] The real challenge is to try to stay open, to learn from the data, but also, at the same time, not drown in the individual detail. We have to keep looking for the forest among all these trees.

## References

Aasness, J. (1983) "Engle Functions, Distribution of Consumption and Errors in Variables". Paper presented at the European Meeting of the Econometric Society in Pisa, Oslo: Institute of Economics.

Aigner, D. J. (1973) "Regression with a Binary Independent Variable Subject to Errors of Observation", *Journal of Econometrics*, 17, 49–59.

[25]An important issue not discussed in this chapter is the testing of models which is a way of staying open and allowing the data to reject our stories about them. There is a wide range of possible tests that models can and should be subjected to. See, e.g. Chapters 5, 13, 14, 15, 18, 19, and 33 and Hausman (1978) and Hendry (1983).

Allison, P. D. (1981) "Maximal Likelihood Estimation in Linear Models When Data Are Missing", *Sociological Methodology*.

Anderson, T. W. and C. Hsiao (1982) "Formulation and Estimation of Dynamic Models Using Panel Data", *Journal of Econometrics*, 18(1), 47–82.

Beale, E. M. L. and R. J. A. Little (1975) "Missing Values in Multivariate Analysis", *Journal of the Royal Statistical Society, Ser. B.*, 37, 129–146.

Berkson, J. (1950) "Are There Two Regressions?", *Journal of the American Statistical Association*, 45, 164–180.

Bhargava, A. and D. Sargan (1983) "Estimating Dynamic Random Effects Models from Panel Data Coverning Short Time Periods", *Econometrica*, 51(6), 1635–1660.

Bielby, W. T., R. M. Hauser and D. L. Featherman (1977) "Response Errors of Non-Black Males in Models of the Stratification Process", in: Aigner and Goldberger, eds., *Latent Variables in Socioeconomic Models*. Amsterdam: North-Holland Publishing Company, 227–251.

Borus, M. E. (1982) "An Inventory of Longitudinal Data Sets of Interest to Economists", *Review of Public Data Use*, 10(1–2), 113–126.

Borus, M. E. and G. Nestel (1973) "Response Bias in Reports of Father's Education and Socioeconomic Status", *Journal of the American Statistical Association*, 68(344), 816–820.

Bound, J., Z. Griliches and B. H. Hall (1984) "Brothers and Sisters in the Family and Labor Market". NBER Working Paper No. 1476. Forthcoming in *International Economic Review*.

Bowles, S. (1972) "Schooling and Inequality from Generation to Generation", *Journal of Political Economy*, Part II, 80(3), S219–S251.

Center for Human Resource Research (1979) *The National Longitudinal Survey Handbook*. Columbus: Ohio State University.

Chamberlain, Gary (1977) "An Instrumental Variable Interpretation of Identification in Variance Components and MIMIC Models", Chapter 7, in: P. Taubman, ed., *Kinometrics*. Amsterdam: North-Holland Publishing Company, 235–254.

Chamberlain, Gary (1980) "Analysis of Covariance with Qualitative Data", *Review of Economic Studies*, 47(1), 225–238.

Chamberlain, Gary (1982) "Multivariate Regression Models for Panel Data", *Journal of Econometrics*, 18(1), 5–46.

Chamberlain, G. and Z. Griliches (1975) "Unobservables with a Variance-Components Structure: Ability, Schooling and the Economic Success of Brothers", *International Economic Review*, 16(2), 422–449.

Chamberlain, Gary (1977) "More on Brothers", in: P. Taubman, ed., *Kinometrics: Determinants of Socioeconomic Success Within and Between Families*. New York: North-Holland Publishing Company, 97–124.

Chow, G. C. (1983) *Econometrics*. New York: McGraw Hill.

Chow, G. C. and A. Lin (1971) "Best Linear Unbiased Interpolation, Distribution and Extrapolation of Time Series by Related Series", *Review of Economics and Statistics*, 53(4), 372–375.

Cole, R. (1969) *Error in Provisional Estimates of Gross National Product*. Studies in Business Cycles #21, New York: NBER.

Council on Wage and Price Stability (1977) *The Wholesale Price Index: Review and Evaluation*. Washington: Executive Office of the President.

Court, A. T. (1939) "Hedonic Price Indexes with Automotive Examples", in: *The Dynamics of Automobile Demand*. New York: General Motors Corporation, 99–117.

de Leeuw, F. and M. J. McKelvey (1983) "A 'True' Time Series and Its Indicators", *Journal of the American Statistical Association*, 78(381), 37–46.

Dempster, A. P., N. M. Laird and D. B. Rubin (1977) "Maximum Likelihood from Incomplete Data via the EM Algorithm", *Journal of the Royal Statistical Society, Ser. B*, 39(1), 1–38.

Dhrymes, P. J. (1974) *Econometrics*. New York: Springer-Verlag.

Dhrymes, P. J. (1978) *Introductory Econometrics*. New York: Springer-Verlag.

Diewert, W. E. (1980) "Aggregation Problems in the Measurement of Capital", in: D. Usher, ed., *The Measurement of Capital, Studies in Income and Wealth*. University of Chicago Press for NBER, 45, 433–538.

Eicker, F. (1967) "Limit Theorems for Regressions with Unequal and Dependent Errors", in: *Proceedings of the Fifth Berkeley Symposium on Mathematical Statistics and Probability*. Berkeley: University of California, Vol. 1.

Feldstein, M. and L. Summers (1977) "Is the Rate of Profit Falling?", *Brookings Papers on Economic Activity*, 211–227.

Ferber, R. (1966) "The Reliability of Consumer Surveys of Financial Holdings: Demand Deposits", *Journal of the American Statistical Association*, 61(313), 91–103.

Fisher, F. M. (1969) "The Existence of Aggregate Production Functions", *Econometrica*, 37(4), 553–577.

Fisher, F. M. (1980) "The Effect of Sample Specification Error on the Coefficients of 'Unaffected' Variables" in: L. R. Klein, M. Nerlove and S. C. Tsiang, eds., *Quantitative Economics and Development*. New York: Academic Press, 157–163.

Freeman, R. B. (1984) "Longitudinal Analyses of the Effects of Trade Unions", *Journal of Labor Economics*, 2(1), 1–26.

Friedman, M. (1957) *A Theory of the Consumption Function*. NBER General Series 63, Princeton: Princeton University Press.

Frisch, R. (1934) *Statistical Confluence Analysis by Means of Complete Regression Systems*. Oslo: University Economics Institute, Publication No. 5.

Gordon, R. J. (1982) "Energy Efficiency, User-Cost Change, and the Measurement of Durable Goods Prices", in: M. Foss, ed., NBER, *Studies in Income and Wealth,The U.S. National Income and Products Accounts*. Chicago: University of Chicago Press, 47, 205–268.

Gordon, R. J. (1985) The Measurement of Durable Goods Prices, unpublished manuscript.

Gourieroux, C. and A. Monfort (1981) "On the Problem of Missing Data in Linear Models", *Review of Economic Studies*, XLVIII(4), 579–586.

Griliches, Z. (1957) "Specification Bias in Estimates of Production Functions", *Journal of Farm Economics*, 39(1), 8–20.

Griliches, Z. (1961) "Hedonic Price Indexes for Automobiles: An Econometric Analysis of Quality Change", in: *The Price Statistics of the Federal Government*, NBER, 173–196.

Griliches, Z. (1963) "Capital Stock in Investment Functions: Some Problems of Concept and Measurement", in: Christ, et al., eds., *Measurement in Economics*. Studies in Mathematical Economics and Econometrics in Memory of Yehuda Grunfeld. Stanford: Stanford University Press, 115–137.

Griliches, Z. (1970) "Notes on the Role of Education in Production Functions and Growth Accounting", in: W. L. Hansen, ed., *Education, Income and Human Capital*. NBER *Studies in Income and Wealth*. 35, 71–127.

Griliches, Z. (1971) *Price Indexes and Quality Change*. Cambridge: Harvard University Press.

Griliches, Z. (1974) "Errors in Variables and Other Unobservables", *Econometrica*, 42(6), 971–998.

Griliches, Z. (1977) "Estimating the Returns to Schooling: Some Econometric Problems", *Econometrica*, 45(1), 1–22.

Griliches, Z. (1979) "Sibling Models and Data in Economics: Beginnings of a Survey", *Journal of Political Economy*, Part 2, 87(5), S37–S64.

Griliches, Z., B. H. Hall and J. A. Hausman (1978) "Missing Data and Self-Selection in Large Panels", *Annales de L'INSEE*, 30–31, 138–176.

Griliches, Z. and J. A. Hausman (1984) "Errors-in-Variables in Panel Data", NBER Technical Paper No. 37, forthcoming in *Journal of Econometrics*.

Griliches, Z. and J. Mairesse (1984) "Productivity and R&D at the Firm Level", in: Z. Griliches, ed., *R&D, Patents and Productivity*. NBER, Chicago: University of Chicago Press, 339–374.

Griliches, Z. and W. M. Mason (1972) "Education, Income and Ability", *Journal of Political Economy*, Part II, 80(3), S74–S103.

Griliches, Z. and V. Ringstad (1970) "Error in the Variables Bias in Non-Linear Contexts", *Econometrica*, 38(2), 368–370.

Griliches, Z. (1971) *Economies of Scale and the Form of the Production Function*. Amsterdam: North-Holland.

Haitovsky, Y. (1968) "Estimation of Regression Equations When a Block of Observations is Missing", ASA, *Proceedings of the Business and Economic Statistics Section*, 454–461.

Haitovsky, Y. (1972) "On Errors of Measurement in Regression Analysis in Economics", *International Statistical Review*, 40(1), 23–35.

Hall, B. H. (1979) *Moments: The Moment Matrix Processor User Manual*. Stanford: California.

Hall, B. H., Z. Griliches and J. A. Hausman (1983) "Patents and R&D: Is There A Lag Structure?". NBER Working Paper No. 1227.

Hamilton, L. C. (1981) "Self Reports of Academic Performance: Response Errors Are Not Well Behaved", *Sociological Methods and Research*, 10(2), 165–185.

Harvey, A. C. and R. G. Pierse (1982) "Estimating Missing Observations in Economic Time Series". London: London School of Economics Econometrics Programme Discussion Paper No. A33.

Hauser, R. M. and A. S. Goldberger (1971) "The Treatment of Unobservable Variables in Path Analysis", in: H. L. Costner, ed., *Sociological Methodology 1971*. San Francisco: Jossey–Bass, 81–117.

Hausman, J. A. (1978) "Specification Tests in Econometrics", *Econometrica*, 46(6), 1251–1271.

Hausman, J. A. (1982) "The Econometrics of Non Linear Budget Constraints", Fisher–Schultz Lecture given at the Dublin Meetings of the Econometric Society, *Econometrica*, forthcoming.

Hausman, J. A., B. H. Hall and Z. Griliches (1984) "Econometric Models for Count Data with Application to the Patents–R&D Relationship", *Econometrica*, 52(4), 909–938.

Hausman, J. A. and W. E. Taylor (1981) "Panel Data and Unobservable Individual Effects", *Econometrica*, 49(6), 1377–1398.

Hausman, J. A. and M. Watson (1983) "Seasonal Adjustment with Measurement Error Present". National Bureau of Economic Research Working Paper No. 1133.

Hausman, J. A. and D. Wise, eds. (1985) *Social Experimentation*. NBER, Chicago: University of Chicago Press, forthcoming.

Hendry, D. F. (1983) "Econometric Modelling: The 'Consumption Function' in Retrospect", *Scottish Journal of Political Economy*, 30, 193–220.

Intriligator, M. D. (1978) *Econometric Models, Techniques and Applications*. Englewood Cliffs: Prentice-Hall.

Joreskog, K. and D. Sorbom (1981) LISRELV, Analysis of Linear Structural Relationships by Maximum Likelihood and Least Squares Method. Chicago: National Educational Resources.

Judge, G. G., W. R. Griffiths, R. C. Hill and T. C. Lee (1980) *The Theory and Practice of Econometrics*. New York: Wiley.

Karni, E. and I. Weissman (1974) "A Consistent Estimator of the Slope in a Regression Model with Errors in the Variables", *Journal of the American Statistical Association*, 69(345), 211–213.

Klepper, S. and E. E. Leamer (1983) "Consistent Sets of Estimates for Regressions with Errors in All Variables", *Econometrica*, 52(1), 163–184.

Kruskal, W. H. and L. G. Telser (1960) "Food Prices and The Bureau of Labor Statistics", *Journal of Business*, 33(3), 258–285.

Kuznets, S. (1954) *National Income and Its Composition 1919–1938*. New York: NBER.

Kuznets, S. (1971) "Data for Quantitative Economic Analysis: Problems of Supply and Demand". Lecture delivered at the Federation of Swedish Industries. Stockholm: Kungl Boktryckeriet P. A. Norsted and Soner.

Little, R. J. A. (1979) "Maximum Likelihood Inference for Multiple Regressions with Missing Values: A Simulation Study", *Journal of the Royal Statistical Society*, Ser. B. 41(1), 76–87.

Little, R. J. A. (1983) "Superpopulation Models for Non-Response", in: Madow, Olkin and Rubin, eds., National Academy of Sciences, *Incomplete Data in Sample Surveys*. New York: Academic Press, Part VI, II, 337–413.

Little, R. J. A. (1982) "Models for Non-Reponse in Sample Surveys", *Journal of the American Statistical Association*, 77(378), 237–250.

MaCurdy, T. E. (1982) "The Use of Time Series Processes to Model the Error Structure of Earnings in Longitudinal Data Analysis", *Journal of Econometrics*, 18(1), 83–114.

Maddala, G. S. (1971) "The Use of Variance Components Models in Pooling Cross Section and Time Series Data", *Econometrica*, 39(2), 341–358.

Maddala, G. S. (1977) *Econometrics*. New York: McGraw Hill.

Maddala, G. S. (1983) *Limited-Dependent and Qualitative Variables in Econometrics*. Cambridge: Cambridge University Press.

Malinvaud, E. (1980) *Statistical Methods of Econometrics*. 3rd revised ed., Amsterdam: North-Holland.

Manski, C. F. and D. MacFadden, eds. (1981) *Structural Analysis of Discrete Data with Econometric Applications*. Cambridge: MIT Press.

Mare, R. D. and W. M. Mason (1980) "Children's Report of Parental Socioeconomic Status: A Multiple Group Measurement Model", *Sociological Methods and Research*, 9, 178–198.

Marini, M. M., A. R. Olsen and D. B. Rubin (1980) "Maximum-Likelihood Estimation in Panel Studies with Missing Data", *Sociological Methodology 1980*, 9, 315–357.

Massagli, M. P. and R. M. Hauser (1983) "Response Variability in Self- and Proxy Reports of Paternal and Filial Socioeconomic Characteristics", *American Journal of Sociology*, 89(2), 420–431.

Medoff, J. and K. Abraham (1980) "Experience, Performance, and Earnings", *Quarterly Journal of Economics*, XVC(4), 703–736.

Morgenstern, O. (1950) *On the Accuracy of Economic Observations*. Princeton: Princeton University Press, 2nd edition, 1963.

Mundlak, Y. (1978) "On the Pooling of Time Series and Cross Section Data", *Econometrica*, 46(1), 69–85.

Mundlak, Y. (1980) "Cross Country Comparisons of Agricultural Productivity". Unpublished manuscript.

National Academy of Sciences (1979) *Measurement and Interpretation of Productivity*. Washington, D.C.

National Academy of Sciences (1983) in: Madow, Olkin and Rubin, eds., *Incomplete Data in Sample Surveys*. New York: Academic Press, Vol. 1–3.

National Bureau of Economic Research (1961) *The Price Statistics of the Federal Government*, Report of the Price Statistic Review Committee, New York: General Series, No. 73.

National Bureau of Economic Research (1957a) *Studies in Income and Wealth, Problems of Capital Formation: Concepts, Measurement, and Controlling Factors*. New York: Arno Press, Vol. 19.

National Bureau of Economic Research (1957b) *Studies in Income and Wealth, Problems in International Comparisons of Economic Accounts*. New York: Arno Press, Vol. 20.

National Bureau of Economic Research (1958) *Studies in Income and Wealth, A Critique of the United States Income and Products Accounts*. New York: Arno Press, Vol. 22.

National Bureau of Economic Research (1961) *Studies in Income and Wealth, Output, Input and Productivity Measurement*. New York: NBER, Vol. 25.

National Bureau of Economic Research (1969) *Studies in Income and Wealth*, V. R. Fuchs, ed., *Production and Productivity in the Service Industries*. New York: Columbia University Press, Vol. 34.

National Bureau of Economic Research (1973) *Studies in Income and Wealth*, M. Moss, ed., *The Measurement of Economic and Social Performance*. New York: Columbia University Press, Vol. 38.

National Bureau of Economic Research (1983a) *Studies in Income and Wealth*, M. Foss, ed., *The U.S. National Income and Product Accounts*. Chicago: University of Chicago Press, Vol. 47.

National Bureau of Economic Research (1983b) *Studies in Income and Wealth*, J. Triplett, ed., *The Measurement of Labor Cost*. Chicago: University of Chicago Press, Vol. 48.

National Commission on Employment and Unemployment Statistics (1979) *Counting the Labor Force*. Washington: Government Printing Office.

Nijman, Th. E. and F. C. Palm (1985) "Consistent Estimation of a Regression Model with Incompletely Observed Exogenous Variable", Netherlands Central Bureau of Statistics, Unpublished paper.

Pakes, A. (1982) "On the Asymptotic Bias of Wald-Type Estimators of a Straight Line When Both Variables Are Subject to Error", *International Economic Review*, 23(2), 491–497.

Pakes, A. (1983) "On Group Effects and Errors in Variables in Aggregation", *Review of Economics and Statistics*, LXV(1), 168–172.

Pakes, A. and Z. Griliches (1984) "Estimating Distributed Lags in Short Panels with An Application to the Specification of Depreciation Patterns and Capital Stock Constructs", *Review of Economic Studies*, LI(2), 243–262.

Palm, F. C. and Th. E. Nijman (1984) "Missing Observations in the Dynamic Regression Model", *Econometrica*, November, 52(6), 1415–1436.

Prakash, V. (1974) "Statistical Indicators of Industrial Development: A Critique of the Basic Data". International Bank for Reconstruction and Development, DES Working Paper No. 189.

President's Committee to Appraise Employment and Unemployment Statistics (1962) *Measuring Employment and Unemployment*. Washington: Government Printing Office.

Rosen, S. (1974) "Hedonic Prices and Implicit Markets: Product Differentiation in Pure Competition", *Journal of Political Economy*, 82(1), 34–55.

Rubin, D. B. (1976) "Inference and Missing Data", *Biometrika*, 63(3), 581–592.

Ruggles, N. D. (1964) Review of O. Morgenstern, *On the Accuracy of Economic Observations*, 2nd edition, *American Economic Review*, LIV(4, part 1), 445–447.

Schultz, H. (1938) *The Theory and Measurement of Demand*. Chicago: University of Chicago Press.

Stewart, M. B. (1983) "The Estimation of Union Wage Differentials from Panel Data: The Problems of Not-So-Fixed Effects". Cambridge: National Bureau of Economic Research Conference on the Economics of Trade Unions, unpublished.

Stigler, G. J. and J. K. Kindahl (1970) *The Behaviour of Industrial Prices*, National Bureau of Economic Research, New York: Columbia University Press.

Theil, H. (1978) *Introduction to Econometrics.* Englewood Cliffs: Prentice Hall.

Triplett, J. E. (1975) "The Measurement of Inflation: A Survey of Research on the Accuracy of Price Indexes", in: P. H. Earl, ed., *Analysis of Inflation.* Lexington: Lexington Books, Chapter 2, 19–82.

Triplett, J. E. (1983) "An Essay on Labor Cost", in: National Bureau of Economic Research, *Studies in Income and Wealth, The Measurement of Labor Cost.* Chicago: University of Chicago Press, 49, 1–60.

U.S. Department of Commerce (1979) *Gross National Product Improvement Report.* Washington: Government Printing Office.

Usher, D., ed. (1980) *The Measurement of Capital*, National Bureau of Economic Research: *Studies in Income and Wealth.* Chicago: University of Chicago Press, Vol. 45.

Van Praag, B. (1983) "The Population-Sample Decomposition in Minimum Distance Estimation". Unpublished paper presented at the Harvard-MIT Econometrics seminar.

Vernon, R. (1983) "The Politics of Comparative National Statistics". Cambridge, Massachusetts, unpublished.

Waugh, F. V. (1928) "Quality Factors Influencing Vegetable Prices", *Journal of Farm Economics*, 10, 185–196.

White, H. (1980) "Using Least Squares to Approximate Unknown Regression Functions", *International Economic Review*, 21(1), 149–170.

Young, A. H. (1974) "Reliability of the Quarterly National Income and Product Accounts in the United States, 1947–1971", *Review of Income and Wealth*, 20(1), 1–39.

*Chapter 26*

# FUNCTIONAL FORMS IN ECONOMETRIC MODEL BUILDING*

LAWRENCE J. LAU

*Stanford University*


## Contents

*The author wishes to thank Kenneth Arrow, Erwin Diewert, Zvi Griliches, Dale Jorgenson and members of the Econometrics Seminar at the Department of Economics, Stanford University, for helpful comments and discussions. Financial support for this research under grant SOC77-11105 from the National Science Foundation is gratefully acknowledged. Responsibility for errors remains with the author.

*Handbook of Econometrics, Volume III, Edited by Z. Griliches and M.D. Intriligator*
© *Elsevier Science Publishers BV, 1986*

## 1.  Introduction

Econometrics is concerned with the estimation of relationships among observable (and sometimes even unobservable) variables. Any relationship to be estimated is almost always assumed to be stochastic. However, the relationship is often specified in such a way that it can be decomposed into a deterministic and a stochastic part. The deterministic part is often represented as a *known* algebraic function of observable variables and unknown parameters. A typical economic relationship to be estimated may take the form:

$$y = f(X; \alpha) + \varepsilon,$$

where $y$ is the observed value of the dependent variable, $X$ is the observed value of the vector of independent variables, $\alpha$ is a finite vector of unknown constant parameters and $\varepsilon$ is a stochastic disturbance term. The deterministic part, $f(X; \alpha)$, is supposed to be a known function. The functional form problem that we consider is the ex ante choice of the algebraic form of the function $f(X; \alpha)$ *prior* to the actual estimation. We ask: What considerations are relevant in the selection of one algebraic functional form over another, using only a priori information not specific to the particular data set?

This problem of ex ante choice of functional forms is to be carefully distinguished from that of ex post choice, that is, the selection of one functional form from among several that have been estimated from the same actual data set on the bases of the estimated results and/or post-sample predictive tests. The ex post choice problem belongs properly to the realm of specification analysis and hypothesis testing, including the testing of nonnested hypotheses.

We do not consider here the choice of functional forms in quantal choice analysis as the topic has been brilliantly covered by McFadden (1984) elsewhere in this *Handbook*. In our discussion of functional forms, we draw our examples largely from the empirical analyses of production and consumer demand because the restrictions implied by the respective theories on functional forms are richer. But the principles that we use are applicable more generally.

Historically, the first algebraic functional forms were chosen because of their ease of estimation. Almost always a functional form chosen is *linear in parameters*, after a transformation of the dependent variable if necessary. Thus, one specializes from

$$y = f(X; \alpha),$$

to

$$y = \sum_i f_i(X)\alpha_i,$$

or

$$g(y) = \sum_i f_i(X)\alpha_i,$$

where $g(\cdot)$ is a known monotonic transformation of a single variable. Moreover, it is often desirable to be able to identify the effect of each independent variable on the dependent variable separately. Thus, one specializes further to:

$$y = \sum_i f_i(X_i)\alpha_i,$$

or

$$g(y) = \sum_i f_i(X_i)\alpha_i,$$

so that $\alpha_i$ can be interpreted as the effect of a change in $X_i$ (or more precisely $f_i(X_i)$). Finally, for ease of computation and interpretation and for aesthetic reasons, the $f_i(\ )$'s are often chosen to be the same $f(\ )$, resulting in:

$$y = \sum_i f(X_i)\alpha_i, \tag{1.1}$$

or

$$g(y) = \sum_i f(X_i)\alpha_i. \tag{1.2}$$

An example of eq. (1.1) is the widely used linear functional form in which $f(X_i) = X_i$. An example of eq. (1.2) is the double-logarithmic functional form in which $g(y) = \ln y$ and $f(X_i) = \ln X_i$. It has the constant-elasticity property with the advantage that the parameters are independent of the units of measurement.

In addition, functional forms of the type in eqs. (1.1) and (1.2) may be interpreted as first-order approximations to any arbitrary function in a neighborhood of some $X = X_0$ and that is one reason why they have such wide currency.

However, linear functions, while they may approximate whatever underlying function reasonably well for small changes in the independent variables, frequently do not work very well for many others purposes. For example, as a production function, it implies perfect substitution among the different inputs and *constant* marginal products. It cannot represent the phenomenon of diminishing marginal returns. Moreover, the perfect substitution property of the linear production function has the unacceptable implication that almost always only a single input will be employed and an ever so slight change in the relative prices of inputs will cause a complete shift from one input to another.

Another linear-in-parameters functional form that was used is that of the Leontief or fixed-coefficients production function in its derived demand functions representation:

$$X_i = \alpha_i Y, \qquad i = 1, \ldots, m,$$

where $X_i$ is the quantity of the $i$th input, $i = 1, \ldots, m$ and $Y$ is the quantity of output. However, this production function implies zero substitution among the different inputs. No matter what the relative prices of inputs may be, the relative proportions of the inputs remain the same. This is obviously not a good functional form to use if one is interested in the study of substitution possibilities among inputs.

The first widely used production function that allows substitution is the Cobb–Douglas (1928) production function, which may be regarded as a special case of eq. (1.2):

$$\ln Y = \alpha_0 + \sum_i \alpha_i \ln X_i, \quad \sum_i \alpha_i = 1. \tag{1.3}$$

However, it should be noted that the Cobb–Douglas production function was discovered *not* from a priori reasoning but through a process of induction from the empirical data. Cobb and Douglas observed that labor's share of national income had been approximately constant over time and independent of the relative prices of capital and labor. They deduced, under the assumptions of constant returns to scale, perfect competition in the output and input markets, and profit maximization by the firms in the economy that the production function must take the form:

$$Y = AK^\alpha L^{(1-\alpha)}, \tag{1.4}$$

where $K$ and $L$ are the quantities of capital and labor respectively. Eq. (1.4) reduces to the form of eq. (1.3) by taking natural logarithms of both sides. The Cobb–Douglas production function became the principal work horse of empirical analyses of production until the early 1960s and is still widely used today.

The next advance in functional forms for production functions came when Arrow, Chenery, Minhas and Solow (1961) introduced the Constant-Elasticity-of-Substitution (C.E.S.) production function:

$$Y = \gamma [(1-\delta) K^{\rho} + \delta L^{\rho}]^{1/\rho}, \tag{1.5}$$

where $\gamma$, $\delta$ and $\rho$ are parameters. This function is not itself linear in parameters. However, it gives rise to average productivity relations which are linear in parameters after a monotonic transformation:

$$\ln \frac{Y}{L} = \alpha + \sigma \ln \frac{w}{p}, \tag{1.6}$$

$$\ln \frac{Y}{K} = \beta + \sigma \ln \frac{r}{p}, \tag{1.7}$$

where $p$, $r$, and $w$ are the prices of output, capital and labor respectively and $\alpha$, $\beta$ and $\sigma$ are parameters. The C.E.S. production function was discovered, again through a process of induction, when the estimated $\sigma$ from eq. (1.6) turned out to be different from one as one would have expected if the production function were actually of the Cobb–Douglas form.

Unfortunately, although the C.E.S. production function is more general than the Cobb–Douglas production function (which is itself a limiting case of the C.E.S. production function), and is perfectly adequate in the two-input case, its generalizations to the three or more-input case impose unreasonably severe restrictions on the substitution possibilities. [See, for example, Uzawa (1962) and McFadden (1963)]. In the meantime, interest in gross output technologies distinguishing such additional inputs as energy and raw materials continued to grow. Almost simultaneously advances in the computing technology lifted any constraint on the number of parameters that could reasonably be estimated. This led to the growth of the so-called "flexible" functional forms, including the generalized Leontief functional form introduced by Diewert (1971) and the transcendental logarithmic functional form introduced by Christensen, Jorgenson and Lau (1973). These functional forms share the common characteristics of linearity-in-parameters and the ability of providing second-order approximations to any arbitrary function. In essence they allow, in addition to the usual linear terms, as in eqs. (1.1) and (1.2), quadratic and interaction terms in the independent variables.

Here we study the problem of the ex ante choice of functional form when the true functional form is unknown. (Obviously, if the true functional form is known, we should use it.) We shall approach this problem by considering the relevant criteria for the selection of functional forms.

## 2.   Criteria for the selection of functional forms

What are some of the criteria that can be used to guide the ex ante selection of an algebraic functional form for a particular economic relationship? Neither economic theory nor available empirical knowledge provide, in general, a sufficiently complete specification of the economic functional relationship so as to determine its precise algebraic form. Consequently the econometrician has wide latitude in deciding which one of many possible algebraic functional forms to use in building an econometric model. Through practice over the years, however, a set of criteria has evolved and developed. These criteria can be broadly classified into five categories:
- (1) Theoretical consistency;
- (2) Domain of applicability;
- (3) Flexibility;
- (4) Computational facility; and
- (5) Factual conformity.

We shall discuss each of these criteria in turn.

### 2.1.   *Theoretical consistency*

Theoretical consistency means that the algebraic functional form chosen must be capable of possessing all of the theoretical properties required of that particular economic relationship for an appropriate choice of parameters. For example, a cost function of a cost-minimizing firm must be homogeneous of degree one, nondecreasing and concave in the prices of inputs, and nondecreasing in the quantity of output. Thus, any algebraic functional form selected to represent a cost function must be capable of possessing these properties for an appropriate choice of the parameters at least in a neighborhood of the prices of inputs and quantity of output of interest. For another example, a complete system of demand functions of a utility-maximizing consumer must be summable,[1] homogeneous of degree zero in the prices of commodities and income or total expenditure and have a Jacobian matrix which gives rise to a negative semidefinite and symmetric Slutsky substitution matrix. Thus, any algebraic functional form selected to represent a complete system of consumer demand functions must be capable of possessing these properties for an appropriate choice of the parameters at least in a neighborhood of the prices of commodities and income of interest.

Obviously, not all functional forms can meet these theoretical requirements, not even in a small neighborhood of the values of the independent variables of

---

[1]Summability means that the sum of expenditures on all commodities must be equal to income or total expenditure.

interest. However, a sufficiently large number of functional forms will satisfy the test of theoretical consistency, at least locally, that other criteria must be used to select one from among them. Moreover, many functional forms, while they may satisfy the theoretical consistency requirement, are in fact readily seen to be rather poor choices. For example, the cost function

$$C(p, Y) = Y \left[ \sum_{i=1}^{m} \alpha_i p_i \right],$$

where $p_i$ is the price of the $i$th input and $Y$ is the quantity of output and $\alpha_i > 0$, $i = 1, \ldots, m$, satisfies all the theoretical requirements of a cost function. It is homogeneous of degree one, nondecreasing and concave in the prices of inputs and nondecreasing in the quantity of output. However, it is not regarded as a good functional form in general because it allows no substitution among the inputs. The cost-minimizing demand functions corresponding to this cost function are given by Hotelling (1932)–Shephard (1953) Lemma as:

$$X_i = \frac{\partial C}{\partial p_i}(p, Y)$$

$$= \alpha_i Y, \qquad i = 1, \ldots, m.$$

Thus all inputs are employed in fixed proportions. While zero substitution or equivalently fixed proportions may be true for certain industries and processes, it is not an assumption that should be imposed a priori. Rather, the *data* should be allowed to indicate whether there is substitution among the inputs, which brings up the question of "flexibility" of a functional form to be considered below.

Yet sometimes considerations of theoretical consistency alone, even locally, can rule out many functional forms otherwise considered acceptable. This is demonstrated by way of the following two examples, one taken from the empirical analysis of producer behavior and one from consumer behavior.

First, we consider the system of derived demand functions of a cost-minimizing, price and output-taking firm with the constant-elasticity property:

$$\ln X_i = \alpha_i + \sum_{j=1}^{m} \beta_{ij} \ln p_j + \beta_{iY} \ln Y, \qquad i = 1, 2, \ldots, m \qquad (2.1)$$

where $X_i$ is the quantity demanded of the $i$th input, $p_j$ is the price of the $j$th input, and $Y$ is the quantity of output. The elasticities of demand with respect to own and cross prices and the quantity of output are all constants:

$$\frac{\partial \ln X_i}{\partial \ln p_j} = \beta_{ij}, \qquad i, j = 1, \ldots, m,$$

$$\frac{\partial \ln X_i}{\partial \ln Y} = \beta_{iY}, \qquad i = 1, \ldots, m.$$

Functional forms with constant elasticities as parameters are often selected over other functional forms with a similar degree of ease of estimation because the values of the parameters are then independent of the units of measurement of the variables. It can be readily verified that in the absence of further restrictions on the values of the parameters $\beta_{ij}$'s and $\beta_{iY}$'s, such a system of derived input demand functions is *flexible*, that is, it is capable of attaining any given value of $X$ (necessarily positive), $\partial X'/\partial p$ and $\partial X/\partial Y$ at any specified positive values of $p = \bar{p}$ and $Y = \bar{Y}$ through a suitable choice of the parameters $\beta_{ij}$'s and $\beta_{iY}$'s.

However, if it were required, in addition, that the system of derived demand functions in eq. (2.1) be consistent with cost-minimizing behavior on the part of the producer, at least in a neighborhood of the prices of input and the quantity of output, then certain restrictions must be satisfied by the parameters $\beta_{ij}$'s and $\beta_{iY}$'s. Specifically, the function:

$$C(p, Y) \equiv \sum_{i=1}^{m} \exp\left\{ \alpha_i + \sum_{j=1}^{m} (\beta_{ij} + \delta_{ij}) \ln p_j + \beta_{iY} \ln Y \right\}, \tag{2.2}$$

where $\delta_{ij} = \begin{cases} 1, & i = j \\ 0, & \text{otherwise} \end{cases}$

must have all the properties of a cost function *and* its partial derivatives with respect to $p_i$:

$$\frac{\partial C}{\partial p_i}(p, Y) = \sum_{k=1}^{m} \frac{(\beta_{ki} + \delta_{ki}) \exp\left\{ \alpha_k + \sum_{j=1}^{m} (\beta_{kj} + \delta_{kj}) \ln p_j + \beta_{kY} \ln Y \right\}}{p_i},$$

$$i = 1, \ldots, m, \tag{2.3}$$

must be identically equal to the original system of derived demand functions in eq. (2.1):

$$X_i = \exp\left\{ \alpha_i + \sum_{j=1}^{m} \beta_{ij} \ln p_j + \beta_{iY} \ln Y \right\}, \qquad i = 1, \ldots, m. \tag{2.4}$$

A cost function is homogeneous of degree one in the prices of inputs and the first-order partial derivative of a cost function with respect to the price of an input is therefore homogeneous of degree zero, implying:

$$\sum_{j=1}^{m} \beta_{ij} = 0, \qquad i = 1, \ldots, m. \tag{2.5}$$

A cost function is also concave in the prices of inputs, which implies:

$$\frac{\partial}{\partial p_i} X_i = \frac{\partial^2 C}{\partial p_i^2}$$

$$= \frac{\beta_{ii} \exp\left\{ \alpha_i + \sum_{j=1}^{m} \beta_{ij} \ln p_j + \beta_{iY} \ln Y \right\}}{p_i} \leq 0, \qquad i = 1, \ldots, m.$$

We conclude that $\beta_{ii} \leq 0$. Moreover, since the value of a second-order cross-partial derivative is independent of the order of differentiation wherever it exists,

$$\frac{\partial}{\partial p_j} \frac{\partial C}{\partial p_i}(p, Y) = \frac{\partial}{\partial p_i} \frac{\partial C}{\partial p_j}(p, Y), \qquad i \neq j, \quad i, j = 1, \ldots, m,$$

which implies:

$$\frac{\partial}{\partial p_j} X_i = \frac{\partial}{\partial p_i} X_j, \qquad i \neq j, \quad i, j = 1, \ldots, m. \tag{2.6}$$

Applying eq. (2.6) to eq. (2.4) yields:

$$\frac{\beta_{ij} \exp\left\{ \alpha_i + \sum_{k=1}^{m} \beta_{ik} \ln p_k + \beta_{iY} \ln Y \right\}}{p_j}$$

$$= \frac{\beta_{ji} \exp\left\{ \alpha_j + \sum_{k=1}^{m} \beta_{jk} \ln p_k + \beta_{jY} \ln Y \right\},}{p_i} \qquad i \neq j, \quad i, j = 1, \ldots, m. \tag{2.7}$$

There are three possible cases. First, $\beta_{ij} = \beta_{ji} = 0$, in which case each of the two inputs has a zero cross elasticity with respect to the price of the other input. Second, $\beta_{ij} > 0$ *and* $\beta_{ji} > 0$ (they cannot have opposite signs because of the positivity of the exponential function and the nonnegativity of prices), in which case the relative expenditures on the two inputs are constants *independent* of the prices of inputs and quantity of output, implying the following restrictions on the parameters:

$$\beta_{ik} - \beta_{jk} = 0, \qquad k \neq i, j; \quad k = 1, \ldots, m;$$
$$\beta_{ii} + 1 - \beta_{ji} = 0;$$
$$\beta_{ij} - (\beta_{jj} + 1) = 0; \tag{2.8}$$
$$\beta_{iY} - \beta_{jY} = 0,$$
$$\beta_{ij} e^{\alpha_i} - \beta_{ji} e^{\alpha_j} = 0.^2$$

We note that for this case, $(\beta_{ii} + 1) > 0$ and $(\beta_{jj} + 1) > 0$, implying that the own-price elasticities of the $i$th and $j$th inputs must be greater than minus unity (or less than unity in absolute value) – a significant restriction. We further note that if $\beta_{ik} \neq 0$ for some $k$, $k \neq i, j$, $\beta_{jk} \neq 0$ for the same $k$. But if $\beta_{ik} \neq 0$ *and* $\beta_{jk} \neq 0$ by eq. (2.7), $\beta_{ki} \neq 0$ and $p_i X_i / p_k X_k = \beta_{ki} / \beta_{ik}$, a constant, and hence the relative expenditures of all three inputs, $i$, $j$ and $k$, are constants. Moreover, the proportionality of expenditures implies that $\beta_{ii} + 1 - \beta_{ki} = 0$ for all $k$ such that $\beta_{ik} \neq 0$, $k \neq i$. Hence all $\beta_{ik}$'s, $k \neq i$, must have the *same* sign – positive, in this case. All $\beta_{ki}$'s, $k \neq i$, must have the *same* positive sign *and* magnitude. And $\beta_{iY} = \beta_{jY} = \beta_{kY}$.

By considering all the $i$'s it can be shown that the inputs are separable into $n$, $n \leq m$, mutually exclusive and jointly exhaustive groups such that

(1) Cross-price elasticities are zero between any two commodities belonging to different groups;
(2) Relative expenditures are constant within each group.

Such a system of derived demand functions corresponds to a cost function of the form:

$$C(p, Y) = \sum_{j=1}^{n} C_j(p^j, Y), \tag{2.9}$$

where $p^j$ is the vector of prices of the $j$th group of inputs and each $C_j(\ )$ has the form:

$$C_j(p^j, Y) = A_j \left( \prod_i p_i^{\alpha^{ji}} \right) Y^{\beta_j},$$

---

[2] This restriction results from setting the prices of all inputs and the quantity of output to unities.

where

$$A_j > 0; \quad \alpha_{ji} > 0, \, i; \quad \sum_i \alpha_{ji} = 1; \quad \beta_j > 0, \quad j = 1, \ldots, n.$$

Third, $\beta_{ij} < 0$ and $\beta_{ij} < 0$, in which case the relative expenditures on the two inputs are again constants *independent* of the prices of inputs and quantity of output, implying the same restrictions on the parameters as those in eq. (2.8). However, as derived earlier, all $\beta_{ik}$'s that are nonzero must have the *same* sign – negative, in this case. But then $\sum_{i=1}^{m} \beta_{ik}$ cannot be zero as required by zero degree homogeneity. We conclude that a cost function of the form in eq. (2.9) is the only possibility, with rather restrictive implications.

From this example we can see that the requirement of theoretical consistency, even locally, may impose very strong restrictions on an otherwise quite flexible functional form.

Second, we consider the complete system of demand functions of a utility-maximizing, budget-constrained consumer with the constant-elasticity property:[3]

$$\ln X_i = \alpha_i + \sum_{j=1}^{m} \beta_{ij} \ln p_j + \beta_{iM} \ln M, \quad i = 1, 2, \ldots, m; \tag{2.10}$$

where $X_i$ is the quantity demanded of the $i$th commodity, $p_j$ is the price of the $j$th commodity, and $M$ is income (or equivalently total expenditure). The elasticities of demand with respect to own and cross prices and to income are all constants:

$$\frac{\partial \ln X_i}{\partial \ln p_j} = \beta_{ij}, \quad i, j = 1, \ldots, m,$$

$$\frac{\partial \ln X_i}{\partial \ln M} = \beta_{iM}, \quad i = 1, \ldots, m.$$

This is also known as the double-logarithmic system of consumer demand functions. It can be readily verified that in the absence of further restrictions on the values of the parameters $\beta_{ij}$'s and $\beta_{iM}$'s, such a system of consumer demand functions is flexible, that is, it is capable of attaining any given value of $X$ (necessarily positive), $\partial X'/\partial p$ and $\partial X/\partial M$ at any specified positive values of $p = \bar{p}$ and $M = \bar{M}$ through a suitable choice of the parameters $\beta_{ij}$'s and $\beta_{iM}$'s.

However, if it were required, in addition, that the system of consumer demand functions in eq. (2.10) be consistent with utility-maximizing behavior on the part of the consumer, at least in a neighborhood of the prices of commodities and

---

[3] Such a system was employed by Schultz (1938), Wold with Jureen (1953) and Stone (1953).

income, it is necessary that the system of consumer demand functions satisfies summability, that is:

$$\sum_{i=1}^{m} p_i X_i = \sum_{i=1}^{m} \exp\left\{\alpha_i + \sum_{j=1}^{m} (\beta_{ij} + \delta_{ij})\ln p_j + \beta_{iM}\ln M\right\}$$

$$= M \qquad\qquad (2.11)$$

identically. It will be shown that (local) summability alone, through eq. (2.11), imposes strong restrictions on the parameters $\beta_{ij}$'s and $\beta_{iM}$'s.

By dividing both sides by $M$, eq. (2.11) can be transformed into:

$$\sum_{i=1}^{m} \exp\left\{\alpha_i + \sum_{j=1}^{m} (\beta_{ij} + \delta_{ij})\ln p_j + (\beta_{iM} - 1)\ln M\right\} = 1. \qquad (2.12)$$

Differentiating eq. (2.12) with respect to $\ln p_k$ twice, we obtain:

$$\sum_{i=1}^{m} (\beta_{ik} + \delta_{ik})^2 \exp\left\{\alpha_i + \sum_{j=1}^{m} (\beta_{ij} + \delta_{ij})\ln p_j + (\beta_{iM} - 1)\ln M\right\} = 0,$$

$$k = 1, \ldots, m. \qquad (2.13)$$

But

$$(\beta_{ik} + \delta_{ik})^2 \geq 0, \qquad i, k = 1, \ldots, m,$$

and

$$\exp\left\{\alpha_i + \sum_{j=1}^{m} (\beta_{ij} + \delta_{ij})\ln p_j + (\beta_{iM} - 1)\ln M\right\} > 0, \qquad i = 1, \ldots, m.$$

Thus, in order for the left-hand side of eq. (2.13) to be zero, one must have:

$$(\beta_{ik} + \delta_{ik}) = 0, \qquad i, k = 1, \ldots, m.$$

Differentiating eq. (2.12) with respect to $\ln M$ twice, we obtain:

$$\sum_{i=1}^{m} (\beta_{iM} - 1)^2 \exp\left\{\alpha_i + \sum_{j=1}^{m} (\beta_{ij} + \delta_{ij})\ln p_j + (\beta_{iM} - 1)\ln M\right\} = 0, \qquad (2.14)$$

which by a similar argument implies

$$(\beta_{iM} - 1) = 0, \qquad i = 1, \ldots, m.$$

We conclude that (local) summability alone implies that the system of consumer demand functions must take the form:

$$\ln X_i = \alpha_i - \ln p_i + \ln M_i, \qquad i = 1, \ldots, m; \qquad \sum_{i=1}^{m} e^{\alpha_i} = 1, \qquad (2.15)$$

which is no longer flexible.[4] For this system, the own-price elasticity is minus unity, the cross-price elasticities are zeroes, and the income elasticity is unity for the demand function of each and every commodity.

We conclude that theoretical consistency, even if applied only locally, can indeed impose strong restrictions on the admissible range of the values of the parameters of an algebraic functional form. It is essential in any empirical application to verify that the algebraic functional form remains reasonably flexible even under all the restrictions imposed by the theory. We shall return to the concept of "flexibility" in Section 2.3 below.

## 2.2. Domain of applicability

The domain of applicability of an algebraic functional form can refer to a number of different concepts. The most common usage of the domain of applicability refers to the set of values of the independent variables over which the algebraic functional form satisfies all the requirements for theoretical consistency. For example, for an algebraic functional form for a unit cost function $C(p; \alpha)$, where $\alpha$ is a vector of parameters, the domain of applicability of the algebraic functional form, for given $\alpha$, consists of the set

$$\{ p | p \geq 0;\ C(p; \alpha) \geq 0;\ \nabla C(p; \alpha) \geq 0;\ \nabla^2 C(p; \alpha) \text{ negative semidefinite} \}.$$

For an algebraic functional form for a complete system of consumer demand functions, $X(p, M; \alpha)$, the domain of applicability, for given $\alpha$, consists of the set

$$\{ p, M | p, M \geq 0;\ X(p, M; \alpha) \geq 0;$$

$$X(\lambda p, \lambda M; \alpha) = X(p, M; \alpha);\ \text{and}$$

the corresponding Slutsky substitution matrix being symmetric

and negative semidefinite $\}$.

---

[4] This result is well known. The proof here follows Jorgenson and Lau (1977) which contains a more general result.

We shall refer to this concept of the domain of applicability as the *extrapolative domain* since it is defined on the space of the *independent variables* with respect to a given value of the vector of parameters $\alpha$.

It would be ideal if the extrapolative domain of applicability consists of all nonnegative (or positive) prices in the case of a unit cost function or of all nonnegative (or positive) prices and incomes in the case of a complete system of consumer demand functions for any value of the vector of parameters $\alpha$. Unfortunately this is in general not the case.

The first question that needs to be examined is thus: for any algebraic functional form $f(X; \alpha)$, what is the set of $\alpha$ such that $f(X; \alpha)$ is theoretically consistent for the whole of the applicable domain? For an algebraic functional form for a unit cost function, the applicable domain is normally taken to be the set of all nonnegative (positive) prices of inputs.[5] For an algebraic functional form for a complete system of consumer demand functions, the applicable domain is normally taken to be the set of all nonnegative (positive) prices of commodities and incomes.[6] If, for given $\alpha$, the algebraic functional form $f(X; \alpha)$ is theoretically consistent over the whole of the applicable domain, it is said to be *globally* theoretically consistent or globally valid. For many functional forms, however, it may turn out that there is no such $\alpha$, such that $f(X; \alpha)$ is globally valid, or that the set of such admissible $\alpha$'s may be quite small relative to the set of possible $\alpha$'s. Only in very rare circumstances does the set of admissible $\alpha$'s coincide with the set of possible $\alpha$'s.

We have already encountered two examples in Section 2.1 in which the set of admissible values of the parameters that satisfy the requirements of theoretical consistency is a significantly reduced subset of the set of possible values of the parameters. For the system of constant-elasticity cost-minimizing input demand functions, the number of independent parameters is reduced from $m$(inputs)$\times (m + 2)(1\alpha_i; \ m\beta_{ij}$'s and $1\beta_{iY})$ parameters to at most $2m$ parameters by the requirements of local theoretical consistency. It may be verified, however, that under the stated restrictions on its parameters, the cost function in eq. (2.9) as well as the system of constant-elasticity input demand functions that may be derived from it, are globally valid. Similarly, for the complete system of constant-elasticity consumer demand functions, the number of independent parameters is reduced from $m$ (commodities)$\times(m + 2)$ $(1\alpha_i; \ m\beta_{ij}$'s and $1\beta_{iM})$ to $(m - 1)$ parameters by the requirements of local summability. It may be verified, however, that under the stated restrictions on its parameters (own-price elasticities of $-1$; cross-price elasticities of 0 and income elasticities of 1), the complete system of constant-elasticity consumer demand functions is globally valid.

---

[5] It is possible, and sometimes advisable, to take the applicable domain to be a compact convex subset of the set of all nonnegative prices.

[6] It is possible, and sometimes advisable, to take the applicable domain to be a compact convex subset of the set of all nonnegative prices and incomes.

These two examples share an interesting property – for given $\alpha$, if the algebraic functional form is locally valid, it is globally valid. This property, however, does not always hold. We shall consider two examples of unit cost functions – the generalized Leontief unit cost function introduced by Diewert (1971) and the transcendental logarithmic unit cost function introduced by Christensen, Jorgenson and Lau (1973).

The generalized Leontief unit cost function for a single-output, two-input technology takes the form:

$$C(p_1, p_2) = \alpha_0 p_1 + \alpha_1 p_1^{1/2} p_2^{1/2} + \alpha_2 p_2. \tag{2.16}$$

Local theoretical consistency requires that in a neighborhood of some price $(\bar{p}_1, \bar{p}_2)$,

$$C(\bar{p}_1, \bar{p}_2) \geq 0;$$

$$\nabla C(\bar{p}_1, \bar{p}_2) \geq 0; \tag{2.17}$$

$$\nabla^2 C(\bar{p}_1, \bar{p}_2) \text{ negative semidefinite.}$$

We note that a change in the units of measurement of the inputs leaves the values of the cost function and the expenditures unchanged. Without loss of generality, the price per unit of any input can be set equal to unity at any specified set of positive prices by a suitable change in the units of measurement. The parameters of the cost function, of course, must be appropriately rescaled. We therefore assume that the appropriate rescaling of the parameters have been done and take $(\bar{p}_1, \bar{p}_2)$ to be $(1,1)$. By direct computation:

$$C(1,1) = \alpha_0 + \alpha_1 + \alpha_2,$$

$$\nabla C(1,1) = \begin{bmatrix} \alpha_0 + \frac{1}{2}\alpha_1 \\ \alpha_2 + \frac{1}{2}\alpha_1 \end{bmatrix},$$

$$\nabla^2 C(1,1) = \begin{bmatrix} \dfrac{-\alpha_1}{4} & \dfrac{\alpha_1}{4} \\ \dfrac{\alpha_1}{4} & \dfrac{-\alpha_1}{4} \end{bmatrix}.$$

It is clear that by choosing $\alpha_1$ to be positive and sufficiently large all three conditions in eq. (2.17) can be strictly satisfied at $(1,1)$. We conclude that for local theoretical consistency $\alpha_1$ positive and sufficiently large is sufficient. (Actually $\alpha_1$ nonnegative is necessary.)

We shall now show that $\alpha_1$ positive and sufficiently large alone is not sufficient for global theoretical consistency. Global theoretical consistency requires that

$$C(p_1, p_2) = \alpha_0 p_1 + \alpha_1 p_1^{1/2} p_2^{1/2} + \alpha_2 p_2 \geq 0; \tag{2.18}$$

$$\nabla C(p_1, p_2) = \begin{bmatrix} \alpha_0 + \tfrac{1}{2}\alpha_1 p_1^{-1/2} p_2^{1/2} \\ \alpha_2 + \tfrac{1}{2}\alpha_1 p_1^{1/2} p_2^{-1/2} \end{bmatrix} \geq 0; \tag{2.19}$$

$$\nabla^2 C(p_1, p_2) = \alpha_1 \begin{bmatrix} -\tfrac{1}{4} p_1^{-3/2} p_2^{1/2} & \tfrac{1}{4} p_1^{-1/2} p_2^{-1/2} \\ \tfrac{1}{4} p_1^{-1/2} p_2^{-1/2} & -\tfrac{1}{4} p_1^{1/2} p_2^{-3/2} \end{bmatrix}, \text{ negative semidefinite;} \tag{2.20}$$

for all $p_1, p_2 \geq 0$.

First, note that as long as $\alpha_1 \geq 0$, negative semidefiniteness of the Hessian matrix of the unit cost function always holds. Second, if $\alpha_0 < 0$, then for sufficiently large $p_1$ and sufficiently small $p_2$, $\nabla C(p_1, p_2)$ will fail to be nonnegative. We conclude that for global monotonicity, $\alpha_0 \geq 0$ and similarly $\alpha_2 \geq 0$. If $\alpha_0$, $\alpha_1$ and $\alpha_2$ are all nonnegative, eq. (2.18) will be nonnegative for all nonnegative prices. We conclude that the restrictions

$$\alpha_0 \geq 0; \qquad \alpha_1 \geq 0; \qquad \alpha_2 \geq 0, \tag{2.21}$$

are necessary and sufficient for global theoretical consistency of the generalized Leontief unit cost function.

The transcendental logarithmic unit cost function for a single-output, two-input technology takes the form:

$$\ln C(p_1, p_2) = \alpha_0 + \alpha_1 \ln p_1 + (1 - \alpha_1) \ln p_2$$

$$+ \frac{\beta_{11}}{2} \ln p_1^2 - \beta_{11} \ln p_1 \ln p_2$$

$$+ \frac{\beta_{11}}{2} \ln p_2^2. \tag{2.22}$$

Local theoretical consistency at $(1,1)$ requires that:

$$C(1,1) = e^{\alpha_0} \geq 0,$$

$$\nabla C(1,1) = \begin{bmatrix} e^{\alpha_0} \alpha_1 \\ e^{\alpha_0}(1 - \alpha_1) \end{bmatrix} \geq 0, \tag{2.23}$$

$$\nabla^2 C(1,1) = e^{\alpha_0} \begin{bmatrix} \alpha_1(\alpha_1 - 1) + \beta_{11} & \alpha_1(1 - \alpha_1) - \beta_{11} \\ \alpha_1(1 - \alpha_1) - \beta_{11} & -(1 - \alpha_1)\alpha_1 + \beta_{11} \end{bmatrix}, \text{ negative semidefinite,}$$

$e^{\alpha_0}$ is always greater than zero. $1 \ge \alpha_1 \ge 0$ is necessary and sufficient for $\nabla C(1,1)$ to be nonnegative. $\alpha_1(\alpha_1 - 1) + \beta_{11} \le 0$ is necessary and sufficient for $\nabla^2 C(1,1)$ to be negative semidefinite. The set of necessary and sufficient restrictions on the parameters for local theoretical consistency at $(1,1)$ is therefore:

$$1 \ge \alpha_1 \ge 0; \qquad \alpha_1(\alpha_1 - 1) + \beta_{11} \le 0. \tag{2.24}$$

We shall now show that the conditions in eq. (2.24) are not sufficient for global theoretical consistency. Global theoretical consistency requires that

$$C(p_1, p_2) = \exp\left\{\alpha_0 + \alpha_1 \ln p_1 + (1 - \alpha)\ln p_2 + \frac{\beta_{11}}{2}\ln p_1^2 - \beta_{11}\ln p_1 \ln p_2\right.$$
$$\left. + \frac{\beta_{11}}{2}\ln p_2^2\right\} \ge 0 \tag{2.25}$$

$$\nabla C(p_1, p_2)' = C\left[\frac{\alpha_1 + \beta_{11}\ln p_1 - \beta_{11}\ln p_2}{p_1}\right.$$
$$\left. \times \frac{(1 - \alpha_1) - \beta_{11}\ln p_1 + \beta_{11}\ln p_2}{p_2}\right] \ge 0 \tag{2.26}$$

$$\frac{\partial^2 C}{\partial p_1^2} = \frac{C}{p_1^2}(\alpha_1 + \beta_{11}\ln p_1 - \beta_{11}\ln p_2)(\alpha_1 - 1 + \beta_{11}\ln p_1 - \beta_{11}\ln p_2)$$
$$+ \beta_{11} \le 0, \tag{2.27}$$

for all $p_1, p_2 > 0$.[7]

Equation (2.27) is necessary and sufficient for the negative semidefiniteness of $\nabla^2 C(p_1, p_2)$ because $C(p_1, p_2)$ is homogeneous of degree one. First, note that eq. (2.25) is always satisfied because of the positivity of the exponential function. Second, because the range of $\ln p_1$ (and $\ln p_2$) for positive prices is from minus infinity to infinity, no matter what the sign of $\beta_{11}$ may be, as long as it is nonzero, one can make $\ln p_1$ arbitrarily large (positive) or small (negative) by choosing $p_1$ to be arbitrarily large or small, and thus causing the nonnegativity of $\nabla C(p_1, p_2)$ to fail. Thus, for global monotonicity, $\beta_{11} = 0$. If $1 \ge \alpha_1 \ge 0$ and $\beta_{11} = 0$, eq. (2.27) reduces to:

$$\frac{\alpha_1(\alpha_1 - 1)}{p_1^2} \le 0,$$

which will always be satisfied. We conclude that the restrictions:

$$1 \ge \alpha_1 \ge 0; \qquad \beta_{11} = 0, \tag{2.28}$$

---

[7] The logarithmic function is not defined at 0.

are necessary and sufficient for global theoretical consistency of the transcendental logarithmic unit cost function.

We shall show later that under the necessary and sufficient restrictions for global theoretical consistency on their parameters both the generalized Leontief unit cost function and the transcendental logarithmic unit cost function lose their flexibility.

Having established that functional forms such as the generalized Leontief unit cost function and the transcendental logarithmic unit cost function can be globally valid only under relatively stringent restrictions on the parameters, but that they can be locally valid under relatively less stringent restrictions we turn our attention to a second question, namely, characterizing the domain of theoretical consistency for a functional form when it fails to be global.

As our first example, we consider again the generalized Leontief unit cost function. We note that $\alpha_1 \geq 0$ is a necessary condition for local theoretical consistency. Given $\alpha_1 \geq 0$, eq. (2.20) is identically satisfied. The set of prices of inputs over which the generalized Leontief unit cost function is theoretically consistent must satisfy:

$$C(p_1, p_2) = \alpha_0 p_1 + \alpha_1 p_1^{1/2} p_2^{1/2} + \alpha_2 p_2 \geq 0. \tag{2.29}$$

$$\nabla C(p_1, p_2) = \begin{bmatrix} \alpha_0 + \frac{1}{2}\alpha_1 p_1^{-1/2} p_2^{1/2} \\ \alpha_2 + \frac{1}{2}\alpha_1 p_1^{1/2} p_2^{-1/2} \end{bmatrix} \geq 0. \tag{2.30}$$

If eq. (2.30) holds, eq. (2.29) must hold because

$$C(p_1, p_2) = \nabla C(p_1, p_2) \cdot p.$$

We conclude that the domain of theoretical consistency consists of the set of prices which satisfy eq. (2.30). Eq. (2.30) can be rewritten as:

$$\begin{bmatrix} \alpha_0 & \frac{1}{2}\alpha_1 \\ \frac{1}{2}\alpha_1 & \alpha_2 \end{bmatrix} \begin{bmatrix} p_1^{1/2} \\ p_2^{1/2} \end{bmatrix} \geq 0. \tag{2.31}$$

Eq. (2.31) thus defines the domain of theoretical consistency of the generalized Leontief unit cost function. If $(1,1)$ were required to be in this domain then the additional restrictions of:

$$\begin{aligned} \alpha_0 + \tfrac{1}{2}\alpha_1 &\geq 0, \\ \alpha_2 + \tfrac{1}{2}\alpha_1 &\geq 0, \end{aligned} \tag{2.32}$$

must also be satisfied.

Next we consider the transcendental logarithmic unit cost function. We note that $1 \geq \alpha_1 \geq 0$ and $\alpha_1(\alpha_1 - 1) + \beta_{11} \leq 0$ are necessary conditions for theoretical consistency if $(1,1)$ were required to be in the domain. If $\beta_{11} \neq 0$, we have seen that the translog unit cost function cannot be globally theoretically consistent. We consider the cases of $\beta_{11} > 0$ and $\beta_{11} < 0$ separately. If $\beta_{11} > 0$, it can be shown that the domain of theoretical consistency is given by:

$$\exp\left\{\frac{1}{\beta_{11}}\left(\frac{1}{2} - \sqrt{\frac{1}{4} - \beta_{11}} - \alpha\right)\right\} \leq \frac{p_1}{p_2} \leq \exp\left\{\frac{1}{\beta_{11}}\left(\frac{1}{2} + \sqrt{\frac{1}{4} - \beta_{11}} - \alpha\right)\right\},$$

(2.33)

where $\frac{1}{4} \geq (1 - \alpha)\alpha \geq \beta_{11} > 0$. If $\beta_{11} < 0$, it can be shown that the domain of theoretical consistency is given by:

$$e^{(1-\alpha)/\beta_{11}} \leq \frac{p_1}{p_2} \leq e^{-\alpha/\beta_{11}}.^{8}$$

(2.34)

Our analysis shows that both the generalized Leontief and the translog unit cost functions cannot be globally theoretically consistent for all choices of parameters. However, even when global theoretical consistency fails, there is still a set of prices of inputs over which theoretical consistency holds and this set may well be large enough for all practical purposes. The question which arises here is that given neither functional form is guaranteed to be globally theoretically consistent, is there any objective criterion for choosing one over the other?

One approach that may provide a basis for comparison is the following: We can imagine each functional form to be attempting to mimic the values of $C$, $\nabla C$ and $\nabla^2 C$ at some arbitrarily chosen set of prices of inputs, say, without loss of generality, $(1,1)$. Once the values of $C$, $\nabla C$ and $\nabla^2 C$ are given, the unknown parameters of each functional form is determined. We can now investigate, holding $C$, $\nabla C$ and $\nabla^2 C$ constant, the domain of theoretical consistency of each functional form. If the domain of theoretical consistency of one functional form always contains the domain of theoretical consistency of the other, no matter what the values of $C$, $\nabla C$ and $\nabla^2 C$ are, we say that the first functional form dominates the second functional form in terms of extrapolative domain of applicability. In general, however, there may not be dominance and one functional form may have a larger domain of theoretical consistency for some values of $C$, $\nabla C$ and $\nabla^2 C$ and a smaller domain for other values.

We shall apply this approach to a comparison of the generalized Leontief and transcendental logarithmic unit cost functions in the single-output, two-input case.

[8] See Lau and Schaible (1984) for a derivation. See also Caves and Christensen (1980).

We choose $(1,1)$ to be the point of interpolation. We let

$$C(1,1) = 1,^9$$

$$\nabla C(1,1) = \begin{bmatrix} k_2 \\ 1 - k_2 \end{bmatrix}, \qquad (2.35)$$

and

$$\nabla^2 C(1,1) = \begin{bmatrix} -k_3 & k_3 \\ k_3 & -k_3 \end{bmatrix},$$

where $1 \geq k_2 \geq 0$ and $k_3 \geq 0$. Eq. (2.35) with $k_2$ and $k_3$ ranging through all of their admissible values represents all the theoretically consistent values that can possibly be attained by a unit cost function, its gradient and its Hessian matrix at $(1,1)$.

We need to establish the rules that relate the values of the parameters to the values of $C$, $\nabla C$, and $\nabla^2 C$ at $(1,1)$. We shall refer to such rules as the *rules of interpolation*. For the generalized Leontief unit cost function, the rules of interpolation are:

$$C(1,1) = 1 = \alpha_0 + \alpha_1 + \alpha_2,$$

$$\nabla C(1,1) = \begin{bmatrix} k_2 \\ 1 - k_2 \end{bmatrix} = \begin{bmatrix} \alpha_0 + \dfrac{\alpha_1}{2} \\ \alpha_2 + \dfrac{\alpha_1}{2} \end{bmatrix},$$

$$\nabla^2 C(1,1) = \begin{bmatrix} -k_3 & k_3 \\ k_3 & -k_3 \end{bmatrix} = \begin{bmatrix} \dfrac{-\alpha_1}{4} & \dfrac{\alpha_1}{4} \\ \dfrac{\alpha_1}{4} & \dfrac{-\alpha_1}{4} \end{bmatrix},$$

which imply:

$$\alpha_1 = 4k_3,$$
$$\alpha_0 = k_2 - 2k_3, \qquad (2.36)$$
$$\alpha_2 = (1 - k_2) - 2k_3.$$

It can be verified that $\alpha_0 + \alpha_1 + \alpha_2$ is indeed equal to unity. Thus, the generalized

---

[9] $C(1,1)$ may be set equal to any positive constant by an appropriate rescaling of all the parameters. We choose $C(1,1) = 1$ for the sake of convenience.

Leontief unit cost function may be rewritten in terms of $k_2$ and $k_3$ as:

$$C(p_1, p_2) = (k_2 - 2k_3)p_1 + 4k_3 p_1^{1/2}p_2^{1/2} + (1 - k_2 - 2k_3)p_2. \tag{2.37}$$

For the translog unit cost function, the rules of interpolation are:

$$C(1,1) = 1 = e^{\alpha_0},$$

$$\nabla C(1,1) = \begin{bmatrix} k_2 \\ 1 - k_2 \end{bmatrix} = \begin{bmatrix} \alpha_1 \\ (1 - \alpha_1) \end{bmatrix},$$

$$\nabla^2 C(1,1) = \begin{bmatrix} -k_3 & k_3 \\ k_3 & -k_3 \end{bmatrix} = \begin{bmatrix} \alpha_1(\alpha_1 - 1) + \beta_{11} & \alpha_1(1 - \alpha_1) - \beta_{11} \\ \alpha_1(1 - \alpha_1) - \beta_{11} & (\alpha_1 - 1)\alpha_1 + \beta_{11} \end{bmatrix},$$

which imply:

$$\alpha_0 = 0,$$
$$\alpha_1 = k_2, \tag{2.38}$$
$$\beta_{11} = -k_3 + k_2(1 - k_2).$$

Thus, the translog unit cost function may be rewritten as:

$$\ln C(p_1, p_2) = k_2 \ln p_1 + (1 - k_2)\ln p_2$$
$$+ \frac{[k_2(1 - k_2) - k_3]}{2}(\ln p_1)^2$$
$$- [k_2(1 - k_2) - k_3]\ln p_1 \ln p_2$$
$$+ \frac{[k_2(1 - k_2) - k_3]}{2}(\ln p_2)^2. \tag{2.39}$$

We can now compare the domains of theoretical consistency of the two functional forms holding $k_2$ and $k_3$ constant. For the generalized Leontief unit cost function, the domain of theoretical consistency is defined by eq. (2.31) as:

$$\begin{bmatrix} \alpha_0 & \dfrac{\alpha_1}{2} \\ \dfrac{\alpha_1}{2} & \alpha_2 \end{bmatrix} \begin{bmatrix} p_1^{1/2} \\ p_2^{1/2} \end{bmatrix} \geq 0,$$

or

$$\begin{bmatrix} k_2 - 2k_3 & 2k_3 \\ 2k_3 & (1 - k_2) - 2k_3 \end{bmatrix} \begin{bmatrix} p_1^{1/2} \\ p_2^{1/2} \end{bmatrix} \geq 0. \tag{2.40}$$

If $k_2 - 2k_3 \geq 0$ and $(1 - k_2) - 2k_3 \geq 0$, then the domain of theoretical consistency is the whole of the nonnegative orthant of $R^2$. If $k_2 - 2k_3 \geq 0$ and $(1 - k_2) - 2k_3 < 0$, then the domain of theoretical consistency is given by:

$$\frac{p_1}{p_2} \geq \left[ \frac{(1 - k_2) - 2k_3}{2k_3} \right]^2. \tag{2.41}$$

If $k_2 - 2k_3 < 0$ and $(1 - k_2) - 2k_3 \geq 0$, then the domain of theoretical consistency is given by:

$$\left( \frac{2k_3}{k_2 - 2k_3} \right)^2 \geq \frac{p_1}{p_2}. \tag{2.42}$$

Finally if $k_2 - 2k_3 < 0$ and $(1 - k_2) = 2k_3 < 0$, then the domain of theoretical consistency is given by:

$$\left( \frac{2k_3}{k_2 - 2k_3} \right)^2 \geq \frac{p_1}{p_2} \geq \left[ \frac{(1 - k_2) - 2k_3}{2k_3} \right]^2. \tag{2.43}$$

For the translog unit cost function, the domain of theoretical consistency is defined by eqs. (2.33) and (2.34). If $\beta_{11} = -k_3 + k_2(1 - k_2) = 0$, the domain of theoretical consistency is the whole of the positive orthant of $R^2$ (and may be uniquely extended to the whole of the nonnegative orthant of $R^2$). If $\beta_{11} = -k_3 + k_2(1 - k_2) > 0$, then the domain of theoretical consistency is given by:

$$\exp\left\{ \left( \tfrac{1}{2} + \sqrt{\tfrac{1}{4} - [k_2(1 - k_2) - k_3]} - k_2 \right) / [k_2(1 - k_2) - k_3] \right\} \geq \frac{p_1}{p_2}$$

$$\geq \exp\left\{ \left( \tfrac{1}{2} - \sqrt{\tfrac{1}{4} - [k_2(1 - k_2) - k_3]} - k_2 \right) / [k_2(1 - k_2) - k_3] \right\}. \tag{2.44}$$

If $\beta_{11} = -k_3 + k_2(1 - k_2) < 0$, then the domain of theoretical consistency is given by:

$$\exp\left\{ -k_2 / [k_2(1 - k_2) - k_3] \right\} \geq \frac{p_1}{p_2} \geq \exp\left\{ (1 - k_2) / [k_2(1 - k_2) - k_3'] \right\}. \tag{2.45}$$

With these formulas we can compare the domains of theoretical consistency for different values of $k_2$ and $k_3$ such that $1 \geq k_2 \geq 0$ and $k_3 \geq 0$. First, suppose $k_3 = 0$, then $k_2 - 2k_3 \geq 0$ and $(1 - k_2) - 2k_3 \geq 0$ and the domain of theoretical consistency for the generalized Leontief unit cost function is the whole of the nonnegative orthant of $R^2$. $k_3 = 0$ implies that $\beta_{11} = k_2(1 - k_2) \geq 0$. Thus, the domain of theoretical consistency for the translog unit cost function is given by:

$$\exp\left\{\left(\tfrac{1}{2} + \sqrt{\tfrac{1}{4} - k_2(1 - k_2)} - k_2\right)/k_2(1 - k_2)\right\} \geq \frac{p_1}{p_2}$$

$$\geq \exp\left\{\left(\tfrac{1}{2} - \sqrt{\tfrac{1}{4} - k_2(1 - k_2)} - k_2\right)/k_2(1 - k_2)\right\},$$

which is clearly smaller than the whole of the nonnegative orthant of $R^2$. We note that the maximum and minimum values of $k_2(1 - k_2)$ over the interval $[0, 1]$ is $\tfrac{1}{4}$ and $0$ respectively. Given $k_3 = 0$, if $k_2(1 - k_2) = 0$, $\beta_{11} = 0$, which implies that the domain of theoretical consistency is the whole of the nonnegative orthant of $R^2$. If $k_2(1 - k_2) = \tfrac{1}{4}$, $\beta_{11} = \tfrac{1}{4}$, and the domain of theoretical consistency reduces to a single ray through the origin defined by $p_1 = p_2$. If $k_2(1 - k_2) = \tfrac{2}{9}$, $(k_2 = \tfrac{1}{3})$, the domain of theoretical consistency is given by:

$$e^{3/2} = 4.48 \geq \frac{p_1}{p_2} \geq 1.$$

Overall, we can say that the domain of theoretical consistency of the translog unit cost function is not satisfactory for $k_3 = 0$.

Next suppose $k_3 = k_2(1 - k_2)$ (which implies that $k_3 \leq \tfrac{1}{4}$), then either

$$k_2 - 2k_3 = k_2 - 2k_2 + 2k_2^2$$
$$= k_2(2k_2 - 1) < 0,$$

or

$$(1 - k_2) - 2k_3 = (1 - k_2) - 2k_2(1 - k_2)$$
$$= (1 - k_2)(1 - 2k_2) < 0,$$

or

$$k_2 = \tfrac{1}{2}.$$

If $k_2 = \tfrac{1}{2}$, $k_3 = \tfrac{1}{4}$, and the domain of theoretical consistency of the generalized Leontief unit cost function remains the whole of the nonnegative orthant of $R^2$. However, if either of the first two cases is true (they cannot both be true), then the domain of theoretical consistency for the generalized Leontief unit cost function will be smaller than the whole of the nonnegative orthant of $R^2$. $k_3 = k_2(1 - k_2)$ implies that $\beta_{11} = 0$. Thus the domain of theoretical consistency for the translog unit cost function is the whole of the positive orthant of $R^2$. We conclude that neither functional form dominates the other. The cases of $k_3 = 0$ and $k_3 = k_2(1 - k_2)$ correspond approximately to the Leontief and Cobb–Douglas production functions respectively.

How do the two functional forms compare at some intermediate values of $k_2$ and $k_3$? Observe that the value of the elasticity of substitution at $(1,1)$ is given by:

$$\sigma(1,1) = \frac{C(1,1)C_{12}(1,1)}{C_1(1,1)C_2(1,1)},$$

$$= k_3 / [k_2(1 - k_2)].$$

If we let $k_2 = \frac{1}{3}$, $(1 - k_2) = \frac{2}{3}$, then $\sigma(1,1) = \frac{3}{4}$ is achieved at $k_3 = \frac{1}{6}$. At these values of $k_2$ and $k_3$, the domain of theoretical consistency of the generalized Leontief unit cost function is still the whole of the nonnegative orthant of $R^2$. At these values of $k_2$ and $k_3$, $\beta_{11} = -\frac{1}{6} + \frac{2}{9} = \frac{1}{18} > 0$. The domain of theoretical consistency of the translog unit cost function is given by:

$$56{,}233 \geq \frac{p_1}{p_2} \geq 0.0072.$$

We see that although it is short of the whole of the nonnegative orthant of $R^2$, for all practical purposes, the domain is large enough. Similarly $\sigma(1,1) = \frac{5}{4}$ is achieved at $k_3 = \frac{5}{18}$. At these values of $k_2$ and $k_3$, the domain of theoretical consistency of the generalized Leontief unit cost function is given by:

$$\left(\frac{25}{4}\right) \geq \frac{p_1}{p_2} \geq 0,$$

or $p_2$ cannot be more than $6\frac{1}{4}$ times greater than $p_1$. The domain of theoretical consistency of the translog unit cost function is given by:

$$e^6 = 403.4 \geq \frac{p_1}{p_2} \geq 0.000006.$$

We see that ignoring extremely small relative prices, the domain of theoretical consistency of the translog unit cost function is much larger than that of the generalized Leontief unit cost function.

The comparison of the domains of theoretical consistency of different functional forms for given values of $k_2$ and $k_3$ is a worthwhile enterprise and should be systematically extended to other functional forms and to the three or more-input cases. The lack of space does not permit an exhaustive analysis here. It suffices to note that the extrapolative domain of applicability does not often provide a clearcut criterion for the choice of functional forms in the absence of a priori information. Of course, if it is known a priori whether the elasticity of substitution is likely to be closer to zero or one a more appropriate choice can be made.

However, it is useful to consider a functional form $f(X; \alpha)$ as in turn a function $g(X; k) \equiv f(X; \alpha(k))$ where $\alpha(k)$ represents the rules of interpolation. If one can prespecify the set of $X$'s of interest, over which theoretical consistency must hold, one can then ask the question: What is the set of $k$'s such that a given functional form $f(X; \alpha(k)) \equiv g(X; k)$ will have a domain of theoretical consistency (in $X$) that contains the prespecified set of $X$'s. We can call this set of $k$'s the "interpolative domain" of the functional form. It characterizes the type of underlying behavior of the data for which a given functional form may be expected to perform satisfactorily.

## 2.3. Flexibility

Flexibility means the ability of the algebraic functional form to approximate arbitrary but theoretically consistent behavior through an appropriate choice of the parameters. The concept of flexibility, first introduced by Diewert (1973, 1974), is best illustrated with examples. First, we consider the cost function:

$$C(p, Y) = Y \left[ \sum_{i=1}^{m} \alpha_i p_i \right], \qquad \alpha_i > 0, \qquad i = 1, \ldots, m.$$

The derived demand functions are given by Hotelling (1932)–Shephard (1953) Lemma as:

$$X_i = \frac{\partial C}{\partial p_i} (p, Y) = \alpha_i Y, \qquad i = 1, \ldots, m.$$

The inputs are always employed in *fixed proportions*, whatever the values of $\alpha$ may be. Moreover, own *and* cross-price elasticities of all inputs are always zero! Thus, although the cost function satisfies the criterion of theoretical consistency, it cannot be considered "flexible" because it is incapable of approximating any theoretically consistent cost function satisfactorily through an appropriate choice of the parameters.[10] If we are interested in estimating the price elasticities of the derived demand for say labor or energy, we would not employ the linear cost function as an algebraic functional form because the price elasticities of demands that can be derived from such a cost function are by a priori assumption always zeroes.

---

[10] There is, of course, the question of what satisfactory approximation means, which is addressed below.

The degree of flexibility required of an algebraic functional form depends on the purpose at hand. In the empirical analysis of producer behavior, flexibility is generally taken to mean that the algebraic functional form used, be it a production function, a profit function, or a cost function, must be capable of generating output supply and input demand functions whose own and cross-price elasticities can assume arbitrary values subject only to the requirements of theoretical consistency at any arbitrarily given set of prices through an appropriate choice of the parameters. We can give a working definition of "flexibility" for an algebraic functional form for a unit cost function as follows:

*Definition*

An algebraic functional form for a unit cost function $C(p; \alpha)$ is said to be flexible if at any given set of nonnegative (positive) prices of inputs the parameters of the cost function, $\alpha$, can be chosen so that the derived unit-output input demand functions and their own and cross-price elasticities are capable of assuming arbitrary values at the given set of prices of inputs subject only to the requirements of theoretical consistency.[11]

More formally, let $C(p; \alpha)$ be an algebraic functional form for a unit cost function where $\alpha$ is a vector of unknown parameters. Then flexibility implies and is implied by the existence of a solution $\alpha(\bar{p}; \bar{C}, \bar{X}, \bar{S})$ to the following set of equations:

$$C(\bar{p},; \alpha) = \bar{C},$$

$$\nabla C(\bar{p}; \alpha) = \bar{X}, \qquad\qquad (2.46)$$

$$\nabla^2 C(\bar{p}; \alpha) = \bar{S},$$

for every nonnegative (positive) value of $\bar{p}$, $\bar{C}$ and $\bar{X}$ and negative semidefinite value of $\bar{S}$[12] such that $\bar{C} = \bar{p}\bar{X}$ and $\bar{S}\bar{p} = 0$. In other words, for every vector of prices of inputs $\bar{p}$, it is possible to choose the vector of parameters $\alpha$ so that at the given $\bar{p}$, the values of the unit cost function, its gradient and its Hessian matrix are equal to prespecified values of $\bar{C}$, $\bar{X}$ and $\bar{S}$ respectively.

An example of a flexible algebraic functional form for a unit cost function is the generalized Leontief cost function. The generalized Leontief unit cost function

[11] This definition of flexibility is sometimes referred to as "second-order" flexibility because it implies that the gradient and the Hessian matrix of the unit cost function with respect to the prices of inputs are capable of assuming arbitrary nonnegative and negative semidefinite values respectively.

[12] Negative semidefiniteness of $S$ follows from homogeneity of degree one and concavity of the unit cost function in the prices of inputs.

is given by:

$$C(p) = \sum_i \sum_j \beta_{ij} p_i^{1/2} p_j^{1/2}, \tag{2.47}$$

where without loss of generality $\beta_{ij} = \beta_{ji}, \forall i, j$. The elements of the gradient and Hessian matrix of the generalized Leontief unit cost function are given by:

$$\frac{\partial C}{\partial p_i} = \beta_{ii} + \frac{1}{2} \sum_{j \neq i} \beta_{ij} p_i^{-1/2} p_j^{1/2}, \qquad i = 1, \dots, m; \tag{2.48}$$

$$\frac{\partial^2 C}{\partial p_i \partial p_j} = \frac{1}{4} \beta_{ij} p_i^{-1/2} p_j^{-1/2}, \qquad i \neq j, \quad i, j = 1, \dots, m; \tag{2.49}$$

$$\frac{\partial^2 C}{\partial p_i^2} = -\frac{1}{4} \sum_{j \neq i} \beta_{ij} p_i^{-3/2} p_j^{1/2}, \qquad i = 1, \dots, m. \tag{2.50}$$

In order to demonstrate the flexibility of the generalized Leontief unit cost function, we need to show that given the left-hand sides of eqs. (2.47) through (2.50) and $\bar{p}$, one can always find a set of parameters $\beta$ that will solve these equations exactly. First, observe that eq. (2.47) can always be solved by an appropriate scaling of the parameters provided that

$$\frac{\partial C}{\partial p_i} = \beta_{ii} + \frac{1}{2} \sum_{j \neq i} \beta_{ij} p_i^{-1/2} p_j^{1/2} \geq 0, \qquad i = 1, \dots, m.$$

Second, eq. (2.48) can always be solved by an appropriate choice of the $\beta_{ii}$'s, $\beta_{ii} \geq 0$, whatever the value of

$$\frac{1}{2} \sum_{j \neq i} \beta_{ij} p_i^{-1/2} p_j^{1/2}, \qquad i = 1, \dots, m.$$

Third, eq. (2.49) can always be solved by setting

$$\beta_{ij} = \frac{4}{p_i^{1/2} p_j^{1/2}} \frac{\partial^2 C}{\partial p_i \partial p_j}, \qquad i \neq j, \quad i, j = 1, \dots, m.$$

Finally, because of homogeneity of degree zero of $\partial C / \partial p_i$,

$$\frac{\partial^2 C}{\partial p_i^2} p_i = -\sum_{j \neq i} \frac{\partial^2 C}{\partial p_i \partial p_j} p_j,$$

so that

$$\frac{\partial^2 C}{\partial p_i^2} = -\frac{1}{p_i} \sum_{j \neq i} \frac{\partial^2 C}{\partial p_i \partial p_j} p_j$$

$$= -\frac{1}{4} \sum_{j \neq i} \beta_{ij} p_i^{-3/2} p_j^{1/2}, \qquad i = 1, \ldots, m,$$

which satisfies eq. (2.50) identically. We note that

$$\sum_{j \neq i} \beta_{ij} p_i^{-3/2} p_j^{1/2} \geq 0, \qquad i = 1, \ldots, m,$$

in order for the Hessian matrix to be negative semidefinite. We conclude that the generalized Leontief unit cost function is flexible.

Another example of a flexible algebraic functional form for a unit cost function is the transcendental logarithmic cost function. The translog unit cost function is given by:

$$\ln C(p) = C_0 + \sum_i \alpha_i \ln p_i + \frac{1}{2} \sum_i \sum_j \beta_{ij} \ln p_i \ln p_j, \qquad (2.51)$$

where $\sum_i \alpha_i = 1$; $\sum_j \beta_{ij} = 0, \forall i$ and without loss of generality $\beta_{ij} = \beta_{ji}, \forall_{i,j}$. The elements of the gradient and Hessian matrix of the translog unit cost function are given by:

$$\frac{\partial C}{\partial p_i} = \frac{C}{p_i} \frac{\partial \ln C}{\partial \ln p_i},$$

$$= \frac{C}{p_i} \left( \alpha_i + \sum_j \beta_{ij} \ln p_j \right), \qquad i = 1, \ldots, m; \qquad (2.52)$$

$$\frac{\partial^2 C}{\partial p_i \partial p_j} = \frac{C}{p_i p_j} \left[ \beta_{ij} + \frac{\partial \ln C}{\partial \ln p_i} \frac{\partial \ln C}{\partial \ln p_j} \right], \qquad \begin{array}{l} i \neq j, \\ i, j = 1, \ldots, m; \end{array} \qquad (2.53)$$

$$\frac{\partial^2 C}{\partial p_i^2} = \frac{C}{p_i^2} \left[ \beta_{ii} + \frac{\partial \ln C}{\partial \ln p_i} \left( \frac{\partial \ln C}{\partial \ln p_i} - 1 \right) \right], \qquad i = 1, \ldots, m. \qquad (2.54)$$

In order to demonstrate the flexibility of the translog unit cost function, we need to show that given the left-hand sides of eqs. (2.51) through (2.54) and $\bar{p}$, one always find a set of parameters $C_0$, $\alpha$ and $\beta$ that will solve these equations exactly. First, we observe that eq. (2.51) can always be satisfied by an appropriate choice of $C_0$. Eq. (2.52) can be rewritten as

$$\frac{p_i}{C}\frac{\partial C}{\partial p_i} = \alpha_i + \sum_j \beta_{ij}\ln p_j, \qquad i=1,\dots,m,$$

which can always be solved by an appropriate choice of the $\alpha_i$'s, $\alpha_i \geq 0$, $i=1,\dots,m$ and $\sum_i \alpha_i = 1$, subject to $\sum_i \beta_{ij} = 0, \forall j$. Eqs. (2.53) and (2.54) combined may be written as:

$$\frac{1}{C}\begin{bmatrix} p_1 & 0 & \cdots & 0 \\ 0 & p_2 & \cdots & 0 \\ 0 & 0 & & p_m \end{bmatrix}\nabla^2 C(p)\begin{bmatrix} p_1 & 0 & \cdots & 0 \\ 0 & p_2 & \cdots & 0 \\ 0 & 0 & & p_m \end{bmatrix}$$
$$= \beta + ww' - \operatorname{diag}[w],$$

or

$$\beta = \frac{1}{C}\begin{bmatrix} p_1 & 0 & \cdots & 0 \\ 0 & p_2 & \cdots & 0 \\ 0 & 0 & \cdots & 0 \\ 0 & 0 & \cdots & p_m \end{bmatrix}\nabla^2 C(p)\begin{bmatrix} p_1 & 0 & & 0 \\ 0 & p_2 & & 0 \\ 0 & 0 & \cdots & 0 \\ 0 & 0 & \cdots & p_m \end{bmatrix}$$
$$- ww' - \operatorname{diag}[w], \tag{2.55}$$

where $w_i \equiv \partial \ln C / \partial \ln p_i$, $i=1,\dots,m$, and $\operatorname{diag}[w]$ is a diagonal matrix with $w_i$'s on the diagonal. Every term on the right-hand side of eq. (2.55) is either known or specified. Thus, $\beta$ can be chosen, subject to $\sum_i \beta_{ij} = 0, \forall j$, to satisfy any negative semidefinite matrix specified for $\nabla^2 C(p)$. We conclude that the translog unit cost function is flexible.

Similarly, we can give a working definition of "flexibility" for an algebraic functional form for a complete system of consumer demand functions as follows:

*Definition*

An algebraic functional form for a complete system of consumer demand functions $F(p, M; \alpha)$, is said to be flexible if at any given set of nonnegative (positive) prices of commodities and income or total expenditure the parameters, $\alpha$, of the complete system of consumer demand functions can be chosen so that the consumer demand functions and their own and cross-price and income elasticities are capable of assuming arbitrary values at the given set of prices of commodities and income subject only to the requirements of theoretical consistency.

More formally, let $F^*(p^*, M^*; \alpha)$ be a vector-valued algebraic functional form for a complete system of consumers demand functions expressed in natural logarithmic form, that is:

$$F_i^*(p^*, M^*; \alpha) = \ln X_i, \qquad i = 1, \ldots, m;$$

$$p_i^* = \ln p_i, \qquad i = 1, \ldots, m;$$

$$M^* = \ln M.$$

Then flexibility implies and is implied by the existence of a solution $\alpha(\bar{p}^*, \bar{M}^*; \bar{F}^*, \overline{\partial F^{*\prime}}/\partial p^*, \overline{\partial F^*}/\partial M^*)$ to the following set of equations:

$$F^*(\bar{p}^*, \bar{M}^*; \alpha) = \bar{F}^*,$$

$$\frac{\partial F^{*\prime}}{\partial p^*}(\bar{p}^*, \bar{M}^*; \alpha) = \frac{\overline{\partial F^{*\prime}}}{\partial p^*}, \qquad (2.56)$$

$$\frac{\partial F^*}{\partial M^*}(\bar{p}^*, \bar{M}^*; \alpha) = \frac{\overline{\partial F^*}}{\partial M},$$

for every positive value of $\bar{p}^*$, $\bar{M}^*$ and $\bar{F}^*$ and symmetric negative semidefinite value of the corresponding Slutsky substitution matrix which depends on $\bar{p}^*$, $\bar{M}^*$, $\overline{\partial F^{*\prime}}/\partial p^*$ and $\overline{\partial F^*}/\partial M^*$.

We note that an equivalent definition may be phrased in terms of the natural derivatives of the demand functions with respect to the prices of commodities and income rather than the logarithmic derivatives or elasticities.

An example of a flexible algebraic functional form for a complete system of consumer demand functions is the transcendental logarithmic demand system introduced by Christensen, Jorgenson and Lau (1975). The transcendental logarithmic demand system is given by:

$$\frac{p_i X_i}{M} = \frac{\alpha_i + \sum_j \beta_{ij}(\ln p_j - \ln M)}{-1 + \sum_j \beta_{jM}(\ln p_j - \ln M)}, \qquad i = 1, \ldots, m, \qquad (2.57)$$

where $\beta_{ij} = \beta_{ji}$, $i, j = 1, \ldots, m$ and $\sum_i \beta_{ij} = \beta_{jM}$, $j = 1, \ldots, m$. It may be verified that this complete system of demand functions can attain at any prespecified positive values of $p = \bar{p}$ and $M = \bar{M}$ and given positive value of $X$ and negative semidefinite value of the Slutsky substitution matrix $S$ such that $S'\bar{p} = 0$, where a typical element of $S$ is given by:

$$S_{ij} = \frac{\partial X_i}{\partial p_j} + X_j \frac{\partial X_i}{\partial M}, \qquad i, j = 1, \ldots, m,$$

through a suitable choice of the parameters $\beta_{ij}$'s and $\beta_{iM}$'s.

Flexibility of a functional form is desirable because it allows the data the opportunity to provide information about critical parameters. An inflexible functional form often prescribes the value, or at least the range of values, of the critical parameters. In general, the degree of flexibility required depends on the application. For most applications involving producer or consumer behavior, the flexibility required is that the own and cross-price derivatives (or equivalently the elasticities) of demand for inputs or commodities be free to attain any set of theoretically consistent values. For other applications, the desired degree of flexibility may be greater or less. Sometimes a knowledge of the sign and/or magnitude of a third-order derivative may be necessary. For example, in the analysis of behavior under uncertainty, the third derivative of the utility function of the decision maker plays a critical role in the comparative statics. In the empirical analysis of such situations, the algebraic functional form should be chosen so that it is "third-order" flexible, that is, it permits the data to inform about the sign and/or magnitude of the third derivative of the utility function (or equivalently, the second-order derivative of the demand function). In other words, we need to know not only the elasticity of demand, but also the rate of change of the elasticity of demand.

## 2.4. Computational facility

The computational facility of a functional form implies one or more of the following properties.

(1) Its unknown parameters are easy to estimate from the data. Usually what this means is that the functional form is, after a known transformation if necessary, linear-in-parameters, and if there are restrictions on the parameters they are linear restrictions. This is called the "Linearity-in-Parameters" property.

(2) The functional form and any functions of interest derived from it are represented in explicit closed form. For example, it is often not enough that the production function is linear in parameters. The input demand functions derivable from it should be representable in explicit closed form and preferably be linear in parameters as well. This property makes it easy to manipulate and calculate the values of different quantities of economic interest and their derivatives with respect to the independent variables. This is called the property of "Explicit Representability".

Explicit representability of a complete system of demand functions for inputs or commodities cannot in general be guaranteed if one begins with an arbitrary production function or utility function. In fact, the only known production functions that give rise to a system of explicitly representable input demand functions are those that are homothetic after a translation of the origin if necessary. Similarly, the only known utility functions that give rise to a complete system of explicitly representable consumer demand functions are those that are homothetic after a translation of the origin if necessary. By contrast, if one beings by specifying a profit or cost function or an indirect utility function, explicit representability is guaranteed. Given a profit or cost function, the system of input demand functions are, by Hotelling–Shephard Lemma, the gradient of the profit or cost function with respect to the vector of prices of inputs. Given an indirect utility function, the complete system of consumer demand functions are given by Roy's (1943) Identity:

$$X_i = \frac{-\dfrac{\partial V}{\partial p_i}(p, M)}{\dfrac{\partial V}{\partial M}(p, M)}, \qquad i = 1, \ldots, m,$$

where $V(p, M)$ is the indirect utility function.

(3) If the functional form pertains to a complete system, say, of either cost-minimizing input demand functions or consumer demand functions, the different functions in the same system should have the same algebraic form but different parameters. This is called the property of "Uniformity".

Uniformity of a functional form is desirable not only for aesthetic reasons but also because it simplifies considerably the statistical estimation and other related computations. In essence the same procedure and computer programming can be applied to all of the different functions in the same complete system if their algebraic forms are the same.

(4) The number of parameters in the functional form should be the minimum possible number required to achieve a given desired degree of flexibility. In many instances the number of observations is quite small and conservation of the degrees of freedom is an important consideration. In addition, the cost of computation for a given problem increases approximately at the rate of $n^2$ where $n$ is the number of parameters to be estimated. This is called the property of "Parsimony".

We may add that both the generalized Leontief and the translog unit cost functions give rise to a system of cost-minimizing input demand functions that satisfies all four of the properties here.

## 2.5. Factual conformity

Factual conformity implies consistency of the functional form with known empirical facts. Fortunately or unfortunately (depending on one's point of view), there are few known, generally accepted and consistently confirmed facts. Perhaps the only generally accepted and consistently confirmed known empirical fact is Engel's Law, which says that the demand for food, or primary commodities in general, has an income elasticity of less than unity.[13] While this fact may seem innocuous enough, it rules out the use of any homothetic direct or indirect utility

---

[13] See Houthakker (1957), (1965).

function as the basis for an empirical study of consumer demand because homotheticity implies that the income elasticity of demand of every commodity is unity.

Less established but still widely accepted empirical facts include:

(1) the six-tenth factor rule between capital cost and output capacity for certain chemical and petrochemical processing industries;

(2) the elasticities of substitution between all pairs of input in the three or more-input case are not all identical;

(3) the proportionality of the quantity of raw material input to the quantity of output (for example, iron ore and steel);

(4) not all Engel curves are linear in income.

Each of these facts has implications on the choice of functional forms. For example, the six-tenth factor rule is inconsistent with the use of functional forms for production functions that are homothetic (unless all other inputs also satisfy the six-tenth factor rule, which is generally not the case). The lack of identity among the elasticities of substitution between all pairs of inputs suggests that the Constant-Elasticity-of-Substitution (and hence the Cobb–Douglas) production function is not an appropriate algebraic functional form. The proportionality of raw material input to output suggests that the production function must have one of the two following forms:

$$Y = \text{Min}\left\{ f(X), \frac{M}{\alpha_M} \right\},$$

where $X$ is the vector of all other inputs, $f(X)$ is a function of $X$ and $M$ is the quantity of raw material input; or

$$Y = f(X)M.$$

The fact that not all Engel curves (of different commodities) are linear suggests that the use of the Gorman (1953) condition for the analysis of aggregate consumer demand can be justified only as an approximation.[14]

In the choice of algebraic functional forms, one should avoid, insofar as possible, the selection of one which has implications that are at variance with established facts.

## 3. Compatibility of the criteria for the selection of functional forms

A natural question that arises is: Are there algebraic functional forms that satisfy all five categories of criteria that we have laid down in Section 2? In other words, does there exist an algebraic functional form that is globally theoretically con-

---

[14] The Gorman condition on the utility function justifies the existence of aggregate demand functions as functions of aggregate income and is widely applied in empirical analyses. See for example Blackorby, Boyce and Russell (1978).

sistent (for all theoretically consistent data), flexible, linear-in-parameters, explicitly representable, uniform (if there are more than one function in the system), parsimonious in the number of parameters and conforms to known facts? Obviously, the answer depends on the specific application. In Section 3.1, we give an example of the incompatibility of a global extrapolative domain of applicability and flexibility. In Section 3.2, we give an example of the incompatibility of computational facility and factual conformity. In Section 3.3, we prove an impossibility theorem which says that there does *not* exist an algebraic functional form for a unit cost function which has a global extrapolative domain of applicability and satisfies the criteria of flexibility and computational facility.

Thus, in general, one should not expect to find an algebraic functional form that satisfies all five categories of criteria. For specific applications, especially in situations in which the relevant theory imposes little or no restriction, it may be possible that such an algebraic functional form can be found.

### 3.1.  *Incompatibility of a global domain of applicability and flexibility*

Consider the generalized Leontief unit cost function for a single-output, two-input technology:

$$C(p_1, p_2) = \alpha_0 p_1 + \alpha_1 p_1^{1/2} p_2^{1/2} + \alpha_2 p_2,$$

which, as shown in Section 2.2, is theoretically consistent over the whole nonnegative orthant of prices of inputs if and only if $\alpha_0 \geq 0$; $\alpha_1 \geq 0$ and $\alpha_2 \geq 0$. We shall show that under these parametric restrictions, the unit cost function is not *flexible*, that is, the parameters cannot be chosen such that it can attain arbitrary but theoretically consistent values of $C$, $\nabla C$ and $\nabla^2 C$ at an arbitrary set of prices of inputs.

Without loss of generality let the set of prices be $(1,1)$, and let the arbitrarily chosen values of $C$, $\nabla C$ and $\nabla^2 C$ at $(1,1)$ be

$$C(1,1) = k_1 \geq 0,$$

$$\nabla C(1,1) = \begin{bmatrix} k_2 \\ k_1 - k_2 \end{bmatrix} \geq 0, \tag{3.1}$$

$$\nabla^2 C(1,1) = \begin{bmatrix} -k_3 & k_3 \\ k_3 & -k_3 \end{bmatrix}, \qquad k_3 \geq 0,$$

where the restrictions on $\nabla C(1,1)$ and $\nabla^2 C(1,1)$ reflect homogeneity of degree one, monotonicity and concavity of the unit cost function in the prices of inputs. Flexibility requires that for arbitrarily given $k_1, k_2, k_3 \geq 0$, with $k_1 - k_2 \geq 0$, the parameters $\alpha_0, \alpha_1, \alpha_2 \geq 0$ can be found such that

$$C(1,1) = \alpha_0 + \alpha_1 + \alpha_2 = k_1,$$

$$\frac{\partial C}{\partial p_1}(1,1) = \alpha_0 + \frac{1}{2}\alpha_1 = k_2, \qquad (3.2)$$

$$\frac{\partial^2 C}{\partial p_1^2}(1,1) = -\frac{1}{4}\alpha_1 = -k_3.$$

The reader can verify that satisfaction of eq. (3.2) is equivalent to the satisfaction of eq. (3.1). It is easy to see that $\alpha_1$ can always be chosen to be $4k_3$ and hence $\geq 0$. However,

$$\alpha_0 + \tfrac{1}{2}\alpha_1 = \alpha_0 + 2k_3 = k_2,$$

cannot hold with $\alpha_0 \geq 0$ if $2k_3 \geq k_2$. Thus, flexibility fails if the generalized Leontief unit cost function is required to be theoretically consistent globally. We note that $2k_3 \geq k_2$ implies that

$$-\frac{p_1}{X_1}\frac{\partial X_1}{\partial p_1} = -p_1\frac{\partial^2 C/\partial p_1^2}{\partial C/\partial p_1} = \frac{k_3}{k_2} \geq \frac{1}{2}.$$

Thus, the generalized Leontief unit cost function, if it were to be required to be valid for all nonnegative prices of inputs, cannot approximate a technology with an elasticity of input demand of greater than $\frac{1}{2}$!

This examples shows that a global extrapolative domain of applicability may be incompatible with flexibility.

The first related question is: Given the rules of interpolation embodied in eq. (3.2), what is the domain of values of $k_1, k_2$ and $k_3$ that will allow the generalized Leontief unit cost function to be globally theoretically consistent? We note from eq. (3.2) that the parameters may be obtained by interpolation as:

$$\alpha_0 = k_2 - 2k_3 \geq 0,$$
$$\alpha_1 = 4k_3 \geq 0,$$
$$\alpha_2 = k_1 - k_2 - 2k_3 \geq 0,$$

which must all be nonnegative. Moreover, by monotonicity, $k_1 - k_2 \geq 0$. The inequalities are, however, all homogeneous of degree one, we may thus arbitrarily normalize $k_1$ to unity. The domain of $k_1, k_2, k_3$'s can then be represented by the following set of inequalities:

$$k_2^* - 2k_3^* \geq 0,$$
$$1 - k_2^* - 2k_3^* \geq 0,$$
$$1 - k_2^* \geq 0,$$
$$k_2^* \geq 0; \quad k_3^* \geq 0.$$

These inequalities can be illustrated graphically in Figure 1. The interpolative domain of the generalized Leontief unit cost function, if it were required to globally theoretically consistent, consists only of the shaded area. The shaded area falls far short of the constraint for theoretical consistency, that is, $1 - k_2^* \geq 0$, $k_2^* \geq 0$ and $k_3^* \geq 0$. It is clear that if the generalized Leontief unit cost function were to be required to be globally theoretically consistent, it can be flexible only for those values of $k_2^*$ and $k_3^*$ in the shaded area.

The elasticity of substitution at $(1,1)$ may be computed as:

$$\sigma = \frac{CC_{12}}{C_1 C_2} = \frac{k_1}{k_2} \frac{k_3}{(k_1 - k_2)}$$

$$= \frac{k_3^*}{k_2^*} \frac{1}{(1 - k_2^*)}.$$

The minimum value of $\sigma$ over the admissible domain of $k^*$'s is of course zero. The maximum value can be shown to occur at $k_2^* = \frac{1}{2}$ and $k_3^* = \frac{1}{4}$, that is, $\sigma = 1$. Thus, the generalized Leontief unit cost function, if it were to be globally theoretically consistent, cannot attain an elasticity of substitution greater than unity.

The own and cross-price elasticities of the input demand functions are given by:

$$\frac{p_j}{X_i} \frac{\partial X_i}{\partial p_j} = \frac{p_j}{C_i} \frac{\partial^2 C}{\partial p_i \partial p_j}, \qquad i, j = 1, 2.$$

At $(1,1)$, they are given by:

$$\frac{\partial \ln X_1}{\partial \ln p_1} = \frac{-k_3^*}{k_2^*},$$

$$\frac{\partial \ln X_1}{\partial \ln p_2} = \frac{k_3^*}{k_2^*},$$

$$\frac{\partial \ln X_2}{\partial \ln p_1} = \frac{k_3^*}{(1 - k_2^*)},$$

$$\frac{\partial \ln X_2}{\partial \ln p_2} = \frac{-k_3^*}{(1 - k_2^*)}.$$

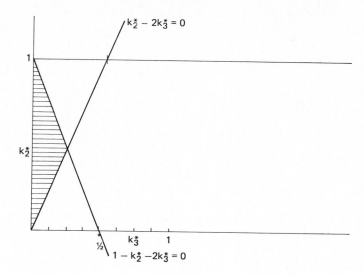

Figure 1

Referring to Figure 1, the maximum absolute value of $\partial \ln X_i / \partial \ln p_j$ within the admissible region is $\frac{1}{2}$, the minimum absolute value is 0.

It should be noted that the incompatibility of a global extrapolative domain of applicability and flexibility is a common problem and not limited to the generalized Leontief unit cost function. It is also true of the translog unit cost function. If the translog unit cost function were required to be globally theoretically consistent, the only value of elasticity of substitution it can take at $(1,1)$ is unity!

The purpose of Section 3.1 is to show that the two criteria of domain of applicability and flexibility are often incompatible. In Section 3.3 we shall show that the two criteria are *never* compatible for any functional form for a unit cost function that is linear in parameters and parsimonious.

## 3.2. Incompatibility of computational facility and factual conformity

In Section 2.5 we pointed out the known fact that some commodities, notably food, have income elasticities less than unity. Thus, any algebraic functional form for a complete system of consumer demand functions that has the property of unitary income elasticity for every commodity must be at variance with the facts and should not be used. This rules out all complete systems of consumer demand functions derived from a homothetic functional form for a direct or indirect utility function.

Unfortunately, all known theoretically consistent (flexible or not) complete system of consumer demand functions of three or more commodities that are linear in parameters,[15] after a known transformation of the dependent variables if necessary, have the property of unitary income elasticities for all commodities.[16] Thus, in the choice of a functional form for a complete system of consumer demand functions, the linearity-in-parameters property has to be abandoned.

It is conjectured that linearity-in-parameters implies unitary income elasticities for all theoretically consistent complete systems of consumer demand functions of three or more commodities. Such a theorem remains to be proved.

### 3.3. *Incompatibility of a global domain of applicability, flexibility and computational facility*

We now proceed to prove a general impossibility theorem which says that a linear-in-parameters and parsimonious functional form for a unit cost function cannot be simultaneously (1) globally theoretically consistent and (2) flexible for all theoretically consistent data. Thus, it is futile to look for a linear-in-parameters functional form for a unit cost function that will satisfy all of our criteria. In Section 3.1 we already demonstrated that a global domain of applicability is incompatible with flexibility as far as the generalized Leontief unit cost function is concerned. Here we show that this incompatibility is true of all linear-in-parameters and parsimonious unit cost functions.

Our presentation is simplified by considering the normalized cost function defined as $C^*(p_2/p_1) \equiv C(1, p_2/p_1)$ instead of the cost function $C(p_1, p_2)$. The two functions are of course equivalent. The properties of the normalized cost function are as follows:

$$C^*(q) - q\frac{dC^*}{dq}(q) \geq 0, \tag{3.3}$$

$$\frac{dC^*}{dq}(q) \geq 0, \tag{3.4}$$

$$\frac{d^2C^*}{dq^2}(q) \leq 0, \tag{3.5}$$

---

[15] Linearity in parameters as used here requires that the restrictions on the parameters, if any, are linear also. Thus, the Linear Expenditure System introduced by Stone (1954) is not a linear-in-parameters functional form.

[16] See, for example, Jorgenson and Lau (1977) and (1979) and Lau (1977).

were $q \equiv p_2/p_1$. We note that eqs. (3.3) and (3.4) together imply that $C*(q) \geq 0$.

*Lemma 1*

Let a normalized unit cost function have the linear-in-parameters and parsimonious form:

$$C(q) = f_0(q)\alpha_0 + f_1(q)\alpha_1 + f_2(q)\alpha_2,[17]$$ (3.6)

where the $f_i(q)$'s are a set of linearly independent twice continuously differentiable functions of $q$. In addition, suppose that the functional form is flexible, that is, for every $\bar{q} > 0$ and every $k \geq 0$, there exists a set of parameters $\alpha_0$, $\alpha_1$ and $\alpha_2$ such that:

$$\sum_{i=0}^{2} (f_i(\bar{q}) - \bar{q}f_i'(\bar{q}))\alpha_i = k_0,$$

$$\sum_{i=0}^{2} f_i'(\bar{q})\alpha_i = k_1,$$

$$-\sum_{i=0}^{2} f_i''(\bar{q})\alpha_i = k_2.$$

Let this system of equations be written as:

$$W(\bar{q})\alpha = k.$$ (3.7)

where

$$W(q) = \begin{bmatrix} f_0(q) - qf_0'(q) & f_1(q) - qf_1'(q) & f_2(q) - qf_2'(q) \\ f_0'(q) & f_1'(q) & f_2'(q) \\ f_0''(q) & f_1''(q) & f_2''(q) \end{bmatrix}.$$

Then $W(\bar{q})$ is nonsingular for all $\bar{q}$.

*Proof*

By hypothesis, for all $\bar{q} > 0$, and for all $k \geq 0$, there is a solution $\alpha$ satisfying

$$W(\bar{q})\alpha = k.$$

---

[17] This functional form is parsimonious because it has the minimum number of independent parameters required for flexibility.

By Gale's (1960) Theorem of the Alternative this implies that there must *not* be a solution $y$ to the equations

$$W(\bar{q})'y = 0, \qquad k'y = 1, \qquad \bar{q} > 0; \qquad k \geq 0.$$

Suppose $W(\bar{q})$ is singular for some $\bar{q}$, then there exists $\bar{y} \neq 0$ such that

$$W(\bar{q})'\bar{y} = 0.$$

Since $\bar{y} \neq 0$ there exists $k \geq 0$ such that $k'\bar{y} \neq 0$. If $k'\bar{y} < 0$, we consider $\bar{y}^* = -\bar{y}$, so that $k'\bar{y}^* > 0$. By defining $k^* \equiv k/k'\bar{y}^*$, $k^*{}'\bar{y}^* = 1$. Then $W(q)'\bar{y}^* = 0$, $k^*{}'\bar{y}^* = 1$, $k^* \geq 0$ which, by Gale's Theorem of the Alternative, implies that

$$W(\bar{q})\alpha = k^*, \ k^* \geq 0,$$

does not have a solution contradicting the hypothesis of flexibility. We conclude that flexibility implies nonsingularity of $W(\bar{q})$ for all $\bar{q} > 0$.                Q.E.D.

We note that if the functions $f_0(q)$, $f_1(q)$ and $f_2(q)$ are linearly dependent, then $W(q)$ is always singular. It is clear that the functional form in eq. (3.6) is parsimonious in the number of parameters since the number of independent unknown parameters is equal to the number of components of $k$ that need to be matched.

### *Lemma 2*

Let $A$ be a real square matrix. Let $x$ be a nonnegative vector of the same dimension. Then

$$Ax \geq 0 \qquad \text{for all} \qquad x \geq 0,$$

if and only if $A$ is nonnegative.

### *Proof*

Sufficiency is straightforward. Necessity is proved by contradiction. Suppose there exist $A$, not nonnegative, such that $Ax \geq 0$ for all $x \geq 0$. Let $A_{ij} < 0$, then let $x$ be a vector with unity as the $j$th element and zero otherwise. The $i$th element of $Ax$ will therefore be negative, contradicting the hypothesis that $Ax \geq 0$. We conclude that $A$ must be nonnegative.                Q.E.D.

### *Lemma 3*

Let $A$ be a real, nonnegative, nonsingular square matrix of finite dimension. The $A^{-1}$ is nonnegative if and only if $A = DP$ where $D$ is a positive diagonal matrix and $P$ is a permutation matrix.

A proof is contained in Appendix 1.[18]

With these three lemmas, we can now proceed to state and prove the main impossibility theorem.

*Theorem*

Let a class of normalized unit cost functions have the linear-in-parameters and parsimonious form:

$$C(q; \alpha) = f_0(q)\alpha_0 + f_1(q)\alpha_1 + f_2(q)\alpha_2,$$

where the $f_i(q)$'s are a set of linearly independent twice continuously differentiable functions of $q$. In addition, suppose that the functional form is *flexible*, that is, for every $\bar{q} > 0$ and every $k \geq 0$, there exists a set of parameters $\alpha_0$, $\alpha_1$ and $\alpha_2$ such that:

$$\sum_{i=0}^{2} \left( f_i(\bar{q}) - \bar{q}f_i'(\bar{q}) \right)\alpha_i = k_0,$$

$$\sum_{i=0}^{2} f_i'(\bar{q})\alpha_i = k_1,$$

$$\sum_{i=0}^{2} f_i''(\bar{q})\alpha_i = -k_2,$$

or equivalently

$$W(\bar{q})\alpha = k.$$

Then $C(q; \alpha)$ cannot be globally theoretically consistent (for all nonnegative prices) for all such $\alpha$'s.

*Proof*

The proof is by contradiction. Global theoretical consistency of $C(q; \alpha)$ implies:

$$W(q)\alpha \geq 0, \qquad \forall q \geq 0.$$

By hypothesis, for every $\bar{q} > 0$ and $k \geq 0$, there exists

$$W(\bar{q})\alpha = k.$$

---

[18]I am grateful to Kenneth Arrow for correcting an error in the original formulation of Lemma 3.

By Lemma 1, $W(\bar{q})$ is nonsingular and hence

$$\alpha = W(\bar{q})^{-1}k.$$

Suppose the theorem is false, then there exists $W(q)$ such that:

$$W(q)\alpha = W(q)W(\bar{q})^{-1}k \geq 0, \forall q > 0, \bar{q} > 0 \text{ and } k \geq 0.$$

By Lemma 2, $W(q)W(\bar{q})^{-1}$ must be nonnegative. Let

$$A(q,\bar{q}) \equiv W(q)W(\bar{q})^{-1},$$

which is nonnegative. Then

$$W(q) = A(q,\bar{q})W(\bar{q}). \tag{3.8}$$

By the symmetry of $q$ and $\bar{q}$,

$$W(\bar{q}) = A(\bar{q},q)W(q),$$

and hence

$$W(q) = A(q,\bar{q})A(\bar{q},q)W(q),$$

which implies that

$$A(q,\bar{q})^{-1} = A(\bar{q},q).$$

Thus, both $A(q,\bar{q})$ and its inverse are nonnegative. By Lemma 3,

$$A(q,\bar{q}) = D(q,\bar{q})P, \tag{3.9}$$

where $D(q,\bar{q})$ is a positive diagonal matrix and $P$ is a permutation matrix.[19] Substituting eq. (3.9) into eq. (3.8), we obtain:

$$W(q) = D(q,\bar{q})PW(\bar{q}).$$

$PW(\bar{q})$ is a nonsingular matrix independent of $q$, so that each element of the $i$th

---

[19]A permutation matrix is a square matrix which can be put into the form of an identity matrix by a suitable reordering of the rows (or columns) if necessary.

row of $W(q)$ is equal to a constant (possibly zero) times $D_{ii}(q)$, a function of $q$. This contradicts the linear independence of the *functions* $f_0(q)$, $f_1(q)$, and $f_2(q)$.

Q.E.D.

The implication of this theorem is that there can be no linear-in-parameters and parsimonious functional form for a normalized unit cost function which can fit arbitrary but theoretically consistent values of a normalized unit cost function and its first and second derivatives at any preassigned value of the normalized price and be itself theoretically consistent for all nonnegative normalized prices. One has to be prepared to give up one or more of the desirable properties of an algebraic functional form.

Since one is not likely to give up theoretical consistency or flexibility, or even computational facility, the logical area for a compromise lies in the domain of applicability. For example, one can be satisfied with an extrapolative domain of a functional form for a unit cost function that excludes, say, unreasonably high values of the elasticity of substitution.

The fact is that requiring the extrapolative domain of a functional form to be global when the data on which the parameters of the functional form are estimated are local does not make too much sense from a practical point of view. In the first place, even assuming that the same functional form and the same parameters hold outside the neighborhood containing the observed data, the confidence band for the estimated function will become so wide for values of independent variables far away from the neighborhood containing the observed data that it will not be very useful at all. Second, values of the parameters and even the functional form itself may be different for values of independent variables far away from the neighborhood containing the observed data.[20] Unfortunately there is no way of knowing a priori. One can only wait until these faraway values are actually experienced and observed. Third, reality is always finite and it is difficult to conceive of any application in which an independent variable, for example, a price or a quantity of an input, becomes arbitrarily large.

For these reasons, it may be just as well that a global extrapolative domain cannot be achieved in general. One should settle for a well-prespecified compact domain of applicability that reflects the actual and potential ranges of data experiences.

The theorem can be generalized in several dimensions: (1) the number of independent variables can be increased; (2) the number of parameters can be increased (but maintained finite); (3) the functional form can be linear-in-parameters after a monotonic transformation.

---

[20]As an example, consider classical Newtonian mechanics and relativistic mechanics. The latter reduces to the former at low velocities. However, an extrapolation of Newtonian mechanics to high-velocity situations would be wrong!

## 4.  Concluding remarks

The most important conclusion that can be drawn from our analysis here is that in general it is not possible to satisfy all five categories of criteria simultaneously. Some trade-offs have to be made. It is however not recommended that one compromises on local theoretical consistency – any algebraic functional form must be capable of satisfying the theoretical consistency restrictions at least in a neighborhood of the values of the independent variables of interest. It is also not recommended, except as a last resort, to give up computational facility, as the burden of and probability of failure in the estimation of nonlinear-in-parameters models is at least one order of magnitude higher than linear-in-parameters models and in many instances the statistical theory is less well developed. It is also not advisable to sacrifice flexibility – inflexibility restricts the sensitivity of the parameter estimates to the data and limits a priori what the data are allowed to tell the econometrician. Unless there is strong a priori information on the true functional form, flexibility should be maintained as much as possible.

This leaves the domain of applicability as the only area where compromises may be made. As argued in Section 3.3, most practical applications can be accommodated even if the functional form is not globally theoretically consistent so long as it is theoretically consistent within a sufficiently large but nevertheless compact subset of the space of independent variables. For example, any extrapolative domain of theoretical consistency which allows the relative price of inputs to vary by factor of one million is plenty large enough. Moreover, by making a compromise on the extrapolative domain of applicability one can also simultaneously reduce the domain over which the functional form has to be flexible. Further, one can also make compromises with regard to the interpolative domain of the functional form, that is, to limit the set of possible values of the derivatives of the function that the functional form has to fit. For example, one may specify that a functional form for a unit cost function $C(p; \alpha(k))$ be theoretically consistent for all prices in a compact subset of positive prices and for all values of $k$ in a compact subset of possible values of its first and second derivatives. This last possibility holds the most promise.

With regard to specific applications, one can say that as far as the empirical analysis of production is concerned, the surest way to obtain a theoretically consistent representation of the technology is to make use of one of the dual concepts such as the profit function, the cost function or the revenue function. There, as we have learned, one has to be prepared to make compromises with regard to the domain of applicability. The impossibility theorem in Section 3.3 applies not only to unit cost functions but to other similar concepts such as profit and revenue functions as well.

As far as the empirical analysis of consumer demand is concerned, the surest way to obtain a theoretically consistent and flexible complete system of demand functions is to specify a theoretically consistent and flexible nonhomothetic indirect utility function and derive the system of consumer demand functions by Roy's Identity. As long as the indirect utility function is theoretically consistent and flexible, the resulting complete system of consumer demand functions will also be theoretically consistent, flexible, and explicitly representable. Unfortunately, linearity-in-parameters of the indirect utility function does not guarantee linearity-in-parameters of the complete systems of consumer demand functions. In fact, the only known linear-in-parameters complete system of consumer demand functions of three or more commodities are derivable from homothetic utility functions with the undesirable implication that the income elasticities of demands of all commodities are unities, an implication that has been repeatedly contradicted by facts. Thus, one has to give up on the linearity-in-parameters property in the choice of a functional form for a complete system of consumer demand functions.

Once linearity-in-parameters is given up, it is not clear what the next best thing may be. However, here one may be guided by parsimony of parameters (and restrictions on parameters). The estimation of nonlinear parameters subject to nonlinear constraints is a considerably more difficult undertaking and the degree of nonlinearity should be kept at a minimum. A device that frequently works is to start with a linear-in-parameters complete system and translate its origin so that the resulting translated system no longer has the property of unitary income elasticities for all commodities.

## Appendix 1

### Lemma 3

Let $A$ be a real, nonnegative, nonsingular square matrix of finite dimension. Then $A^{-1}$ is nonnegative if and only if $A = DP$ where $D$ is a positive diagonal matrix and $P$ is a permutation matrix.[21]

### Proof

Sufficiency follows from the fact that the inverse of a permutation matrix is its transpose, which is also a permutation matrix. The proof of necessity is by induction on the order of the matrix $n$. First, we verify the necessity of the lemma

---

[21]A permutation matrix is a square matrix which can be put into the form of an identity matrix by a suitable reordering of the rows (or columns) if necessary.

for $n = 2$. The elements of $A$ and $A^{-1}$, both nonnegative, must satisfy the following equations:

$$A_{11}A_{11}^{-1} + A_{12}A_{21}^{-1} = 1, \tag{A.1}$$

$$A_{11}A_{12}^{-1} + A_{12}A_{22}^{-1} = 0, \tag{A.2}$$

$$A_{21}A_{11}^{-1} + A_{22}A_{21}^{-1} = 0, \tag{A.3}$$

$$A_{21}A_{12}^{-1} + A_{22}A_{22}^{-1} = 1, \tag{A.4}$$

where $A \geq 0$; $A^{-1} \geq 0$. First suppose $A_{11} \neq 0$. Then by eq. (A.2) $A_{12}^{-1} = 0$ which in turn implies that $A_{11}^{-1} \neq 0$ and $A_{22}^{-1} \neq 0$ (otherwise $A^{-1}$ is singular). $A_{11}^{-1} \neq 0$ implies by eq. (A.3) that $A_{21} = 0$, $A_{22}^{-1} \neq 0$ implies by eq. (A.2) that $A_{12} = 0$. Thus $A$ is a diagonal matrix and nonsingularity implies that $A$ is a positive diagonal matrix. Next suppose $A_{11} = 0$, then $A_{12} \neq 0$ and $A_{21} \neq 0$ (otherwise $A$ is singular) and by eq. (A.1) $A_{11}^{-1} \neq 0$. $A_{21}^{-1} \neq 0$ implies by eq. (A.3) $A_{22} = 0$. Thus, $A$ can be expressed as

$$A = \begin{bmatrix} 0 & A_{12} \\ A_{21} & 0 \end{bmatrix} = \begin{bmatrix} A_{12} & 0 \\ 0 & A_{21} \end{bmatrix} \begin{bmatrix} 0 & 1 \\ 1 & 0 \end{bmatrix},$$

the product of a positive diagonal matrix and a permutation matrix.

Now suppose the lemma is true for all real, nonnegative nonsingular square matrices for all orders up to $n$, we shall show that it is true for order $(n+1)$. Let the matrices $A$ and its inverse $A^{-1}$ be partitioned conformably as

$$A = \begin{bmatrix} A_{11} & a_{1n} \\ a_{n1} & A_n \end{bmatrix}; \qquad A^{-1} = \begin{bmatrix} B_{11} & b_{1n} \\ b_{n1} & B_n \end{bmatrix},$$

where $A_{11}$ and $B_{11}$ are scalars. The elements of $A$ and $A^{-1}$ must satisfy the following equations:

$$A_{11}B_{11} + a_{1n}b_{n1} = 1, \tag{A.5}$$

$$A_{11}b_{1n} + a_{1n}B_n = 0, \tag{A.6}$$

$$a_{n1}B_{11} + A_n b_{n1} = 0, \tag{A.7}$$

$$a_{n1}b_{1n} + A_n B_n = I_n. \tag{A.8}$$

First, suppose $A_{11} \neq 0$, then by eq. (A.6) $b_{1n} = 0$ which implies that $B_{11} \neq 0$ and $B_n$ is nonsingular (otherwise $A^{-1}$ is singular). $B_{11} \neq 0$ implies by eq. (A.7) $a_{n1} = 0$. $B_n$ is nonsingular implies by eq. (A.6) $a_{1n} = 0$. By eq. (A.8) $B_n = A_n^{-1}$. By eq.

(A.5) $B_{11} = A_{11}^{-1}$. Thus the matrices $A$ and $A^{-1}$ have the following forms:

$$A = \begin{bmatrix} A_{11} & 0 \\ 0 & A_n \end{bmatrix}, \qquad A^{-1} = \begin{bmatrix} A_{11}^{-1} & 0 \\ 0 & A_n^{-1} \end{bmatrix}.$$

But $A_n$ and $A_n^{-1}$ are both nonnegative, implying, by the lemma that

$$A_n = D_n P_n.$$

We conclude that

$$A = \begin{bmatrix} A_{11} & 0 \\ 0 & D_n \end{bmatrix} \begin{bmatrix} 1 & 0 \\ 0 & P_n \end{bmatrix},$$

the product of a positive diagonal matrix and a permutation matrix.

Next suppose $A_{11} = 0$, then $a_{1n} \neq 0$ and $a_{n1} \neq 0$ (otherwise $A$ is singular), which in turn imply:

(1) by eq. (A.5),

$$a_{1n} b_{n1} = 1.$$

(2) by eq. (A.6),

$$a_{1n} B_n = 0.$$

(3) by eq. (A.7),

$$B_{11} = 0 \text{ and } A_n b_{n1} = 0.$$

We note, first of all, that eq. (A.8) implies that $a_{n1} b_{1n}$ must be a diagonal matrix. A typical element of $a_{n1} b_{1n}$ is $a_{n1,i} b_{1n,j}$. In order for this to be identically zero for $i \neq j$, all $i$, $j$, it is necessary and sufficient that $a_{n1}$ and $b_{1n}$ be nonzero in only one element which is common to both $a_{n1}$ and $b_{1n}$. Let this element be the $k$th element of $a_{n1}$ (and $b_{1n}$). Moreover, since $a_{n1} b_{1n}$ is then a diagonal matrix with the $k$th element on the diagonal nonzero, $I_n - a_{n1} b_{1n}$ is also a diagonal matrix. However, it must have a rank equal to $A_n B_n$ and hence less than or equal to $n - 1$. We conclude that the nonzero diagonal element of $a_{n1} b_{1n}$ must be equal to unity. The product $A_n B_n$ is then equal to an identity matrix with the $k$th element on the diagonal replaced by a zero. The ranks of $A_n$ and $B_n$ must be equal to $(n - 1)$. If either of them were less than $(n - 1)$, then the matrix $A$ (or $A^{-1}$) would be singular.

Second, we note that because

$$A_n b_{n1} = 0,$$

whenever an element of $b_{n1}$ is nonzero, the corresponding column of $A_n$ must be zero. The rank condition on $A_n$ implies that there can only be one such zero column. Hence $b_{n1}$ can only have one nonzero element, say, the $l$th. Similarly, because

$$a_{1n} B_n = 0,$$

$a_{1n}$ can have only one nonzero element. Moreover, because $a_{1n} b_{n1} = 1$, the same element in $a_{1n}$ and $b_{n1}$ must be nonzero. Thus, the matrix $A$ has the form:

$$A = \begin{bmatrix} 0 & 0 \cdots 0 & a_{1n,l} & 0 \cdots 0 \\ 0 & & & \\ \vdots & & & \\ 0 & & & \\ a_{n1,k} & & A_n & \\ 0 & & & \\ \vdots & & & \\ 0 & & & \end{bmatrix},$$

where the $l$th column of $A_n$ is a column of zeros. Similarly, $A^{-1}$ has the form:

$$A^{-1} = \begin{bmatrix} 0 & 0 \cdots 0 & b_{1n,k} & 0 \cdots 0 \\ \vdots & & & \\ 0 & & & \\ b_{n1,l} & & B_n & \\ 0 & & & \\ \vdots & & & \\ 0 & & & \end{bmatrix},$$

where the $l$th row of $B_n$ is a row of zeros.

Moreover, the product of the $k$th row of $A_n$ and $B_n$ must be identically zero by eq. (A.8). This means that the $k$th row of $A_n$ must be proportional to $a_{1n}$ (with the constant of proportionality being possibly zero). But the $l$th element of the $k$th row of $A_n$ is zero, whereas the $l$th element of $a_{1n}$ is nonzero. We conclude that the $k$th row of $A_n$ is identically zero. Similarly, the product of $A_n$ and the $k$th column of $B_n$ must also be identically zero. This means, by a similar argument, that the $k$th column of $B_n$ is identically zero. Thus, the matrices $A$ and $A^{-1}$ have the following forms.

$$
A = \begin{bmatrix}
0 & 0 \cdots 0 & a_{1n,l} & 0 \cdots 0 \\
\vdots & & 0 & \\
& & \vdots & \\
0 & A^*_{k-1,l-1} & & A^*_{k-1,n-l} \\
a_{n1,k} & 0 \quad 0 \cdots & 0 & \cdots 0 \\
0 & & \vdots & \\
\vdots & & & \\
0 & A^*_{n-k,l-1} & 0 & A^*_{n-k,n-l}
\end{bmatrix},
\tag{A.9}
$$

$$
A^{-1} = \begin{bmatrix}
0 & 0 \cdots 0 & b_{1n,k} & 0 \cdots 0 \\
0 & & 0 & \\
\vdots & & \vdots & \\
0 & B^*_{l-1,k-1} & 0 & B^*_{l-1,n-k} \\
b_{n1,l} & 0 \cdots 0 & 0 & 0 \cdots 0 \\
0 & & 0 & \\
\cdot & & 0 & \\
\cdot & B^*_{n-l,k-1} & 0 & B^*_{n-l,n-k} \\
\cdot & & \vdots & \\
0 & & 0 &
\end{bmatrix},
$$

where $A^*_{ij}$ and $B^*_{ij}$ are conformable partitions of $A$ and $A^{-1}$ respectively. Further, by direct multiplication,

$$
A_n B_n = \begin{bmatrix}
& 0 & \\
I_{k-1} & \vdots & 0 \\
& \vdots & \\
0 & \cdots \quad 0 & \cdots 0 \\
& 0 \quad \vdots & I_{n-k} \\
& 0 &
\end{bmatrix}.
$$

Let $A^*_n$ be the matrix formed by deleting the $k$th row and $l$th column of $A_n$ and $B^*_n$ be the matrix formed by deleting the $l$th row and $k$th column of $B_n$, it can be shown that the resulting product of the two square matrices $A^*_n$ and $B^*_n$ is:

$$
A^*_n B^*_n = I_{n-1},
$$

so that $B_n^* = A_n^{*-1}$. But $A_n^*$ is of order $n-1$. Thus, applying the lemma,

$$A_n^* = D_{n-1}P_{n-1},$$

where $D_{n-1}$ is a positive diagonal matrix and $P_{n-1}$ is a permutation matrix. Substituting this result into eq. (A.9) we obtain:

$$A = \begin{bmatrix} a_{1n,l} & & & & & & & \\ & D_{11} & & & & 0 & & \\ & & \ddots & & & & & \\ & & & D_{k-1,k-1} & & & & \\ & & & & a_{n1,k} & & & \\ 0 & & & & & D_{kk} & & \\ & & & & & & \ddots & \\ & & & & & & & D_{n-1,n-1} \end{bmatrix}$$

$$\cdot \begin{bmatrix} 0 & 0 \cdots & & 1 & \cdots & & 0 \\ 0 & & & 0 & & & \\ 0 & P_{k-1,l-1}^* & \vdots & & P_{k-1,n-l}^* & & \\ \vdots & & 0 & & & & \\ \vdots & & & & & & \\ 1 & 0 \cdots & & 0 & \cdots & & 0 \\ 0 & & & 0 & & & \\ \vdots & P_{n-k,l-1}^* & \vdots & & P_{n-k,n-l}^* & & \\ 0 & & & 0 & & & \end{bmatrix}, \qquad \text{(A.10)}$$

where the $D_{ii}$'s are the elements of the positive diagonal matrix $D_{n-1}$ and $P_{ij}^*$'s are conformable partitions of the permutation matrix $P_{n-1}^*$. It can be verified that the second matrix of the product in eq. (A.10) is a permutation matrix.     Q.E.D.

# References

Arrow, K. J., H. B. Chenery, B. S. Minhas and R. M. Solow (1961) "Capital-Labor Substitution and Economic Efficiency", *Review of Economics and Statistics*, 43, 225–250.

Barten, A. P. (1967) "Evidence on the Slutsky Conditions for Demand Equations", *Review of Economics and Statistics*, 49, 77–84.

Barten, A. P. (1977) "The Systems of Consumer Demand Functions Approach: A Review", in: M. D. Intriligator, ed., *Frontiers of Quantitative Economics*. IIIA, Amsterdam: North-Holland, 23–58.

Berndt, E. R., M. N. Darrough and W. E. Diewert (1977) "Flexible Functional Forms and Expenditure Distributions: An Application to Canadian Consumer Demand Functions", *International Economic Review*, 18, 651–676.

Blackorby, C., R. Boyce and R. R. Russell (1978) "Estimation of Demand Systems Generated by the Gorman Polar Form: A Generalization of the S-branch Utility Tree", *Econometrica*, 46, 345–364.

Caves, D. W. and L. R. Christensen (1980) "Global Properties of Flexible Functional Forms", *American Economic Review*, 70, 422–432.

Christensen, L. R., D. W. Jorgenson and L. J. Lau (1973) "Transcendental Logarithmic Production Frontiers", *Review of Economics and Statistics*, 55, 28–45.

Christensen, L. R., D. W. Jorgenson and L. J. Lau (1975) "Transendental Logarithmic Utility Functions", *American Economic Review*, 65, 367–383.

Cobb, C. W. and P. C. Douglas (1928) "A Theory of Production", *American Economic Review*, 18, 139–165.

Deaton, A. and J. S. Muellbauer (1980a) "An Almost Ideal Demand System", *American Economic Review*, 70, 312–326.

Deaton, A. and J. S. Muellbauer (1980b) *Economics and Consumer Behavior*. Cambridge: Cambridge University Press.

Diewert, W. E. (1971) "An Application of the Shephard Duality Theorem, A Generalized Leontief Production Function", *Journal of Political Economy*, 79, 481–507.

Diewert, W. E. (1973) "Functional Forms for Profit and Transformation Functions", *Journal of Economic Theory*, 6, 284–316.

Diewert, W. E. (1974) "Functional Forms for Revenue and Factor Requirement Functions", *International Economic Review*, 15, 119–130.

Fuss, M. A., D. L. McFadden and Y. Mundlak (1978) "Functional Forms in Production Theory", in: M. A. Fuss and D. L. McFadden, eds., *Production Economics: A Dual Approach to Theory and Applications*. Amsterdam: North-Holland, 1, 219–268.

Gale, D., (1960) *The Theory of Linear Economic Models*. New York: McGraw-Hill.

Gorman, W. M. (1953) "Community Preference Fields", *Econometrica*, 21, 63–80.

Gorman, W. M. (1981) "Some Engel Curves", in: A. S. Deaton, ed., *Essays in the Theory and Measurement of Consumer Behavior: In Honor of Sir Richard Stone*. New York: Cambridge University Press, 7–29.

Griliches, Z. and V. Ringstad (1971) *Economies of Scale and the Form of the Production Function*. Amsterdam: North-Holland.

Hanoch, G. (1971) "CRESH Production Functions", *Econometrica*, 39, 695–712.

Heady, E. O. and J. L. Dillon (1961) *Agricultural Production Functions*. Ames: Iowa State University Press.

Hotelling, H. S. (1932) "Edgeworth's Taxation Paradox and the Nature of Demand and Supply Functions", *Journal of Political Economy*, 40, 577–616.

Houthakker, H. S. (1957) "An International Comparison of Household Expenditure Patterns, Commemorating the Centenary of Engel's Law", *Econometrica*, 25, 532–551.

Houthakker, H. S. (1960) "Additive Preferences", *Econometrica*, 28, 244–257.

Houthakker, H. S. (1965) "New Evidence on Demand Elasticities", *Econometrica*, 33, 277–288.

Jorgenson, D. W. and L. J. Lau (1977) "Statistical Tests of the Theory of Consumer Behavior", in: H. Albach, E. Helmstadter and R. Hemm, eds., *Quantitative Wirtschaftforschung*. Tübingen: J. C. B. Mohr, 384–394.

Jorgenson, D. W. and L. J. Lau (1979) "The Integrability of Consumer Demand Functions", *European Economic Review*, 12, 115–147.

Jorgenson, D. W., L. J. Lau and T. M. Stoker (1980) "Welfare Comparison Under Exact Aggregation", *American Economic Review*, 70, 268–272.

Jorgenson, D. W., L. J. Lau and T. M. Stoker (1982) "The Transcendental Logarithmic Model of Aggregate Consumer Behavior", in: R. L. Basmann and G. F. Rhodes, eds., *Advances in Econometrics*. Greenwhich: JAI Press, Vol. 1.

Klein, L. R. and H. Rubin (1947–1948) "A Constant-Utility Index of the Cost of Living", *Review of Economic Studies*, 15, 84–87.

Lau, L. J. (1977) "Complete Systems of Consumer Demand Functions Through Duality", in: M. D. Intriligator, ed., *Frontiers of Quantitative Economics*. IIIA, Amsterdam: North-Holland, 59–86.

Lau, L. J. (1978) "Applications of Profit Functions", in: M. A. Fuss and D. L. McFadden, eds., *Production Economics: A Dual Approach to Theory and Applications*. Amsterdam: North-Holland, 1, 133–216.

Lau, L. J. (1982) "A Note on the Fundamental Theorem of Exact Aggregation", *Economics Letters*, 9, 119–126.

Lau, L. J., W. L. Lin and P. A. Yotopoulos (1978) "The Linear Logarithmic Expenditure System: An Application to Consumption-Leisure Choice", *Econometrica*, 46, 843–868.

Lau, L. J. and S. Schaible (1984) "A Note on the Domain of Monotonicity and Concavity of the Transcendental Logarithmic Unit Cost Function", Department of Economics, Stanford: Stanford University, mimeographed.

Lau, L. J. and B. A. Van Zummeren (1980) "The Choice of Functional Forms when Prior Information is Diffused". Paper presented at the Fourth World of Congress of the Econometric Society, Aix-en-Provence, France, August 28–September 2, 1980.

McFadden, D. L. (1963) "Further Results on C.E.S. Production Functions", *Review of Economic Studies*, 30, 73–83.

McFadden, D. L. (1964) "Existence Conditions for Theil-Type Preferences", Department of Economics, Berkeley: University of California, mimeographed.

McFadden, D. L. (1978) "Cost, Revenue, and Profit Functions", in: M. A. Fuss and D. L. McFadden, eds., *Production Economics: A Dual Approach to Theory and Applications*. Amsterdam: North-Holland, 1, 3–109.

McFadden, D. L. (1984) "Econometric Analysis of Qualitative Response Models", in: Z. Griliches and M. D. Intriligator, eds., *Handbook of Econometrics*. Amsterdam: North-Holland, Vol. 2.

Muellbauer, J. S. (1975) "Aggregation, Income Distribution, and Consumer Demand", *Review of Economic Studies*, 42, 525–543.

Muellbauer, J. S. (1976) "Community Preferences and the Representative Consumer", *Econometrica*, 44, 979–999.

Nerlove, M. (1963) "Returns to Scale in Electricity Supply", in: C. F. Christ, et al., eds., *Measurement in Economics: Studies in Mathematical Economics and Econometrics in Memory of Yehuda Grunfeld*. Stanford: Stanford University Press, Vol. I.

Pollak, R. A. and T. J. Wales (1978) "Estimation of Complete Demand Systems from Household Budget Data: The Linear and Quadratic Expenditure Systems", *American Economic Review*, 68, 348–359.

Pollak, R. A. and T. J. Wales (1980) "Comparison of the Quadratic Expenditure System and Translog Demand Systems with Alternative Specifications of Demographic Effects", *Econometrica*, 48, 595–612.

Roy, R. (1943) *De l'utilité*. Paris: Hermann.

Schultz, H. (1938) *The Theory and Measurement of Demand*. Chicago: University of Chicago Press.

Shephard, R. W. (1953) *Cost and Production Functions*. Princeton: Princeton University Press.

Shephard, R. W. (1970) *Theory of Cost and Production Functions*. Princeton: Princeton University Press.

Stone, J. R. N. (1953) *The Measurement of Consumer's Expenditure and Behavior in the United Kingdom, 1820–1938*. Cambridge: Cambridge University Press Vol. 1.

Stone, J. R. N. (1954) "Linear Expenditure Systems and Demand Analysis: An Application to the Pattern of British Demand", *Economic Journal*, 64, 511–527.

Theil, H. (1967) *Economics and Information Theory*. Amsterdam: North-Holland.

Uzawa, H. (1962) "Production Functions with Constant Elasticities of Substitution", *Review of Economic Studies*, 29, 291–299.

Wold, H. with L. Jureen (1953) *Demand Analysis*. New York: Wiley.

*Chapter 27*

# LIMITED DEPENDENT VARIABLES

PHOEBUS J. DHRYMES

*Columbia University*

## Contents

*Handbook of Econometrics, Volume III, Edited by Z. Griliches and M.D. Intriligator*
© *Elsevier Science Publishers BV, 1986*

## 0.  Introduction

This is intended to be an account of certain salient themes of the Limited Dependent Variable (LDV) literature. The object will be to acquaint the reader with the nature of the basic problems and the major results rather than recount just who did what when. An extended bibliography is given at the end, that attempts to list as many papers as have come to my attention – even if only by title.

By LDV we will mean instances of (dependent) variables – i.e. variables to be explained in terms of some economic model or rationalizing scheme for which (a) their range is intrinsically a finite discrete set and any attempt to extend it to the real line (or the appropriate multivariable generalization) not only does not lead to useful simplification, but befouls any attempt to resolve the issues at hand; (b) even though their range may be the real (half) line (or the appropriate multivariable generalization) their behavior is conditioned on another process(es).

Examples of the first type are models of occupational choice, entry into labor force, entry into college upon high school graduation, utilization of recreational facilities, utilization of modes of transport, childbearing, etc.

Examples of the latter are models of housing prices and wages in terms of the relevant characteristics of the housing unit or the individual – what is commonly referred to as hedonic price determination. Under this category we will also consider the case of truncated dependent observations.

In examining these issues we shall make an attempt to provide an economic rationalization for the model considered, but our main objective will be to show why common procedures such as least squares fail to give acceptable results; how one approaches these problems by maximum likelihood procedures and how one can handle problems of inference – chiefly by determining the limiting distributions of the relevant estimators. An attempt will be made to handle all problems in a reasonably uniform manner and by relatively elementary means.

## 1.  Logit and probit

### 1.1.  Generalities

Consider first the problem faced by a youth completing high school; or by a married female who has attained the desired size of her family. In the instance of the former the choice to be modelled is going to college or not; in the case of the latter we need to model the choice of entering the labor force or not.

Suppose that as a result of a properly conducted survey we have observations on $T$ individuals, concerning their socioeconomic characteristics and the choices they have made.

In order to free ourselves from dependence on the terminology of a particular subject when discussing these problems, let us note that, in either case, we are dealing with binary choice; let us denote this by

Alternative 1    Going to College or Entering Labor Force

Alternative 2    Not Going to College or Not Entering Labor Force

Since the two alternatives are exhaustive we may make alternative 1 correspond to an abstract event $\mathscr{E}$ and alternative 2 correspond to its complement $\bar{\mathscr{E}}$. In this context it will be correct to say that what we are interested in is the set of factors affecting the occurrence or nonoccurrence of $\mathscr{E}$. What we have at our disposal is some information about the *attributes of these alternatives* and the (*socioeconomic*) *attributes of the individual exercising choice*. Of course we also observe the choices of the individual agent in question. Let

$$y_t = 1 \quad \text{if individual } t \text{ chooses in accordance with event } \mathscr{E},$$
$$\quad = 0 \quad \text{otherwise.}$$

Let

$$w = (w_1, w_2, \ldots, w_s),$$

be a vector of characteristics relative to the alternatives corresponding to the events $\mathscr{E}$ and $\bar{\mathscr{E}}$; finally, let

$$r_t = (r_{t1}, \ldots, r_{tm}),$$

be the vector describing the socioeconomic characteristics of the $t$th individual economic agent.

We may be tempted to model this phenomenon as

$$y_t = x_t.\beta + \varepsilon_t, \qquad t = 1, 2, \ldots, T, \tag{1}$$

where

$$x_t. = (w, r_t.).$$

$\beta$ is a vector of unknown constants and

$$\varepsilon_t : t = 1, 2, \ldots, T,$$

is a sequence of suitably defined error terms.

The formulation in (1) and subsequent estimation by least squares procedures was a common occurrence in the empirical research of the sixties.

## 1.2. Why a general linear model (GLM) formulation is inappropriate

Although the temptation to think of LDV problems in a GLM context is enormous a close examination will show that this is also fraught with considerable problems. At an intuitive level, we seek to approximate the dependent variable by a linear function of some other observables; the notion of approximation is based on ordinary Euclidean distance. That is quite sensible, in the usual GLM context, since no appreciable violence is done to the essence of the problem by thinking of the dependent variable as ranging without restriction over the real line – perhaps after suitably centering it first.

Since the linear function by which we approximate it is similarly unconstrained, it is not unreasonable to think of Euclidean distance as a suitable measure of proximity. Given these considerations we proceed to construct a logically consistent framework in which we can optimally apply various inferential procedures.

In the present context, however, it is not clear whether the notion of Euclidean distance makes a great deal of sense as a proximity measure. Notice that the dependent variable can only assume two possible values, while no comparable restrictions are placed on the first component of the right hand side of (1). Second, note that if we insist on putting this phenomenon in the GLM mold, then for observations in which

$$y_t = 1,$$

we must have

$$\varepsilon_t = 1 - x_t.\beta, \tag{2}$$

while for observations in which

$$y_t = 0,$$

we must have

$$\varepsilon_t = - x_t.\beta. \tag{3}$$

Thus, the error term can only assume two possible values, and we are immediately led to consider an issue that is important to the proper conceptualization of such models, viz., that what we need is *not* a linear model "explaining" the choices individuals make, but rather a model of the probabilities corresponding to the choices in question. Thus, if we ask ourselves: what is the expectation of $\varepsilon_t$, we shall be forced to think of the probabilities attaching to the relations described in (2) and (3) and thus conclude that

$$\varepsilon_t = 1 - x_t.\beta,$$

with probability equal to

$$p_{t1} = P(y_t = 1),\tag{4}$$

and

$$\varepsilon_t = - x_t.\beta,$$

with probability

$$p_{t2} = P(y_t = 0) = 1 - p_{t1}.\tag{5}$$

What we really should be asking is: what determines the probability that the $t$th economic agent chooses in accordance with event $\mathscr{E}$, and eq. (1) should be viewed as a clumsy way of going about it. We see that putting

$$p_{t1} = F(x_t.\beta) = \int_{-\infty}^{x_t.\beta} f(\xi)\,\mathrm{d}\xi,\tag{6}$$

$$p_{t2} = 1 - F(x_t.\beta) = \int_{x_t.\beta}^{\infty} f(\xi)\,\mathrm{d}\xi,\tag{7}$$

where $f(\cdot)$ is a suitable density function with known parameters, formalizes the dependence of the probabilities of choice on the observable characteristics of the individual and/or the alternatives.

To complete the argumentation about why the GLM is inapplicable in the present context we note further

$$E(\varepsilon_t) = F(x_t.\beta)(1 - x_t.\beta) + [1 - F(x_t.\beta)](- x_t.\beta) = F(x_t.\beta) - x_t.\beta,\tag{8}$$

$$\mathrm{Var}(\varepsilon_t) = F(x_t.\beta)[1 - F(x_t.\beta)].\tag{9}$$

Hence, prima facie, least squares techniques are not appropriate, even if the formulations in (1) made intuitive sense.

We shall see that similar situations arise in other LDV contexts in which the absurdity of least squares procedures is not as evident as it is here.

Thus, to recapitulate, least squares procedures are inapplicable

i. because we should be interested in estimating the probability of choice; however, we are using a linear function to predict actual choices, without ensuring that the procedure will yield "predictions" satisfying the conditions that probabilities ought to satisfy

ii. on a technical level the conditions on the error term that are compatible with the desirable properties of least squares estimators in the context of the GLM are patently false in the present case.

### 1.3. A utility maximization motivation

As before, consider an individual, $t$, who is faced with the choice problem as in the preceding section but who is also hypothesized to behave so as to maximize his utility in choosing between the two alternatives. In the preceding it is assumed that the individual's utility contains a random component. It involves little loss in relevance to write the utility function as

$$U_t = u(w, r_{t.}; \theta) + \varepsilon_t, \qquad t = 1, 2, \ldots, T,$$

where

$$u(w, r_{t.}; \theta) \equiv E(U | w, r_{t.}), \qquad \varepsilon_t \equiv U_t - u(w, r_{t.}; \theta).$$

For the moment we shall dispense with the subscript $t$ referring to the $t$th individual.

If the individual chooses according to event $\mathscr{E}$, his utility is (where now any subscripts refer to alternatives),

$$U_1 = u(w, r; \theta_1) + \varepsilon_1. \tag{10}$$

The justification for the parameter vector $\theta$ being subscripted is that, since $w$ is constant across alternatives, $\theta$ must vary. While this may seem unnatural to the reader it is actually much more convenient, as the following development will make clear.

If the individual chooses in accordance with $\bar{\mathscr{E}}$, then

$$U_2 = u(w, r; \theta_2) + \varepsilon_2. \tag{11}$$

Hence, choice is in accordance with event $\mathscr{E}$ if, say,

$$U_1 \geq U_2. \tag{12}$$

But (12) implies

Alternative 1 is chosen or choice is made in accordance with event $\mathscr{E}$ if

$$\varepsilon_2 - \varepsilon_1 \leq u(w, r; \theta_1) - u(w, r; \theta_2), \tag{13}$$

which makes it abundantly clear that we can speak unambiguously *only* about the probabilities of choice. To "predict" choice we need an additional "rule" – such as, for example,

Alternative 1 is chosen when the probability attaching to event $\mathscr{E}$ is 0.5 or higher.

If the functions $u(\cdot)$ in (13) are linear, then the $t$th individual will choose Alternative 1 if

$$\varepsilon_{t2} - \varepsilon_{t1} \leq x_t . \beta, \tag{14}$$

where

$$x_t. = (w, r_t.), \qquad \beta = \theta_1 - \theta_2. \tag{15}$$

Hence, in the notation of the previous section

$$P(y_t = 1) = P(\varepsilon_{t2} - \varepsilon_{t1} \leq x_t.\beta) = \int_{-\infty}^{x_t.\beta} f_t(\xi) \, d\xi = F_t(x_t.\beta), \tag{16}$$

where now $f_t$ is the density function of $\varepsilon_{t2} - \varepsilon_{t1}$.

If

$$f_t(\cdot) = f_{t'}(\cdot), \qquad t \neq t',$$

then we have a basis for estimating the parametric structure of our model. Before we examine estimation issues, however, let us consider some possible distribution for the errors, i.e. the random variables $\varepsilon_{t1}, \varepsilon_{t2}$.

Thus, suppose

$$\varepsilon_{t'}. \sim N(0, \Sigma), \qquad \Sigma = \begin{bmatrix} \sigma_{11} & \sigma_{12} \\ \sigma_{21} & \sigma_{22} \end{bmatrix}, \qquad \Sigma > 0,$$

and the $\varepsilon_{t}.$'s are independent identically distributed (i.i.d.). We easily find that

$$\varepsilon_{t2} - \varepsilon_{t1} \sim N(0, \sigma^2), \qquad \sigma^2 = \sigma_{22} - 2\sigma_{12} + \sigma_{11}.$$

Hence

$$\Pr\{y_t = 1\} = \frac{1}{\sqrt{2\pi\sigma^2}} \int_{-\infty}^{x_t.\beta} e^{-1/2\sigma^2 \xi^2} \, d\xi = \frac{1}{\sqrt{2\pi}} \int_{-\infty}^{v_t} e^{-(1/2)\zeta^2} \, d\zeta = F(v_t),$$

where

$$v_t = \frac{x_t . \beta}{\sqrt{\sigma^2}},$$

and $F(\cdot)$ is the c.d.f. of the unit normal. Notice that in this context it is not possible to identify separately $\beta$ and $\sigma^2$ by observing solely the choices individuals make; we can only identify $\beta/\sigma$.

For reasons that we need not examine here, analysis based on the assumption that errors in (10) and (11) are normally distributed is called *Probit Analysis*.

We shall now examine another specification that is common in applied research, which is based on the logistic distribution. Thus, let $q$ be an exponentially distributed random variable so that its density is

$$g(q) = e^{-q} \qquad q \in (0, \infty), \tag{17}$$

and consider the distribution of

$$v = \ln(q)^{-1} = -\ln q. \tag{18}$$

The Jacobian of this transformation is

$$J(r \to q) = e^{-v}.$$

Hence, the density of $v$ is

$$h(v) = \exp - v \exp - e^{-v} \qquad v \in (-\infty, \infty). \tag{19}$$

If the $\varepsilon_{ti}$, $i = 1, 2$ of (14) are mutually independent with density as in (19), then the joint density is

$$h(\varepsilon_1, \varepsilon_2) = \exp - (\varepsilon_1 + \varepsilon_2) \exp - (e^{-\varepsilon_1} + e^{-\varepsilon_2}). \tag{20}$$

Put

$$v_1 + v_2 = \varepsilon_2,$$
$$v_2 = \varepsilon_1. \tag{21}$$

The Jacobian of this transformation is 1; hence the joint density of the $v_i$, $i = 1, 2$, is given by

$$\exp - (v_1 + 2v_2) \exp - (e^{-v_2} + e^{-v_1 - v_2}).$$

Since

$$v_1 = \varepsilon_2 - \varepsilon_1,$$

the desired density is found as

$$g(v_1) = \int_{-\infty}^{\infty} \exp - (v_1 + 2v_2) \exp - (e^{-v_2} + e^{-v_1 - v_2}) \, \mathrm{d}v_2.$$

To evaluate this put

$$1 + e^{-v_1} = t, \qquad s = te^{-v_2},$$

to obtain

$$g(v_1) = \frac{e^{-v_1}}{t^2} \int_0^{\infty} s e^{-s} \, \mathrm{d}s = \frac{e^{-v_1}}{(1 + e^{-v_1})^2}.$$

Hence, in this case the probability of choosing Alternative 1 is given by

$$P(y_t = 1) = \int_{-\infty}^{x_t \cdot \beta} \frac{e^{-\zeta}}{(1 - e^{-\zeta})^2} \, \mathrm{d}\zeta = \left. \frac{1}{1 + e^{-\zeta}} \right|_{-\infty}^{x_t \cdot \beta} = \frac{1}{1 + e^{-x_t \cdot \beta}},$$

$$P(y_t = 0) = 1 - F(x_t \cdot \beta) = \frac{e^{-x_t \cdot \beta}}{1 + e^{-x_t \cdot \beta}}.$$

This framework of binary or *dichotomous* choice easily generalizes to the case of *polytomous* choice, without any appreciable complication – see, e.g. Dhrymes (1978a).

## 1.4. Maximum likelihood estimation

Although alternative estimation procedures are available we shall examine only the maximum likelihood (ML) estimator, which appears to be the most appropriate, given the sorts of data typically available to economists.

To recapitulate: we have the problem of estimating the parameters in a dichotomous choice context, characterized by a density function $f(\cdot)$; we shall deal with the case where $f(\cdot)$ is the *unit normal* and the *logistic*.

As before we define

$$y_t = 1 \qquad \text{if choice corresponds to event } \mathscr{E}$$
$$= 0 \qquad \text{if choice corresponds to event } \bar{\mathscr{E}}$$

The event $\mathscr{E}$ may correspond to entering the labor force or going to college in the examples considered earlier.

$$P(y_t = 1) = F(x_t.\beta),$$

where

$$x_t. = (w, r_t.),$$

$w$ is the $s$-element row vector describing the relevant attributes of the alternatives and $r_t.$ is the $m$-element row vector describing the relevant socioeconomic characteristics of the $t$th individual.

We recall that a likelihood function may be viewed in two ways: for purposes of estimation we take the sample as given (here the $y_t$'s and $x_t.$'s) and regard it as a function of the unknown parameters (here the vector $\beta$) with respect to which it is to be maximized; for purposes of deriving the limiting distribution of estimators it is appropriate to think of it as a function of the dependent variable(s) – and hence as one that encompasses the probabilistic structure imposed on the model. This dual view of the likelihood function (LF) will become evident below.

The LF is easily determined to be

$$L^* = \prod_{t=1}^{T} F(x_t.\beta)^{y_t} [1 - F(x_t.\beta)]^{1-y_t}. \tag{22}$$

As usual, we find it more convenient to operate with its logarithm

$$\ln L^* = L = \sum_{t=1}^{T} \{ y_t \ln F(x_t.\beta) + (1 - y_t) \ln[1 - F(x_t.\beta)] \}. \tag{23}$$

For purposes of estimation, this form is unduly complicated by the presence of the random variables, $y_t$'s. Given the sample, we will know that some of the $y_t$'s assume the value one and others assume the value zero. We can certainly rearrange the observations so that the first $T_1 \leq T$ observations correspond to

$$y_t = 1, \qquad t = 1, 2, \ldots, T_1,$$

while the remaining $T_2 < T$ correspond to

$$y_{T_1 + t} = 0, \qquad t = 1, 2, \ldots, T_2,$$

If we give effect to these statements the log likelihood function becomes

$$L = \sum_{t=1}^{T_1} \ln F(x_t.\beta) + \sum_{t=T_1+1}^{T_1+T_2} \ln[1 - F(x_t.\beta)], \tag{24}$$

and as such it does not contain any random variables[1] – *even symbolically*! Thus, it is rather easy for a beginning scholar to become confused as to how, solving

$$\frac{\partial L}{\partial \beta} = 0,$$

will yield an estimator, say $\hat{\beta}$, with any probabilistic properties. At least the analogous situation in the GLM

$$y = X\beta + u,$$

using the standard notation yields

$$\hat{\beta} = (X'X)^{-1}X'y,$$

and $y$ is recognized to be a random variable with a probabilistic structure induced by our assumption on the structural error vector $u$.

Thus, we shall consistently avoid the use of the form in (24) and use instead the form in (23). As is well known, the ML estimator is found by solving

$$\frac{\partial L}{\partial \beta} = \sum_{t=1}^{T} \left[ y_t \frac{f(x_t.\beta)}{F(x_t.\beta)} - (1 - y_t) \frac{f(x_t.\beta)}{1 - F(x_t.\beta)} \right] x_t. = 0. \tag{25}$$

We note that, in general, (25) is a highly nonlinear function of the unknown parameter $\beta$ and, hence, can only be solved by iteration.

Since by definition a ML estimator, $\hat{\beta}$, is one obeying

$$L(\hat{\beta}) \geq L(\beta), \text{ for all admissible } \beta, \tag{26}$$

it is important to ensure that solving (25) does, indeed, yield a maximum in the form of (26) and not merely a local stationary point – at least asymptotically.

The assumptions under which the properties of the ML estimator may be established are partly motivated by the reason stated above. These assumptions are

*Assumption A.1.1.*

The explanatory variables are uniformly bounded, i.e. $x_t. \varepsilon H_*$, for all $t$, where $H_*$ is a closed bounded subset of $R_{s+m}$, i.e. the $(s + m)$-dimensional Euclidean space.

*Assumption A.1.2.*

The (admissible) parameter space is, similarly, a closed bounded subset of $R_{s+m}$, say, $P_*$ such that $P_* \supset N(\beta^0)$, where $N(\beta^0)$ is an open neighborhood of the true parameter point $\beta^0$.

---

[1] For any sample, of course, the choice of $T_1$ is random.

*Remark 1*

Assumption (A.1.1.) is rather innocuous and merely states that the socioeconomic variables of interest are bounded. Assumption (A.1.2.) is similarly innocuous. The technical import of these assumptions is to ensure that, at least asymptotically, the maximum maximorum of (24) is properly located by the calculus methods of (25) and to also ensure that the equations in (25) are well defined by precluding a singularity due to

$$F(x_t.\beta) = 0 \qquad \text{or} \qquad 1 - F(x_t.\beta) = 0.$$

Moreover, these assumptions also play a role in the argument demonstrating the consistency of the ML estimator.

   To the above we add another condition, well known in the context of the general linear model (GLM).

*Assumption A.1.3.*

Let

$$X = (x_t.) \qquad t = 1, 2, \ldots, T,$$

where the elements of $x_t.$ are nonstochastic. Then

$$\text{rank}\,(X) = s + m, \qquad \lim_{T \to \infty} \frac{X'X}{T} = M > 0.$$

   With the aid of these assumptions we can easily demonstrate (the proof will not be given here) the validity of the following

*Theorem 1*

Given assumption A.1.1. through A.1.3. the log likelihood function, $L$ of (24), is concave in $\beta$, whether $F(\cdot)$ is the unit normal or the logistic c.d.f..

*Remark 2*

The practical implication of Theorem 1 is that, at any sample size, if we can satisfy ourselves that the LF of (24) does not attain its maximum on the boundary of the parameter space, then a solution to (25), say $\hat{\beta}$, obeys

$$L(\hat{\beta}) \geq L(\beta) \qquad \text{for all admissible } \beta.$$

On the other hand as the sample size tends to infinity then with probability one the condition above is satisfied.

The (limiting) properties of the ML estimator necessary for carrying out tests of hypotheses are given in

*Theorem 2*

The ML estimator, $\hat{\beta}$, in the logistic as well as the normal case is strongly consistent and moreover it obeys

$$\sqrt{T}(\hat{\beta} - \beta) \sim N(0, C),$$

when

$$C^{-1} = - \lim_{T \to \infty} \frac{1}{T} E\left[ \frac{\partial^2 L}{\partial \beta \partial \beta}(\beta^0) \right].$$

*Corollary 1*

A consistent estimator of the covariance matrix of the limiting distribution is given, in the case of normal density, $f$, and c.d.f., $F$, by

$$\hat{C} = \left\{ \frac{1}{T} \Sigma \left[ \frac{f^2(x_t.\hat{\beta})}{F(x_t.\hat{\beta})[1 - F(x_t.\hat{\beta})]} x_t'.x_t. \right] \right\}^{-1}, \tag{27}$$

For the logistic c.d.f. (logit) this reduces to

$$\hat{C} = \left[ \frac{1}{T} \sum_{t=1}^{T} f(x_t.\hat{\beta}) x_t'.x_t. \right]^{-1}. \tag{28}$$

## 1.5.  Goodness of fit

In the context of the GLM the coefficient of determination of multiple regression ($R^2$) has at least three useful interpretations.

i. it stands in a one-to-one relation to the $F$-statistic for testing the hypothesis that the coefficients of the *bona fide* explanatory variables are zero;
ii. it is a measure of the reduction of the variability of the dependent variable through the *bona fide* explanatory variables;
iii. it is the square of the simple correlation coefficient between predicted and actual values of the dependent variable within the sample.

Unfortunately, in the case of the discrete choice models under consideration we do not have a statistic that fits all three characterizations above. We can, on the other hand, define one that essentially performs the first two functions.

In order to demonstrate these facts it will be convenient to represent the maximized (log) LF more informatively. Assuming that the ML estimator corresponds to an interior point of the admissible parameter space we can write

$$L(\hat{\beta}) = L(\beta^0) + \frac{\partial L}{\partial \beta}(\beta^0)(\hat{\beta} - \beta^0) + \frac{1}{2}(\hat{\beta} - \beta^0)' \frac{\partial^2 L}{\partial \beta \partial \beta}(\beta^0)(\hat{\beta} - \beta^0)$$

$$+ \text{ third order terms.} \tag{29}$$

The typical third order term involves

$$\phi_T = \frac{1}{6} \frac{1}{T^{3/2}} \frac{\partial^3 L}{\partial \beta_i \partial \beta_j \partial \beta_k}(\beta^*)\sqrt{T}(\hat{\beta}_i - \beta_i^0)\sqrt{T}(\hat{\beta}_j - \beta_j^0)\sqrt{T}(\hat{\beta}_k - \beta_k^0).$$

It is our contention that

$$\plim_{T \to \infty} \phi_T = 0. \tag{30}$$

Now,

$$\plim_{T \to \infty} \frac{1}{T} \frac{\partial^3 L}{\partial \beta_i \partial \beta_j \partial \beta_k}(\beta^*) = \bar{L}_{ijk},$$

is a well defined, finite quantity, where

$$|\beta^* - \beta^0| \le \hat{\beta} - \beta^0|.$$

But then, (30) is obvious since it can be readily shown that

$$\plim_{T \to \infty} \frac{1}{T^{3/2}} \frac{\partial^3 L}{\partial \beta_i \partial \beta_j \partial \beta_k} = 0,$$

and moreover that

$$\sqrt{T}(\hat{\beta}_i - \beta_i^0), \ \sqrt{T}(\hat{\beta}_j - \beta_j^0), \ \sqrt{T}(\hat{\beta}_k - \beta_k^0),$$

are a.c. finite. Hence, for large samples, approximately

$$L(\hat{\beta}) \sim L(\beta^0) + \frac{\partial L}{\partial \beta}(\beta^0)(\hat{\beta} - \beta^0) + \frac{1}{2}(\hat{\beta} - \beta^0)' \frac{\partial^2 L}{\partial \beta \partial \beta}(\beta^0)(\hat{\beta} - \beta^0).$$

On the other hand, expanding $\dfrac{\partial L}{\partial \beta}$ by Taylor series we find

$$\frac{1}{\sqrt{T}} \frac{\partial L}{\partial \beta}(\beta^0) \sim -\left[\frac{1}{T} \frac{\partial^2 L}{\partial \beta \partial \beta}(\beta^0)\right] \sqrt{T}(\hat{\beta} - \beta^0).$$

Thus,

$$\frac{\partial L}{\partial \beta}(\beta^0)(\hat{\beta} - \beta^0) \sim -(\hat{\beta} - \beta^0)' \frac{\partial^2 L}{\partial \beta \partial \beta}(\beta^0)(\hat{\beta} - \beta^0),$$

and, consequently, for large samples

$$L(\hat{\beta}) \sim L(\beta^0) - \frac{1}{2}(\hat{\beta} - \beta^0)' \frac{\partial^2 L}{\partial \beta \partial \beta}(\beta^0)(\hat{\beta} - \beta^0).$$

Hence

$$2[L(\hat{\beta}) - L(\beta^0)] \sim -(\hat{\beta} - \beta^0)' \frac{\partial^2 L}{\partial \beta \partial \beta}(\beta^0)(\hat{\beta} - \beta^0) \sim \chi^2_{s+m}. \tag{31}$$

Consider now the hypothesis

$$H_0: \qquad \beta^0 = 0, \tag{32}$$

as against

$$H_1: \qquad \beta^0 \neq 0.$$

Under $H_0$

$$L(\beta^0) = \sum_{t=1}^{T} \{y_t \ln F(0) + (1 - y_t)\ln[1 - F(0)]\} = T \ln(\tfrac{1}{2}),$$

and

$$2[L(\hat{\beta}) - T \ln\tfrac{1}{2}] \sim \chi^2_{s+m},$$

is a test statistic for testing the null hypothesis in (32). On the other hand, this is not a useful basis for defining an $R^2$ statistic, for it implicitly juxtaposes the economically motivated model that defines the probability of choice as a function of

$$x_t.\beta,$$

and the model based on the *principle of insufficient reason* which states that the probability to be assigned to choice corresponding to the event $\mathscr{E}$ and that corresponding to its complement $\bar{\mathscr{E}}$ are both $\frac{1}{2}$. It would be far more meaningful to consider the null hypothesis to be

$$\beta^0 = \begin{pmatrix} \beta^0_0 \\ 0 \end{pmatrix},$$

i.e. to follow for a nonzero constant term, much as we do in the case of the GLM. The null hypothesis as above would correspond to assigning a probability to choice corresponding to event $\mathscr{E}$ by

$$\bar{y} = F(\tilde{\beta}_0) \quad \text{or} \quad \tilde{\beta}_0 = F^{-1}(\bar{y}),$$

where

$$\bar{y} = \frac{1}{T} \sum_{t=1}^{T} y_t.$$

Thus, for some null hypothesis $H_0$, let

$$L(\tilde{\beta}) = \sup_{H_0} L(\beta).$$

By an argument analogous to that leading to (31) we conclude that

$$2[L(\hat{\beta}) - L(\tilde{\beta})] \sim -(\hat{\beta} - \beta^0)' \frac{\partial^2 L}{\partial \beta \partial \beta}(\beta^0)(\hat{\beta} - \beta^0)$$

$$+ (\tilde{\beta} - \beta^0)' \frac{\partial^2 L}{\partial \beta \partial \beta}(\beta^0)(\tilde{\beta} - \beta^0). \tag{33}$$

In fact, (33) represents a transform of the likelihood ratio (LR) and as such it is a LR test statistic. We shall now show that in the case where

$$H_0: \quad \beta^0_{(2)} = 0, \quad \beta^0 = \begin{pmatrix} \beta^0_{(1)} \\ \beta^0_{(2)} \end{pmatrix},$$

the quantity in the right member of (33) reduces to a test[2] based on the marginal (limiting) distribution of

$$\sqrt{T}\left(\hat{\beta}_{(2)}-\beta_{(2)}^{0}\right).$$

To this effect put

$$\tilde{C}_{*} = \frac{1}{T}\frac{\partial^{2}L}{\partial\beta\,\partial\beta}(\beta^{0}),$$

and note that

$$\frac{1}{\sqrt{T}}\frac{\partial L}{\partial\beta}(\beta^{0}) \sim -\tilde{C}_{*}\sqrt{T}(\hat{\beta}-\beta^{0}). \tag{34}$$

Partitioning

$$\tilde{C}_{*} = \begin{bmatrix} C_{*11} & C_{*12} \\ C_{*21} & C_{*22} \end{bmatrix},$$

conformably with

$$(\hat{\beta}-\beta^{0}) = \begin{bmatrix} \hat{\beta}_{(1)}-\beta_{(1)}^{0} \\ \hat{\beta}_{(2)}-\beta_{(2)}^{0} \end{bmatrix},$$

we find

$$\frac{1}{\sqrt{T}}\frac{\partial L}{\partial\beta_{(1)}}(\beta^{0}) \sim -\left[\tilde{C}_{*11}\sqrt{T}\left(\hat{\beta}_{(1)}-\beta_{(1)}^{0}\right)+\tilde{C}_{*12}\sqrt{T}\left(\hat{\beta}_{(2)}-\beta_{(2)}^{0}\right)\right],$$

$$\frac{1}{\sqrt{T}}\frac{\partial L}{\partial\beta_{(2)}}(\beta^{0}) \sim -\left[\tilde{C}_{*21}\sqrt{T}\left(\hat{\beta}_{(1)}-\beta_{(1)}^{0}\right)+\tilde{C}_{*22}\sqrt{T}\left(\hat{\beta}_{(2)}-\beta_{(2)}^{0}\right)\right]. \tag{35}$$

Using (34) we can rewrite (33) as

$$-2[L(\hat{\beta})-L(\tilde{\beta})] \sim -(\hat{\beta}-\beta^{0})'\frac{\partial L}{\partial\beta}(\beta^{0})+(\tilde{\beta}-\beta^{0})'\frac{\partial L}{\partial\beta}(\beta^{0})$$

$$= -\left\{\left[(\hat{\beta}_{(1)}-\beta_{(1)}^{0})-(\tilde{\beta}_{(1)}-\beta_{(1)}^{0})\right]'\frac{\partial L}{\partial\beta_{(1)}}(\beta^{0})\right.$$

$$\left.+(\hat{\beta}_{(2)}-\beta_{(2)}^{0})'\frac{\partial L}{\partial\beta_{(2)}}(\beta^{0})\right\}.$$

---

[2] It should be remarked that a similar result in the context of the GLM is called, somewhat redundantly, a Chow test.

From (34) we find, bearing in mind that under $H_0$ we estimate

$$\tilde{\beta} = \begin{pmatrix} \tilde{\beta}_{(1)} \\ 0 \end{pmatrix},$$

$$\sqrt{T}\left(\tilde{\beta}_{(1)} - \beta_{(1)}^0\right) \sim -\tilde{C}_{*11}^{-1}\frac{1}{\sqrt{T}}\frac{\partial L}{\partial \beta_{(1)}}(\beta^0). \tag{36}$$

From (35) we find

$$-\sqrt{T}\left(\hat{\beta}_{(1)} - \beta_{(1)}^0\right) \sim \tilde{C}_{*11}^{-1}\left[\frac{1}{\sqrt{T}}\frac{\partial L}{\partial \beta_{(1)}}(\beta^0) + \tilde{C}_{*12}\sqrt{T}\left(\hat{\beta}_{(2)} - \beta_{(2)}^0\right)\right]. \tag{37}$$

Hence

$$\left[\sqrt{T}\left(\tilde{\beta}_{(1)} - \beta_{(1)}^0\right) - \sqrt{T}\left(\hat{\beta}_{(1)} - \beta_{(1)}^0\right)\right] \sim \tilde{C}_{*11}^{-1}\tilde{C}_{*12}\sqrt{T}\left(\hat{\beta}_{(2)} - \beta_{(2)}^0\right),$$

and thus (33) may be further rewritten as

$$-2[L(\hat{\beta}) - L(\tilde{\beta})] \sim \left[\hat{\beta}_{(2)} - \beta_{(2)}^0\right]'\left[\tilde{C}_{*21}C_{*11}^{-1}\frac{\partial L}{\partial \beta_{(1)}}(\beta^0) - \frac{\partial L}{\partial \beta_{(2)}}(\beta^0)\right]. \tag{38}$$

Again, from (35) we see that

$$\tilde{C}_{*21}\tilde{C}_{*11}^{-1}\frac{1}{\sqrt{T}}\frac{\partial L}{\partial \beta_{(1)}}(\beta^0) - \frac{1}{\sqrt{T}}\frac{\partial L}{\partial \beta_{(2)}}(\beta^0)$$

$$\sim \left[\tilde{C}_{*22} - \tilde{C}_{*21}\tilde{C}_{*11}^{-1}\tilde{C}_{*12}\right] \cdot \sqrt{T}\left(\hat{\beta}_{(2)} - \beta_{(2)}^0\right),$$

and thus (38) reduces to

$$-2[L(\hat{\beta}) - L(\tilde{\beta})] \sim T\left(\hat{\beta}_{(2)} - \beta_{(2)}^0\right)'\left[\tilde{C}_{*22} - \tilde{C}_{*21}\tilde{C}_{*11}^{-1}\tilde{C}_{*12}\right]\left(\hat{\beta}_{(2)} - \beta_{(2)}^0\right). \tag{39}$$

But under $H_0$, (39) is exactly the test statistic based on the (limiting) marginal distribution of

$$\sqrt{T}\left(\hat{\beta}_{(2)} - \beta_{(2)}^0\right) \sim N(0, 1 - C_{22}), \tag{40}$$

where

$$C_{22} = \plim_{T \to \infty}\left[\tilde{C}_{*22} - \tilde{C}_{*21}\tilde{C}_{*11}^{-1}\tilde{C}_{*12}\right]^{-1}. \tag{41}$$

In the special case where

$$\beta_{(1)} = \beta_0,$$

i.e. it is the constant term in the expression

$$x_t.\beta,$$

so that no bona fide explanatory variables "explain" the probability of choice, we can define $R^2$ by

$$R^2 = 1 - \frac{L(\hat{\beta})}{L(\tilde{\beta})}. \tag{42}$$

The quantity in (42) has the property

i. $R^2 \in [0,1)$
ii. the larger the contribution of the bona fide variables to the maximum of the LF the closer is $R^2$ to 1
iii. $R^2$ stands in a one-to-one relation to the chi-square statistic for testing the hypothesis that the coefficients of the bona fide variables are zero. In fact, under $H_0$

$$-2L(\tilde{\beta})R^2 \sim \chi^2_{s+m-1}.$$

It is desirable, in empirical practice, that a statistic like $R^2$ be reported and that a constant term be routinely included in the specification of the linear functional

$$x_t.\beta,$$

Finally, we should also stress that $R^2$ as in (42) *does not* have the interpretation as the square of the correlation coefficient between "predicted" and "actual" observations.

## 2. Truncated dependent variables

### 2.1. Generalities

Suppose we have a sample conveying information on consumer expenditures; in particular, suppose we are interested in studying household expenditures on consumer durables. In such a sample survey it would be routine that many households report zero expenditures on consumer durables. This was, in fact, the situation faced by Tobin (1958) and he chose to model household expenditure on consumer durables as

$$
\begin{aligned}
y_t &= x_t.\beta + u_t, && \text{if } x_t.\beta + u_t > 0 \\
&= 0 && \text{otherwise}
\end{aligned} \tag{43}
$$

The same model was later studied by Amemiya (1973). We shall examine below the inference and distributional problem posed by the manner in which the model's dependent variable is truncated.

## 2.2.  Why simple OLS procedures fail

Let us append to the model in (43) the standard assumptions that

(A.2.1.)  The $\{u_t:\ t=1,2,\dots\}$ is a sequence of i.i.d. random variables with

$$
u_t \sim N(0,\sigma^2), \qquad \sigma^2 \in (0,\infty).
$$

(A.2.2.)  The elements of $x_t.$ are bounded for all $t$, i.e.

$$
|x_{ti}| < k_i, \qquad \text{for all } t, \qquad i=1,2,\dots,n,
$$

are linearly independent and

$$
(p)\ \lim_{T\to\infty} \frac{X'X}{T} = M,
$$

exists as a nonsingular nonstochastic matrix.

(A.2.3.)  If the elements of $x_t.$ are stochastic, then $x_t., u_t,$ are mutually independent for all $t, t'$, i.e. the error and data generating processes are mutually independent.

(A.2.4.)  The parameter space, say $H \subset R_{n+2}$, is compact and it contains an open neighborhood of the true parameter point $(\beta^{0\prime}, \sigma_0^2)'$.

The first question that occurs is why not use the entire sample to estimate $\beta$? Thus, defining

$$
X = (x_t.), \qquad t=1,2,\dots,T,
$$

$$
u = (u_1, u_2,\dots, u_T)', \qquad y^{(1)} = (y_1, y_2,\dots, y_{T1})', \qquad y^{(2)} = (0,\dots,0)',
$$

$$
y = \left(y^{(1)\prime},\ y^{(2)\prime}\right)',
$$

we may write

$$y = X\beta + u,$$

and estimate $\beta$ by

$$\tilde{\beta} = (X'X)^{-1}X'y. \tag{44}$$

A little reflection will show, however, that this leads to serious and palpable specification error *since in (43) we do not assert that the zero observations are generated by the same process that generates the positive observations.* Indeed, a little further reflection would convince us that it would be utterly inappropriate to insist that the same process that generates the zero observations should also generate the nonzero observations, since for the zero observations we should have that

$$u_t = -x_t.\beta, \qquad t = T_{1+1},\ldots,T_1+T_2,$$

and this would be inconsistent with assumption (A.1.1.).

We next ask, why not confine our sample solely to the nonzero observations,

$$y^{(1)} = X_1\beta + u_{(1)},$$

and thus estimate $\beta$ by

$$\tilde{\beta} = (X_1'X_1)^{-1}X_1'y^{(1)}.$$

This may appear quite reasonable at first, even though it is also apparent that we are ignoring some (perhaps considerable) information. Deeper probing, however, will disclose a much more serious problem. After all, ignoring some sample elements would affect only the degrees of freedom and the *t*- and *F*-statistics alone. If we already have a large sample, throwing out even a substantial part of it will not affect matters much. But now it is in order to ask: What is the process by which some dependent variables are assigned the value zero? A look at (43) convinces us that it is a random process governed by the behavior of the error process and the characteristics relevant to the economic agent, $x_t.$. Conversely, the manner in which the sample on the basis of which we shall estimate $\beta$ is selected is governed by some aspects of the error process. In particular we note that for us to observe a positive $y_t$, according to

$$y_t = x_t.\beta + u_t, \tag{45}$$

the error process should satisfy

$$u_t > - x_t.\beta. \tag{46}$$

Thus, for the positive observations we should be dealing with the *truncated* distribution function of the error process. But, what is the mean of the truncated distribution? We have, if $f(\cdot)$ is the density and $F(\cdot)$ the c.d.f. of $u_t$

$$E(u_t|u_t > - x_t.\beta) = \frac{1}{1 - F(- x_t.\beta)} \int_{-x_t.\beta}^{\infty} \xi f(\xi) \, d\xi.$$

If $f(\cdot)$ is the $N(0, \sigma^2)$ density the integral can be evaluated as

$$f(x_t.\beta),$$

and, in addition, we also find

$$1 - F(- x_t.\beta) = F(x_t.\beta).$$

Moreover, if we denote by $\phi(\cdot)$, $\Phi(\cdot)$ the $N(0,1)$ density and c.d.f., respectively, and by

$$\nu_t = \frac{x_t.\beta}{\sigma}, \tag{47}$$

then

$$E(u_t|u_t > + x_t.\beta) = \sigma \frac{\phi(\nu_t)}{\Phi(\nu_t)} = \sigma\psi_t. \tag{48}$$

Since the mean of the error process in (45) is given by (48) we see that we are committing a misspecification error by leaving out the "variable" $\phi(\nu_t)/\Phi(\nu_t)$ [see Dhrymes (1978a)].

Defining

$$v_t = u_t - \sigma \frac{\phi(\nu_t)}{\Phi(\nu_t)}, \tag{49}$$

we see that $\{v_t: t = 1, 2, \ldots\}$ is a sequence of independent *but non-identically distributed random variables*, since

$$\text{Var}(v_t) = \sigma^2(1 - \nu_t\psi_t - \psi_t^2). \tag{50}$$

Thus, there is no simple procedure by which we can obtain efficient and/or consistent estimators by confining ourselves to the positive subsample; consequently, we are forced to revert to the entire sample and employ ML methods.

## 2.3. Estimation of parameters with ML methods

We are operating with the model in (43), subject to (A.2.1.) through (A.2.4.) and the convention that the first $T_1$ observations correspond to positive dependent variables, while the remaining $T_2$, $(T_1 + T_2 = T)$, correspond to zero observations.

Define

$$
\begin{aligned}
c_t &= 1 \quad\quad \text{if } y_t > 0,\\
&= 0 \quad\quad \text{otherwise,}
\end{aligned}
\tag{51}
$$

and note that the (log) LF can be written as

$$
L = \sum_{t=1}^{T} \left\{ (1 - c_t)\ln \Phi(-\nu_t) - c_t\left[ \frac{1}{2}\ln(2\pi) + \frac{1}{2}\ln\sigma^2 + \frac{1}{2\sigma^2}(y_t - x_t.\beta)^2 \right] \right\}.
\tag{52}
$$

Differentiating with respect to $\gamma = (\beta', \sigma^2)'$, we have

$$
\frac{\partial L}{\partial \beta} = -\frac{1}{\sigma} \sum_{t=1}^{T} \left\{ (1 - c_t)\frac{\phi(\nu_t)}{\Phi(-\nu_t)} - c_t\left( \frac{y_t - x_t.\beta}{\sigma} \right) \right\} x_t. = 0,
\tag{53}
$$

$$
\frac{\partial L}{\partial \sigma^2} = -\frac{1}{2\sigma^2} \sum_{t=1}^{T} \left\{ c_t\left[ 1 - \frac{1}{\sigma^2}(y_t - x_t.\beta)^2 \right] - (1 - c_t)\frac{\nu_t\phi(\nu_t)}{\Phi(-\nu_t)} \right\} = 0,
$$

and these equations have to be solved in order to obtain the ML estimator. It is, first, interesting to examine how the conditions in (53) differ from the equations to be satisfied by simple OLS estimators applied to the positive component of the sample. By simple rearrangement we obtain, using the convention alluded to above,

$$
X_1'X_1\beta = X_1'y^{(1)} - \sigma \sum_{t=T_1+1}^{T} \psi(-\nu_t)x_t'.,
\tag{54}
$$

$$
\sigma^2 = \frac{1}{T_1}(y^{(1)} - X_1\beta)'(y^{(1)} - X_1\beta) + \frac{\sigma^2}{T_1} \sum_{t=T_1+1}^{T} \psi(-\nu_t)\nu_t,
\tag{55}
$$

where

$$
\psi(\nu_t) = \frac{\phi(\nu_t)}{\Phi(\nu_t)}, \quad\quad \psi(-\nu_t) = \frac{\phi(\nu_t)}{\Phi(-\nu_t)}.
\tag{56}
$$

Since these expressions occur very frequently, we shall often employ the abbreviated notation

$$\psi_t = \psi(\nu_t), \qquad \psi_t^* = \psi(-\nu_t).$$

Thus, if in some sense

$$z_{T\cdot}' = \sum_{t=T_1+1}^{T} \psi_t^* x_{t\cdot}', \tag{57}$$

is negligible, the ML estimator, say $\hat{\beta}$, could yield results that are quite similar, from an applications point of view, to those obtained through the simple OLS estimator, say $\tilde{\beta}$, as applied to the positive component of the sample. From (54) it is evident that if $z_{T\cdot}'$ of (57) is small then

$$\sigma^2 \sum_{t=T_1+1}^{T} \psi_t^* \nu_t = \sigma z_{T\cdot}\beta,$$

is also small. Hence, under these circumstances

$$\hat{\beta} = \tilde{\beta}, \quad \hat{\sigma}^2 = \tilde{\sigma}^2$$

which explains the experience occasionally encountered in empirical applications.

The eqs. (53) or (54) and (55) are highly nonlinear and can only be solved by iterative methods. In order to ensure that the root of

$$\frac{\partial L}{\partial \gamma} = 0, \qquad \gamma = (\beta', \sigma^2)',$$

so located is the ML estimator it is necessary to show either that the equation above has only one root – which is difficult – or that we begin the iteration with an initial consistent estimator.

### 2.4.   An initial consistent estimator

Bearing in mind the development in the preceding section we can rewrite the model describing the positive component of the sample as

$$y_t = x_{t\cdot}\beta + \sigma\psi_t + v_t = \sigma(\nu_t + \psi_t) + v_t, \tag{58}$$

such that

$$\{v_t : t = 1, 2, \dots\},$$

is a sequence of mutually independent random variables with

$$E(v_t) = 0, \qquad \text{Var}(v_t) = \sigma^2(1 - \nu_t \psi_t - \psi_t^2),$$

(59)

and such that they are independent of the explanatory variables $x_t$.

The model in (58) cannot be estimated by simple means owing to the fact that $\psi_t$ is not directly observable; thus, we are forced into nonstandard procedures.

We shall present below a modification and simplification of a consistent estimator due to Amemiya (1973). First we note that, confining our attention to the positive component of the sample

$$y_t^2 = \sigma^2(\nu_t + \psi_t)^2 + v_t^2 + 2v_t(\nu_t + \psi_t)\sigma.$$

(60)

Hence

$$E(y_t^2|x_{t\cdot}, u_t > -x_{t\cdot}\beta) = \sigma^2(\nu_t^2 + \nu_t\psi_t) + \sigma^2$$

$$= x_{t\cdot}\beta E(y_t|x_{t\cdot}, u_t > -x_{t\cdot}\beta) + \sigma^2.$$

(61)

Defining

$$\varepsilon_t \equiv y_t^2 - E(y_t^2|x_{t\cdot}, u_t > -x_{t\cdot}\beta),$$

(62)

we see that $\{\varepsilon_t: t = 1, 2, \dots\}$ is a sequence of independent random variables with mean zero and, furthermore, we can write

$$w_t = y_t^2 = x_{t\cdot}\beta y_t + \sigma^2 + \varepsilon_t, \qquad t = 1, 2, \dots, T_1.$$

(63)

The problem, of course, is that $y_t$ is correlated with $\varepsilon_t$ and hence simple regression will not produce a consistent estimator for $\beta$ and $\sigma^2$.

However, we can employ an instrumental variables (I.V.) estimator[3]

$$\bar{\gamma} = (\tilde{X}'_* X_*)^{-1} X'_* w, \qquad w = (w_1, w_2, \dots, w_{T_1})',$$

(64)

[3] It is here that the procedure differs from that suggested by Amemiya (1973). He defines

$$\tilde{y}_t = x_{t\cdot}(X_1'X_1)^{-1}X_1'y^{(1)},$$

while we define

$$\tilde{y}_t = x_{t\cdot}a,$$

for nontrivial vector $a$.

where

$$X_* = (D_y X_1, e), \qquad \tilde{X}_* = (D_{\tilde{y}} X_1, e), \tag{65}$$

and

$$\tilde{y}_t = x_{t.} a, \qquad D_{\tilde{y}} = \operatorname{diag}(\tilde{y}_1, \tilde{y}_2, \ldots, \tilde{y}_{T_1}), \qquad D_y = (y_1, y_2, \ldots, y_{T_1}), \tag{66}$$

for an arbitrary nontrivial vector $a$.

It is clear that by substitution we find

$$\tilde{\gamma} = \gamma + (\tilde{X}'_* X_*)^{-1} \tilde{X}'_* \varepsilon. \tag{67}$$

We easily establish that

$$\tilde{X}'_* X_* = \begin{bmatrix} X_1' D_{\tilde{y}} D_y X_1 & X_1' \tilde{y} \\ y' X_1 & e'e \end{bmatrix}.$$

Clearly

$$\lim_{T \to \infty} \frac{e'e}{T_1} = 1, \qquad \lim_{T \to \infty} \frac{1}{T_1} X_1' \tilde{y} = \left( \lim_{T \to \infty} \frac{X_1' X_1}{T} \right) a,$$

$$\operatorname*{plim}_{T \to \infty} \frac{1}{T} X_1' y = \left( \lim_{T \to \infty} \frac{X_1' X_1}{T_1} \right) \beta + \operatorname*{plim}_{T \to \infty} \frac{X_1' u_{.1}}{T_1}.$$

Now

$$\frac{1}{T_1} X_1' u_{.1} = \frac{1}{T_1} \sum_{t=1}^{T_1} x_{t.}' u_t,$$

and

$$\{ x_{t.}' u_t : t = 1, 2, \ldots \},$$

is a sequence of independent random variables with mean

$$E(x_{t.}' u_t) = \sigma x_{t.}' \psi_t, \tag{68}$$

and covariance matrix

$$\operatorname{Cov}(x_{t.}' u_t) = \sigma^2 (1 - \nu_t \psi_t - \psi_t) x_{t.}' x_{t.} = \omega_t x_{t.}' x_{t.}, \tag{69}$$

where

$$\omega_t = \sigma^2 \left( 1 - \nu_t \psi_t - \psi_t^2 \right),$$

is uniformly bounded by assumption (A.2.2) and (A.2.4). Hence, by (A.2.2)

$$\lim_{T_1 \to \infty} \frac{1}{T_1} \sum_{t=1}^{T_1} \omega_t x'_{t.} x_{t.},$$

converges to a matrix with finite elements. Further and similar calculations will show that

$$\frac{\tilde{X}'_* X_*}{T},$$

converges a.c. to a nonsingular matrix. Thus, we are reduced to examining the limiting behavior of

$$\frac{1}{\sqrt{T_1}} \tilde{X}'_* \varepsilon = \frac{1}{\sqrt{T_1}} \sum_{t=1}^{T_1} \begin{pmatrix} x_{t.} a x'_{t.} \\ 1 \end{pmatrix} \varepsilon_t. \tag{70}$$

But this is a sequence of independent nonidentically distributed random variables with mean zero and uniformly bounded (in $x_{t.}$ and $\beta$) moments to any finite order. Now for any arbitrary $(n+2 \times 1)$ vector $\alpha^*$ consider

$$\frac{1}{\sqrt{T_1}} \alpha^{*\prime} \tilde{X}'_* \varepsilon = \frac{1}{\sqrt{T_1}} \sum_{t=1}^{T_1} \alpha_t \varepsilon_t, \tag{71}$$

where

$$\alpha_t = (x_{t.} \alpha)(x_{t.} a), \qquad \alpha^* = (\alpha', \alpha_{n+2}),$$

and note that

$$\lim_{T_1 \to \infty} \frac{S_{T_1}^{*2}}{T_1} = S,$$

is well defined where

$$S_{T_1}^{*2} = \sum_{t=1}^{T_1} \alpha_t^2 \text{Var}(\varepsilon_t). \tag{72}$$

Define, further

$$S_{T_1}^2 = \frac{S_{T_1}^{*2}}{T_1},$$

and note that

$$S_{T_1}^* = T_1^{1/2} S_{T_1}.$$

But then it is evident that Liapounov's condition is satisfied, i.e. with $K$ a uniform bound on $E|\alpha_t \varepsilon_t|^{2+\delta}$

$$\lim_{T_1 \to \infty} \frac{\sum\limits_{t=1}^{T_1} E|\alpha_t \varepsilon_t|^{2+\delta}}{S_{T_1}^{*2+\delta}} \le K \lim_{T \to \infty} \frac{T_1}{T_1^{1+\delta/2} S_{T_1}^{2+\delta}} = \lim_{T_1 \to \infty} \frac{K}{T_1^{\delta/2} S^{2+\delta}} = 0.$$

By a theorem of Varadarajan, see Dhrymes (1970), we conclude that

$$\frac{1}{\sqrt{T_1}} \tilde{X}_*' \varepsilon \sim N(0, H),$$

where

$$H = \lim_{T \to \infty} \frac{1}{T_1} \begin{bmatrix} \sum\limits_{t=1}^{T_1} (x_t.a)^2 x_t'.x_t. \mathrm{Var}(\varepsilon_t) & \sum\limits_{t=1}^{T_1} (x_t.a) x_t'. \mathrm{Var}(\varepsilon_t) \\ \sum\limits_{t=1}^{T_1} (x_t.a) x_t. \mathrm{Var}(\varepsilon_t) & \sum\limits_{t=1}^{T_1} \mathrm{Var}(\varepsilon_t) \end{bmatrix}. \tag{73}$$

Consequently we have shown that

$$\sqrt{T_1} (\tilde{\gamma} - \gamma) \sim N(0, Q^{-1}HQ^{-1}),$$

where

$$Q = \lim_{\mathrm{a.c.}} \frac{(\tilde{X}_*' X_*)}{T_1}. \tag{74}$$

Moreover since

$$\sqrt{T_1} (\tilde{\gamma} - \gamma) \sim \zeta,$$

where $\zeta$ is an a.c. finite random vector it follows that

$$\tilde{\gamma} - \gamma_0 \sim \frac{\zeta}{\sqrt{T_1}},$$

which shows that $\tilde{\gamma}$ converges a.c. to $\gamma_0$.

We may summarize the development above in

*Lemma 1*

Consider the model in (43) subject to assumptions (A.2.1.) through (A.2.4.); further consider the I.V. estimator of the parameter vector $\gamma$ in

$$w_t = (x_t. y_t, 1)\gamma + \varepsilon_t, \qquad w_t = y_t^2,$$

given by

$$\tilde{\gamma} = (\tilde{X}'_* X_*)^{-1} \tilde{X}'_* w,$$

where $\tilde{X}_*$, $X_*$ and $w$ are as defined in (65) and (66). Then

  i.  $\tilde{\gamma}$ converges to $\gamma_0$ almost certainly,

  ii.  $\sqrt{T_1}(\tilde{\gamma} - \gamma_0) \sim N(0, Q^{-1}HQ'^{-1})$,

where $Q$ and $H$ are as defined in (74) and (73) respectively.

## 2.5. Limiting properties and distribution of the ML estimator

Returning now to eqs. (53) or (54) and (55) we observe that since the initial estimator, say $\tilde{\gamma}$, is strongly consistent, at each step of the iterative procedure we get a (strongly) consistent estimator. Hence, at convergence, the estimator so determined, say $\hat{\gamma}$, is guaranteed to be (strongly) consistent.

The perceptive reader may ask: Why did we not use the apparatus of Section 1.d. instead of going through the intermediate step of obtaining the initial consistent estimator? The answer is, essentially, that Theorem 1 (of Section 1.d.) does not hold in the current context. To see that, recall the (log) LF of our problem and write it as

$$L_T(\gamma) = \frac{1}{T} \sum_{t=1}^{T} \left\{ (1 - c_t)\ln \Phi(-v_t) - c_t \right.$$

$$\left. \cdot \left[ \frac{1}{2}\ln(2\pi) + \frac{1}{2}\ln \sigma^2 + \frac{1}{2\sigma^2}(y_t - x_t.\beta)^2 \right] \right\}. \tag{75}$$

Since $L_T$ is at least twice differentiable it is concave if and only if its Hessian is negative (semi)definite over the space of admissible $\gamma$-parameters. After some manipulation we can show that

$$\frac{\partial^2 L_T}{\partial \sigma^2 \partial \sigma^2} = -\frac{1}{4\sigma^4} \frac{1}{T} \left\{ \sum_{t=1}^{T} (1-c_t) \frac{\nu_t \phi(\nu_t)}{\Phi(-\nu_t)} \left( 3 - \nu_t^2 + \frac{\nu_t \phi(\nu_t)}{\Phi(-\nu_t)} \right) \right.$$
$$\left. + c_t \left[ 4 \frac{(y_t - x_t.\beta)^2}{\sigma^2} - 2 \right] \right\}.$$

When $\beta = 0$ the entire first term in brackets is null so that the derivative reduces to

$$-\frac{1}{\sigma^4} \frac{1}{T} \sum_{t=1}^{T} c_t \left( \frac{y_t^2}{\sigma^2} - \frac{1}{2} \right),$$

which could well be positive for some realizations. Hence, we cannot unambiguously assert that over some (large) compact subset of $R_{n+2}$, the Hessian of the (log) LF is negative semidefinite. Consequently, we have no assurance that, if we attempted to solve

$$\frac{\partial L_T}{\partial \gamma}(\gamma) = 0, \tag{76}$$

beginning with an arbitrary initial point, say $\tilde{\gamma}$, upon convergence we should arrive at the consistent root of (93). On the other hand, from the general theory of ML estimation we know that if the true parameter point is in the interior of the $\gamma$-admissible space then (76) has at most one consistent root. Of course, it may have many roots if the function $L_T$ is nonconcave and herein lies the problem. In the previous Section, however, because of Theorem 1 we knew that the (log) likelihood function was concave and hence starting from an arbitrary point we could locate, upon convergence, the global maximizer and hence the ML estimator.

Many of the other results of Section 1.d., however, are available to us in virtue of

*Lemma 2*

The (log) LF of the problem of this section as exhibited in (75) converges a.c. uniformly in $\gamma$. In particular

$$L_T(\gamma) \overset{\text{a.c.}}{\to} \lim_{T \to \infty} E[L_T(\gamma)],$$

uniformly in $\gamma$.

*Proof*

Consider the log LF of (75) and in particular its $t$th term

$$\xi_t = (1 - c_t)\ln \Phi(-\nu_t) - c_t\left[\frac{1}{2}\ln(2\pi) + \frac{1}{2}\ln\sigma^2 + \frac{1}{2\sigma^2}(y_t - x_t.\beta)^2\right],$$

$$t = 1, 2, \ldots. \quad (77)$$

For any $x$-realization

$$\{\xi_t: t = 1, 2, \ldots\},$$

is a sequence of independent random variables with uniformly bounded moments in virtue of assumption (A.2.1) through (A.2.3). Thus, there exists a constant, say $k$, such that

$$\text{Var}(\xi_t) < k, \qquad \text{for all } t.$$

Consequently, by Kolmogorov's criterion, for all admissible $\gamma$,

$$\{L_T(\gamma) - E[L_T(\gamma)]\} \overset{\text{a.c.}}{\rightarrow} 0. \qquad\qquad \text{Q.E.D.}$$

*Remark 3*

The device of beginning the iterative process for solving (76) with a consistent estimator ensures that for sufficiently large $T$ we will be locating the estimator, say $\hat{\gamma}_T$, satisfying

$$L_T(\hat{\gamma}_T) = \sup_{\gamma} L_T(\gamma).$$

Lemma 2, can be shown to imply that

$$L_T(\hat{\gamma}_T) \overset{\text{a.c.}}{\rightarrow} \overline{L}(\overline{\gamma}, \gamma^0), \qquad \overline{L}(\overline{\gamma}, \gamma^0) = \sup_{\gamma} \overline{L}(\gamma, \gamma^0).$$

Moreover, we can also show that

$$\overline{\gamma} = \gamma^0.$$

On the other hand, it is not possible to show routinely that $\hat{\gamma}_T \overset{\text{a.c.}}{\rightarrow} \gamma^0$. Essentially, the problem is the term corresponding to $\sigma^2$ which contains expressions like

$$c_t\frac{(y_t - x_t.\beta)^2}{\sigma^2},$$

which cannot be (absolutely) bounded. This does not prevent us from showing convergence a.c. of $\hat{\gamma}_T$ to $\gamma^0$. By the iterative process we have shown that $\hat{\gamma}_T$ converges to $\gamma^0$ at least in probability. Convergence a.c. is shown easily once we obtain the limiting distribution of $\hat{\gamma}_T$ – a task to which we now turn.

Thus, as before, consider the expansion

$$\frac{\partial L_T}{\partial \gamma}(\hat{\gamma}_T) = \frac{\partial L_T}{\partial \gamma}(\gamma^0) + \frac{\partial^2 L_T}{\partial \gamma \partial \gamma}(\gamma^*)(\hat{\gamma}_T - \gamma^0), \tag{78}$$

where $\gamma^0$ is the true parameter point and

$$|\hat{\gamma}_T - \gamma^0| \le |\gamma^* - \gamma^0|.$$

We already have an explicit expression in eq. (53) for the derivative $\partial L_T / \partial \gamma$. So let us obtain the Hessian of the LF. We find

$$\frac{\partial^2 L_T}{\partial \beta \partial \beta}(\gamma) = -\frac{1}{\sigma^2}\frac{1}{T}\sum_{t=1}^{T}\left\{(1-c_t)\psi_t^*(\psi_t^* - \nu_t) + c_t\right\}x_{t.}'x_{t.},$$

$$\frac{\partial^2 L_T}{\partial \beta \partial \sigma^2} = -\frac{1}{2\sigma^3}\frac{1}{T}\sum_{t=1}^{T}\left\{2c_t\left(\frac{y_t - x_{t.}\beta}{\sigma}\right) - (1-c_t)\psi_t^*(1 + \nu_t\psi_t^* - \nu_t^2)\right\}x_{t.},$$

$$\tag{79}$$

$$\frac{\partial^2 L_T}{\partial \sigma^2 \partial \sigma^2} = -\frac{1}{4\sigma^4}\frac{1}{T}\sum_{t=1}^{T}\left\{c_t^2\left(\frac{y_t - x_{t.}\beta}{\sigma}\right)^2 + (1-c_t)\nu_t\psi_t^*(1 + \nu_t\psi_t^* - \nu_t^2)\right\}$$

$$+ \frac{1}{2\sigma^4}\frac{1}{T}\sum_{t=1}^{T}\left\{c_t\left[1 - \left(\frac{y_t - x_{t.}\beta}{\sigma}\right)^2\right] - (1-c_t)\nu_t\psi_t^*\right\}.$$

We may now define

$$\xi_{1t} = (1-c_t)\frac{\phi(\nu_t^0)}{\Phi(-\nu_t^0)} - c_t\left(\frac{y_t - x_{t.}\beta^0}{\sigma_0}\right), \tag{80}$$

$$\xi_{2t} = c_t\left[1 - \left(\frac{y_t - x_{t.}\beta^0}{\sigma_0}\right)^2\right] - (1-c_t)\frac{\nu_t^0\phi(\nu_t^0)}{\Phi(-\nu_t^0)},$$

and

$$\xi_{11t} = (1-c_t)\psi_t^{*0}(\psi_t^{*0} - \nu_t^0) + c_t,$$

$$\xi_{12t} = \xi_{21t} = (1-c_t)\psi_t^{*0}(1 + \nu_t^0 - \nu_t^0\psi_t^{*0}), \tag{81}$$

$$\xi_{22t} = c_t^2\left(\frac{y_t - x_{t.}\beta^0}{\sigma_0}\right)^2 + (1-c_t)\nu_t^0\psi_t^{*0}(1 + \nu_t^0\psi_t^{*0} - \nu_t^{02}),$$

where, evidently,

$$\psi_t^{*0} = \frac{\phi(v_t^0)}{\Phi(-v_t^0)}, \qquad v_t^0 = \frac{x_t. \beta^0}{\sigma_0}, \qquad \psi_t^0 = \frac{\phi(v_t^0)}{\Phi(v_t^0)}.$$

With the help of the notation in (80) and (81) we find

$$\frac{\partial L_T}{\partial \gamma'}(\gamma^0) = -\frac{1}{\sigma_0}\frac{1}{T}\sum_{t=1}^{T}\begin{bmatrix} x_t'. & 0 \\ 0 & \frac{1}{2\sigma_0} \end{bmatrix}\begin{bmatrix} \xi_{1t} \\ \xi_{2t} \end{bmatrix}, \tag{82}$$

and

$$\frac{\partial^2 L_T}{\partial \gamma \partial \gamma}(\gamma^0) = -\frac{1}{\sigma_0^2}\frac{1}{T}\sum_{t=1}^{T}\begin{bmatrix} \xi_{11t}x_t'.x_t. & \frac{1}{2\sigma_0}\xi_{12t}x_t'. \\ \frac{1}{2\sigma_0}\xi_{21t}x_t. & \frac{1}{4\sigma^2}\xi_{22t} \end{bmatrix} + \Omega_{*T}, \tag{83}$$

where $\Omega_{*T}$ is a matrix all of whose elements are zero except the last diagonal element, which is

$$\frac{1}{T}\sum_{t=1}^{T}\frac{1}{2\sigma^4}\xi_{2t}..$$

Thus, for every $T$ we have

$$E(\Omega_{*T}) = 0. \tag{84}$$

Consequently, we are now ready to prove

**Theorem 3**

Consider the model of eq. (43) subject to assumption (A.2.1.) through (A.2.4.); moreover, consider the ML estimator, $\hat{\gamma}_T$, obtained by iteration from an initial consistent estimator as a solution of (76). Then

$$\sqrt{T}(\hat{\gamma}_T - \gamma^0) \sim N(0, \sigma_0^2 C^{-1}),$$

where

$$C = \lim_{T \to \infty}\frac{1}{T}\sum_{t=1}^{T}\begin{bmatrix} \omega_{11t}x_t'.x_t. & \frac{1}{2\sigma_0}\omega_{12t}x_t'. \\ \frac{1}{2\sigma_0}\omega_{21t}x_t. & \frac{1}{4\sigma_0^2}\omega_{22t} \end{bmatrix},$$

and

$$\omega_{ijt} = E(\xi_{ijt}) \qquad i, j = 1, 2.$$

*Proof*

From the expansion in (78) and the condition under which the ML estimator is obtained we find

$$\sqrt{T}(\hat{\gamma}_T - \gamma^0) = -\left[\frac{\partial^2 L_T}{\partial \gamma \partial \gamma}(\gamma^*)\right]^{-1} \frac{1}{\sqrt{T}} \frac{\partial L}{\partial \gamma}(\gamma^0).$$

But

$$\frac{1}{\sqrt{T}} \frac{\partial L}{\partial \gamma}(\gamma^0) = -\frac{1}{\sigma_0} \frac{1}{\sqrt{T}} \sum_{t=1}^{T} \begin{bmatrix} x'_{t\cdot} & 0 \\ 0 & \frac{1}{2\sigma_0} \end{bmatrix} \begin{bmatrix} \xi_{1t} \\ \xi_{2t} \end{bmatrix}. \tag{85}$$

The right member of (85) involves the sum of a sequence of independent random variables with mean zero. Moreover, it is easily verified that such variables have uniformly bounded moments to order at least four. Hence, a Liapounov condition holds. Since the covariance matrix of each term is

$$\frac{1}{\sigma_0^2} \begin{bmatrix} x'_{t\cdot} x_{t\cdot} \omega_{11t} & \frac{1}{2\sigma_0} x'_{t\cdot} \omega_{12t} \\ \frac{1}{2\sigma_0} x_{t\cdot} \omega_{21t} & \frac{1}{4\sigma_0^2} \omega_{22t} \end{bmatrix},$$

with

$$\begin{aligned}
\omega_{11t} &= E(\xi_{1t}^2) = \Phi(\nu_t^0) + \psi(\nu_t^0)[\psi_t^{*0} - \nu_t^0], \\
\omega_{12t} &= \omega_{21t} = E(\xi_{1t}\xi_{2t}) = \phi(\nu_t^0)[1 - \nu_t^0 \psi_t^{*0} + \nu_t^{02}], \\
\omega_{22t} &= E(\xi_{2t}^2) = 2\Phi(\nu_t^0) - \nu_t^0 \phi(\nu_t^0)[1 - \nu_t^0 \psi_t^{*0} + \nu_t^{02}].
\end{aligned} \tag{86}$$

Thus we see that

$$\frac{1}{\sqrt{T}} \frac{\partial L}{\partial \gamma}(\gamma^0) \sim N\left(0, \frac{1}{\sigma_0^2} C\right).$$

From (79) we also verify that

$$\frac{\partial^2 L_T}{\partial \gamma \partial \gamma}(\gamma^*) - \frac{\partial^2 L_T}{\partial \gamma \partial \gamma}(\gamma^0),$$

converges in probability to the null matrix, element by element. But the elements of

$$\frac{\partial^2 L_T}{\partial \gamma \partial \gamma}(\gamma^0),$$

are seen to be sums of independent random variables with finite means and bounded variances; hence, they obey a Kolmogorov criterion and thus

$$\frac{\partial^2 L_T}{\partial \gamma \partial \gamma}(\gamma^*) \overset{\text{a.c.}}{\rightarrow} \lim_{T \to \infty} E\left[\frac{\partial^2 L_T}{\partial \gamma \partial \gamma}(\gamma^0)\right].$$

We easily verify that

$$E(\xi_{11t}) = \omega_{11t}, \; E(\xi_{12t}) = E(\xi_{21t}) = \phi(\nu_t^0)\left[1 - \nu_t^0 \psi_t^{*0} + \nu_t^{02}\right]$$

$$= \omega_{12t} = \omega_{21t},$$

$$E(\xi_{22t}) = \omega_{22t}.$$

Hence

$$\underset{T \to \infty}{\text{plim}} \frac{\partial^2 L_T}{\partial \gamma \partial \gamma}(\gamma^*) = \lim_{T \to \infty} -\frac{1}{\sigma_0^2}\frac{1}{T}\sum_{t=1}^{T}\begin{bmatrix} \omega_{11t}x_t'.x_t. & \dfrac{1}{2\sigma_0}\omega_{12t}x_t'. \\[2ex] \dfrac{1}{2\sigma_0}\omega_{21t}x_t. & \dfrac{1}{4\sigma_0^2}\omega_{22t} \end{bmatrix}$$

$$= -\frac{1}{\sigma_0^2}C,$$

and, moreover,

$$\sqrt{T}(\hat{\gamma} - \gamma^0) \sim N(0, \sigma_0^2 C^{-1}) \tag{Q.E.D.}$$

*Corollary 2*

The estimator $\hat{\gamma}_T$ converges a.c. to $\gamma_0$.

*Proof*

From Theorem 3

$$\sqrt{T}\left(\hat{\gamma}_T - \gamma^0\right) \sim \zeta,$$

where $\zeta$ is an a.c. finite random variable; hence,

$$\hat{\gamma}_T - \gamma^0 \sim \frac{\zeta}{\sqrt{T}},$$

and thus

$$\hat{\gamma}_T \overset{\text{a.c.}}{\to} \gamma^0.$$

*Corollary*

The marginal (limiting) distribution of $\hat{\beta}_T$ is given by

$$\sqrt{T}\left(\hat{\beta}_T - \beta\right) \sim N\left(0, \sigma_0^2 P^{-1}\right),$$

where

$$P = \lim_{T \to \infty} \frac{1}{T} \sum_{t=1}^{T} \left(\omega_{11t} - \frac{\omega_{21t}^2}{\omega_{22t}}\right) x_t'.x_t. \tag{87}$$

*Proof*

Evident from the definition of $C$ in Theorem 3.

*Remark 4*

The unknown parameters of the limiting distribution of $\hat{\gamma}_T$ can be estimated by the standard procedure as

$$-\frac{\partial^2 L_T(\hat{\gamma}_T)}{\partial \gamma \partial \gamma}.$$

However, it would be much preferable to estimate $C$ as

$$
\hat{C} = \frac{1}{\hat{\sigma}^2} \frac{1}{T} \sum_{t=1}^{T}
\begin{bmatrix}
\hat{\omega}_{11t} x'_{t.} x_{t.} & \dfrac{1}{2\hat{\sigma}} \hat{\omega}_{12t} x'_{t.} \\[2ex]
\dfrac{1}{2\hat{\sigma}} \hat{\omega}_{21t} x_{t.} & \dfrac{1}{4\hat{\sigma}^2} \hat{\omega}_{22t}
\end{bmatrix},
$$

with $\hat{\omega}_{ijt}$ given as in (86) evaluated at $\hat{\gamma}_T$.

## 2.6. Goodness of fit

In the context of the truncated dependent variable model the question arises as to what we would want to mean by a "goodness of fit" statistic.

As analyzed in the Section on discrete choice models the usual $R^2$, in the context of the GLM, serves a multiplicity of purposes; when we complicate the process in which we operate it is not always possible to define a single statistic that would be meaningful in all contexts.

Since the model is

$$
\begin{aligned}
y_t &= x_{t.}\beta + u_t && \text{if } x_{t.}\beta + u_t > 0, \\
&= 0 && \text{if } u_t \le -x_{t.}\beta,
\end{aligned}
$$

the fitted model may "describe well" the first statement but poorly the second or vice versa. A useful statistic for the former would be the square of the simple correlation coefficient between predicted and actual $y_t$. Thus, e.g. suppose we follow our earlier convention about the numbering of observations; then for the positive component of the sample we put

$$
\hat{y}_t = x_{t.}\hat{\beta} + \hat{\sigma}\hat{\psi}_t, \qquad t = 1, 2, \ldots, T_1. \tag{88}
$$

An intuitively appealing statistic is

$$
r^2 = \frac{\left[ \displaystyle\sum_{t=1}^{T_1} (\hat{y}_t - \bar{\hat{y}})(y_t - \bar{y}) \right]^2}{\left[ \displaystyle\sum_{t=1}^{T_1} (\hat{y}_t - \bar{\hat{y}})^2 \right]\left[ \displaystyle\sum_{t=1}^{T_1} (y_t - \bar{y})^2 \right]}, \tag{89}
$$

where

$$
\bar{y} = \frac{1}{T_1} \sum_{t=1}^{T_1} y_t, \qquad \bar{\hat{y}} = \frac{1}{T_1} \sum_{t=1}^{T_1} \hat{y}_t. \tag{90}
$$

As to how well it discriminates between the zero and positive (dependent variable) observations we may compute $\Phi(-\nu_t)$ for all $t$; in the perfect discrimination case

$$\Phi(-\hat{\nu}_{t_2}) > \Phi(-\hat{\nu}_{t_1}), \qquad t_1 = 1, 2, \ldots, T_1, \qquad t_2 = T_1 + 1, \ldots, T. \tag{91}$$

The relative frequency of the reversal of ranks would be another interesting statistic, as would the average probability difference, i.e.

$$\frac{1}{T_2} \sum_{t_2 = T_1 + 1}^{T} \Phi(-\hat{\nu}_{t_2}) - \frac{1}{T_1} \sum_{t_1 = 1}^{T_1} \Phi(-\hat{\nu}_{t_1}) = d. \tag{92}$$

We have a "right" to expect as a minimum that

$$d > 0. \tag{93}$$

## 3. Sample selectivity

### 3.1. Generalities

This is another important class of problems that relate specifically to the issue of how observations on a given economic phenomenon are generated. More particularly, we hypothesize that whether a certain variable, say $y_{t1}^*$, is *observed or not depends on another variable, say $y_{t2}^*$*. Thus, the observability of $y_{t1}^*$ depends on the probability structure of the stochastic process that generates $y_{t2}^*$, as well as on that of the stochastic process that governs the behavior of $y_{t1}^*$. The variable $y_{t2}^*$ may be inherently unobservable although we assert that we know the variables that enter its "systematic part."

To be precise, consider the model

$$y_{t1}^* = x_{t1}^* . \beta_{.1}^* + u_{t1}^*,$$

$$t = 1, 2, \ldots, T, \tag{94}$$

$$y_{t2}^* = x_{t2}^* . \beta_{.2}^* + u_{t2}^*,$$

where $x_{t1}^*.$, $x_{t2}^*.$ are $r_1$, $r_2$-element row vectors of observable "exogenous" variables which may have elements in common. The vectors

$$u_{t\cdot}^* = \left( u_{t1}^*, u_{t2}^* \right), \qquad t = 1, 2, \ldots,$$

form a sequence of i.i.d. random variables with distribution

$$u_{t\cdot}^{*\prime} \sim N(0, \Sigma^*), \qquad \Sigma^* > 0.$$

The variable $y_{t2}^*$ is inherently unobservable, while $y_{t1}^*$ is observable if and only if

$$y_{t1}^* \geq y_{t2}^*.$$

An example of such a model is due to Heckman (1979) where $y_{t1}^*$ is an observed wage for the $t$th worker and $y_{t2}^*$ is his reservation wage. Evidently, $y_{t1}^*$ is the "market valuation" of his skills and other pertinent attributes, represented by the vector $x_{t1}^*$, while $y_{t2}^*$ represents, through the vector $x_{t2}^*$, those personal and other relevant attributes that lead him to seek employment at a certain wage or higher.

Alternatively, in the market for housing $y_{t1}^*$ would represent the "market valuation" of a given structure's worth while $y_{t2}^*$ would represent the current owner's evaluation.

Evidently a worker accepts a wage for employment or a structure changes hands if and only if

$$y_{t1}^* \geq y_{t2}^*.$$

If the covariance matrix, $\Sigma^*$, is diagonal, then there is no correlation between $y_{t1}^*$ and $y_{t2}^*$ and hence in view of the assumption regarding the error process

$$\{ u_{t\cdot}^{*\prime} : t = 1, 2, \ldots \},$$

we could treat the sample

$$\{ ( y_{t1}^*, x_{t1}^* ) : t = 1, 2, \ldots, T \},$$

as one of i.i.d. observations; consequently, we can estimate consistently the parameter vector $\beta_{\cdot 1}^*$ by OLS given the sample, irrespectively of the second relation in (94).

On the other hand, if the covariance matrix, $\Sigma^*$, is not diagonal, then the situation is far more complicated, since now there does exist a stochastic link between $y_{t1}^*$ and $y_{t2}^*$. The question then becomes: If we apply OLS to the first equation in (94) do we suffer more than just the usual loss in efficiency?

## 3.2.   *Inconsistency of least squares procedures*

In the current context, it would be convenient to state the problem in canonical form before we attempt further analysis. Thus, define

$$y_{t1} = y_{t1}^*, \qquad y_{t2} = y_{t1}^* - y_{t2}^*, \qquad x_{t1\cdot} = x_{t1\cdot}^*, \qquad x_{t2\cdot} = \left( x_{t1\cdot}^*, x_{t2\cdot}^* \right),$$

$$\beta_{\cdot 1} = \beta_{\cdot 1}^*, \qquad \beta_{\cdot 2} = \begin{pmatrix} \beta_{\cdot 1}^* \\ -\beta_{\cdot 2}^* \end{pmatrix}, \qquad u_{t1} = u_{t1}^*, \qquad u_{t2} = u_{t1}^* - u_{t2}^*, \tag{95}$$

with the understanding that if $x_{t1\cdot}^*$ and $x_{t2\cdot}^*$ have elements in common, say,

$$x_{t1\cdot}^* = \left( z_{t1}, z_{t1}^* \right), \qquad x_{t2\cdot}^* = \left( z_{t1\cdot}, z_{t2\cdot}^* \right),$$

then

$$x_{t2\cdot} = \left( z_{t1\cdot}, z_{t1\cdot}^*, z_{t2\cdot}^* \right), \qquad \beta_{\cdot 2} = \begin{pmatrix} \beta_{\cdot 11}^* - \beta_{\cdot 12}^* \\ \beta_{\cdot 21}^* \\ -\beta_{\cdot 22}^* \end{pmatrix}, \tag{96}$$

where $\beta_{\cdot 11}^*, \beta_{\cdot 12}^*$ are the coefficients of $z_{t1\cdot}$ in $x_{t1\cdot}^*$ and $x_{t2\cdot}^*$ respectively, $\beta_{21}^*$ is the coefficient of $z_{t1}^*$ and $\beta_{\cdot 22}^*$ is the coefficient of $z_{t2}^*$.

Hence, the model in (94) can be stated in the canonical form

$$\begin{cases} y_{t1} = x_{t1\cdot}\beta_{\cdot 1} + u_{t1}, \\ y_{t2} = x_{t2\cdot}\beta_{\cdot 2} + u_{t2}, \end{cases} \tag{97}$$

*such that $x_{t2\cdot}$ contains at least as many elements as $x_{t1\cdot}$,*

$$\left\{ u_{t\cdot}' = (u_{t1}, u_{t2})' : t = 1, 2, \dots \right\},$$

is a sequence of i.i.d. random variables with distribution

$$u_{t\cdot}' \sim N(0, \Sigma), \qquad \Sigma > 0,$$

and subject to the condition that $y_{t1}$ is observable (observed) if and only if $y_{t2} \geq 0$.

If we applied OLS methods to the first equation in (97) do we obtain, at least, consistent estimators for its parameters? The answer hinges on whether that question obeys the standard assumptions of the GLM.

Clearly, and solely in terms of the system in (97),

$$\left\{ u_{t1} : t = 1, 2, \dots \right\}, \tag{98}$$

is a sequence of i.i.d. random variables and if in (94) we are prepared to assert that the standard conditions of the typical GLM hold, nothing in the subsequent discussion suggests a correlation between $x_{t1.}$ and $u_{t1}$; hence, if any problem should arise it ought to be related to the probability structure of the sequence in (98) insofar as it is associated with observable $y_{t1}$ – a problem to which we now turn. We note that the conditions hypothesized by the model imply that (potential) realizations of the process in (98) are conditioned on[4]

$$u_{t2} \geq - x_{t2}.\beta_{.2}. \tag{99}$$

Or, perhaps more precisely, we should state that (implicit) realizations of the process in (98) associated with *observable realizations*

$$\{ y_{t1}: t = 1, 2, \ldots \},$$

are conditional on (99). Therefore, in dealing with the error terms of (potential) samples the marginal distribution properties of (98) are not relevant; what *are relevant* are its conditional properties – as conditioned by (99).

We have

*Lemma 3*

The distribution of realizations of the process in (98) as conditioned by (99) has the following properties:

i. The elements $\{ u_{t1}, u_{t2} \}$ are mutually independent for $t \neq t'$.

ii. The density of $u_{t1}$, given that the corresponding $y_{t1}$ is observable (observed) is

$$f(u_{t1}|u_{t2} > - x_{t2}.\beta_{.2}) = \frac{\Phi(\pi_t)}{\Phi(\nu_{t2})} \frac{1}{\sqrt{2\pi\sigma_{11}}} \exp - \frac{1}{2\sigma_{11}} u_{t1}^2, \tag{100}$$

where

$$\nu_{t2} = \frac{x_{t2}.\beta_{.2}}{\sigma_{22}^{1/2}}, \qquad \pi_t = \frac{1}{\alpha^{1/2}} \left( \nu_{t2} + \frac{\rho_{12}}{\sigma_{11}^{1/2}} u_{t1} \right),$$

$$\rho_{12}^2 = \frac{\sigma_{12}^2}{\sigma_{11}\sigma_{22}} \qquad \alpha = 1 - \rho_{12}^2, \tag{101}$$

and $\Phi(\cdot)$ is the c.d.f. of a $N(0,1)$.

---

[4] Note that in terms of the original variables (99) reads

$$u_{t1}^* \geq u_{2t}^* + x_{t2}^*.\beta_{.2}^* - x_{t1}^*.\beta_{.1}^*.$$

We shall not use this fact in subsequent discussion, however.

*Proof*

i. is quite evidently valid since by the standard assumptions of the GLM we assert that $(x_{t1}^*, x_{t2}^*)$ and $u_{t\cdot}^* = (u_{t1}^*, u_{t2}^*)$ are mutually independent and that

$$\{u_{t\cdot}^{*\prime}: t = 1, 2, \ldots\},$$

is a sequence of i.i.d. random variables.

As for part ii. we begin by noting that since the conditional density of $u_{t1}$ given $u_{t2}$ is given by

$$u_{t1} | u_{t2} \sim N\left(\frac{\sigma_{12}}{\sigma_{22}} u_{t2}, \sigma_{11}\alpha\right),$$

and since the restriction in (99) restricts us to the space

$$u_{t2} \geq -x_{t2} . \beta_{.2},$$

the required density can be found as

$$f(u_{t1} | u_{t2} \geq -x_{t2} . \beta_{.2}) = \frac{1}{\Phi(\nu_{t2})} \frac{1}{\sqrt{2\pi\alpha\sigma_{11}}} \frac{1}{\sqrt{2\pi\sigma_{22}}}$$

$$\cdot \int_{-x_{t2} . \beta_{.2}}^{\infty} \exp - \frac{1}{2\sigma_{11}}\left(u_{t1} - \frac{\sigma_{12}}{\sigma_{22}}\xi\right)^2 \exp - \frac{1}{2\sigma_{22}}\xi^2 \, \mathrm{d}\xi.$$

Completing the square (in $\xi$) and making the change in variable

$$\zeta = \left(\xi - \frac{\sigma_{12}}{\sigma_{11}} u_{t1}\right) / (\sigma_{22}\alpha)^{1/2},$$

we find

$$f(u_{t1} | u_{t2} \geq -x_{t2} . \beta_{.2}) = \frac{\Phi(\pi_t)}{\Phi(\nu_{t2})} \frac{1}{\sqrt{2\pi\sigma_{11}}} \exp - \frac{1}{2\sigma_{11}} u_{t1}^2.$$

To verify that this is, indeed, a density function we note that it is everywhere nonnegative and

$$\int_{-\infty}^{\infty} f(\xi_1 | u_{t2} \geq -x_{t2} . \beta_{.2}) \, \mathrm{d}\xi_1$$

$$= \frac{1}{\Phi(\nu_{t2})} \int_{-\infty}^{\infty} \frac{1}{\sqrt{2\pi\sigma_{11}}} \left[\int_{-\infty}^{\pi_t} \frac{1}{\sqrt{2\pi}} \exp - \frac{1}{2}\xi_2^2 \, \mathrm{d}\xi_2\right] \cdot \exp - \frac{1}{2\sigma_{11}}\xi_1^2 \, \mathrm{d}\xi_1.$$

Making the transformation

$$\zeta_1 = \frac{1}{\sigma_{11}^{1/2}} \xi_1, \qquad \zeta_2 = \alpha^{1/2} \xi_2 - \rho_{12} \zeta_1,$$

the integral is reduced to

$$\frac{1}{\Phi(\nu_{t2})} \int_{-\infty}^{\infty} \frac{1}{\sqrt{2\pi\alpha}} \frac{1}{\sqrt{2\pi}} \int_{-\infty}^{\nu_{t2}} \exp - \frac{1}{2\alpha} (\zeta_2 + \rho_{12}\zeta_1)^2 \exp - \tfrac{1}{2}\zeta_1^2 \, \mathrm{d}\zeta_2 \mathrm{d}\zeta_1$$

$$= \frac{1}{\Phi(\nu_{t2})} \int_{-\infty}^{\nu_{t2}} \frac{1}{\sqrt{2\pi}} \left[ \int_{-\infty}^{\infty} \frac{1}{\sqrt{2\pi\alpha}} \exp - \frac{1}{2\alpha} (\zeta_1 + \rho_{12}\zeta_2)^2 \, \mathrm{d}\zeta_1 \right]$$

$$\cdot \exp - \tfrac{1}{2}\zeta_2^2 \, \mathrm{d}\zeta_2 = 1. \hspace{4cm} \text{Q.E.D.}$$

*Lemma 4*

The $k$th moment of realizations of the process in (98) corresponding to observable realizations $\{ y_{t1}: t = 1, 2, \ldots \}$ is given, for $k$ even $(k = 2, 4, 6, \ldots)$, by

$$I_{k,t} = \sigma_{11}(k-1)I_{k-2,t} - \sigma_{11}^{k/2}\alpha^{(k-2)/2}\rho_{12}^2\nu_{t2}\psi(\nu_{t2}) \sum_{r=0}^{(k-2)/2} \binom{k-1}{2r+1}\left(\frac{\rho_{12}^2\nu_{t2}^2}{\alpha}\right)^r$$

$$\cdot \frac{\left[2\left(\dfrac{k-2}{2} - r\right)\right]!}{2^{\frac{k-2}{2} - r}\left(\dfrac{k-2}{2} - r\right)!}, \hspace{3cm} (102)$$

while for $k$ odd $(k = 3, 5, 7, \ldots)$ it is given by

$$I_{k,t} = \sigma_{11}(k-1)I_{k-2,t} + \sigma_{11}^{k/2}\alpha^{(k-1)/2}\rho_{12}\psi(\nu_{t2}) \sum_{r=0}^{(k-1)/2} \binom{k-1}{2r}\left(\frac{\rho_{12}^2\nu_{t2}^2}{\alpha}\right)^r$$

$$\cdot \frac{\left[2\left(\dfrac{k-1}{2} - r\right)\right]!}{2^{\frac{k-1}{2} - r}\left(\dfrac{k-1}{2} - r\right)!}, \hspace{3cm} (103)$$

where

$$\psi(\nu_{t2}) = \frac{\phi(\nu_{t2})}{\Phi(\nu_{t2})}, \qquad I_{0,t} = 1, \qquad I_{1,t} = \sigma_{11}^{1/2}\rho_{12}\psi(\nu_{t2}). \hspace{2cm} (104)$$

## Remark 5

It is evident, from the preceding discussion, that the moments of the error process corresponding to observable $y_{t1}$ are uniformly bounded in $\beta_{\cdot 1}$, $\beta_{\cdot 2}$, $\sigma_{11}$, $\sigma_{12}$, $\sigma_{22}$, $x_{t1\cdot}$ and $x_{t2\cdot}$ – provided the parameter space is compact and the elements of $x_{t1\cdot}, x_{t2\cdot}$ are bounded.

## Remark 6

It is also evident from the preceding that for (potential) observations from the model

$$y_{t1} = x_{t1\cdot}\beta_{\cdot 1} + u_{t1},$$

we have that

$$E(y_{t1}|x_{t1\cdot}) = x_{t1\cdot}\beta_{\cdot 1} + \sigma_{11}^{1/2}\rho_{12}\psi(\nu_{t2}). \tag{105}$$

We are now in a position to answer the question, raised earlier, whether OLS methods applied to the first equation in (97) will yield at least consistent estimators. In this connection we observe that the error terms of *observations* on the first equation of (97) obey

$$E(u_{t1}|u_{t2} \geq -x_{t2\cdot}\beta_{\cdot 2}) = I_{1t} = \sigma_{11}^{1/2}\rho_{12}\psi(\nu_{t2}),$$

$$\mathrm{Var}(u_{t1}|u_{t2} \geq -x_{t2\cdot}\beta_{\cdot 2}) = I_{2t} - I_{1t}^2 = \sigma_{11} - \sigma_{11}\rho_{12}^2\nu_{t2}\psi(\nu_{t2})$$

$$- \sigma_{11}\rho_{12}^2\psi^2(\nu_{t2})$$

$$= \sigma_{11} - \sigma_{11}\rho_{12}^2\psi(\nu_{t2})[\nu_{t2} + \psi(\nu_{t2})].$$

As is well known, the second equation shows the errors to be *heteroskedastic* – whence we conclude that OLS estimators *cannot be efficient*. The first equation above shows the errors to have a nonzero mean. As shown in Dhrymes (1978a) a nonconstant (nonzero) mean implies misspecification due to left out variables and *hence inconsistency*.

Thus, OLS estimators are inconsistent; hence, we must look to other methods for obtaining suitable estimators for $\beta_{\cdot 1}, \sigma_{11}$, etc. On the other hand, *if, in* (105), $\rho_{12} = 0$, *then OLS estimators would be* consistent but inefficient.

## 3.3.   The LF and ML estimation

We shall assume that in our sample we have entities for which $y_{t1}$ is observed and entities for which it is not observed; if $y_{t1}$ is not observable, then we know that

$y_{t2} < 0$, hence that

$$u_{t2} < - x_{t2}.\beta_{.2}.$$

Consequently, the probability attached to that event is

$$\Phi(- \nu_{t2}).$$

Evidently, the probability of observing $y_{t1}$ is $\Phi(\nu_{t2})$ and given that $y_{t1}$ is observed the probability it will assume a value in some internal $\Delta$ is

$$\frac{1}{\Phi(\nu_{t2})} \frac{1}{\sqrt{2\pi\sigma_{11}}} \int_{\Delta} \Phi(\pi_t) \exp - \frac{1}{2\sigma_{11}} \xi^2 \, d\xi.$$

Hence, the unconditional probability that $y_{t1}$ will assume a value in the interval $\Delta$ is

$$\frac{1}{\sqrt{2\pi\sigma_{11}}} \int_{\Delta} \Phi(\pi_t) \exp - \frac{1}{2\sigma_{11}} \xi^2 \, d\xi.$$

Define

$$c_t = 1 \qquad \text{if } y_{t1} \text{ is observed,}$$
$$= 0 \qquad \text{otherwise,}$$

and note that the LF is given by

$$L^* = \prod_{t=1}^{T} \left[ \Phi(\nu_{t2}) f(y_{t1} - x_{t1}.\beta_{.1} | u_{t2} \geq - x_{t2}.\beta_{.2}) \right]^{c_t} \left[ \Phi(- \nu_{t2}) \right]^{1 - c_t}. \qquad (106)$$

Thus, e.g. if for a given sample we have no observations on $y_{t1}$ the LF becomes

$$\prod_{t=1}^{T} \Phi(- \nu_{t2}),$$

while, if all sample observations involve observable $y_{t1}$'s the LF becomes

$$\prod_{t=1}^{T} \left\{ \frac{1}{\sqrt{2\pi\sigma_{11}}} \Phi\left[ \frac{1}{\alpha^{1/2}} \left( \nu_{t2} + \rho_{12} \left( \frac{y_{t1} - x_{t1}.\beta_{.1}}{\sigma_{11}^{1/2}} \right) \right) \right] \exp - \frac{1}{2\sigma_{11}} (y_{t1} - x_{t1}.\beta_{.1})^2 \right\}.$$

Finally, if the sample contains entities for which $y_{t1}$ is observed as well as entities for which it is not observed, then we have the situation in (106). We shall examine the estimation problems posed by (106) in its general form.

*Remark 7*

It is evident that we can parametrize the problem in terms of $\beta_{.1}, \beta_{.2}, \sigma_{11}, \sigma_{22}, \sigma_{12}$; it is further evident that $\beta_{.2}$ and $\sigma_{22}$ appear only in the form $(\beta_{.2}/\sigma_{22}^{1/2})$ – hence, that $\sigma_{22}$ cannot be, separately, identified. We shall, thus, adopt the convention

$$\sigma_{22} = 1. \tag{107}$$

A consequence of (107) is that (105) reduces to

$$E(y_{t1}|x_{t1.}, u_{t2} \geq -x_{t2}.\beta_{.2}) = x_{t1}.\beta_{.1} + \sigma_{12}\psi(\nu_{t2}). \tag{108}$$

The logarithm of the LF is given by

$$L = \sum_{t=1}^{T} \left\{ (1 - c_t)\ln \Phi(-\nu_{t2}) \right.$$

$$+ c_t\left[ -\frac{1}{2}\ln(2\pi) - \frac{1}{2}\ln \sigma_{11} - \frac{1}{2\sigma_{11}}(y_{t1} - x_{t1}.\beta_{.1})^2 \right]$$

$$\left. + \ln \Phi\left[ \frac{1}{\alpha^{1/2}}\left( \nu_{t2} + \rho_{12}\left( \frac{y_{t1} - x_{t1}.\beta_{.1}}{\sigma_{11}^{1/2}} \right) \right) \right] \right\}. \tag{109}$$

*Remark 8*

We shall proceed to maximize (109) treating $\beta_{.1}, \beta_{.2}$ as free parameters. As pointed out in the discussion following eq. (95) the two vectors will, generally, have elements in common. While we shall ignore this aspect here, for simplicity of exposition, we can easily take account of it by considering as the vector of unknown parameters $\gamma$ whose elements are the distinct elements of $\beta_{.1}, \beta_{.2}$ and $\sigma_{11}, \rho_{12}$.

The first order conditions yield

$$\frac{\partial L}{\partial \beta_{.1}} = \frac{1}{\sigma_{11}^{1/2}} \sum_{t=1}^{T} c_t\left[ \frac{y_{t1} - x_{t1}.\beta_{.1}}{\sigma_{11}^{1/2}} - \frac{\rho_{12}}{\alpha^{1/2}}\frac{\phi(\pi_t)}{\Phi(\pi_t)} \right] x_{t1.}, \tag{110}$$

$$\frac{\partial L}{\partial \beta_{.2}} = \sum_{t=1}^{T} \left[ c_t\frac{\phi(\pi_t)}{\Phi(\pi_t)}\frac{1}{\alpha^{1/2}} - (1 - c_t)\frac{\phi(\nu_{t2})}{\Phi(-\nu_{t2})} \right] x_{t2.}, \tag{111}$$

$$\frac{\partial L}{\partial \sigma_{11}} = \frac{1}{2\sigma_{11}} \sum_{t=1}^{T} c_t\left[ -1 + \left( \frac{y_{t1} - x_{t1}.\beta_{.1}}{\sigma_{11}^{1/2}} \right)^2 - \frac{\phi(\pi_t)}{\Phi(\pi_t)}\frac{\rho_{12}}{\alpha^{1/2}}\left( \frac{y_{t1} - x_{t1}.\beta_{.1}}{\sigma_{11}^{1/2}} \right) \right], \tag{112}$$

$$\frac{\partial L}{\partial \rho_{12}} = \frac{1}{\alpha^{3/2}} \sum_{t=1}^{T} c_t\frac{\phi(\pi_t)}{\Phi(\pi_t)}\left[ \rho_{12}\nu_{t2} + \left( \frac{y_{t1} - x_{t1}.\beta_{.1}}{\sigma_{11}^{1/2}} \right) \right]. \tag{113}$$

Putting

$$\gamma = \left( \beta'_{\cdot 1}, \beta'_{\cdot 2}, \sigma_{11}, \sigma_{12} \right)', \tag{114}$$

we see that the ML estimator, say $\hat{\gamma}$, is defined by the condition

$$\frac{\partial L}{\partial \gamma} (\hat{\gamma}) = 0. \tag{115}$$

Evidently, this is a highly nonlinear set of relationships which can be solved only by iteration, from an initial consistent estimator, say $\tilde{\gamma}$.

### 3.4. An initial consistent estimator

To obtain an initial consistent estimator we look at the sample solely from the point of view of whether information is available on $y_{t1}$, i.e. whether $y_{t1}$ is observed with respect to the economic entity in question. It is clear that this, at best, will identify only $\beta_{\cdot 2}$, since absent any information on $y_{t1}$ we cannot possibly hope to estimate $\beta_{\cdot 1}$. Having estimated $\beta_{\cdot 2}$ by this procedure we proceed to construct the variable

$$\tilde{\psi}_t = \tilde{\psi}_t(\tilde{\nu}_{t2}) = \frac{\phi\left( x_{t2} . \tilde{\beta}_{\cdot 2} \right)}{\Phi\left( x_{t2} . \tilde{\beta}_{\cdot 2} \right)}, \qquad t = 1, 2, \ldots, T. \tag{116}$$

Then, turning our attention to that part of the sample which contains observations on $y_{t1}$, we simply regress $y_{t1}$ on $(x_{t1}., \tilde{\psi}_t)$. In this fashion we obtain estimators of

$$\delta = \left( \beta'_{\cdot 1}, \sigma_{12} \right)' \tag{117}$$

as well as of $\sigma_{11}$.

Examining the sample from the point of view first set forth at the beginning of this section we have the log likelihood function

$$L_1 = \sum_{t=1}^{T} \left[ c_t \ln \Phi(\nu_{t2}) + (1 - c_t) \ln \Phi(-\nu_{t2}) \right], \tag{118}$$

which is to be maximized with respect to the unknown vector $\beta_{\cdot 2}$. In Section 1.d. we noted that $L_1$ is strictly concave with respect to $\beta_{\cdot 2}$; moreover, the matrix of its second order derivatives is given by

$$\frac{\partial^2 L_1}{\partial \beta_{\cdot 2} \partial \beta_{\cdot 2}} = -\sum_{t=1}^{T} \phi(x_{t2}.\beta_{\cdot 2}) \left[ c_t \frac{S_1(x_{t2}.\beta_{\cdot 2})}{\Phi^2(x_{t2}.\beta_{\cdot 2})} + (1 - c_t) \frac{S_2(x_{t2}.\beta_{\cdot 2})}{\Phi^2(-x_{t2}.\beta_{\cdot 2})} \right]$$
$$\cdot x'_{t2}.x_{t2}. \tag{119}$$

where

$$S_1(x_{t2}.\beta_{\cdot 2}) = \phi(x_{t2}.\beta_{\cdot 2}) + (x_{t2}.\beta_{\cdot 2})\Phi(x_{t2}.\beta_{\cdot 2}), \tag{120}$$
$$S_2(x_{t2}.\beta_{\cdot 2}) = \phi(x_{t2}.\beta_{\cdot 2}) - (x_{t2}.\beta_{\cdot 2})\Phi(-x_{t2}.\beta_{\cdot 2}). \tag{121}$$

It is also shown in the discussion of Section 1.d. that asymptotically

$$\sqrt{T}(\hat{\beta}_{\cdot 2} - \beta_2^0) \sim N\left(0, -\lim_{T \to \infty} \left\{ \frac{1}{T} E\left[ \frac{\partial^2 L}{\partial \beta_{\cdot 2} \partial \beta_{\cdot 2}}(\beta_{\cdot 2}^0) \right] \right\}^{-1} \right), \tag{122}$$

where $\hat{\beta}_{\cdot 2}$ is the ML estimator, i.e. it solves

$$\frac{\partial L_1}{\partial \beta_{\cdot 2}}(\hat{\beta}_{\cdot 2}) = 0, \tag{123}$$

and $\beta_{\cdot 2}^0$ is the true parameter point. It is evident from (122) that $\hat{\beta}_{\cdot 2}$ converges a.c. to $\beta_{\cdot 2}^0$. Define now

$$\tilde{\psi}_t = \frac{\phi(x_{t2}.\hat{\beta}_{\cdot 2})}{\Phi(x_{t2}.\hat{\beta}_{\cdot 2})}, \qquad t = 1, 2, \ldots, T, \tag{124}$$

and consider the estimator

$$\tilde{\delta} = (X_1^{*\prime} X_1^*)^{-1} X_1^{*\prime} y_{\cdot 1'} \qquad \delta = (\beta'_{\cdot 1'} \sigma_{12})', \tag{125}$$

where we have written

$$y_{t1} = x_{t1}.\beta_1 + \sigma_{12}\psi_t + v_{t1}, \qquad v_{t1} = u_{t1} - \sigma_{12}\psi_t, \tag{126}$$
$$X_1^* = (X_1, \tilde{\psi}), \qquad X_1 = (x_{t1}.), \qquad \tilde{\psi} = (\tilde{\psi}_t), \qquad t = 1, 2, \ldots, T. \tag{127}$$

We observe that

$$(\tilde{\delta} - \delta^0) = (X_1^{*\prime} X_1^*)^{-1} X_1^{*\prime} [v_{\cdot 1} - \sigma_{12}(\tilde{\psi} - \psi)]. \tag{128}$$

*It is our contention that the estimator in (125) is consistent for $\beta_{.1}$ and $\sigma_{12}$; moreover that it naturally implies a consistent estimator for $\sigma_{11}$, thus yielding the initial consistent estimator, say*

$$\tilde{\gamma} = \left( \tilde{\beta}'_{.1}, \hat{\beta}'_{.2}, \tilde{\sigma}_{11}, \tilde{\rho}_{12} \right)' \tag{129}$$

*which we require for obtaining the LM estimator.*

Formally, we will establish that

$$\sqrt{T}(\hat{\delta} - \delta^0) = \left( \frac{X_1^{*'}X_1^{*}}{T} \right) \frac{1}{\sqrt{T}} X_1^{*'} \left[ v_{.1} - \sigma_{12}(\tilde{\psi} - \psi) \right] \sim N(0, F), \tag{130}$$

for suitable matrix $F$, thus showing that $\hat{\delta}$ converges to $\delta^0$ with probability one (almost surely).

In order that we may accomplish this task it is imperative that we must specify more precisely the conditions under which we are to consider the model[5] in (94), as expressed in (97). We have:

(A.3.1.)  The basic error process

$$\{ u'_{t.} : t = 1, 2, \ldots \}, \qquad u_{t.} = (u_{t1}, u_{t2}),$$

is one of i.i.d. random variables with

$$u'_{t.} \sim N(0, \Sigma), \qquad \Sigma > 0, \qquad \sigma_{22} = 1,$$

and is independent of the process generating the exogenous variables $x_{t1.}, x_{t2.}$.

(A.3.2.)  The admissible parameter space, say $H \subset R_{n+3}$, is closed and bounded and contains an open neighborhood of the true parameter point

$$\gamma^0 = \left( \beta^0_{.1}, \beta^{0\prime}_{.2}, \sigma^0_{11}, \rho^0_{12} \right)'.$$

(A.3.3.)  The exogenous variables are nonstochastic and are bounded, i.e.

$$|x_{t2i}| < k_i, \qquad i = 0, 1, 2, \ldots n$$

for all $t$.[6]

---

[5]As pointed out earlier, it may be more natural to state conditions in terms of the basic variables $x^*_{t1.}, x^*_{t2.}, u^*_{t1}$ and $u^*_{t2}$; doing so, however, will disrupt the continuity of our discussion; for this reason we state conditions in terms of $x_{t1.}, x_{t2.}, u_{t1.}$ and $u_{t2}$.

(A.3.4.) The matrix

$$X_2 = (x_{t2\cdot}) \qquad t = 1, 2, \ldots, T,$$

is of rank $n + 1$ and moreover

$$\lim_T \frac{1}{T} X_2' X_2 = P, \qquad P > 0.$$

*Remark 9*

It is a consequence of the assumptions above that, for any $x_{t2\cdot}$ and admissible $\beta_{\cdot 2}$, there exists $k$ such that

$$-r \le x_{t2\cdot}\beta_{\cdot 2} \le r, \qquad 0 < r < k, \qquad k < \infty,$$

so that, for example,

$$\begin{aligned}
&\phi(x_{t2\cdot}\beta_{\cdot 2}) > \phi(-k) > 0, \\
&\Phi(x_{t2\cdot}\beta_{\cdot 2}) < \Phi(k) < 1, \\
&\Phi(x_{t2\cdot}\beta_{\cdot 2}) > \Phi(-k) > 0.
\end{aligned} \tag{131}$$

Consequently,

$$\psi(\nu_t) = \frac{\phi(x_{t2\cdot}\beta_{\cdot 2})}{\Phi(x_{t2\cdot}\beta_{\cdot 2})}, \qquad \psi^*(\nu_t) = \frac{\phi(x_{t2\cdot}\beta_{\cdot 2})}{\Phi(-x_{t2\cdot}\beta_{\cdot 2})},$$

are both bounded continuous functions of their argument.

   To show the validity of (130) we proceed by a sequence of Lemmata.

*Lemma 5*

The probability limit of the matrix to be inverted in is given by

$$\operatorname*{plim}_{T \to \infty} \frac{1}{T} X_1^{*\prime} X_1^* = \lim_{T \to \infty} \frac{1}{T} X_1^{0\prime} X_1^0 = Q_0, \qquad Q_0 > 0,$$

---

[6] We remind the reader that in the canonical representation of (97), the vector $x_{t1\cdot}$ is a subvector of $x_{t2\cdot}$; hence the boundedness assumptions on $x_{t2\cdot}$ imply similar boundedness conditions on $x_{t1\cdot}$.

   Incidentally, note that $\beta_{\cdot 1}^0$ is not necessarily a subvector of $\beta_{\cdot 2}^0$, since the latter would contain $\beta_{\cdot 11}^* - \beta_{\cdot 12}^*$ and in addition $\beta_{\cdot 21}^*, -\beta_{\cdot 22}^*$, while the former will contain $\beta_{\cdot 11}^*, \beta_{\cdot 21}^*$.

where

$$X_1^0 = (X_1, \psi^0), \qquad \psi^0 = (\psi_t^0), \qquad \psi_t^0 = \frac{\phi(x_{t2}.\beta_2^0)}{\Phi(x_{t2}.\beta_{.2}^0)}.$$

*Proof*

We examine

$$S_T = \frac{1}{T}[X_1^{*\prime}X_1^* - X_1^{0\prime}X_1^0] = \frac{1}{T}\begin{bmatrix} 0 & X_1'(\tilde{\psi} - \psi^0) \\ (\tilde{\psi} - \psi^0)'X_1 & (\tilde{\psi} + \psi^0)(\tilde{\psi} - \psi^0) \end{bmatrix}, \quad (132)$$

and the problem is reduced to considering

$$\tilde{\psi}_t - \psi_t^0 = \alpha_t^0 x_{t2}.(\hat{\beta}_{.2} - \beta_2^0) + s_t^*(\hat{\beta}_{.2} - \beta_{.2}^0)'x_{t2}'.x_{t2}.(\hat{\beta}_{.2} - \beta_{.2}^0), \quad (133)$$

where

$$\alpha_t^0 = \frac{\partial \psi(\nu_{t2})}{\partial \nu_{t2}} \qquad \text{evaluated at } \beta_{.2} = \beta_{.2}^0,$$

$$s_t^* = \frac{\partial^2 \psi(\nu_{t2})}{\partial \nu_{t2}^2} \qquad \text{evaluated at } \beta_{.2} = \beta_{.2}^*,$$

$$|\beta_{.2}^* - \beta_{.2}^0| < |\beta_{.2} - \beta_{.2}^0|.$$

It is evident that, when the expansion in (133) is incorporated in (132) quadratic terms in $(\hat{\beta}_{.2} - \beta_{.2}^0)$ will vanish with $T$.

Hence we need be concerned only with the terms of the form

$$\frac{1}{T}\sum_{t=1}^{T} x_{t1}'.(\tilde{\psi}_t - \psi_t^0) \sim \frac{1}{T^{3/2}}\sum_{t=1}^{T}[\alpha_t^0 x_{t1}'.x_{t2}]\sqrt{T}(\hat{\beta}_{.2} - \beta_{.2}^0),$$

or of the form

$$\frac{1}{T}\sum_{t=1}^{T}(\tilde{\psi}_t + \psi_t^0)(\tilde{\psi}_t - \psi_t^0) \sim \frac{1}{T^{3/2}}\sum_{t=1}^{T}[\alpha_t^0(\tilde{\psi}_t + \psi_t^0)x_{t2}.]\sqrt{T}(\hat{\beta}_{.2} - \beta_{.2}^0).$$

In either case we note that by assumption (A.3.4.) and Remark 9

$$\lim_{T \to \infty} \frac{1}{T} \sum_{t=1}^{T} \alpha_t^0 x'_{t1} . x_{t2}.,$$

has bounded elements; similarly, for

$$\lim_{T \to \infty} \frac{1}{T} \sum_{t=1}^{T} \alpha_t^0 (\tilde{\psi}_t + \psi_t^0) x_{t2}..$$

Consequently, in view of (122) and (132) we conclude

$$\plim_{T \to \infty} S_T = 0,$$

which implies

$$\plim_{T \to \infty} \frac{1}{T} X_1^{*\prime} X_1^{*} = \lim_{T \to \infty} \frac{1}{T} X_1^{0\prime} X_1^{0} = Q_0. \tag{134}$$

*Corollary 4*

The limiting distribution of the left member of (130) is obtainable through

$$\sqrt{T} (\tilde{\delta} - \delta^0) \sim Q_0^{-1} X_1^{*\prime} [v_{.1} - \sigma_{12}(\tilde{\psi} - \psi^0)], \quad v_{.1} = (v_{11}, v_{21} \cdots v_{T1})'.$$

Indeed, by standard argumentation we may establish

*Theorem 4*

Under assumption (A.3.1) through (A.3.4) the initial (consistent) estimator of this section has the limiting distribution

$$\sqrt{T} (\tilde{\delta} - \delta^0) \sim N(0, F), \qquad F = Q_0^{-1} P Q_0^{-1},$$

where

$$P = \sigma_{11} \lim_{T \to \infty} \frac{1}{T} \begin{bmatrix} \sum\limits_{t=1}^{T} \omega_{11t} x'_{t1} . x_{t1} . & \sum\limits_{t=1}^{T} \omega_{11t} \psi_t^0 x'_{t1} . \\ \sum\limits_{t=1}^{T} \omega_{11t} \psi_t^0 x_{t1} . & \sum\limits_{t=1}^{T} \omega_{11t} \psi_t^{0^2} \end{bmatrix}$$

$Q_0$ is defined in (134) and

$$E(v_{t1}^2) = \sigma_{11}\omega_{11t} = \sigma_{11}\left[1 - \rho_{12}^{02}v_{t2}^0\psi_t^0 - \rho_{12}^{02}\psi_t^{02}\right].$$

*Corollary 5*

The initial estimator above is strongly consistent.

*Proof*

From the theorem above

$$\sqrt{T}(\tilde{\delta} - \delta^0) \sim \zeta,$$

where $\zeta$ is an a.c. finite random vector.
Thus

$$\tilde{\delta} - \delta^0 \sim \frac{1}{\sqrt{T}}\zeta,$$

$\tilde{\delta}$ converges to $\delta^0$ a.c.
Evidently, the parameter $\sigma_{11}$ can be estimated (at least consistently) by

$$\tilde{\sigma}_{11} = \frac{1}{T}\left[\tilde{v}_{t1}^2 + \tilde{\sigma}_{12}\tilde{\psi}_t\tilde{v}_t 2 + \tilde{\sigma}_{12}\tilde{\psi}_t^2\right].$$

## 3.5. *Limiting distribution of the ML estimator*

In the previous section we outlined a procedure for obtaining an initial estimator, say

$$\tilde{\gamma} = (\tilde{\beta}_1, \tilde{\beta}_2', \tilde{\sigma}_{11}, \tilde{\sigma}_{12})',$$

and have shown that it converges to the true parameter point, say $\gamma^0$, with probability one (a.c.).
We now investigate the properties of the ML estimator, say $\hat{\gamma}$, obtained by solving

$$\frac{\partial L}{\partial \gamma}(\hat{\gamma}) = 0$$

through iteration, beginning with $\tilde{\gamma}$. The limiting distribution of $\hat{\gamma}$ may be found by examining

$$\sqrt{T}\left(\hat{\gamma} - \gamma^0\right) = \left[-\frac{1}{T}\frac{\partial^2 L}{\partial\gamma\partial\gamma}(\gamma^*)\right]^{-1}\frac{1}{\sqrt{T}}\frac{\partial L}{\partial\gamma}(\gamma^0),\tag{135}$$

where $\gamma^0$ is the true parameter point, $\gamma^*$ obeys

$$|\gamma^* - \gamma^0| \le |\hat{\gamma} - \gamma^0|,$$

and

$$\frac{\partial L}{\partial\gamma'}(\gamma^0) = \sum_{t=1}^{T} A_t \xi_{\cdot t},\tag{136}$$

where

$$A_t = \begin{bmatrix} \dfrac{1}{\sigma_{11}^{1/2}}x'_{t1}. & 0 & 0 & 0 \\[2ex] 0 & x'_{t2}. & 0 & 0 \\[2ex] 0 & 0 & \dfrac{1}{2\sigma_{11}} & 0 \\[2ex] 0 & 0 & 0 & \dfrac{1}{\alpha^{3/2}} \end{bmatrix}\tag{137}$$

$$\xi_{1t} = c_t\left[\frac{u_{t1}}{\sigma_{11}^{1/2}} - \frac{\rho_{12}}{\alpha^{1/2}}\frac{\phi(\pi_t)}{\Phi(\pi_t)}\right],$$

$$\xi_{2t} = c_t\left[\frac{1}{\alpha^{1/2}}\frac{\phi(\pi_t)}{\Phi(\pi_t)}\right] - (1 - c_t)\frac{\phi(\nu_{t2})}{\Phi(\nu_{t2})},$$

$$\xi_{3t} = c_t\left[\left(\frac{u_{t1}}{\sigma_{11}^{1/2}}\right)^2 - \frac{\rho_{12}}{\alpha^{1/2}}\frac{\phi(\pi_t)}{\phi(\pi_t)}\left(\frac{u_{t1}}{\sigma_{11}^{1/2}}\right) - 1\right],\tag{138}$$

$$\xi_{4t} = c_t\frac{\phi(\pi_t)}{\Phi(\pi_t)}\left[\rho_{12}\nu_{t2} + \frac{u_{t1}}{\sigma_{11}^{1/2}}\right],$$

$$\xi_{\cdot t} = (\xi_{1t}, \xi_{2t}, \xi_{3t}, \xi_{4t})'.$$

In order to complete the problem we also require an expression for the Hessian of the likelihood function in addition to the expressions in (110) through (113).

To this effect define

$$\xi_{11t} = c_t \left[ 1 + \frac{\rho_{12}^2}{\alpha} \pi_t \frac{\phi(\pi_t)}{\Phi(\pi_t)} + \frac{\rho_{12}^2}{\alpha} \frac{\phi^2(\pi_t)}{\Phi^2(\pi_t)} \right],$$

$$\xi_{21t} = - c_t \left[ \frac{\rho_{12}}{\alpha} \frac{\phi(\pi_t)}{\Phi(\pi_t)} \pi_t + \frac{\rho_{12}}{\alpha} \frac{\phi^2(\pi_t)}{\Phi^2(\pi_t)} \right],$$

$$\xi_{31t} = c_t \left[ 2 \left( \frac{u_{t1}}{\sigma_{11}^{1/2}} \right) - \frac{\rho_{12}}{\alpha} \frac{\phi(\pi_t)}{\Phi(\pi_t)} + \frac{\rho_{12}^2}{\alpha} \frac{\phi(\pi_t)}{\Phi(\pi_t)} \pi_t \left( \frac{u_{t1}}{\sigma_{11}^{1/2}} \right) \right.$$
$$\left. + \frac{\rho_{12}^2}{\alpha} \frac{\phi^2(\pi_t)}{\Phi^2(\pi_t)} \left( \frac{u_{t1}}{\sigma_{11}^{1/2}} \right) \right],$$

$$\xi_{41t} = c_t \left[ \frac{\phi(\pi_t)}{\Phi(\pi_t)} - \frac{\rho_{12}}{\alpha^{1/2}} \frac{\phi(\pi_t)}{\Phi(\pi_t)} \pi_t \left( \rho_{12} v_{t2} + \frac{u_{t1}}{\sigma_{11}^{1/2}} \right) \right.$$
$$\left. - \frac{\rho_{12}}{\alpha^{1/2}} \frac{\phi^2(\pi_t)}{\Phi^2(\pi_t)} \left( \rho_{12} v_{t2} + \frac{u_{t1}}{\sigma_{11}^{1/2}} \right) \right],$$

$$\xi_{22t} = \frac{1}{\alpha} c_t \left[ \frac{\phi(\pi_t)}{\Phi(\pi_t)} \pi_t + \frac{\phi^2(\pi_t)}{\Phi^2(\pi_t)} \right] + (1 - c_t) \psi^*(v_{t2}) \left[ \psi^*(v_{t2}) - v_{t2} \right], \qquad (139)$$

$$\xi_{32t} = - c_t \left[ \frac{\rho_{12}}{\alpha} \frac{\phi(\pi_t)}{\Phi(\pi_t)} \pi_t \left( \frac{u_{t1}}{\sigma_{11}^{1/2}} \right) + \frac{\rho_{12}}{\alpha} \frac{\phi^2(\pi_t)}{\Phi^2(\pi_t)} \left( \frac{u_{t1}}{\sigma_{11}^{1/2}} \right) \right],$$

$$\xi_{42t} = - c_t \frac{1}{\alpha^{1/2}} \left[ \frac{\phi^2(\pi_t)}{\Phi^2(\pi_t)} \left( \rho_{12} v_{t2} + \frac{u_{t1}}{\sigma_{11}^{1/2}} \right) \right],$$

$$\xi_{42t}^* = c_t \left[ \rho_{12} \frac{\phi(\pi_t)}{\Phi(\pi_t)} - \frac{1}{\alpha^{1/2}} \frac{\phi(\pi_t)}{\Phi(\pi_t)} \pi_t \left( \rho_{12} v_{t2} + \frac{u_{t1}}{\sigma_{11}^{1/2}} \right) \right],$$

$$\xi_{33t} = c_t \left[ 2 \left( \frac{u_{t1}}{\sigma_{11}^{1/2}} \right)^2 - \frac{\rho_{12}}{\alpha^{1/2}} \frac{\phi(\pi_t)}{\Phi(\pi_t)} \frac{u_{t1}}{\sigma_{11}^{1/2}} + \frac{\rho_{12}^2}{\alpha} \frac{\phi(\pi_t)}{\Phi(\pi_t)} \pi_t \left( \frac{u_{t1}}{\sigma_{11}^{1/2}} \right)^2 \right.$$
$$\left. + \frac{\rho_{12}^2}{\alpha} \frac{\phi^2(\pi_t)}{\Phi^2(\pi_t)} \left( \frac{u_{t1}}{\sigma_{11}^{1/2}} \right)^2 \right],$$

$$\xi_{33t}^{*} = \xi_{3t},$$

$$\xi_{43t} = c_t \left[ \frac{\phi(\pi_t)}{\Phi(\pi_t)} \left( \frac{u_{t1}}{\sigma_{11}^{1/2}} \right) - \frac{\rho_{12}}{\alpha^{1/2}} \frac{\phi(\pi_t)}{\Phi(\pi_t)} \pi_t \left( \rho_{12} \nu_{t2} \left( \frac{u_{t1}}{\sigma_{11}^{1/2}} \right) + \frac{u_{t1}^2}{\sigma_{11}} \right) \right.$$

$$\left. - \frac{\rho_{12}}{\alpha^{1/2}} \frac{\phi^2(\pi_t)}{\Phi^2(\pi_t)} \left( \rho_{12} \nu_{t2} \left( \frac{u_{t1}}{\sigma_{11}^{1/2}} \right) + \frac{u_{t1}^2}{\sigma_{11}} \right) \right],$$

$$\xi_{44t} = \xi_{4t}^2,$$

$$\xi_{44t}^{*} = \frac{3\rho_{12}}{\alpha} \xi_{4t} + c_t \left[ \frac{\phi(\pi_t)}{\Phi(\pi_t)} \nu_{t2} - \frac{1}{\alpha^{3/2}} \frac{\phi(\pi_t)}{\Phi(\pi_t)} \pi_t \left( \rho_{12} \nu_{t2} + \frac{u_{t1}}{\sigma_{11}^{1/2}} \right)^2 \right].$$

In the expressions of (138) and (139) we have replaced, for reasons of notational economy only,

$$\left( \frac{y_{t1} - x_t \cdot \beta_{\cdot 1}}{\sigma_{11}^{1/2}} \right),$$

by

$$\left( \frac{u_{t1}}{\sigma_{11}^{1/2}} \right).$$

## Remark 10

The starred symbols, for example, $\xi_{42t}^{*}$, $\xi_{33t}^{*}$, $\xi_{44t}^{*}$, all correspond to components of the Hessian of the log LF *having mean zero*. Hence, such components can be ignored both in determining the limiting distribution of the ML estimator and in its numerical derivation, given a sample. We can, then, represent the Hessian of the log of the LF as

$$\frac{\partial^2 L}{\partial \gamma \partial \gamma} = \sum_{t=1}^{T} \Omega_t + \sum_{t=1}^{T} \Omega_t^{*},$$

where $\Omega_t^{*}$ contains only zeros or elements having mean zero. It is also relatively straightforward to verify that

$$A_t \text{Cov}(\xi_{\cdot t}) A_t' = E(\Omega_t),$$

where the elements of $A_t$, $\xi_{\cdot t}$ and $\Omega_t$ have been evaluated at the true parameter point $\gamma^0$.

To determine the limiting distribution of the ML estimator (i.e. the converging iterate beginning with an initial consistent estimator) we need

*Lemma 6*

Let $A_t$, $\xi._t$ be as defined in (139) and (138); then,

$$\frac{1}{\sqrt{T}} \frac{\partial L}{\partial \gamma'}(\gamma^0) = \frac{1}{\sqrt{T}} \sum_{t=1}^{T} A_t \xi._t \sim N(0, C_*),$$

where

$$C_* = \lim_{T \to \infty} \frac{1}{T} \sum_{t=1}^{T} A_t \text{Cov}(\xi._t) A'_t = \lim_{T \to \infty} \frac{1}{T} \sum_{t=1}^{T} E(\Omega_t). \tag{141}$$

*Proof*

The sequence

$$\{ A_t \xi._t : t = 1, 2, \ldots \},$$

is one of independent nonidentically distributed random vectors with mean zero and uniformly bounded moments to any finite order; moreover, the sequence obeys a Liapounov condition. Consequently

$$\frac{1}{\sqrt{T}} \frac{\partial L}{\partial \gamma}(\gamma^0) \sim N(0, C_*). \tag{Q.E.D.}$$

An explicit representation of $\Omega_t$ or $C_*$ is omitted here because of its notational complexity. To complete the argument concerning the limiting distribution of the ML estimator we obtain the limit of

$$\frac{1}{T} \frac{\partial^2 L}{\partial \gamma \partial \gamma}(\gamma), \quad \gamma \in H.$$

Again for the sake of brevity of exposition we shall only state the result without proof

*Lemma 7*

Under assumptions (A.3.1) through (A.3.4)

$$\frac{1}{T} \frac{\partial^2 L}{\partial \gamma \partial \gamma}(\gamma) \overset{\text{a.c.}}{\to} \lim_{T \to \infty} \frac{1}{T} E\left[ \frac{\partial^2 L}{\partial \gamma \partial \gamma}(\gamma) \right],$$

uniformly in $\gamma$.

*Remark 11*

We note that for $\gamma = \gamma^0$

$$\lim_{T \to \infty} \frac{1}{T} E\left[\frac{\partial^2 L}{\partial\gamma\partial\gamma}(\gamma)\right] = -C_*.$$

We finally have

*Theorem 5*

(Asymptotic Normality): Consider the model of (97) subject to the condition in (107) and assumptions (A.3.1.) through (A.3.4.). Let $\hat{\gamma}$ be the ML estimator of $\gamma^0$ – the true parameter point. Then

$$\sqrt{T}(\hat{\gamma} - \gamma^0) \sim N(0, C),$$

where

$$C = C_*^{-1},$$

and $C_*$ is as is defined in (141).

*Proof*

From the expansion in (135)

$$\sqrt{T}(\hat{\gamma} - \gamma^0) \sim C_*^{-1} \frac{1}{\sqrt{T}} \frac{\partial L}{\partial\gamma'}(\gamma^0).$$

From Lemma 6 we then conclude

$$\sqrt{T}(\hat{\gamma} - \gamma^0) \sim N(0, C). \qquad\qquad\qquad\qquad\qquad\text{Q.E.D.}$$

*Corollary 6*

The ML estimator $\hat{\gamma}$ obeys

$$\hat{\gamma} \overset{a.c.}{\to} \gamma^0.$$

*Proof*

By Theorem 5

$$\sqrt{T}(\hat{\gamma} - \gamma^0) \sim \zeta,$$

where $\zeta$ is a well defined a.c. finite random variable. Hence,

$$\tilde{\gamma} - \gamma^0 \sim \frac{\zeta}{\sqrt{T}} \xrightarrow{\text{a.c.}} 0.$$

*Corollary 7*

The matrix in the expansion of (135) obeys

$$\frac{1}{T} \frac{\partial^2 L}{\partial \gamma \partial \gamma} (\gamma^*) \xrightarrow{\text{a.c.}} \lim_{T \to \infty} \frac{1}{T} E \left[ \frac{\partial^2 L}{\partial \gamma \partial \gamma} (\gamma^0) \right].$$

*Proof*

Lemma 7 and Corollary 6.

### 3.6. *A test for selectivity bias*

A test for selectivity bias is formally equivalent to the test of

$$H_0: \rho_{12} = 0 \quad \text{or} \quad \gamma = \left( \beta'_{\cdot 1}, \beta'_{\cdot 2}, \sigma_{11}, 0 \right)'$$

as against the alternative

$H_1$: $\gamma$ unrestricted (except for the obvious conditions, $\sigma_{11} > 0$, $\rho_{12}^2 \varepsilon [0,1]$). From the likelihood function in eq. (109) the (log) LF under $H_0$ becomes

$$L(\gamma | H_0) = \sum_{t=1}^{T} \left\{ (1 - c_t) \ln \Phi(-v_{t2}) + c_t \ln \Phi(v_{t2}) \right.$$

$$\left. - \frac{1}{2} c_t \left[ \ln(2\pi) + \ln \sigma_{11} + \frac{1}{\sigma_{11}} (y_{t1} - x_{t1}.\beta_{\cdot 1})^2 \right] \right\}. \tag{142}$$

We note that (142) is separable in the parameters $(\beta'_{\cdot 1}, \sigma_{11})'$ and $\beta_{\cdot 2}$. Indeed, the ML estimator of $\beta_{\cdot 2}$ is the "probit" estimator, $\hat{\beta}_{\cdot 2}$, obtained in connection with eq. (118) in Section 3.d.; the ML estimator of $(\beta'_{\cdot 1}, \sigma_{11})'$ is the usual one obtained by least squares except that $\sigma_{11}$ is estimated with bias – as all maximum likelihood procedures imply in the normal case. Denote the estimator of $\gamma$ obtained under $H_0$, by $\tilde{\gamma}$. Denote by $\hat{\gamma}$ the ML estimator whose limiting distribution was obtained in the preceding section.

Thus

$$\lambda = L(\tilde{\gamma}|H_1) - L(\hat{\gamma}|H_0).\tag{143}$$

is the usual likelihood rationtest statistic. It may be shown that

$$-2\lambda \sim \chi_1^2.\tag{144}$$

We have thus proved

*Theorem 6*

In the context of the model of this section a test for the absence of selectivity bias can be carried out by the likelihood ratio (LR) principle. The test statistic is

$$-2\lambda \sim \chi_1^2,$$

where

$$\lambda = \sup_{H_0} L(\gamma) - \sup_{H_1} L(\gamma).$$

# References

Aitchison, J. and J. Bennett (1970) "Polychotomous Quantal Response by Maximum Indicant", *Biometrika*, 57, 253–262.

Aitchison, J. and S. Silvey (1957) "The Generalization of Probit Analysis to the Case of Multiple Responses", *Biometrika*, 37, 131–140.

Amemiya, T. (1973) "Regression Analysis When the Dependent Variable Is Truncated Normal", *Econometrica*, 41, 997–1016.

Amemiya, T. (1974) "Bivariate Probit Analysis: Minimum Chi-Square Methods", *Journal of the American Statistical Association*, 69, 940–944.

Amemiya, T. (1974) "Multivariate Regression and Simultaneous Equation Models When the Dependent Variables Are Truncated Normal", *Econometrica*, 42, 999–1012.

Amemiya, T. (1974) "A Note on the Fair and Jaffee Model", *Econometrica*, 42, 759–762.

Amemiya, T. (1975) "Qualitative Response Models", *Annals of Economic and Social Measurement*, 4, 363–372.

Amemiya, T. (1976) "Tne Maximum Likelihood, the Minimum Chi-Square, and the Non-linear Weighted Least Squares Estimator in the General Qualitative Response Model", *JASA*, 71.

Amemiya, T. (1978) "The Estimation of a Simultaneous Equation Generalized Probit Model", *Econometrica*, 46, 1193–1205.

Amemiya, T. (1978) "On a Two-Step Estimation of a Multivariate Logit Model", *Journal of Econometrics*, 8, 13–21.

Amemiya, T. and F. Nold (1975) "A Modified Logit Model", *Review of Economics and Statistics*, 57, 255–257.

Anscombe, E. J. (1956) "On Estimating Binomial Response Relations", *Biometrika*, 43, 461–464.

Ashford, J. R. and R. R. Sowden (1970) "Multivariate Probit Analysis", *Biometrics*, 26, 535–546.

Ashton, W. (1972) *The Logit Transformation*. New York: Hafner.

Bartlett, M. S. (1935) "Contingent Table Interactions", *Supplement to the Journal of the Royal Statistical Society*, 2, 248–252.

Berkson, J. (1949) "Application of the Logistic Function to Bioassay", *Journal of the American Statistical Association*, 39, 357–365.

Berkson, J. (1951) "Why I Prefer Logits to Probits", *Biometrika*, 7, 327–339.

Berkson, J. (1953) "A Statistically Precise and Relatively Simple Method of Estimating the Bio-Assay with Quantal Response, Based on the Logistic Function", *Journal of the American Statistical Association*, 48, 565–599.

Berkson, J. (1955) "Estimate of the Integrated Normal Curve by Minimum Normit Chi-Square with Particular Reference to Bio-Assay", *Journal of the American Statistical Association*, 50, 529–549.

Berkson, J. (1955) "Maximum Likelihood and Minimum Chi-Square Estimations of the Logistic Function", *Journal of the American Statistical Association*, 50, 130–161.

Bishop, T., S. Feiberg and P. Hollan (1975) *Discrete Multivariate Analysis*. Cambridge: MIT Press.

Block, H. and J. Marschak (1960) "Random Orderings and Stochastic Theories of Response", in: I. Olkin, ed., *Contributions to Probability and Statistics*. Stanford: Stanford University Press.

Bock, R. D. (1968) "Estimating Multinomial Response Relations", in: *Contributions to Statistics and Probability: Essays in Memory of S. N. Roy*. Chapel Hill: University of North Carolina Press.

Bock, R. D. (1968) *The Measurement and Prediction of Judgment and Choice*. San Francisco: Holden-Day.

Boskin, M. (1974) "A Conditional Logit Model of Occupational Choice", *Journal of Political Economy*, 82, 389–398.

Boskin, M. (1975) "A Markov Model of Turnover in Aid to Families with Dependent Children", *Journal of Human Resources*, 10, 467–481.

Chambers, E. A. and D. R. Cox (1967) "Discrimination between Alternative Binary Response Models", *Biometrika*, 54, 573–578.

Cosslett, S. (1980) "Efficient Estimators of Discrete Choice Models", in: C. Manski and D. McFadden, eds., *Structural Analysis of Discrete Data*. Cambridge: MIT Press.

Cox, D. (1970) *Analysis of Binary Data*. London: Methuen.

Cox, D. (1972) "The Analysis of Multivariate Binary Data", *Applied Statistics*, 21, 113–120.

Cox, D. (1958) "The Regression Analysis of Binary Sequences", *Journal of the Royal Statistical Society*, Series B, 20, 215–242.

Cox, D. (1966) "Some Procedures Connected with the Logistic Response Curve", in: F. David, ed., *Research Papers in Statistics*. New York: Wiley.

Cox, D. and E. Snell (1968) "A General Definition of Residuals", *Journal of the Royal Statistical Society*, Series B, 30, 248–265.

Cragg, J. G. (1971) "Some Statistical Models for Limited Dependent Variables with Application to the Demand for Durable Goods", *Econometrica*, 39, 829–844.

Cragg, J. and R. Uhler (1970) "The Demand for Automobiles", *Canadian Journal of Economics*, 3, 386–406.

Cripps, T. F. and R. J. Tarling (1974) "An Analysis of the Duration of Male Unemployment in Great Britain 1932–1973", *The Economic Journal*, 84, 289–316.

Daganzo, C. (1980) *Multinomial Probit*. New York: Academic Press.

Dagenais, M. G. (1975) "Application of a Threshold Regression Model to Household Purchases of Automobiles", *The Review of Economics and Statistics*, 57, 275–285.

Debreu, G. (1960) "Review of R. D. Luce Individual Choice Behavior", *American Economic Review*, 50, 186–188.

Dhrymes, P. J. (1970) *Econometrics: Statistical Foundations and Applications*. Harper & Row, 1974, New York: Springer-Verlag.

Dhrymes, P. J. (1978a) *Introductory Econometrics*. New York: Springer-Verlag.

Dhrymes, P. J. (1978b) *Mathematics for Econometrics*. New York: Springer-Verlag.

Domencich, T. and D. McFadden (1975) *Urban Travel Demand: A Behavioral Analysis*. Amsterdam: North-Holland.

Efron, B. (1975) "The Efficiency of Logistic Regression Compared to Normal Discriminant Analysis", *Journal of the American Statistical Association*, 70, 892–898.

Fair, R. C. and D. M. Jaffee (1972) "Methods of Estimation for Markets in Disequilibrium", *Econometrica*, 40, 497–514.

Finney, D. (1964) *Statistical Method in Bio-Assay*. London: Griffin.

Finney, D. (1971) *Probit Analysis*. New York: Cambridge University Press.

Gart, J. and J. Zweifel (1967) "On the Bias of Various Estimators of the Logit and Its Variance", *Biometrika*, 54, 181–187.

Gillen, D. W. (1977) "Estimation and Specification of the Effects of Parking Costs on Urban Transport Mode Choice", *Journal of Urban Economics*, 4, 186–199.

Goldberger, A. S. (1971) "Econometrics and Psychometrics: A Survey of Communalities", *Psychometrika*, 36, 83–107.

Goldberger, A. S. (1973) "Correlations Between Binary Outcomes and Probabilistic Predictions", *Journal of American Statistical Association*, 68, 84.

Goldfeld, S. M. and R. E. Quandt (1972) *Nonlinear Methods on Econometrics*. Amsterdam: North-Holland.

Goldfeld, S. M. and R. E. Quandt (1973) "The Estimation of Structural Shifts by Switching Regressions", *Annals of Economic and Social Measurement*, 2, 475–485.

Goldfeld, S. M. and R. E. Quandt (1976) "Techniques for Estimating Switching Regressions", in: S. Goldfeld and R. Quandt, eds., *Studies in Non-Linear Estimation*. Cambridge: Ballinger.

Goodman, I. and W. H. Kruskal (1954) "Measures of Association for Cross Classifications", *Journal of the American Statistical Association*, 49, 732–764.

Goodman, I. and W. H. Kruskal (1954) "Measures of Association for Cross Classification II, Further Discussion and References", *Journal of the American Statistical Association*, 54, 123–163.

Goodman, L. A. (1970) "The Multivariate Analysis of Qualitative Data: Interactions Among Multiple Classifications", *Journal of the American Statistical Association*, 65, 226–256.

Goodman, L. A. (1971) "The Analysis of Multidimensional Contingency Tables: Stepwise Procedures and Direct Estimation Methods for Building Models for Multiple Classifications", *Technometrics*, 13, 33–61.

Goodman, L. A. (1972) "A Modified Multiple Regression Approach to the Analysis of Dichotomous Variables", *American Sociological Review*, 37, 28–46.

Goodman, L. A. (1972) "A General Model for the Analysis of Surveys", *American Journal of Sociology*, 77, 1035–1086.

Goodman, L. A. (1973) "Causal Analysis of Panel Study Data and Other Kinds of Survey Data", *American Journal of Sociology*, 78, 1135–1191.

Griliches, Z., B. H. Hall and J. A. Hausman (1978) "Missing Data and Self-Selection in Large Panels", *Annals de l'Insee*, 30–31, 137–176.

Grizzle, J. (1962) "Asymptotic Power of Tests of Linear Hypotheses Using the Probit and Logit Transformations", *Journal of the American Statistical Association*, 57, 877–894.

Grizzle, J. (1971) "Multivariate Logit Analysis", *Biometrics*, 27, 1057–1062.

Gronau, R. (1973) "The Effect of Children on the Housewife's Value of Time", *Journal of Political Economy*, 81, 168–199.

Gronau, R. (1974) "Wage Comparisons: A Selectivity Bias", *Journal of Political Economy*, 82, 1119–1143.

Gurland, J., I. Lee and P. Dahm (1960) "Polychotomous Quantal Response in Biological Assay", *Biometrics*, 16, 382–398.

Haberman, S. (1974) *The Analysis of Frequency Data*. Chicago: University of Chicago Press.

Haldane, J. (1955) "The Estimation and Significance of the Logarithm of a Ratio of Frequencies", *Annals of Human Genetics*, 20, 309–311.

Harter, J. and A. Moore (1967) "Maximum Likelihood Estimation, from Censored Samples, of the Parameters of a Logistic Distribution", *Journal of the American Statistical Association*, 62, 675–683.

Hausman, J. (1979) "Individual Discount Rates and the Purchase and Utilization of Energy Using Durables", *Bell Journal of Economics*, 10, 33–54.

Hausman, J. A. and D. A. Wise (1976) "The Evaluation of Results from Truncated Samples: The New Jersey Negative Income Tax Experiment", *Annals of Economic and Social Measurement*, 5, 421–445.

Hausman, J. A. and D. A. Wise (1977) "Social Experimentation, Truncated Distributions and Efficient Estimation", *Econometrica*, 45, 319–339.

Hausman, J. A. and D. A. Wise (1978) "A Conditional Probit Model for Qualitative Choice: Discrete Decisions Recognizing Interdependence and Heterogeneous Preferences", *Econometrica*, 46, 403–426.

Hausman, J. A. and D. A. Wise (1980) "Stratification on Endogenous Variables and Estimation: The Gary Experiment", in: C. Manski and D. McFadden, eds., *Structural Analysis of Discrete Data*. Cambridge: MIT Press.

Heckman, J. (1974) "Shadow Prices, Market Wages, and Labor Supply", *Econometrica*, 42, 679–694.

Heckman, J. (1976) "Simultaneous Equations Model with Continuous and Discrete Endogenous Variables and Structural Shifts", in: S. M. Goldfeld and E. M. Quandt, eds., *Studies in Non-Linear Estimation*. Cambridge: Ballinger.

Heckman, J. (1976) "The Common Structure of Statistical Models of Truncation, Sample Selection and Limited Dependent Variables and a Simple Estimation for Such Models", *Annals of Economic and Social Measurement*, 5, 475–492.

Heckman, J. (1978) "Dummy Exogenous Variables in a Simultaneous Equation System", *Econometrica*, 46, 931–959.

Heckman, J. (1978) "Simple Statistical Models for Discrete Panel Data Developed and Applied to Test the Hypothesis of True State Dependence Against the Hypothesis of Spurious State Dependence", *Annals de l'Insee*, 30–31, 227–270.

Heckman, J. (1979) "Sample Selection Bias as a Specification Error", *Econometrica*, 47, 153–163.

Heckman, J. (1980) "Statistical Models for the Analysis of Discrete Panel Data", in: C. Manski and D. McFadden, eds., *Structural Analysis of Discrete Data*. Cambridge: MIT Press.

Heckman, J. (1980) "The Incidental Parameters Problem and the Problem of Initial Conditions in Estimating a Discrete Stochastic Process and Some Monte Carlo Evidence on Their Practical Importance", in: C. Manski and D. McFadden, eds., *Structural Analysis of Discrete Data*. Cambridge: MIT Press.

Heckman, J. and R. Willis (1975) "Estimation of a Stochastic Model of Reproduction: An Econometric Approach", in: N. Terleckyj, ed., *Household Production and Consumption*. New York: National Bureau of Economic Research.

Heckman, J. and R. Willis (1977) "A Beta Logistic Model for the Analysis of Sequential Labor Force Participation of Married Women", *Journal of Political Economy*, 85, 27–58.

Joreskog, K. and A. S. Goldberger (1975) "Estimation of a Model with Multiple Indicators and Multiple Causes of a Single Latent Variable Model", *Journal of the American Statistical Association*, 70, 631–639.

Kiefer, N. (1978) "Discrete Parameter Variation: Efficient Estimation of a Switching Regression Model", *Econometrica*, 46, 427–434.

Kiefer, N. (1979) "On the Value of Sample Separation Information", *Econometrica*, 47, 997–1003.

Kiefer, N. and G. Neumann (1979) "An Empirical Job Search Model with a Test of the Constant Reservation Wage Hypothesis", *Journal of Political Economy*, 87, 89–107.

Kohn, M., C. Manski and D. Mundel (1976), "An Empirical Investigation of Factors Influencing College Going Behavior", *Annals of Economic and Social Measurement*, 5, 391–419.

Ladd, G. (1966) "Linear Probability Functions and Discriminant Functions", *Econometrica*, 34, 873–885.

Lee, L. F. (1978) "Unionism and Wage Rates: A Simultaneous Equation Model with Qualitative and Limited Dependent Variables", *International Economic Review*, 19, 415–433.

Lee, L. F. (1979) "Identification and Estimation in Binary Choice Models with Limited (Censored) Dependent Variables", *Econometrica*, 47, 977–996.

Lee, L. F. (1980) "Simultaneous Equations Models with Discrete and Censored Variables", in: C. Manski and D. McFadden, eds., *Structural Analysis of Discrete Data*. Cambridge: MIT Press.

Lee, L. F. and R. P. Trost (1978) "Estimation of Some Limited Dependent Variable Models with Applications to Housing Demand", *Journal of Econometrics*, 8, 357–382.

Lerman, S. and C. Manski (1980) "On the Use of Simulated Frequencies to Approximate Choice Probabilities", in: C. Manski and D. McFadden, eds., *Structural Analysis of Discrete Data*. Cambridge: MIT Press.

Li, M. (1977) "A Logit Model of Home Ownership", *Econometrica*, 45, 1081–1097.

Little, R. E. (1968) "A Note on Estimation for Quantal Response Data", *Biometrika*, 55, 578–579.

Luce, R. D. (1959) *Individual Choice Behavior: A Theoretical Analysis*. New York: Wiley.

Luce, R. D. (1977) "The Choice Axiom After Twenty Years", *Journal of Mathematical Psychology*, 15, 215–233.

Luce, R. D. and P. Suppes (1965) "Preference, Utility, and Subjective Probability", in: R. Luce, R.

Bush and E. Galanter, eds., *Handbook of Mathematical Psychology III*. New York: Wiley.

Maddala, G. S. (1977) "Self-Selectivity Problem in Econometric Models", in: P. Krishniah, ed., *Applications of Statistics*. Amsterdam: North-Holland.

Maddala, G. S. (1977) "Identification and Estimation Problems in Limited Dependent Variable Models", in: A. S. Blinder and P. Friedman, eds., *Natural Resources, Uncertainty and General Equilibrium Systems: Essays in Memory of Rafael Lusky*. New York: Academic Press.

Maddala, G. S. (1978) "Selectivity Problems in Longitudinal Data", *Annals de l'INSEE*, 30–31, 423–450.

Maddala, G. S. and L. F. Lee (1976) "Recursive Models with Qualitative Endogenous Variables", *Annals of Economic and Social Measurement*, 5.

Maddala, G. and F. Nelson (1974) "Maximum Likelihood Methods for Markets in Disequilibrium", *Econometrica*, 42, 1013–1030.

Maddala, G. S. and R. Trost (1978) "Estimation of Some Limited Dependent Variable Models with Application to Housing Demand", *Journal of Econometrics*, 8, 357–382.

Maddala, G. S. and R. Trost (1980) "Asymptotic Covariance Matrices of Two-Stage Probit and Two-Stage Tobit Methods for Simultaneous Equations Models with Selectivity", *Econometrica*, 48, 491–503.

Manski, C. (1975) "Maximum Score Estimation of the Stochastic Utility Model of Choice", *Journal of Econometrics*, 3, 205–228.

Manski, C. (1977) "The Structure of Random Utility Models", *Theory and Decision*, 8, 229–254.

Manski, C. and S. Lerman (1977) "The Estimation of Choice Probabilities from Choice-Based Samples", *Econometrica*, 45, 1977–1988.

Manski, C. and D. McFadden (1980) "Alternative Estimates and Sample Designs for Discrete Choice Analysis", in: C. Manski and D. McFadden, eds., *Structural Analysis of Discrete Data*. Cambridge: MIT Press.

Marshak, J. "Binary-Choice Constraints and Random Utility Indicators", in: K. Arrow, S. Karlin and P. Suppes, eds., *Mathematical Methods in the Social Sciences*. Stanford University Press.

McFadden, D. "Conditional Logit Analysis of Qualitative Choice Behavior", in: P. Zarembka, ed., *Frontiers in Econometrics*. New York: Academic Press.

McFadden, D. (1976) "A Comment on Discriminant Analysis 'Versus' Logit Analysis", *Annals of Economics and Social Measurement*, 5, 511–523.

McFadden, D. (1976) "Quantal Choice Analysis: A Survey", *Annals of Economic and Social Measurement*, 5, 363–390.

McFadden, D. (1976) "The Revealed Preferences of a Public Bureaucracy", *Bell Journal*, 7, 55–72.

Miller, L. and R. Radner (1970) "Demand and Supply in U.S. Higher Education", *American Economic Review*, 60, 326–334.

Moore, D. H. (1973) "Evaluation of Five Discrimination Procedures for Binary Variables", *Journal of American Statistical Association*, 68, 399–404.

Nelson, F. (1977) "Censored Regression Models with Unobserved Stochastic Censoring Thresholds", *Journal of Econometrics*, 6, 309–327.

Nelson, F. S. and L. Olsen (1978) "Specification and Estimation of a Simultaneous Equation Model with Limited Dependent Variables", *International Economic Review*, 19, 695–710.

Nerlove, M. (1978) "Econometric Analysis of Longitudinal Data: Approaches, Problems and Prospects", *Annales de l'Insee*, 30–31, 7–22.

Nerlove, M. and J. Press (1973) "Univariable and Multivariable Log-Linear and Logistic Models", RAND Report No. R-1306-EDA/NIH.

Oliveira, J. T. de (1958) "Extremal Distributions", *Revista de Faculdada du Ciencia, Lisboa, Serie A*, 7, 215–227.

Olsen, R. J. (1978) "Comment on 'The Effect of Unions on Earnings and Earnings on Unions: A Mixed Logit Approach'", *International Economic Review*, 259–261.

Plackett, R. L. (1974) *The Analysis of Categorical Data*. London: Charles Griffin.

Poirier, D. J. (1976) "The Determinants of Home Buying in the New Jersey Graduated Work Incentive Experiment", in: H. W. Watts and A. Rees, eds., *Impact of Experimental Payments on Expenditure, Health and Social Behavior, and Studies on the Quality of the Evidence*. New York: Academic Press.

Poirier, D. J. (1980) "A Switching Simultaneous Equation Model of Physician Behavior in Ontario", in: D. McFadden and C. Manski, eds., *Structural Analysis of Discrete Data: With Econometric Applications.* Cambridge: MIT Press.

Pollakowski, H. (1980) *Residential Location and Urban Housing Markets.* Lexington: Heath.

Quandt, R. (1956) "Probabilistic Theory of Consumer Behavior", *Quarterly Journal of Economics,* 70, 507–536.

Quandt, R. (1970) *The Demand for Travel.* London: Heath.

Quandt, R. (1972) "A New Approach to Estimating Switching Regressions", *Journal of the American Statistical Association,* 67, 306–310.

Quandt, R. (1978) "Tests of the Equilibrium vs. Disequilibrium Hypothesis", *International Economic Review,* 19, 435–452.

Quandt, R. and W. Baumol (1966) "The Demand for Abstract Travel Modes: Theory and Measurement", *Journal of Regional Science,* 6, 13–26.

Quandt, R. E. and J. B. Ramsey (1978) "Estimating Mixtures of Normal Distributions and Switching Regressions", *Journal of the American Statistical Association,* 71, 730–752.

Quigley, J. M. (1976) "Housing Demand in the Short-Run: An Analysis of Polytomous Choice", *Explorations in Economic Research,* 3, 76–102.

Radner, R. and L. Miller (1975) *Demand and Supply in U.S. Higher Education.* New York: McGraw-Hill.

Sattath, S. and A. Tversky (1977) "Additive Similarity Trees", *Psychometrika,* 42, 319–345.

Shakotko, Robert A. and M. Grossman (1981) "Physical Disability and Post-Secondary Educational Choices", in: V. R. Fuchs, ed., *Economic Aspects of Health.* National Bureau of Economic Research, Chicago: University of Chicago Press.

Sickles, R. C. and P. Schmidt (1978) "Simultaneous Equation Models with Truncated Dependent Variables: A Simultaneous Tobit Model", *Journal of Economics and Business,* 31, 11–21.

Theil, H. (1969) "A Multinomial Extension of the Linear Logit Model", *International Economic Review,* 10, 251–259.

Theil, H. (1970) "On the Estimation of Relationships Involving Qualitative Variables", *American Journal of Sociology,* 76, 103–154.

Thurstone, L. (1927) "A Law of Comparative Judgement", *Psychological Review,* 34, 273–286.

Tobin, J. (1958) "Estimation of Relationships for Limited Dependent Variables", *Econometrica,* 26, 24–36.

Tversky, A. (1972) "Choice by Elimination", *Journal of Mathematical Psychology.* 9, 341–367.

Tversky, A. (1972) "Elimination by Aspects: A Theory of Choice", *Psychological Review,* 79, 281–299.

Walker, S. and D. Duncan (1967) "Estimation of the Probability of an Event as a Function of Several Independent Variables", *Biometrika,* 54, 167–179.

Westin, R. (1974) "Predictions from Binary Choice Models", *Journal of Econometrics,* 2, 1–16.

Westin, R. B. and D. W. Gillen (1978) "Parking Location and Transit Demand: A Case Study of Endogenous Attributes in Disaggregate Mode Choice Functions", *Journal of Econometrics,* 8, 75–101.

Willis, R. and S. Rosen (1979) "Education and Self-Selection", *Journal of Political Economy,* 87, 507–536.

Yellot, J. (1977) "The Relationship Between Luce's Choice Axiom, Thurstone's Theory of Comparative Judgment, and the Double Exponential Distribution", *Journal of Mathematical Psychology,* 15, 109–144.

Zellner, A. and T. Lee (1965) "Joint Estimation of Relationships Involving Discrete Random Variables", *Econometrica,* 33, 382–394.

*Chapter 28*

# DISEQUILIBRIUM, SELF-SELECTION, AND SWITCHING MODELS*

G. S. MADDALA

*University of Florida*

## Contents

*This chapter was first prepared in 1979. Since then Quandt (1982) has presented a survey of disequilibrium models and Maddala (1983a) has treated self-selection and disequilibrium models in two chapters of the book. The present paper is an updated and condensed version of the 1979 paper. If any papers are not cited, it is just through oversight rather than any judgment on their importance. Financial support from the NSF is gracefully acknowledged.

*Handbook of Econometrics, Volume III, Edited by Z. Griliches and M.D. Intriligator*
© *Elsevier Science Publishers BV, 1986*

## 1. Introduction

The title of this chapter stems from the fact that there is an underlying similarity between econometric models involving disequilibrium and econometric models involving self-selection, the similarity being that both of them can be considered switching structural systems. We will first consider the switching regression model and show how the simplest models involving disequilibrium and self-selection fit in this framework. We will then discuss switching simultaneous equation models, disequilibrium models and self-selection models.

A few words on the history of these models might be in order at the outset. Disequilibrium models have a long history. In fact all the "partial adjustment" models are disequilibrium models.[1] However, the disequilibrium models considered here are different in the sense that they add the extra element of 'quantity rationing'. The differences will be made clear later (in Section 6). As for self-selection models, one can quote an early study by Roy (1951) who considers an example of two occupations: Hunting and fishing and individuals self-select based on their comparative advantage. This example and models of self-selection are discussed later (in Section 9). Finally, as for switching models, almost all the models with discrete parameter changes fall in this category and thus they have a long history. The models considered here are of course different in the sense that we consider also "endogenous" switching. We will first start with some examples of switching regression models. Switching simultaneous equations models are considered later (in Section 5).

Suppose the observations on a dependent variable $y$ can be classified into two regimes and are generated by different probability laws in the two regimes. Define

$$y_1 = X\beta_1 + u_1. \tag{1.1}$$
$$y_2 = X\beta_2 + u_2. \tag{1.2}$$

and

$$y = y_1 \quad \text{iff } Z\alpha - u > 0. \tag{1.3}$$
$$y = y_2 \quad \text{iff } Z\alpha - u \le 0. \tag{1.4}$$

$X$ and $Z$ are (possibly overlapping) sets of explanatory variables. $\beta_1$, $\beta_2$ and $\alpha$ are sets of parameters to be estimated. $u_1$, $u_2$ and $u$ are residuals that are only contemporaneously correlated. We will assume that $(u_1, u_2, u)$ are jointly nor-

---

[1] The disequilibrium model in continuous time analyzed by Bergstrom and Wymer (1976) is also a partial adjustment model except that it is formulated in continuous time.

mally distributed with mean vector 0, and covariance matrix

$$\Sigma = \begin{pmatrix} \sigma_1^2 & \sigma_{12} & \sigma_{1u} \\ \sigma_{12} & \sigma_2^2 & \sigma_{2u} \\ \sigma_{1u} & \sigma_{2u} & 1 \end{pmatrix}.$$

We have set $\text{var}(u) = 1$ because, by the nature of the conditions (1.3) and (1.4) $\alpha$ is estimable only up to a scale factor.

The model given by eqs. (1.1) to (1.4) is called a switching regression model. If $\sigma_{1u} = \sigma_{2u} = 0$ then we have a model with exogenous switching. If $\sigma_{1u}$ or $\sigma_{2u}$ is non-zero, we have a model with endogenous switching. This distinction between switching regression models with exogenous and endogenous switching has been discussed at length in Maddala and Nelson (1975).

We will also distinguish between two types of switching regression models.

Model A: Sample separation known.

Model B: Sample separation unknown.

In the former class we know whether each observed $y$ is generated by (1.1) or (1.2). In the latter class we do not have this information. Further, in the models with known sample separation we can consider two categories of models:

Model A-1: $y$ observed in both regimes.

Model A-2: $y$ observed in only one of the two regimes.

We will discuss the estimation of this type of models in the next section. But first, we will given some examples for the three different types of models.

*Example 1:* Disequilibrium market model

Fair and Jaffee (1972) consider a model of the housing market. There is a demand function and a supply function but demand is not always equal to supply. (As to why this happens is an important question which we will discuss in a later section.) The specification of the model is:

Demand function: $D = X\beta_1 + u_1$
Supply function: $S = X\beta_2 + u_2$

The quantity transacted, $Q$, is given by

$Q = \text{Min}(D, S)$ (the points on the thick lines in Figure 1).

Thus $Q = X\beta_1 + u_1$ if $D < S$,

$Q = X\beta_2 + u_2$    if $D > S$.

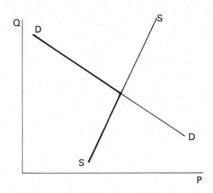

Figure 1

The condition $D < S$ can be written as:

$$X\left(\frac{\beta_2 - \beta_1}{\sigma}\right) - \left(\frac{u_1 - u_2}{\sigma}\right) > 0,$$

where $\sigma^2 = \text{Var}(u_1 - u_2) = \sigma_1^2 + \sigma_2^2 - 2\sigma_{12}$. Thus the model is the same as the switching regression model in eqs. (1.1) to (1.4) with $Z = X$, $\alpha = (\beta_2 - \beta_1)/\sigma$ and $u = (u_1 - u_2)/\sigma$. If sample separation is somehow known, i.e. we know which observations correspond to excess demand and which correspond to excess supply, then we have Model A-1. If sample separation is not known, we have Model B.

*Example 2:*     Model with self-selection

Consider the labor supply model considered by Gronau (1974) and Lewis (1974). The wages offered $W_0$ to an individual, and the reservation wages $W_r$ (the wages at which the individual is willing to work) are given by the following equations:

$$W_0 = X\beta_1 + u_1 \qquad W_r = X\beta_2 + u_2.$$

The individual works and the observed wage $W = W_0$ if $W_0 \geq W_r$. If $W_0 < W_r$, the individual does not work and the observed wages are $W = 0$. This is an example of Model A-2. The dependent variable is observed in only one of the two regimes. The observed distribution of wages is a truncated distribution–it is the distribution of wage offers truncated by the "Self-selection" of individuals–each individual choosing to be 'in the sample' of working individuals or not, by comparing his (or her) wage offer with his (or her) reservation wage.

*Example 3:*    Demand for durable goods

This example is similar to the labor-force participation model in Example 2. Let $y_1$ denote the expenditures the family can afford to make, and $y_2$ denote the value of the minimum acceptable car to the family (the threshold value). The actual expenditures $y$ will be defined as $y = y_1$ iff $y_1 \geq y_2$ and $= 0$ otherwise.

*Example 4:*    Needs vs. reluctance hypothesis

Banks are reluctant to frequent the discount window too often for fear of adverse sanctions from the Federal Reserve. One can define:

$y_1 =$ Desired borrowings
$y_2 =$ Threshhold level below which banks will not use the discount window.

The structure of this model is somewhat different from that given in examples 2 and 3, because we observe $y_1$ all the time. We do not observe $y_2$ but we know for each observation whether $y_1 \leq y_2$ (the bank borrows in the Federal funds market) or $y_1 > y_2$ (the bank borrows from the discount window).

Some other examples of the type of switching regression model considered here are the unions and wages model by Lee (1978), the housing demand model by Lee and Trost (1978), and the education and self-selection model of Willis and Rosen (1979).

## 2.   Estimation of the switching regression model: Sample separation known

Returning to the model given by eqs. (1.1) to (1.4), we note that the likelihood function is given by (dropping the $t$ subscripts on $u$, $X$, $Z$, $y$ and $I$)

$$L\left(\beta_1, \beta_2, \alpha, \sigma_1^2, \sigma_2^2, \sigma_{12}, \sigma_{1u}, \sigma_{2u} | y, I, X\right)$$

$$\propto \prod \left[ g_1(y - X\beta_1) \int_{-\infty}^{Z\alpha} f_1(u | y - X\beta_1) \, du \right]^I$$

$$\cdot \left[ g_2(y - X\beta_2) \int_{Z\alpha}^{\infty} f_2(u | y - X\beta_2) \, du \right]^{1-I}. \tag{2.1}$$

where

$I = 1$    iff $Z\alpha - u > 0$,
$\quad = 0$    otherwise.

and the bivariate normal density of $(u_1, u)$ has been factored into the marginal density $g_1(u_1)$ and the conditional density $f_1(u|u_1)$, with a similar factorization of the bivariate normal density of $(u_2, u)$. Note that $\sigma_{12}$ does not occur at all in the likelihood function and thus is not estimable in this model. Only $\sigma_{1u}$ and $\sigma_{2u}$ are estimable. In the special case $u = (u_1 - u_2)/\sigma$ where $\sigma^2 = \text{Var}(u_1 - u_2)$ as in the examples in the previous section, it can be easily verified that from the consistent estimates of $\sigma_1^2$, $\sigma_2^2$, $\sigma_{1u}$ and $\sigma_{2u}$ we can get a consistent estimate of $\sigma_{12}$.

The maximum likelihood estimates can be obtained by an iterative solution of the likelihood equations using the Newton-Raphson method or the Berndt et al. (1974) method. The latter involves obtaining only the first derivatives of the likelihood function and has better convergence properties. In Lee and Trost (1978) it is shown that the log-likelihood function for this model is uniformly bounded from above. The maximum likelihood estimates of this model can be shown to be consistent and asymptotically efficient following the lines of proof that Amemiya (1973) gave for the Tobit model. To start the iterative solution of the likelihood equations, one should use preliminary consistent estimates of the parameters which can be obtained by using a two-stage estimation method which is described in Lee and Trost (1978),[2] and will not be reproduced here.

There are some variations of this switching regression model that are of considerable interest. The first is the case of the labor supply model where $y$ is observed in only one of the two regimes (Model A-2). The model is given by the following relationships:

$$y = y_1 \quad \text{if } y_1 \geq y_2$$
$$= 0 \quad \text{otherwise.}$$

For the group $I = 1$, we know $y_1 = y$ and $y_2 \leq y$
For the group $I = 0$, all we know is $y_1 < y_2$

Hence the likelihood function for this model can be written as:

$$L(\beta_1, \beta_2, \sigma_1^2, \sigma_2^2, \sigma_{12}) = \prod_{t=1}^{T} \left[ \int_{-\infty}^{g_{2t}} f(g_{1t}, u_{2t}) \, du_{2t} \right]^{I_t} [1 - \Phi_t]^{1 - I_t} \qquad (2.2)$$

where

$$g_{1t} = y_t - X_t \beta_1,$$
$$g_{2t} = y_t - X_t \beta_2,$$
$$\Phi_t = \Phi\left[ \frac{X_t(\beta_2 - \beta_1)}{\sigma} \right],$$
$$\sigma^2 = \text{Var}(u_1 - u_2) = \sigma_1^2 + \sigma_2^2 - 2\sigma_{12},$$

[2] This procedure first used by Heckman (1976) for the labor supply model was extended to a wide class of models by Lee (1976).

$\Phi(\cdot)$ is the distribution function of the standard normal and $f$ is the joint density of $(u_{1t}, u_{2t})$. Since $y$ is observed only in one of the regimes, we need to impose some identifiability restrictions on the parameters of the model. These restrictions are:

(a) There should be at least one explanatory variable in (1.1) not included in (1.2)

or

(b) $\text{Cov}(u_1, u_2) = 0$.

These conditions were first derived in Nelson (1975) and since then have been re-derived by others.

The second variation of the switching regression model that has found wide application is where the criterion function determining the switching also involves $y_1$ and $y_2$ i.e. eqs. (1.3) and (1.4) are replaced by

$$y = y_1 \qquad \text{iff } I^* > 0,$$

$$y = y_2 \qquad \text{iff } I^* \leq 0.$$

Where

$$I^* = \gamma_1 y_1 + \gamma_2 y_2 + Z\alpha - u. \tag{2.3}$$

Examples of this model are the unions and wages model by Lee (1978) and the education and self-selection model by Willis and Rosen (1979). In both cases, the choice function (2.3) determining the switching involves the income differential $(y_1 - y_2)$. Thus $\gamma_2 = -\gamma_1$. Interest centers on the sign and significance of the coefficient of $(y_1 - y_2)$.

The estimation of this model proceeds as before. We first write the criterion function in its reduced form and estimate the parameters by the probit method. Note that, for normalization purposes, instead of imposing the condition $\text{Var}(u) = 1$, it is more convenient to impose the condition that the variance of the residual $u^*$ in the reduced form for (2.3) is unity.

$$\text{i.e. } \text{Var}(u^*) = \text{Var}(\gamma_1 u_1 + \gamma_2 u_2 - u) = 1. \tag{2.4}$$

This means that $\text{Var}(u) = \sigma_u^2$ is a parameter to be estimated. But, in the switching regression model, the parameters that are estimable are: $\beta_1$, $\beta_2$, $\sigma_1^2$, $\sigma_2^2$, $\sigma_{1u*}$, and $\sigma_{2u*}$ where $\sigma_{1u}^* = \text{Cov}(u_1, u^*)$ and $\sigma_{2u}^* = \text{Cov}(u_2, u^*)$. The estimates of $\sigma_{1u*}$ and $\sigma_{2u*}$ together with the normalization eq. (2.4) give us only 3 equations from which we still have to estimate four parameters $\sigma_{12}$, $\sigma_{1u}$, $\sigma_{2u}$ and $\sigma_u^2$. Thus, in this model we have to impose the condition that one of the covariances $\sigma_{12}, \sigma_{1u}, \sigma_{2u}$ is zero. The most natural assumption is $\sigma_{12} = 0$.

As for the estimation of the parameters in the choice function (2.3), again we have to impose some conditions on the explanatory variables in $y_1$ and $y_2$. After obtaining estimates of the parameters $\beta_1$ and $\beta_2$, we get the estimated values $\hat{y}_1$ and $\hat{y}_2$ or $y_1$ and $y_2$ respectively and estimate the parameters in (2.3) by the probit method using these estimated values of $y_1$ and $y_2$. The condition for the estimability of the parameters in (2.3) is clearly that there be no perfect multicollinearity between $\hat{y}_1$, $\hat{y}_2$ and $z$.

This procedure, called the "two-stage probit method" gives consistent estimates of the parameters of the choice function. Note that since $(y_1 - \hat{y}_1)$ and $(y_2 - \hat{y}_2)$ are heteroscedastic, the residuals in this two-stage probit method are heteroscedastic. But this heteroscedasticity exists only in small samples and the residuals are homoscedastic asymptotically, thus preserving the consistency properties of the two-stage probit estimates. For a proof of this proposition and the derivation of the asymptotic covariance matrix of the two-stage probit estimates, see Lee (1979).

## 3.   Estimation of the switching regression model: Sample separation unknown

In this case we do not know whether each observation belongs to Regime 1 or Regime 2. The labor supply model clearly does not fall in this category because the sample separation is known automatically. In the disequilibrium market model, where the assumption of unknown sample separation has been often made, what this implies is that given just the data on quantity transacted and the explanatory variables, we have to estimate the parameters of both the demand and supply functions. Once we estimate these parameters, we can estimate the probability that each observation belongs to the demand and the supply function.

Consider the simplest disequilibrium model with sample separation unknown:

$$D_t = X_{1t}\beta_1 + u_{1t} \text{ (Demand function)},$$

$$S_t = X_{2t}\beta_2 + u_{2t} \text{ (Supply function)},$$

$$Q_t = \text{Min}(D_t, S_t).$$

The probability that observation $t$ belongs to the demand function is:

$$\lambda_t = \text{Prob}(D_t < S_t),$$

$$= \text{Prob}(u_{1t} - u_{2t} < X_{2t}\beta_2 - X_{1t}\beta_1). \tag{3.1}$$

Let $f(u_1, u_2)$ be the joint density of $(u_1, u_2)$ and $g(D, S)$ the joint density of $D$ and $S$ derived from it. If observation $t$ is on the demand function, we know that

$D_t = Q_t$ and $S_t > Q_t$. Hence,

$$h(Q_t|Q_t = D_t) = \int_{Q_t}^{\infty} g(Q_t, S_t)\,\mathrm{d}S_t/\lambda_t. \tag{3.2}$$

The denominator $\lambda_t$ in (3.2) is the normalizing constant. It is equal to the numerator integrated over $Q_t$ over its entire range. Similarly, if observation $t$ is on the supply function, we know that $S_t = Q_t$ and $D_t > Q_t$. Hence,

$$h(Q_t|Q_t = S_t) = \int_{Q_t}^{\infty} g(D_t, Q_t)\,\mathrm{d}D_t/(1-\lambda_t). \tag{3.3}$$

The unconditional density of $Q_t$ is:

$$
\begin{aligned}
h(Q_t) &= \lambda_t h(Q_t|Q_t = D_t) + (1-\lambda_t)h(Q_t|Q_t = S_t) \\
&= \int_{Q_t}^{\infty} g(Q_t, S_t)\,\mathrm{d}S_t + \int_{Q_t}^{\infty} g(D_t, Q_t)\,\mathrm{d}D_t.
\end{aligned} \tag{3.4}
$$

The likelihood function is:

$$L = \prod_t h(Q_t). \tag{3.5}$$

As will be shown later, the likelihood function for this model is unbounded for certain parameter values.

Once the parameters in the model have been estimated, we can estimate the probability that each observation is on the demand function or the supply function. Maddala and Nelson (1974) suggest estimating the expressions $\lambda_t$ in (3.1). These were the probabilities calculated in Sealy (1979) and Portes and Winter (1980). Kiefer (1980a) and Gersovitz (1980) suggest calculating:

$$P(D_t < S_t|Q_t), \tag{3.6}$$

and classifying an observation as belonging to the demand function if this probability is $> 0.5$ and belonging to the supply function if this probability is $< 0.5$.

For the model we are considering, we have

$$\mathrm{Prob}(D_t < S_t|Q_t) = \int_{Q_t}^{\infty} g(Q_t, S_t)\,\mathrm{d}S_t/h(Q_t), \tag{3.7}$$

where $h(Q_t)$ is defined in (3.4). Lee (1983b) treats the classification of sample observations to periods of excess demand or excess supply as a problem in discriminant analysis. He shows that the classification rule suggested by Kiefer and Gersovitz is optimal in the sense that it minimizes the total probability of misclassification. Even in a complicated model, these relationships hold good. Note that in a more complicated model (say with stochastic price adjustment equations) to calculate $\lambda_t$ as in (3.1) or to compute (3.7) we need to derive the marginal distribution of $D_t$ and $S_t$.

There are two major problems with the models with unknown sample separation, one conceptual and the other statistical. The conceptual problem is that we are asking too much from the data when we do not know which observations are on the demand function and which are on the supply function. The results cannot normally be expected to be very good though the frequency with which 'good' results are reported with this method are indeed surprising. For instance, in Sealey (1979) the standard errors for the disequilibrium model (with sample separation unknown) are in almost all cases lower than the corresponding standard errors for the equilibrium model! Goldfeld and Quandt (1975) analyze the value of sample separation information by Monte-Carlo methods and Kiefer (1979) analyzes analytically the value of such information by comparing the variances of the parameter estimates in a switching regression model from a joint density of $(y, D)$ and the marginal density of $y$ (where $y$ is a continuous variable and $D$ is a discrete variable). These results show that there is considerable loss of information if sample separation is not known. In view of this, some of the empirical results being reported from the estimation of disequilibrium models with unknown sample separation are surprisingly good. Very often, if we look more closely into the reasons why disequilibrium exists, then we might be able to say something about the sample separation itself. This point will be discussed later in our discussion of disequilibrium models.

The statistical problem is that the likelihood functions for this class of models are usually unbounded unless some restrictions (usually unjustifiable) are imposed on the error variances. As an illustration, consider the model in eqs. (1.1) to (1.4):

Define

$$\text{Prob}(y = y_1) = \pi,$$

$$\text{Prob}(y = y_2) = 1 - \pi.$$

The conditional density of $y$ given $y = y_1$ is:

$$f(y|y = y_1) = f_1(y - X\beta_1)/\pi.$$

Similarly,

$$f(y|y = y_2) = f_2(y - X\beta_2/(1 - \pi).$$

Hence, the unconditional density of $y$ is:

$$f(y) = \pi \cdot \frac{f_1}{\pi} + (1 - \pi)\frac{f_2}{\pi} = [f_1(y - X\beta_1) + f_2(y - X\beta_2)].$$

Where $f_1$ and $f_2$ are the density functions of $u_1$ and $u_2$ respectively. Thus, the distribution of $y$ is the mixture of two normal distributions. Given $n$ observations $y_i$, we can write the likelihood functions as:

$$L = (A_1 + B_1)(A_2 + B_2) - (A_n + B_n),$$

where

$$A_i = \frac{1}{\sigma_1} \exp\left[ -\frac{1}{2\sigma_1{}^2} (Y_i - X_i\beta_1)^2 \right],$$

and

$$B_i = \frac{1}{\sigma_2} \exp\left[ -\frac{1}{2\sigma_2{}^2} (Y_i - X_i\beta_2)^2 \right].$$

Take $\sigma_2 \neq 0$ and consider the behaviour of $L$ as $\sigma_1 \to 0$. If $X_1\hat{\beta}_1 = y_1$, then $A_1 \to \infty$ and $A_2$, $A_3, -A_n$ all $\to 0$. But $B_1, B_2, -B_n$ are finite. Hence $L \to \infty$. Thus, as $\sigma_1 \to 0$ the likelihood function tends to infinity if $X_i\hat{\beta}_1 = y_i$ for any value of $i$. Similarly, if $\sigma_1 \neq 0$, then as $\sigma_2 \to 0$ the likelihood function tends to infinity if $X_i\hat{\beta}_2 = y_i$ for any value of $i$.

In more complicated models, this proof gets more complicated, but the structure of the proof is the same as in the simple model above. [See Goldfeld and Quandt (1975) and Quandt (1983, pp. 13–16) for further discussion of the problem of unbounded likelihood functions in such models.]

Another problem in this model, pointed out by Goldfeld and Quandt (1978) is the possibility of convergence to a point where the correlation between the residuals is either $+1$ or $-1$. This problem, of course, does not arise if one assumes $\sigma_{12} = 0$ to start with.

The disequilibrium model with unknown sample separation that we have been discussing is a switching regression model with endogenous switching. The case of a switching regression model with exogenous switching and unknown sample separation has been extensively discussed in Quandt and Ramsay (1978) and the discussion that followed their paper.

The model in this case is:

Regime 1: $y_i = X'_{1i}\beta_1 + \varepsilon_{1i}$ with probability $\lambda$
Regime 2: $y_i = X_{2i}\beta_2 + \varepsilon_{2i}$ with probability $(1 - \lambda)$

$$\varepsilon_{1i} \sim IN(0, \sigma_1{}^2) \varepsilon_{2i} \sim IN(0, \sigma_2{}^2).$$

As noted earlier, the likelihood function for this model becomes unbounded for certain parameter values. However, Kiefer (1978) has shown that a root of the likelihood equations corresponding to a *local* maximum is consistent, asymptotically normal and efficient.[3]

Quandt and Ramsay (1978) suggest an MGF (moment generating function) estimator for this model. Note that the moment generating function of $y$ is:

$$E(e^{\theta y}) = \lambda \exp\left[x_1'\beta_1\theta + \theta^2\frac{\sigma_1^2}{2}\right] + (1-\lambda)\exp\left[x_2'\beta_2\theta + \frac{\theta^2\sigma_2^2}{2}\right]. \qquad (3.8)$$

Select a set of $\theta_j$ ($j=1,2\ldots k$) and replace in eq. (3.8).

$$E(e^{\theta_j y}) \qquad \text{by} \quad \frac{1}{n}\sum_{i=1}^{n} e^{\theta_j y_i},$$

$$\exp(\theta_j x_1'\beta_1) \qquad \text{by} \quad \frac{1}{n}\sum_{i=1}^{n} \exp(\theta_j x_{1i}'\beta_1),$$

and

$$\exp(\theta_j x_2'\beta_2) \qquad \text{by} \quad \frac{1}{n}\sum_{i=1}^{n} \exp(\theta_j x_{2i}'\beta_2).$$

Quandt and Ramsay's MGF method is to estimate the parameters $\gamma = (\lambda, \beta_1, \beta_2, \sigma_1^2, \sigma_2^2)$ by minimizing

$$\sum_{j=1}^{k}\left[\frac{1}{n}\sum_{i=1}^{n} Z_i(\theta_j) - \frac{1}{n}\sum_{i=1}^{n} G(\gamma, x_i, \theta_j)\right]^2, \qquad (3.9)$$

where

$$Z_i(\theta_j) = \exp(\theta_j y_i),$$

and $G(\gamma, x_i, \theta_j)$ is the value of the expression on the right hand side of (3.8) for $\theta = \theta_j$ and the $i$th observation.

The normal equations obtained by minimizing (3.9) with respect to $\gamma$ are the same as those obtained by minimizing

$$\sum_{i=1}^{n}\sum_{j=1}^{k} \varepsilon_{ij}^2, \qquad (3.10)$$

---

[3] Hartley and Mallela (1977) prove the strong consistency of the maximum likehood estimator but on the assumption that $\sigma_1$ and $\sigma_2$ are bounded away from zero. Amemiya and Sen (1977) show that even if the likelihood function is unbounded, a consistent estimator of the true parameter value in this model corresponds to a local maximum of the likelihood function rather than a global maximum.

where

$$\varepsilon_{ij} = Z_i(\theta_j) - G(\gamma, x_i, \theta_j).$$

The normal equations in both cases are:

$$\sum_i \sum_j \varepsilon_{ij} \frac{\partial G}{\partial \gamma} = 0.$$

Schmidt (1982) shows that we get more efficient estimates if we minimize weighted sum of squares rather than the simple sum of squares (3.10), making use of the covariance matrices $\Omega_i$ of $(\varepsilon_{i1}, \varepsilon_{i2}, \ldots \varepsilon_{ik})$ for $i = 1, 2 \ldots n$.

Two major problems with the MGF estimator is the choice of the number of $\theta$'s to be chosen (the choice of $k$) and the choice of the particular values of $\theta_j$ for a given choice of $k$. Schmidt (1982) shows that the asymptotic efficiency of the modified MGF estimator (the estimator corresponding to generalized least squares) is a non-decreasing function of $k$ and conjectures that the lower bound of the asymptotic variance is the asymptotic variance of the ML estimator. Thus, the larger the $k$ the better. As for the choice of the particular values of $\theta_j$ for given $k$, Kiefer, in his comment on Quandt and Ramsay's paper notes that the $\theta$'s determine the weights given to the moments of the raw data by the MGF estimator. Small $\theta$'s imply heavy weight attached to low order moments. He also suggests choosing $\theta$'s by minimizing some measure of the size of the asymptotic covariance matrix (say the generalized variance). But this depends on the values of the unknown parameters, though some preliminary estimates can be substituted. Schmidt (1982) presents some Monte-Carlo evidence on this but it is inconclusive.

The discussants of the Quandt and Ramsay paper pointed out that the authors had perhaps exaggerated the problems with the ML method, that they should compare their method with the ML method, and perhaps use the MGF estimates as starting values for the iterative solution of likelihood equations.

In summary, there are many problems with the estimation of switching models with unknown sample separation and much more work needs to be done before one can judge either the practical usefulness of the model or the empirical results already obtained in this area. The literature on self-selection deals with switching models with known sample separation but the literature on disequilibrium models contains several examples of switching models with unknown sample separation [see Sealey (1979), Rosen and Quandt (1979) and Portes and Winter (1980)]. Apart from the computational problems mentioned above, there is also the problem that these studies are all based on the hypothesis of the minimum condition holding on the aggregate so that the aggregate quantity transacted switches between being on the demand curve and the supply curve. The validity of this assumption could be as much a problem in the interpretation of the empirical results as the estimation problems discussed above. Though the "minimum condition" can be justified at the micro-level, it would no longer be valid at the macro-level. Muellbauer (1978) argues that at the macro-level a more reasonable assumption is that

$$Q_t \leq \text{Min}(D_t, S_t).$$

The problems of aggregation are as important as the problems of estimation with unknown sample separation discussed at length above. The econometric problems posed by aggregation have also been discussed in Batchelor (1977), Kooiman and Kloek (1979), Malinvaud (1982) and Muellbauer and Winter (1980).

## 4.  Estimation of the switching regression model with imperfect sample separation information

The discussion in the previous two sections is based on two polar cases: sample separation completely known or unknown. In actual practice there may be many cases where information about sample separation is imperfect rather than perfect or completely unavailable. Lee and Porter (1984) consider the case. They consider the model:

Regime 1: $Y_{1t} = X_{1t}\beta_1 + \varepsilon_{1t}$                                              (4.1)

Regime 2: $Y_{2t} = X_{2t}\beta_2 + \varepsilon_{2t}$,                                             (4.2)

for $t = 1, 2, \ldots, T$. There is a dichotomous indicator $W_t$ for each $t$ which provides sample separation information for each $t$. We define a latent dichotomous variable $I_t$ where

$I_t = 1$ if the sample observation $Y_t = Y_{1t}$

     $= 0$ if the sample observation $Y_t = Y_{2t}$.

The relation between $I_t$ and $W_t$ can be best described by a transition probability matrix

|            | $W_t = 1$ | $W_t = 0$ |
|------------|-----------|-----------|
| $I_t = 1$  | $p_{11}$  | $p_{10}$  |
| $I_t = 0$  | $p_{01}$  | $p_{00}$  |

where

$$p_{ij} = \text{Prob}(W_t = j | I_t = i) \qquad \text{for } i, j = 1, 2.$$

$$p_{11} + p_{10} = 1 \text{ and } p_{01} + p_{00} = 1.$$

Let

$$\text{Prob}(W_t = 1) = p.$$

Then

$$p = \lambda p_{11} + (1 - \lambda) p_{01},$$

where

$$\lambda = \text{prob}(I_t = 1).$$

If we assume $\varepsilon_{1t}$ and $\varepsilon_{2t}$ to be normally distributed as $N(0, \sigma_1^2)$ and $N(0, \sigma_2^2)$ respectively and define

$$f_i = \frac{1}{(2\pi)^{1/2} \sigma_i} \exp\left[ -\frac{1}{2\sigma_i^2} (Y_t - X_{it}\beta_i)^2 \right] \qquad \text{for } i = 1, 2,$$

then the joint density of $Y_t$ and $W_t$ is

$$f(Y_t, W_t) = \left[ f_1 \lambda p_{11} + f_2 (1 - \lambda) p_{01} \right]^{W_t}$$
$$\cdot \left[ f_1 \lambda (1 - p_{11}) + f_2 (1 - \lambda)(1 - p_{01}) \right]^{1 - W_t}, \tag{4.3}$$

and the marginal density of $Y_t$ is

$$g(Y_t) = \lambda f_1 + (1 - \lambda) f_2. \tag{4.4}$$

If $p_{11} = p_{01}$, then the joint density $f(Y_t, W_t)$ can be factored as:

$$f(Y_t, W_t) = g(Y_t) \left[ p_{11}^{W_t} (1 - p_{11})^{1 - W_t} \right],$$

and hence the indicators $W_t$ do not contain any information on the sample separation. One can test the hypothesis $p_{11} = p_{01}$ in any actual empirical case, as shown by Lee and Porter. Also, if $p_{11} = 1$ and $p_{01} = 0$, the indicator $W_t$ provides perfect sample separation, and

$$f(Y_t, W_t) = \left( f_1^{W_t} f_2^{1-W_t} \right) \left[ \lambda^{W_t} (1-\lambda)^{1-W_t} \right].$$

Thus, both the cases considered earlier – sample separation known and sample separation unknown are particular cases of the model considered here.

Lee and Porter also show that if $p_{11} \neq p_{01}$, then there is a gain in efficiency by using the indicator $W_t$. Lee and Porter show that the problem of unbounded likelihood functions encountered in switching regression models with unknown sample separation also exists in this case of imperfect sample separation. As for ML estimation, they suggest a suitable modification of the EM algorithm suggested by Hartley (1977, 1978) and Kiefer (1980b) for the switching regression model with unknown sample separation.

The paper by Lee and Porter is concerned with a switching regression model with exogenous switching but it can be readily extended to a switching regression model with endogenous switching. For instance, in the simple disequilibrium market model

$$D_t = X_{1t}\beta_1 + \varepsilon_{1t},$$
$$S_t = X_{2t}\beta_2 + \varepsilon_{2t},$$
$$Q_t = \text{Min}(D_t, S_t),$$

the joint density of $Q_t$ and $W_t$ can be derived by a procedure analogous to that in (4.3) and it is

$$f(Q_t, W_t) = \left[ p_{11}G_{1t} + p_{01}G_{2t} \right]^{W_t} \left[ (1-p_{11})G_{1t} + (1-p_{01})G_{2t} \right]^{1-W_t},$$

where

$$G_{1t} = \int_{Q_t}^{\infty} g(Q_t, S_t)\, dS_t,$$

$$G_{2t} = \int_{Q_t}^{\infty} G(D_t, Q_t)\, dD_t,$$

and $g(D, S)$ is the joint density of $D$ and $S$. The marginal density $h(Q_t)$ of $Q_t$ is given by the eq. (3.4). As before, if $p_{11} = p_{01}$ then the joint density $f(Q_t, W_t)$ can be written as

$$f(Q_t, W_t) = h(Q_t)\left[ p_{11}^{W_t}(1-p_{11})^{1-W_t} \right].$$

One can use the sign of $\Delta P_t$ for $W_t$. The procedure would then be an extension of the 'directional method' of Fair and Jaffee (1972) in the sense that the sign of $\Delta P_t$ is taken to be a noisy indicator rather than a precise indicator as in Fair and Jaffee. Further discussion of the estimation of disequilibrium models with noisy indicators can be found in Maddala (1984).

## 5. Switching simultaneous systems

We now consider generalizations of the model (1.1) to (1.4) to a simultaneous equation system. Suppose the set of endogenous variables $Y$ are generated by the following two probability laws:

$$B_1 Y_1 + \Gamma_1 X = U_1. \tag{5.1}$$
$$B_2 Y_2 + \Gamma_2 X = U_2. \tag{5.2}$$

and

$$Y = Y_1 \qquad \text{iff } Z\alpha - v > 0. \tag{5.3}$$
$$Y = Y_2 \qquad \text{iff } Z\alpha - v < 0. \tag{5.4}$$

If $v$ is uncorrelated with $U_1$ and $U_2$, we have switching simultaneous systems with exogenous switching. Goldfeld and Quandt (1976) consider models of this kind. Davidson (1978) and Richard (1980) consider switching simultaneous systems where the number of endogenous variables could be different in the two regimes. The switching is still exogenous. An example of this type of model mentioned by Davidson is the estimation of a simultaneous equation model where exchange rates are fixed part of the time and floating the rest of the time. Thus the exchange rate is endogenous in one regime and exogenous in the other regime.

If the residual $v$ is correlated with $U_1$ and $U_2$ we have endogenous switching. The analysis of such models proceeds the same way as Section 2 and the details, which merely involve algebra, will not be pursued here. [See Lee (1979) for the details.] Problems arise, however, when the criterion function in (5.3) and (5.4) involves some of the endogenous variables in the structural system. In this case we have to write the criterion function in its reduced form and make sure that the two reduced form expressions amount to the same condition. As an illustration, consider the model

$$Y_t = \gamma_1 Y_2 + \beta_1' X_1 + u_1,$$
$$Y_2 = \gamma_2 Y_1 + \beta_2' X_2 + u_2 \qquad \text{if } Y_1 < c,$$
$$\quad = \gamma_2' Y_1 + \beta_2' X_2 + u_2 \qquad \text{if } Y_1 > c.$$

Unless $(1 - \gamma_1 \gamma_2)$ and $(1 - \gamma_1 \gamma_2')$ are of the same sign, there will be an inconsistency in the conditions $Y_1 < c$ and $Y_1 > c$ from the two reduced forms. Such conditions for logical consistency have been pointed out by Amemiya (1974), Maddala and Lee (1976) and Heckman (1978). They need to be imposed in switching simultaneous systems where the switch depends on some of the endogenous variables. Gourieroux et al. (1980b) have derived some general conditions which they call "coherency conditions" and illustrate them with a number of examples. These conditions are derived from a theorem by Samelson et al. (1958) which gives a necessary and sufficient condition for a linear space to be partitioned in cones. We will not go into these conditions in detail here. In the case of the switching simultaneous system considered here, the condition they derive is that the determinants of the matrices giving the mapping from the endogenous variables $(Y_1, Y_2, \ldots, Y_k)$ to the residuals $(u_1, u_2, \ldots, u_k)$ are of the same sign, in the different regimes. The two determinants under consideration are $(1 - \gamma_1\gamma_2)$ and $(1 - \gamma_1\gamma_2')$. The condition for logical consistency of the model is that they are of the same sign or $(1 - \gamma_1\gamma_2)(1 - \gamma_1\gamma_2') > 0$. A question arises about what to do with these conditions. One can impose them and then estimate the model. Alternatively, since the condition is algebraic, if it cannot be given an economic interpretation, it is important to check the basic structure of the model. An illustration of this is the dummy endogenous variable model in Heckman (1976a). The model discusses the problem of estimation of the effect of fair employment laws on the wages of blacks relative to whites, when the passage of the law is endogenous. The model as formulated by Heckman is a switching simultaneous equations model for which we have to impose a condition for "logical consistency". However, the condition does not have any meaningful economic interpretation and as pointed out in Maddala and Trost (1981), a careful examination of the arguments reveals that there are two sentiments, not one as assumed by Heckman, that lead to the passage of the law, and when the model is reformulated, there is no condition for logical consistency that needs to be imposed.

The simultaneous equations models with truncated dependent variables considered by Amemiya (1974) are also switching simultaneous equations models which require conditions for logical consistency. Again, one needs to examine whether these conditions need to be imposed exogenously or whether a more logical formulation of the problem leads to a model where these conditions are automatically satisfied. For instance, Waldman (1981) gives an example of time allocation of young men to school and work where the model is formulated in terms of underlying behavioural relations and the conditions derived by Amemiya follow naturally from economic theory. On the other hand, these conditions have to be imposed exogenously (and are difficult to give an economic interpretation) if the model is formulated in a mechanical fashion where time allocated to work was modelled as a linear function of school time and exogenous variables and time allocated to school was modelled as a linear function of work time and exogenous variables.

The point of this lengthy discussion is that in switching simultaneous equation models, we often have to impose some conditions for the logical consistency of the model. If these conditions cannot be given a meaningful economic interpretation, it is worthwhile checking the original formulation of the model rather than imposing these conditions exogenously and estimating the parameters in the model subject to these conditions.

An interesting feature of the switching simultaneous systems is that it is possible to have underidentified systems in one of the regimes. As an illustration, consider the following model estimated by Avery (1982):

$$D = \beta_1' X_1 + \alpha_1 Y + u_1 \text{ Demand for Durables.}$$

$$Y_1 = \beta_2' X_1 + \alpha_2 D + u_2 \text{ Demand for Debt.}$$

$$Y_2 = \beta_3' X_3 + \alpha_3 D + u_3 \text{ Supply of Debt.}$$

$$Y = \min(Y_1, Y_2) \text{ Actual quantity of Debt.}$$

$D, Y_1, Y_2$ are the endogenous variables and $X_1$ and $X_3$ are sets of exogenous variables. Note that the exogenous variables in the demand for durables equation and the demand for debt equation are the same.

The model is a switching simultaneous equations model with endogenous switching. We can write the model as follows:

| Regime 1: $Y_1 < Y_2$ | Regime 2: $Y_2 < Y_1$ |
|---|---|
| $D = \beta X_1 + \alpha_1 Y + u_1$ | $D = \beta_1' X_1 + \alpha_1 Y + u_1$ |
| $Y = \beta_2' X_1 + \alpha_2 D + u_2$ | $Y = \beta_3' X_3 + \alpha_3 D + u_3$ |

If we get the reduced forms for $Y_1$ and $Y_2$ in the two regimes and simplify the expression $Y_1 - Y_2$, we find that:

$$(Y_1 - Y_2) \text{ in Regime 2} = \frac{1 - \alpha_1 \alpha_3}{1 - \alpha_1 \alpha_2} \{(Y_1 - Y_2) \text{ in Regime 1}\}.$$

Thus, the condition for the logical consistency of this model is that $(1 - \alpha_1 \alpha_2)$ and $(1 - \alpha_1 \alpha_3)$ are of the same sign – a condition that can also be derived by using the theorems in Gourieroux et al. (1980b).

The interesting thing to note is that the simultaneous equation system in Regime 1 is under-identified. However, if the system of equations in Regime 2 is identified, the fact that we can get consistent estimates of the parameters in the demand equation for durables from Regime 2, enables us to get consistent estimates of the parameters in the $Y_1$ equation. Thus the parameters in the simultaneous equations system in Regime 1 are identified. One can construct a formal and rigorous proof but this will not be attempted here. Avery (1982) found that he could not estimate the parameters of the structural equation for $Y_1$ but this is possibly due to the estimation methods used.

In summary, switching simultaneous equations models often involve the imposition of constraints on parameters so as to avoid some internal inconsistencies in the model. But it is also very often the case that such logical inconsistencies arise when the formulation of the model is mechanical. In many cases, it has been found that a re-examination and a more careful formulation leads to an alternative model where such constraints need not be imposed.

There are also some switching simultaneous equations models where a variable is endogenous in one regime and exogenous in another and, unlike the cases considered by Richard (1980) and Davidson (1978), the switching is endogenous. An example is the disequilibrium model in Maddala (1983b).

## 6. Disequilibrium models: Different formulations of price adjustment

Econometric estimation of disequilibrium models has a long history. The partial adjustment models are all disequilibrium models and in fact this is the type of model that the authors had in mind when they talked of "disequilibrium model." Some illustrative examples of this are Rosen and Nadiri (1974), and Jonson and Taylor (1977).

The recent literature on disequilibrium econometrics considers a different class of models and has a different structure. These models are more properly called "rationing models." This literature started with the paper by Fair and Jaffee (1972). The basic equation in their models is

$$Q_t = \text{Min}(D_t, S_t), \tag{6.1}$$

where

$Q_t$ = quantity transacted
$D_t$ = quantity demanded
$S_t$ = quantity supplied.

Fair and Jaffee considered two classes of models
(i) Directional models: In these we infer whether $Q_t$ is equal to $D_t$ or $S_t$ based on the direction of price movement, i.e.

$$D_t > S_t \text{ and hence } Q_t = S_t \qquad \text{if } \Delta P_t > 0,$$
$$D_t < S_t \text{ and hence } Q_t = D_t \qquad \text{if } \Delta P_t < 0,$$

where $\Delta P_t = P_t - P_{t-1}$,
and
(ii) Quantitative models: In these the price change is proportional to excess demand (or supply), i.e.

$$P_t - P_{t-1} = \gamma(D_t - S_t). \tag{6.2}$$

The maximum likelihood estimation of the quantitative model is discussed in Amemiya (1974a). The maximum likelihood estimation of the directional model, and models with stochastic sample separation (i.e. where only (6.1) is used or (6.2) is stochastic) is discussed in Maddala and Nelson (1974).

The directional method is logically inconsistent since the condition that $\Delta P_t$ gives information on sample separation implies that $P_t$ is endogenous, in which case there are not enough equations to determine the endogenous variables $Q_t$ and $P_t$.[4] We will, therefore, discuss only models with the price determination eq. (6.2) included.

There are three important problems with the specification of this model that need some discussion. These are:
(i) The meaning of the price adjustment eq. (6.2)
(ii) The modification in the specification of the demand and supply functions that need to be made because of the existence of the disequilibrium, and
(iii) The validity of the min. condition (6.1).
We will discuss these problems in turn.

### 6.1. The meaning of the price adjustment equation

The disequilibrium market model usually considered consists of the following demand and supply functions:

$$D_t = X_{1t}\beta_1 + \alpha_1 P_t + u_{1t} \tag{6.3}$$

$$S_t = X_{2t}\beta_2 + \alpha_2 P_t + u_{2t}. \tag{6.4}$$

and the eqs. (6.1) and (6.2). To interpret the "price adjustment" eq. (6.2) we have to ask the basic question of why disequilibrium exists. One interpretation is that prices are fixed by someone. The model is thus a *fix-price model*. The disequilibrium exists because price is fixed at a level different from the market equilibrating level (as is often the case in centrally planned economies). In this case the

---

[4] The directional method makes sense only for the estimation of the reduced form equations for $D_t$ and $S_t$ in a model with a price adjustment equation. There are cases where this is needed. The likelihood function for the estimation of the parameters in this model is derived in Maddala and Nelson (1974). It is:

$$L \propto \prod_{\Delta P < 0} \int_Q^\infty g(Q,S)\,\mathrm{d}s \cdot \prod_{\Delta P > 0} \int_Q^\infty g(D,Q)\,\mathrm{d}D.$$

where $g(D,S)$ is the joint density of $D$ and $S$ (from the reduced form equations). When $\Delta P < 0$ we have $D = Q$ and $S > Q$ and when $\Delta P > 0$ we have $S = Q$ and $D > Q$. Note that the expression given in Fair and Kelejian (1974) as the likelihood function for this model is not correct though it gives consistent estimates of the parameters.

price adjustment eq. (6.2) can be interpreted as the rule by which the price-fixing authority is changing the price. However, there is the problem that the price-fixing authority does not know $D_t$ and $S_t$ since they are determined only after $P_t$ is fixed. Thus, the eq. (6.2) cannot make any sense in the fix-price model. Laffont and Garcia (1977) suggested a modification of the price adjustment equation which is:

$$P_{t+1} - P_t = \gamma(D_t - S_t). \tag{6.2'}$$

In this case the price fixing authority uses information on the past period's demand and supply to adjust prices upwards or downwards. In this case the price-fixing rule is an operational one but one is still left wondering why the price-fixing authority follows such a dumb rule as (6.2'). A more reasonable thing to do is to fix the price at a level that equates expected demand and supply. One such rule is to determine price by equating the components of (6.3) and (6.4) after ignoring the stochastic disturbance terms. This gives

$$P_t = \frac{X_{1t}\beta_1 - X_{2t}\beta_2}{\alpha_2 - \alpha_1}. \tag{6.5}$$

This is the procedure suggested by Green and Laffont (1981) under the name of "anticipatory pricing".

As mentioned earlier, the meaning of the price adjustment equation depends on the source of disequilibrium. An alternative to the fix-price model as an explanation of disequilibrium is the *partial adjustment model* (see Bowden, 1978 a, b). The source of disequilibrium in this formulation is stickiness of prices (due to some institutional constraints or other factors). Let $P_t^*$ be the market equilibrating price. However, prices do not adjust fully to the market equilibrating level and we specify the "partial adjustment" model:

$$\begin{aligned} P_t - P_{t-1} &= \lambda(P_t^* - P_{t-1}) \qquad 0 < \lambda < 1 \\ &= \lambda(P_t^* - P_t + P_t - P_{t-1}). \end{aligned} \tag{6.6}$$

Hence

$$P_t - P_{t-1} = \frac{\lambda}{1-\lambda}(P_t^* - P_t). \tag{6.7}$$

If $P_t < P_t^*$ there will be excess demand and if $P_t > P_t^*$ there will be excess supply. Hence, if $\Delta P_t < 0$ we have a situation of excess supply.

Note that in this case it is $\Delta P_t$ (not $\Delta P_{t+1}$ as in the Laffont–Garcia case) that gives the sample separation. *But the interpretation is not that prices rise in response to excess demand* (as implicitly argued by Fair and Jaffee) *but that there is excess demand* (or excess supply) *because prices do not fully adjust to the equilibrating values.*[5]

Equation (6.7) can also be written as

$$P_t - P_{t-1} = \gamma(D_t - S_t),\tag{6.8}$$

if we assume that the excess demand $(D_t - S_t)$ is proportional to the difference $(P_t^* - P_t)$, i.e. the difference between the equilibrating price and the actual price. The interpretation of the coefficient $\gamma$ in (6.8) is of course different from what Fair and Jaffee gave to the same equation.

One can also allow for different speeds of upward and downward partial adjustment. Consider the following formulation:

$$
\begin{aligned}
P_t - P_{t-1} &= \lambda_1\big(P_t^* - P_{t-1}\big) && \text{if } P_t^* > P_{t-1}, \\
&= \lambda_2\big(P_t^* - P_{t-1}\big) && \text{if } P_t^* < P_{t-1}.
\end{aligned}\tag{6.9}
$$

These equations imply

$$
\begin{aligned}
P_t - P_{t-1} &= \frac{\lambda_1}{1-\lambda_1}\big(P_t^* - P_t\big) && \text{if } P_t^* > P_t. \\
&= \frac{\lambda_2}{1-\lambda_2}\big(P_t^* - P_t\big) && \text{if } P_t^* < P_t
\end{aligned}\tag{6.10}
$$

Note first that the conditions $P_t^* > P_{t-1}$, $P_t > P_{t-1}$, $P_t^* > P_t$ and $D_t > S_t$ are all equivalent. Also assuming that excess demand is proportional to $P_t^* - P_t$ we can write eqs. (6.10) as

$$
\begin{aligned}
\Delta P_t &= \gamma_1(D_t - S_t) && \text{if } D_t > S_t \\
&= \gamma_2(D_t - S_t) && \text{if } D_t < S_t.
\end{aligned}\tag{6.11}
$$

Again note that we get $\Delta P_t$ and not $\Delta P_{t+1}$ in these equations.

Ito and Ueda (1979) use Bowden's formulation with different speeds of adjustment as given by (6.9) to estimate the rates of adjustment in interest rates for business loans in the U.S. and Japan. They prefer this formulation to that of Fair and Jaffee or Laffont and Garcia because in eq. (6.9), $\lambda_1$ and $\lambda_2$ are pure numbers which can be compared across countries. The same cannot be said about the parameters $\gamma_1$ and $\gamma_2$ in eq. (6.11).

---

[5] The formulation in terms of partial adjustment towards $P^*$ was suggested by Bowden (1978a) though he does not use the interpretation of the Fair–Jaffee equation given here. Bowden (1978b) discusses this approach in greater detail under the title: "The PAMEQ Specification".

There is still one disturbing feature about the partial adjustment eq. (6.6) that Bowden adopts and under which we have given a justification for the Fair and Jaffee directional and quantitative methods. This is that $\Delta P_t$ unambiguously gives us an idea about whether there is excess demand or excess supply. As mentioned earlier this does not make intuitive sense. On closer examination one sees that the problem is with eq. (6.6), in particular the assumption that $\lambda$ lies between 0 and 1. This is indeed a very strong assumption and implies that prices are sluggish but never change to overshoot $P_t^*$ the equilibrium prices. There is, however, no a priori reason why this should happen.[6] Once we drop the assumption that $\lambda$ should lie between 0 and 1, it is no longer true that we can use $\Delta P_t$ to classify observations as belonging to excess demand or excess supply. As noted earlier the assumption $0 < \lambda < 1$ implies that the conditions $P_t^* > P_{t-1}$, $P_t > P_{t-1}$, $P_t^* > P_t$ and $D_t > S_t$ are all equivalent. With $\lambda > 1$, this no longer holds good.

In summary, we considered two models of disequilibrium – the fix-price model and the partial adjustment model. In the fix-price model the price adjustment eq. (6.2) is non-operational. The modification (6.2′) suggested by Laffont and Garcia is an operational rule but really does not make much sense. A more reasonable formula for a price-setting rule is the anticipatory pricing rule (6.5). But this implies that a price-adjustment equation like (6.2) or (6.2′) is not valid.

In the case of the partial adjustment model one can derive an equation of the form (6.2) though its meaning is different from the one given by Fair and Jaffee and many others using this price adjustment equation. The meaning is not that prices adjust in response to excess demand or supply but that excess demand and supply exist because prices do not adjust to the market equilibrating level. However, as discussed earlier, eq. (6.2) can be derived from the partial adjustment model (6.6) only under a restrictive set of assumptions.

The preceding arguments hold good when eq. (6.2) is made stochastic with the addition of a disturbance term. In this case there is not much use for the price-adjustment equation. The main use of eq. (6.2) is that it gives a sample separation, and estimation with sample separation known is much simpler than estimation with sample separation unknown. If one is anyhow going to estimate a model with sample separation unknown, then one can as well eliminate eq. (6.2). For fix-price models, one substitutes the anticipatory price eq. (6.5) and for partial adjustment models one uses eq. (6.6) directly.

## 6.2. Modifications in the specification of the demand and supply functions

The preceding discussion refers to alternative formulations of the price adjustment equation. One can also question the specification of the other equations as

---

[6] Since no economic model has been specified, there is no reason to make any alternative assumption either.

well. We will now discuss alternative specifications of the demand and supply functions.

The probability that there would be rationing should affect the demand and supply functions. There are two ways of taking account of this. One procedure suggested by Eaton and Quandt (1983) is to introduce the probability of rationing as an explanatory variable in the demand and/or supply functions (6.3) and (6.4). A re-specification of eq. (6.3), they consider is

$$D_t = X_{1t}\beta_1 + \alpha_1 P_t + \gamma_1 \pi_t + u_{1t}, \tag{6.3$'$}$$

where

$$\pi_t = \text{Prob}(D_t > S_t),$$

$\gamma_1$ is expected to be $< 0$. Eaton and Quandt show that the solution for $\pi_t$ is unique.[7] In their empirical work they include $(1 - \pi_t)$ as an explanatory variable in the supply function. They also include a price adjustment equation in their model.

An alternative procedure to take account of the probability of rationing is to re-formulate the demand and supply functions in terms of expected prices and incorporate the probability of disequilibrium as a determining factor in the formation of expectations. This is the approach followed in Chanda (1984). Since price expectations anyhow need to be introduced into the model and since stickiness in price movement or other limitations on price movement are the sources of disequilibrium, this procedure of incorporating probability of rationing into price expectations is the logical one and is more meaningful than introducing the probability of disequilibrium as an explanatory variable, as done by Eaton and Quandt. The approach adopted by Eaton and Quandt does not say what disequilibrium is due to, whereas the approach based on price expectations depends on what the sources of disequilibrium are.

As an illustration of this approach we will re-formulate the supply function by introducing expected prices. We leave eqs. (6.1), (6.2) and (6.3) as they are and re-define (6.4) as

$$S_t = X_{2t}\beta_2 + \alpha_2 P_t^e + u_{2t}, \tag{6.4$'$}$$

where $P_t^e$ is the expected price, i.e. the price the suppliers expect to prevail in period $t$, the expectation being formed at time $t-1$ (we will assume a one period lag between production decisions and supply). Regarding the expected price $P_t^e$, if we use some naive extrapolative or the adaptive expectations formulae, then the estimation proceeds as in earlier models with no price expectations, with minor modifications. For instance, with the adaptive expectations formula, one would

[7]Though the analysis is similar, the computations are more complex because of the presence of $\pi_t$ in the demand function.

first get the ML estimates conditional on a value $\lambda$ of the weighting parameter and then choose the value of $\lambda$ for which the likelihood is maximum.

An alternative procedure is to use the rational expectations hypothesis

$$P_t^e = E(P_t | I_{t-1}), \tag{6.12}$$

where $P_t^e$ is the expected price and $I_{t-1}$ represents the information set the economic agents are assumed to have.

Equation (6.12) implies that we can write

$$P_t = P_t^e + v_t,$$

where $v_t$ is uncorrelated with all the variables in the information set $I_{t-1}$. If the information set $I_{t-1}$ includes the exogenous variables $X_{1t}$ and $X_{2t}$, i.e. if these exogenous variables are known at time $t-1$, then we can substitute $P_t^e = P_t - v_t$ in eq. (6.12). We can re-define a residual $u_{2t}^* = u_{2t} - \alpha_2 v_t$ and $u_{2t}^*$ has the same properties as $u_{2t}$. Thus the estimation of the model simplifies to the case considered by Fair and Jaffee.

If, on the other hand, $X_{1t}$ and $X_{2t}$ are not known at time $(t-1)$ we cannot treat $v_t$ the same way as we treat $u_{2t}$ since $v_t$ can be correlated with $X_{1t}$ and $X_{2t}$. In this case we proceed as follows.

From eqs. (6.2), (6.3), and (6.4') we have

$$\Delta P_t = \gamma(D_t - S_t),$$

or

$$P_t - P_{t-1} = \gamma\left[(\beta_1' X_{1t} - \beta_2' X_{2t}) + (\alpha_1 P_t - \alpha_2 P_t^e) + (u_{1t} - u_{2t})\right].$$

Taking expectations of both sides conditional on the information set $I_{t-1}$,

$$P_t^e - P_{t-1} = \gamma\left[\beta_1' X_{1t}^* - \beta_2' X_{2t}^*) + (\alpha_1 - \alpha_2)P_t^e\right],$$

or

$$P_t^e = \lambda_1(\beta_1' X_{1t}^* - \beta_2' X_{2t}^*) + \lambda_2 P_{t-1}, \tag{6.13}$$

where

$$\lambda_1 = \frac{\gamma}{[1 + \gamma(\alpha_2 - \alpha_1)]},$$

$$\lambda_2 = \frac{1}{[1 + \gamma(\alpha_2 - \alpha_1)]},$$

and $X_{1t}^*$ and $X_{2t}'$ are the expected values of $X_{1t}$ and $X_{2t}$. (Note that this equation is valid even if the price adjustment eq. (6.2) is made stochastic.)

To obtain $X_{1t}^*$ and $X_{2t}^*$ we have to make some assumptions about how these exogenous variables are generated. A common assumption is that they follow vector autoregressive processes. Let us for the sake of simplicity of notation assume a first order autoregressive process.

$$X_{1t} = \phi_{11} X_{1,t-1} + \phi_{12} X_{2,t-1} + \varepsilon_{1t},$$
$$X_{2t} = \phi_{21} X_{1,t-1} + \phi_{22} X_{2,t-1} + \varepsilon_{2t}. \qquad (6.14)$$

Then

$$X_{1t}^* = \phi_{11} X_{1,t-1} + \phi_{12} X_{2,t-1},$$

and

$$X_{2t}^* = \phi_{21} X_{1,t-1} + \phi_{22} X_{2,t-1}.$$

We substitute these equations in (6.13) and substitute the resulting expression for $P_t^e$ in eq. (6.4').

The estimation of the model will proceed as with the usual disequilibrium model. The likelihood function in this model is derived in exactly the same way as with the Fair and Jaffee model, as derived in Amemiya (1974). The only extra complication is the existence of cross-equation restrictions as implied by eqs. (6.14), as discussed in Wallis (1980). The two-stage least squares estimation suggested in Amemiya (1974) can also be easily adapted to the above model. For details of this, see Chanda (1984).

Yet another modification in the specification of the demand and supply function that one needs to make is that of 'spillovers'. The unsatisfied demand and excess supply from the previous period will spill over to current demand and supply. The demand and supply functions (6.3) and (6.4) are now reformulated respectively as:

$$D_t = X_{1t}\beta_1 + \alpha_1 P_t + \delta_1(D_{t-1} - Q_{t-1}) + u_{1t},$$
$$S_t = X_{2t}\beta_2 + \alpha_2 P_t + \delta_2(S_{t-1} - Q_{t-1}) + u_{2t}, \qquad (6.15)$$

with $\delta_1 > 0$, $\delta_2 > 0$, and $\delta_1\delta_2 < 1$. [See Orsi (1982) for this last condition.]

At time $(t-1)$, $Q_{t-1}$ is equal to $D_{t-1}$ or $S_{t-1}$. Thus, one of these is not observed. However, if the price adjustment eq. (6.2) is not stochastic, one has a four-way regime classification depending on excess demand or excess supply at time periods $(t-1)$ and $t$. Thus, the method of estimation suggested by Amemiya (1974a) for the Fair and Jaffee model can be extended to this case. Such extension is done in Laffont and Monfort (1979). Orsi (1982) applied this model to the Italian labor market but the estimates of the spill-over coefficients were not significantly different from zero. This method is further extended by Chanda (1984) to the case where the supply function depends on expected prices and expectations are formed rationally.

## 6.3. The validity of the 'Min' condition

As mentioned in the introduction, the main element that distinguishes the recent econometric literature on disequilibrium models from the earlier literature is the 'Min' condition' (6.1). This condition has been criticized on the grounds that:
  (a) Though it can be justified at the micro-level, it cannot be valid at the aggregate level where it has been very often used.
  (b) It introduces unnecessary computational problems which can be avoided by replacing it with

$$Q = \text{Min}[E(D), E(S)] + \varepsilon. \tag{6.1'}$$

  (c) In some disequilibrium models, the appropriate condition for the transacted quantity is

$$Q = 0 \quad \text{if } D \neq S.$$

Criticism (a) made by Muellbauer (1978) is a valid one. The appropriate modifications depend on the assumptions made about the aggregation procedure. These problems have been discussed in Batchelor (1977), Kooiman and Kloek (1979), Malinvaud (1982) and Muellbauer and Winter (1980). Bouisson, Laffont and Vuong (1982) suggests using survey data to analyze models of disequilibrium at the aggregate level.

Regarding criticism (b), Richard (1980b) and Hendry and Spanos (1980) argue against the use of the 'Min' condition as formulated in (6.1). Sneessens (1981, 1983) adopts the condition (6.1'). However, eq. (6.1') is hard to justify as a behavioural equation. Even the computational advantages are questionable [see Quandt (1983) pp. 25–26]. The criticism of Hendry and Spanos is also not valid on closer scrutiny [see Maddala (1983a), pp. 34–35 for details].

Criticism (c) is elaborated in Maddala (1983a, b), where a distinction is made between "Rationing models" and "Trading Models", the former term applying to models for which the quantity transacted is determined by the condition (6.1), and the latter term applying to models where no transaction takes place if $D_t \neq S_t$. Condition (6.1) is thus replaced by

$$Q_t = 0 \quad \text{if } D_t \neq S_t. \tag{6.1''}$$

The term 'trading model' arose by analogy with commodity trading where trading stops when prices hit a floor or a ceiling (where there is excess demand or excess supply respectively). However, in commodity trading, a sequence of trades takes place and all we have at the end of the day is the total volume of trading and the opening, high, low and closing prices.[8] Thus, commodity trading models do not necessarily fall under the category of 'trading' models defined here. On the other hand models that involve 'rationing' at the aggregate level might fall into the class of 'trading' models defined here at the micro-level. Consider, for instance, the loan demand problem with interest rate ceilings. At the aggregate level there would be an excess demand at the ceiling rate and there would be rationing. The question is how rationing is carried out. One can argue that for each individual there is a demand schedule giving the loan amounts $L$ the individual would want to borrow at different rates of interest $R$. Similarly, the bank would also have a supply schedule giving the amounts $L$ it would be willing to lend at different rates of interest $R$. If the rate of interest at which these two schedules intersect is $\leq \overline{R}$ the ceiling rate, then a transaction takes place. Otherwise no transaction takes place. This assumption is perhaps more appropriate in mortgage loans rather than consumer loans. In this situation $Q$ is not $\text{Min}(D, S)$. In fact $Q = 0$ if $D \neq S$. The model would be formulated as:

$$\left. \begin{array}{ll} \text{Loan Demand} & L_i = \alpha_1 R_i + \beta_1' X_{1i} + u_{1i} \\ \text{Loan Supply} & L_i = \alpha_2 R_i + \beta_2' X_{2i} + u_i \end{array} \right\} \quad \text{if } R_i^e \leq \overline{R},$$

$$L_i = 0 \text{ otherwise.}$$

$R_i^e$ is the rate of interest that equilibrates demand and supply. If the assumption is that the individual borrows what is offered at the ceiling rate $\overline{R}$, an assumption more appropriate for consumer loans, we have

$$L_i = \alpha_2 \overline{R} + \beta_2' X_{2i} + u_{2i} \quad \text{if } R_i^e > \overline{R}.$$

In this case of course $Q = \text{Min}(D, S)$, but there is never a case of excess-supply. Further discussion of this problem can be found in Maddala and Trost (1982).

---

[8]Actually, in studies on commodity trading, the total number of contracts is treated as $Q_t$ and the closing price for the day as $P_t$. The closing price is perhaps closer to an equilibrium price than the opening, low and high prices. But it cannot be treated as an equilibrium price. There is the question of what we mean by 'equilibrium' price in a situation where a number of trades take place in a day. One can interpret it as the price that would have prevailed if there was to be a Walrasian auctioneer and a single trade took place for a day. If this is the case, then the closing price would be an equilibrium price only if a day is a long enough period for prices at the different trades to converge to some equilibrium. These problems need further work. See Monroe (1981).

The important situations where this sort of disequilibrium model arises is where there are exogenous controls on the movement of prices. There are essentially three major sources of disequilibrium that one can distinguish.

(1) Fixed prices

(2) Imperfect adjustment of prices

(3) Controlled prices

We have till now discussed the case of fixed prices and imperfect adjustment to the market equilibrating price. The case of controlled prices is different from the case of fixed prices. The disequilibrium model considered earlier in example 1, Section 1 is one with fix ed prices. With fixed prices, the market is almost always in disequilibrium. With controlled prices, the market is sometimes in equilibrium and sometimes in disequilibrium.[9]

Estimation of disequilibrium models with controlled prices is discussed in Maddala (1983a, pp. 327–34 and 1983b) and details need not be presented here. Gourieroux and Monfort (1980) consider endogenously controlled prices and Quandt (1984) discusses switching between equilibrium and disequilibrium.

In summary, not all situations of disequilibrium involve the 'Min' condition (6.1). In those formulations where there is some form of rationing, the alternative condition (6.1′), that has been suggested on grounds of ccmputational simplicity, is not a desirable one to use and is difficult to justify conceptually.

What particular form the 'Min' condition takes depends on how the rationing is carried out and whether we are analyzing micro or macro data. The discussion of the loan problem earlier shows how the estimation used depends on the way customers are rationed. This analysis applies at the micro level. For analysis with macro data Goldfeld and Quandt (1983) discuss alternative decision criteria by which the Federal Home Loan Bank Board (FHLBB) rations its advances to savings and loan institutions. The paper based on earlier work by Goldfeld, Jaffee and Quandt (1980) discusses how different targets and loss functions lead to different forms of the 'Min' condition and thus call for different estimation methods. This approach of deriving the appropriate rationing condition from explicit loss functions is the appropriate thing to do, rather than writing down the demand and supply functions (6.3), and (6.4), and saying that since their is disequilibrium (for some unknown and unspecified reasons) we use the 'Min' condition (6.1).

## 7.  Some other problems of specification in disequilibrium models

We will now discuss some problems of specifications in disequilibrium models that need further work.

---

[9]Mackinnon (1978) discusses this problem but the likelihood functions he presents are incorrect. The correct analysis of this model is presented in Maddala (1983b).

## 7.1. *Problems of serial correlation*

The econometric estimation of disequilibrium models is almost exclusively based on the assumption that the error terms are serially independent. If they are serially correlated, the likelihood functions are intractable since they involve integrals of a very high dimension. One can, however, derive a test for serial correlation based on the Lagrangian multiplier principle that does not involve the evaluation of multiple integrals. (See Lee and Maddala, 1983a.) Quandt (1981) discusses the estimation of a simple disequilibrium model with autocorrelated errors but the likelihood function maximized by him is $L = \prod_t h(Q_t)$ which is not correct since $Q_t$ and $Q_{t-1}$ are correlated. The only example till now where estimation is done with autocorrelated errors is the paper by Cosslett and Lee (1983) who analyze the model

$$Y_t = X_t \beta + \alpha I_t - u_t,$$

where $u_t$ are first-order autocorrelated, $Y_t$ is a continuous indicator and $I_t$ is a discrete indicator measured with error. The model they consider is thus, a switching regression model with exogenous switching and imperfect sample separation. Cosslett and Lee derive a test statistic for detecting serial correlation in such a model and show that the likelihood function can be evaluated by a recurrence relation, and thus maximum likelihood estimation is computationally feasible.

For the disequilibrium model with *known* sample separation, one can just transform the demand and supply eqs. (6.3) and (6.4). For instance, if the residuals in the two equations are first-order autocorrelated, we have

$$u_{1t} = \rho_1 u_{1,t-1} + e_{1t},$$

$$u_{2t} = \rho_2 u_{2,t-1} + e_{2t}. \tag{7.1}$$

Then we have

$$D_t = \rho_1 D_{t-1} + \beta_1 X_{1t} - \beta_1 \rho_1 X_{1,t-1} + e_{1t},$$

and

$$S_t = \rho_2 S_{t-1} + \beta_2 X_{2t} - \beta_2 \rho_2 X_{2,t-1} + e_{2t}. \tag{7.2}$$

Since sample separation is available, the procedure in Laffont and Monfort (1979) can be used with the modification that there are nonlinear restrictions on the parameters in (7.2). The same procedure holds good if, instead of (7.1) we specify an equation where $u_{1t}$ and $u_{2t}$ depended on lagged values of both $u_{1t}$ and $u_{2t}$.

Thus, serially correlated errors can be handled if the sample separation is known and in models with exogenous switching even if the sample separation is imperfect.

## 7.2. Tests for distributional assumptions

The econometric estimation of disequilibrium models is entirely based on the assumption of normality of the disturbances. It would be advisable to devise tests of the normality assumption and suggest methods of estimation that are either distribution-free or based on distributions more general than the normal distribution. Lee (1982b) derives a test for the assumption of normality in the disequilibrium market model from the Lagrangean multiplier principle. The test is based on some measures of cumulants. He finds that for the data used by Fair and Jaffee (1972) the normality assumption is strongly rejected. More work, therefore, needs to be done in devising methods of estimation based on more general distributions, or deriving some distribution-free estimators [see Cosslett (1984), and Heckman and Singer (1984) for some work in this direction].

## 7.3. Tests for disequilibrium

There have been many tests suggested for the "disequilibrium hypothesis", i.e. to test whether the data have been generated by an equilibrium model or a disequilibrium model. Quandt (1978) discusses several tests and says that there does not exist a uniformly best procedure for testing the hypothesis that a market is in equilibrium against the alternative that it is not.

A good starting point for "all" tests for disequilibrium is to ask the basic question of what the disequilibrium is due to. In the case of the partial adjustment model given by eq. (6.7), the disequilibrium is clearly due to imperfect adjustment of prices. In this case the proper test for the equilibrium vs. disequilibrium hypothesis is to test whether $\lambda = 1$. See Ito and Ueda (1981). This leads to a test that $1/\gamma = 0$ in the Fair and Jaffee quantitative model, since $\gamma$ is proportional to $1/1 - \lambda$. This is the procedure Fair and Jaffee suggest. However, if the meaning of the price adjustment equation is that prices adjust in response to either excess demand or excess supply, then as argued in Section 6, the price adjustment equation should have $\Delta P_{t+1}$ not $\Delta P_t$, and also it is not clear how one can test for the equilibrium hypothesis in this case. The intuitive reason is that now the price adjustment equation does not give any information about the source of the disequilibrium.

Quandt (1978) argues that there are two classes of disequilibrium models which are;

(a) Models where it is known for which observations $D_t < S_t$ and for which $D_t > S_t$, i.e. the sample separation is known, and

(b) Models in which such information is not available.

He says that in case (a) the question of testing for disequilibrium does not arise at all. It is only in case (b) that it makes sense.

The example of the partial adjustment model (6.7) is a case where we have sample separation given by $\Delta P_t$. However, it still makes sense to test for the disequilibrium hypothesis which in this case merely translates to a hypothesis about the speed of adjustment of prices to levels that equilibrate demand and supply. Adding a stochastic term $u_{3t}$ to the price adjustment equation does not change the test. When $\lambda = 1$ this says $P_t = P_t^* + u_{3t}$.

There is considerable discussion in Quandt's paper on the question of nested vs. non-nested hypothesis. Quandt argues that very often the hypothesis of equilibrium vs. disequilibrium is non-nested, i.e. the parameter set under the null hypothesis that the model is an equilibrium model is not a subset of the parameter set for the disequilibrium model. The problem in these cases may be that there is no adequate explanation of why disequilibrium exists in the first place.

Consider for instance, the disequilibrium model: with the demand and supply functions specified by eqs. (6.3) and (6.4).

Quandt argues that if one takes the limit of the likelihood function for this model with price adjustment equation as:

$$\Delta P_t = \gamma(D_t - S_t) + u_{3t} \tag{7.3}$$

and

$$\sigma_{23} = \text{Cov}(u_2, u_3) = 0,$$

$$\sigma_{13} = \text{Cov}(u_1, u_3) = 0,$$

$$\sigma_3^2 \neq 0,$$

and

$$\gamma \to \infty,$$

then we get the likelihood function for the equilibrium model ($Q_t = D_t = S_t$) and thus the hypothesis is "nested"; but that if $\sigma_3^2 = 0$, the likelihood function for the disequilibrium model does not tend to the likelihood function for the equilibrium model even if $\gamma \to \infty$ and thus the hypothesis is not tested. The latter conclusion, however, is counter-intuitive and if we consider the correct likelihood function for this model derived in Amemiya (1974) and if we take the limits as $\gamma \to \infty$, we get the likelihood function for the equilibrium model.

Quandt also shows that if the price adjustment equation is changed to

$$\Delta P_{t+1} = \gamma(D_t - S_t) + u_{3t}, \tag{7.4}$$

then the limit of the likelihood function of the disequilibrium model as $\gamma \to \infty$ is not the likelihood function for the equilibrium model. This makes intuitive sense and is also clear when we look at the likelihood functions derived in Section 5. In this case the hypothesis is nonnested, but the problem is that as discussed earlier, this price adjustment equation does not tell us anything about what disequilibrium is due to. As shown in Section 6, the price adjustment eq. (7.3) follows from the partial adjustment eq. (6.7) and thus throws light on what disequilibrium is due to, but the price adjustment eq. (7.4) says nothing about the source of disequilibrium. If we view the equation as a forecast equation, then the disequilibrium is due to imperfect forecasts of the market equilibrating price. In this case it is clear that as $\gamma \to \infty$, we do not get perfect forecasts. What we need to have for a nested model is a forecasting equation which for some limiting values of some parameters yields perfect forecasts at the market equilibrating prices.

Consider now the case where we do not have a price adjustment equation and the model merely consists of a demand equation and a supply equation. Now, clearly the source of the disequilibrium is that $P_t$ is exogenous. Hence the test boils down to testing whether $P_t$ is exogenous or endogenous. The methods developed by Wu (1973) and Hausman (1978) would be of use here.

As mentioned earlier, if a disequilibrium is due to partial adjustment of prices, then a test for disequilibrium is a test for $\lambda = 0$ in eq. (6.7) or a test that $1/\gamma = 0$ in eq. (6.2). The proper way to test this hypothesis is to re-parameterize the equations in terms of $\eta = 1/\lambda$ before the estimation is done. This re-parameterization is desirable in all models (models with expectational variables, spillovers, inventories etc.) where the price adjustment eq. (6.2) or its stochastic version is used.

There is only one additional problem and it is that the model is instable for $\eta < 0$. Thus the null hypothesis $\eta = 0$ lies on the boundary of the set of admissible values of $\eta$. In this case one can use the upper $2\alpha$ percentage point of the $\chi^2$ distribution in order that the test may have a significance level of $\alpha$ in large samples. Upcher (1980) developed a Lagrange multiplier or score statistic. The score test is not affected by the boundary problem and only requires estimation of the constrained model, i.e. the model under the hypothesis of equilibrium. This test is therefore computationally much simpler than either LR or Wald test and in case the null hypothesis of equilibrium is accepted, one can avoid the burdensome method of estimating the disequilibrium model. Upcher's analysis shows that the score test statistic is identical for both stochastic and non-stochastic specification of the price-adjustment equation. The advantage of this result is that it encompasses a broad spectrum of alternatives. But, in case the null hypothesis of equilibrium is rejected, a range of alternative specifications for the disequilibrium model is possible.

However, a major objection to the use of Lagrange multiplier procedure is that it ignores the one-sided nature of the alternative and, therefore, is likely to result in a test with low power compared to the LR or Wald test procedures.

This issue has been recently addressed by Rogers (1983) who has proposed a test statistic that is asymptotically equivalent under the null hypothesis and a sequence of local alternatives to the LR and Wald statistics, and which has the same computational advantage over these statistics as does the Lagrange multiplier statistic over the LR and Wald statistics in the case of the usual two-sided alternatives.

An alternative test for disequilibrium developed by Hwang (1980) relies on the idea of deriving an equation of the form

$$Q_t = \pi_1 X_{1t} + \pi_2 X_{2t} + \pi_3 P_t + v_t, \tag{7.5}$$

from the equilibrium and disequilibrium model. The difference between the two models is that $\pi_1, \pi_2, \pi_3$ are stable over time in the equilibrium model and varying over time in the disequilibrium model. Hwang, therefore, proposes to use stability tests available in the literature for testing the hypothesis of equilibrium. In the case of the equilibrium model $P_t$ is endogenous. Eq. (7.5) is derived from the conditional distribution of $Q_t$ given $P_t$ and hence can be estimated by ordinary least squares. The only problem with the test suggested by Hwang is that parameter instability can arise from a large number of sources and if the null hypothesis is rejected, we do not know what alternative model to consider.

In summary, it is always desirable to base a test for disequilibrium on a discussion of the source for disequilibrium.

## 7.4. *Models with inventories*

In Section 6 we considered modifications of the demand and supply functions taking account of spillovers. However, spillovers on the supply side are better accounted for by introducing inventories explicitly. Dagenais (1980) considers inventories and spillovers in the demand function and suggests a limited information method. Chanda (1984) extends this analysis to take into account expected prices in the supply function.

Green and Laffont (1981) consider inventories in the context of a disequilibrium model with anticipatory pricing. Laffont (1983) presents a survey of the theoretical and empirical work on inventories in the context of fixed-price models.

The issues of how to formulate the desired inventory holding and how to formulate inventory behaviour in the presence of disequilibrium are problems that need further study.

## 8. Multimarket disequilibrium models

The analysis in the preceding sections on single market disequilibrium models has been extended to multimarket disequilibrium models by Gourieroux et al. (1980) and Ito (1980). Quandt (1978) first considered a two-market disequilibrium model of the following form: (the exogenous variables are omitted):

$$D_{1t} = \alpha_1 Q_{2t} + U_{1t},$$
$$S_{1t} = \beta_1 Q_{2t} + U_{2t},$$
$$D_{2t} = \alpha_2 Q_{1t} + V_{1t},$$
$$S_{2t} = \beta_2 Q_{1t} + V_{2t}, \tag{8.1}$$
$$Q_{1t} = \text{Min}(D_{1t}, S_{1t}),$$
$$Q_{2t} = \text{Min}(D_{2t}, S_{2t}). \tag{8.2}$$

Quandt did not consider the logical consistency of the model. This is considered in Amemiya (1977) and Gourieroux et al. (1980a).

Consider the regimes:

$$R_1: D_1 \geq S_1 \cdot D_2 \geq S_2,$$
$$R_2: D_1 \geq S_1 \cdot D_2 < S_2,$$
$$R_3: D_1 < S_1 \cdot D_2 < S_2,$$
$$R_4: D_1 < S_2 \cdot D_2 \geq S_2. \tag{8.3}$$

In regime 1, we have $Q_1 = S_1$, $Q_2 = S_2$ and substituting these in (8.1) we have

$$A_1 \begin{bmatrix} D_1 \\ S_1 \\ D_2 \\ S_2 \end{bmatrix} = \begin{bmatrix} 1 & 0 & 0 & -\alpha_1 \\ 0 & 1 & 0 & -\beta_1 \\ 0 & -\alpha_2 & 1 & 0 \\ 0 & -\beta_2 & 0 & 1 \end{bmatrix} \begin{bmatrix} D_1 \\ S_1 \\ D_2 \\ S_2 \end{bmatrix} = \begin{bmatrix} u_1 \\ u_2 \\ u_3 \\ u_4 \end{bmatrix}.$$

Similarly, we can define the corresponding matrices $A_2, A_3, A_4$ in regimes

$R_2, R_3, R_4$ respectively that give the mapping from $(D_1, S_1, D_2, S_2)$ to $(u_1, u_2, u_3, u_4)$.

$$A_2 = \begin{bmatrix} 1 & 0 & -\alpha_1 & 0 \\ 0 & 1 & -\beta_1 & 0 \\ 0 & -\alpha_2 & 1 & 0 \\ 0 & -\beta_2 & 0 & 1 \end{bmatrix} \qquad A_3 = \begin{bmatrix} 1 & 0 & -\alpha_1 & 0 \\ 0 & 1 & -\beta_1 & 0 \\ -\alpha_2 & 0 & 1 & 0 \\ -\beta_2 & 0 & 0 & 1 \end{bmatrix},$$

and

$$A_4 = \begin{bmatrix} 1 & 0 & 0 & -\alpha_1 \\ 0 & 1 & 0 & -\beta_1 \\ -\alpha_2 & 0 & 1 & 0 \\ -\beta_2 & 0 & 0 & 1 \end{bmatrix}.$$

The logical consistency or 'coherency' conditions derived by Gourieroux et al. are that the determinants of these four matrices, i.e. $(1 - \beta_1\beta_2)$, $(1 - \alpha_2\beta_1)$, $(1 - \alpha_1\alpha_2)$, $(1 - \alpha_1\beta_2)$ must be the same sign.

The major problem that the multimarket disequilibrium models are supposed to throw light on (which the models in eqs. (8.1) and (8.2) does not) refers to the "spill-over effects" – the effects of unsatisfied demand or supply in one market on the demand and supply in other markets. Much of this discussion on spill-over effects has been in the context of macro-models, the two markets considered are the commodity market and the labor market. The commodity is supplied by producers and consumed by households. Labor is supplied by households and used by producers. The quantities actually transacted are given by

$$C = \text{Min}(C^d, C^s),$$

$$L = \text{Min}(L^d, L^s). \qquad (8.4)$$

The demands and supplies actually presented in each market are called "effective" demands and supplies and these are determined by the exogenous variables and the endogenous quantity constraints (8.4). By contrast, the "notional" demands and supplies refer to the unconstrained values. Denote these by $\overline{C}^d, \overline{C}^s, \overline{L}^d, \overline{L}^s$. The different models of multi-market disequilibrium differ in the way 'effective' demands and "spill-over effects" are defined. Gourieroux et al. (1980a) define the effective demands and 'spill-over effects' as follows:

*Model I*

$$
\begin{aligned}
C^d &= \bar{C}^d & &\text{if } L = L^s \le L^d, \\
&= \bar{C}^d + \alpha_1(L - \bar{L}^s) & &\text{if } L = L^d < L^s,
\end{aligned}
\tag{8.5}
$$

$$
\begin{aligned}
C^s &= \bar{C}^s & &\text{if } L = L^d \le L^s, \\
&= \bar{C}^s + \alpha_2(L - \bar{L}^d) & &\text{if } L = L^s < L^d,
\end{aligned}
\tag{8.6}
$$

$$
\begin{aligned}
L^d &= \bar{L}^d & &\text{if } C = C^s \le C^d, \\
&= \bar{L}^d + \beta_1(C - \bar{C}^s) & &\text{if } C = C^d < C^s,
\end{aligned}
\tag{8.7}
$$

$$
\begin{aligned}
L^s &= \bar{L}^s & &\text{if } C = C^d \le C^s, \\
&= \bar{L}^s + \beta_2(C - \bar{C}^d) & &\text{if } C = C^s < C^d.
\end{aligned}
\tag{8.8}
$$

This specification is based on Malinvaud (1977) and assumes that agents on the short-side of the market present their notional demand as their effective demand in the other market. For instance eq. (8.5) says that if households are able to sell all the labor they want to, then their effective demand for goods is the same as their 'notional' demand. On the other hand, if they cannot sell all the labor they want to, there is a "spill-over effect" but note that this is proportional to $L - \bar{L}^s$ not $L - L^s$. (I.e. it is proportional to the difference between actual labor sold and the 'notional' supply of labor.)

The model considered by Ito (1980) is as follows:

*Model II*

$$
C^d = \bar{C}^d + \alpha_1(L - \bar{L}^s),
\tag{8.5'}
$$

$$
C^s = \bar{C}^s + \alpha_2(L - \bar{L}^d),
\tag{8.6'}
$$

$$
L^d = \bar{L}^d + \beta_1(C - \bar{C}^s),
\tag{8.7'}
$$

$$
L^s = \bar{L}^s + \beta_2(C - \bar{C}^d).
\tag{8.8'}
$$

An alternative model suggested by Portes (1977) based on work by Benassy is the following:

*Model III*

$$
C^d = \bar{C}^d + \alpha_1(L - L^s),
\tag{8.5''}
$$

$$
C^s = \bar{C}^s + \alpha_2(L - L^d),
\tag{8.6''}
$$

$$L^d = \overline{L}^d + \beta_1(C - C^s), \tag{8.7''}$$

$$L^s = \overline{L}^s + \beta_2(C - C^d). \tag{8.8''}$$

Portes compares the reduced forms for these three models and argues that econometrically, there is little to choose between the alternative definitions of effective demand.

The conditions for logical consistency (or coherency) are the same in all these models viz: $0 < \alpha_i\beta_j < 1$ for $i, j = 1, 2$. Both Gourieroux et al. (1980a) and Ito (1980) derive these conditions, suggest price and wage adjustment equations similar to those considered in Section 6, and discuss the maximum likelihood estimation of their models. Ito also discusses two-stage estimation similar to that proposed by Amemiya for the Fair and Jaffee model, and derives sufficient conditions for the uniqueness of a quantity-constrained equilibrium in his model. We cannot go into the details of all these derivations here. The details involve more of algebra than any new conceptual problems in estimation. In particular, the problems mentioned in Section 6 about the different price adjustment equations apply here as well.

Laffont (1983) surveys the empirical work on multi-market disequilibrium models. Quandt (1982, pp. 39–54) also has a discussion of the multi-market disequilibrium models.

The applications of multi-market disequilibrium models all seem to be in the macro area. However, here the problems of aggregation are very important and it is not true that the whole economy switches from a regime of excess demand to one of excess supply or vice versa. Only some segments might behave that way. The implications of aggregation for econometric estimation have been studied in some simple models by Malinvaud (1982).

The problems of spillovers also tend to arise more at a micro-level rather than a macro-level. For instance, consider two commodities which are substitutes in consumption (say natural gas and coal) one of which has price controls. We can define the demand and supply functions in the two markets (omitting the exogenous variables) as follows:

$$D_1 = \alpha_1 P_1 + \beta_1 P_2 + u_1,$$
$$S_1 = \alpha_2 P_1 + u_2 P_1 \le \overline{P},$$
$$Q_1 = \text{Min}(D_1, S_1),$$
$$D_2 = \gamma_1 P_2 + \delta_1 P_1 + \lambda(D_1 - S_1) + V_1,$$
$$S_2 = \gamma_2 P_2 + V_2,$$
$$Q_2 = D_2 = S_2, \text{ i.e. the second market is always in equilibrium.}$$

If $P_1 \le \overline{P}$, we have the usual simultaneous equations model with the two quantities and two prices as the endogenous variables. If $P_1 > \bar{P}$, then there is excess demand in the first market and a spill-over of this into the second market. This model is still in a "partial equilibrium" framework but would have interesting empirical applications. It is at least one step forward from the single-market disequilibrium model which does not say what happens to the unsatisfied demand or supply.

## 9. Models with self-selection

As mentioned in the introduction, there is an early discussion of the self-selection problem in Roy (1951) who discussed the case of individuals choosing between two occupations, hunting and fishing, on the basis of their comparative advantage. See Maddala (1983a) pp. 257–8 for a discussion of this model.

The econometric discussion of the consequences of self-selectivity started with the papers by Gronau (1974), Lewis (1974) and Heckman (1974). In this case the problem is about women choosing to be in the labor force or not. The observed distribution of wages is a truncated distribution. It is the distribution of wage offers truncated by reservation wages. The Gronau–Lewis model consisted of two equations:

$$
\begin{aligned}
W_0 &= X\beta_1 + u_1, \\
W_r &+ X\beta_2 + u_2.
\end{aligned}
\tag{9.1}
$$

We observe $W = W_0$ iff $W_0 \geq W_r$. Otherwise $W = 0$. We discussed the estimation of this model in Section 2 and we will not repeat it here. The term 'selectivity bias' refers to the fact that if we estimate eq. (9.1) by OLS based on the observations for which we have wages $W$, we get inconsistent estimates of the parameters.

Note that

$$
E(u_1 | W_0 \geq W_r) = -\sigma_{1u} \frac{\phi(Z)}{\Phi(Z)},
$$

where

$$
Z = \frac{X\beta_1 - X\beta_2}{\sigma}, \quad u = \frac{u_2 - u_1}{\sigma}, \quad \sigma = \mathrm{Var}(u_2 - u_1) \quad \text{and} \quad \sigma_{1u} = \mathrm{Cov}(u, u_1).
$$

Hence we can write (9.1) as:

$$
W = X\beta_1 - \sigma_{1u} \frac{\phi(Z)}{\Phi(Z)} + V,
\tag{9.2}
$$

where $E(V) = 0$.

A test for selectivity bias is a test for $\sigma_{1u} = 0$. Heckman (1976) suggested a two-stage estimation method for such models. First get consistent estimates for the parameters in $Z$ by the probit method applied to the dichotomous variable (in the labor force or not). Then estimate eq. (9.2) by OLS using the estimated values $\hat{Z}$ for $Z$.

The self-selectivity problem has since been analyzed in different contexts by several people. Lee (1978) has applied it to the problem of unions and wages. Lee and Trost (1978) have applied it to the problem of housing demand with choices of owning and renting. Willis and Rosen (1979) have applied the model to the problem of education and self-selection. These are all switching regression models. Griliches et al. (1979) and Kenny et al. (1979) consider models with both selectivity and simultaneity. These models are switching simultaneous equations models. As for methods of estimation, both two-stage and maximum likelihood methods have been used. For two-stage methods, the paper by Lee et al. (1980) gives the asymptotic covariance matrices when the selectivity criterion is of the probit and tobit types.

In the literature of self-selectivity a major concern has been with testing for selectivity bias. These are tests for $\sigma_{1u} = 0$ and $\sigma_{2u} = \text{Cov}(u, u_2) = 0$. However, a more important issue is the sign and magnitude of these covariances and often not much attention is devoted to this. In actual practice we ought to have $\sigma_{2u} - \sigma_{1u} > 0$ but $\sigma_{1u}$ and $\sigma_{2u}$ can have any signs.[10] It is also important to estimate the mean values of the dependent variables for the alternate choice. For instance, in the case of college education and income, we should estimate the mean income of college graduates had they chosen not to go to college, and the mean income of non-college graduates had they chosen to go to college. In the example of hunting and fishing we should compute the mean income of hunters had they chosen to be fishermen and the mean income of fishermen had they chosen to be hunters. Such computations throw light on the effects of self-selection and also reveal difficiencies in the model which simple tests for the existence of selectivity bias do not. See Björlund and Moffitt (1983) for such calculations.

In the literature on labor supply, there has been considerable discussion of "individual heterogeneity", i.e. the observed self-selection is due to individual characteristics not captured by the observed variables (some women want to work no matter what and some women want to sit at home no matter what). Obviously, these individual specific effects can only be analyzed if we have panel data. This problem has been analyzed by Heckman and Chamberlain, but since these problems will be discussed in the chapters on labor supply models by Heckman

---

[10] This is pointed out in Lee (1978b). Trost (1981) illustrates this with an empirical example on returns to college education.

and analysis of cross-section and time-series data by Chamberlain they will not be elaborated here.

One of the more important applications of the procedures for the correction of selectivity bias is in the evaluation of programs.

In evaluating the effects of several social programs, one has to consider the selection and truncation that can occur at different levels. We can depict the situation by a decision tree as follows.

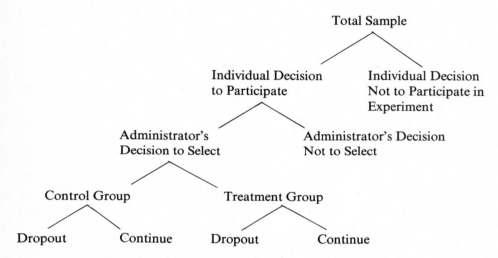

Figure 2   A decision tree for the evaluation of social experiments.

In practical situations, one would have to assume randomness at certain levels or else the model can get too unwieldy to be of any use. As to the level at which selection and truncation bias needs to be introduced, this is a question that depends on the nature of the problem. Further, in Figure 2 the individual's decision to participate preceded the administrator's decision to select. This situation can be reversed or both the decisions could be simultaneous. Another problem is that caused by the existence of multiple categories such as no participation, partial or full participation or different types of treatment. These cases fall in the class of models with polychotomous choice and selectivity. The selectivity problem with polychotomous choice has been analyzed in Hay (1980), Dubin and McFadden (1984) and Lee (1981). A summary of these methods can be found in Maddala (1983a, pp. 275–278). An empirical application illustrating the approach suggested by Lee is in Trost and Lee (1984).

One further problem is that of truncated samples. Very often we do not have data on all the individuals – participants and non-participants. If the data consists of only participants in a program and we know nevertheless that there is self-selection and we have data on the variables determining the participation decision function, then we can still correct for selectivity bias. The methodology for this problem is discussed in the next section. The important thing to note is that though, theoretically, truncation does not change the identifiability of the parameters, there is, nevertheless a loss of information.

There is a vast amount of literature on program evaluation. Some important references are: Goldberger (1972) and Barnow, Cain and Goldberger (1981). These papers and the selectivity problem in program evaluation have been surveyed in Maddala (1983a, pp. 260–267).

One other problem is that of correcting for selectivity bias when the explanatory variables are measured with error. An example of this occurs in problems of measuring wage discrimination, particularly a comparison between the Federal and non-Federal sectors. A typical regression equation considered is one of regressing earnings on productivity and a dummy variable depicting race or sex or ethnic group. Since productivity cannot be measured, some proxies are used. When such equations are estimated, say for the Federal (or non-Federal) sectors, one has to take account of individual choices to belong to one or the other sector. To avoid the selection bias we have to model not only the determinants of wage offers but also the process of self-selection by which individuals got into that sector. An analysis of this problem is in Abowd, Abowd and Killingsworth (1983).

Finally, there is the important problem that most of the literature on selectivity bias adjustment is based on the assumption of normality. Consider the simple two equation model to analyze the selectivity problem.

$$Y = X\beta + u,$$

$$I^* = Z\gamma - \varepsilon,$$

$X$ and $Z$ are exogenous variables. $I^*$ is never observed. All we observe is $I = 1$ if $I^* > 0$, $I = 0$ otherwise. Also $Y$ is not observed unless $I^* > 0$.

Olsen (1980) shows how the only assumption we need to make a correction for selection bias in the estimation of $\beta$, is that $\varepsilon$ is normal and that the conditional expectation of $u$ given $\varepsilon$ is linear. If $u$ and $\varepsilon$ are bivariate normal, this condition follows automatically. Goldberger (1980) made some calculations with alternative error distributions and showed that the normal selection bias adjustment is quite sensitive to departures from normality. Lee (1982a, 1983a) suggests some general transformations to normality. The transformations suggested by him can be done using some methods outlined in Hildebrand (1956) and Appendix II, c in Bock and Jones (1968). This approach permits the analysis of selection bias with any distributional assumptions. Details can be found in the papers by Lee, and a summary in Maddala (1983a, pp. 272–275).

## 10. Multiple criteria for selectivity

There are several practical instances where selectivity could be due to several sources rather than just one as considered in the examples in the previous Section. Griliches et al. (1979) cite several problems with the NLS young men data set that could lead to selectivity bias. Prominent among these are attrition and (other) missing data problems. In such cases we would need to formulate the model as switching regression or switching simultaneous equations models where the switch depends on more than one criterion function.

During recent years there have been many applications involving multiple criteria of selectivity. Abowd and Farber (1982) consider a model with two decisions: the decision of individuals to join a queue for union jobs and the decision of employers to draw from the queue. Poirier (1980) discusses a model where the two decisions are those of the employee to continue with the sponsoring agency after training and the decisions of the employer to make a job offer after training. Fishe et al. (1981) consider a two-decision model: whether to go to college or not and whether to join the labor force or not. Ham (1982) examines the labor supply problem by classifying individuals into four categories according to their unemployment and under-employment status. Catsiapis and Robinson (1982) study the demand for higher education and the receipt of student aid grants. Tunali (1983) studies migration decisions involving single and multiple moves. Danzon and Lillard (1982) analyze a sequential process of settlement of malpractice suits. Venti and Wise (1982) estimate a model combining student preferences for colleges and the decision of the university to admit the student.

All these problems can be classified into different categories depending on whether the decision rules are joint or sequential. This distinction, however, is not made clear in the literature and the studies all use the multivariate normal distribution to specify the joint probabilities.

With a two decision model, the specification is as follows:

$$Y_1 = X_1\beta_1 + u_1, \tag{10.1}$$

$$Y_2 = X_2\beta_2 + u_2, \tag{10.2}$$

$$I_1^* = Z_1\gamma_1 - \varepsilon_1, \tag{10.3}$$

$$I_2^* = Z_2\gamma_2 - \varepsilon_2. \tag{10.4}$$

We also have to consider whether the choices are completely observed or they are partially observed. Define the indicator variables

$$I_1 = 1 \quad \text{iff } I_1^* > 0$$
$$\phantom{I_1} = 0 \quad \text{otherwise,}$$
$$I_2 = 1 \quad \text{iff } I_2^* > 0$$
$$\phantom{I_2} = 0 \quad \text{otherwise.}$$

The question is whether we observe $I_1$ and $I_2$ separately or only as a single indicator variable $I = I_1 I_2$. The latter is the case with the example of Abowd and Farber. Poirier (1980) also considers a bivariate probit model with partial observability but his model is a joint model–not a sequential model as in the example of Abowd and Farber. In the example Poirier considers, the employer must decide whether or not to give a job offer and the applicant must decide whether or not to seek a job offer. We do not observe these individual decisions. What we observe is whether the trainee continues to work after training. If either the employer or the employee makes the decision first, then the model would be a sequential model.

The example considered by Fishe et al. (1981) is a joint decision model but both indicators $I_1$ and $I_2$ are observed. Similar is the case considered by Ham (1982) though it is hard to see how unemployment and underemployment could be considered as two decisions. Workers do not choose to be unemployed and underemployed. Rather both unemployment and underemployment are consequences of more basic decisions of employers and employees. The example considered by Catsiapis and Robinson (1982) is a sequential decision, though one can also present arguments that allow it to be viewed as a joint decision model.

In the joint decision model with partial observability, i.e. where we observe $I = I_1 \cdot I_2$ only and not $I_1$ and $I_2$ individually, the parameters $\gamma_1$ and $\gamma_2$ in eqs. (10.3) and (10.4) are estimable only if there is at least one non-overlapping variable in either one of $Z_1$ and $Z_2$. Since $V(\varepsilon_1) = V(\varepsilon_2) = 1$ by normalization, let us define $\text{Cov}(\varepsilon_1, \varepsilon_2) = \rho$. Also write

$$\text{Prob}(I_1^* > 0,\ I_2^* > 0)$$
$$= \text{Prob}(\varepsilon_1 < Z_1\gamma_1,\ \varepsilon_2 < Z_2\gamma_2)$$
$$= F(Z_1\gamma_1, Z_2\gamma_2, \rho).$$

Then the ML estimates of $\gamma_1$, $\gamma_2$ and $\rho$ are obtained by maximizing the likelihood function

$$L_1 = \prod_{I=1} F(Z_1\gamma_1, Z_2\gamma_2, \rho) \cdot \prod_{I=0} \left[ 1 - F(Z_1\gamma_1, Z_2\gamma_2, \rho) \right]. \tag{10.5}$$

With the assumption of bivariate normality of $\varepsilon_1$ and $\varepsilon_2$, this involves the use of bivariate probit analysis.

In the sequential decision model with partial observability, if we assume that the function (10.4) is defined only on the subpopulation $I_1 = 1$, then since the distribution of $\varepsilon_2$ that is assumed is considered on $\varepsilon_1 < Z_1\gamma_1$, the likelihood function to be maximized would be

$$L_2 = \prod_{I=1} \left[ \Phi(Z_1\gamma_1)\Phi(Z_2\gamma_2) \right] \cdot \prod_{I=0} \left[ 1 - \Phi(Z_1\gamma_1)\Phi(Z_2\gamma_2) \right]. \tag{10.6}$$

Again, the parameters $\gamma_1$ and $\gamma_2$ are estimable only if there is at least one non-overlapping variable in either one of $Z_1$ and $Z_2$ (otherwise we would not know which estimates refer to $\gamma_1$ and which refer to $\gamma_2$). In their example on job queues and union status of workers, Abowd and Farber (1982) obtain their parameter estimates using the likelihood function (10.6). One can, perhaps, argue that even in the sequential model, the appropriate likelihood function is still (10.5) and not (10.6). It is possible that there are persons who do not join the queue ($I_1 = 0$) but for whom employers would want to give a union job. The reason we do not observe these individuals in union jobs is because they had decided not to join the queue. But we do not also observe in the union jobs all those with $I_2 = 0$. Thus, we can argue that $I_2^*$ exists and is, in principle, defined even for the observations $I_1 = 0$. If the purpose of the analysis is to examine what factors influence the employers' choice of employees for union jobs, then possibly the parameter estimates should be obtained from (10.5). The difference between the two models is in the definition of the distribution of $\varepsilon_2$. In the case of (10.5), the distribution of $\varepsilon_2$ is defined over the whole population. In the case of (10.6), it is defined over the subpopulation $I_1 = 1$. The latter allows us to make only conditional inferences.[11] The former allows us to make both conditional and marginal inferences. To make marginal inferences, we need estimates of $\gamma_2$. To make conditional inferences we consider the conditional distribution $f(\varepsilon_2|\varepsilon_1 < Z_1\gamma_1)$ which involves $\gamma_1$, $\gamma_2$, and $\rho$.

Yet another type of partial observability arises in the case of truncated samples. An example is that of measuring discrimination in loan markets. Let $I_1^*$ refer to the decision of an individual on whether or not to apply for a loan, and let $I_2^*$ refer to the decision of the bank on whether or not to grant the loan.

$I_1 = 1$     if the individual applies for a loan

   $= 0$     otherwise,

$I_2 = 1$     if the applicant is given a loan

   $= 0$     otherwise.

---

[11] The conditional model does not permit us to allow for the fact that changes in $Z_2$ also might affect the probability of being in the queue. Also, the decision of whether or not to join the queue can be influenced by the perception of the probability of being drawn from the queue.

Rarely do we have data on the individuals for whom $I_1 = 0$. Thus what we have is a truncated sample. We can, of course, specify the distribution of $I_2^*$ only for the subset of observations $I_1 = 1$ and estimate the parameters $\gamma_2$ by say the probit ML method and examine the significance of the coefficients of race, sex, age, etc. to see whether there is discrimination by any of these variables. This does not, however, allow for self-selection at the application stage, say for some individuals not applying because they feel they will be discriminated against. For this purpose we define $I_2^*$ over the whole population and analyze the model from the truncated sample. The argument is that, in principle $I_2^*$ exists even for the non-applicants. The parameters $\gamma_1$, $\gamma_2$ and $\rho$ can be estimated by maximizing the likelihood function

$$ L_3 = \prod_{I_2 = 1} \frac{F(Z_1\gamma_1, Z_2\gamma_2, \rho)}{\Phi(Z_1\gamma_1)} \cdot \prod_{I_2 = 0} \frac{\Phi(Z_1\gamma_1) - F(Z_1\gamma_1, Z_2\gamma_2, \rho)}{\Phi(Z_1\gamma_1)}. \tag{10.7} $$

In this model the parameters $\gamma_1$, $\gamma_2$ and $\rho$ are, in principle, estimable even if $Z_1$ and $Z_2$ are the same variables. In practice, however, the estimates are not likely to be very good. Muthén and Jörekog (1981) report the results of some Monte-Carlo experiments on this. Bloom et al. (1981) report that attempts at estimating this model did not produce good estimates. However, the paper by Bloom and Killingsworth (1981) shows that correction for selection bias can be done even with truncated samples. Wales and Woodland (1980) also present some encouraging Monte-Carlo evidence. Since the situation of truncated samples is of frequent occurrence (see Bloom and Killingsworth for a number of examples) more evidence on this issue will hopefully accumulate in a few years.

The specification of the distributions of $\varepsilon_1$ and $\varepsilon_2$ in (10.3) and (10.4) depends on whether we are considering a joint decision model or a sequential decision model. For problems with sequential decisions, the situation can diagrammatically be described as follows:

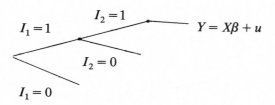

In a sequential decision model, the disturbance $\varepsilon_2$ can be defined only on the subpopulation for which $I_1 = 1$. The specification of the joint distribution for $(\varepsilon_1, \varepsilon_2)$ over the whole population will not be appropriate in principle and will introduce unnecessarily complicated functional forms for the conditional probabilities. This point is emphasized in Lee and Maddala (1983b). On the other hand, if we specify the marginal distribution of $\varepsilon_1$ and the conditional distribution of $\varepsilon_2$ given $I_1 = 1$ then there is no way we can allow for the correlations among the decisions. Lee and Maddala (1983b) and Lee (1984) suggest the following: Let $F_1(\varepsilon_1)$ be the marginal distribution of $\varepsilon_1$ defined on the subpopulation $(I_1 = 1)$ which is, of course, implied by the marginal distribution of $\varepsilon_1$ on the whole population. $F_2(\varepsilon_2)$ is the marginal distribution of $\varepsilon_2$ defined on the subpolation $I_1 = 1$. Given the marginal distributions $F_1(\varepsilon_1)$ and $F_2(\varepsilon_2)$ defined on the common measurable space, there are infinitely many ways of generating joint distributions with given marginals. Lee (1983a) discusses some computable methods of generating these distributions. This procedure can be applied to correct for selectivity bias in sequential decision models with any specifications of the marginal distributions of $\varepsilon_1$ on the whole population and of $\varepsilon_2$ on the subpopulation and allowing for correlations in the decisions. See Lee and Maddala (1983b) and Lee (1984) for details.

## 11. Concluding remarks

In the preceding sections we have reviewed the recent literature on disequilibrium and selectivity models. We will now go through some deficiencies of these models and examine future areas of research.

The cornerstone of the "disequilibrium" models discussed in this chapter is the "minimum condition." One of the most disturbing points in the empirical applications is that the models have been mechanically applied with no discussion of what disequilibrium is due to and what the consequences are. In spite of all the limitations mentioned in Section 3, the model discussed there (with slight variation) has been the model with the most empirical applications. For instance, Sealy (1979) used the model to study credit rationing in the commercial loan market. Portes and Winter (1978) used it to estimate demand for money and savings functions in centrally planned economies (Czechoslovakia, East Germany, Hungary and Poland). Portes and Winter (1980) used it to study the demand for consumption goods in centrally planned economies. Chambers et al. (1978) used it to study the effects of import quotas on the U.S. beef market.

The reason for the popularity of this model is that it needs us to specify very little. The authors of the above papers specify the demand and supply functions as usual, and then say there is "rationing" and disequilibrium because of regulations. But even if the regulations control prices, it does not imply that prices are fixed at certain levels continuously which is what the model says. Further, there is no discussion of how the rationing is carried out and in almost all cases the data used are macro-data and the implications of aggregation are ignored.

The main application of the methodology discussed in this chapter is to regulated markets and centrally planned economies, where there are price and quantity regulations. In Section 6 we discussed the case of controlled prices and showed how the analysis can be applied to credit markets with interest rate ceilings (or equivalently, labor markets with minimum-wage laws). The interest rate ceiling problem has been analyzed in Maddala and Trost (1982). The minimum wage problem has been analyzed in Meyer and Wise (1983a, b). An analysis of price supports is in Maddala and Shonkwiler (1984). The case of centrally planned economies has been analyzed by Charemza and Quandt (1982).

Another major criticism of the disequilibrium models appears in two papers by Richard (1980) and Hendry and Spanos (1980). These criticisms are also elaborated in the comments by Hendry and Richard as the survey paper by Quandt (1982). Hendry and Spanos point out that the "minimum condition" was actually discussed by Frisch (1949) but that he suggested formulation of "market pressures" that are generated by the inequality between the unobserved latent variables $D_t$ and $S_t$. These pressures were formulated in the price adjustment eqs. discussed in Section 6 but we also saw the serious limitations of this eq. in the presence of the "minimum condition". Hendry and Spanos suggest dropping the "minimum condition" (which is the main source of all the headaches in estimation), concentrating on the "pressures" and dynamic adjustment processes, and modelling the observables directly. Though there is some merit in their argument, as mentioned earlier, the main application of the methodology described in this chapter is to the analysis of regulated markets and planned economies and the methods suggested by Hendry and Spanos are not applicable to such problems. Since the Hendry–Spanos paper is discussed in detail in Maddala (1983a, pp. 343–345) we will not repeat the criticism here.

Finally, mention must be made of the criticism of the switching regression models with endogenous switching (of which the disequilibrium and selection models are particular cases) by Poirier and Rudd (1981). These authors argue that there has been substantial confusion in the econometrics literature over switching regression models with endogenous switching and that this confusion can cause serious interpretation problems when the model is employed in empirical work. Fortunately, however, the arguments presented by these authors are incorrect. Since their paper has been discussed in detail in Maddala (1983a, pp. 283–287) we will not repeat the criticism here.

The literature on self-selection contains interesting empirical applications in the areas of labor supply, unions and wages, education and self-selection, program evaluation, measuring discrimination and so on. However, the literature on disequilibrium models lacks interesting empirical applications. Part of the problem here is that not much thought is often given to the substantive question of what the sources of disequilibrium are and also there are few micro data sets to which the methods have been applied. Almost all applications [Avery (1982), Maddala and Trost (1982), Meyer and Wise (1983a, b) are perhaps some exceptions] are based on aggregate time-series data and there is not enough discussion of problems of aggregation. The Fair and Jaffee example on the housing market as well as the different models of "credit rationing" are all based on aggregate data and there is much to be desired in the detailed specification of these models.

Perhaps the most interesting application of the disequilibrium models are in the areas of regulated industries. After all, it is regulation that produces disequilibrium in these markets. Estimation of some disequilibrium models with micro-data sets for regulated industries and estimation of the effects of regulation would make the disequilibrium literature more intellectually appealing than it has been. There are also some issues that need to be investigated regarding the appropriate formulation of the demand and supply functions under disequilibrium. The expectation of disequilibrium can itself be expected to change the demand and supply functions. Thus, one needs to incorporate expectations into the modelling of disequilibrium.

The literature on self-selection, by contrast to the disequilibrium literature, has several interesting empirical applications. However, even here a lot of work remains to be done. The case of selectivity being based on several criteria rather than one has been mentioned in Section 10. Here one needs a clear distinction to be made between joint decision and sequential decision models. Another problem is that of correcting for selectivity bias when the explanatory variables are measured with error. Almost all the usual problems in the single equation regression model need to be analyzed in the presence of the selection (self-selection) problem.

# References

Abowd, A. M., J. M. Abowd and M. R. Killingsworth (1983) "Race, Spanish Origin and Earnings Differentials Among Men: The Demise of Two Stylized Facts". Discussion Paper #83-11, Economics Research Center/NORC, University of Chicago.

Abowd, J. M. and H. S. Farber (1982) "Job Queues and Union Status of Workers", *Industrial and Labor Relations Review*, 35(4), 354–367.

Amemiya, T. (1973) "Regression Analysis When the Dependent Variable is Truncated Normal", *Econometrica*, 41(6), 997–1016.

Amemiya, T. (1974a) "A Note on a Fair and Jaffee Model", *Econometrica*, 42(4), 759–762.

Amemiya, T. (1974b) "Multivariate Regression and Simultaneous Equations Models When the Dependent Variables are Truncated Normal", *Econometrica*, 42(6), 999–1012.

Amemiya, T. (1977) "The Solvability of a Two-Market Disequilibrium Model", Working Paper 82, IMSSS, Stanford University, August 1977.

Amemiya, T. and Sen G. (1977) "The Consistency of the Maximum Likelihood Estimator in a Disequilibrium Model", Technical Report No. 238, IMSSS, Stanford University.

Avery, R. B. (1982) "Estimation of Credit Constraints by Switching Regressions", in: C. Manski and D. McFadden, eds., *Structural Analysis of Discrete Data: With Econometric Applications*. MIT Press.

Barnow, B. S., G. G. Cain and A. S. Goldberger (1980) "Issues in the Analysis of Selectivity Bias", in: E. W. Stromsdorder and G. Farkas, eds., *Evaluation Studies – Review Annual*, 5, 43–59.

Batchelor, R. A. (1977) "A Variable-Parameter Model of Exporting Behaviour", *Review of Economic Studies*, 44(1), 43–58.

Bergstrom, A. R. and C. R. Wymer (1976) "A Model for Disequilibrium Neoclassical Growth and its

Application to the United Kingdom", in: A. R. Bergstrom, ed., *Statistical Inference in Continuous Time Economic Models*. Amsterdam, North-Holland Publishing Co.

Berndt, E. R., Hall, B. H., Hall, R. E. and J. A. Hausman (1974) "Estimation and Inference in Non-Linear Structural Models", *Annals of Economic and Social Measurement*, 3(4), 653–665.

Björklund, A. and R. Moffitt (1983) "The Estimation of Wage Gains and Welfare Gains From Self-Selection Models". Manuscript, Institute for Research on Poverty, University of Wisconsin.

Bloom, D. E. and M. R. Killingsworth (1981) "Correcting for Selection Bias in Truncated Samples: Theory, With an Application to the Analysis of Sex Salary Differentials in Academe". Paper presented at the Econometric Society Meetings, Washington, D.C., Dec. 1981.

Bloom, D. E., B. J. Preiss and J. Trussell (1981) "Mortgage Lending Discrimination and the Decision to Apply: A Methodological Note". Manuscript, Carnegie Mellon University.

Bock, R. D. and L. V. Jones (1968) *The Measurement and Prediction of Judgement and Choice*. San Francisco: Holden-Day.

Bouissou, M. B., J. J. Laffont and Q. H. Vuong (1983) "Disequilibrium Econometrics on Micro Data". Paper presented at the European Meeting of the Econometric Society, Pisa, Italy.

Bowden, R. J. (1978a) "Specification, Estimation and Inference for Models of Markets in Disequilibrium", *International Economic Review*, 19(3), 711–726.

Bowden, R. J. (1978b) *The Econometrics of Disequilibrium*. Amsterdam: North Holland Publishing Co.

Catsiapis, G. and C. Robinson (1982) "Sample Selection Bias With Multiple Selection Rules", *Journal of Econometrics*, 18, 351–368.

Chambers, R. G., R. E. Just, L. J. Moffitt and A. Schmitz (1978) "International Markets in Disequilibrium: A Case Study of Beef". Berkeley: California Agricultural Experiment Station.

Chanda, A. K. (1984) *Econometrics of Disequilibrium and Rational Expectations*. Ph.D. Dissertation, University of Florida.

Charemza, W. and R. E. Quandt (1982) "Models and Estimation of Disequilibrium for Centrally Planned Economies", *Review of Economic Studies*, 49, 109–116.

Cosslett, S. R. (1984) "Distribution-Free Estimation of a Model with Sample Selectivity". Discussion Paper, Center for Econometrics and Decision Sciences, University of Florida.

Cosslett, S. R. and Long-Fei Lee (1983) "Serial Correlation in Latent Discrete Variable Models". Discussion Paper, University of Florida, forthcoming in *Journal of Econometrics*.

Dagenais, M. G. (1980) "Specification and Estimation of a Dynamic Disequilibrium Model", *Economics Letters*, 5, 323–328.

Danzon, P. M. and L. A. Lillard (1982) *The Resolution of Medical Malpractice Claims: Modelling the Bargaining Process*. Report #R-2792-ICJ, California: Rand Corporation.

Davidson, J. (1978) "FIML Estimation of Models with Several Regimes". Manuscript, London School of Economics, October 1978.

Dempster, A. P., Laird, N. M. and Rubin, D. B. (1977) "Maximum Likelihood from Incomplete Data via the EM Algorithm", *Journal of the Royal Statistical Society*, Series B, 39, 1–38 with discussion.

Dubin, J. and D. McFadden (1984) "An Econometric Analysis of Residential Electrical Appliance Holdings and Consumption", *Econometrica*, 52(2), 345–62.

Eaton, J. and R. E. Quandt (1983) "A Model of Rationing and Labor Supply: Theory and Estimation", *Econometrica*, 50, 221–234.

Fair, R. C. and D. M. Jaffee (1972) "Methods of Estimation for Markets in Disequilibrium", *Econometrica*, 40, 497–514.

Fair, R. C. and H. H. Kelejian (1974) "Methods of Estimation for Markets in Disequilibrium: A Further Study", *Econometrica*, 42(1), 177–190.

Fishe, R. P. H., R. P. Trost and P. Lurie (1981) "Labor Force Earnings and College Choice of Young Women: An Examination of Selectivity Bias and Comparative Advantage", *Economics of Education Review*, 1, 169–191.

Frisch, R. (1949) "Prolegomena to a Pressure Analysis of Economic Phenomena", *Metroeconomica*, 1, 135–160.

Gersovitz, M. (1980) "Classification Probabilities for the Disequilibrium Model", *Journal of Econometrics*, 41, 239–246.

Goldberger, A. S. (1972) "Selection Bias in Evaluating Treatment Effects: Some Formal Illustrations". Discussion Paper #123-72, Institute for Research on Poverty, University of Wisconsin.

Goldberger, A. S. (1981) "Linear Regression After Selection", *Journal of Econometrics*, 15, 357–66.

Goldberger, A. S. (1980) "Abnormal Selection Bias". Workshop Series #8006, SSRI, University of Wisconsin.

Goldfeld, S. M., D. W. Jaffee and R. E. Quandt (1980) " A Model of FHLBB Advances: Rationing or Market Clearing?", *Review of Economics and Statistics*, 62, 339–347.

Goldfeld, S. M. and R. E. Quandt (1975) "Estimation in a Disequilibrium Model and the Value of Information", *Journal of Econometrics*, 3(3), 325–348.

Goldfeld, S. M. and R. E. Quandt (1978) "Some Properties of the Simple Disequilibrium Model with Covariance", *Economics Letters*, 1, 343–346.

Goldfeld, S. M. and R. E. Quandt (1983) "The Econometrics of Rationing Models". Paper presented at the European Meetings of the Econometric Society, Pisa, Italy.

Gourieroux, C., J. J. Laffont and A. Monfort (1980a) "Disequilibrium Econometrics in Simultaneous Eqs. Systems", *Econometrica*, 48(1), 75–96.

Gourieroux, C., J. J. Laffont and A. Monfort (1980b) "Coherency Conditions in Simultaneous Linear Eqs. Models with Endogenous Switching Regimes", *Econometrica*, 48(3), 675–695.

Gourieroux, C. and A. Monfort (1980) "Estimation Methods for Markets with Controlled Prices". Working Paper #8012, INSEE, Paris, October 1980.

Green, J. and J. J. Laffont (1981) "Disequilibrium Dynamics with Inventories and Anticipatory Price Setting", *European Economic Review*, 16(1), 199–223.

Griliches, Z., B. H. Hall and J. A. Hausman (1978) "Missing Data and Self-Selection in Large Panels", *Annales de L'INSEE*, 30–31, *The Econometrics of Panel Data*, 137–176.

Gronau, R. (1974) "Wage Comparisons: A Selectivity Bias", *Journal of Political Economy*, 82(6), 1119–1143.

Ham, J. C. (1982) "Estimation of a Labor Supply Model with Censoring Due to Unemployment and Underemployment", *Review of Economic Studies*, 49, 335–354.

Hartley, M. J. (1977) "On the Estimation of a General Switching Regression Model via Maximum Likelihood Methods". Discussion Paper #415, Department of Economics, State University of New York at Buffalo.

Hartley, M. J. (1979) "Comment", *Journal of the American Statistical Association*, 73(364), 738–741.

Hartley, M. J. and P. Mallela (1977) "The Asymptotic Properties of a Maximum Likelihood Estimator for a Model of Markets in Disequilibrium", *Econometrica*, 45(5), 1205–1220.

Hausman, J. A. (1978) "Specification Tests in Econometrics", *Econometrica*, 46(6), 1251–1272.

Hay, J. (1980) "Selectivity Bias in a Simultaneous Logit-OLS Model: Physician Specialty Choice and Specialty Income". Manuscript, University of Connecticut Health Center.

Heckman, J. J. (1974) "Shadow Prices, Market Wages and Labor Supply", *Econometrica*, 42(4), 679–694.

Heckman, J. J. (1967a) "Simultaneous Equations Models with Continuous and Discrete Endogenous Variables and Structural Shifts", in: Goldfeld and Quandt, eds., *Studies in Nonlinear Estimation*. Cambridge: Ballinger Publishing.

Heckman, J. J. (1976b) "The Common Structure of Statistical Models of Truncation, Sample Selection, and Limited Dependent Variables, and a Simple Estimator for Such Models", *Annals of Economic and Social Measurement*, 5(4), 475–492.

Heckman, J. J. (1978) "Dummy Endogenous Variables in a Simultaneous Equations System", *Econometrica*, 46(6), 931–959.

Heckman, J. J. (1979) "Sample Selection Bias as a Specification Error", *Econometrica*, 47(1), 153–161.

Heckman, J. and B. Singer (1984) "A Method for Minimizing the Impact of Distributional Assumptions in Econometric Models for Duration Data", *Econometrica*, 52(2), 271–320.

Hendry, D. F. and A. Spanos (1980) "Disequilibrium and Latent Variables". Manuscript, London School of Economics.

Hildebrand, F. B. (1956) *Introduction to Numercial Analysis*. New York: McGraw-Hill.

Hwang, H. (1980) "A Test of a Disequilibrium Model", *Journal of Econometrics*, 12, 319–333.

Ito, T. (1980) "Methods of Estimation for Multi-Market Disequilibrium Models", *Econometrica*, 48(1), 97–125.

Ito, T. and K. Ueda (1981) "Tests of the Equilibrium Hypothesis in Disequilibrium Econometrics: An International Comparison of Credit Rationing", *International Economic Review*, 22(3), 691–708.

Johnson, N. L. and S. Kotz (1972) *Distributions in Statistics: Continuous Multivariate Distributions*. Wiley: New York.

Johnson, P. D. and J. C. Taylor (1977) "Modelling Monetary Disequilibrium", in: M. G. Porter, ed., *The Australian Monetary System in the 1970's*. Australia: Monash University.

Kenny, L. W., L. F. Lee, G. S. Maddala and R. P. Trost (1979) "Returns to College Education: An Investigation of Self-Selection Bias Based on the Project Talent Data", *International Economic Review*, 20(3), 751–765.

Kiefer, N. (1978) "Discrete Parameter Variation: Efficient Estimation of a Switching Regression Model", *Econometrica*, 46(2), 427–434.

Kiefer, N. (1979) "On the Value of Sample Separation Information", *Econometrica*, 47(4), 997–1003.

Kiefer, N. (1980a) "A Note on Regime Classification in Disequilibrium Models", *Review of Economic Studies*, 47(1), 637–639.

Kiefer, N. (1980b) "A Note on Switching Regression and Logistic Discrimination", *Econometrica*, 48, 637–639.

King, M. (1980) "An Econometric Model of Tenure Choice and Housing as a Joint Decision", *Journal of Public Economics*, 14(2), 137–159.

Kooiman, T. and T. Kloek (1979) "Aggregation and Micro-Markets in Disequilibrium: Theory and Application to the Dutch Labor Market: 1948–1975". Working Paper, Rotterdam: Econometric Institute, April 1979.

Laffont, J. J. (1983) "Fix-Price Models: A Survey of Recent Empirical Work". Discussion Paper #8305, University of Toulouse.

Laffont, J. J. and R. Garcia (1977) "Disequilibrium Econometrics for Business Loans", *Econometrica*, 45(5), 1187–1204.

Laffont, J. J. and A. Monfort (1979) "Disequilibrium Econometrics in Dynamic Models", *Journal of Econometrics*, 11, 353–361.

Lee, L. F. (1976) *Estimation of Limited Dependent Variable Models by Two-Stage Methods*. Ph.D. Dissertation, University of Rochester.

Lee, L. F. (1978a) "Unionism and Wage Rates: A Simultaneous Equations Model with Qualitative and Limited Dependent Variables", *International Economic Review*, 19(2), 415–433.

Lee, L. F. (1978b) "Comparative Advantage in Individuals and Self-Selection". Manuscript, University of Minnesota.

Lee, L. F. (1979) "Identification and Estimation in Binary Choice Models with Limited (Censored) Dependent Variables", *Econometrica*, 47(4), 977–996.

Lee, L. F. (1982a) "Some Approaches ot the Correction of Selectivity Bias", *Review of Economic Studies*, 49, 355–372.

Lee, L. F. (1982b) "Test for Normality in the Econometric Disequilibrium Markets Model", *Journal of Econometrics*, 19, 109–123.

Lee, L. F. (1983a) "Generalized Econometric Models with Selectivity", *Econometrica*, 51(2), 507–512.

Lee, L. F. (1983b) "Regime Classification in the Disequilibrium Market Models". Discussion Paper #93, Center for Econometrics and Decision Sciences, University of Florida.

Lee, L. F. (1984) "Sequential Discrete Choice Econometric Models With Selectivity". Discussion Paper, University of Minnesota.

Lee, L. F. and R. P. Trost (1978) "Estimation of Some Limited Dependent Variable Models with Application to Housing Demand", *Journal of Econometrics*, 8, 357–382.

Lee, L. F., G. S. Maddala and R. P. Trost (1980) "Asymptotic Covariance Matrices of Two-Stage Probit and Two-Stage Tobit Methods for Simultaneous Equations Models with Selectivity", *Econometrica*, 48(2), 491–503.

Lee, L. F. and G. S. Maddala (1983a) "The Common Structure of Tests for Selectivity Bias, Serial Correlation, Heteroscedasticity and Normality in the Tobit Model". Manuscript, Center for Econometrics and Decision Sciences, University of Florida. Forthcoming in the *International Economic Review*.

Lee, L. F. and G. S. Maddala (1983b) "Sequential Selection Rules and Selectivity in Discrete Choice Econometric Models". Manuscript, Center for Econometrics and Decision Sciences, University of Florida.

Lee, L. F. and R. H. Porter (1984) "Switching Regression Models with Imperfect Sample Separation Information: With an Application on Cartel Stability", *Econometrica*, 52(2), 391–418.

Lewis, H. G. (1974) "Comments on Selectivity Biases in Wage Comparisons", *Journal of Political Economy*, 82(6), 1145–1155.

Mackinnon, J. G. (1978) "Modelling a Market Which is Sometimes in Disequilibrium". Discussion Paper #287, Canada: Queens University, April 1978.

Mackinnon, J. F. and N. D. Olewiler (1980) "Disequilibrium Estimation of the Demand for Copper", *The Bell Journal of Economics*, 11, 197–211.

Maddala, G. S. (1977a) "Self-Selectivity Problems in Econometrica Models", in: P. R. Krishnan, ed., *Applications of Statistics*. North-Holland Publishing, 351–366.

Maddala, G. S. (1977b) "Identification and Estimation Problems in Limited Dependent Variable Models", in: A. S. Blinder and P. Friedman, eds., *Natural Resources, Uncertainty and General Equilibrium Systems: Essays in Memory of Rafael Lusky*. New York: Academic Press, 219–239.

Maddala, G. S. (1983a) *Limited Dependent and Qualitative Variables in Econometrics*. New York: Cambridge University Press.

Maddala, G. S. (1983b) "Methods of Estimation for Models of Markets with Bounded Price Variation", *International Economic Review*, 24(2), 361–378.

Maddala, G. S. (1984) "Estimation of the Disequilibrium Model with Noisy Indicators". Manuscript, University of Florida.

Maddala, G. S. and F. D. Nelson (1974) "Maximum Likelihood Methods for Models of Markets in Disequilibrium", *Econometrica*, 42(6), 1013–1030.

Maddala, G. S. and L. F. Lee (1976) "Recursive Models with Qualitative Endogenous Variables", *Annals of Economic and Social Measurement*, 5(4), 525–545.

Maddala, G. S. and F. D. Nelson (1975) "Switching Regression Models with Exogenous and Endogenous Switching", *Proceedings of the Business and Economic Statistics Section*, American Statistical Association, 423–426.

Maddala, G. S. and J. S. Shonkwiler (1984) "Estimation of a Disequilibrium Model Under Rational Expectations and Price Supports: The Case of Corn in the U.S". Manuscript, University of Florida.

Maddala, G. S. and R. P. Trost (1981) "Alternative Formulations of the Nerlove-Press Models", *Journal of Econometrics*, 16, 35–49.

Maddala, G. S. and R. P. Trost (1982) "On Measuring Discrimination in Loan Markets", *Housing Finance Review*, 1(1), 245–268.

Malinvaud, E. (1977) *The Theory of Unemployment Reconsidered*. Oxford: Blackwell.

Malinvaud, E. (1982) "An Econometric Model for Macro-Disequilibrium Analysis", in: M. Hazewinkel and A. H. G. Rinnoy Kan, eds., *Current Developments in the Interface: Economics, Econometrics, Mathematics*. D. Reidel Publishing Co., 239–258.

Melino, A. (1982) "Testing for Sample Selection Bias", *Review of Economic Studies*, 49(1), 151–153.

Meyer, R. H. and D. A. Wise (1983a) "The Effect of Minimum Wage on the Employment and Earnings of Youth", *Journal of Labor Economics*, 1(1), 66–100.

Meyer, R. H. and D. A. Wise (1983b) "Discontinuous Distributions and Missing Persons: The Minimum Wage and Unemployed Youth", *Econometrica*, 51(6), 1677–1698.

Monroe, Margaret A. (1981) *A Disequilibrium Econometric Analysis of Interest Rate Futures Markets*. Ph.D. Dissertation, University of Florida.

Muelbauer, J. and Winter D. (1980) "Unemployment, Employment and Exports in British Manufacturing: A Non-clearing Markets Approach", *European Economic Review*, 13(2), 383–409.

Muthen, B. and K. G. Joreskog (1981) "Selectivity Problems in Quasi-experimental Studies". Paper presented at the Conference on "Experimental Research in Social Sciences". University of Florida, January 1981.

Nelson, F. D. (1975) *Estimation of Economic Relationships with Censored, Truncated and Limited Dependent Variables*. Ph.D. Dissertation, University of Rochester.

Nelson, F. D. (1977) "Censored Regression Models with Unobserved Stochastic Censoring Thresholds", *Journal of Econometrics*, 6, 309–327.

Olsen, R. J. (1980) "A Least Squares Correction for Selectivity Bias", *Econometrica*, 48(6), 1815–1820.

Olsen, R. J. (1982) "Distribution Tests for Selectivity Bias and a More Robust Likelihood Estimator", *International Economic Review*, 23(1), 223–240.

Orsi, R. (1982) "On the Dynamic Specification of Disequilibrium Econometrics: An Analysis of Italian Male and Female Labor Markets". CORE Discussion Paper #8228, Louvain, Belgium.

Poirier, D. J. (1980) "Partial Observability in Bivariate Probit Models", *Journal of Econometrics*, 12, 209–217.

Poirier, D. J. and P. A. Rudd (1981) "On the Appropriateness of Endogenous Switching", *Journal of*

*Econometrics*, 16(2), 249–256.

Portes, R. D. (1978) "Effective Demand and Spillovers in Empirical Two-Market Disequilibrium Models". Discussion Paper #595, Harvard Institute of Economic Research, November 1977.

Portes, R. D. and D. Winter (1978) "The Demand for Money and for Consumption Goods in Centrally Planned Economies", *The Review of Economics and Statistics*, 60(1), 8–18.

Portes, R. D. and D. Winter (1980) "Disequilibrium Estimates for Consumption Goods Markets in Centrally Planned Economies", *Review of Economic Studies*, 47(1), 137–159.

Quandt, R. E. (1978) "Maximum Likelihood Estimation of Disequilibrium Models", *Pioneering Economics*, Italy: Padova.

Quandt, R. E. (1978) "Tests of the Equilibrium vs. Disequilibrium Hypothesis", *International Economic Review*, 19(2), 435–452.

Quandt, R. E. and J. D. Ramsey (1978) "Estimating Mixtures of Normal Distributors and Switching Regressions", with discussion, *Journal of the American Statistical Association*, 73, 730–752.

Quandt, R. E. (1981) "Autocorrelated Errors in Simple Disequilibrium Models", *Economics Letters*, 7, 55–61.

Quandt, R. E. (1982) "Econometric Disequilibrium Models". With comments by D. F. Hendry, A. Monfort and J. F. Richard, *Econometric Reviews*, 1(1), 1–63.

Quandt, R. E. (1983) "Bibliography of Quantity Rationing and Disequilibrium Models". Princeton University, Dec. 1983, updated every 3–6 months.

Quandt, R. E. (1984) "Switching Between Equilibrium and Disequilibrium", *Review of Economics and Statistics*, forthcoming.

Richard, J. F. (1980a) "Models with Several Regime Changes and Changes in Exogeneity", *Review of Economic Studies*, 47(1), 1–20.

Richard, J. F. (1980b) "C-Type Distributions and Disequilibrium Models". Paper presented in the Toulouse Conference on "Economics and Econometrics of Disequilibrium".

Rogers, A. J. (1983) "Generalized Lagrange Multiplier Tests for Problems of One-Sided Alternatives". Manuscript, Princeton University.

Rosen, S. and M. I. Nadiri (1974) "A Disequilibrium Model of Demand for Factors of Production", *American Economic Review*, papers and proceedings, 64(2), 264–270.

Rosen, H. and R. E. Quandt (1978) "Estimation of a Disequilibrium Aggregate Labor Market", *Review of Economics and Statistics*, 60, 371–379.

Roy, A. D. (1951) "Some Thoughts on the Distribution of Earnings", *Oxford Economic Papers*, 3, 135–146.

Samelson, H., R. M. Thrall and O. Wesler (1958) "A Partition Theorem for Euclidean *n* Space", *Proceedings of the American Mathematical Society*, 9, 805–807.

Schmidt, P. (1982) "An Improved Version of the Quandt-Ramsay MGF Estimator for Mixtures of Normal Distributions and Switching Regressions", *Econometrica*, 50(2), 501–516.

Sealy, C. W., Jr. (1979) "Credit Rationing in the Commercial Loan Market: Estimates of a Structural Model Under Conditions of Disequilibrium", *Journal of Finance*, 34(2), 689–702.

Sneessens, H. (1981) *Theory and Estimation of Macroeconomic Rationing Models*. New York: Springer-Verlag, 1981.

Sneessens, H. (1983) "A Macro-Economic Rationing Model of the Belgian Economy", *European Economic Review*, 20, 193–215.

Tishchler, A. and I. Zang (1979) "A Switching Regression Model Using Inequality Conditions", *Journal of Econometrics*, 11, 259–274.

Trost, R. P. (1981) "Interpretation of Error Covariances With Non-Random Data: An Empirical Illustration of Returns of College Education", *Atlantic Economic Journal*, 9(3), 85–90.

Trost, R. P. and L. F. Lee (1984) "Technical Training and Earnings: A Polychotomous Choice Model with Selectivity", *The Review of Economics and Statistics*, 66(1), 151–156.

Tunali, I. (1983) "A Common Structure for Models of Double Selection". Report #8304, Social Systems Research Institute, University of Wisconsin.

Upcher, M. R. (1980) *Theory and Applications of Disequilibrium Econometrics*. Ph.D. Dissertation, Canberra: Australian National University.

Venti, S. F. and D. A. Wise (1982) "Test Scores, Educational Opportunities, and Individual Choice", *Journal of Public Economics*, 18, 35–63.

Waldman, D. M. (1981) "An Economic Interpretation of Parameter Constraints in a Simultaneous

Equations Model with Limited Dependent Variables", *International Economic Review*, 22(3), 731–730.

Wales, T. J. and A. D. Woodland (1980) "Sample Selectivity and the Estimation of Labor Supply Functions", *International Economic Review*, 21, 437–468.

Wallis, K. F. (1980) "Econometric Implications of the Rational Expectations Hypothesis", *Econometrica*, 48(1), 49–72.

Willis, R. J. and S. Rosen (1979) "Education and Self-Selection", *Journal of Political Economy*, Part 2, 87(5), 507–526.

Wu, De-Min (1973) "Alternative Tests of Independence Between Stochastic Regressors and Disturbances", *Econometrica*, 41(3), 733–750.

*Chapter 29*

# ECONOMETRIC ANALYSIS OF LONGITUDINAL DATA*

JAMES J. HECKMAN

*University of Chicago and NORC*

BURTON SINGER

*Yale University and NORC*

## Contents

*This research was supported by NSF Grant SES-8107963 and NIH Grant NIH-1-RO1-HD16846-01 to the Economics Research Center, NORC, 6030 S. Ellis, Chicago, Illinois 60637. We thank Takeshi Amemiya and Aaron Han for helpful comments.

*Handbook of Econometrics, Volume III, Edited by Z. Griliches and M.D. Intriligator*
© *Elsevier Science Publishers BV, 1986*

## 0. Introduction

In analyzing discrete choices made over time, two arguments favor the use of continuous time models. (1) In most economic models there is no natural time unit within which agents make their decisions and take their actions. Often it is more natural and analytically convenient to characterize the agent's decision and action processes as operating in continuous time. (2) Even if there were natural decision periods, there is no reason to suspect that they correspond to the annual or quarterly data that are typically available to empirical analysts, or that the discrete periods are synchronized across individuals. Inference about an underlying stochastic process that is based on interval or point sampled data may be very misleading especially if one falsely assumes that the process being investigated operates in discrete time. Conventional discrete choice models such as logit and probit when defined for one time interval are of a different functional form when applied to another time unit, if they are defined at all. Continuous time models are invariant to the time unit used to record the available data. A common set of parameters can be used to generate probabilities of events occurring in intervals of different length. For these reasons the use of continuous time duration models is becoming widespread in economics.

This paper considers the formulation and estimation of continuous time econometric duration models. Research on this topic is relatively new and much of the available literature has borrowed freely and often uncritically from reliability theory and biostatistics. As a result, most papers in econometric duration analysis present statistical models only loosely motivated by economic theory and assume access to experimental data that are ideal in comparison to the data actually available to social scientists.

This paper is in two parts. Part I – which is by far the largest – considers single spell duration models which are the building blocks for the more elaborate multiple spell models considered in Part II. Many issues that arise in multiple spell models are more easily discussed in a single spell setting and in fact many of the available duration data sets only record single spells.

Our discussion of single spell duration models is in six sections. In Section 1.1 we present some useful definitions and statistical concepts. In Section 1.2 we present a short catalogue of continuous time duration models that arise from choice theoretic economic models. In Section 1.3 we consider conventional methods for introducing observed and unobserved variables into reduced form versions of duration models. We discuss the sensitivity of estimates obtained from single spell duration models to inherently ad hoc methods for controlling for observed and unobserved variables.

The extreme sensitivity to ad hoc parameterizations of duration models that is exhibited in this section leads us to ask the question "what features of duration models can be identified nonparametrically?" Our answer is the topic of Section 1.4. There we present nonparametric procedures for assessing qualitative features of conditional duration distributions in the presence of observed and unobserved variables. We discuss nonparametric identification criteria for a class of duration models (proportional hazard models) and discuss tradeoffs among criteria required to secure nonparametric identification. We also discuss these questions for a more general class of duration models. The final topic considered in this section is nonparametric estimation of duration models.

In Section 1.5 we discuss the problem of initial conditions. There are few duration data sets for which the beginning of the sample observation period coincides with the start of a spell. More commonly, the available data for single spell models consist of interrupted spells or portions of spells observed after the sample observation period begins. The problem raised by this sort of sampling frame and its solution are well known for duration models with no unobservables in time homogeneous environments. We present these solutions and then discuss this problem for the more difficult but empirically relevant case of models with unobservable variables in time inhomogeneous environments. In Section 1.6 we return to the structural duration models discussed in Section 1.2 and consider new econometric issues that arise in attempting to recover explicit economic parameters.

Part II on multiple spells is divided into two sections. The first (Section 2.1) presents a general framework which contains many interesting multiple spell models as a special case. The second (Section 2.2) presents a multiple spell event history model and considers conditions under which access to multiple spell data aids in securing model identification. This paper concludes with a brief summary.

# 1. Single spell models

## 1.1. Statistical preliminaries

There are now a variety of excellent textbooks on duration analysis that discuss the formulation of duration models so that a lengthy introduction to standard survival models is unnecessary.[1] In an effort to make this chapter self-contained, however, this section sets out the essential ideas that we need from this literature in the rest of the chapter.

---

[1] See especially, Kalbfleisch and Prentice (1980), Lawless (1982) and Cox and Oakes (1984).

A nonnegative random variable $T$ with absolutely continuous distribution function $G(t)$ and density $g(t)$ may be uniquely characterized by its hazard function. The hazard for $T$ is the conditional density of $T$ given $T > t \geq 0$, i.e.

$$h(t) = f(t|T > t) = \frac{g(t)}{1 - G(t)} \geq 0. \tag{1.1.1}$$

Knowledge of $G$ determines $h$.

Conversely, knowledge of $h$ determines $G$ because by integration of (1.1.1)

$$\int_0^t h(u)\,du = -\ln(1 - G(x))\Big|_0^t + c,$$

$$G(t) = 1 - \exp\left(-\int_0^t h(u)\,du\right); \tag{1.1.2}$$

$c = 0$ since $G(0) = 0$. The density of $T$ is

$$g(t) = h(t)\exp\left(-\int_0^t h(u)\,du\right). \tag{1.1.3}$$

For the rest of this paper we assume that the distribution of $T$ is absolutely continuous, and we associate $T$ with spell duration.[2] In this case it is also natural to interpret $h(t)$ as an *exit rate* or *escape rate* from the state because it is the limit (as $\Delta \to 0$) of the probability that a spell terminates in interval $(t, t + \Delta)$ given that the spell has lasted $t$ periods, i.e.

$$h(t) = \lim_{\Delta \to 0} \Pr(t < T < t + \Delta|T > t)\Delta$$

$$= \lim_{\Delta \to 0}\left[\frac{G(t + \Delta) - G(t)}{\Delta}\right]\frac{1}{(1 - G(t))}$$

$$= \frac{g(t)}{1 - G(t)}. \tag{1.1.4}$$

Equation (1.1.4) constitutes an alternative definition of the hazard that links the models discussed in Part I to the more general multistate models discussed in Part II.

---

[2] For a treatment of duration distributions that are not absolutely continuous see, e.g. Lawless (1982).

The survivor function is the probability that a duration exceeds $t$. Thus

$$S(t) = P(T > t) = 1 - G(t) = \exp\left(-\int_0^t h(u)\,du\right). \tag{1.1.5}$$

In terms of the survivor function we may write the density $g(t)$ as

$$g(t) = h(t)S(t).$$

Note that there is no requirement that

$$\lim_{t \to \infty} \int_0^t h(u)\,du \to \infty, \tag{1.1.6}$$

or equivalently that

$$S(\infty) = 0.$$

If (1.1.6) is satisfied, the duration distribution is termed nondefective. Otherwise, it is termed *defective*.

The technical language here creates the possibility of confusion. There is nothing wrong with defective distributions. In fact they emerge naturally from many optimizing models. For example, Jovanovic (1979) derives an infinite horizon worker-firm matching model with a defective job tenure distribution. Condition (1.1.6) is violated in his model so $S(\infty) > 0$ because some proportion of workers find that their current match is so successful that they never wish to leave their jobs.

Duration dependence is said to exist if

$$\frac{dh(t)}{dt} \neq 0.$$

The only density with no duration dependence almost everywhere is the exponential distribution. For in this case $h(t) \equiv h$, a constant, and hence from (1.1.2), $T$ is an exponential random variable. Obviously if $G$ is exponential, $h(t) \equiv h$.

If $dh(t)/dt > 0$, at $t = t_0$, there is said to be *positive duration dependence at $t_0$*. If $dh(t)/dt < 0$, at $t = t_0$, there is said to be *negative duration dependence at $t_0$*. In job search models of unemployment, positive duration dependence arises in the case of a "declining reservation wage" (see, e.g. Lippman and McCall, 1976). In this case the exit rate from unemployment is monotonically increasing in $t$. In job turnover models negative duration dependence (at least asymptotically) is associated with worker-firm matching models (see, e.g. Jovanovic, 1979).

For many econometric duration models it is natural to analyze conditional duration distributions where the conditioning is with respect to observed $(x(t))$ and unobserved $(\theta(t))$ variables. Indeed, by analogy with conventional regression analysis, much of the attention in many duration analyses focuses on the effect of regressors $(x(t))$ on durations.

We define the conditional hazard as

$$h(t|x(t),\theta(t)) = \lim_{\Delta \to 0} \frac{\Pr(t < T < t + \Delta | T > t, x(t), \theta(t))}{\Delta}. \tag{1.1.7}$$

The dating on regressor vector $x(t)$ is an innocuous convention. $x(t)$ may include functions of the entire past or future or the entire paths of some variables, e.g.

$$x_1(t) = \int_t^\infty k_1(z_1(u)) \, du,$$

$$x_2(t) = \int_{-\infty}^t k_2(z_2(u)) \, du,$$

$$x_3(t) = \int_{-\infty}^\infty k_3(z_3(u), t) \, du,$$

where the $z_i(u)$ are underlying time dated regressor variables.

We make the following assumptions about these conditioning variables.

(A.1) $\theta(t)$ is distributed independently of $x(t')$ for all $t, t'$. The distribution of $\theta$ is $\mu(\theta)$. The distribution of $x$ is $D(x)$.

(A.2) There are no functional restrictions connecting the conditional distribution of $T$ given $\theta$ and $x$ and the marginal distributions of $\theta$ and $x$.

Speaking very loosely, $x$ is assumed to be "weakly exogenous" with respect to the duration process. More precisely $x$ is ancillary for $T$.[3]

By analogy with the definitions presented for the raw duration models, we may integrate (1.1.7) to produce the conditional duration distribution

$$G(t|x,\theta) = 1 - \exp\left(-\int_0^t h(u|x(u), \theta(u)) \, du\right), \tag{1.1.8}$$

the conditional survivor function

$$S(t|\theta, x) = P(T > t|x, \theta)$$

$$= \exp\left(-\int_0^t h(u|x(u), \theta(u)) \, du\right), \tag{1.1.9}$$

[3] See, e.g. Cox and Hinkley (1974) for a discussion of ancillarity.

and the conditional density

$$g(t|x,\boldsymbol{\theta}) = h(t|x(t),\boldsymbol{\theta}(t))S(t|x,\boldsymbol{\theta}). \tag{1.1.10}$$

One specification of conditional hazard (1.1.7) that has received much attention in the literature is the *proportional hazard specification* [see Cox (1972)]

$$h(t|x(t),\boldsymbol{\theta}(t)) = \psi(t)\varphi(x(t))\eta(\boldsymbol{\theta}(t)), \tag{1.1.11}$$

which postulates that the log of the conditional hazard is linear in functions of $t$, $x$ and $\boldsymbol{\theta}$ and that

$$\psi(t) \geq 0, \eta(\boldsymbol{\theta}(t)) \geq 0, \varphi(x(t)) \geq 0 \qquad \text{for all } t,$$

where $\eta$ is a monotonic continuous increasing function of $\boldsymbol{\theta}(t)$.

## 1.2. Examples of duration models produced by economic theory

In this section of the paper, we present three examples of duration models produced by economic choice models. These examples are (A) a continuous time labor supply, (B) a continuous time search unemployment model, and (C) a continuous time consumer purchase model that generalizes conventional discrete choice models in a straightforward way.

Examples A and B contain most of the essential ideas. We demonstrate how a continuous time formulation avoids the need to specify arbitrary decision periods as is required in conventional discrete time models (see, e.g. Heckman, 1981a). We also discuss a certain identification problem that arises in single spell models that is "solved" by assumption in conventional discrete time formulations.

### 1.2.1. Example A: A dynamic model of labor force participation

The one period version of this model is the workhorse of labor economics. Consumers at age $a$ are assumed to possess a concave twice differentiable one period utility function defined over goods ($X(a)$) and leisure ($L(a)$). Denote this utility function by $U(X(a), L(a))$. Define leisure hours so that $0 \leq L(a) \leq 1$. The consumer is free to choose his hours of work at parametric wage $W(a)$. There are no fixed costs of work, and for convenience taxes are ignored. At each age the consumer receives unearned income $Y(a)$. There is no saving or borrowing. Decisions are assumed to be made under perfect certainty.

The consumer works at age $a$ if the marginal rate of substitution between goods and leisure evaluated at the no work position (also known as the nonmarket wage)

$$M(Y(a)) = \frac{U_2(Y(a),1)}{U_1(Y(a),1)},\tag{1.2.1}$$

is less than the market wage $W(a)$. For if this is so, his utility is higher in the market than at home. The subscripts on $U$ denote partial derivatives with respect to the appropriate argument. It is convenient to define an index function $I(a)$ written as

$$I(a) = W(a) - M(Y(a)).$$

If $I(a) \geq 0$, the consumer works at age $a$, and we record this event by setting $d(a) = 1$. If $I(a) < 0$, $d(a) = 0$.

In a discrete time model, a spell of employment begins at $a_1$ and ends at $a_2 + 1$ provided that $I(a_1 - 1) < 0$, $I(a_1 + j) \geq 0$, $j = 0, \ldots, a_2 - a_1$, $I(a_2 + 1) < 0$. Reversing the direction of the inequalities generates a characterization of a nonwork spell that begins at $a_1$ and ends at $a_2$.

To complete the econometric specification, error term $\varepsilon(a)$ is introduced. Under an assumption of perfect certainty, the error term arises from variables observed by the consumer but not observed by the econometrician. In the current context, $\varepsilon(a)$ can be interpreted as a shifter of household technology and tastes. For each person successive values of $\varepsilon(a)$ may be correlated, but it is assumed that $\varepsilon(a)$ is independent of $Y(a)$ and $W(a)$. We define the index function inclusive of $\varepsilon(a)$ as

$$I^*(a) = W(a) - M(Y(a)) + \varepsilon(a).\tag{1.2.2}$$

If $I^*(a) \geq 0$, the consumer works at age $a$.

The distribution of $I^*(a)$ induces a distribution on employment spells. To demonstrate this point in a simple way we assume that (i) the $\varepsilon(a)$ are serially independent, (ii) the environment is time homogeneous so $W(a)$ and $Y(a)$ remain constant over time for the individual, (iii) the probability that a new value of $\varepsilon$ is received in an interval is $P$, and (iv) that the arrival times of new values of $\varepsilon(a)$ are independent of $W, Y$, and other arrival times. We denote the c.d.f. of $\varepsilon$ by $F$. By virtue of the perfect certainty assumption, the individual knows when new values of $\varepsilon$ will arrive and what they will be. The econometrician, however, does not have this information at his disposal. He never directly observes $\varepsilon(a)$ and only knows that a new value of nonmarket time has arrived if the consumer actually changes state.

The probability that an employed person does not leave the employed state is

$$1 - F(\psi),\tag{1.2.3}$$

where $\psi = M(Y) - W$. The probability of receiving $j$ new values of $\varepsilon$ in interval $t_e$ is

$$P_j = \binom{t_e}{j} P^j (1 - P)^{t_e - j}.$$

The probability that a spell is longer than $t_e$ is the sum over $j$ of the products of the probability of receiving $j$ innovations in $t_e(P_j)$ and the probability that the person does not leave the employed state on each of the $j$ occasions $(1 - F(\psi))^j$. Thus

$$P(T_e > t_e) = \sum_{j=0}^{t_e} \binom{t_e}{j} P^j (1 - P)^{t_e - j} (1 - F(\psi))^j$$

$$= (1 - PF(\psi))^{t_e}.\tag{1.2.4}$$

Thus the probability that an employment spell starting at calendar time $t_e = 0$ terminates *at* $t_e$ is

$$P(T_e = t_e) = P(T_e > t_e - 1) - P(T_e > t_e)$$

$$= (1 - PF(\psi))^{t_e - 1}(PF(\psi)).\tag{1.2.5}$$

By similar reasoning it can be shown that the probability that a nonemployment spell lasts $t_n$ periods is

$$P(T_n = t_n) = [(1 - P(1 - F(\psi)))]^{t_n - 1} P(1 - F(\psi)).\tag{1.2.6}$$

In conventional models of discrete choice over time [see, e.g. Heckman (1981a)] $P$ is implicitly set to one. Thus in these models it is assumed that the consumer receives a new draw of $\varepsilon$ each period. The model just presented generalizes these models to allow for the possibility that $\varepsilon$ may remain constant over several periods of time. Such a generalization creates an identification problem because from a single employment or nonemployment spell it is only possible to estimate $PF(\psi)$ or $P(1 - F(\psi))$ respectively. This implies that any single spell model of the duration of employment or nonemployment is consistent with the model of eq. (1.2.2) with $P = 1$ or with another model in which (1.2.2) does not characterize behavior but in which the economic variables determine the arrival time of new values of $\varepsilon$. However, access to both employment and nonemployment spells solves this problem because $P = PF(\psi) + P(1 - F(\psi))$, and hence $F(\psi)$ and $P$ are separately identified.

The preceding model assumes that there are natural periods of time within which innovations in $\varepsilon$ may occur. For certain organized markets there may be well-defined trading intervals, but for the consumer's problem considered here no such natural time periods exist. This suggests the following continuous time reformulation.

In place of the Bernoulli assumption for the arrival of fresh values of $\varepsilon$, suppose instead that a Poisson process governs the arrival of shocks. As is well known [see, e.g. Feller (1970)] the Poisson distribution is the limit of a Bernoulli trial process in which the probability of success in each interval $\eta = \Delta/n$, $P_\eta$, goes to zero in such a way that $\lim_{n \to 0} nP_\eta \to \lambda \neq 0$. Thus in the reformulated continuous time model it is assumed that an infinitely large number of very low probability Bernoulli trials occur within a specified interval of time.

For a time homogeneous environment the probability of receiving $j$ offers in time period $t_e$ is

$$P(j|t_e) = \exp(-\lambda t_e) \frac{(\lambda t_e)^j}{j!}. \tag{1.2.7}$$

Thus for the continuous time model the probability that a person who begins employment at $a = a_1$ will stay in the employed state at least $t_e$ periods is, by reasoning analogous to that used to derive (1.2.6),

$$\Pr(T_e > t_e) = \sum_{j=0}^{\infty} \exp(-\lambda t_e) \frac{(\lambda t_e)^j}{j!} (1 - F(\psi))^j$$

$$= \exp(-\lambda F(\psi) t_e), \tag{1.2.8}$$

so the density of spell lengths is

$$g(t_e) = \lambda F(\psi) \exp(-\lambda F(\psi) t_e).$$

A more direct way to derive (1.2.8) notes that from the definition of a Poisson process, the probability of receiving a new value of $\varepsilon$ in interval $(a, a + \Delta)$ is

$$p = \lambda \Delta + o(\Delta),$$

where $\lim_{\Delta \to 0} (o(\Delta)/\Delta) \to 0$, and the probability of exiting the employment state conditional on an arrival of $\varepsilon$ is $F(\psi)$. Hence the exit rate or hazard rate from the employment state is

$$h_e = \lim_{\Delta \to 0} \frac{\lambda \Delta F(\psi)}{\Delta} + o(\Delta),$$
$$= \lambda F(\psi).$$

Using (1.1.4) relating the hazard function and the survivor function we conclude that

$$\Pr(T_e > t_e) = \exp\left(-\int_0^{t_e} h_e(u) \, du\right) = \exp(-\lambda F(\psi) t_e).$$

By similar reasoning, the probability that a person starting in the nonemployed state will stay on in that state for at least duration $t_n$ is

$$\Pr(T_n > t_n | \lambda) = \exp(-\lambda(1 - F(\psi)) t_n).$$

Analogous to the identification result already presented for the discrete time model, it is impossible using single spell employment or nonemployment data to separate $\lambda$ from $F(\psi)$ or $1 - F(\psi)$ respectively. However, access to data on both employment and nonemployment spells makes it possible to identify both $\lambda$ and $F(\psi)$.

The assumption of time homogeneity of the environment is made only to simplify the argument. Suppose that nonmarket time arrives via a nonhomogeneous Poisson process so that the probability of receiving one nonmarket draw in interval $(a, a + \Delta)$ is

$$p(a) = \lambda(a)\Delta + o(\Delta). \tag{1.2.9}$$

Assuming that $W$ and $Y$ remain constant, the hazard rate for exit from employment at time period $a$ for a spell that begins at $a_1$ is

$$h_e(a | a_1) = \lambda(a) F(\psi), \tag{1.2.10}$$

so that the survivor function for the spell is

$$P(T_e > t_e | a_1) = \exp\left(-F(\psi) \int_{a_1}^{a_1 + t_e} \lambda(u) \, du\right).^4 \tag{1.2.11}$$

---

[4] As first noted by Lundberg (1903), it is possible to transform this model to a time homogeneous Poisson model if we redefine duration time to be

$$\Omega^*(t_e, a_1) = \int_{a_1}^{a_1 + t_e} \lambda(u) \, du.$$

Allowing for time inhomogeneity in $Y(a)$ and $W(a)$ raises a messy, but not especially deep problem. It is possible that the values of these variables would change at a point in time in between the arrival of $\varepsilon$ values and that such changes would result in a reversal of the sign of $I^*(a)$ so that the consumer would cease working at points in time when $\varepsilon$ did not change. Conditioning on the paths of $Y(a)$ and $W(a)$ formally eliminates the problem.

By similar reasoning

$$P(T_n > t_n | a_1) = \exp\left(-(1 - F(\psi)) \int_{a_1}^{a_1 + t_n} \lambda(u)\,du\right).$$

### 1.2.2.  Example B: A one state model of search unemployment

This model is well exposited in Lippman and McCall (1976). The environment is assumed to be time homogeneous. Agents are assumed to be income maximizers. If an instantaneous cost $c$ is incurred, job offers arrive from a Poisson process with parameter $\lambda$ independent of the level of $c(c > 0)$. The probability of receiving a wage offer in time interval $\Delta t$ is $\lambda \Delta t + o(\Delta t)$.[5] Thus the probability of two or more job offers in interval $\Delta t$ is negligible.[6]

Successive wage offers are independent realizations from a known absolutely continuous wage distribution $F(w)$ with finite mean that is assumed to be common to all agents. Once refused, wage offers are no longer available. Jobs last forever, there is no on the job search, and workers live forever. The instantaneous rate of interest is $r(> 0)$.

$V$ is the value of search. Using Bellman's optimality principle for dynamic programming [see, e.g. Ross (1970)], $V$ may be decomposed into three components plus a negligible component [of order $o(\Delta t)$].

$$V = -\frac{c\Delta t}{1 + r\Delta t} + \frac{(1 - \lambda \Delta t)}{1 + r\Delta t} V$$

$$+ \frac{\lambda \Delta t}{1 + r\Delta t} E \max[w/r; V] + o(\Delta t), \qquad \text{for } V > 0,$$

$$= 0 \text{ otherwise.} \tag{1.2.12}$$

The first term on the right of (1.2.12) is the discounted cost of search in interval $\Delta t$. The second term is the probability of not receiving an offer $(1 - \lambda \Delta t)$ times the discounted value of search at the end of interval $\Delta t$. The third term is the probability of receiving a wage offer, $(\lambda \Delta t)$, times the discounted value of the expectation [computed with respect to $F(w)$] of the maximum of the two options confronting the agent who receives a wage offer: to take the offer (with present value $w/r$) or to continue searching (with present value $V$). Note that eq. (1.2.12)

---

[5] $o(\Delta t)$ is defined as a term such that $\lim_{\Delta t \to 0} o(\Delta t)/\Delta t \to 0$.
[6] For one justification of the Poisson wage arrival assumption, see, e.g. Burdett and Mortensen (1978).

is defined only for $V > 0$. If $V = 0$, we may define the agent as out of the labor force [see Lippman and McCall (1976)]. As a consequence of the time homogeneity of the environment, once out the agent is always out. Sufficient to ensure the existence of an optimal reservation wage policy in this model is $E(|W|) < \infty$ [Robbins (1970)].

Collecting terms in (1.2.12) and passing to the limit, we reach the familiar formula [Lippman and McCall (1976)]

$$c + rV = (\lambda/r) \int_{rV}^{\infty} (w - rV) \, dF(w) \qquad \text{for } V > 0, \tag{1.2.13}$$

where $rV$ is the reservation wage, which is implicitly determined from (1.2.13). For any offered wage $w \geq rV$, the agent accepts the offer. The probability that an offer is unacceptable is $F(rV)$.

To calculate the probability that an unemployment spell $T_u$ exceeds $t_u$, we may proceed as in the preceding discussion of labor supply models and note that the probability of receiving an offer in time interval $(a, a + \Delta)$ is

$$p = \lambda \Delta + o(\Delta), \tag{1.2.14}$$

and further note that the probability that an offer is accepted is $(1 - F(rV))$ so

$$h_u = \lambda (1 - F(rV)), \tag{1.2.15}$$

and

$$P(T_u > t_u) = \exp\left(-\lambda(1 - F(rV))t_u\right). \tag{1.2.16}$$

For discussion of the economic content of this model, see, e.g Lippman and McCall (1976) or Flinn and Heckman (1982a).

Accepted wages are truncated random variables with $rV$ as the lower point of truncation. The density of accepted wages is

$$g(w|w > rV) = \frac{f(w)}{1 - F(rV)}, \qquad w \geq rV. \tag{1.2.17}$$

Thus the one spell search model has the same statistical structure for accepted wages as other models of self selection in labor economics [Lewis (1974), Heckman (1974), and the references in Amemiya (1984)].

From the assumption that wages are distributed independently of wage arrival times, the joint density of duration times $t_u$ and accepted wages $(w)$ is the product of the density of each random variable,

$$m(t_u, w) = \{\lambda(1 - F(rV))\exp - (\lambda(1 - F(rV)t_u))\} \frac{f(w)}{1 - F(rV)}$$

$$= (\lambda \exp - \lambda(1 - F(rV))t_u)f(w), \qquad w \geq rV. \qquad (1.2.18)$$

Time homogeneity of the environment is a strong assumption to invoke especially for the analysis of data on unemployment spells. Even if the external environment were time homogeneous, finiteness of life induces time inhomogeneity in the decision process of the agent. We present a model for a time inhomogeneous environment.

For simplicity we assume that a reservation wage property characterizes the optimal policy noting that for general time inhomogeneous models it need not.[7] We denote the reservation wage at time $\tau$ as $rV(\tau)$.

The probability that an individual receives a wage offer in time period $(\tau, \tau + \Delta)$ is

$$p(\tau) = \lambda(\tau)\Delta + o(\Delta). \qquad (1.2.19)$$

The probability that it is accepted is $(1 - F(rV(\tau)))$. Thus the hazard rate at time $\tau$ for exit from an unemployment spell is

$$h(\tau) = \lambda(\tau)(1 - F(rV(\tau))), \qquad (1.2.20)$$

so that the probability that a spell that began at $\tau_1$ lasts at least $t_u$ is

$$P(T_u > t_u | \tau_1) = \exp\left(-\int_{\tau_1}^{\tau_1 + t_u} \lambda(z)(1 - F(rV(z)))\,dz\right). \qquad (1.2.21)$$

The associated density is

$$g(t_u | \tau_1) = \lambda(\tau_1 + t_u)(1 - F(rV(\tau_1 + t_u)))$$

$$\cdot \exp\left(-\int_{\tau_1}^{\tau_1 + t_u} \lambda(z)(1 - F(rV(z)))\,dz\right).^{8}$$

---

[7] For time inhomogeneity induced solely by the finiteness of life, the reservation wage property characterizes an optimal policy (see, e.g. De Groot, 1970).

[8] Note that in this model it is trivial to introduce time varying forcing variables because by assumption the agent cannot accept a job in between arrival of job offers. Compare with the discussion in footnote 4.

### 1.2.3.  *Example C: A dynamic McFadden model*

As in the marketing literature (see, e.g. Hauser and Wisniewski, 1982a, b, and its nonstationary extension in Singer, 1982), we imagine consumer choice as a sequential affair. An individual goes to a grocery store at randomly selected times. Let $\lambda(\tau)$ be the hazard function associated with the density generating the probability of the event that the consumer goes to the store at time $\tau$. We assume that the probability of two or more visits to the store in interval $\Delta$ is $o(\Delta)$. Conditional on arrival at the store, he may purchase one of $J$ items. Denote the purchase probability by $P_j(\tau)$. Choices made at different times are assumed to be independent, and they are also independent of arrival times. Then the probability that the consumer purchases good $j$ at time $\tau$ is

$$h(j|\tau) = \lambda(\tau)P_j(\tau), \qquad (1.2.22)$$

so that the probability that the next purchase is item $j$ at a time $t = \tau + \tau_1$ or later is

$$P(t, j|\tau_1) = \exp\left(-\int_{\tau_1}^{\tau_1 + t} \lambda(u)P_j(u)\,\mathrm{d}u\right). \qquad (1.2.23)$$

The $P_j$ may be specified using one of the many discrete choice models discussed in Amemiya's survey (1981). For the McFadden random utility model with Weibull errors (1974), the $P_j$ are multinominal logit. For the Domencich–McFadden (1975) random coefficients preference model with normal coefficients the $P_j$ are specified by multivariate probit.

In the dynamic McFadden model few new issues of estimation and specification arise that have not already been discussed above or in Amemiya's survey article (1984). For concreteness, we consider the most elementary version of this model.

Following McFadden (1974), the utility associated with each of $J$ possible choices at time $\tau$ is written as

$$U_j(\tau) = V\big(s, x_j(\tau)\big) + \varepsilon\big(s, x_j(\tau)\big), \qquad j = 1, \ldots, J$$

where $s$ denotes vectors of measured attributes of individuals, $x(\tau)$ represents vectors of attributes of choices, $V$ is a nonstochastic function and the $\varepsilon(s, x_j(\tau))$ are i.i.d. Weibull, i.e.

$$P\big(\varepsilon(s, x_j(\tau)) \le \varphi\big) = \exp[-\exp(-\varphi)].$$

Then as demonstrated by McFadden (p. 110),

$$P_j\big(s, x_j(\tau)\big) = \frac{\exp\big(V\big(s, x_j(\tau)\big)\big)}{\displaystyle\sum_{l=1}^{J} \exp\big(V\big(s, x_l(\tau)\big)\big)}.$$

Adopting a linear specification for $V$ we write

$$V\big(s, x_j(\tau)\big) = x_j'(\tau)\boldsymbol{\beta}(s),$$

so

$$P_j\big(s, x_j(\tau)\big) = \frac{\exp\big(x_j'(\tau)\boldsymbol{\beta}(s)\big)}{\displaystyle\sum_{l=1}^{J} \exp\big(x_l'(\tau)\boldsymbol{\beta}(s)\big)}.$$

In a model without unobservable variables, the methods required to estimate this model are conventional.

The parameter $\boldsymbol{\beta}(s)$ can be estimated by standard logit analysis using data on purchases made at purchase times. The estimation of the times between visits to stores can be conducted using the conventional duration models described in Section 1.3. More general forms of Markovian dependence across successive purchases can be incorporated (see Singer, 1982, for further details).

## 1.3.    Conventional reduced form models

The most direct approach to estimating the economic duration models presented in Section 1.2 is to specify functional forms for the economic parameters and their dependence on observed and unobserved variables. This approach is both costly and controversial. It is controversial because economic theory usually does not produce these functional forms – at best it specifies potential lists of regressor variables some portion of which may be unobserved in any data set. Moreover in many areas of research such as in the study of unemployment durations, there is no widespread agreement in the research community about the correct theory. The approach is costly because it requires nonlinear optimization of criterion functions that often can be determined only as implicit functions. We discuss this point further in Section 1.6.

Because of these considerations and because of a widespread belief that it is useful to get a "feel for the data" before more elaborate statistical models are fit, reduced form approaches are common in the duration analysis literature. Such an approach to the data is inherently ad hoc because the true functional form of the duration model is unknown. At issue is the robustness of the qualitative inferences obtained from these models with regard to alternative ad hoc specifications. In this section of the paper we review conventional approaches and reveal their lack of robustness. Section 1.4 presents our response to this lack of robustness.

The problem of nonrobustness arises solely because regressors and unobservables are introduced into the duration model. If unobservables were ignored and the available data were sufficiently rich, it would be possible to estimate a duration model by a nonparametric Kaplan–Meier procedure [see, e.g. Lawless (1982) or Kalbfleisch and Prentice (1980)]. Such a general nonparametric approach is unlikely to prove successful in econometrics because (a) the available samples are small especially after cross classification by regressor variables and (b) empirical modesty leads most analysts to admit that some determinants of any duration decision may be omitted from the data sets at their disposal.

Failure to control for unobserved components leads to a well known bias toward negative duration dependence. This is the content of the following proposition:

*Proposition 1*

Uncontrolled unobservables bias estimated hazards towards negative duration dependence. □

The proof is a straightforward application of the Cauchy–Schwartz theorem. Let $h(t|x, \theta)$ be the hazard conditional on $x, \theta$ and $h(t|x)$ is the hazard conditional only on $x$. These hazards are associated respectively with conditional distributions $G(t|x, \theta)$ and $G(t|x)$.

From the definition,

$$h(t|x) = \frac{\int g(t|x, \theta)\,d\mu(\theta)}{\int (1 - G(t|x, \theta))\,d\mu(\theta)}.$$

Thus[9]

$$\frac{\partial h(t|x)}{\partial t} = \frac{\int (1 - G(t|x, \theta))\frac{\partial h(t|x, \theta)}{\partial t}\,d\mu(\theta)}{\int (1 - G(t|x, \theta))\,d\mu(\theta)}$$

$$+ \frac{\left[\int g(t|x, \theta)\,d\mu(\theta)\right]^2 - \int \frac{g^2(t|x, \theta)}{1 - G(t|x, \theta)}\,d\mu(\theta)\int (1 - G(t|x, \theta)\,d\mu(\theta)}{\left[\int (1 - G(t|x, \theta))\,d\mu(\theta)\right]^2}.$$

(1.3.1)

---

[9] We use the fact that

$$\frac{\partial h(t|x, \theta)}{\partial t} = \frac{\frac{\partial g(t|x, \theta)}{\partial t}}{1 - G(t|x, \theta)} + \left[\frac{g(t|x, \theta)}{1 - G(t|x, \theta)}\right]^2.$$

The second term on the right-hand side is always nonpositive as a consequence of the Cauchy–Schwartz theorem. □

Intuitively, more mobility prone persons are the first to leave the population leaving the less mobile behind and hence creating the illusion of stronger negative duration dependence than actually exists.

To ignore unobservables is to bias estimated hazard functions in a known direction. Ignoring observables has the same effect. So in response to the limited size of our samples and in recognition of the myriad of plausible explanatory variables that often do not appear in the available data, it is unwise to ignore observed or unobserved variables. The problem is how to control for these variables.

There are many possible conditional hazard functions [see, e.g. Lawless (1982)]. One class of proportional hazard models that nests many previous models as a special case and therefore might be termed "flexible" is the Box–Cox conditional hazard

$$h(t|x,\boldsymbol{\theta}) = \exp\left( x'(t)\boldsymbol{\beta} + \left( \frac{t^{\lambda_1} - 1}{\lambda_1} \right) \gamma_1 + \left( \frac{t^{\lambda_2} - 1}{\lambda_2} \right) \gamma_2 + \boldsymbol{\theta}(t) \right), \qquad (1.3.2)$$

where $\lambda_1 \neq \lambda_2$, $x(t)$ is a $1 \times k$ vector of regressors and $\beta$ is a $k \times 1$ vector of parameters, and $\theta$ is assumed to be scalar. (See Flinn and Heckman, 1982b.) Exponentiating ensures that the hazard is nonnegative as is required for a conditional density.

Setting $\gamma_2 = 0$ and $\lambda_1 = 0$ produces a Weibull hazard; setting $\gamma_2 = 0$ and $\lambda_1 = 1$ produces a Gompertz hazard. Setting $\gamma_1 = \gamma_2 = 0$ produces an exponential model. Conditions under which this model is identified for the case $\gamma_2 = 0$ are presented in Section 1.4.

The conventional approach to single spell econometric duration analysis assumes a specific functional form known up to a finite set of parameters for the conditional hazard and a specific functional form known up to a finite set of parameters for the distribution of unobservables. $\theta(t)$ is assumed to be a time invariant scalar random variable $\theta$. An implicit assumption in most of this literature is that the origin date of the sample is also the start date of the spells being analyzed so that initial conditions or left censoring problems are ignored. We question this assumption in Section 1.5 below.

The conventional approach does, however, allow for right censored spells assuming independent censoring mechanisms. We consider two such schemes.

Let $V(t)$ be the probability that a spell is censored at duration $t$ or later. If

$$\begin{aligned} V(t) &= 0 & t < L, \\ V(t) &= 1 & t \geq L, \end{aligned} \qquad (1.3.3)$$

there is censoring at fixed duration $L$. This type of censoring is common in many economic data sets. More generally, for continuous censoring times let $v(t)$ be the density associated with $V(t)$. In an independent censoring scheme, the censoring time is assumed to be independent of the survival time, and the censoring distribution is assumed to be functionally independent of the survival distribution, and does not depend on $\theta$.

Let $d = 1$ if a spell is not right censored and $d = 0$ if it is. Let $t$ denote an observed spell length. Then the joint frequency of $(t, d)$ conditional on $x$ for the case of absolutely continuous distribution $V(t)$ is

$$f(t, d|x) = v(t)^{1-d} \int_{\theta} [h(t|x(t), \theta) V(t)]^d S(t|x(t), \theta) \, d\mu(\theta),$$

$$= \{v(t)^{1-d} V(t)^d\} \int_{\theta} [h(t|x(t), \theta)]^d S(t|x(t), \theta) \, d\mu(\theta). \quad (1.3.4)$$

By the assumption of functional independence between $V(t)$ and $G(t|x)$, we may ignore the $V$ and $v$ functions in estimating $\mu(\theta)$ and $h(t|x(t), \theta)$ via maximum likelihood.

For the Dirac censoring distribution (1.3.3), the density of observed durations is

$$f(t, d|x) = \int_{\theta} [h(t|x(t), \theta)]^d S(t|x(t), \theta) \, d\mu(\theta). \quad (1.3.5)$$

It is apparent from (1.3.4) or (1.3.5) that without further restrictions, a variety of $h(t|x, \theta)$ and $\mu(\theta)$ pairs will be consistent with any $f(t, d|x)$.[10] Conditions under which a unique pair is determined are presented in Section 1.4. It is also apparent from (1.3.4) or (1.3.5) that given the functional form of either $h(t|x, \theta)$ or $\mu(\theta)$ and the data $(f(t, d|x))$ it is possible, at least in principle, to appeal to the theory of integral equations and solve for either $\mu(\theta)$ or $h(t|x, \theta)$. Current practice thus *overparameterizes* the duration model by specifying the functional form of both $h(t|x, \theta)$ and $\mu(\theta)$. In Section 1.4, we discuss methods for estimating $\mu(\theta)$ nonparametrically given that the functional form of $h(t|x, \theta)$ is specified up to a finite number of parameters. In the rest of this section we demonstrate consequences of incorrectly specifying either $h(t|x, \theta)$ or $\mu(\theta)$.

First consider the impact of incorrect treatment of time varying regressor variables. Many conventional econometric duration analyses are cavalier about

---

[10] Heckman and Singer (1982) present some examples. They are not hard to generate for anyone with access to tables of integral transforms.

such variables because introducing them into the analysis raises computational problems. Except for special time paths of variables the term

$$\int_0^t h(u|x(u),\theta)\,du,$$

which appears in survivor function (1.1.8) does not have a closed form expression. To evaluate it requires numerical integration.

To circumvent this difficulty, one of two expedients is often adopted (see, e.g. Lundberg, 1981, Cox and Lewis, 1966):

(i) Replacing time trended variables with their within spell average

$$\bar{x}(t) = \frac{1}{t}\int_0^t x(u)\,du \qquad t > 0,$$

(ii) Using beginning of spell values

$$x(0).$$

Expedient (i) has the undesirable effect of building spurious dependence between duration time $t$ and the manufactured regressor variable. To see this most clearly, suppose that $x$ is a scalar and $x(u) = a + bu$. Then clearly

$$\bar{x}(t) = a + \frac{b}{2}t,$$

and $t$ and $\bar{x}(t)$ are clearly linearly dependent. Expedient (ii) ignores the time inhomogeneity in the environment.[11]

To illustrate the potential danger from adopting these expedients consider the numbers presented in Table 1. These record Weibull hazards ((1.3.2) with $\gamma_2 = 0$ and $\lambda_1 = 0$) estimated on data for employment to nonemployment transitions using the CTM program described by Hotz (1983). In these calculations, unobservables are ignored. A job turnover model estimated using expedient (i) indicates weak negative duration dependence (column one row two) and a strong *negative* effect of high national unemployment rates on the rate of exiting jobs. The same model estimated using expedient (ii) now indicates (see column two) strong negative duration dependence and a strong *positive* effect of high national

---

[11] Moreover, in the multistate models with heterogeneity that are presented in Part II of this paper, treating $x(0)$ as exogenous is incorrect because the value of $x(0)$ at the start of the current spell depends on the lengths of outcomes of preceding spells. See the discussion in Section 2.2. This problem is also discussed in Flinn and Heckman (1982b, p. 62).

Table 1
Weibull model – Employment to nonemployment transitions
(absolute value of normal statistics in parentheses)[a]

| | Regressors fixed at average value over the spell (expedient i) | Regressors fixed at value as of start of spell (expedient ii) | Regressors vary freely |
|---|---|---|---|
| Intercept | 0.971 (1.535) | −3.743 (12.074) | −3.078 (8.670) |
| ln duration ($\gamma_1$) | −0.137 (1.571) | −0.230 (2.888) | −0.341 (3.941) |
| Married with spouse present? ( =1 if yes; = 0 otherwise) | −1.093 (2.679) | −0.921 (2.310) | −0.610 (1.971) |
| National unemployment rate | −1.800 (6.286) | 0.569 (3.951) | 0.209 (1.194) |

[a]*Source*: See Flinn and Heckman (1982b, p. 69).

unemployment rates on the rate of exiting employment. Allowing regressors to vary freely reveals that the strong negative duration dependence effect remains, but now the effect of the national unemployment rate on exit rates from employment is weak and statistically insignificant.

These empirical results are typical. Introducing time varying variables into single spell duration models is inherently dangerous, and ad hoc methods for doing so can produce wildly misleading results. More basically, separating the effect of time varying variables from duration dependence is only possible if there is "sufficient" independent variation in $x(t)$. To see this, consider hazard (1.3.2) with $\gamma_2 = 0$ and $x(t)$ scalar. Taking logs, we reach

$$\ln(h(t|x,\theta)) = x(t)\beta + \left( \frac{t^{\lambda_1} - 1}{\lambda_1} \right) \gamma_1 + \theta(t).$$

If

$$x(t) = \frac{t^{\lambda_1} - 1}{\lambda_1},$$

it is obviously impossible to separately estimate $\beta$ and $\gamma_1$. There is a classical multicollinearity problem. For a single spell model in a time inhomogeneous environment with general specifications for duration dependence, the analyst is at the mercy of the data to avoid such linear dependence problems. Failure to

Table 2
Sensitivity to misspecification of the mixing distribution $\mu(\theta)$.[a,b]

|  | Normal heterogeneity | Log normal heterogeneity | Gamma heterogeneity |
|---|---|---|---|
| Intercept | −3.92 | −13.2 | 5.90 |
|  | (2.8)[b] | (4.7) | (3.4) |
| ln duration ($\gamma$) | −0.066 | −0.708 | −0.576 |
|  | (0.15) | (0.17) | (0.17) |
| Age | 0.0036 | −0.106 | −0.202 |
|  | (0.048) | (0.03) | (0.06) |
| Education | 0.0679 | −0.322 | −0.981 |
|  | (0.233) | (0.145) | (0.301) |
| Tenure on previous job | −0.0512 | 0.00419 | −0.034 |
|  | (0.0149) | (0.023) | (0.016) |
| Unemployment benefits | −0.0172 | 0.0061 | −0.003 |
|  | (0.0036) | (0.0051) | (0.004) |
| Married | 0.833 | 0.159 | −0.607 |
| (0.1) | (0.362) | (0.30) | (0.496) |
| Unemployment rate | −26.12 | 25.8 | −17.9 |
|  | (9.5) | (10.3) | (11.2) |
| Ed. × age | −0.00272 | 0.00621 | 0.0152 |
|  | (0.0044) | (0.034) | (0.0053) |

[a] Sample size is 456.
[b] Standard errors in parentheses.
*Source*: See Heckman and Singer (1982) for further discussion of these numbers.

control for time varying regressor variables may mislead, but introducing such variables may create an identification problem.

Next we consider the consequence of misspecifying the distribution of unobservables. Table 2 records estimates of a Weibull duration model with three different specifications for $\mu(\theta)$ as indicated in the column headings. The estimates and inference vary greatly depending on the functional form selected for the mixing distribution. Trussell and Richards (1983) report similar results and exhibit similar sensitivity to the choice of the functional form of the conditional hazard $h(t|x,\theta)$ for a fixed $\mu(\theta)$.

## 1.4. Identification and estimation strategies

In our experience the rather vivid examples of the sensitivity of estimates of duration models to changes in specification presented in the previous section of the paper are the rule rather than the exception. This experience leads us to address the following three questions in this section of the paper:

(A) What features, if any, of $h(t|x,\theta)$ and/or $\mu(\theta)$ can be identified from the "raw data", i.e. $G(t|x)$?

(B) Under what conditions are $h(t|x, \theta)$ and $\mu(\theta)$ identified? i.e. how much a priori information has to be imposed on the model before these functions are identified?

(C) What empirical strategies exist for estimating $h(t|x, \theta)$ and/or $\mu(\theta)$ non-parametrically and what is their performance?

We assume a time homogeneous environment throughout. Little is known about the procedure proposed below for general time inhomogeneous environments.

### 1.4.1. Nonparametric procedures to assess the structural hazard $h(t|x, \theta)$

This section presents criteria that can be used to test the null hypothesis of no structural duration dependence and that can be used to assess the degree of model complexity that is required to adequately model the duration data at hand. The criteria to be set forth here can be viewed in two ways: As identification theorems and as empirical procedures to use with data.

We consider the following problem: $G(t|x)$ is estimated. We would like to infer properties of $G(t|x, \theta)$ without adopting any parametric specification for $\mu(\theta)$ or $h(t|x, \theta)$. We ignore any initial conditions problems. We further assume that $x(t)$ is time invariant.[12]

As a consequence of Proposition 1 proved in the preceding section, if $G(t|x)$ exhibits positive duration dependence for some intervals of $t$ values, $h(t|x, \theta)$ must exhibit positive duration dependence for some interval of $\theta$ values in those intervals of $t$. As noted in Section 1.3, this is so because the effect of scalar heterogeneity is to make the observed conditional duration distribution exhibit more negative duration dependence (more precisely, never less negative duration dependence) than does the structural hazard $h(t|x, \theta)$.

In order to test whether or not an empirical $G(t|x)$ exhibits positive duration dependence, it is possible to use the *total time on test statistic* (Barlow et al. 1972, p. 267). This statistic is briefly described here. For each set of $x$ values, constituting a sample of $I_x$ durations, order the first $k$ durations starting with the smallest

$$t_1 \leq t_2 \leq \cdots \leq t_k, \qquad 1 \leq k \leq I_x.$$

Let $D_{i:I_x} = [I_x - (i+1)](t_i - t_{i-1})$, where $t_0 \equiv 0$. Define

$$V_k = k^{-1} \sum_{i=1}^{k-1} \left[ \sum_{j=1}^{i} D_{j:I_x} \right] \Big/ k^{-1} \sum_{i=1}^{k} D_{i:I_x}.$$

---

[12] If $x(t)$ is not time invariant, additional identification problems arise. In particular, nonparametric estimation of $G(t|x(t))$ becomes much more difficult.

$V_k$ is called the cumulative total time on test statistic. If the observations are from a distribution with an increasing hazard rate, $V_k$ tends to be large. Intuitively, if $G(t|x)$ is a distribution that exhibits positive duration dependence, $D_{1:I_x}$ stochastically dominates $D_{2:I_x}$, $D_{2:I_x}$ stochastically dominates $D_{3:I_x}$, and so forth. Critical values for testing the null hypothesis of no duration dependence have been presented by Barlow and associates (1972, p. 269). This test can be modified to deal with censored data (Barlow et al. 1972, p. 302). The test is valuable because it enables the econometrician to test for positive duration dependence without imposing any arbitrary parametric structure on the data.

Negative duration dependence is more frequently observed in economic data. That this should be so is obvious from eq. (1.3.1) in the proof of Proposition 1. Even when the structural hazard has a positive derivative $\partial h(t|x, \theta)/\partial t > 0$, it often occurs that the second term on the right-hand side of (1.3.1) outweighs the first term. It is widely believed that it is impossible to distinguish structural negative duration dependence from a pure heterogeneity explanation of observed negative duration dependence when the analyst has access only to single spell data. To investigate duration distributions exhibiting negative duration dependence, it is helpful to distinguish two families of distributions.

Let $\mathscr{G}_1 = \{ G: -\ln[1 - G(t|x)] \text{ is concave in } t \text{ holding } x \text{ fixed} \}$. Membership in this class can be determined from the total time on test statistic. If $G_1$ is log concave, the $D_{i:I_x}$ defined earlier are stochastically increasing in $i$ for fixed $I_x$ and $x$. Ordering the observations from the largest to the smallest and changing the subscripts appropriately, we can use $V_k$ to test for log concavity.

Next let $\mathscr{G}_2 = \{ G: G(t|x) = \int (1 - \exp(-t\phi(x)\eta(\theta))) \, d\mu(\theta) \text{ for some probability measure } \mu \text{ on } [0, \infty] \}$. It is often erroneously suggested that $\mathscr{G}_1 = \mathscr{G}_2$, i.e. that negative duration dependence by a homogeneous population ($G \in \mathscr{G}_1$) cannot be distinguished from a pure heterogeneity explanation ($G \in \mathscr{G}_2$).

In fact, by virtue of Bernstein's theorem (see, e.g. Feller, 1971, p. 439–440) if $G \in \mathscr{G}_2$ it is completely monotone, i.e.

$$(-1)^n \frac{\partial^n}{\partial t^n} (1 - G(t|x)) \geq 0 \qquad \text{for } n \geq 1 \qquad \text{and all } t \geq 0 \tag{1.4.1}$$

and if $G(t|x)$ satisfies (1.4.1), $G(t|x) \in \mathscr{G}_2$.

Setting $n = 3$, (4.1) is violated if $(-1)^3 [\partial^3/\partial t^3](1 - G(t|x)) < 0$, i.e. if for some $t = t_0$

$$\left[ -\frac{\partial^2 h(t|x)}{\partial t^2} + 3h(t|x) \frac{\partial h(t|x)}{\partial t} - h^3(t|x) \right]_{t = t_0} > 0,$$

[see Heckman and Singer (1982) and Lancaster and Nickell (1980)].

Formal verification of (1.4.1) requires uncensored data sufficiently rich to support numerical differentiation twice. Note that if the data are right censored at $t = t^*$, we may apply (1.4.1) over the interval $0 < t \leq t^*$ provided that we define

$$1 - G^*(t|x) = \frac{\int_\theta [1 - \exp(-t\varphi(x)\eta(\theta))] \, d\mu(\theta)}{\int_\theta [1 - \exp(-t^*\varphi(x)\eta(\theta))] \, d\mu(\theta)},$$

and test whether

$$(-1)^n \frac{\partial^n}{\partial t^n}(1 - G^*(t|x)) \geq 0 \qquad \text{for } n \geq 1 \qquad \text{and} \qquad 0 < t \leq t^*. \quad (1.4.2)$$

Satisfaction of (1.4.2) for all $0 < t < t^*$ is only a necessary condition. It is sufficient only if $t^* \to \infty$.

Chamberlain (1980) has produced an alternative test of the *necessary* conditions that must be satisfied for a distribution to belong to $\mathcal{G}_2$ that does not require numerical differentiation of empirical distribution functions and that can be applied to censored data.

The key insight in his test is as follows. For $G \in \mathcal{G}_2$, the probability that $T > k$ is the survivor function

$$S(k|x) = \int_0^\infty [\exp - (k\varphi(x)\eta(\theta))] \, d\mu(\theta). \quad (1.4.3)$$

By a transformation of variables $z = \exp(-\phi(x)\eta(\theta))$, we may transform (1.4.3) for fixed $x$ to

$$S(k|x) = \int_0^1 z^k \, d\mu^*(z), \quad (1.4.4)$$

i.e. as the $k$th moment of a random variable defined on the unit interval.

From the solution to the classical Hausdorff moment problem (see, e.g. Shohat and Tamarkin, 1943, p. 9) it is known that there exists a $\mu^*(z)$ that satisfies (1.4.4) if

$$\Delta^k S(l|x) \geq 0 \qquad k, l = 0.1, \ldots, \infty \quad (1.4.5)$$

where

$$\Delta^0 S(l|x) = S(l|x),$$

$$\Delta^1 S(l|x) = S(l|x) - S(l+1|x),$$

$$\Delta^k S(l|x) = S(l|x) - \binom{k}{1} S(l|x)$$

$$+ \binom{k}{2} S(l+2|x) + \cdots$$

$$+ (-1)^k S(l+k|x).$$

Choosing equispaced intervals $(0,1,\ldots,[t^*])$ where $[t^*]$ is the nearest whole integer less than $t^*$, form the $S(l|x)$ functions $l = 0,\ldots,[t^*]$. Compute the survivor functions so defined and test a subset of the necessary conditions $(l = 1,\ldots,k)$. The estimated survivor functions are asymptotically normally distributed as the number of independent observations becomes large, and thus the asymptotic distribution of the subset of survivor functions is straightforward to compute. Failure of these necessary conditions implies that (1.4.4) and hence (1.4.3) cannot represent the underlying duration distribution. Thus it is possible to reject $G \in \mathscr{G}_2$ if some subset of conditions (1.4.5) is not satisfied. Note that if $x$ are i.i.d., the same test procedure can be applied to the full sample based on the unconditional survivor functions $S(l)$, $l = 1,\ldots,[t^*]$. In an important paper Robb (1984) extends this analysis by presenting a larger set of necessary conditions and by producing finite sample test statistics for the strengthened conditions.

It is important to note that (1.4.5) or (1.4.1) are rejection criteria. There are other models that may satisfy (1.4.1). For example

$$S(t) = \int_0^\infty [\exp(-t^\alpha \theta)] \, d\mu(\theta), \tag{1.4.6}$$

for $\alpha < 1$ is completely monotone. By Bernstein's theorem this distribution has a representation in $\mathscr{G}_2$.

### 1.4.2. *Nonparametric procedures to assess the mixing distribution*

In this subsection we consider some procedures that enable us to assess the modality of the mixing distribution. For expositional simplicity we suppress the dependence of $x$. Let $\mathscr{G}_3 = \{G : G(t) = \int_0^t g(u) \, du, \ g(t) = \int g(t|\theta) m(\theta) \, d\theta$ for some probability density $m(\theta)$ and $g(t|\theta) = k(t|\theta) v(t)$, where $k(t|\theta)$ is sign regular of order 2 $(SR_2)\}$.

Sign regularity means that if $t_1 < t_2$ and $\theta_1 < \theta_2$, then

$$\varepsilon_2 \det\begin{pmatrix} k(t_1|\theta_1) & k(t_1|\theta_2) \\ k(t_2|\theta_1) & k(t_2|\theta_2) \end{pmatrix} \geq 0,$$

where $\varepsilon_2$ is either $+1$ or $-1$. If $\varepsilon_2 = +1$, then $k(t|\theta)$ is called totally positive of order 2, abbreviated $TP_2$. From the point of view of inferring properties about the density of a mixing measure from properties of $g$, models with $SR_2$ conditional densities allow us to obtain lower bounds on the number of modes in $m(\theta)$ from knowledge of the number of modes in $g(v)$. Models for which $k(t|\theta) = g(t|\theta)/v(t)$ satisfies $SR_2$ include all members of the exponential family. In fact, for the exponential family, $k(t|\theta)$ is $TP_2$. Thus an assessment of modality of an estimated density, using, for example, the procedure of Hartigan and Hartigan (1985), is an important guide to specifying the characteristics of the density of $\Theta$.

Sign regular (particularly totally positive) kernels include many examples that are central to model specification in economics. In particular, if $d\nu(t)$ is any measure on $[0, +\infty)$ such that $\int_0^\infty [\exp(t\theta)] d\nu(t) < +\infty$ for $\theta \in \Theta$ (an ordered set), let

$$\beta(\theta) = \frac{1}{\displaystyle\int_0^\infty [\exp(t\theta)] d\nu(t)},$$

and, in what follows, set $d\nu(t) = v(t)dt$ and $g(t|\theta) = \beta(\theta)[\exp t\theta]\nu(t)$. Then the density $g(t) = \int \beta(\theta)\exp(t\theta)v(t)m(\theta)d\theta$ governs the observable durations of spells, $g(t|\theta)$ is a member of the exponential family, and $k(t|\theta) = \beta(\theta)\exp(t\theta)$ is $TP_2$ (Karlin, 1968). The essential point in isolating this class of duration densities is that knowledge of the number and character of the modes of $g/v$ implies that the density, $m$, of the mixing measure must have at least as many modes. In particular, if $g/v$ is unimodal, $m$ cannot be monotonic; it must have at least one mode. More generally, if $c$ is an arbitrary positive level and $(g(t)/v(t)) - c$ changes sign $k$ times as $t$ increases from 0 to $+\infty$, then $m(\theta) - c$ must change sign at least $k$ times as $\theta$ traverses the parameter set $\Theta$ from left to right (Karlin, 1968, p. 21).

The importance of this variation-diminishing character of the transformation $\int k(t|\theta)m(\theta)d\theta$ for modeling purposes is that if we assess the modality of $g$ using, for example, the method of Hartigan and Hartigan (1985), then because $v$ is given a priori, we know the modality of $g/v$, which in turn, implies restrictions on $m$ in fitting mixing densities to data. In terms of a strategy of fitting finite mixtures, a bimodal $g/v$ suggests fitting a measure with support at, say, five

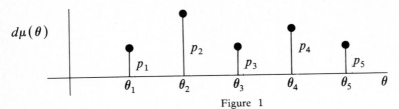

Figure 1

points to the data, but subject to the constraints that $p_1 < p_2$, $p_2 > p_3$, $p_3 < p_4$, and $p_4 > p_5$, as shown in Figure 1.

Subsequent specification of a mixing density $m(\theta)$ to describe the same data could proceed by fitting spline polynomials with knots at $\theta_1, \ldots, \theta_5$ to the estimated discrete mixing distribution.

### 1.4.3. Identifiability

In the preceding section, nonparametric procedures were proposed to assess qualitative features of conditional hazards and mixing distributions. These procedures aid in model selection but provide only qualitative guidelines to aid in model specification. In this section we consider conditions under which the conditional hazards and mixing distributions are identified. Virtually all that is known about this topic is for proportional hazard models (1.1.10) with scalar time invariant heterogeneity ($\theta(t) = \theta$) and time invariant regressors ($x(t) = x$). Thus identification conditions are presented for the model

$$h(t|x, \theta) = \psi(t)\varphi(x)\theta. \tag{1.4.7}$$

Before stating identifiability conditions, it is useful to define

$$Z(t) = \int_0^t \psi(u)\,du.$$

Then for the proportional hazard model (1.4.7) we have the following proposition due to Elbers and Ridder (1982).

*Proposition 2*

If (i) $E(\Theta) = 1$, (ii) $Z(t)$ defined on $[0, \infty)$ can be written as the integral of a nonnegative integrable function $\psi(t)$ defined on $[0, \infty)$, $Z(t) = \int_0^t \psi(u)\,du$, (iii) the set $S, x \in S$, is an open set in $R^k$ and the function $\varphi$ is defined on $S$ and is nonnegative, differentiable and nonconstant on $S$, then $Z$, $\varphi$, and $\mu(\theta)$ are identified. $\square$

The important point to note about Proposition 2 is that the identification analysis is completely nonparametric. No restrictions are imposed on $\psi$, $\varphi$ or $\mu(\theta)$ except for those stated in the proposition. Condition (iii) requires the existence of at least one continuous valued regressor variable defined on an interval. In the Appendix to their paper, Elbers and Ridder modify their proof to establish identifiability in models with only discrete valued regressor variables. However, the existence of at least one regressor variable in the model is essential in securing identification. Condition (i) requires the existence of a mean for the distribution of $\theta$. This assumption excludes many plausible fat tailed mixing distributions. Defining $\eta$ by $\theta = e^{\eta}$, condition (i) is not satisfied for distributions of $\eta$ which do not possess a moment generating function. For example Pareto $\Theta$ with finite mean, Cauchy distributed $\eta$ and certain members of the Gamma family fail condition (i).[13]

The requirement that $E(\Theta) < \infty$ is essential to the Elbers-Ridder proof. If this condition is not satisfied, and if no further restrictions are placed on $\mu(\theta)$, the duration model is not identified.

As an example of this point, suppose that the true model is Weibull with $Z_0(t) = t^{\alpha_0}$, $\varphi_0(x) = e^{x'\beta_0}$ and $\mu_0$ such that $E(\Theta) < \infty$. The survivor function for this model is

$$S_0(t|x) = 1 - G_0(t|x) = \int_0^{\infty} \left[\exp\left(-t^{\alpha_0}\exp(x'\beta_0)\right)\right] d\mu_0(\theta).$$

Define $\omega = t^{\alpha} \exp(x'\beta)$. Then

$$S_0(\omega) = \int_0^{\infty} \left[\exp(-\omega\theta)\right] d\mu_0(\theta) = L_0(\omega),$$

is the Laplace transform $(L)$ of random variable $\Omega$.

We have already noted in the discussion surrounding (4.6) that by virtue of Bernstein's theorem, if $0 < c < 1$,

$$L_1(\omega^c) = \int_0^{\infty} \left[\exp(-\omega^c\theta^*)\right] d\mu_1(\theta^*),$$

is completely monotone and is the Laplace transform of some random variable $\Theta^*$ where $E(\Theta^*) = \infty$. Thus

$$1 - G_1(t|x) = L_1(\omega^c) = L_0(\omega) = 1 - G_0(t|x),$$

[13] For

$$f(\eta) = \left[\exp(-\lambda\eta)\right]\frac{(\lambda\eta)^{r-1}}{\Gamma(r)} \qquad \text{and} \qquad \lambda < 1, \ E(\exp\eta) = \infty.$$

so a model with

$$Z_1(t) = t^{\alpha_0 c}, \varphi_1(x) = \exp(cx'\beta_0) \qquad \text{and } \mu_1 \text{ such that } E(\Theta^*) = \infty$$

explains the data as well as the original model ($\alpha = \alpha_0$, $\beta = \beta_0$ and $\mu = \mu_0$ with $E(\Theta) < \infty$).

The requirement that $E(\Theta) < \infty$ is overly strong. Heckman and Singer (1984a) establish identifiability when $E(\Theta) = \infty$ by restricting the tail behavior of the admissible mixing distribution. Their results are recorded in the following proposition.

*Proposition 3*

If

(i) The random variable $\Theta$ is nonnegative with a nondefective distribution $\mu$. For absolutely continuous $\mu$, the density $m(\theta)$ is restricted so that

$$m(\theta) \sim \frac{c}{(\ln \theta)^\gamma \theta^{(1+\varepsilon)} L(\theta)}, \tag{1.4.8}$$

as $\theta \to \infty$ where $c > 0$, $0 < \varepsilon < 1$ and $\gamma \geq 0$ where $L(\theta)$ is slowly varying in the sense of Karamata.[14] $\varepsilon$ is assumed known.

(ii) $Z \in \mathbf{Z} = \{ Z(t), t \geq 0 : Z(t)$ is a nonnegative increasing function with $Z(0) = 0$ and $\exists c > 0$ and $t_+$ not depending on the function $Z(t)$ such that $Z(t_+) = c$ where $c$ is a known constant$\}$.

(iii) $\phi \in \mathbf{\Phi} = \{\phi(x), x \in S : \phi$ is nonconstant on $S$, $\exists$ at least one coordinate $x_i$ defined on $(-\infty, \infty)$ such that $\phi(0, 0, \ldots, x_i, 0, \ldots)$ traverses $(0, \infty)$ as $x_i$ traverses $(-\infty, \infty)$, $0 \in S$, and $\phi(0) = 1\}$.

Then $Z$, $\varphi$ and $\mu$ are identified. $\square$ For proof, see Heckman and Singer (1984a).

Condition (i) is weaker than the Elbers and Ridder condition (i). $\Theta$ need not possess moments of any order nor need the distribution function $\mu$ have a density. However, in order to satisfy (i) the tails of the true distribution are assumed to die off at a fast enough rate and the rate is assumed known. The condition that $Z(t_+) = c$ for some $c > 0$ and $t_+ > 0$ for all admissible $Z$ plays an important role. This condition is satisfied, for example, by a Weibull integrated hazard since for all $\alpha$, $Z(1) = 1$. The strengthened condition (ii) substitutes for the weakened (i) in our analysis. Condition (iii) has identical content in both analyses. The essential idea in both is that $\phi$ varies continuously over an interval. In the

---

[14] Heckman and Singer (1984a) also present conditions for $\mu(\theta)$ that are not absolutely continuous. For a discussion of slowly varying functions see Feller (1971, p. 275).

absence of a finiteness of first moment assumption, Proposition 3 gives a conditional identification result. Given $\varepsilon$, it is possible to estimate $\psi$, $\mu$, $\varphi$ provided cross over condition (ii) is met.

A key assumption in the Heckman-Singer proof and in the main proof of Elbers and Ridder is the presence in the model of at least one exogenous variable that takes values in an interval of the real line. In duration models with no regressors or with only categorical regressors both proofs of identifiability break down. This is so because both proofs require exogenous variables that trace out the Laplace transform of $\Theta$ over some interval in order to uniquely identify the functions of interest.[15]

The requirement that a model possess at least one regressor is troublesome. It explicitly rules out an interaction detection strategy that cross-classifies the data on the basis of exogenous variables and estimates separate $Z(t)$ and $\mu(\theta)$ functions for different demographic groups. It rules out interactions between $x$ and $\theta$ and $x$ and $Z$.

In fact some widely used parametric hazard models can be identified together with the mixing distribution $\mu(\theta)$ even when no regressors appear in the model. Identification is secured under these conditions by specifying the functional form of the hazard function up to a finite number of unknown parameters and placing some restrictions on the moments of admissible $\mu$ distributions.

A general strategy of proof for this case is as follows [for details see Heckman and Singer (1984a)]. Assume that $Z'_\alpha(t)$ is a member of a parametric family of nonnegative functions and that the pair $(\alpha, \mu)$ is not identified. Assuming that $Z'_\alpha$ is differentiable to order $j$, nonidentifiability implies that the identities

$$1 = \frac{g_1(t)}{g_0(t)} = \frac{Z'_{\alpha_1}(t)\int_0^\infty \left[\theta\exp\left(-Z_{\alpha_1}(t)\theta\right)\right]d\mu_1(\theta)}{Z'_{\alpha_0}(t)\int_0^\infty \theta\left[\exp\left(-Z_{\alpha_0}(t)\right)\right]d\mu_0(\theta)},$$

$$\cdots$$

$$1 = \frac{g_1^{(j)}(t)}{g_0^{(j)}(t)},$$

for all $t \geq 0$ must hold for at least two distinct pairs $(\alpha_0, \mu_0)$, $(\alpha_1, \mu_1)$. We then derive contradictions. We demonstrate under certain stated conditions that these identities cannot hold unless $\alpha_1 = \alpha_0$. Then $\mu$ is identified by the uniqueness theorem for Laplace transforms.

---

[15]As previously noted, in their Appendix Elbers and Ridder (1982) generalize their proofs to a case in which all of the regressors are discrete valued. However, a regressor is required in order to secure identification.

To illustrate this strategy consider identifiability for the class of Box-Cox hazards (see eq. 1.3.2 with $\gamma_2 = 0$):

$$Z'_\alpha = \exp\left(\gamma\left(\frac{t^\lambda - 1}{\lambda}\right)\right).$$

For this class of hazard models there is an interesting tradeoff between the interval of admissible $\lambda$ and the number of bounded moments that is assumed to restrict the admissible $\mu(\theta)$. More precisely, the following propositions are proved in our joint work.

*Proposition 4*

For the true value of $\lambda$, $\lambda_0$, defined so that $\lambda_0 \leq 0$, if $E(\Theta) < \infty$ for all admissible $\mu$, and for all bounded $\gamma$, then the triple $(\gamma_0, \lambda_0, \mu_0)$ is uniquely identified. □ [For proof, see Heckman and Singer (1984a).]

*Proposition 5*

For the true value of $\lambda$, $\lambda_0$, such that $0 < \lambda_0 < 1$, if all admissible $\mu$ are restricted to have a common finite mean that is assumed to be known a priori ($E(\Theta) = m_1$) and a bounded (but not necessarily common) second moment $E(\Theta^2) < \infty$, and all admissible $\gamma$ are bounded, then the triple $(\gamma_0, \lambda_0, \mu_0)$ is uniquely identified. □ (For proof see Heckman and Singer, 1984a.)

*Proposition 6*

For the true value of $\lambda$, $\lambda_0$, restricted so that $0 < \lambda_0 < j$, $j$ a positive integer, if all admissible $\mu$ are restricted to have a common finite mean that is assumed to be known a priori ($E(\Theta) = m_1$) and $\varepsilon$ bounded (but not necessarily common) $(j+1)^{st}$ moment ($E(\Theta^{j+1}) < \infty$), and all admissible $\gamma$ are bounded, then the triple $(\gamma_0, \lambda_0, \mu_0)$ is uniquely identified. □ (For proof see Heckman and Singer, 1984a.)

It is interesting that each integer increase in the value of $\lambda_0 > 0$ requires an integer increase in the highest moment that must be assumed finite for all admissible $\mu$.

The general strategy of specifying a flexible functional form for the hazard and placing moment restrictions on the admissible $\mu$ works in other models besides the Box-Cox class of hazards. For example consider a nonmonotonic log logistic model used by Trussell and Richards (1985).

$$Z'(t) = \frac{(\lambda\alpha)(\lambda t)^{\alpha-1}}{1+(\lambda t)^\alpha}, \qquad \infty > \lambda, \qquad \alpha > 0. \tag{1.4.9}$$

*Proposition 7*

For hazard model (1.4.9), the triple $(\lambda_0, \alpha_0, \mu_0)$ is identified provided that the admissible $\mu$ are restricted to have a common finite mean $E(\Theta) = m_1 < \infty$. □ (For proof, see Heckman and Singer, 1984a.)

An interesting and more direct strategy of proof of identifiability which works for some of the hazard model specifications given above is due to Arnold and Brockett (1983). To illustrate their argument, consider the Weibull hazard

$$h(t|\theta) = \alpha t^{\alpha-1}\theta,$$

and mixing distributions restricted to those having a finite mean. Then a routine calculation shows that $\alpha$ may be calculated directly in terms of the observed survivor function via the recipe

$$\alpha = \lim_{t \to 0} \frac{\ln(tS'(t)/S(t))}{\ln t}.$$

The mixing distribution is then identified using the uniqueness theorem for Laplace transforms. Their proof of identifiability is constructive in that it also provides a direct procedure for estimation of $\mu(\theta)$ and $\alpha$ that is distinct from the procedure discussed below.

Provided that one adopts a parametric position on $h(t|\theta)$ these propositions show that it is possible to completely dispense with regressors. Another way to interpret these results is to note that since for each value of $x$, we may estimate $Z_\alpha$ and $\mu(\theta)$, it is not necessary to adopt proportional hazards specification (1.4.7) in order to secure model identification. All that is required is a conditional (on $x$) proportional hazards specification. $Z$ and $\mu$ may be arbitrary functions of $x$.

Although we have no theorems yet to report, it is obvious that it should be possible to reverse the roles of $\mu(\theta)$ and $h(t|\theta)$: i.e. if $\mu(\theta)$ is parameterized it should be possible to specify conditions under which $h(t|\theta)$ is identified nonparametrically.

The identification results reported here are quite limited in scope. First, as previously noted in Section 1.3, the restriction that the regressors are time invariant is crucial. If the regressors contain a common (to all observations) time trended variable, $\varphi$ can be identified from $\psi$ only if strong functional form assumptions are maintained so that $\ln \psi$ and $\ln \varphi$ are linearly independent. Since one cannot control the external environment, it is always possible to produce a $\psi$ function that fails this linear independence test. Moreover, even when $x(t)$ follows a separate path for each person, so that there is independent variation between $\ln \psi(t)$ and $\ln \varphi(t)$, at least for some observations, a different line of proof is required than has been produced in the literature.

Second, and more important, the proportional hazard model is not derived from an economic model. It is a statistically convenient model. As is implicit from the models presented in Section 1.2 and as will be made explicit in Section 1.6 duration models motivated by economic theory cannot in general be cast into a proportional hazards mold. Accordingly, the identification criteria discussed in this section are of limited use in estimating explicitly formulated economic models. In general, the hazard functions produced by economic theory are not separable as is assumed in (1.4.7).

Research is underway on identifiability conditions for nonseparable hazards. As a prototype we present the following identification theorem for a specific nonseparable hazard.

*Proposition 8*

Nonseparable model with (i) $Z_\alpha(t) = t^{(\alpha x)^2 + \theta}$, (ii) density $w_\beta(x|\theta) = (\theta + \beta)[\exp-(\theta + \beta)x]$ and (iii) $\int \theta \, d\mu(\theta) < \infty$ is identified. $\square$ For proof, see Heckman and Singer (1983).

Note that not only is the hazard nonseparable in $x$ and $\theta$ but the density of $x$ depends on $\theta$ so that $x$ is not weakly exogenous with respect to $\theta$.

Before concluding this discussion of identification, it is important to note that the concept of identifiability employed in this and other papers is the requirement that the mapping from a space of (conditional hazards) $X$ (a restricted class of probability distributions) to (a class of joint frequency functions for durations and covariates) be one to one and onto. This formulation of identifiability is standard. In this literature there is no requirement of a metric on the spaces or of completeness. Such requirements are essential if consistency of an estimator is desired. In this connection, Kiefer and Wolfowitz (1956) propose a definition of identifiability in a metric space whereby the above-mentioned mapping is $1:1$ on the completion (with respect to a given metric) of the original spaces. Without some care in defining the original space, undesirable distributions can appear in the completions.

As an example, consider a Weibull hazard model with conditional survivor function given an observed $k$-dimensional covariate $x$ defined as

$$1 - G(t|x, \beta) = S(t|x) = \int_0^\infty \exp\left(- t^{\alpha_0}\left(\exp x'\beta_0\right)v\right) dF_0(v),$$

where

$$0 < \alpha_0 \leq A < +\infty,$$

$\beta \in$ compact subset of $k$-dimensional Euclidean space, and $F_0$ is restricted to be a probability distribution on $[0, +\infty)$ with $\int_0^\infty v \, dF_0(v) = 1$. As a specialization of Elbers and Ridder's (1982) general proof, $\alpha_0$, $\beta_0$ and $F_0$ are identified. Now consider the completion with respect to the Kiefer–Wolfowitz (1956) metric of the

Cartesian product of the parameter space of allowed $\alpha$ and $\beta$ values and the probability distributions on $[0, +\infty)$ satisfying $\int_0^\infty v\,\mathrm{d}F_0(v) = 1$. The completion contains distributions $F_1$ on $[0, +\infty)$ satisfying $\int_0^\infty v\,\mathrm{d}F_1(v) = \infty$. Now observe that if $S(t|x)$ has a representation as defined above for some $\alpha \in (0,1)$ and $F_0$ with mean 1, then it is also a completely monotone function of $t$. Thus we also have the representation

$$ S(t|x) = \int_0^\infty \left[ \exp\!\left( -t\!\left( \exp\!\left( x'\beta_1 \right) \right) v \right) \right] \mathrm{d}F_1(v), $$

but now $F_1$ must have an infinite mean. This implies that $(\alpha_0, \beta_0, F_0)$ and $(1, \beta_1, F_1)$ generate the same survivor function. Hence the model is not identifiable on the completion of a space where probability distributions are restricted to have a finite mean.

This difficulty can be eliminated by further restricting $F_0$ to belong to a uniformly integrable family of distribution functions. Then all elements in the completion with respect to the Kiefer–Wolfowitz and a variety of other metrics will also have a finite mean and identifiability is again ensured. The comparable requirement for the case when $E_0(V) = \infty$ is that (1.4.8) converges uniformly to its limit.

The a priori restriction of identifiability considerations to complete metric spaces is not only central to establishing consistency of estimation methods but also provides a link between the concept of identifiability as it has developed in econometrics and notions of identifiability which are directly linked to consistency as in the engineering literature on control theory.

### 1.4.4. Nonparametric estimation

Securing identifiability of a nonparametric model is only the first step toward estimating the model. At the time of this writing, no nonparametric estimator has been devised that consistently estimates the general proportional hazard model (1.4.7).

In Heckman and Singer (1984b) we consider consistent estimation of the proportional hazard model when $\psi(t)$ and $\varphi(x)$ are specified up to a finite number of parameters but $\mu(\theta)$ is unrestricted except that it must have either a finite mean and belong to a uniformly integrable family or satisfy tail condition (1.4.8) with uniform convergence. We verify sufficiency conditions due to Kiefer and Wolfowitz (1956) which, when satisfied, guarantee the existence of a consistent nonparametric maximum likelihood estimator. We analyze a Weibull model for censored and uncensored data and demonstrate how to verify the sufficiency conditions for more general models. The analysis only ensures the existence of a consistent estimator. The asymptotic distribution of the estimator is unknown.

Drawing on results due to Laird (1978) and Lindsey (1983a, b), we characterize the computational form of the nonparametric maximum likelihood estimator.[16] To state these results most succinctly, we define

$$t^* = \varphi(x) \int_0^t \psi(u) \, du = \varphi(x) Z(t).$$

For any fixed value of the parameters determining $\varphi(x)$ and $Z(t)$ in (1.4.7), $t^*$ conditional on $\theta$ is an exponential random variable i.e.

$$f(t^*|\theta) = \theta \exp(-t^*\theta) \qquad \theta \geq 0. \tag{1.4.10}$$

For this model, the following propositions can be established for the Nonparametric Maximum Likelihood Estimator (NPMLE).

*Proposition 9*

Let $I^*$ be the number of distinct $t^*$ values in the sample of $I(\geq I^*)$ observations. Then the NPMLE of $\mu(\theta)$ is a finite mixture with at most $I^*$ points of increase, i.e. for censored and uncensored data (with $d=1$ for uncensored observations)

$$f(t^*) = \sum_{i=1}^{I^*} \theta_i^d \exp(-t^*\theta_i) P_i,$$

where $P_i \geq 0$, $\sum_{i=1}^{I^*} P_i = 1$. $\square$

Thus the NPMLE is a finite mixture but in contrast to the usual finite mixture model, $I^*$ is estimated along with the $P_i$ and $\theta_i$. Other properties of the NPMLE are as follows.

*Proposition 10*

Assuming that no points of support $\{\theta_i\}$ come from the boundary of $\theta$ the NPMLE is unique. $\square$ (See Heckman and Singer, 1984b.)

*Proposition 11*

For uncensored data, $\hat{\theta}_{\min} = 1/t^*_{\max}$ and $\hat{\theta}_{\max} = 1/t^*_{\min}$ where "ˆ" denotes the NPMLE estimate, and $t^*_{\max}$ and $t^*_{\min}$ are, respectively, the sample maximum and

---

[16] In computing the estimator it is necessary to impose all of the identifiability conditions in order to secure consistent estimators. For example, in a Weibull model with $E(\Theta) < \infty$, it is important to impose this requirement in securing estimates. As our example in the preceding subsection indicated, there are other models with $E(\Theta) = \infty$ that will explain the data equally well. In large samples, this condition is imposed, for example, by picking estimates of $\mu(\theta)$ such that $|\int (1 - \hat{\mu}(\theta)) \, d\theta| < \infty$ or equivalently $|\int (1 - \hat{\mu}(\theta) \, d\theta|^{-1} > 0$. Similarly, if identification is secured by tail condition (1.4.8), this must be imposed in selecting a unique estimator. See also the discussion at the end of Section 1.4.3.

minimum values for $t^*$. For censored data, $\hat{\theta}_{min} = 0$ and $\hat{\theta}_{max} = 1/t^*_{min}$. □ (See Heckman and Singer, 1984b.)

These propositions show that the NPMLE for $\mu(\theta)$ in the proportional hazard model is in general unique and the estimated points of support lie in a region with known bounds (given $t^*$). In computing estimates one can confine attention to this region. Further characterization of the NPMLE is given in Heckman and Singer (1984b).

It is important to note that all of these results are for a given $t^* = Z(t)\varphi(x)$. The computational strategy we use fixes the parameters determining $Z(t)$ and $\varphi(x)$ and estimates $\mu(\theta)$. For each estimate of $\mu(\theta)$ so achieved $Z(t)$ and $\varphi(x)$ are estimated by traditional parametric maximum likelihood methods. Then fresh $t^*$ are generated and a new $\mu(\theta)$ is estimated until convergence occurs. There is no assurance that this procedure converges to a global optimum.

In a series of Monte Carlo runs reported in Heckman and Singer (1984b) the following results emerge.

(i) The NPMLE recovers the parameters governing $Z(t)$ and $\varphi(x)$ rather well.
(ii) The NPMLE does not produce reliable estimates of the underlying mixing distribution.
(iii) The estimated c.d.f. for duration times $G(t|x)$ produced via the NPMLE predicts the sample c.d.f. of durations quite well even in fresh samples of data with different distributions for the $x$ variables.

A typical run is reported in Table 3. The structural parameters $(\alpha_1, \alpha_2)$ are estimated rather well. The mixing distribution is poorly estimated but the within sample agreement between the estimated c.d.f. of $T$ and the observed c.d.f. is good. Table 4 records the results of perturbing a model by changing the mean of the regressors from 0 to 10. There is still close agreement between the estimated model (with parameters estimated on a sample where $X \sim N(0,1)$) and the observed durations (where $X \sim N(10,1)$).

The NPMLE can be used to check the plausibility of any particular parametric specification of the distribution of unobserved variables. If the estimated parameters of a structural model achieved from a parametric specification of the distribution of unobservables are not "too far" from the estimates of the same parameters achieved from the NPMLE, the econometrician would have much more confidence in adopting a particular specification of the mixing distribution. Development of a formal test statistic to determine how far is "too far" is a topic for the future. However, because of the consistency of the nonparametric maximum likelihood estimator a test based on the difference between the parameters of $Z(t)$ and $\varphi(x)$ estimated via the NPMLE and the same parameters estimated under a particular assumption about the functional form of the mixing distribution would be consistent.

Table 3
Results from a typical estimation

| $d\mu(\theta) = [\exp(\Delta\theta)\exp-(e^{\theta}/\beta)\,d\theta]/\Gamma(1/2)$ with $\Delta = 1/2$ $\beta = 1$ | | |
|---|---|---|
| True model | $\alpha_1 = 1$ | $\alpha_2 = 1$ |
| Estimated model | 0.9852 | 0.9846 |
|  | (0.0738)* | (0.1022)* |

where $Z(t) = t^{\alpha_1}$ and $\varphi(x) = \exp(\alpha_2 x)$
Sample size $L = 500$
Log likelihood $-1886.47$

|  |  | Estimated mixing distribution | | |
|---|---|---|---|---|
| Estimated $\theta_i$ | Estimated $P_i$ | Estimated c.d.f. | True c.d.f | Observed c.d.f. |
| $-12.9031$ | 0.008109 | 0.008109 | 0.001780 | 0.0020 |
| $-7.0938$ | 0.06524 | 0.07335 | 0.03250 | 0.0400 |
| $-4.0107$ | 0.1887 | 0.2621 | 0.1510 | 0.1620 |
| $-1.7898$ | 0.3681 | 0.6302 | 0.4366 | 0.4280 |
| $-0.0338$ | 0.3698 | 1.000 | 0.8356 | 0.8320 |

|  | Estimated cumulative distribution of duration vs. actual ($\hat{G}(t)$ vs. $G(t)$) | |
|---|---|---|
| Value of $t$ | Estimated $t$ c.d.f. | Observed c.d.f. |
| 0.25 | 0.1237 | 0.102 |
| 0.50 | 0.2005 | 0.186 |
| 1.00 | 0.3005 | 0.296 |
| 3.00 | 0.4830 | 0.484 |
| 5.00 | 0.5661 | 0.556 |
| 10.00 | 0.6675 | 0.660 |
| 20.00 | 0.7512 | 0.754 |
| 40.00 | 0.8169 | 0.818 |
| 99.00 | 0.8800 | 0.880 |

*The numbers reported below the estimates are standard errors from the estimated information matrix for $(\alpha, P, \theta)$ given $I^*$. As noted in the text these have no rigorous justification.

Table 4
Predictions on a fresh sample, $X \sim N(10,1)$
(The model used to fit the parameters is $X \sim N(0,1)$).

| Estimated cumulative distribution of duration vs. actual $\hat{G}(t)$ vs. $G(t)$ | | |
|---|---|---|
| Value of $t$ ($\times 10^5$) | Estimated $t$ c.d.f. | Observed c.d.f. |
| 1.0 | 0.1118 | 0.1000 |
| 4.0 | 0.2799 | 0.2800 |
| 8.0 | 0.3924 | 0.3920 |
| 10.0 | 0.4300 | 0.4360 |
| 25.0 | 0.5802 | 0.5740 |
| 100.0 | 0.7607 | 0.7640 |
| 300.0 | 0.8543 | 0.8620 |
| 5000.0 | 0.9615 | 0.9660 |

The fact that we produce a good estimator of the structural parameters while producing a poor estimator for $\mu$ suggests that it might be possible to protect against the consequences of misspecification of the mixing distribution by fitting duration models with mixing distributions from parametric families, such as members of the Pearson system, with more than the usual two parameters. Thus the failure of the NPMLE to estimate more than four or five points of increase for $\mu$ can be cast in a somewhat more positive light. A finite mixture model with five points of increase is, after all, a nine (independent) parameter model for the mixing distribution. Imposing a false, but very flexible, mixing distribution may not cause much bias in estimates of the structural coefficients. Morever, for small $I^*$, computational costs are *lower* for the NPMLE than they are for traditional parametric maximum likelihood estimators of $\mu(\theta)$. The computational costs of precise evaluation of $\mu(\theta)$ over "small enough" intervals of $\theta$ are avoided by estimating a finite mixtures model.

We conclude this section noting that the Arnold and Brockett (1983) estimator for $\alpha$ discussed in Section 1.4.3 circumvents the need to estimate $d\mu(\theta)$ and so in this regard is more attractive than the estimator discussed in this subsection. Exploiting the fact that $t^*$ is independent of $x$, it is possible to extend their estimator to accommodate models with regressors. (The independence conditions provide orthogonality restrictions from which it is possible to identify $\beta$.) However, it is not obvious how to extend their estimator to deal with censored data. Our estimator can be used without modification on censored data.

## 1.5.  Sampling plans and initial conditions problems

There are few duration data sets for which the start date of the sample coincides with the origin date of all sampled spells. Quite commonly the available data are random samples of interrupted spells or else are spells that begin after the start date of the sample. For interrupted spells one of the following duration times may be observed: (1) time in the state up to the sampling date ($T_b$), (2) time in the state after the sampling date ($T_a$), or (3) total time in a completed spell observed at the origin of the sample ($T_c = T_a + T_b$). Durations of spells that begin after the origin date of the sample are denoted $T_d$.

In this section we derive the density of each of these durations for time homogeneous and time inhomogeneous environments and for models with and without observed and unobserved explanatory variables. The main message of this section is that in general the distributions of each of the random variables $T_a$, $T_b$, $T_c$ and $T_d$ differ from the population duration distribution $G(t)$. Estimators based on the wrong duration distribution in general produce invalid estimates of the parameters of $G(t)$ and will lead to incorrect inference about the population duration distribution.

*1.5.1. Time homogeneous environments and models without observed and unobserved explanatory variables*[17]

We first consider the analytically tractable case of a single spell duration model without regressors and unobservables in a time homogeneous environment. To simplify notation we assume that the sample at our disposal begins at calendar time 0. Looking backward, a spell of length $t_b$ interrupted at 0 began $t_b$ periods ago. Looking forward, the spell lasts $t_a$ periods after the sampling date. The completed spell is $t_c = t_b + t_a$ in length. We ignore right censoring and assume that the underlying distribution is nondefective. (These assumptions are relaxed in Subsection 1.5.2 below.)

Let $k(-t_b)$ be the intake rate; i.e. $t_b$ periods before the sample begins, $k(-t_b)$ is the proportion of the population that enters the state of interest at time $\tau = -t_b$. The time homogeneity assumption implies that

$$k(-t_b) = k, \qquad \forall t_b. \tag{1.5.1}$$

Let $g(t) = h(t)\exp(-\int_0^t h(u)\,du)$ be the density of completed durations *in the population*. The associated survivor function is

$$S(t) = 1 - G(t) = \exp\left(-\int_0^t h(u)\,du\right).$$

The proportion of the population experiencing a spell at calendar time $\tau = 0$, $P_0$, is obtained by integrating over the survivors from each cohort i.e.

$$P_0 = \int_0^\infty k(-t_b)(1-G(t_b))\,dt_b = \int_0^\infty k(-t_b)\exp\left(-\int_0^{t_b} h(u)\,du\right)dt_b.$$

Thus the density of an interrupted spell of length $t_b$ is the ratio of the proportion surviving from those who entered $t_b$ periods ago to the total stock

$$f(t_b) = \frac{k(-t_b)(1-G(t_b))}{P_0} = \frac{k(-t_b)\exp\left(-\int_0^{t_b} h(u)\,du\right)}{P_0}. \tag{1.5.2}$$

Assuming $m = \int_0^\infty x g(x)\,dx < \infty$ (and so ruling out defective distributions) and

---

[17]See Cox (1962), Cox and Lewis (1966), Sheps and Menken (1973), Salant (1977) and Baker and Trevedi (1982) for useful presentations of time homogeneous models.

integrating the denominator of the preceding expression by parts, we reach the familiar expression [see, e.g. Cox and Lewis (1966)]

$$f(t_b) = \frac{(1 - G(t_b))}{m} = \frac{S(t_b)}{m} = \frac{1}{m} \exp\left( -\int_0^{t_b} h(u)\, du \right). \tag{1.5.3}$$

The density of sampled interrupted spells is *not* the same as the population density of completed spells.

The density of sampled completed spells is obtained by the following straightforward argument. In the population, the conditional density of $t_c$ given $0 < t_b < t_c$ is

$$g(t_c|t_b) = \frac{g(t_c)}{(1 - G(t_b))} = h(t_c)\exp\left( -\int_{t_b}^{t_c} h(u)\, du \right), \qquad t_c > t_b. \tag{1.5.4}$$

Using (1.5.3), the marginal density of $t_c$ in the sample is

$$f(t_c) = \int_0^{t_c} g(t_c|t_b) f(t_b)\, dt_b = \int_0^{t_c} \frac{g(t_c)}{m}\, dt_b \tag{1.5.5}$$

so

$$f(t_c) = \frac{t_c g(t_c)}{m}.$$

The density of the forward time $t_a$ can be derived from (1.5.4). Substitute for $t_c$ using $t_c = t_a + t_b$ and integrate out $t_b$ using density (1.5.3). Thus

$$f(t_a) = \int_0^{\infty} g(t_a + t_b|t_b) f(t_b)\, dt_b = \int_0^{\infty} \frac{g(t_a + t_b)}{m}\, dt_b$$

$$= \frac{1}{m} \int_{t_a}^{\infty} g(z)\, dz = \frac{(1 - G(t_a))}{m}$$

$$= \frac{S(t_a)}{m} = \frac{\exp\left( -\int_0^{t_a} h(u)\, du \right)}{m}. \tag{1.5.6}$$

So in a time homogeneous environment the functional form of $f(t_a)$ is identical to $f(t_b)$.

The following results are well known about the distributions of the random variables $T_a$, $T_b$ and $T_c$.

(i) If $g(t)$ is exponential with parameter $\theta$ (i.e. $g(t) = \theta \exp(-t\theta)$) then so are $f(t_a)$ and $f(t_b)$. The proof is immediate.

(ii) $E(T_a) = (m/2)(1 + (\sigma^2/m^2))$[18]

   where $\sigma^2 = E(T - m)^2 = \int_0^\infty (t - m)^2 g(t) \, dt$.

(iii) $E(T_b) = (m/2)(1 + (\sigma^2/m^2))$

   (since $T_a$ and $T_b$ have the same density)

(iv) $E(T_c) = m(1 + (\sigma^2/m^2))$[19]

   so $E(T_c) = 2E(T_a) = 2E(T_b)$

   and $E(T_c) > m$ unless $\sigma^2 = 0$.

(v) If $(-\ln(1 - G(t))/t)\uparrow$ in $t$, $\sigma^2/m^2 > 1$.

   (This condition is implied if $h(t) = g(t)/1 - G(t)$ is increasing in $t$ i.e. $h'(t) > 0$). In this case, $E(T_a) = E(T_b) > m$. (See Barlow and Proschan, 1975 for proof).

(vi) If $(-\ln(1 - G(t))/t)\downarrow t$, $\sigma^2/m^2 < 1$.

   (This condition is implied if $h'(t) < 0$.) In this case $E(T_a) = E(T_b) < m$. (See Barlow and Proschan, 1975 for proof).

Result (i) restates the classical result (see Feller, 1970) that if the population distribution of durations is exponential so are the sample distributions of $T_a$ and $T_b$. Result (iii) coupled with result (v) indicates that if the population distribution of durations exhibits positive duration dependence, the mean of interrupted spells ($T_b$) *exceeds* the population mean duration. Result (iii) coupled with (vi) reverses this ordering for duration distributions with negative duration dependence. Result

---

[18] Proof: $E(T_a) = \dfrac{1}{m} \int_0^\infty t_a (1 - G(t_a)) \, dt_a$.

Integrating by parts assuming that $E(T^2) = \int_0^\infty t^2 g(t) \, dt < \infty$, we obtain

$$E(T_a) = \frac{1}{m} \int_0^\infty \frac{t_a^2}{2} g(t_a) \, dt_a + \frac{1}{m}(1 - G(t_a)) \frac{t_a^2}{2} \Big|_0^\infty$$

$$= \frac{1}{m} \frac{(\sigma^2 + m^2)}{2} = \frac{m}{2}\left(1 + \frac{\sigma^2}{m^2}\right).$$

[19] $E(T_c) = \dfrac{1}{m} \int_0^\infty x^2 g(x) \, dx = \dfrac{\sigma^2 + m^2}{m} = m\left(1 + \dfrac{\sigma^2}{m^2}\right).$

(iv) indicates that sampled completed spells have a mean in excess of the population mean unless $\sigma^2 = 0$ (hence the term "length biased sampling") and that completed spells have a mean twice that of interrupted $(T_b)$ or partially completed forward spells $(T_a)$.

We next present the distribution of $T_d$, the duration time for spells that begin after the origin date of the sample. Let $\mathcal{T}$ denote the time a spell begins. The density of $\mathcal{T}$ is $k(\tau)$. Assuming that $\mathcal{T}$ and $T_d$ are independent the joint probability that a spell begins at $\mathcal{T} = \tau$ and lasts less than $t_d$ periods is

$$\Pr\{\mathcal{T} = \tau \text{ and } T_d < t_d\} = k(\tau)G(t_d).$$

Thus the density of $T_d$ in a time homogeneous environment is

$$f(t_d) = g(t_d). \tag{1.5.7}$$

The distributions of $T_a$, $T_b$ and $T_c$ are of a different functional form than the distribution of $T$. The only exception is the case in which $T$ is an exponential random variable with parameter $\lambda$; in this case $T_a$ and $T_b$ are also exponential with parameter $\lambda$. The distribution of $T_d$ has the same functional form as the distribution of $T$.

Thus in a typical longitudinal sample in which data are available for the completed portions of durations of spells in progress $(T_a)$ and on durations initiated after the origin date of the sample $(T_d)$, two different distributions are required to analyze the data.

It is common to "solve" the left censoring problem by assuming that $G(t)$ is exponential. The bias that results from invoking this assumption when it is false can be severe. As an example suppose that the population distribution of $t$ is Weibull so

$$g(t) = \alpha\varphi t^{\alpha-1}\exp(-\varphi t^\alpha) \qquad \varphi > 0, \qquad \alpha > 0. \tag{1.5.8}$$

Suppose that the sample data are on the completed portions of interrupted spells and that there is no right censoring so that using formula (1.5.6)

$$f(t_a) = \frac{\exp(-t^\alpha\varphi)}{\left(\dfrac{\Gamma\left(\dfrac{1}{\alpha}+1\right)}{\varphi^{1/\alpha+1}}\right)},$$

If it is falsely assumed that $g^*(t) = \lambda e^{-t\lambda}$, the maximum likelihood estimator of

$\lambda$ for a random sample of durations is

$$\hat{\lambda} = \left( \frac{\Sigma t_i}{I} \right)^{-1},$$

which has probability limit

$$\text{plim } \hat{\lambda} = \varphi^{(1/\alpha)} \frac{\Gamma\left( \dfrac{1}{\alpha} \right)}{\Gamma\left( \dfrac{2}{\alpha} \right)}.$$

For $\alpha = 2$,

$$\text{plim } \hat{\lambda} = (\varphi)^{1/2} \Gamma(1/2).$$

As another example, suppose the sample being analyzed consists of complete spells sampled at time zero (i.e. $T_c$) generated by an underlying population exponential density

$$g(t) = \lambda \exp(-t\lambda).$$

Then from (1.5.5)

$$f(t_c) = \lambda^2 t_c \exp(-\lambda t_c).$$

If it is falsely assumed that $g(t)$ characterizes the duration data and $\lambda$ is estimated by maximum likelihood

$$\text{plim } \hat{\lambda} = 2\lambda.$$

This is an immediate consequence of results (i) and (v) previously stated.

Continuing this example, suppose instead that a Weibull model is falsely assumed, i.e.

$$g^*(t) = \alpha t^{\alpha - 1} \varphi \exp(-t^\alpha \varphi)$$

and the parameters $\alpha$ and $\varphi$ are estimated by maximum likelihood. The maximum likelihood estimator solves the following equations,

$$\frac{1}{\hat{\varphi}} = \frac{\sum\limits_{i=1}^{I} t_i^{\hat{\alpha}}}{I},$$

$$\frac{1}{\hat{\alpha}} + \frac{\sum\limits_{i=1}^{I} \ln t_i}{I} = \frac{\hat{\varphi} \sum\limits_{i=1}^{I} (\ln t_i) t_i^{\hat{\alpha}}}{I},$$

so

$$\frac{1}{\hat{\alpha}} + \frac{\sum\limits_{i=1}^{I} \ln t_i}{I} = \frac{\sum\limits_{i=1}^{I} t_i^{\hat{\alpha}} \ln t_i}{\sum\limits_{i=1}^{I} t_i^{\hat{\alpha}}}. \tag{1.5.9}$$

Using the easily verified result that

$$\int_0^\infty t^{P-1}(\ln t)\exp(-t\lambda)\,dt = \lambda^{-P}\left\{\frac{\partial \Gamma(P)}{\partial P} - \ln \lambda \Gamma(P)\right\},$$

and the fact that in large samples plim $\hat{\alpha} = \alpha*$ is the value of $\alpha*$ that solves (1.5.9), $\alpha*$ is the solution to

$$\frac{1}{\alpha*} + E(\ln t) = \frac{E(t^{\alpha*}\ln t)}{E(t^{\alpha*})},$$

and we obtain the equation

$$\frac{1}{\alpha*} + \left(\frac{\partial \Gamma(P)}{\partial P}\bigg|_{P=2} - \ln \lambda\right) = \left(\frac{\partial \ln \Gamma(P)}{\partial P}\bigg|_{P=\alpha*+2} - \ln \lambda\right). \tag{1.5.10}$$

Using the fact that

$$\frac{\Gamma'(P+1)}{\Gamma(P+1)} = \frac{1}{P} + \frac{\Gamma'(P)}{\Gamma(P)},$$

and collecting terms, we may rewrite (1.5.10) as

$$\frac{1}{\alpha^*(\alpha^*+1)} + \frac{\partial \Gamma(P)}{\partial P}\bigg|_{P=2} = \frac{1}{\Gamma(P)} \frac{\partial \Gamma(P)}{\partial P}\bigg|_{P=\alpha^*+1}. \tag{1.5.11}$$

Since $\Gamma(2)=1$, it is clear that $\alpha^*=1$ is never a solution of this equation. In fact, since the left hand side is monotone decreasing in $\alpha^*$ and the right hand side is monotone increasing in $\alpha^*$, and since at $\alpha^*=1$, the left hand side exceeds the right hand side, the value of $\alpha^*$ that solves (1.5.11) exceeds unity. Thus if a Weibull model is fit by maximum likelihood to length biased completed spells generated by an exponential population model, in large samples positive duration dependence will always be found, i.e. $\alpha^*>1$.

It can also be shown that

$$\text{plim } \hat{\varphi} = \frac{\lambda^{\alpha^*-1}}{\Gamma(\alpha^*+2)}.$$

If the Weibull is fit to data on $T_a$ and $T_b$ generated from an exponential population, $\alpha^*=1$.

These examples dramatically illustrate the importance of recognizing the impact of the sampling plan on the distribution of observed durations. As a general proposition only the distribution of $T_d$ – the length of spells initiated after the origin date of the sample – is invariant to the sampling plan. As a short cut, one can obtain inefficient but consistent estimates of $G(t)$ by confining an empirical analysis to such spells.

However, in the presence of unobserved variables this strategy will in general produce inconsistent parameter estimates. We turn next to consider initial conditions problems in models with observed and unobserved explanatory variables.

### 1.5.2. The densities of $T_a$, $T_b$, $T_c$, and $T_d$ in time inhomogeneous environments for models with observed and unobserved explanatory variables

We define $k(\tau|x(\tau),\theta)$ to be the intake rate into a given state at calendar time $\tau$. We assume that $\theta$ is a scalar heterogeneity component and $x(\tau)$ is a vector of explanatory variables. It is convenient and correct to think of $k(\tau|x(\tau),\theta)$ as the density associated with the random variable $\mathcal{T}$ for a person with characteristics $(x(\tau),\theta)$. We continue the useful convention that spells are sampled at $\mathcal{T}=0$. The densities of $T_a$, $T_b$, $T_c$ and $T_d$ are derived for two cases: (a) conditional on a sample path $\{x(u)\}_{-\infty}^t$ and (b) marginally on the sample path $\{x(u)\}_{-\infty}^t$, (i.e. integrating it out). We denote the distribution of $\{x(u)\}_{-\infty}^t$ as $D(x)$ with associated density $dD(x)$.

The derivation of the density of $T_b$ conditional on $\{x(u)\}^0_{-\infty}$ is as follows. The proportion of the population in the state at time $\tau = 0$ is obtained by integrating over the survivors of each cohort of entrants. Thus

$$P_0(x) = \int_0^\infty \int_\theta k(-t_b|x(-t_b), \theta) \exp\left(-\int_0^{t_b} h(u|x(u-t_b), \theta) \, du\right) d\mu(\theta) \, dt_b.$$

$$(1.5.12)$$

Note that, unlike the case in the models analyzed in Section 1.5.1, this integral may exist even if the underlying distribution is defective provided that the $k(\cdot)$ factor damps the survivor function. We require

$$\lim_{\substack{\sup \\ \tau \to -\infty}} |\tau|^{1+\varepsilon} k(\tau) S(\tau) = 0 \qquad \text{for } \varepsilon > 0.$$

The proportion of people in the state with sample path $\{x(u)\}^0_{-\infty}$ whose spells are exactly of length $t_b$ is the set of survivors from a spell that initiated at $\tau = -t_b$ or

$$\int_\theta k(-t_b|x(-t_b), \theta) \exp\left(-\int_0^{t_b} h(u|x(u-t_b), \theta) \, du\right) d\mu(\theta).$$

Thus the density of $T_b$ conditional on $\{x(u)\}^0_{-\infty}$ is

$$f\left(t_b|\{x(u)\}^0_{-\infty}\right)$$

$$= \frac{\int_\theta k(-t_b|x(-t_b), \theta) \exp\left(-\int_0^{t_b} h(u|x(u-t_b), \theta) \, du\right) d\mu(\theta)}{P_0(x)}.$$

$$(1.5.13)$$

The marginal density of $T_b$ (integrating out $x$) is obtained by an analogous argument: Divide the marginal flow rate as of time $\mathcal{T} = -t_b$ (the integrated flow rate) by the marginal (integrated) proportion of the population in the state at $\tau = 0$.

Thus defining

$$P_0 = \int_X P_0(x) \, dD(x),$$

where $X$ is the domain of integration for $x$, we write

$$f(t_b) = \frac{\int_X \int_\theta k[-t_b|x(-t_b), \theta] \times \exp\left(-\int_0^{t_b} h(u|x(u-t_b), \theta) \, du\right) d\mu(\theta) \, dD(x)}{P_0}. \tag{1.5.14}$$

Note that we use a function space integral to integrate out $\{x(u)\}^0_{-\infty}$. [See Kac (1959) for a discussion of such integrals.] Note further that one obtains an incorrect expression for the marginal density of $T_b$ if one integrates (5.13) against the population density of $x(dD(x))$. The error in this procedure is that the appropriate density of $x$ against which (1.5.13) should be integrated is a density of $x$ conditional on the event that an observation is in the sample at $\tau = 0$. By Bayes' theorem this density is

$$f(x|T_b > 0) = \left(\int_0^\infty f\left(t_b|\{x(u)\}^0_{-\infty}\right) dD(x) \, dt_b\right) \frac{P_0(x)}{P_0},$$

which is not in general the same as the density $dD(x)$. For proper distributions for $T_b$,

$$f(x|T_b > 0) = dD(x) \frac{P_0(x)}{P_0}.$$

The derivation of the density of $T_c$, the completed length of a spell sampled at $\mathcal{T} = 0$ is equally straightforward. For simplicity we ignore right censoring problems so that we assume that the sampling frame is of sufficient length that all spells are not censored and further assume that the underlying duration distribution is not defective. (But see the remarks at the conclusion of this section.) Conditional on $\{x(u)\}^\tau_{-\infty}$ and $\theta$ the probability that the spell began at $\tau$ is

$$k(\tau|x(\tau), \theta).$$

The conditional density of a completed spell of length $t$ that begins at $\tau$ is

$$h(t|x(\tau + t), \theta) \exp\left(-\int_0^t h(u|x(\tau + u), \theta) \, du\right).$$

For any fixed $\tau \leq 0$, $t_c$ by definition exceeds $-\tau$. Conditional on $x$, the probability that $T_c$ exceeds $\tau$ is $P_0(x)$. Thus, integrating out $\tau$, respecting the fact that $t_c > -\tau$

$$f\left(t_c | \{x(u)\}^{t_c}_{-\infty}\right) = \frac{\int_{-t_c}^0 \int_{\theta} k(\tau | x(\tau), \theta) h(t | x(\tau + t_c), \theta) \times \exp\left(-\int_0^{t_c} h(u | x(\tau + u), \theta) \, du\right) d\mu(\theta) \, d\tau}{P_0(x)}.$$ (1.5.15)

The marginal density of $T_c$ is

$$f(t_c) = \frac{\int_{-t_c}^0 \int_{X} \int_{\theta} k(\tau | x(\tau), \theta) h(t_c | x(\tau + t), \theta) \times \exp\left(-\int_0^{t_c} h(u | x(\tau + u), \theta) \, du\right) d\mu(\theta) \, dD(x) \, d\tau}{P_0}.$$ (1.5.16)

Ignoring right censoring, the derivation of the density of $T_a$ proceeds by recognizing that $T_a$ conditional on $\mathscr{T} \leq 0$ is the right tail portion of random variable $-\mathscr{T} + T_a$, the duration of a completed spell that begins at $\mathscr{T} = \tau$. The probability that the spell is sampled is $P_0(x)$. Thus the conditional density of $T_a = t_a$ given $\{x(u)\}^{t_a}_{-\infty}$ is obtained by integrating out $\tau$ and correctly conditioning on the event that the spell is sampled, i.e.

$$f\left(t_a | \{x(u)\}^{t_a}_{-\infty}\right) = \frac{\int_{-\infty}^0 \int_{\theta} k(\tau | x(\tau), \theta) h(t_a - \tau | x(t_a), \theta) \times \exp\left(-\int_0^{t_a - \tau} h(u | x(u + \tau), \theta) \, du\right) d\mu(\theta) \, d\tau}{P_0(x)},$$

(1.5.17)

and the corresponding marginal density is

$$f(t_a) = \frac{\int_{-\infty}^0 \int_{X} \int_{\theta} k(\tau | x(\tau), \theta) h(t_a - \tau | x(t_a), \theta) \times \exp\left(-\int_0^{t_a - \tau} h(u | x(u + \tau), \theta) \, du\right) d\mu(\theta) \, dD(x) \, d\tau}{P_0}.$$ (1.5.18)

Of special interest is the case $k(\tau | x, \theta) = k(x)$ in which the intake rate does not depend on unobservables and is constant for all $\tau$ given $x$, and in which $x$ is time invariant. Then (1.5.13) specializes to

$$f(t_b|x) = \frac{1}{m(x)} \int_{\theta} \exp\left(-\int_0^{t_b} h(u|x,\theta) \, du\right) d\mu(\theta),$$ (1.5.13′)

where

$$m(x) = \int_0^{\infty} \int_{\theta} \exp\left(-\int_0^z h(u|x,\theta) \, du\right) d\mu(\theta) \, dz.$$

This density is very similar to (1.5.3). Under the same restrictions on $k$ and $x$, (1.5.15) and (1.5.17) specialize respectively to

$$f(t_c|x) = \frac{t_c \int_{\theta} h(t_c|x,\theta) \exp\left(-\int_0^{t_c} h(u|x,\theta) \, du\right) d\mu(\theta)}{m(x)},$$ (1.5.15′)

which is to be compared to (1.5.5), and

$$f(t_a|x) = \frac{\int_{\theta} \exp\left(-\int_0^{t_a} h(u|x,\theta) \, du\right) d\mu(\theta)}{m(x)},$$ (1.5.17′)

which is to be compared to (1.5.6). For this special case all of the results (i)-(vi) stated in Section 1.5.1 go through with obvious redefinition of the densities to account for observed and unobserved variables.

*It is only for the special case of $k(\tau|x,\theta) = k(\tau|x)$ with time invariant regressors that the densities of $T_a$, $T_b$ and $T_c$ do not depend on the parameters of $k$.*

In order to estimate the parameters of $h(t|x,\theta)$ from data on $T_a$, $T_b$ or $T_c$ gathered in a time inhomogeneous environment for a model with unobservables, knowledge of $k$ is required. As long as $\theta$ appears in the conditional hazard and $k$ depends on $\theta$ or $\tau$ or if $x$ is not time invariant, $k$ must be specified along with $\mu(\theta)$ and $h(t|x,\theta)$.

The common expedient for "solving" the initial conditions problem for the density of $T_a$ – assuming that $G(t|x,\theta)$ is exponential – does not avoid the dependence of the density of $T_a$ on $k$ even if $k$ does not depend on $\theta$ as long as it depends on $\tau$ or $x(\tau)$ where $x(\tau)$ is not time invariant. Thus in the exponential case in which $h(u|x(u+\tau),\theta)=h(x(u+\tau),\theta)$, we may write (1.5.17) for the case $k=k(\tau|x(\tau))$ as

$$f\left(t_a|\{x(u)\}_{-\infty}^{t_a}\right)=\frac{\displaystyle\int_{-\infty}^{0}\int_{\theta}k(\tau|x(\tau))\exp\left(-\int_{0}^{-\tau}h(x(u+\tau),\theta)\,du\right)h(x(t_a),\theta)}{\displaystyle\int_{-\infty}^{0}\int_{\theta}k(\tau|x(\tau))\exp\left(-\int_{0}^{-\tau}h(x(u+\tau),\theta)\,du\right)d\mu(\theta)\,d\tau}.$$
$$\times\exp\left(-\int_{0}^{t_a}h(x(u),\theta)\,du\right)d\mu(\theta)\,d\tau$$

Only if $h(x(u+\tau),\theta)=h(x(u+\tau))$, so that unobservables do not enter the model (or equivalently that the distribution of $\Theta$ is degenerate), does $k$ cancel in the expression. In that case the numerator factors into two components, one of which is the denominator of the density. "$k$" also disappears if it is a time invariant constant that is functionally independent of $\theta$.[20]

At issue is the plausibility of alternative specifications of $k$. Although nothing can be said about this matter in a general way, for a variety of economic models, it is plausible that $k$ depends on $\theta, \tau$ and $x(\tau)$ and that the $x$ are not time invariant. For example, in a study of unemployment spells over the business cycle, the onset of a spell of unemployment is the result of prior job termination or entry into the workforce. So $k$ is the density of the length of a spell resulting from a prior economic decision. The same unobservables that determine unemployment are likely to determine such spells as well. In addition, it is odd to assume a time invariant general economic and person specific environment in an analysis of unemployment spells: Aggregate economic conditions change, and person specific variables like age, health, education and wage rates change over time. Similar arguments can be made on behalf of a more general specification of $k$ for most economic models.

---

[20] We note that one "short cut" procedure frequently used does not avoid these problems. The argument correctly notes that conditional on $\theta$ and the start date of the sample

$$(*)f\left(t_a|\{x(u)\}_{0}^{t_a},\theta\right)=h(x(t_a),\theta)\exp\left(-\int_{0}^{t_a}h(x(u),\theta)\,du\right).$$

This expression obviously does not depend on $k$. The argument runs astray by integrating this expression against $d\mu(\theta)$ to get a marginal (with respect to $\theta$) density. The correct density of $\theta$ is not $d\mu(\theta)$ and depends on $k$ by virtue of the fact that sample $\theta$ are generated by the selection mechanism that an observation must be in the sample at $\tau=0$. Precisely the same issue arises with regard to the distribution of $x$ in passing from (1.5.13) to (1.5.14). However, density $(*)$ can be made the basis of a simpler estimation procedure in a multiple spell setting as we note below in Section 2.2.

The initial conditions problem for the general model has two distinct components.

(i)   The functional form of $k(\tau|x(\tau), \theta)$ is not in general known. This includes as a special case the possibility that for some unknown $\tau^* < 0$, $k(\tau|x(\tau), \theta) \equiv 0$ for $\tau < \tau^*$. In addition, the value of $\tau^*$ may vary among individuals so that if it is unknown it must be treated as another unobservable.

(ii)  If $x(\tau)$ is not time invariant, its value may not be known for $\tau < 0$ so that even if the functional form of $k$ is known, the correct conditional duration densities cannot be constructed.

These problems exacerbate the problem of securing model identification. Assumptions made about the functional form of $k$ and the presample values of $x(\tau)$ inject a further source of arbitrariness into single spell model specification. Even if $x(\tau)$ is known for $\tau \leq 0$, $k$, $\mu$ and $h$ cannot all be identified nonparametrically.

The initial conditions problem stated in its most general form is intractable. However, various special cases of it can be solved. For example, suppose that the functional form of $k$ is known up to some finite number of parameters, but presample values of $x(\tau)$ are not. If the distribution of these presample values is known or can be estimated, one method of solution to the initial conditions problem is to define duration distributions conditional on past sample values of $x(\tau)$ but marginal on presample values, i.e. to integrate out presample $x(\tau)$ from the model using the distribution of their values. This suggests using (1.5.14) rather than (1.5.13) for the density of $T_b$. In place of either (1.5.15) or (1.5.16) for the density of $T_c$, this approach suggests using

$$
f\big(t_c|\{x(u)\}_0^{t_c}\big) = \frac{\displaystyle\int_{-t_c}^0 \int_\theta \int_{\{x(\tau):\,\tau<0\}} k\big(\tau|x(\tau),\theta\big) h\big(t_c|x(t_c+\tau),\theta\big)}{\times \exp\left(-\int_0^{t_c} h(u|x(\tau+u),\theta)\,\mathrm{d}u\right)\mathrm{d}D(x)\,\mathrm{d}\mu(\theta)\,\mathrm{d}\tau}{P_0},
$$

$$(1.5.19)$$

with a similar modification in the density of $T_a$.

This procedure requires either that the distribution of presample $x(\tau)$ be known or else that it be estimated along with the other functions in the model. The latter suggestion complicates the identification problem one further fold. The former suggestion requires either access to another sample from which it is possible to estimate the distribution of presample values of $x$ or else that it be possible to use within sample data on $x$ to estimate the distribution of the presample data, as would be possible, for example, if presample and within sample data distributions differed only by a finite order polynomial time trend.

Recall, however, that the distribution of $x$ within the sample is *not* the distribution of $x$ in the population, $D(x)$. This is a consequence of the impact of the sample selection rule on the joint distribution of $x$ and $T$.[21] The distribution of the $x$ within sample depends on the distribution of $\theta$, and the parameters of $h(t|x, \theta)$ and the presample distribution of $x$. Thus, for example, the joint density of $T_a$ and $x$ for $\tau > 0$ is

$$
f(t_a, x(\tau)|\tau \geq 0) = \frac{dD(x) \int_{-t_a}^{0} \int_{\theta} \int\int_{\{x:\tau<0\}} k(\tau|x(\tau), \theta) h(t_a - \tau|x(t_a), \theta)}{P_0} \times \exp\left(-\int_{0}^{t_a-\tau} h(u|x(u+\tau), \theta) \, du\right) dD(x) \, d\mu(\theta) \, d\tau ,
$$

(1.5.20)

so, the density of within sample $x(\tau)$ is

$$
f(x(\tau)|\tau \geq 0) = \int_{0}^{\infty} f(t_a, x(\tau)) \, dt_a = \frac{dD(x)}{P_0}
$$

$$
\times \int_{0}^{\infty} \int_{-t_a}^{0} \int_{\theta} \int\int_{\{x:\tau<0\}} k(\tau|x(\tau), \theta) h(t_a - \tau|x(t_a), \theta)
$$

$$
\times \exp\left(-\int_{0}^{t_a-\tau} h(u|x(u+\tau), \theta) \, du\right) dD(x) \, d\mu(\theta) \, d\tau \, dt_a.
$$

It is *this* density and not $dD(x)$ that is estimated using within sample data on $x$.

This insight suggests two further points. (1) By direct analogy with results already rigorously established in the choice based sampling literature (see, e.g. Manski and Lerman, 1977; Manski and McFadden, 1981, and Cosslett, 1981) more efficient estimates of the parameters of $h(t|x, \theta)$, and $\mu(\theta)$ can be secured using the joint densities of $T_a$ and $x$ since the density of within sample data depends on the structural parameters of the model as a consequence of the sample selection rule. (2) Access to other sources of data on the $x$ will be essential in order to "integrate out" presample $x$ via formulae like (1.5.19).

A partial avenue of escape from the initial conditions problem exploits $T_d$, i.e. durations for spells initiated after the origin date of the sample. The density of $T_d$

---

[21] Precisely the same phenomenon appears in the choice based sampling literature (see, e.g. Manski and Lerman, 1977, Manski and McFadden, 1981 and Cosslett, 1981). In fact the suggestion of integrating out the missing data is analogous to the suggestions offered in Section 1.7 of the Manski and McFadden paper.

conditional on $\{x(u)\}_0^{t_d+\tau_d}$ where $\tau_d > 0$ is the start date of the spell is

$$f\left(t_d|\{x(u)\}_0^{t_d+\tau_d}\right) = \frac{\int_0^\infty \int_\theta k(\tau|x(\tau),\theta)h(t_d|x(\tau+t_d),\theta) \times \exp\left(-\int_0^{t_d} h(u|x(\tau+u),\theta)\,du\right)d\mu(\theta)\,d\tau}{\int_0^\infty \int_\theta k(\tau|x(\tau),\theta)\,d\mu(\theta)\,d\tau}. \quad (1.5.21)$$

The denominator is the probability that $\mathcal{T} \geq 0$. Only if $k$ does not depend on $\theta$ will the density of $T_d$ not depend on the parameters of $k$. More efficient inference is based on the joint density of $\mathcal{T}$ and $t_d$

$$f\left(t_d,\tau|\{x(u)\}_0^{t_d+\tau_d}\right) = \frac{\int_\theta k(\tau|x(\tau),\theta)h(t_d|x(\tau+t_d),\theta) \times \exp\left(-\int_0^{t_d} h(u|x(\tau+u),\theta)\,du\right)d\mu(\theta)\,d\tau}{\int_0^\infty \int_\theta k(\tau|x(\tau),\theta)\,d\mu(\theta)\,d\tau}.$$

$$(1.5.22)$$

Inference based on (1.5.21) or (1.5.22) requires *fewer* a priori assumptions than are required to use data on $T_a$, $T_b$, or $T_c$. Unless $x$ is specified to depend on lagged values of explanatory variables, presample values of $x$ are not required. Since the start dates of spells are known, it is now in principle possible to estimate $k$ nonparametrically. Thus in samples with spells that originate after the origin date of the sample, inference is more robust.

As previously noted, the densities of the durations of $T_a$, $T_b$, $T_c$ and $T_d$ are in general different. However they depend on a common set of parameters. In samples with spells that originate after the start date of the sample, these cross density restrictions aid in solving the initial conditions problem because the parameters estimated from the relatively more informative density of $T_d$ can be exploited to estimate parameters from the other types of duration densities.

Before concluding this section, it is important to recall that we have abstracted from the problems raised by a finite length sample frame and the problems of right censoring. If the sampling frame is such that $\tau^* > \mathcal{T} > 0$, for example, the formulae for the durations of $T_a$, $T_c$ and $T_d$ presented above must be modified to account for this data generation process.

For example, the density of measured completed spells that begin after the start date of the sample incorporates the facts that $0 \leq \mathcal{T} \leq \tau^*$ and $T_d \leq \tau^* - \mathcal{T}$, i.e. that the onset of the spell occurs after $\tau = 0$ and that all completed spells must be of length $\tau^* - \mathcal{T}$ or less. Thus in place of (5.21) we write (recalling that $\tau_d$ is the start date of the spell)

$$
\begin{aligned}
&f\left(t_d | \{x(u)\}_0^{t_d+\tau_d}, T_d \le \tau^* - \mathcal{T}, \mathcal{T} \ge 0\right) \\
&= \frac{\displaystyle\int_0^{\tau^*-t_d} \int_\theta k(\tau|x(\tau),\theta) h(t_d|x(\tau+t_d),\theta) \times \exp\left(-\int_0^{t_d} h(u|x(\tau+u),\theta)\,du\right) d\mu(\theta)\,d\tau}{\displaystyle\int_0^{\tau^*} \int_0^{\tau^*-t_d} \int_\theta k(\tau|x(\tau),\theta) h(t_d|x(\tau+t_d),\theta) \times \exp\left(-\int_0^{t_d} h(u|x(\tau+u),\theta)\,du\right) d\mu(\theta)\,d\tau\,dt_d}.
\end{aligned}
$$

The denominator is the joint probability of the events $0 < \mathcal{T} < \tau^* - T_d$ and $0 < T_d < \tau^*$ which must occur if we are to observe a completed spell that begins during the sampling frame $0 < \mathcal{T} < \tau^*$. As $\tau^* \to \infty$, this expression is equivalent to the density in (5.21).

The density of right censored spells that begin after the start date of the sample is simply the joint probability of the events $0 < \mathcal{T} < \tau^*$ and $T_d > \tau^* - \mathcal{T}$, i.e.

$$
\begin{aligned}
&P\left(0 < \mathcal{T} < \tau^* \wedge T_d > \tau^* - \mathcal{T} | \{x(u)\}_0^{\tau^*}\right) \\
&= \int_0^{\tau^*} \int_{\tau^*-\tau}^\infty \int_\theta k(\tau|x(\tau),\theta) \exp\left(-\int_0^{\tau^*-t_d} h(u|x(\tau+u),\theta)\,du\right) d\mu(\theta)\,dt_d\,d\tau.
\end{aligned}
$$

The modifications required in the other formulae presented in this subsection to account for the finiteness of the sampling plan are equally straightforward. For spells sampled at $\tau = 0$ for which we observe presample values of the duration and post sample *completed* durations ($T_c$), it must be the case that (a) $\mathcal{T} \le 0$ and (b) $\tau^* - \mathcal{T} \ge T_c \ge -\mathcal{T}$ where $\tau^* > 0$ is the length of the sampling plan. Thus in place of (1.5.15) we write

$$
\begin{aligned}
&f\left(t_c | \{x(u)\}_{-\infty}^{t_c}, -\mathcal{T} \le T_c \le \tau^* - \mathcal{T}, \mathcal{T} \le 0\right) \\
&= \frac{\displaystyle\int_{-t_c}^{\tau^*-t_c} \int_\theta k(\tau|x(\tau),\theta) h(t_c|x(\tau+t_c),\theta) \times \exp\left(-\int_0^{t_c} h(u|x(\tau+u),\theta)\,du\right) d\mu(\theta)\,d\tau}{\displaystyle\int_{-\infty}^0 \int_{-\tau}^{\tau^*-\tau} \int_\theta k(\tau|x(\tau),\theta) h(t_c|x(\tau+t_c),\theta) \times \exp\left(-\int_0^{t_c} h(u|x(\tau+u),\theta)\,du\right) d\mu(\theta)\,dt_c\,d\tau}.
\end{aligned}
$$

The denominator of this expression is the joint probability of the events that $-\mathscr{T} < T_c < \tau^* - \mathscr{T}$ and $\mathscr{T} \leq 0$. For spells sampled at $\tau = 0$ for which we observe presample values of the duration and post-sample *right censored durations*, it must be the case that (a) $\mathscr{T} < 0$ and (b) $T_c \geq \tau^* - \mathscr{T}$ so the density for such spells is

$$f\left(t_c | \{x(u)\}_{-\infty}^{t_c}, T_c \geq \tau^* - \mathscr{T}, \mathscr{T} \leq 0\right)$$

$$= \int_{-\infty}^{0} \int_{\tau^* - \tau}^{\infty} \int_{\theta} k(\tau | x(\tau), \theta) h(t_c | x(\tau + t_c), \theta)$$

$$\times \exp\left(-\int_{0}^{t_c} h(u | x(\tau + u), \theta) \, du\right) d\mu(\theta) \, dt_c \, d\tau.$$

The derivation of the density for $T_a$ in the presence of a finite length sample frame is straightforward and for the sake of brevity is deleted. It is noted in Sheps–Menken (1973) (for models without regressors) and Flinn–Heckman (1982b) (for models with regressors) that failure to account for the sampling frame produces the wrong densities and inference based on such densities may be seriously misleading.

## 1.6.   *New issues that arise in formulating and estimating choice theoretic duration models*

In this section we briefly consider new issues that arise in the estimation of choice theoretic duration models. For specificity, we focus on the model of search unemployment in a time homogeneous environment that is presented in Section 1.2.2. Our analysis of this model serves as a prototype for a broad class of microeconomic duration models produced from optimizing theory.

We make the following points about this model assuming that the analyst has access to longitudinal data on $I$ independent spells of unemployment.

(A) Without data on accepted wages, the model of eqs. (1.2.10)–(1.2.21) is hopelessly underidentified even if there are no regressors or unobservables in the model.

(B) Even with data on accepted wages, the model is not identified unless the distribution of wage offers satisfies a recoverability condition to be defined below.

(C) For models without unobserved variables, the asymptotic theory required to analyze the properties of the maximum likelihood estimator of the model is nonstandard.

(D) Allowing for individuals to differ in observed and unobserved variables injects an element of arbitrariness into model specification, creates new identification and computational problems, and virtually guarantees that the hazard is not of the proportional hazards functional form.

(E) A new feature of duration models with unobservables produced by optimizing theory is that the support of $\Theta$ now depends on parameters of the model.

We consider each of the points in turn.

### 1.6.1. Point A

From a random sample of durations of unemployment spells in a model without observed or unobserved explanatory variables, it is possible to estimate $h_u$ (in eq. (1.2.15)) via maximum likelihood or Kaplan–Meier procedures (see, e.g. Kalbfleisch and Prentice, 1980, pp. 10–16). It is obviously not possible using such data alone to separate $\lambda$ from $(1 - F(rV))$ much less to estimate the reservation wage $rV$.

### 1.6.2. Point B

Given access to data on accepted wage offers it is possible to estimate the reservation wage $rV$. A consistent estimator of $rV$ is the minimum of the accepted wages observed in the sample

$$\widehat{rV} = \min\{W_i\}_{i=1}^I. \tag{1.6.1}$$

For proof see Flinn and Heckman (1982a).

Access to accepted wages does not secure identification of $F$. Only the truncated wage offer distribution can be estimated:

$$F(w|w \geq rV) = \frac{F(w) - F(rV)}{1 - F(rV)}, \qquad w \geq rV.$$

To recover an untruncated distribution from a truncated distribution with a known point of truncation requires further conditions. If $F$ is normal, such recovery is possible. If it is Pareto, it is not.[22] A sufficient condition that ensures recoverability is that $F(w)$ be real analytic over the support of $W$ so that by an analytic continuation argument, $F(w)$ can be continued outside of the region of truncation.[23] In the Pareto example, the support of $W$ is unknown.

---

[22] Thus if $F(w) = \varphi w^\beta$, $c_2 \leq w \leq \infty$, $\beta \leq -2$, where $\varphi = -(\beta + 1)/(c_2)^{\beta+1}$, $F(w|w \geq rV) = -(\beta + 1)w^\beta/(rV)^{\beta+1}$ so $\varphi$ (or $c_2$) does not appear in the conditional distribution.

[23] For a good discussion of real analytic functions, see Rudin (1974). If a function is real analytic, knowledge of the function over an interval is sufficient to determine the function over its entire domain of definition.

If the recoverability condition is not satisfied, it is not possible to determine $F$ even if $rV$ can be consistently estimated. Hence it is not possible to decompose $h_u$ in (1.2.15) into its constituent components.

If the recoverability condition is satisfied, it is possible to consistently estimate $F$, $\lambda$ and $rV$. From (1.2.13), it is possible to estimate a linear relationship between $r$ and $c$. The model is identified only by restricting $r$ or $c$ in some fashion. The most commonly used assumption fixes $r$ at a prespecified value.

### 1.6.3. *Point C*

Using density (1.2.18) in a maximum likelihood procedure creates a non-standard statistical problem. The range of random variable $W$ depends on a parameter of the model ($W \geq rV$). For a model without observed or unobserved explanatory variables, the maximum likelihood estimator of $rV$ is in fact the order statistic estimator (1.6.1). The likelihood based on (1.2.18) is monotonically increasing in $rV$, so that imposing the restriction that $W \geq rV$ is essential in securing maximum likelihood estimates of the model. Assuming that the density of $W$ is such that $f(rV) \neq 0$, the consistent maximum likelihood estimator of the remaining parameters of the model can be obtained by inserting $r\hat{V}$ in place of $rV$ everywhere in (1.2.18) and the *sampling distribution of this estimator is the same whether or not $rV$ is known* a priori *or estimated.* For a proof, see Flinn and Heckman (1982a). In a model with observed explanatory variables but without unobserved explanatory variables, a similar phenomenon occurs. However, at the time of this writing, a rigorous asymptotic distribution theory is available only for models with discrete valued regressor variables which assume a finite number of values.

### 1.6.4. *Point D*

Introducing observed and unobserved explanatory variables into a structural duration model raises the same sort of issues about ad hoc model specifications already discussed in the analysis of reduced form models in Section 1.3. However, there is the additional complication that structural restriction (1.2.13) produced by economic theory must be satisfied. One is not free to arbitrarily specify the parameters of the model.

It is plausible that $c$, $r$, $\lambda$ and $F$ in (1.2.13) all depend on observed and unobserved explanatory variables. Introducing such variables into the econometric search model raises three new problems.

(i) Economic theory provides no guidance on the functional form of the $c$, $r$, $\lambda$ and $F$ functions (other than the restriction given by (1.2.13)).[24] Estimates

---

[24]As discussed in Flinn and Heckman (1982a), some equilibrium search models place restrictions on the functional form of $F$.

secured from these models are very sensitive to the choice of these functional forms. Model identification is difficult to check and is very functional form dependent.

(ii) In order to impose the restrictions produced by economic theory to secure estimates, it is necessary to solve nonlinear eq. (1.2.13). Of special importance is the requirement that $V > 0$. If this restriction is not satisfied, the model cannot explain the data. If $V < 0$, an unemployed individual will not search. Closed form solutions exist only for special cases and in general numerical algorithms must be developed to impose or test these restrictions. Such numerical analysis procedures are costly even for a simple one spell search model and for models with more economic content often become computationally intractable. (One exception is a dynamic McFadden model with no restrictions between the choice and interarrival time distributions.)

(iii) Because of restrictions like (1.2.13), proportional hazard specifications (1.1.10) are rarely produced by economic models.

### 1.6.5. Point E

In the search model without unobserved variables, the restriction that $W \geq rV$ is an essential piece of identifying information. In a model with unobservable $\Theta$ introduced in $c$, $r$, $\lambda$ or $F$, $rV = rV(\theta)$ as a consequence of functional restriction (1.2.13). In this model, the restriction that $W \geq rV$ is replaced with an implicit equation restriction on the support of $\Theta$; i.e. for an observation with accepted wage $W$ and reservation wage $rV(\theta)$, the admissible support set for $\Theta$ is

$$\{ \theta : 0 \leq rV(\theta) \leq W \}.$$

This set is not necessarily connected.

The left hand side of the inequality states the requirement that must be satisfied if search is undertaken ($rV > 0$ for $r > 0$). The right hand side of the inequality states the requirement that accepted wages must exceed reservation wages. Unless this restriction is imposed on the support of $\Theta$, the structural search model is not identified. (See Flinn and Heckman, 1982a.)[25]

Thus in a duration model produced from economic theory not only is the conditional hazard $h(t|x(t), \theta)$ unlikely to be of the proportional hazard functional form, but the support of $\Theta$ will depend on parameters of the model. The mixing distribution representations presented in Section 1.4.4 above are unlikely to characterize structural duration models. Accordingly, the nonparametric iden-

---

[25] Kiefer and Neumann (1981) fail to impose this requirement in their discrete time structural search model so their proposed estimator is inconsistent. See Flinn and Heckman, 1982c.

tification and estimation strategies presented in Section 1.4 require modification before they can be applied to explicit economic models.

## 2.   Multiple spell models

The single spell duration models discussed in Part I are the principal building blocks for the richer, more behaviorally interesting models presented in this part of the paper. Sequences of birth intervals, work histories involving movements among employment states, the successive issuing of patents to firms and individual criminal victimization histories are examples of multiple spell processes which require a more elaborate statistical framework than the one presented in Part I.

In this part of the paper we confine our attention to new issues that arise in the analysis of multiple spell data. Issues such as the sensitivity of empirical estimates to ad hoc specifications of mixing distributions and initial conditions problems which also arise in multiple spell models are not discussed except in cases where access to multiple spell data aid in their resolution.

This part of the paper is in two sections. In Section 2.1 we present a unified statistical framework within which a rich variety of discrete state continuous time processes can be formulated and analyzed. We indicate by example how specializations of this framework yield a variety of models, some of which already appear in the literature. We do not present a complete analysis of multiple spell processes including their estimation and testing on data generated by various sampling processes because at the time of this writing too little is known about this topic.

Section 2.2 considers in somewhat greater detail a class of multiple spell duration models that have been developed for the analysis of event history data. In this Section we also consider some alternative approaches to initial conditions problems and some alternative approaches to controlling for unobserved variables that are possible if the analyst has access to multiple spell data.

### 2.1.   A unified framework

#### 2.1.1.   A general construction

To focus on main issues, in this section we ignore models with unobserved variables. We retain the convention that the sample at our disposal starts at calendar time $\tau = 0$.

Let $\{Y(\tau), \tau > 0\}$, $Y(\tau) \in \overline{N}$ where $\overline{N} = \{1, \dots, C\}$, $C < \infty$, be a finite state continuous time stochastic process. We define random variable $R(j)$, $j \in$

$\{1,\ldots,\infty\}$ as the value assumed by $Y$ at the $j$th transition time. $Y(\tau)$ or $R(j)$ is generated by the following sequence.

(i) An individual begins his evolution in a state $Y(0) = R(0) = r(0)$ and waits there for a random length of time $T_1$ governed by a conditional survivor function

$$P(T_1 > t_1 | r(0)) = \exp\left(-\int_0^{t_1} h(u|x(u), r(0)) \, du\right).$$

As before $h(u|x(u), r(0))$ is a calendar time (or age) dependent function and we now make explicit the origin state of the process.

(ii) At time $T(1) = \tau(1)$, the individual moves to a new state $R(1) = r(1)$ governed by a conditional probability law

$$P(R(1) = r(1) | \tau(1), r(0)),$$

which may also be age dependent.

(iii) The individual waits in state $R(1)$ for a random length of time $T_2$ governed by

$$P(T_2 > t_2 | \tau(1), r(1), r(0)) = \exp\left(-\int_0^{t_2} h(u|x(u + \tau(1)), r(1), r(0)) \, du\right).$$

Note that one coordinate of $x(u)$ may be $u + \tau(1)$, and that $\mathcal{T}(2) - \mathcal{T}(1) = T_2$. At the transition time $\mathcal{T}(2) = \tau(2)$ he switches to a new state $R(2) = r(2)$ where the transition probability

$$P(R(2) = r(2) | \tau(1), \tau(2), r(1), r(0)),$$

may be calendar time dependent.

Continuing this sequence of waiting times and moves to new states gives rise to a sequence of random variables

$$R(0) = r(0), \mathcal{T}(1) = \tau(1), R(1) = r(1), \mathcal{T}(2) = \tau(2), R(2) = r(2),\ldots$$

and suggests the definitions

$$Y(\tau) = R(k) \qquad \text{for} \qquad \tau(k) \le \tau < \tau(k+1),$$

where $R(k), k = 0, 1, 2,\ldots$ is a discrete time stochastic process governed by the conditional probabilities

$$P(R(k) = r(k) | t_k, r_{k-1}),$$

where

$$t_k = (t_1, \ldots, t_k) \qquad \text{and} \qquad r_{k-1} = (r(0), \ldots, r(k-1)).$$

$T_k = \mathscr{T}(k) - \mathscr{T}(k-1)$ is governed by the conditional survivor function,

$$P(T_k \geq t_k | t_{k-1}, r_{k-1}) = \exp\left(-\int_0^{t_k} h(u|x(u + \tau(k-1)), t_{k-1}, r_{k-1}) \, du\right).$$

### 2.1.2.  Specializations of interest

We now present a variety of special cases to emphasize the diversity of models encompassed by the preceding construction.

*2.1.2.1.  Repeated events of the same kind.*  This is a one state process, e.g. births in a fertility history. $R(\cdot)$ is a degenerate process and attention focuses on the sequence of waiting times $T_1, T_2, \ldots$.

One example of such a process writes

$$P(T_k > t_k | t_{k-1}) = \exp\left(-\int_0^{t_k} h_k(u|x(u + \tau(k-1))) \, du\right).$$

The hazard for the $k$th interval depends on the number of previous spells. This special form of dependence is referred to as *occurrence dependence*. In a study of fertility, $k-1$ corresponds to birth parity for a woman at risk. Heckman and Borjas (1980) consider such models for the analysis of unemployment.

Another variant writes the hazard of a current spell as a function of the mean duration of previous spells, i.e. for spell $j > 1$

$$h(u|x(u + \tau(j-1)), t_{j-1}) = h\left(u \bigg| \frac{1}{j-1} \sum_{i=1}^{j-1} t_i, \tau(j-1) + u\right).$$

[See, e.g. Braun and Hoem (1979).]

Yet another version of the general model writes for the $j$th spell

$$h(u|x(u + \tau(j-1)), t_{j-1}) = h_j(u|x, t_1, t_2, \ldots, t_{j-1}).$$

This is a model with both occurrence dependence and lagged duration dependence, where the latter is defined as dependence on lengths of preceding spells.

A final specification writes

$$h(u|x(u + \tau(j-1)), t_{j-1}) = h(x(u + \tau(j-1))).$$

For spell $j$ this is a model for independent non-identically distributed durations; and $Y(\tau)$ is a nonstationary renewal process.

*2.1.2.2. Multistate processes.* Let

$$P(R(k) = r(k)|t_k, r_{k-1}) = m_{r(k-1), r(k)},$$

where

$$\|m_{ij}\| = M,$$

is a finite stochastic matrix

$$P(T_k > t_k|t_{k-1}, r_{k-1}) = \exp(-\lambda_{r(k-1)}t_k),$$

where the elements of $\{\lambda_i\}$ are positive constants. Then $Y(\tau)$ is a time homogeneous Markov chain with constant intensity matrix

$$Q = \Lambda(M - I)$$

where

$$\Lambda = \begin{pmatrix} \lambda_1 & & \boldsymbol{O} \\ & \ddots & \\ \boldsymbol{O} & & \lambda_C \end{pmatrix},$$

and $C$ is the number of states in the chain.[26]

In the dynamic McFadden model for a stationary environment presented in Section 1.2.3, $M$ has the special structure $m_{ij} = m_{lj} = P_j$ for all $i$ and $l$; i.e. the origin state is irrelevant in determining the destination state. This restricted model can be tested against a more general specification.[27]

A time inhomogeneous semi-Markov process emerges as a special case of the general model if we let

$$P(R(k) = r(k)|t_k, r_{k-1}, \tau(k-1)) = \pi_{r(k-1), r(k)}(\tau(k), t_k),$$

---

[26] Note that without further restrictions on the elements of $M$, it is not possible to separate $\lambda_i$ from $(m_{ii} - 1)$ so that one might as well normalize $m_{ii} = 0$.

[27] Note that in the McFadden model it is not necessary to normalize $m_{ii} = 0$ to identify $\lambda_i$ because of the cross row restrictions on $M$.

where

$$\|\pi_{ij}(\tau, u)\| = \Pi(\tau, u),$$

is a two parameter family of time ($\tau$) and duration ($u$) dependent stochastic matrices with each element a function $\tau$ and $u$ and

$$m_{ii} = 0.$$

We further define

$$P(T_k > t_k | t_{k-1}, r_{k-1}, \tau(k-1)) = \exp\left(-\int_0^{t_k} h(u | r_{(k-1)}, \tau(k-1)) \, du\right).$$

With this restricted form of dependence, $Y(\tau)$ is a time inhomogeneous semi-Markov process. (Hoem, 1972, provides a nice expository discussion of such processes.)

Moore and Pyke (1968) consider the problem of estimating a time inhomogeneous semi-Markov model without observed or unobserved explanatory variables. The natural estimator for a model without restrictions connecting the parameters of $P(R(k) = r(k) | t_k, r_{k-1}, \tau(k-1))$ and $P(\dot{T}_k > t_k | t_{k-1}, r_{k-1}, \tau(k-1))$ breaks the estimation into two components.

(i) Estimate $\Pi$ by using data on transitions from $i$ to $j$ for observations with transitions having identical (calendar time $\tau$, duration $u$) pairs. A special case of this procedure for a model with no duration dependence in a time homogeneous environment pools $i$ to $j$ transitions for all spells to estimate the components of $M$ (see also Billingsley, 1961). Another special case for a model with duration dependence in a time homogeneous environment pools $i$ to $j$ transitions for all spells of a given duration.

(ii) Estimate $P(T_k > t_k | t_{k-1}, r_{k-1}, \tau(k-1))$ using standard survival methods (as described in Section 1.3 or in Lawless (1982)) on times between transitions.

These two estimators are consistent, asymptotically normal, and efficient and are *independent* of each other as the number of persons sampled becomes large. There is no efficiency gain from joint estimation. The same results carry over if $\Pi$ and $P(T_k > t_k | t_{k-1}, r_{k-1}, \tau(k-1))$ are parameterized (e.g. elements of $\Pi$ as a logit, $P(T_k > t_k | \cdot)$ as a general duration model) provided, for example, the regressors are bounded iid random variables. The two component procedure is efficient. However, if there are parameter restrictions connecting $\Pi$ and the conditional survivor functions, the two component estimation procedure produces inefficient estimators. If $\Pi$ and the conditional survivor functions depend on a common unobservable, a joint estimation procedure is required to secure a consistent random effect estimator.

## 2.2. *General duration models for the analysis of event history data*

In this section we present a multistate duration model for event history data, i.e. data that give information on times at which people change state and on their transitions. We leave for another occasion the analysis of multistate models designed for data collected by other sampling plans. This is a major area of current research.

An equivalent way to derive the densities of duration times and transitions for the multistate processes described in Section 2.1 that facilitates the derivation of the likelihoods presented below is based on the exit rate concept introduced in Part I. An individual event history is assumed to evolve according to the following steps.

(i) At time $\tau = 0$, an individual is in state $r_{(0)} = (i)$, $i = 1, \ldots, C$. Given occupancy of state $i$, there are $N_i \leq C - 1$ possible destinations.[28] The limit (as $\Delta t \to 0$) of the probability that a person who starts in $i$ at calendar time $\tau = 0$ leaves the state in interval $(t_1, t_1 + \Delta t)$ given regressor path $\{x(u)\}_0^{t_1 + \Delta t}$ and unobservable $\theta$ is the conditional hazard or escape rate

$$\lim_{\Delta t \to 0} \frac{P\left(t_1 < T_1 < t_1 + \Delta t \mid r_{(0)} = (i), \mathcal{T}(0) = 0, x(t_1), \theta, T_1 > t_1\right)}{\Delta t}$$

$$= h\left(t_1 \mid r_{(0)} = (i), \mathcal{T}(0) = 0, x(t_1), \theta\right). \tag{2.2.1}$$

This limit is assumed to exist.

The limit (as $\Delta t \to 0$) of the probability that a person starting in $r_{(0)} = (i)$ at time $\tau(0)$ leaves to go to $j \neq i$, $j \in N_i$ in interval $(t_1, t_1 + \Delta t)$ given regressor path $\{x(u)\}_0^{t_1 + \Delta t}$ and $\theta$ is

$$\lim_{\Delta t \to 0} \frac{P\left(t_1 < T_1 < t_1 + \Delta t, R(1) = j \mid r_{(0)} = (i), \mathcal{T}(0) = 0, x(t_1), \theta, T_1 \geq t_1\right)}{\Delta t}$$

$$= h\left(t_1, j \mid r_{(0)} = (i), \mathcal{T}(0) = 0, x(t_1), \theta\right). \tag{2.2.2}$$

---

[28] If some transitions are prohibited then $N_i < C - 1$.

From the laws of conditional probability

$$\sum_{j=1}^{N_i} h\big(t_1, j | \mathbf{r}_{(0)} = (i), \mathcal{T}(0) = 0, x(t_1), \boldsymbol{\theta}\big)$$
$$= h\big(t_1 | \mathbf{r}_{(0)} = (i), \mathcal{T}(0) = 0, x(t_1), \boldsymbol{\theta}\big).$$

(ii) The probability that a person starting in state $i$ at calendar time $\tau = 0$ survives to $T_1 = t_1$ is (from the definition of the survivor function in (1.8) and from hazard (2.2.1))

$$P\big(T_1 > t_1 | \mathbf{r}_{(0)} = (i), \mathcal{T}(0) = 0, \{x(u)\}_0^{t_1}, \boldsymbol{\theta}\big)$$
$$= \exp\bigg(-\int_0^{t_1} h\big(u | \mathbf{r}_{(0)} = (i), \mathcal{T}(0) = 0, x(u), \boldsymbol{\theta}\big) du\bigg).$$

Thus the density of $T_1$ is

$$f\big(t_1 | \mathbf{r}_{(0)} = (i), \mathcal{T}(0) = 0, \{x(u)\}_0^{t_1}, \boldsymbol{\theta}\big)$$
$$= -\frac{\partial P\big(T_1 > t_1 | \mathbf{r}_{(0)} = (i), \mathcal{T}(0) = 0, \{x(u)\}_0^{t_1}, \boldsymbol{\theta}\big)}{\partial t_1}$$
$$= h\big(t_1 | \mathbf{r}_{(0)} = i, \mathcal{T}(0) = 0, x(t_1), \boldsymbol{\theta}\big)$$
$$\times P\big(T_1 > t_1 | \mathbf{r}_{(0)} = (i), \mathcal{T}(0) = 0, \{x(u)\}_0^{t_1}, \boldsymbol{\theta}\big).$$

The density of the joint event $R(1) = j$ and $T_1 = t_1$ is

$$f\big(t_1, j | \mathbf{r}_{(0)} = (i), \mathcal{T}(0) = 0, \{x(u)\}_0^{t_1}, \boldsymbol{\theta}\big)$$
$$= h\big(t_1 | \mathbf{r}_{(0)} = (i), \mathcal{T}(0) = 0, x(t_1), \boldsymbol{\theta}\big)$$
$$\times P\big(T_1 > t_1 | \mathbf{r}_{(0)} = (i), \mathcal{T}(0) = 0, \{x(u)\}_0^{t_1}, \boldsymbol{\theta}\big).$$

This density is sometimes called a subdensity. Note that

$$\sum_{j=1}^{N_i} f\big(t_1, j | \mathbf{r}_{(0)} = (i), \mathcal{T}(0) = 0, \{x(u)\}_0^{t_1}, \boldsymbol{\theta}\big)$$
$$= f\big(t_1 | \mathbf{r}_{(0)} = (i), \mathcal{T}(0) = 0, \{x(u)\}_0^{t_1}, \boldsymbol{\theta}\big).$$

Proceeding in this fashion, one can define densities corresponding to each duration in the individual's event history. Thus, for an individual who starts in state $\mathbf{r}_{(m)}$ after his $m$th transition, the subdensity for $T_{m+1} = t_{m+1}$ and $R(m+1)$

$= j, \; j = 1, \ldots, N_m$ is

$$f\left(t_{m+1}, j | r_{(m)}, \mathscr{T}(m) = \tau(m), \{x(u)\}_0^{\tau(m+1)}, \theta\right),$$

where

$$\tau(m+1) = \sum_{n=1}^{m+1} t_n. \tag{2.2.3}$$

As in Part I we assume an independent censoring mechanism. The most commonly encountered form of such a mechanism is upper limit truncation on the final spell. As noted in Part I, in forming the likelihood we can ignore the censoring densities.

The conditional density of completed spells $T_1, \ldots, T_k$ and right censored spell $T_{k+1}$ given $\{x(u)\}_0^{\tau(k)+t_{k+1}}$ assuming that $\mathscr{T}(0) = 0$ is the exogenous start date of the event history (and so corresponds to the origin date of the sample) is, allowing for more general forms of dependence,

$$g\left(t_1, r(1), t_2, r(2), \ldots, t_k, r(k), t_{k+1} | \{x(u)\}_0^{\tau(k)+t_{k+1}}\right)$$

$$= \int_\theta \left\{ \prod_{i=1}^k f\left(t_i, r(i) | r_{(i-1)}, t_{(i-1)}, \tau(i-1), \{x(u)\}_{\tau(i-1)}^{\tau(i)}, \theta\right) \right. \tag{2.2.4}$$

$$\left. \left\{ P\left(T_{k+1} > t_{k+1} | r_{(k)}, t_k, \tau(k), \{x(u)\}_{\tau(k)}^{\tau(k)+t_k}, \theta\right) \right\} \right\} d\mu(\theta).$$

As noted in Section 1.5, it is unlikely that the origin date of the sample coincides with the start date of the event history. Let

$$\varphi\left(r(0), \mathscr{T}(0) = 0, r(1), t_{1a}, \{x(u)\}_{-\infty}^{\tau(1)}, \theta\right),$$

be the probability density for the random variables describing the events that a person is in state $R(0) = r(0)$ at time $\mathscr{T}(0) = 0$ with a spell of length $t_{1a}$ (measured after the start of the sample) that ends with an exit to state $R(1) = r(1)$ given $\{x(u)\}_0^{\tau(1)}$ and $\theta$. The derivation of this density in terms of the intake density $k$ appears in Section 1.5 (see the derivation of the density of $T_a$). The only new point to notice is that the $h$ in Section 1.5 should be replaced with the appropriate $h$ as defined in (2.2.2). The joint density of $(r(0), t_{1c}, r(1))$ the *completed spell* density sampled at $\mathscr{T}(0) = 0$ terminating in state $r(1)$ is defined analogously. For such spells we write the density as

$$\varphi\left(r(0), \mathscr{T}(0) = 0, t_{1c}, r(1), \{x(u)\}_{-\infty}^{\tau(1)}, \theta\right).$$

In a multiple spell model setting in which it is plausible that the process has been in operation prior to the origin data of the sample, intake rate $k$ introduced in Section 1.5 is the density of the random variable $\mathcal{T}$ describing the event "entered the state $r(0)$ at time $\mathcal{T} = \tau \leq 0$ and did not leave the state until $\mathcal{T} > 0$." The expression for $k$ in terms of exit rate (2.2.2) depends on (i) presample values of $x$ and (ii) the date at which the process began. Thus in principle given (i) and (ii) it is possible to determine the functional form of $k$. In this context it is plausible that $k$ depends on $\theta$.

The joint likelihood for $r(0), t_{1l}(l = a, c), r(1), t_2, \ldots, r(k), t_{k+1}$ conditional on $\theta$ and $\{x(u)\}_{-\infty}^{\tau(k) + t_{k+1}}$ for a right censored $k + 1$st spell is

$$g\left(r(0), t_{1l}, r(1), t_2, r(2), \ldots, t_k, r(k), t_{k+1}|\{x(u)\}_{-\infty}^{\tau(k) + t_{k+1}}, \theta\right)$$

$$= \varphi\left(r(0), \mathcal{T}(0) = 0, t_{1l}, r(1)|\{x(u)\}_{-\infty}^{\tau(1)}, \theta\right)$$

$$\cdot \left[\prod_{i=2}^{k} f\left(t_i, r(i)|r_{(i-1)}, t_{(i-1)}, \tau(i-1), \{x(u)\}_{\tau(i-1)}^{\tau(i)}, \theta\right)\right].$$

$$P\left(T_{k+1} > t_{k+1}|r_k, t_{(k)}, \tau(k-1), \{x(u)\}_{\tau(k)}^{\tau(k) + t_{k+1}}, \theta\right).$$

$$(2.2.5)$$

The marginal likelihood obtained by integrating out $\theta$ is,

$$g\left(r(0), t_{1l}, r(1), t_2, \ldots, t_k, r(k), t_{k+1}|\{x(u)\}_{-\infty}^{\tau(k) + t_{k+1}}\right)$$

$$= \int_{\theta} g\left(r(0), t_{1l}, r(1), t_2, \ldots, t_k, r(k), t_{k+1}|\{x(u)\}_{-\infty}^{\tau(k) + t_{k+1}}, \theta\right) d\mu(\theta).$$

$$(2.2.6)$$

Equation (2.2.6) makes explicit that the date of onset of spell $m + 1$ ($\mathcal{T}(m+1)$) depends on the durations of the preceding spells. Accordingly, in a model in which the exit rates (2.2.2) depend on $\theta$, the distribution of time varying $x$ variables (including date of onset of the spell) *sampled at the start of each spell* depends on $\theta$. Such variables are not (weakly) exogenous or ancillary in duration regression equations, and least squares estimators of models that include such variables are, in general, inconsistent. (See Flinn and Heckman, 1982b.) Provided that in the population $X$ is distributed independently of $\Theta$, time varying variables create no econometric problem for maximum likelihood estimators based on density (2.2.6) which accounts for the entire history of the process. However, a maximum likelihood estimator based on a density of the *last $n < k + 1$ spells* that conditions on $\tau(k + 1 - n)$ or $\{x(u)\}_{-\infty}^{\tau(k+1-n)}$ assuming they are independent of $\Theta$ is inconsistent.

Using (2.2.5) and conditioning on $T_{1l} = t_{1l}$ produces conditional likelihood

$$g\left(r(0), t_{1l}, r(1), t_2, \ldots, t_k, r(k), t_{k+1} | \{x(u)\}_{-\infty}^{\tau(k) + t_{k+1}}, \theta, T_{1l} = t_{1l}\right)$$

$$= \prod_{i=2}^{k} f\left(t_i, r(i) | r_{(i-1)}, t_{(i-1)}, \tau(i-1), \{x(u)\}_{\tau(i-1)}^{\tau(i)}, \theta\right). \qquad (2.2.7)$$

$$P\left(T_{k+1} > t_{k+1} | r_k, t_k, \tau(k), \{x(u)\}_{\tau(k)}^{\tau(k) + t_{k+1}}, \theta\right).$$

For three reasons, inference based on conditional likelihood (2.2.7) appears to be attractive (see Heckman, 1981b). (1) With this likelihood it is not necessary to specify or estimate the distribution $\mu(\theta)$. It thus appears possible to avoid one element of arbitrariness in model specification. (2) With this likelihood we avoid the initial conditions problem because $\varphi$ and $\{x(u)\}_{-\infty}^{\tau(1)}$ do not appear in density (2.2.7). (3) Treating $\theta$ as a parameter allows for arbitrary dependence between $\theta$ and $x$. These three reasons demonstrate the potential gains that arise from having multiple spell data.[29]

However for general duration distributions, inference based on (2.2.7) fit on panel data produces inconsistent estimators. This is so because the conditional likelihood function depends on person specific component $\theta$. Estimating $\theta$ as a parameter for each person along with the other parameters of the model produces inconsistent estimators of all parameters if $k < \infty$ in the available panel data because the likelihood equations are not in general separable (see Neyman and Scott, 1948). In most panel data sets, $k$ is likely to be small.

No Monte Carlo study of the performance of the inconsistent estimator has been performed. By analogy with the limited Monte Carlo evidence reported in Heckman (1981b) for a fixed effect discrete choice model if $x$ does not contain lagged values of the dependent variable, the inconsistency is likely to be negligible even if the likelihood is fit on short panels. The inconsistency issue may be a matter of only theoretical concern.

Chamberlain (1984) drawing on results due to Andersen (1973, 1980) presents a class of multiple spell duration models for which it is possible to find sufficient or ancillary statistics for $\theta$. Estimation within this class of models avoids the inconsistency problem that arises in likelihoods based on (2.2.7). The class of exponential family distributions for which the Andersen–Chamberlain procedures are valid is very special and does not provide arbitrarily close approximations to a general duration density. Most economically motivated duration models are not likely to be members of the exponential family. With these procedures it is not

---

[29] The conditional likelihood cannot be used to analyze single spell data. Estimating $\theta$ as a person specific parameter would explain each single spell observation perfectly and no structural parameters of the model would be identified.

possible to estimate duration dependence parameters. These procedures avoid the need to specify or estimate $\mu(\theta)$ and solve the problem of initial conditions by making very strong and nonrobust assumptions about the functional form of the conditional hazard $h(t|x, \theta)$.

The random effect maximum likelihood estimator based on density (2.2.6) is the estimator that is likely to see the greatest use in multispell models that control for unobservables. Flinn and Heckman (1982b) and Hotz (1983) have developed a general computational algorithm called CTM for a likelihood based on (2.2.6) that has the following features.

(i) It allows for a flexible Box–Cox hazard for (2.2.2) with scalar heterogeneity.

$$h(t|x, \theta) = \exp\left( x(t)\beta + \left( \frac{t^{\lambda_1} - 1}{\lambda_1} \right)\lambda_1 + \left( \frac{t^{\lambda_2} - 1}{\lambda_2} \right)\gamma_2 + c\theta \right),$$

$$\lambda_1 < \lambda_2. \quad (1.3.2)'$$

where $\beta$, $\gamma_1$, $\gamma_2$, $\lambda_1$, $\lambda_2$ and $c$ are permitted to depend on the origin state, the destination state and the serial order of the spell. Lagged durations may be included among the $x$. Using maximum likelihood procedures it is possible to estimate all of these parameters except for one normalization of $c$.

(ii) It allows for general time varying variables and right censoring. The regressors may include lagged durations.[30]

(iii) $\mu(\theta)$ can be specified as either normal, log normal or gamma or the NPMLE procedure discussed in Section 1.4.1 can be used.[31]

(iv) It solves the left censoring or initial conditions problem by assuming that the functional form of the initial duration distribution for each origin state is different from that of the other spells.[32]

The burden of computing likelihoods based on (2.2.6) is lessened by the following recursive estimation strategy. (1) Integrate out $T_2, \ldots, T_{k+1}$ from (2.2.6) and estimate the parameters of the reduced likelihood. (2) Then integrate out $T_3, \ldots, T_{k+1}$ from (2.2.6) and estimate the parameters of the reduced likelihood fixing the parameters estimated from stage one. (3) Proceed in this fashion until all parameters are estimated. One Newton step from these parameter values produces efficient maximum likelihood estimators.

---

[30] The random effect maximum likelihood estimator based on (2.2.6) can be shown to be consistent in the presence of $\theta$ with lagged durations included on $x$.

[31] The NPMLE procedure of Heckman and Singer (1984b) can be shown to be consistent for multiple spell data.

[32] This procedure is identical to the procedure discussed in Section 1.5.2, using spells that originate after the origin of the sample.

For more details on the CTM program see Hotz (1983). For further details on the CTM likelihood function and its derivatives, see Flinn and Heckman (1983).[33] For examples of structural multispell duration models see Coleman (1983) and Flinn and Heckman (1982a).

## 3. Summary

This paper considers the formulation and estimation of continuous time social science duration models. The focus is on new issues that arise in applying statistical models developed in biostatistics to analyze economic data and formulate economic models. Both single spell and multiple spell models are discussed. In addition, we present a general time inhomogeneous multiple spell model which contains a variety of useful models as special cases.

Four distinctive features of social science duration analysis are emphasized:

(1) Because of the limited size of samples available in economics and because of an abundance of candidate observed explanatory variables and plausible omitted explanatory variables, standard nonparametric procedures used in biostatistics are of limited value in econometric duration analysis. It is necessary to control for observed and unobserved explanatory variables to avoid biasing inference about underlying duration distributions. Controlling for such variables raises many new problems not discussed in the available literature.

(2) The environments in which economic agents operate are not the time homogeneous laboratory environments assumed in biostatistics and reliability theory. Ad hoc methods for controlling for time inhomogeneity produce badly biased estimates.

(3) Because the data available to economists are not obtained from the controlled experimental settings available to biologists, doing econometric duration analysis requires accounting for the effect of sampling plans on the distributions of sampled spells.

(4) Econometric duration models that incorporate the restrictions produced by economic theory only rarely can be represented by the models used by biostatisticians. The estimation of structural econometric duration models raises new statistical and computational issues.

---

[33] In Flinn and Heckman (1983), the likelihood is derived using a "competing risks" framework. [See, e.g. Kalbfleisch and Prentice (1980) for a discussion of competing risks models.] This framework is in fact inessential to their approach. A more direct approach starts with hazards (2.2.1) and (2.2.2) that are not based on "latent failure times." This direct approach, given hazard specification (1.3.2), produces exactly the same estimating equations as are given in their paper.

Because of (1) it is necessary to parameterize econometric duration models to control for both observed and unobserved explanatory variables. Economic theory only provides qualitative guidance on the matter of selecting a functional form for a conditional hazard, and it offers no guidance at all on the matter of choosing a distribution of unobservables. This is unfortunate because empirical estimates obtained from econometric duration models are very sensitive to assumptions made about the functional forms of these model ingredients.

In response to this sensitivity we present criteria for inferring qualitative properties of conditional hazards and distributions of unobservables from raw duration data sampled in time homogeneous environments; i.e. from unconditional duration distributions. No parametric structure need be assumed to implement these procedures.

We also note that current econometric practice *overparameterizes* duration models. Given a functional form for a conditional hazard determined up to a finite number of parameters, it is possible to consistently estimate the distribution of unobservables nonparametrically. We report on the performance of such an estimator and show that it helps to solve the sensitivity problem.

We demonstrate that in principle it is possible to identify both the conditional hazard and the distribution of unobservables without assuming parametric functional forms for either. Tradeoffs in assumptions required to secure such model identification are discussed. Although under certain conditions a fully nonparametric model can be identified, the development of a consistent fully nonparametric estimator remains to be done.

We also discuss conditions under which access to multiple spell data aids in solving the sensitivity problem. A superficially attractive conditional likelihood approach produces inconsistent estimators, but the practical significance of this inconsistency is not yet known. Conditional inference schemes for eliminating unobservables from multiple spell duration models that are based on sufficient or ancillary statistics require unacceptably strong assumptions about the functional forms of conditional hazards and so are not robust. Contrary to recent claims, they offer no general solution to the model sensitivity problem.

The problem of controlling for time inhomogeneous environments (Point (2)) remains to be solved. Failure to control for time inhomogeneity produces serious biases in estimated duration models. Controlling for time inhomogeneity creates a potential identification problem.

For a single spell data it is impossible to separate the effect of duration dependence from the effect of time inhomogeneity by a fully nonparametric procedure. Although it is intuitively obvious that access to multiple spell data aids in the solution of this identification problem, the development of precise conditions under which this is possible is a topic left for future research.

We demonstrate how sampling schemes distort the functional forms of sample duration distributions away from the population duration distributions that are the usual object of econometric interest (Point (3)). Inference based on mis-specified duration distributions is in general biased. New formulae for the densities of commonly used duration measures are produced for duration models with unobservables in time inhomogeneous environments. We show how access to spells that begin after the origin date of a sample aids in solving econometric problems created by the sampling schemes that are used to generate economic duration data.

We also discuss new issues that arise in estimating duration models explicitly derived from economic theory (Point (4)). For a prototypical search unemployment model we discuss and resolve new identification problems that arise in attempting to recover structural economic parameters. We also consider non-standard statistical problems that arise in estimating structural models that are not treated in the literature. Imposing or testing the restrictions implied by economic theory requires duration models that do not appear in the received literature and often requires numerical solution of implicit equations derived from optimizing theory.

# References

Amemiya, T. (1981) "Qualitative Response Models: A Survey", *Journal of Economic Literature*, 19, 1483–1536.

Amemiya, T. (1984) "Tobit Models: A Survey", *Journal of Econometrics*, 24, 1–63.

Andersen, E. B. (1973) *Conditional Inference and Models for Measuring*. Copenhagen: Mentalhygiejnisk Forlag.

Andersen, E. B. (1980) *Discrete Statistical Models with Social Science Applications*. Amsterdam: North-Holland.

Arnold, Barry and P. Brockett (1983) "Identifiability For Dependent Multiple Decrement/Competing Risks Models", *Scandanavian Actuarial Journal*, 10, 117–127.

Baker, G. and P. Trevedi (1982) "Methods for Estimating the Duration of Periods of Unemployment". Australian National University Working Paper.

Barlow, R. E. and F. Proschan (1975) *Statistical Theory of Reliability and Life Testing*. New York: Holt, Rinehart and Winston.

Barlow, R. E., D. J. Bartholomew, J. M. Bremner and H. D. Brunk (1972) *Statistical Inference Under Order Restrictions*. London: Wiley.

Billingsley, P. (1961) *Statistical Inference for Markov Processes*. Chicago: University of Chicago Press.

Braun, H. and J. Hoem (1979) "Modelling Cohabitational Birth Intervals in the Current Danish Population: A Progress Report". Copenhagen University, Laboratory of Actuarial Mathematics, working paper no. 24.

Burdett, K. and D. Mortensen (1978) "Labor Supply under Uncertainty", in: R. Ehrenberg, ed., *Research in Labor Economics*. London: JAI Press, 2, 109–157.

Chamberlain, G. (1985) "Heterogeneity, Duration Dependence and Omitted Variable Bias", in: J. Heckman and B. Singer, eds., *Longitudinal Analysis of labor Market Data*. New York: Cambridge University Press.

Chamberlain, G. (1980) "Comment on Lancaster and Nickell", *Journal of Royal Statistical Society*, Series A, 160.

Coleman, T. (1983) "A Dynamic Model of Labor Supply under Uncertainty". U. of Chicago, presented at 1983 Summer Meetings of the Econometric Society, Evanston, Ill., unpublished manuscript.

Cosslett, S. (1981) "Efficient Estimation of Discrete Choice Models", in: C. Manski and D. McFadden, eds., *Structural Analysis of Discrete Data with Econometric Applications*. Cambridge: MIT Press, 41–112.

Cox, D. R. (1962) *Renewal Theory*. London: Methuen.

Cox, D. R. (1972) "Regression Models and Lifetables", *Journal of the Royal Statistical Society*, Series B, 34, 187–220.

Cox, D. R. and D. Hinkley (1974) *Theoretical Statistics*. London: Chapman and Hall.

Cox, D. R. and P. A. W. Lewis (1966) *The Statistical Analysis of a Series of Events*. London: Chapman and Hall.

Cox, D. R. and D. O. Oakes (1984) *Analysis of Survival Data*. London: Chapman and Hall.

DeGroot, M. (1970) *Optimal Statistical Decisions*. New York: McGraw-Hill.

Domencich, T. and D. McFadden (1975) *Urban Travel Demand*. Amsterdam: North-Holland.

Elbers, C. and G. Ridder (1982) "True and Spurious Duration Dependence: The Identifiability of the Proportional Hazard Model", *Review of Economic Studies*, 49, 403–410.

Feller, W. (1970) *An Introduction to Probability Theory and Its Applications*. New York: Wiley, Vol. I, third edition.

Feller, W. (1971) *An Introduction to Probability Theory and Its Applications*. New York: Wiley, Vol. II.

Flinn, C. and J. Heckman (1982a) "New Methods for Analyzing Structural Models of Labor Force Dynamics", *Journal of Econometrics*, 18, 115–168.

Flinn, C. and J. Heckman (1982b) "Models for the Analysis of Labor Force Dynamics", in: R. Basmann and G. Rhodes, eds., *Advances in Econometrics*, 1, 35–95.

Flinn, C. and J. Heckman (1982c) "Comment on 'Individual Effects in a Nonlinear Model: Explicit Treatment of Heterogeneity in the Empirical Job Search Literature'", unpublished manuscript, University of Chicago.

Flinn, C. and J. Heckman (1983) "The Likelihood Function for the Multistate–Multiepisode Model in 'Models for the Analysis of Labor Force Dynamics'", in: R. Basmann and G. Rhodes, eds., *Advances in Econometrics*. Greenwich: JAI Press, 3.

Hartigan, J. and P. Hartigan (1985) "The Dip Test for Unimodalities", *The Annals of Statistics*, 13(1), 70–84.

Hauser, J. R. and K. Wisniewski (1982a) "Dynamic Analysis of Consumer Response to Marketing Strategies", *Management Science*, 28, 455–486.

Hauser, J. R. and K. Wisniewski (1982b) "Application, Predictive Test and Strategy Implications for a Dynamic Model of Consumer Response", *Marketing Science*, 1, 143–179.

Heckman, J. (1981a) "Statistical Models for Discrete Panel Data", in: C. Manski and D. McFadden eds., *The Structural Analysis of Discrete Data*. Cambridge: MIT Press.

Heckman, J. (1981b) "The Incidental Parameters Problem and the Problem of Initial Conditions in Estimating a Discrete Time-Discrete Data Stochastic Process", in: C. Manski and D. McFadden, eds., *Structural Analysis of Discrete Data with Economic Applications*. Cambridge: MIT Press, 179–197.

Heckman, J. (1974) "Shadow Prices, Market Wages and Labor Supply", *Econometrica*, 42(4), 679–694.

Heckman, J. and G. Borjas (1980) "Does Unemployment Cause Future Unemployment? Definitions, Questions and Answers from a Continuous Time Model of Heterogeneity and State Dependence", *Economica*, 47, 247–283.

Heckman, J. and B. Singer (1982) "The Identification Problem in Econometric Models for Duration Data", in: W. Hildenbrand, ed., *Advances in Econometrics*. Proceedings of World Meetings of the Econometric Society, 1980. Cambridge: Cambridge University Press.

Heckman, J. and B. Singer (1983) "The Identifiability of Nonproportional Hazard Models". University of Chicago, unpublished manuscript.

Heckman, J. and B. Singer (1984a) "The Identifiability of the Proportional Hazard Model", *Review of Economic Studies*, 51(2), 231–243.

Heckman, J. and B. Singer (1984b) "A Method for Minimizing the Impact of Distributional Assumptions in Econometric Models for duration Data", *Econometrica*, 52(2), 271–320.

Hoem, J. (1972) "Inhomogeneous Semi-Markov Processes, Select Actuarial Tables and Duration Dependence in Demography", in: T. Greville, ed., *Population Dynamics*. New York: Academic Press, 251–296.

Hotz, J. (1983) "Continuous Time Models (CTM): A Manual". GSIA, Pittsburgh: Carnegie-Melon University.

Jovanovic, B. (1979) "Job Matching and the Theory of Turnover", *Journal of Political Economy*, October, 87, 972–990.

Kac, M. (1959) Probability and Related Topics in the Physical Science. New York: Wiley.

Kalbfleisch, J. and R. Prentice (1980) *The Statistical Analysis of Failure Time Data*. New York: Wiley.

Kieter, N. and G. Neumann (1981) "Individual Effects in a Nonlinear Model", *Econometrica*, 49(4), 965–980.

Lancaster, T. and S. Nickell (1980) "The Analysis of Reemployment Probabilities for the Unemployed", *Journal of the Royal Statistical Society*, Series A, 143, 141–165.

Lawless, J. F. (1982) *Statistical Models and Methods for Lifetime Data*. New York: Wiley.

Lewis, H. G. (1974) "Comments on Selectivity Biases in Wage Comparisons", *Journal of Political Economy*, November, 82(6), 1145–1156.

Lindsey, B. (1983a) "The Geometry of Mixture Likelihoods, Part I", *Annals of Statistics*, 11, 86–94.

Lindsey, B. (1983b) "The Geometry of Mixture Likelihoods, Part II", *Annals of Statistics*, 11(3), 783–792.

Lippman, S. and J. McCall (1976) "The Economics of Job Search: A Survey", *Economic Inquiry*, September, 14, 113–126.

Lundberg, F. (1903) "I. Approximerad Framställning af Sannolikhetsfunktionen II. Aterforsakring af Kollektivrisker". Uppsala: Almquist und Wicksell.

Lundberg, S. (1981) "The Added Worker: A Reappraisal". NBER Working Paper no. 706, Cambridge, Mass.

Manski, C. and D. McFadden (1981) "Alternative Estimators and Sample Designs for Discrete Choice Analysis", in: C. Manski and D. McFadden, *Structural Analysis of Discrete Data with Econometric Applications*. Cambridge: MIT Press, 2–50.

Manski, C. and S. Lerman (1977) "The Estimation of Choice Probabilities from Choice Based Samples", *Econometrica*, 45, 1977–1988.

McFadden, D. (1974) "Conditional Logit Analysis of Qualitative Choice Behavior", in: P. Zarembka, ed., *Frontiers in Econometrics*. New York: Academic Press.

Moore, E. and R. Pyke (1968) "Estimation of the Transition Distributions of a Markov Renewal Process", *Annals of the Institute of Statistical Mathematics*. Tokyo, 20(3), 411–424.

Neyman, J. and E. Scott (1948) "Consistent Estimates Based on Partially Consistent Observations", *Econometrica*, 16, 1–32.

Robb, R. (1984) "Two Essays on the Identification of Economic Models". University of Chicago, May, unpublished manuscript.

Robbins, H. (1970) "Optimal Stopping", *American Mathematical Monthly*, 77, 333–43.

Ross, S. M. (1970) *Applied Probability Models with Optimization Applications*. San Francisco: Holden-Day.

Rudin, W. (1974) *Real and Complex Analysis*. New York: McGraw Hill.

Salant, S. (1977) "Search Theory and Duration Data: A Theory of Sorts", *Quarterly Journal of Economics*, February, 91, 39–57.

Sheps, M. and J. Menken (1973) *Mathematical Models of Conception and Birth*. Chicago: University of Chicago Press.

Shohat, J. and J. Tamarkin (1943) *The Problem of Moments*. New York: American Mathematical Society.

Singer, B. (1982) "Aspects of Nonstationarity", *Journal of Econometrics*, 18(1), 169–190.

Trussell, J. and T. Richards (1985) "Correcting for Unobserved Heterogeneity in Hazard Models: An Application of the Heckman-Singer Procedure, in N. Tuma, *Sociological Methodology*. San Francisco: Jossey Bass.

# PART 8

# SELECTED APPLICATIONS AND USES OF ECONOMETRICS

Chapter 30

# DEMAND ANALYSIS

ANGUS DEATON*

*Princeton University*

## Contents

*I am grateful to E. Berndt, J. Muellbauer, A. Powell, H. Theil, H. Varian, T. Wales, R. Williams and the editors for helpful comments and criticism. The views expressed, as well as the remaining errors, remain my own. This is an inadequately revised version of a paper written in 1980. Given the very limited time available for revision, I am conscious that I have not been able to do justice to the considerable literature that has appeared in the intervening five years.

*Handbook of Econometrics, Volume III, Edited by Z. Griliches and M.D. Intriligator*
© *Elsevier Science Publishers BV, 1986*

## 0.  Introduction

The empirical analysis of consumer behavior has always held a central position in econometrics and many of what are now standard techniques were developed in response to practical problems in interpreting demand data. An equally central position in economic analysis is held by the theory of consumer behavior which has provided a structure and language for model formulation and data analysis. Demand analysis is thus in the rare position in econometrics of possessing long interrelated pedigrees on both theoretical and empirical sides. And although the construction of models which are both theoretically and empirically satisfactory is never straightforward, no one who reads the modern literature on labor supply, on discrete choice, on asset demands, on transport, on housing, on the consumption function, on taxation or on social choice, can doubt the current vigor and power of utility analysis as a tool of applied economic reasoning. There have been enormous advances towards integration since the days when utility theory was taught as a central element in microeconomic courses but then left unused by applied economists and econometricians.

Narrowly defined, demand analysis is a small subset of the areas listed above, referring largely to the study of commodity demands by consumers, most usually based on aggregate data but occasionally, and more so recently, on cross-sections or even panels of households. In this chapter, I shall attempt to take a somewhat broader view and discuss, if only briefly, the links between conventional demand analysis and such topics as labor supply, the consumption function, rationing, index numbers, equivalence scales and consumer surplus. Some of the most impressive recent econometric applications of utility theory are in the areas of labor supply and discrete choice, and these are covered in other chapters. Even so, a very considerable menu is left for the current meal. Inevitably, the choice of material is my own, is partial (in both senses), and does not pretend to be a complete survey of recent developments. Nor have I attempted to separate the economic from the statistical aspects of the subject. The strength of consumer demand analysis has been its close articulation of theory and evidence and the theoretical advances which have been important (particularly those concerned with duality) have been so precisely because they have permitted a more intimate contact between the theory and the interpretation of the evidence. It is not possible to study applied demand analysis without keeping statistics and economic theory simultaneously in view.

The layout of the chapter is as follows. Section 1 is concerned with utility and the specification of demand functions and attempts to review the theory from the point of view of applied econometrics. Duality aspects are particularly emphasized. Section 2 covers what I shall call 'naive' demand analysis, the estimation and testing, largely on aggregate time series data, of 'complete' systems of demand equations linking quantities demanded to total expenditure and prices. The label "naive" implies simplicity neither in theory nor in econometric technique. Instead, the adjective refers to the belief that, by itself, the simple, static, neoclassical model of the individual consumer could (or should) yield an adequate description of aggregate time-series data. Section 3 is concerned with microeconomic or cross-section analysis including the estimation of Engel curves, the treatment of demographic variables, and the particular econometric problems which arise in such contexts. There is also a brief discussion of the econometric issues that arise when consumers face non-linear budget constraints. Sections 4 and 5 discuss two theoretical topics of considerable empirical importance, separability and aggregation. The former provides the analysis underpinning econometric analysis of subsystems on the one hand and of aggregates, or supersystems, on the other. The latter provides what justification there is for grouping over different consumers. Econometric analysis of demand under conditions of rationing or quantity constraints is discussed in Section 6. Section 7 provides a brief overview of three important topics which, for reasons of space, cannot be covered in depth, namely, intertemporal demand analysis, including the analysis of the consumption function and of durable goods, the choice over qualities, and the links between demand analysis and welfare economics, particularly as concerns the measurement of consumer surplus, cost-of-living index numbers and the costs of children. Many other topics are inevitably omitted or dealt with less fully than is desirable; some of these are covered in earlier surveys by Goldberger (1967), Brown and Deaton (1972) and Barten (1977).

# 1. Utility and the specification of demand

## 1.1. Assumptions for empirical analysis

As is conventional, I begin with the specification of preferences. The relationship "is at least as good as", written $\geq$, is assumed to be reflexive, complete, transitive and continuous. If so, it may be represented by a utility function, $v(q)$ say, defined over commodity vector $q$ with the property that the statement $q^A \geq q^B$ for vectors $q^A$ and $q^B$ is equivalent to the statement $v(q^A) \geq v(q^B)$. Clearly, for most purposes, it is more convenient to work with a utility function than with a preference ordering. There seem few prior empirical grounds for objecting to reflexivity, completeness, transitivity or continuity, nor indeed to the assumption that $v(q)$ is monotone increasing in $q$. Again, for empirical work, there is little objection to the assumption that preferences are *convex*, i.e. that for $q^A \geq q^B$, and for $0 \leq \lambda \leq 1$, $\lambda q^A + (1-\lambda)q^B \geq q^B$. This translates immediately into quasi-concavity of the utility function $v(q)$, i.e. for $q^A$, $q^B$, $0 \leq \lambda \leq 1$,

$$v(q^A) \geq v(q^B) \quad \text{implies} \quad v(\lambda q^A + (1-\lambda)q^B) \geq v(q^B). \tag{1}$$

Henceforth, I shall assume that the consumer acts so as to maximise the monotone, continuous and quasi-concave utility function $v(q)$.

It is common, in preparation for empirical work, to assume, in addition to the above properties, that the utility function is *strictly* quasi-concave (so that for $0 < \lambda < 1$ the second inequality in (1) is strict), *differentiable*, and that all goods are *essential*, i.e. that in all circumstances all goods are bought. All these assumptions are convenient in particular situations. But they are all restrictive and all rule out phenomena that are likely to be important in some empirical situations. Figure 1 illustrates in two dimensions. All of the illustrated indifference curves are associated with quasi-concave utility functions, but only $A$ is either differentiable or strictly quasi-concave. The flat segments on $B$ and $C$ would be ruled out by strict quasi-concavity; hence, strictness ensures single-val-

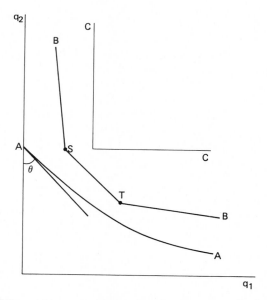

Figure 1   Indifference curves illustrating quasi-concavity, differentiability and essential goods.

ued demand functions. Empirically, flats are important because they represent perfect substitutes; for example, between $S$ and $T$ on $B$, the precise combination of $q_1$ and $q_2$ makes no difference and this situation is likely to be relevant, say, for two varieties of the same good. Non-differentiabilities occur at the kink points on the curves $B$ and $C$. With a linear budget constraint, kinks imply that for relative prices within a certain range, two or more goods are bought in fixed proportions. Once again, this may be practically important and fixed relationships between complementary goods are often a convenient and sensible modelling strategy. The $n$-dimensional analogue of the utility function corresponding to $C$ is the fixed coefficient or Leontief utility function

$$v(q) = \min\{\alpha_1 q_1, \alpha_2 q_2, \ldots, \alpha_n q_n\}. \tag{2}$$

For positive parameters $\alpha_1, \ldots, \alpha_n$. Finally curve $A$ illustrates the situation where $q_2$ is essential but $q_1$ is not. As $q_2$ tends to zero, its marginal value relative to that of $q_1$ tends to infinity along any given indifference curve. Many commonly used utility functions impose this condition which implies that $q_2$ is always purchased in positive amounts. But for many goods, the behavior with respect to $q_1$ is a better guide; if $p_1 > p_2 \theta$, the consumer on indifference curve $A$ buys none of $q_1$. Data on individual households always show that, even for quite broad commodity groups, many households do not buy all goods. It is therefore necessary to have models that can deal with this fact.

### 1.2. Lagrangians and matrix methods

If $v(q)$ is strictly quasi-concave and differentiable, the maximization of utility subject to the budget constraint can be handled by Lagrangian techniques. Writing the constraint $p \cdot q = x$ for price vector $p$ and *total expenditure x*, the first-order conditions are

$$\frac{\partial v(q)}{\partial q_i} = \lambda p_i, \tag{3}$$

which, under the given assumptions, solve for the demand functions

$$q_i = g_i(x, p). \tag{4}$$

For example, the *linear expenditure system* has utility function

$$u = \Pi(q_i - \gamma_i)^{\beta_i}, \tag{5}$$

for parameters $\gamma$ and $\beta$, the first-order conditions of which are readily solved to give the demand functions

$$p_i q_i = p_i \gamma_i + \beta_i (x - p \cdot \gamma). \tag{6}$$

In practice, the first-order conditions are rarely analytically soluble even for quite simple formulations (e.g. Houthakker's (1960) "direct addilog" $u = \sum \alpha_i q_i^{\beta_i}$), nor is it at all straightforward to pass back from given demand functions to a closed form expression for the utility function underlying them, should it indeed exist.

The generic properties of demands are frequently derived from (3) by total differentiation and matrix inversion to express $dq$ as a function of $dx$ and $dp$, the so-called "fundamental matrix equation" of consumer demand analysis, see Barten (1966) originally and its frequent later exposition by Theil, e.g. (1975b, pp. 14ff), also Phlips (1974, 1983, p. 47), Brown and Deaton (1972, pp. 1160–2). However, such an analysis requires that $v(q)$ be twice-differentiable, and it is usually assumed in addition that utility has been monotonically transformed so that the Hessian is non-singular and negative definite. Neither of these last assumptions follows in any natural way from reasonable axioms; note in particular that is is *not* always possible to transform a quasi-concave function by means of a monotone increasing function into a concave one, see Kannai (1977), Afriat (1980). Hence, the methodology of working through first-order conditions involves an expansive and complex web of restrictive and unnatural assumptions, many of which preclude consideration of phenomena requiring analysis. Even in the hands of experts, e.g. the survey by Barten and Bohm (1980), the analytical apparatus becomes very complex. At the same time, the difficulty of solving the conditions in general prevents a close connection between preferences and demand, between the a priori and the empirical.

## 1.3. Duality, cost functions and demands

There are many different ways of representing preferences and great convenience can be obtained by picking that which is most appropriate for the problem at hand. For the purposes of generating empirically useable models in which quantities are a function of prices and total expenditure, dual representations are typically most convenient. In this context, duality refers to a switch of variables, from quantities to prices, and to the respecification of preferences in terms of the latter. Define the *cost function*, sometimes *expenditure function*, by

$$c(u, p) = \left\{ \min_q p \cdot q; \, v(q) \geq u \right\}. \tag{7}$$

If $x$ is the total budget to be allocated, then $x$ will be the cheapest way of reaching whatever $u$ can be reached at $p$ and $x$, so that

$$c(u, p) = x. \tag{8}$$

The function $c(u, p)$ can be shown to be continuous in both its arguments, monotone increasing in $u$ and monotone non-decreasing in $p$. It is linearly homogeneous and *concave* in prices, and first and second differentiable almost everywhere. It is *strictly* quasi-concave if $v(q)$ is *differentiable* and *everywhere* differentiable if $v(q)$ is *strictly* quasi-concave. For proofs and further discussions see McFadden (1978), Diewert (1974a), (1980b) or, less rigorously, Deaton and Muellbauer (1980a, Chapter 2).

The empirical importance of the cost function lies in two features. The first is the 'derivative property', often known as Shephard's Lemma, Shephard (1953). By this, whenever the derivative exists

$$\frac{\partial c(u, p)}{\partial p_i} \equiv h_i(u, p) = q_i. \tag{9}$$

The functions $h_i(u, p)$ are known as *Hicksian* demands, in contrast to the *Marshallian* demands $g_i(x, p)$. The second feature is the Shephard–Uzawa duality theorem [again see McFadden (1978) or Diewert (1974a), (1980b)] which given convex preferences, allows a constructive recovery of the utility function from the cost function. Hence, all the information in $v(q)$ which is relevent to behavior and empirical analysis is encoded in the function $c(u, p)$. Or put another way, any function $c(u, p)$ with the correct properties can serve as an alternative to $v(q)$ as a basis for empirical analysis. The direct utility function need never be explicitly evaluated or derived; if the cost function is correctly specified, corresponding preferences always exist. The following procedure is thus suggested in empirical work. Starting from some linearly homogeneous concave cost function $c(u, p)$, derive the Hicksian demand functions $h_i(u, p)$ by differentiation. These can be converted into Marshallian demands by substituting for $u$ from the inverted form of (8); this is written

$$u = \psi(x, p), \tag{10}$$

and is known as the *indirect* utility function. (The original function $v(q)$ is the *direct* utility function and the two are linked by the identity $\psi(x, p) \equiv v\{g(x, p)\}$ for utility maximizing demands $g(x, p)$). Substituting (10) into (9) yields

$$q_i = h_i(u, p) = h_i\{\psi(x, p), p\} = g_i(x, p), \tag{11}$$

which can then be estimated. Of course, the demands corresponding to the original cost function may not fit the data or may have other undesirable properties for the purpose at hand. To build this back into preferences, we must be able to go from $g_i(x, p)$ back to $c(u, p)$. But, from Shephard's Lemma, $q_i = g_i(x, p)$ may be rewritten as

$$\frac{\partial c(u, p)}{\partial p_i} = g_i \{ c(u, p), p \}, \tag{12}$$

which may be solved for $c(u, p)$ provided the mathematical integrability conditions are satisfied. These turn out to be equivalent to Slutsky symmetry, so that demand functions displaying symmetry always imply some cost function, see, for example, Hurwicz and Uzawa (1971) for further details. If the Slutsky matrix is also negative semi-definite (together with symmetry, the 'economic' integrability condition), the cost function will be appropriately concave which it must be to represent preferences. This possibility, of moving relatively easily between preferences and demands, is of vital importance if empirical knowledge is to be linked to economic theory.

An alternative and almost equally straightforward procedure is to start from the indirect utility function $\psi(x, p)$. This must be zero degree homogeneous in $x$ and $p$ and quasi-convex in $p$ and Shephard's Lemma takes the form

$$q_i = g_i(x, p) = \frac{- \partial \psi(x, p)/\partial p_i}{\partial \psi(x, p)/\partial x}, \tag{13}$$

a formula known as *Roy's identity*, Roy (1942). This is sometimes done in "normalized" form. Clearly, $\psi(x, p) = \psi(1, p/x) = \psi^*(r)$ where $r = p/x$ is the vector of normalized prices. Hence, using $\psi^*$ instead of $\psi$, Roy's identity can be written in the convenient form

$$w_i = \frac{p_i q_i}{x} = \frac{\partial \psi^*/\partial \log r_i}{\sum_k \partial \psi^*/\partial \log r_k} = \frac{\partial \log c(u, p)}{\partial \log p_i}, \tag{14}$$

where the last equality follows from rewriting (9).

One of the earliest and best practical examples of the use of these techniques is Samuelson's (1947–8) derivation of the utility function (5) from the specification of the linear expenditure system suggested earlier by Klein and Rubin (1947–8). A more recent example is provided by the following. In 1943, Holbrook Working suggested that a useful form of Engel curve was given by expressing the budget share of good $i$, $w_i$, as a linear function of the logarithm of total expenditure.

Hence,

$$w_i = \alpha_i + \beta_i \ln x, \tag{15}$$

for parameters $\alpha$ and $\beta$, generally functions of prices, and this form was supported in later comparative tests by Leser (1963). From (14), the budget shares are the logarithmic derivatives of the cost function, so that (15) corresponds to differential equations of the form

$$\frac{\partial \ln c(u, p)}{\partial \ln p_i} = \alpha_i(p) + \beta_i(p) \ln c(u, p), \tag{16}$$

which give a solution of the general form

$$\ln c(u, p) = u \ln b(p) + (1 - u) \ln a(p), \tag{17}$$

where $\alpha_i(p) = (a_i \ln b - b_i \ln a)/(\ln b - \ln a)$ and $\beta_i(p) = b_i/(\ln b - \ln a)$ for $a_i = \partial \ln a / \partial \ln p_i$ and $b_i = \partial \ln b / \partial \ln p_i$. The form (17) gives the cost function as a utility-weighted geometric mean of the linear homogeneous functions $a(p)$ and $b(p)$ representing the cost functions of the very poor ($u = 0$) and the very rich ($u = 1$) respectively. Such preferences have been called the PIGLOG class by Muellbauer (1975b), (1976a), (1976b). A full system of demand equations within the Working–Leser class can be generated by suitable choice of the functions $b(p)$ and $a(p)$. For example, if

$$\ln a(p) = a_o + \sum \alpha_k \ln p_k + \tfrac{1}{2} \sum_k \sum_m \gamma_{km}^* \ln p_k \ln p_m,$$
$$\ln b(p) = \ln a(p) + \beta_0 \Pi p_k^{\beta_k}, \tag{18}$$

we reach the "almost ideal demand system" (AIDS) of Deaton and Muellbauer (1980b) viz

$$w_i = \alpha_i + \beta_i \ln(x/P) + \sum_j \gamma_{ij} \ln p_j,$$
$$\ln P = \alpha_0 + \sum \alpha_k \ln p_k + \tfrac{1}{2} \sum_k \sum_m \gamma_{km} \ln p_k \ln p_m, \tag{19}$$

and $\gamma_{ij} = \tfrac{1}{2}(\gamma_{ij}^* + \gamma_{ji}^*)$. A variation on the same theme is to replace the geometric mean (17) by a mean of order $\varepsilon$

$$c(u, p) = \{ ub(p)^\varepsilon + (1 - u) a(p)^\varepsilon \}^{1/\varepsilon}, \tag{20}$$

with Engel curves

$$w_i = \alpha_i + \beta_i x^{-\varepsilon}. \tag{21}$$

This is Muellbauer's PIGL class; equation (21), in an equivalent Box–Cox form, has recently appeared in the literature as the "generalized Working model", see Tran van Hoa, Ironmonger, and Manning (1983) and Tran van Hoa (1983).

I shall return to these and similar models below, but for the moment note how the construction of these models allows empirical knowledge of demands to be built into the specification of preferences. This works at a less formal level too. For example, prior information may relate to the shape of indifference curves, say that two goods are poor substitutes or very good substitutes as the case may be. This translates directly into curvature properties of the cost function; 'kinks' in quantity space turn into 'flats' in price space and vice versa so that the specification can be set accordingly. For further details, see the elegant diagrams in McFadden (1978).

The duality approach also provides a simple demonstration of the generic properties of demand functions which have played such a large part in the testing of consumer rationality, see Section 2 below. The budget constraint implies immediately that the demand functions *add-up* (trivially) and that they are zero-degree *homogeneous* in prices and total expenditure together (since the budget constraint is unaffected by proportional changes in $p$ and $x$). Shephard's Lemma (9) together with the mild regularity conditions required for Young's Theorem implies that

$$\frac{\partial h_i}{\partial p_j} = \frac{\partial^2 c}{\partial p_j \, \partial p_i} = \frac{\partial^2 c}{\partial p_i \, \partial p_j} = \frac{\partial h_j}{\partial p_i}, \tag{22}$$

so that, if $s_{ij}$, the Slutsky substitution term is $\partial h_i / \partial p_j$, the matrix of such terms, $S$, is *symmetric*. Furthermore, since $c(u, p)$ is a concave function of $p$, $S$ must be *negative semi-definite*. (Note that the homogeneity of $c(u, p)$ implies that $p$ lies in the nullspace of $S$). Of course, $S$ is not directly observed, but it can be evaluated using (12); differentiating with respect to $p_j$ gives the *Slutsky equation*.

$$s_{ij} = \frac{\partial g_i}{\partial p_j} + \frac{\partial g_i}{\partial x} q_j. \tag{23}$$

Hence to the extent that $\partial g_i / \partial p_j$ and $\partial g_i / \partial x$ can be estimated econometrically, symmetry and negative semi-definiteness can be checked. I shall come to practical attempts to do so in the next section.

## 1.4. Inverse demand functions

In practical applications, it is occasionally necessary to estimate prices as a function of quantities rather than the other way round. An approach to specification exists for this case which is precisely analogous to that suggested above. From the direct utility function and the first-order conditions (10), apply the budget constraint $p \cdot q = x$ to give

$$\frac{p_i q_i}{x} = \frac{\partial v / \partial \ln q_i}{\sum_k \partial v / \partial \ln q_k}, \tag{24}$$

which is the dual analogue of (14), though now determination goes from the quantities $q$ to the normalized prices $p/x$. Alternatively, define the distance function $d(u, q)$, dual to the cost function, by

$$d(u, q) = \min_p \{ p \cdot q; \ \psi(1, p) \leq u \}. \tag{25}$$

The distance function has properties analogous to the cost function and, in particular,

$$p_i / x = \partial d(u, q) / \partial q_i = a_i(u, q), \tag{26}$$

are the inverse compensated demand functions relating an indifference curve $u$ and a quantity ray $q$ to the price to income ratios at the intersection of $q$ and $u$. See McFadden (1978), Deaton (1979) or Deaton and Muellbauer (1980a, Chapter 2.7) for fuller discussions.

Compensated and uncompensated inverse demand functions can be used in exactly the same way as direct demand functions and are appropriate for the analysis of situations when quantities are predetermined and prices adjust to clear the market. Hybrid situations can also be analysed with some prices fixed and some quantities fixed; again see McFadden (1978) for discussion of "restricted" preference representation functions. Note one final point, however. The Hessian matrix of the distance function $d(u, q)$ is the *Antonelli* matrix $A$ with elements

$$a_{ij} = \frac{\partial^2 d}{\partial q_i \partial q_j} = a_{ji} = \frac{\partial a_i(u, q)}{\partial q_j}, \tag{27}$$

which can be used to define $q$-substitutes and $q$-complements just as the Slutsky matrix defines $p$-substitutes and $p$-complements, see Hicks (1956) for the original discussion and derivations. Unsurprisingly the Antonelli and Slutsky matrices are intimately related and given the close parallel been duality and matrix inversion, it is appropriate that they should be generalised inverses of one another. For example, using $\nabla$ to denote the vector of price or quantity partial derivatives, (9) and (26) combine to yield

$$q = \nabla c \{ u, \nabla d \{ u, \nabla c(u, p) \} \}. \tag{28}$$

Hence, differentiating with respect to $p/x$ and repeatedly applying the chain rule, we obtain at once

$$S^* = S^* A S^*. \tag{29}$$

Similarly,

$$A = A S^* A, \tag{30}$$

where $S^* = xS$. Note that the homogeneity restrictions imply $Aq = S^*p = 0$ which together with (29) and (30) complete the characterization as generalized inverses. These relationships also allow passage from one type of demand function to another so that the Slutsky matrix can be calculated from estimates of indirect demand functions while the Antonelli matrix may be calculated from the usual demands. The explicit formula for the latter is easily shown to be

$$A = \left( xS + qq' \right)^{-1} - x^{-2} pp', \tag{31}$$

with primes denoting transposition, see Deaton (1981a). The Antonelli matrix has important applications in measuring quantity index numbers, see, e.g. Diewert (1981, 1983) and in optimal tax theory, see Deaton (1981a). Formula (31) allows its calculation from an estimate of the Slutsky matrix.

  This brief review of the theory is sufficient to permit discussion of a good deal of the empirical work in the literature. Logically, questions of aggregation and separability ought to be treated first, but since they are not required for an understanding of what follows, I shall postpone their discussion to Section 4.

## 2.   Naive demand analysis

Following Stone's first empirical application of the linear expenditure system in 1954, a good deal of attention was given in the subsequent literature to the problems involved in estimating complete, and generally nonlinear, systems of demand equations. Although the issues are now reasonably well understood, they deserve brief review. I shall use the linear expenditure system as representative of the class

$$p_{it}q_{it} = f_i(p_t, x_t; b) + u_{it}, \tag{32}$$

for commodity $i$ on observation $t$, parameter vector $b$, and error $u_{it}$. For the linear expenditure system the function takes the form

$$f_i(p_t, x_t; b) = \gamma_i p_{it} + \beta_i(x_t - p_t \cdot \gamma). \tag{33}$$

## 2.1. Simultaneity

The first problem of application is to give a sensible interpretation to the quantity $x_t$. In loose discussion of the theory $x_t$ is taken as "income" and is assumed to be imposed on the consumer from outside. But, if $q_t$ is the vector of commodity purchases in period $t$, then (a) only exceptionally is any real consumer given a predetermined and inflexible limit for total commodity expenditure and (b) the only thing which expenditures add up to is total expenditure *defined* as the sum of expenditures. Clearly then, $x_t$ is in general jointly endogenous with the expenditures and ought to be treated as such, a point argued, for example, by Summers (1959), Cramer (1969) and more recently by Lluch (1973), Lluch and Williams (1974). The most straightforward solution is to instrument $x_t$ and there are no shortages of theories of the consumption function to suggest exogenous variables. However, in the spirit of demand analysis this can be formalized rather neatly using any intertemporally separable utility function. For example, loosely following Lluch, an intertemporal or extended linear expenditure system can be proposed of the form

$$p_{it}q_{it} = p_{it}\gamma_{it} + \beta_{it}\left\{W - \sum_{\tau=t}^{L}\sum_k p_{\tau k}^* \gamma_{\tau k}\right\} + v_{it}, \tag{34}$$

where the $\gamma_{it}$ and $\beta_{it}$ parameters are now specific to periods (needs vary over the life-cycle), $W$ is the current present discounted value of present and future income and current financial assets, and $p_{\tau k}^*$ is the current discounted price of good $k$ in future period $\tau$($p_{tk}^* = p_{tk}$ since $t$ is the present). As with any such system based on intertemporally separable preferences, see Section 4 below, (34) can be solved for $x_t$ by summing the left-hand side over $i$ and the result, i.e. the consumption function, used to substitute for $W$. Hence (34) *implies* the familiar static linear expenditure system, i.e.

$$p_{it}q_{it} = p_{it}\gamma_{it} + \frac{\beta_{it}}{\beta_t}\left\{x_t - \sum_k p_{kt}\gamma_{kt}\right\} + \left\{v_{it} - \frac{\beta_{it}}{\beta_t}v_t\right\}, \tag{35}$$

where $v_t = \sum v_{it}$, $\beta_t = \sum \beta_{it}$ and it is assumed, as is reasonable, that $\beta_t \neq 0$. This not only relates the parameters in the static version (33) to their intertemporal counterparts, but it also gives valuable information about the structure of the error term in (32). Given this, the bias introduced by ignoring the simultaneity between $x_t$ and $p_{it}q_{it}$ can be studied. For the usual reasons, it will be small if the equations fit well, as Prais (1959) argued in his reply to Summers (1959). But there is a rather more interesting possibility. It is easily shown, on the basis of (35), that

$$\text{Cov}(x_t, u_{it}) = \sum_k \sigma_{ik} - \frac{\beta_{it}}{\beta_t}\sum_k \sum_m \sigma_{km}, \tag{36}$$

where $\sigma_{ij}$ is the (assumed constant) covariance between $v_{it}$ and $v_{jt}$, i.e.

$$\text{Cov}(v_{it}, v_{js}) = \delta_{ts}\sigma_{ij}. \tag{37}$$

where $\sigma_{ts}$ is the Kronecker delta. Clearly, the covariance in (36) is zero if $\sum_k \sigma_{ik}/\sum \sigma_{km} = \beta_{it}/\beta_t$. One specialized theory which produces exactly this relationship is Theil's (1971b, 1974, 1975a, 1975b, pp. 56–90, 1979) "rational random behaviour" under which the variance, covariance matrix of the errors $v_{it}$ is rendered proportional to the Slutsky matrix by consumers' trading-off the costs of exact maximization against the utility losses of not doing so. If this model is correct, there is no simultaneity bias, see Deaton (1975a, pp. 161–8) and Theil (1976, pp. 4–6, 80–82) for applications. However, most econometricians would tend to view the error terms as reflecting, at least in part, those elements *not* allowed for by the theory, i.e. misspecifications, omitted variables and the like. Even so, it is not implausible that (36) should be close to zero since the requirement is that error covariances between each category and total expenditure should be proportional to the marginal propensity to spend for that good. This is a type of "error separability" whereby omitted variables influence demands in much the same way as does total outlay.

In general, simultaneity will exist and the issue deserves to be taken seriously; it is likely to be particularly important in cross-section work, where occasional large purchases affect both sides of the Engel curve. Ignoring it may also bias the other tests discussed below, see Altfield (1985).

## 2.2. Singularity of the variance – covariance matrix

The second problem arises from the fact that with $x_t$ *defined* as the sum of expenditures, expenditures automatically add-up to total expenditure identically, i.e. without error. Hence, provided $f_i$ in (32) is properly chosen, we must have

$$\sum_i p_{it} q_{it} = x_t; \qquad \sum f_i(p_t, x_t; b) = x_t; \qquad \sum_i u_{it} = 0. \tag{38}$$

Writing $\Omega$ as the $n \times n$ contemporaneous variance–covariance matrix of the $u_{it}$'s with typical element $\omega_{ij}$, i.e.

$$E(u_{it}, u_{js}) = \delta_{ts}\omega_{ij}, \tag{39}$$

then the last part of (38) clearly implies

$$\sum_j \omega_{ij} = \sum_i \omega_{ij} = 0, \tag{40}$$

so that the variance–covariance matrix is singular. If (32) is stacked in the usual way as an $nT$ observation regression, its covariance matrix is $\Omega \otimes I$ which cannot have rank higher than $(n-1)T$. Hence, the usual generalized least squares estimator or its non-linear analogue is not defined since it would require the non-existent inverse $\Omega^{-1} \otimes I$.

This non-existence is, however, a superficial problem. For a set of equations such as (32) satisfying (38), one equation is essentially redundant and all of its parameters can be inferred from knowledge of those in the other equations. Hence, attempting to estimate all the parameters in all equations is equivalent to including some parameters more than once and leads to exactly the same problems as would arise if, for example, some independent variables were included more than once on the right hand side of an ordinary single-variable regression. The solution is obviously to drop one of the equations and estimate the resulting $(n-1)$ equations by GLS, Zellner's (1962) seemingly unrelated regressions estimator (SURE), or similar technique. Papers by McGuire, Farley, Lucas and Winston (1968) and by Powell (1969) show that the estimates are invariant to the particular equation which is selected for omission. Barten (1969) also considered the maximum-likelihood estimation of such systems when the errors follow the multivariate normal assumption. If $\Omega_n$ is the variance–covariance matrix of the system (32) excluding the $n$th equation, a sample of $T$ observations has a log-likelihood conditional on normality of

$$\ln L = -\frac{T}{2}(n-1)\ln 2\pi - \frac{T}{2}\ln \det \Omega_n - \frac{1}{2}\sum_{t=1}^{T} u'_{(n)t}\Omega_n^{-1}u_{(n)t}, \tag{41}$$

where $u_{(n)}$ is the $(n-1)$-vector of $u_{it}$ excluding element $n$. Barten defines a new non-singular matrix $V$ by

$$V = \Omega + \kappa ii, \qquad (42)$$

where $i$ is the normalized vector of units, i.e. $i_i = 1/n$, and $0 < \kappa < \infty$. Then (41) may be shown to be equal to

$$\ln L = \frac{T}{2} \{\ln \kappa + \ln n - (n-1)\ln 2\pi - \ln \det V\} - \frac{1}{2} \sum_{t=1}^{T} u_t' V^{-1} u_t. \qquad (43)$$

This formulation establishes that the likelihood is independent of the equation deleted (and incidentally of $\kappa$ since (41) does not depend on it) and also returns the original symmetry to the problem. However, in practice, the technique of dropping one equation is usually to be preferred since it reduces the dimension of the parameter vector to be estimated which tends to make computation easier.

Note two further issues associated with singularity. First, if the system to be estimated is a "subsystem" of commodities that does not exhaust the budget, the variance covariance matrix of the residuals need not, and usually will not be singular. In consequence, SURE or FIML (see below) can be carried out directly on the subsystem. However, it is still necessary to assume a non-diagonal variance–covariance matrix; overall singularity precludes *all* goods from having orthogonal errors and there is usually no good reason to implicitly confine all the off-diagonal covariances to the omitted goods. Second, there are additional complications if the residuals are assumed to be serially correlated. For example, in (32), it might be tempting to write

$$u_{it} = \rho_i u_{it-1} + \varepsilon_{it}, \qquad (44)$$

for serially uncorrelated errors $\varepsilon_{it}$. If $R$ is the diagonal matrix of $\rho_i$'s, (44) implies that

$$\Omega = R\Omega R + \Sigma, \qquad (45)$$

where $\Sigma$ is the contemporaneous variance–covariance matrix of the $\varepsilon$'s. Since $\Omega i = \Sigma i = 0$, we must have $\Omega \rho = 0$, which, since $i$ spans the null space of $\Omega$, implies that $\rho \propto i$, i.e. that all the $\rho_i$'s are the same, a result first established by Berndt and Savin (1975). Note that this does *not* mean that (44) with $\rho_i = \rho$ for all $i$ is a sensible specification for autocorrelation in singular systems. It would seem better to allow for autocorrelation at an earlier stage in the modeling, for example by letting $v_{it}$ be autocorrelated in (34) and following through the consequences for the compound errors in (35). In general, this will imply vector autoregressive structures, as, for example, in Guilkey and Schmidt (1973) and Anderson and Blundell (1982). But provided autocorrelation is handled in a way that respects the singularity (as it should be), so that the omitted equation is not implicitly treated differently from the others, then it will always be correct to estimate by dropping one equation since all the relevant information is contained in the other $(n-1)$.

## 2.3. Estimation

For estimation purposes, rewrite (32) in the form

$$y_{ti} = f_{ti}(\beta) + u_{ti}, \tag{46}$$

with $t = 1, \ldots, T$ indexing observations and $i = 1, \ldots, (n-1)$ indexing goods. I shall discuss only the case where $u_{ti}$ are independently and identically distributed as multivariate normal with zero mean and nonsingular covariance matrix $\Omega$. [For other specifications, see, e.g. Woodland (1979)]. Since $\Omega$ is not indexed on $t$, homoskedasticity is being assumed; this is always more likely to hold if the $y_{ti}$'s are the budget *shares* of the goods, not quantities or expenditures. Using budget shares as dependent variables also ensures that the $R^2$ statistics mean something. Predicting better than $w_{it} = \alpha_i$ is an achievement (albeit a mild one), while with quantities or expenditures, $R^2$ tend to be extremely high no matter how poor the model.

Given the variance–covariance matrix $\Omega$, typical element $\omega_{ij}$, the MLE's of $\beta$, $\tilde{\beta}$ say, satisfy the first-order conditions, for all $i$,

$$\sum_t \sum_l \sum_k \frac{\partial f_{tk}}{\partial \beta_i} \omega^{kl} \{ y_{tl} - f_{tl}(\tilde{\beta}) \} = 0, \tag{47}$$

where $\omega^{kl}$ is the $(\cdot k, l)$th element of $\Omega^{-1}$. These equations also define the linear or non-linear GLS estimator. Since $\Omega$ is usually unknown, it can be replaced by its maximum likelihood estimator,

$$\tilde{\omega}_{ij} = \frac{1}{T} \sum_t \{ y_{ti} - f_{ti}(\tilde{\beta}) \} \{ y_{tj} - f_{tj}(\tilde{\beta}) \}. \tag{48}$$

If $\tilde{\omega}_{ij}$ replaces $\omega_{ij}$ in (47) and (47) and (48) are solved simultaneously, $\tilde{\beta}$ and $\tilde{\Omega}$ are the full-information maximum likelihood estimators (FIML). Alternatively, some consistent estimator of $\beta$ can be used in place of $\tilde{\beta}$ in (48) and the resulting $\tilde{\Omega}$ used in (47); the resulting estimates of $\beta$ will be asymptotically equivalent to FIML. Zellner's (1962) seemingly unrelated regression technique falls in this class, see also Gallant (1975) and the survey by Srivastava and Dwivedi (1979) for variants. *Consistency* of estimation of $\tilde{\beta}$ in (47) is unaffected by the choice of $\Omega$; the MLE's of $\beta$ and $\Omega$ are asymptotically independent, as calculation of the information matrix will show. All this is standard enough, except possibly for computation, but the use of standard algorithms such as those of Marquardt (1963), scoring, Berndt, Hall, Hall and Hausman (1974), Newton–Raphson, Gauss–Newton all work well for these models, see Quandt (1984) in this Handbook for a survey. Note also Byron's (1982) technique for estimating very large symmetric systems.

Nevertheless, there are a number of problems, particularly concerned with the estimation of the covariance matrix $\Omega$, and these may be severe enough to make the foregoing estimators undesirable, or even infeasible. Taking feasibility first, note that the estimated covariance matrix $\tilde{\Omega}$ given by (48) is the mean of $T$ matrices each of rank 1 so that its rank cannot be greater than $T$. In consequence, systems for which $(n-1) > T$ cannot be estimated by FIML or SURE if the inverse of the estimated $\tilde{\Omega}$ is required. Even this underestimates the problem. In the linear case (e.g. the Rotterdam system considered below) the demand system becomes the classical multivariate regression model

$$Y = XB + U, \tag{49}$$

with $Y$ a $(T \times (n-1))$ matrix, $X$ a $(T \times K)$ matrix, $B(k \times (n-1))$ and $U(T \times (n-1))$. (The $n$th equation has been dropped). The estimated variance–covariance matrix from (48) is then

$$\tilde{\Omega} = \frac{1}{T} Y' \left( I - X(X'X)^{-1} X' \right) Y. \tag{50}$$

Now the idempotent matrix in backets has rank $(T - k)$ so that the inverse will not exist if $n - 1 > T - k$. Since $X$ is likely to contain at least $n + 2$ variables (prices, the budget and a constant), an eight commodity system would require *at least* 19 observations. Non-linearities and cross-section restrictions can improve matters, but they need not. Consider the following problem, first pointed out to me by Teun Kloek. The AIDS system (19) illustrates most simply, though the problem is clearly a general one. Combine the two parts of (19) into a single set of equations,

$$w_{it} = (\alpha_i - \beta_i \alpha_0) + \beta_i \ln x_t + \sum_j (\gamma_{ij} - \beta_i \alpha_j) \ln p_{jt}$$

$$- \tfrac{1}{2} \beta_i \sum_k \sum_m \gamma_{km} \ln p_{kt} \ln p_{mt} + u_{it}. \tag{51}$$

Not counting $\alpha_0$, which is unidentified, the system (without restrictions) has a total of $(2 + n)(n - 1)$ parameters – $(n - 1)$ $\alpha$'s and $\beta$'s, and $n(n - 1)$ $\gamma$'s – or $(n + 2)$ per equation as in the previous example. But now, each equation has $2 + (n - 1)n$ parameters since all $\gamma$'s always appear. In consequence, if the constant, $\ln x$, $\ln p$, and the cross-terms are linearly independent in the sample, and if $T < 2 + (n - 1)n$, it is possible to choose parameters such that the calcu-lated residuals for any one (arbitrarily chosen) equation will be exactly zero for all sample points. For these parameters, one row and one column of the estimated $\tilde{\Omega}$ will also be zero, its determinant will be zero and the log likelihood (41) or (43) will be infinite. Hence full information MLE's do not exist. In such a case, at least 56 observations would be necessary to estimate an 8 commodity disaggregation. All these cases are variants of the familiar "undersized sample" problem in FIML estimation of simultaneous equation systems and they set upper limits to the amount of commodity disaggregation that can be countenanced on any given time-series data.

Given a singular variance–covariance matrix, for whatever reason, the log likelihood (41) which contains the term $-T/2 \log \det \tilde{\Omega}$, will be infinitely large and FIML estimates do not exist. Nor, in general, can (47) be used to calculate GLS or SURE estimators if a singular estimate of $\Omega$ is employed. However, there are a number of important special cases in which (47) has solutions that can be evaluated even when $\Omega$ is singular (though it is less than clear what is the status of these estimators). For example, in the classical multivariate regression model (49), the solution to (47) is the OLS matrix estimator $\hat{B} = (X'X)^{-1}X'Y$ which does not involve $\Omega$, see e.g. Goldberger (1964, pp. 207–12). Imposing identical *within equation* restrictions on (49), e.g. homogeneity, produces another (restricted) classical model with the same property. With *cross-equation* restrictions of the form $R\beta = r$, e.g. symmetry, for stacked $\beta$, $\tilde{\beta}$, the solution to (47) is

$$\tilde{\beta} = \hat{\beta} + \left\{ \Omega \otimes (X'X)^{-1} \right\} R' \left[ R \left\{ \Omega \otimes (X'X)^{-1} \right\} R' \right]^{-1} (r - R\hat{\beta}), \qquad (52)$$

which, though involving $\Omega$, can still be calculated with $\Omega$ singular provided the matrix in square brackets is non-singular. I have not been able to find the general conditions on (47) that allow solutions of this form, nor is it clear that it is important to do so. General non-linear systems will not be estimable on under-sized samples, and except in the cases given where closed-form solutions exist, attempts to solve (47) and (48) numerically will obviously fail.

The important issue, of course, is the small sample performance of estimators based on near-singular or singular estimates of $\Omega$. In most time series applications with more than a very few commodities, $\tilde{\Omega}$ is likely to be a poor estimator of $\Omega$ and the introduction of very poor estimates of $\Omega$ into the procedure for parame-ter estimation is likely to give rise to extremely inefficient estimates of the latter. Paradoxically, the search for (asymptotic) efficiency is likely to lead, in this case, to much greater (small-sample) inefficiency than is actually obtainable. Indeed it may well be that estimation techniques which do not depend on estimating $\Omega$ will give better estimates in such situations. One possibility is the minimization of the *trace* of the matrix on the right-hand side of (48) rather than its *determinant* as required by FIML. This is equivalent to (non-linear) least squares applied to the sum of the residual sums of squares over each equation and can be shown to be ML if (the true) $\Omega = \sigma^2(I - ii')$ for some $\sigma^2$, see Deaton (1975a, p. 39). There is some general evidence that such methods can dominate SURE and FIML in small samples, see again Srivastava and Dwivedi (1979). Fiebig and Theil (1983) and Theil and Rosalsky (1984) have carried out Monte Carlo simulations of symmetry constrained linear systems, i.e. with estimators of the form (52). The system used has 8 commodities, 15 observations and 9 explanatory variables so that their estimate of $\tilde{\Omega}$ from (50) based on the unconstrained regressions is singular. Fiebig and Theil find that replacing $\Omega$ by $\tilde{\Omega}$ yielded "estimates with greatly reduced efficiency and standard errors which considerably underestimate the true variability of these estimates". A number of alternative specifications for were examined and Theil and Rosalsky found good performance in terms of MSE for Deaton's (1975a) specification $\Omega = \sigma^2(\hat{v} - vv')$ where $v$ is the sample mean of the vector of budget shares and $\hat{v}$ is the diagonal matrix of $v$'s. Their results also give useful information on procedures for evaluating standard errors. Define the matrix $A(\Sigma)$, element $a_{ij}$ by

$$a_{ij}(\Sigma) = \sum_t \sum_l \sum_k \frac{\partial f_{tk}}{\partial \beta_i} \sigma^{kl} \frac{\partial f_t}{\partial \beta_j}, \tag{53}$$

where $\sigma^{kl}$ is the $(k, l)$th element of $\Sigma^{-1}$, so that $\{A(\tilde{\Omega})\}^{-1}$ is the conventionally used (asymptotic) variance–covariance matrix of the FIML estimates $\tilde{\beta}$ from (47). Define also $B(\Sigma, \Omega)$ by

$$b_{ij}(\Sigma, \Omega) = \sum_t \sum_k \sum_l \sum_m \sum_n \frac{\partial f_{tk}}{\partial \beta_i} \sigma^{kl} \omega_{lm} \sigma^{mn} \frac{\partial f_{tn}}{\partial \beta_j}. \tag{54}$$

Hence, if $\beta^*$ is estimated from (47) using some assumed variance–covariance matrix $\bar{\Omega}$ say (as in the experiments reported above), then the variance–covariance matrix $V^*$ is given by

$$V^* = A(\bar{\Omega}) B(\bar{\Omega}, \Omega) A(\bar{\Omega}). \tag{55}$$

Fiebig and Theil's experiments suggest good performance if $\Omega$ in $B(\bar{\Omega}, \Omega)$ is replaced by $\tilde{\Omega}$ from (48).

## 2.4. *Interpretation of results*

It is perhaps not surprising that authors who finally surmounted the obstacles in the way of estimating systems of demand equations should have professed themselves satisfied with their hard won results. Mountaineers are not known for criticising the view from the summit. And certainly, models such as the linear expenditure system, or which embody comparably strong assumptions, yield very high $R^2$ statistics for expenditures or quantities with $t$-values that are usually closer to 10 than to unity. Although there are an almost infinite number of studies using the linear expenditure system from which to illustrate, almost certainly the most comprehensive is that by Lluch, Powell and Williams (1977) who fit the model (or a variant) to data from 17 developed and developing countries using an eightfold disaggregation of commodities. Of the 134 $R^2$ statistics reported (for 2 countries 2 of the groups were combined) 40 are greater than 0.99, 104 are greater than 0.95 and only 14 are below 0.90. (Table 3.9 p. 49). The parameter estimates nearly all "look sensible" and conform to theoretical restrictions, i.e. marginal propensities to consume are positive yielding, in the case of the linear expenditure system, a symmetric negative semi-definite Slutsky matrix. However, as is almost invariably the case with the linear expenditure system, the estimated residuals display substantial positive autocorrelation. Table 3.10 in Lluch, Powell and Williams displays Durbin–Watson statistics for all countries and commodities: of the 134 ratios, 60 are less than 1.0 and only 15 are greater than 2.0. Very similar results were found in my own, Deaton (1975a), application of the linear expenditure system to disaggregated expenditures in post-war Britain. Such results suggest that the explanatory power of the model reflects merely the common upward time trends in individual and total expenditures. The estimated $\beta$ parameters in (33), the marginal propensities to consume, will nevertheless be sensible, since the model can hardly fail to reflect the way in which individual expenditures evolve relative to their sum over the sample as a whole. Obtaining sensible estimates of marginal propensities to spend on time-series data is not an onerous task. Nevertheless, the model singularly fails to account for variations around trend, the high $R^2$ statistics could be similarly obtained by replacing total expenditure by virtually any trending variable, and the $t$-values are likely to be grossly overestimated in the presence of the very severe autocorrelation, see, e.g. Malinvaud (1970, pp. 521–2) and Granger and Newbold (1974). In such circumstances, the model is almost certainly a very poor approximation to whatever process actually generated the data and should be abandoned in favor of more appropriate alternatives. It makes little sense to "treat" the autocorrelation by transforming the residuals by a Cochrane–Orcutt type technique, either based on (44) with a common parameter, or using a full vector autoregressive specification. [See Hendry (1980) for some of the consequences of trying to do so in similar situations.]

In spite of its clear misspecifications, there may nevertheless be cases where the linear expenditure system or a similar model may be the best that can be done. Because of its very few parameters, $(2n-1)$ for an $n$ commodity system, it can be estimated in situations (such as the LDC's in Lluch, Powell and Williams book) where data are scarce and less parsimonious models cannot be used. In such situations, it will at the least give a theoretically consistent interpretation of the data, albeit one that is probably wrong. But in the absence of alternatives, this may be better than nothing. Even so, it is important that such applications be seen for what they are, i.e. untested theory with "sensible" parameters, and not as fully-tested data-consistent models.

## 2.5. *Flexible functional forms*

The immediately obvious problem with the linear expenditure system is that it has too few parameters to give it a reasonable chance of fitting the data. Referring back to (33) and dividing through by $p_i$, it can be seen that the $\gamma_i$ parameters are essentially intercepts and that, apart from them, there is only one free parameter per equation. Essentially, the linear expenditure system does little more than fit bivariate regressions between individual expenditures and their total. Of course, the prices also enter the model but all own- and cross-price effects must also be allowed for within the two parameters per equation, one of which is an intercept. Clearly then, in interpreting the results from such a model, for example, total expenditure elasticities, own and cross-price elasticities, substitution matrices, and so on, there is no way to sort out which numbers are determined by measurement and which by assumption. Certainly, econometric analysis requires the application of prior reasoning and theorizing. But it is not helped if the separate influences of measurement and assumption cannot be practically distinguished.

Such difficulties can be avoided by the use of what are known as "flexible functional forms," Diewert (1971). The basic idea is that the choice of functional form should be such as to allow at least one free parameter for the measurement of each effect of interest. For example, the basic linear regression with intercept is a flexible functional form. Even if the true data generation process is not linear, the linear model without parameter restrictions can offer a first-order Taylor approximation around at least one point. For a system of $(n-1)$ independent demand functions, $(n-1)$ intercepts are required, $(n-1)$ parameters for the total expenditure effects and $n(n-1)$ for the effects of the $n$ prices. Barnett (1983b) offers a useful discussion of how Diewert's definition relates to the standard mathematical notions of approximation.

Flexible functional form techniques can be applied either to demand functions or to preferences. For the former, take the differential of (9) around some convenient point, i.e.

$$dq_i = h_{i0} + h_{iu}\,du + \sum_j s_{ij}\,dp_j. \tag{56}$$

But from (10) and (14)

$$d\ln u = \left(d\ln x - \sum_k w_k\,d\ln p_k\right)\cdot(\partial\ln c/\partial\ln u)^{-1}, \tag{57}$$

so that writing $dq_i = q_i\,d\ln q_i$ and multiplying (56) by $p_i/x$, the approximation becomes

$$w_i\,d\ln q_i = a_i + b_i(d\ln x - w\cdot d\ln p) + \sum_j c_{ij}\,d\ln p_j, \tag{58}$$

where

$$a_i = p_i h_{i0}/x$$
$$b_i = \frac{up_i h_{iu}}{x}\left(\frac{\partial\ln c}{\partial\ln u}\right)^{-1} = p_i\frac{\partial q_i}{\partial x} \tag{59}$$
$$c_{ij} = p_i s_{ij} p_j/x.$$

Eq. (58), with $a_i$, $b_i$ and $c_{ij}$ parametrized, is the *Rotterdam* system of Barten (1966), (1967), (1969) and Theil (1965), (1975b), (1976). It clearly offers a local first-order approximation to the underlying relationship between $q$, $x$ and $p$.

There is, of course, no guarantee that a function $h_i(u, p)$ exists which has $a_i$, $b_i$ and $c_{ij}$ constant. Indeed, if it did, Young's theorem gives $h_{iuj} = h_{iju}$ which, from (59), is easily seen to hold only if $c_{ij} = -(\delta_{ij}b_i - b_i b_j)$. If imposed, this restriction would remove the system's ability to act as a flexible functional form. (In fact, the restriction implies unitary total expenditure and own-price elasticities). Contrary to assertions by Phlips (1974, 1983), Yoshihara (1969), Jorgenson and Lau (1976) and others, this only implies that it is not sensible to impose the restriction; it does not affect the usefulness of (58) for approximation and study of the true demands via the approximation, see also Barten (1977) and Barnett (1979b).

Flexible functional forms can also be constructed by approximating *preferences* rather than demands. By Shephard's Lemma, an order of approximation in prices (or quantities) – but *not* in utility – is lost by passing from preferences to demands, so that in order to guarantee a first-order linear approximation in the latter, *second-order* approximation must be guaranteed in preferences. Beyond that, one can freely choose to approximate the direct utility function, the indirect utility function, the cost-function or the distance function provided only that the appropriate quasi-concavity, quasi-convexity, concavity and homogeneity restrictions are observed. The best known of these approximations is the *translog*, Sargan (1971), Christensen, Jorgenson and Lau (1975) and many subsequent applications. See in particular Jorgenson, Lau and Stoker (1982) for a comprehensive treatment. The *indirect translog* gives a quadratic approximation to the indirect function $\psi^*(r)$ for normalized prices, and then uses (14) to derive the system of share equations. The forms are

$$\psi^*(r) = \alpha_0 + \sum_k \alpha_k \ln r_k + \tfrac{1}{2} \sum_k \sum_j \beta_{kj}^* \ln r_k \ln r_j \tag{60}$$

$$w_i = \frac{\alpha_i + \sum_j \beta_{ij} \ln r_j}{\sum_k \alpha_k + \sum_k \sum_j \beta_{kj} \ln r_j}, \tag{61}$$

where $\beta_{ij} = \tfrac{1}{2}(\beta_{ij}^* + \beta_{ji}^*)$. In estimating (61), some normalization is required, e.g. that $\sum_k \alpha_k = 1$. The *direct translog* approximates the direct utility function as a quadratic in the vector $q$ and it yields an equation of the same form as (61) with $w_i$ on the left-hand side but with $q_i$ replacing $r_i$ on the right. Hence, while (61) views the budget share as being determined by quantity adjustment to exogenous price to outlay ratios, the direct translog views the share as adapting by prices adjusting to exogenous quantities. Each could be appropriate under its own assumptions, although presumably not on the same set of data. Yet another flexible functional form with close affinities to the translog is the second-order approximation to the cost function offered by the AIDS, eqs. (17), (18) and (19) above. Although the translog considerably predates the AIDS, the latter is a good deal simpler to estimate, at least if the price index $\ln P$ can be adequately approximated by some fixed pre-selected index.

The AIDS and translog models yield demand functions that are first-order flexible subject to the theory, i.e. they automatically possess symmetric substitution matrices, are homogeneous, and add up. However, trivial cases apart, the AIDS cost function will not be *globally* concave nor the translog indirect utility function globally convex, though they can be so over a restricted range of $r$ (see below). The functional forms for both systems are such that, by relaxing certain restrictions, they can be made first-order flexible without theoretical restrictions, as is the Rotterdam system. For example, in the AIDS, eq. (19), the restrictions $\gamma_{ij} = \gamma_{ji}$ and $\sum_j \gamma_{ij} = 0$ can be relaxed while, in the indirect translog, eq. (61), $\beta_{ij} = \beta_{ji}$ can be relaxed and $\ln x$ included as a separate variable without necessarily assuming that its coefficient equals $-\sum \beta_{ij}$. Now, if the theory is correct, and the flexible functional form is an adequate representation of it over the data, the restrictions should be satisfied, or at least not significantly violated. Similarly, for the Rotterdam system, if the underlying theory is correct, it might be expected that its approximation by (58) would estimate derivatives conforming to the theoretical restrictions. From (59), homogeneity requires $\sum c_{ij} = 0$ and symmetry $c_{ij} = c_{ji}$. Negative semi-definiteness of the Slutsky matrix can also be imposed (globally for the Rotterdam model and at a point for the other models) following the work of Lau (1978) and Barten and Geyskens (1975).

The AIDS, translog, and Rotterdam models far from exhaust the possibilities and many other flexible functional forms have been proposed. Quadratic logarithmic approximations can be made to distance and cost functions as well as to utility functions. The direct quadratic utility function $u = (q - a)'A(q - a)$ is clearly flexible, though it suffers from other problems such as the existence of "bliss" points, see Goldberger (1967). Diewert (1973b) suggested that $\psi^*(r)$ be approximated by a "Generalized Leontief" model

$$
\psi^*(r) = \left\{ \delta_0 + 2\sum_i \delta_i r_i^{1/2} + \sum_i \sum_j \gamma_{ij} r_i^{1/2} r_j^{1/2} \right\}^{-1}. \tag{62}
$$

This has the nice property that it is globally quasi-convex if $\delta_i \geq 0$ and $\gamma_{ij} \geq 0$ for all $i$, $j$; it also generalizes Leontief since with $\delta_0 = \delta_i = 0$ and $\gamma_{ij} = 0$ for $i \neq j$, $\psi^*(r)$ is the indirect utility function corresponding to the Leontief preferences (2). Berndt and Khaled (1979) have, in the production context, proposed a further generalization of (62) where the $\frac{1}{2}$ is replaced by a parameter, the "generalized Box–Cox" system.

There is now a considerable body of literature on testing the symmetry and homogeneity restrictions using the Rotterdam model, the translog, or these other approximations, see, e.g. Barten (1967), (1969), Byron (1970a), (1970b), Lluch (1971), Parks (1969), Deaton (1974a), (1978), Deaton and Muellbauer (1980b), Theil (1971a), (1975b), Christensen, Jorgensen and Lau (1975), Christensen and Manser (1977), Berndt, Darrough and Diewert (1977), Jorgenson and Lau (1976), and Conrad and Jorgenson (1979). Although there is some variation in results through different data sets, different approximating functions, different estimation and testing strategies, and different commodity disaggregations, there is a good deal of accumulated evidence rejecting the restrictions. The evidence is strongest for homogeneity, with less (or perhaps no) evidence against symmetry over and above the restrictions embodied in homogeneity. Clearly, for any one model, it is impossible to separate failure of the model from failure of the underlying theory, but the results have now been replicated frequently using many different functional forms, so that it seems implausible that an inappropriate specification is at the root of the difficulty. There are many possible substantive reasons why the theory as presented might fail, and I shall discuss several of them in subsequent sections. However, there are a number of arguments questioning this sort of procedure for testing. One is a statistical issue, and questions have been raised about the appropriateness of standard statistical tests in this context; I deal with these matters in the next subsection. The other arguments concern the nature of flexible functional forms themselves.

Empirical work by Wales (1977), Thursby and Lovell (1978), Griffin (1978), Berndt and Khaled (1979), and Guilkey and Lovell (1980) cast doubt on the ability of flexible functional forms both to mimic the properties of actual preferences and technologies, and to behave "regularly" at points in price-outlay space other than the point of local approximation (i.e. to generate non-negative, downward sloping demands). Caves and Christensen (1980) investigated theoretically the global properties of the (indirect) translog and the generalized Leontief forms. For a number of two and three commodity homothetic and non-homothetic systems, they set the parameters of the two systems to give the same pattern of budget shares and substitution elasticities at a point in price space, and then mapped out the region for which the models remained regular. Note that regularity is a mild requirement; it is a minimal condition and does not by itself suggest that the system is a good approximation to true preferences or behavior. It is not possible here to reproduce Caves and Christensen's diagrams, nor do the authors give any easily reproducible summary statistics. Nevertheless, although both systems *can* do well (e.g. when substitutability is low so that preferences are close to Leontief, the GL is close to globally regular, and similarly for the translog when preferences are close to Cobb–Douglas), there are also many cases where the regular regions are worryingly small. Of course, these results apply only to the translog and the GL systems, but I see no reason to suppose that similar problems would not occur for the other flexible functional forms discussed above.

These results raise questions as to whether Taylor series approximations, upon which most of these functional forms are based, are the best type of approximations to work with, and there has been a good deal of recent activity in exploring alternatives. Barnett (1983a) has suggested that Laurent series expansions are a useful avenue to explore. The Laurent expansion of a function $f(x)$ around the point $x_0$ takes the form

$$f(x) = \sum_{n=-\infty}^{+\infty} a_n(x-x_0)^n, \tag{63}$$

and Barnett has suggested generalizing the GL form (62) to

$$\{\psi^*(r)\}^{-1} = a_0 + 2a'v + v'Av - 2b'\bar{v} - \bar{v}'B\bar{v}, \tag{64}$$

where $v_i = r_i^{1/2}$ and $\bar{v}_i = r_i^{-1/2}$. The resulting demand system has too many parameters to be estimated in most applications, and has more than it needs to be a second-order flexible functional form. To overcome this, Barnett suggests setting $b = 0$, the diagonal elements of $B$ to zero, and forcing the off-diagonal elements of both $A$ and $B$ to be non-negative (the Laurent model (64) like the GL model (62) is globally regular if all the parameters are non-negative). The resulting budget equations are

$$w_i = \left( a_i v_i + a_{ii} v_i + \sum_{j \neq i} a_{ij}^2 v_i v_j + \sum_{j \neq i} b_{ij}^2 \bar{v}_j \bar{v}_i \right) / D, \tag{65}$$

where $D$ is the sum over $i$ of the bracketed expression. Barnett calls this the miniflex Laurent model. The squared terms guarantee non-negativity, but are likely to cause problems with multiple optima in estimation. Barnett and Lee (1983) present results comparable to those of Caves and Christensen's which suggest that the miniflex Laurent has a substantially larger regular region than either translog or GL models.

A more radical approach has been pioneered by Gallant, see Gallant (1981), and Gallant and Golub (1983), who has shown how to approximate indirect utility functions using Fourier series. Interestingly, Gallant replicates the Christensen, Jorgenson and Lau (1975) rejection of the symmetry restriction, suggesting that their rejection is not caused by the approximation problems of the translog. Fourier approximations are superior to Taylor approximations in a number of ways, not least in their ability to keep their approximating qualities in the face of the separability restrictions discussed in Section 4 below. However, they are also heavily parametrized and superior approximation may be being purchased at the expense of low precision of estimation of key quantities. Finally, many econometricians are likely to be troubled by the sinusoidal behavior of fitted demands when projected outside the region of approximation. There is something to be said for using approximating functions that are themselves plausible for preferences and demands.

The whole area of flexible functional forms is one that has seen enormous expansion in the last five years and perhaps the best results are still to come. In particular, other bases for spanning function space are likely to be actively explored, see, e.g. Barnett and Jones (1983).

## 2.6. Statistical testing procedures

The principles involved are most simply discussed within a single model and for convenience I shall use the Rotterdam system written in the form, $i = 1, \ldots, (n-1)$

$$w_i \, \mathrm{d} \ln q_i = a_i + b_i \, \mathrm{d} \ln \bar{x}_t + \sum_j \gamma_{ij} \, \mathrm{d} \ln p_j + u_{it}, \tag{66}$$

where $d \ln \bar{x}_t$ is an abbreviated form of the term in (58) and, in practice, the differentials would be replaced by finite approximations, see Theil (1975b, Chapter 2) for details. I shall omit the $n$th equation as a matter of course so that $\Omega$ stands for the $(n-1) \times (n-1)$ variance–covariance matrix of the $u$'s.

The $u_t$ vectors are assumed to be identically and independently distributed as $N(0, \Omega)$. I shall discuss the testing of two restrictions: *homogeneity* $\sum_j \gamma_{ij} = 0$, and symmetry, $\gamma_{ij} = \gamma_{ji}$.

Equation (66) is in the classical multivariate regression form (49), so equation by equation OLS yields SURE and FIML estimates. Let $\hat{\beta}$ be the stacked vector of OLS estimates and $\hat{\Omega}$ for the unrestricted estimate of the variance–covariance matrix (50). If the matrix of unrestricted residuals $Y - X\hat{B}$ is denoted by $\hat{E}$, (50) takes the form

$$\hat{\Omega} = T^{-1} \hat{E}' \hat{E}. \tag{67}$$

Testing homogeneity is relatively straightforward since the restrictions are *within* equation restrictions. A simple way to proceed is to substitute $\gamma_{in} = -\sum_1^{n-1} \gamma_{ij}$ into (66) to obtain the restricted model

$$w_i \, d \ln q_i = a_i + b_i \, d \ln \bar{x}_t + \sum_{j=1}^{n-1} \gamma_{ij} (d \ln p_j - d \ln p_n), \tag{68}$$

and re-estimate. Once again OLS is SURE is FIML and the restriction can be tested equation by equation using standard text-book $F$-tests. These are *exact* tests and no problems of asymptotic approximation arise. For examples, see Deaton and Muellbauer's (1980b) rejections of homogeneity using AIDS. If an overall test is desired, a Hotelling $T^2$ test can be constructed for the system as a whole, see Anderson (1958 pp. 207–10) and Laitinen (1978). Laitinen also documents the divergence between Hotelling's $T^2$ and its limiting $\chi^2$ distribution when the sample size is small relative to the number of goods, see also Evans and Savin (1982). In consequence, homogeneity should *always* be tested using exact $F$ or $T^2$ statistics and *never* using asymptotic test statistics such as uncorrected Wald, likelihood ratio, or Lagrange multiplier tests. However, my reading of the literature is that the rejection of homogeneity in practice tends to be confirmed using exact tests and is not a statistical illusion based on the use of inappropriate asymptotics.

Testing *symmetry* poses much more severe problems since the presence of the cross-equation restrictions makes estimation more difficult, separates SUR from FIML estimators and precludes exact tests. Almost certainly the simplest testing procedure is to use a Wald test based on the unrestricted (or homogeneous) estimates. Define $R$ as the $\frac{1}{2}n(n-1) \times (n-1)(n+2)$ matrix representing the symmetry (and homogeneity) restrictions on $\beta$, so that

$$(R\beta)' = (\gamma_{12} - \gamma_{21}, \gamma_{13} - \gamma_{31}, \ldots, \gamma_{(n-1)n} - \gamma_{n(n-1)}). \tag{69}$$

Then, under the null hypothesis of homogeneity and symmetry combined,

$$W_1 = \hat{\beta}' R' \left[ R \left\{ \hat{\Omega} \otimes (X'X)^{-1} \right\} R' \right]^{-1} R\hat{\beta}, \tag{70}$$

is the Wald test statistic which is asymptotically distributed as $\chi^2_{1/2n(n-1)}$. Apart from the calculation of $W_1$ itself, computation requires no more than OLS estimation. Alternatively, the symmetry constrained estimator $\tilde{\beta}$ given by (52) with $r = 0$, can be calculated. From this, restricted residuals $E$ can be derived, and a new (restricted) estimate of $\Omega$, $\tilde{\Omega}$, i.e.

$$\tilde{\Omega} = T^{-1}\tilde{E}'\tilde{E}. \tag{71}$$

The new estimate of $\tilde{\Omega}$ can be substituted into (52) and iterations continued to convergence yielding the FIML estimators of $\beta$ and $\Omega$. Assume that this process has been carried out and that (at the risk of some notational confusion) $\tilde{\beta}$ and $\tilde{\Omega}$ are the final estimates. A likelihood ratio test can then be computed according to

$$W_2 = T \ln\{\det \tilde{\Omega}/\det \hat{\Omega}\}, \tag{72}$$

and $W_2$ is also asymptotically distributed as $\chi^2_{1/2n(n-1)}$. Finally, there is the Lagrange multiplier, or score test, which is derived by replacing $\hat{\Omega}$ in (70) by $\tilde{\Omega}$, so that

$$W_3 = \hat{\beta}' R' \left[ R \left\{ \tilde{\Omega} \otimes (X'X)^{-1} \right\} R' \right]^{-1} R\hat{\beta}, \tag{73}$$

with again the same limiting distribution.

From the general results of Berndt and Savin (1977), it is known that $W_1 \geq W_2 \geq W_3$; these are mechanical inequalities that always hold, no matter what the configuration of data, parameters, and sample size. In finite samples, with inaccurate and inefficient estimates of $\Omega$, the asymptotic theory may be a poor approximation and the difference between the three statistics may be very large. In my own experience I have encountered a case with 8 commodities and 23 observations where $W_1$ was more than a hundred times greater than $W_3$. Meisner (1979) reports experiments with the Rotterdam system in which the null hypothesis was correct. With a system of 14 equations and 31 observations, $W_1$ rejected symmetry at 5% 96 times out of 100 and at 1% 91 times out of 100. For 11 equations the corresponding figures were 50 and 37. Bera, Byron and Jarque (1981) carried out similar experiments for $W_2$ and $W_3$. From the inequalities, we know that rejections will be less frequent, but it was still found that, with $n$ large relative to $(T - k)$ both $W_2$ and $W_3$ grossly over-rejected.

These problems for testing symmetry are basically the same as those discussed for estimation in (2.3) above; typical time series are not long enough to give reliable estimates of the variance–covariance matrix, particularly for large systems. For estimation, and for the testing of within equation restrictions, the difficulties can be circumvented. But for testing cross-equation restrictions, such as symmetry, the problem remains. For the present, it is probably best to suspend judgment on the existing tests of symmetry (positive or negative) and to await theoretical or empirical developments in the relevant test statistics. [See Byron and Rosalsky (1984) for a suggested ad hoc size correction that appears to work well in at least some situations.]

## 2.7.  Non-parametric tests

All the techniques of demand analysis so far discussed share a common approach of attempting to fit demand functions to the observed data and then enquiring as to the compatibility of these fitted functions with utility theory. If unlimited experimentation were a real possibility in economics, demand functions could be accurately determined. As it is, however, what is observed is a *finite* collection of pairs of quantity and price vectors. It is thus natural to argue that the basic question is whether or not these observed pairs are consistent with any preference ordering whatever, bypassing the need to specify particular demands or preferences. It may well be true that a given set of data is perfectly consistent with utility maximization and yet be very poorly approximated by AIDS, the translog, the Rotterdam system or any other functional form which the limited imagination of econometricians is capable of inventing.

Non-parametric demand analysis takes a direct approach by searching over the price-quantity vectors in the data for evidence of inconsistent choices. If these do exist, a utility function exists and algorithms exist for constructing it (or at least one out of the many possible). The origins of this type of analysis go back to Samuelson's (1938) introduction of revealed preference analysis. However, the recent important work on developing test criteria is due to Hanoch and Rothschild (1972) and especially to Afriat (1967), (1973), (1976), (1977) and (1981). Unfortunately, some of Afriat's best work has remained unpublished and the published work has often been difficult for many economists to understand and assimilate. However, as the techniques involved have become more widespread in economics, other workers have taken up the topic, see the interpretative essays by Diewert (1973a) and Diewert and Parkan (1978) – the latter contains actual test results – and also the recent important work by Varian (1982, 1983).

Afriat proposes that a finite set of data be described as cyclically consistent if, for any "cycle", $a, b, c, \ldots, r, a$ of indices, $p^a \cdot q^a \geq p^a \cdot q^b$, $p^b \cdot q^b \geq p^b \cdot q^c$,

$\ldots, p^r q^r \geq p^r q^a$, then it must be true that $p^a \cdot q^a = p^a \cdot q^b$, $p^b q^b = p^b q^c, \ldots, p^r q^r = p^r q^a$. He then shows that cyclical consistency is necessary and sufficient for the finite set of points to be consistent with the existence of a continuous, non-satiated, concave and monotonic utility function. Afriat also provides a constructive method of evaluating such a utility function. Varian (1982) shows that cyclical consistency is equivalent to a "generalized axiom of revealed preference" (GARP) that is formulated as follows. Varian defines $q^i$ as *strictly directly revealed preferred* to $q$, written $q^i P^0 q$ if $p^i q^i > p^i q$, i.e. $q^i$ was bought at $p^i$ even though $q$ cost less. Secondly $q^i$ is *revealed preferred* to $q$, written $q^i R q$, if $p^i q^i \geq p^i q^j$, $p^j q^j \geq p^j q^k, \ldots, p^m q^m \geq p^m q$, for some sequence of observations $(q^i, q^j, \ldots, q^m)$, i.e. $q^i$ is indirectly or directly (weakly) revealed preferred to $q$. GARP then states that $q^i R q^j$ implies not $q^j P^0 q^i$, and all the nice consequences follow. Varian has also supplied an efficient and easily used algorithm for checking GARP, and his methods have been widely applied. Perhaps not surprisingly, the results show few conflicts with the theory, since on aggregate time series data, most quantities consumed increase over time so that contradictions with revealed preference theory are not possible; each new bundle was unobtainable at the prices and incomes of all previous periods.

Since these methods actually allow the construction of a well-behaved utility function that accounts exactly for most aggregate time-series data, the rejections of the theory based on parametric models (and on semi-parametric models like Gallant's Fourier system) must result from rejection of functional form and not from rejection of the theory per se. Of course, one could regard the non-parametric utility function as being a very profligately parametrized parametric utility function, so that if the object of research is to find a reasonably parsimonious theory-consistent formulation, the non-parametric results are not very helpful.

Afriat's and Varian's work, in particular see Afriat (1981) and Varian (1983), also allows testing of restricted forms of preferences corresponding to the various kinds of separability discussed in Section 4. Varian has also shown how to handle goods that are rationed or not freely chosen, as in Section 6 below. Perhaps most interesting are the tests for homotheticity, a condition that requires the utility function to be a monotone increasing transform of a linearly homogeneous function and which implies that all total expenditure elasticities are unity. Afriat (1977) showed that for two periods, 0 and 1, the necessary and sufficient condition for consistency with a homothetic utility function is that the Laspeyres price index be no less than the Paasche price index, i.e. that

$$\frac{p^1 \cdot q^0}{p^0 \cdot q^0} \geq \frac{p^1 \cdot q^1}{p^0 \cdot q^1}. \tag{74}$$

For many periods simultaneously, Afriat (1981) shows that the Laspeyres index between any two periods $i$ and $j$, say, should be no less than the chain-linked Paasche index obtained by moving from $i$ to $j$ in any number of steps. Given that no one using *any* parametric form has ever suggested that all total expenditure elasticities are unity, it comes as something of a surprise that the Afriat condition appears to be acceptable for an 111 commodity disaggregation of post-war U.S. data, see Manser and McDonald (1984).

Clearly, more work needs to be done on reconciling parametric and non-parametric approaches. The non-parametric methodology has not yet been successfully applied to cross-section data because it provides no obvious way of dealing with non-price determinants of demand. There are also difficulties in allowing for "disturbance terms" so that failures of, e.g. GARP, can be deemed significant or insignificant, but see the recent attempts by Varian (1984) and by Epstein and Yatchew (1985).

## 3. Cross-section demand analysis

Although the estimation of complete sets of demand functions on time-series data has certainly been the dominant concern in demand analysis in recent years, a much older literature is concerned with the analysis of "family budgets" using sample-survey data on cross-sections of households. Until after the Second World War, such data were almost the only sources of information on consumer behavior. In the last few years, interest in the topic has once again become intense as more and more such data sets are being released in their individual microeconomic form, and as computing power and econometric technique develop to deal with them. In the United Kingdom, a regular Family Expenditure Survey with a sample size of 7000 households has been carried out annually since 1954 and the more recent tapes are now available to researchers. The United States has been somewhat less forward in the area and until recently, has conducted a Consumer Expenditure Survey only once every decade. However, a large rotating panel survey has recently been begun by the B.L.S. which promises one of the richest sets of data on consumer behavior ever available and it should help resolve many of the long-standing puzzles over differences between cross-section and time-series results. For example, most very long-run time-series data sets which are available show a rough constancy of the food share, see Kuznets (1962), (1966), Deaton (1975c). Conversion to farm-gate prices, so as to exclude the increasing component of transport and distribution costs and built in services, gives a food share which declines, but does so at a rate which is insignificant in comparison to its rate of decline with income in cross-sections [for a survey of cross-section results, see Houthakker (1957)]. Similar problems exist with other categories of expenditure as well as with the relationship between total expenditure and income.

There are also excellent cross-section data for many less developed countries, in particular from the National Sample Survey in India, but also for many other South–East Asian countries and for Latin America. These contain a great wealth of largely unexploited data, although the pace of work has recently been increasing, see, for example, the survey paper on India by Bhattacharrya (1978), the work on Latin America by Musgrove (1978), Howe and Musgrove (1977), on Korea by Lluch, Powell and Williams (1977, Chapter 5) and on Sri Lanka by Deaton (1981c).

In this section, I deal with four issues. The first is the specification and choice of functional form for Engel curves. The second is the specification of how expenditures vary with household size and composition. Third, I discuss a group of econometric issues arising particularly in the analysis of micro data with particular reference to the treatment of zero expenditures, including a brief assessment of the Tobit procedure. Finally, I give an example of demand analysis with a non-linear budget constraint.

## 3.1. Forms of Engel curves

This is very much a traditional topic to which relatively little has been added recently. Perhaps the classic treatment is that of Prais and Houthakker (1955) who provide a list of functional forms, the comparison of which has occupied many manhours on many data sets throughout the world. The Prais–Houthakker methodology is unashamedly pragmatic, choosing functional forms on grounds of fit, with an attempt to classify particuiar forms as typically suitable for particular types of goods, see also Tornqvist (1941), Aitchison and Brown (1954–5), and the survey by Brown and Deaton (1972) for similar attempts. Much of this work is not very edifying by modern standards. The functional forms are rarely chosen with any theoretical model in mind, indeed all but one of Prais and Houthakker's Engel curves are incapable of satisfying the adding-up requirement, while, on the econometric side, satisfactory methods for comparing different (non-nested) functional forms are very much in their infancy. Even the apparently straightforward comparison between a double-log and a linear specification leads to considerable difficulties, see the simple statistic proposed by Sargan (1964) and the theoretically more satisfactory (but extremely complicated) solution in Aneuryn–Evans and Deaton (1980).

More recent work on Engel curves has reflected the concern in the rest of the literature with the theoretical plausibility of the specification. Perhaps the most general results are those obtained in a paper by Gorman (1981), see also Russell (1983) for alternative proofs. Gorman considers Engel curves of the general form

$$w_i = \sum_{r \in R} a_{ir}(p)\phi_r(\ln x),\tag{75}$$

where $R$ is some finite set and $\phi_r(\ )$ are a series of functions. If such equations are to be theory consistent, there must exist a cost function $c(u, p)$ such that

$$\frac{\partial \ln c(u, p)}{\partial \ln p_i} = \sum_{r \in R} a_{ir}(p)\phi_r\{\ln c(u, p)\}. \tag{76}$$

Gorman shows that for these partial differential equations to have a solution, (a) the rank of the matrix formed from the coefficients $a_{ir}(p)$ can be no larger than 3 and (b), the functions $\phi_n(\ )$ must take specific restricted forms. There are three generic forms for (75), two of which are reproduced below

$$w_i = a_i(p) + b_i(p)\ln x + d_i(p)\sum_{m=1}^{M} \gamma_m(p)(\ln x)^m \tag{77}$$

$$w_i = a_i(p) + b_i(p)\sum_{\sigma_m \in S_-} \mu_m(p)x^{\sigma_m} + d_i(p)\sum_{\sigma_m \in S_+} \theta_m(p)x^{\sigma_m}, \tag{78}$$

where $S$ is a finite set of elements $\sigma_i$, $S_-$ its negative elements and $S_+$ its positive elements. A third form allows combinations of trigonometrical functions of $x$ capable of approximating a quite general function of $x$. However, note that the $\gamma_m$, $\mu_m$ and $\theta_m$ functions in (77) and (78) are *not* indexed on the commodity subscript $i$, otherwise the rank condition on $a_{ir}$ could not hold.

Equations (77) and (78) provide a rich source of Engel curve specifications and contain as special cases a number of important forms. From (77), with $m = 1$, the form proposed by Working and Leser and discussed above, see (15), is obtained. In econometric specifications, $a_i(p)$ adds to unity and $b_i(p)$ to zero, as will their estimates if OLS is applied to each equation separately. The log quadratic form

$$w_i = a_i(p) + b_i(p)\ln x + d_i(p)(\ln x)^2, \tag{79}$$

was applied in Deaton (1981c) to Sri Lankan micro household data for the food share where the quadratic term was highly significant and a very satisfactory fit was obtained (an $R^2$ of 0.502 on more than 3,000 observations.) Note that, while for a single commodity, higher powers of $\ln x$ could be added, doing so in a complete system would require cross-equation restrictions since, according to (77), the ratios of coefficients on powers beyond unity should be the same for all commodities. Testing such restrictions (and Wald tests offer a very simple method – see Section 4(a) below) provides yet another possible way of testing the theory.

Equation (78) together with $S = \{-1, 1, 2, ..., r, ...\}$ gives general polynomial Engel curves. Because of the rank condition, the quadratic with $S = \{-1, 1\}$ is as general as any, i.e.

$$p_i q_i = b_i^*(p) + a_i(p)x + d_i^*(p)x^2,  \tag{80}$$

where $b_i^*(p) = b_i(p)\mu_m(p)$ and $d_i^*(p) = d_i(p)\theta_m(p)$. This is the "quadratic expenditure system" independently derived by Howe, Pollak and Wales (1979), Pollak and Wales (1978) and (1980). The cost function underlying (80) may be shown to be

$$c(u, p) = \alpha(p) - \frac{\beta(p)}{u + \gamma(p)},  \tag{81}$$

where the links between the $a_i$, $b_i^*$ and $d_i^*$ on the one hand and the $\alpha$, $\beta$ and $\gamma$ on the other are left to the interested reader. (With $\ln c(u, p)$ on the left hand side, (81) also generates the form (79)). This specification, like (79), is also of considerable interest for time-series analysis since, in most such data, the range of variation in $x$ is much larger than that in relative prices and it is to be expected that a higher order of approximation in $x$ than in $p$ would be appropriate. Indeed, evidence of failure of linearity in time-series has been found in several studies, e.g. Carlevaro (1976). Nevertheless, in Howe, Pollak and Wales' (1979) study using U.S. data from 1929–1975 for four categories of expenditure, tests against the restricted version represented by the linear expenditure system yielded largely insignificant results. On grouped British cross-section data pooled for two separate years and employing a threefold categorization of expenditures, Pollak and Wales (1978) obtain a $\chi^2$ values of 8.2 (without demographics) and 17.7 (with demographics) in likelihood ratio tests against the linear expenditure system. These tests have 3 degrees of freedom and are notionally significant at the 5% level (the 5% critical value of a $\chi_3^2$ variate is 7.8) but the study is based on only 32 observations and involves estimation of a $3 \times 3$ unknown covariance matrix. Hence, given the discussion in Section 2.6 above, a sceptic could reasonably remain unconvinced of the importance of the quadratic terms for this particular data set.

Another source of functional forms for Engel curves is the study of conditions under which it is possible to aggregate over consumers and I shall discuss the topic in Section 5 below.

## 3.2. Modelling demographic effects

In cross-section studies, households typically vary in much more than total expenditure; age and sex composition varies from household to household, as do the numbers and ages of children. These demographic characteristics have been the object of most attention and I shall concentrate the discussion around them, but other household characteristics can often be dealt with in the same way, (e.g. race, geographical region, religion, occupation, pattern of durable good owner-ship, and so on). If the vector of these characteristics is a, and superscripts denote individual households, the general model becomes

$$q_i^h = g_i(x^h, p, a^h),$$

(82)

with $g_i$ taken as common and, in many studies, with $p$ assumed to be the same across the sample and suppressed as an argument in the function.

The simplest methodology is to estimate a suitable linearization of (82) and one question which has been extensively investigated in this way is whether there are economies of scale to household size in the consumption of some or all goods. A typical approach is to estimate

$$\ln q_i^h = \alpha_i + \beta_i \ln x^h + \gamma_i \ln n^h + u_i,$$

(83)

where $n^h$ is the (unweighted) number of individuals in the household. Tests are then conducted for whether $(\gamma_i + \beta_i - 1)$ is negative (economies of scale), zero (no economies or diseconomies) or positive (diseconomies of scale), since this magni-tude determines whether, at a given level of per capita outlay, quantity per head decreases, remains constant, or increases. For example, Iyengar, Jain and Srinivasan (1968), using (83) on data from the 17th round of the Indian N.S.S. found economies of scale for cereals and for fuel and light, with roughly constant returns for milk and milk products and for clothing.

A more sophisticated approach attempts to relate the effects of characteristics on demand to their role in preferences, so that the theory of consumer behavior can be used to suggest functional forms for (82) just as it is used to specify relationships in terms of prices and outlay alone. Such models can be used for welfare analysis as well as for the interpretation of demand; I deal with the latter here leaving the welfare applications to Section 7 below. A fairly full account of the various models is contained in Deaton and Muellbauer (1980a, Chapter 8) so that the following is intended to serve as only a brief summary.

Fully satisfactory models of household behavior have to deal both with the specification of needs or preferences at the individual level and with the question of how the competing and complementary needs of different individuals are reconciled within the overall budget constraint. The second question is akin to the usual question of social choice, and Samuelson (1956) suggested that family utility $u$, might be written as

$$u^h = V\{u^1(q^1),\ldots,u^{n^h}(q^{n^h})\},$$

(84)

for the $n^h$ individuals in household $h$. Such a form allows decentralized budgeting over members subject to central (parental) control over members' budgets. Presumably the problems normally inherent in making interpersonal comparisons of welfare are not severe within a family since, typically, such allocations seem to be made in a satisfactory manner. Building on this idea, Muellbauer (1976c) has suggested that utility is equalised within the family (e.g. for a maximin social welfare function), so that if $\gamma^r(u, p)$ is the cost function for individual $r$, the family cost function is given by

$$c^h(u, p) = \sum_{r=1}^{n^h} \gamma^r(u, p) = x, \qquad (85)$$

which, if needs can be linked to, say, age through the $\gamma$ functions, would yield an applicable specification with strong restrictions on behavior. However, such models are somewhat artificial in that they ignore the 'public' or shared goods in family consumption, though suitable modifications can be made. They also lack empirical sharpness in that the consumption vectors of individual family members are rarely observed. The exception is in the case of family labor supply, see Chapter 32 of this volume.

Rather more progress has been made in the specification of needs under the assumption that the family acts as a homogeneous unit. The simplest possibility is that, for a given welfare level, costs are affected multiplicatively by some index depending on characteristics and welfare, i.e.

$$c^h(u^h, p, a^h) = m(a^h, u^h)c(u^h, p), \qquad (86)$$

where $c(u^h, p)$ is the cost function for some reference household type, e.g. one with a single adult. The index $m(a^h, u^h)$ can then be thought of as the number of adult equivalences generated by $a^h$ at the welfare level $u^h$. Taking logarithms and differentiating (86) with respect to $\ln p_i$ gives

$$w_i^h = \frac{\ln c(u^h, p)}{\partial \ln p_i}, \qquad (87)$$

which is independent of $a^h$. Hence, if households face the same prices, those with the same consumption patterns $w_i$ have the same $u^h$, so that by comparing their outlays the ratio of their costs is obtained. By (86), this ratio is the equivalence scale $m(a^h, u^h)$. This procedure derives directly from Engel's (1895) pioneering work, see Prais and Houthakker (1955). In practice, a single good, food, is usually used although there is no reason why the model cannot be applied more generally under suitable specification of the $m$ and $c$ functions in (86), see e.g. Muellbauer

(1977). For examples of the usual practice, see Jackson (1968), Orshansky (1965), Seneca and Taussig (1971) and Deaton (1981c).

Although the Engel model is simple to apply, it has the long recognised disadvantage of neglecting any commodity specific dimension to needs. Common observation suggests that changes in demographic composition cause substitution of one good for another as well as the income effects modelled by (86) and (87). In a paper of central importance to the area, Barten (1964) suggested that household utility be written

$$u^h = v(q^*),\qquad(88)$$

$$q_i^* = q_i/m_i(a^h).\qquad(89)$$

So that, using Pollak and Wales' (1981) later terminology, the demographic variables generate indices which "scale" commodity consumption levels. The Barten model is clearly equivalent to writing the cost function in the form

$$c^h(u^h, p, a^h) = c(u^h\ p^*),\qquad(90)$$

$$p_i^* = p_i m_i(a^h),\qquad(91)$$

for a cost function $c(u, p)$ for the reference household. Hence, if $g_i(x, p)$ are the Marshallian demands for the household, household $h$'s demands are given by

$$q_i^h = m_i(a^h)g_i(x^h, p^*).\qquad(92)$$

Differentiation with respect to $a_j$ gives

$$\frac{\partial \ln q_i}{\partial a_j} = \frac{\partial \ln m_i}{\partial a_j} + \sum_{k=1}^{n} e_{ik}\frac{\partial \ln m_k}{\partial a_j},\qquad(93)$$

where $e_{ik}$ is the cross-price elasticity between $i$ and $k$. Hence, a change in demographic composition has a direct affect through the change in needs (on $m_i$) and an indirect effect through the induced change in the "effective" price structure. It is this recognition of the quasi-price substitution effects of demographic change, that "a penny bun costs threepence when you have a wife and child" that is the crucial contribution of the Barten model. The specification itself may well neglect other important aspects of the problem, but this central insight is of undeniable importance.

The main competition to the Barten specification comes from the model originally due to Sydenstricker and King (1921) but rediscovered and popularized by Prais and Houthakker (1955). This begins from the empirical specification, apparently akin to (89)

$$q_i / m_i(a^h) = f_i(x^h / m_0(a^h)),$$ (94)

where $m_i(a^h)$ is the *specific* commodity scale, and $m_0(a^h)$ is some *general* scale. In contrast to (93), we now have the relationship

$$\frac{\partial \ln q_i}{\partial a_j} = \frac{\partial \ln m_i}{\partial a_j} - e_i \frac{\partial \ln m_0}{\partial a_j},$$ (95)

so that the substitution effects embodied in (93) are no longer present. Indeed, if $x^h / m_0(a^h)$ is interpreted as a welfare indicator (which is natural in the context) (94) can only be made consistent with (88) and (89) if indifference curves are Leontief, ruling out all substitution in response to relative price change, see Muellbauer (1980) for details, and Pollak and Wales (1981) for a different interpretation.

On a single cross-section, neither the Barten model nor the Prais–Houthakker model are likely to be identifiable. That there were difficulties with the Prais–Houthakker formulation has been recognized for some time, see Forsyth (1960) and Cramer (1969) and a formal demonstration is given in Muellbauer (1980). In the Barten model, (93) may be rewritten in matrix notation as

$$F = (I + E)M,$$ (96)

and we seek to identify $M$ from observable information on $F$. In the most favorable case, $E$ may be assumed to be known (and suitable assumptions may make this practical even on a cross-section, see Section 4.2 below). The problem lies in the budget constraint, $p \cdot q = x$ which implies $w'[I + E] = 0$ so that the matrix $(I + E)$ has at most rank $n-1$. Hence, for any given $F$ and $E$, both of which are observable, there exist an infinite number of $M$ matrices satisfying (96). In practice, with a specific functional form, neither $F$ nor $E$ may be constant over households so that the information matrix of the system could conceivably not be singular. However, such identification, based on choice of functional form and the existence of high nonlinearities, is inherently controversial. A much better solution is the use of several cross-sections between which there is price variation and, in a such a case, several quite general functional forms are fully identified. For the Prais–Houthakker model, (95) may be written as

$$F = M - em',$$ (97)

where $m = \partial \ln m^0 / \partial a$. From the budget constraint, $w'F = 0$ so that $m' = w'M$

which yields

$$F = (I - ew')M. \tag{98}$$

Once again $(I - ew')$ is singular, and the identification problem recurs. Here price information is likely to be of less help since, with Leontief preferences, prices have only income effects. Even so, it is not difficult to construct Prais–Houthakker models which identified given sufficient variation in prices.

Since Prais and Houthakker, the model has nevertheless been used on a number of occasions, e.g by Singh (1972), (1973), Singh and Nagar (1973), and McClements (1977) and it is unclear how identification was obtained in these studies. The use of a double logarithmic formulation for $f_i$ helps; as is well-known, such a function cannot add up even *locally*, see Willig (1976), Varian (1978), and Deaton and Muellbauer (1980a, pp 19–20) so that the singularity arguments given above cannot be used. Nevertheless, it seems unwise to rely upon a clear misspecification to identify the parameters of the model. Coondoo (1975) has proposed using an assumed independence of $m_0$ on $x$ as an identifying restriction; this is ingenious but, unfortunately, turns out to be inconsistent with the model. There are a number of other possible means of identification, see Muellbauer (1980), but essentially the only practical method is the obvious one of assuming a priori a value for one of the $m_i$'s. By this means, the model can be estimated and its results compared with those of the Barten model. Some results for British data are given in Muellbauer (1977) (1980) and are summarized in Deaton and Muellbauer (1980a, pp 202–5). In brief, these suggest that each model is rather extreme, the Prais–Houthakker with its complete lack of substitution and the Barten with its synchronous equivalence of demographic and price substitution effects. If both models are normalized to have the same food scale, the Prais–Houthakker model also tends to generate the higher scales for other goods since, unless the income effects are very large, virtually all variations with composition must be ascribed directly to the $m_i$'s. The Barten scales are more plausible but evidence suggests that price effects and demographic effects are not linked as simply as is suggested by (93).

Gorman (1976) has proposed an extension to (90) which appears appropriate in the light of this evidence. In addition to the Barten substitution responses he adds fixed costs of children $\gamma_i(a^h)$ say; hence (90) becomes

$$c^h(u^h, p, a^h) = p \cdot \gamma(a^h) + c(u^h, p^*), \tag{99}$$

with (94) retained as before. Clearly, (99) generates demands of the form

$$q_i^h = \gamma_i(a^h) + g_i(x^h - p \cdot \gamma(a^h), p^*). \tag{100}$$

Pollak and Wales (1981) call the addition of fixed costs "demographic translating" as opposed to "demographic scaling" of the Barten model; the Gorman model (99) thus combines translating and scaling. In their paper, Pollak and Wales test various specifications of translating and scaling. Their results are not decisive but tend to support scaling; with little additional explanatory power from translating once scaling has been allowed for. Note, however, that the translating term in (99) might itself form the starting point for the modelling, just as did the multiplicative term in the Engel model. If the scaling terms in (99) are dropped, so that $p$ replaces $p^*$, and if it is recognized that the child cost term $p \cdot \gamma(a^h)$ is likely to be zero for certain "adult" goods, then for $i$ an adult good, we have

$$q_i^h = h_i(u^h, p),$$
(101)

independent of $a^h$. For all such goods, additional children exert only income effects, a proposition that can be straightforwardly tested by comparing the ratios of child to income derivatives across goods, while families with the same outlay on adult goods can be identified as having the same welfare level. This is the model first proposed by Rothbarth (1943) and later implemented by Henderson (1949–50a) (1949–50b) and Nicholson (1949), see also Cramer (1969). Deaton and Muellbauer (1983) have recently tried to reestablish it as a simply implemented model that is superior to the Engel formulation for applications where computational complexity is a problem.

## 3.3. Zero expenditures and other problems

In microeconomic data on consumers expenditure, it is frequently the case that some units do not purchase some of the commodities, alcohol and tobacco being the standard examples. This is of course entirely consistent with the theory of consumer behavior; for example, two goods (varieties) may be very close to being perfect substitutes so that (sub) utility for the two might be

$$u = \alpha_1 q_1 + \alpha_2 q_2,$$
(102)

so that, if outlay is $x$, the demand functions are

$$q_i = x_i/p_i \quad \text{if } p_i/p_j < \alpha_i/\alpha_j$$
$$= 0 \quad \text{otherwise,}$$
(103)

for $i, j = 1, 2$ and for $p_1\alpha_2 \neq p_2\alpha_1$. It is not difficult to design more complex (and more realistic) models along similar lines. For a single commodity, many of these models can be made formally equivalent to the Tobit, Tobin (1958) model

$$y_i^* = x_i'\beta + u_i$$

$$\left.\begin{aligned} y_i &= y_i^* \\ &= 0 \end{aligned}\right\} \quad \begin{aligned} &\text{if } y_i^* \geq 0 \\ &\text{otherwise,} \end{aligned} \qquad (104)$$

and the estimation of this is well-understood.

However, there are a number of extremely difficult problems in applying the Tobit model to the analysis of consumer behavior. First, there is typically more than one good and whenever the demand for one commodity switches regime (i.e. becomes positive having been zero, or vice versa), there are, in general, regime changes in all the other demands, if only to satisfy the budget constraint. In fact, the situation is a good deal more complex since, as will be discussed in Section 6 below, non-purchase is formally equivalent to a zero ration and the imposition of such rations changes the functional form for other commodities in such a way as to generate both income and substitution effects. With a $n$ goods in the budget, and assuming at least one good purchased, there are $2^{n-1}$ possible regimes, each with its own particular set of functional forms for the non-zero demands. Wales and Woodland (1983) have shown how, in principle, such a problem can be tackled and have estimated such a system for a three good system using a quadratic (direct) utility function. Even with these simplifying assumptions, the estimation is close to the limits of feasibility. Lee and Pitt (1983) have demonstrated that a dual approach is as complicated. An alternative approach may be possible if only a small number (one or two) commodities actually take on zero values in the sample. This is to condition on non-zero values, omitting all observations where a zero occurs, and to allow specifically for the resulting sample selection bias in the manner suggested, for example, by Heckman (1979). This technique has been used by Blundell and Walker (1982) to estimate a system of commodity demands simultaneously with an hours worked equation for secondary workers.

The second problem is that it is by no means obvious that the Tobit specification is correct, even for a single commodity. In sample surveys, zeros frequently occur simply because the item was not bought over a relatively short enumeration period (usually one or two weeks, and frequently less in developing countries). Hence, an alternative to (104) might be

$$y_i^* = x_i'\beta + u_i,$$

$$\left.\begin{aligned} y_i &= y_i^*/\pi_i \\ y_i &= 0 \end{aligned}\right\} \quad \begin{aligned} &\text{with probability } \pi_i, \\ &\text{with probability } (1-\pi_i). \end{aligned} \qquad (105)$$

Hence, if, $p(u_i)$ is the p.d.f. of $u_i$ the likelihood for the model is

$$L = \prod_0 (1 - \pi_i) \prod_+ \pi_i p(\pi_i y_i - x_i'\beta). \tag{106a}$$

This can be maximized directly to estimate $\beta$ and $\pi_i$ given some low parameter specification for $\pi_i$. But note in particular that for $\pi_i = \pi$ for all $i$ and $u_i$ taken as i.i.d. $N(0, \sigma^2)$ the likelihood is, for $n_0$ the number of zero $y_i$'s,

$$L = (1 - \pi)^{n_0} \prod_+ \phi\left( \frac{x_i'\beta}{\pi}, \frac{\sigma}{\pi} \right). \tag{106b}$$

Hence OLS on the positive $y_i$'s alone is consistent and fully efficient for $\beta/\pi$ and $\sigma/\pi$. The MLE of $\pi$ is simply the ratio of the number of positive $y_i$'s to the sample size, so that, in this case, all parameters are easily estimated. If this is the true model, Tobit will not generally be consistent. However, note that (105) allows $y_i$ to be negative (although this may be very improbable) and ideally the Tobit and the binary model should be combined. A not very successful attempt to do this is reported in Deaton and Irish (1984). See also Kay, Keen and Morris (1984) for discussion of the related problem of measuring total expenditure when there are many zeroes.

In my view, the problem of dealing appropriately with zero expenditures is currently one of the most pressing in applied demand analysis. We do not have a theoretically satisfactory and empirically implementable method for modelling zeroes for more than a few commodities at once. Yet all household surveys show large fractions of households reporting zero purchases for some goods. Since household surveys typically contain several thousands observations, it is important that procedures be developed that are also computationally inexpensive.

There are also a number of other problems which are particularly acute in cross-section analysis and are not specific to the Tobit specification. *Heteroscedasticity* tends to be endemic in work with micro data and, in my own practical experience, is extremely difficult to remove. The test statistics proposed by Breusch and Pagan (1979) and by White (1980) are easily applied, and White has proposed an estimator for the variance-covariance matrix which is consistent under heteroscedasticity and does not require any specification of its exact form. Since an adequate specification seems difficult in practice, and since in micro studies efficiency is rarely a serious problem, White's procedure is an extremely valuable one and should be applied routinely in large cross-section regressions. Note, however, that with Tobit-like models, untreated heteroscedasticity generates inconsistency in the parameter estimates, see Chapter 27, thus presenting a much more serious problem. The heteroscedasticity introduced by *grouping* has become less important as grouped data has given way to the analysis of the original micro observations, but see Haitovsky (1973) for a full discussion.

Finally, there are a number of largely unresolved questions about the way in which survey design should be taken into account (if at all) in econometric analysis. One topic is whether or not to use inverse probability weights in regression analysis, see e.g. DuMouchel and Duncan (1983) for a recent discussion. The other concerns the possible implications for regression analysis of Godambe's (1955) (1966) theorem on the non-existence of uniformly minimum variance or maximum likelihood estimators for means in finite populations, see Cassel, Sarndal and Wretman (1977) for a relatively cool discussion.

### 3.4. *Non-linear budget constraints*

Consumer behavior with non-linear budget constraints has been extensively discussed in the labor supply literature, where tax systems typically imply a non-linear relationship between hours worked and income received, see Chapter 32 in this Handbook and especially Hausman (1985). I have little to add to Hausman's excellent treatment, but would nevertheless wish to emphasize the potential for these techniques in demand analysis, particularly in "special"

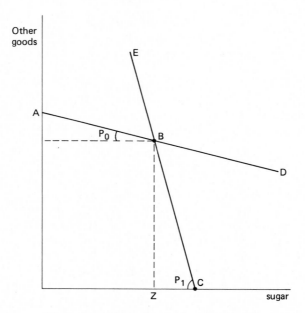

Figure 2.   Budget constraint for a fair price shop.

markets. Housing is the obvious example, but here I illustrate with a simple case based on Deaton (1984). In many developing countries, the government operates so-called "fair price" shops in which certain commodities, e.g. sugar or rice, are made available in limited quantities at subsidized prices. Typically, consumers can buy more than the fair price allocation in the free market at a price $p_1$, with $p_1 > p_0$ the fair price price. Figure 2 illustrates for "sugar" versus a numeraire good with unit price. $Z$ is the amount available in the fair price shop and the budget constraint assumes that resale of surplus at free market prices is impossible.

There are two interrelated issues here for empirical modelling. At the micro level, using cross-section data, we need to know how to use utility theory to generate Engel curves. At the macro-level, it is important to know how the two prices $p_0$ and $p_1$ and the quantity $Z$ affect total demand. As usual, we begin with the indirect utility function, though the form of this can be dictated by prior beliefs about demands (e.g. there has been heavy use of the indirect utility function associated with a linear demand function for a single good – for the derivation, see Deaton and Muellbauer (1980a, p. 96) (1981) and Hausman (1980)). Maximum utility along AD is $u_0 = \psi(x, p, p_0)$ with associated demand, by Roy's identity, of $s_0 = g(x, p, p_0)$. Now, by standard revealed preference, if $s_0 < Z$, $s_0$ is optimal since BC is obtainable by a consumer restricted to being within AD. Similar, maximum utility along EC is $u_1 = \psi(x + (p_1 - p_0)Z, p, p_1)$ with $s = g(x + (p_1 - p_0)Z, p, p_1)$. Again, if $s_1 > Z$, then $s_1$ is optimal. The remaining case is $s_0 > Z$ and $s_1 < Z$ (both of which are infeasible), so that sugar demand is exactly $Z$ (at the kink $B$). Hence, for individual $h$ with expenditure $x^h$ and quota $Z^h$, the demand functions are given by

$$s^h = g^h(x^h, p, p_0) \quad \text{if } g^h(x^h, p, p_0) < Z^h \tag{107}$$

$$s^h = g^h(x^h + (p_1 - p_0)Z^h, p, p_1) \quad \text{if } g^h(x^h + (p_1 - p_0)Z^h, p, p_1) > Z^h \tag{108}$$

$$s^h = Z^h \quad \text{if } g^h(x^h + (p_1 - p_0)Z^h, p, p_1) \leq Z^h \leq g^h(x^h, p, p_0) \tag{109}$$

Figure 3 gives the resulting Engel curve. Estimation on cross-section data is straightforward by an extension of the Tobit method; the demand functions $g^h$ are endowed with taste variation in the form of a normally distributed random term, and a likelihood with three "branches" corresponding to $s^h < Z^h$, $s^h = Z^h$, and $s^h > Z^h$ is constructed. The middle branch corresponds to the zero censoring for Tobit; the outer two are analogous to the non-censored observations in Tobit.

The aggregate free-market demand for sugar can also be analysed using the model. To simplify, assume that households differ only in outlay, $x^h$. Define $x_T$ by $g\{x_T + (p_1 - p_0)Z, p, p_1\} = Z$, so that consumers with $x > x_T$ enter the free

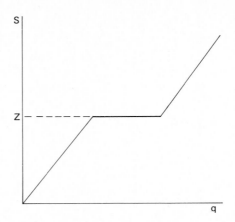

Figure 3.   Engel curve with a non-linear budget constraint.

market. Hence per capita free market demand is

$$s = \int_{x_T}^{\infty} \{ g(x + (p_1 - p_0)Z, p, p_1) - Z \} dF(x) \tag{110}$$

$$\frac{\partial s}{\partial Z} = \int_{x_T}^{\infty} \left\{ \frac{\partial s}{\partial x}(p_1 - p_0) - 1 \right\} dF(x) \tag{111}$$

$$- \{ g(x_T + (p_1 - p_0)Z, p, p_1) - Z \} f(x_T)$$

which, from the definition of $x_T$ is simply

$$\frac{\partial s}{\partial Z} = \int_{x_T}^{\infty} \left\{ \frac{\partial s}{\partial x}(p_1 - p_0) - 1 \right\} dF(x). \tag{112}$$

Since, at the entensive margin, consumers buy nothing in the free market, only the intensive margin is of importance. Note that *all* of these estimations and calculations take a particularly simple form if the Marshallian demand functions are assumed to be linear, so that, even in this non-standard situation, linearity can still greatly simplify.

    The foregoing is a very straightforward example but is illustrates the flavor of the analysis. In practice, non-linear budget constraints may have several kink points and the budget set may be non-convex. While such things can be dealt with, e.g. see King (1980), or Hausman and Wise (1980) for housing, and Reece and Zieschang (1984) for charitable giving, the formulation of the likelihood becomes increasingly complex and the computations correspondingly more burdensome. While virtually all likelihood functions can be maximized *in principle*, doing so for real applied examples with several thousand observations can be prohibitively expensive.

## 4. Separability

In the conventional demand analysis discussed so far, a number of important assumptions have not been justified. First, demand within each period is analysed conditional on total expenditure and prices for that period alone, with no mention of the broader determinants of behavior, wealth, income, other prices and so on. Second, considerations of labor supply were completely ignored. Third, no attention was given to questions of consumption and saving or to the problems arising for goods which are sufficiently durable to last for more than one period. Fourth, the practical analysis has used, not the elementary goods of the theory, but rather aggregates such as food, clothing, etc., each with some associated price index. Separability of one sort or another is behind each of these assumptions and this section gives the basic results required for applied analysis. No attempt is made to give proofs, for more detailed discussion the reader may consult Blackorby, Primont and Russell (1978), Deaton and Muellbauer (1980a Chapter 5) or the original creator of much of the material given here, Gorman (1959) (1968) as well as many unpublished notes.

### 4.1. Weak separability

Weak separability is the central concept for much of the analysis. Let $q^A$ be some subvector of the commodity vector $q$ so that $q = (q^A, q^{\bar{A}})$ without loss of generality. $q^A$ is then said to be (weakly) *separable* if the direct utility function takes the form

$$u = v\left(v_A(q^A), q^{\bar{A}}\right),$$  (113)

$v_A(q^A)$ is the subutility (or felicity) function associated with $q^A$. This equation is equivalent to the existence of a preference ordering over $q^A$ alone; choices over the $q^A$ bundles are consistent independent of the vector $q^{\bar{A}}$. More symmetrically, preferences as a whole are said to be separable if $q$ can be partitioned into $(q^A, q^B, \ldots, q^N)$ such that

$$u = v\left(v_A(q^A), v_B(q^B), \ldots, v_N(q^N)\right).$$  (114)

Since $v$ is increasing in the subutility levels, it is immediately obvious that maximization of overall $u$ implies maximization of the subutilities subject to whatever is optimally spent on the groups. Hence, (113) implies the existence of subgroup demands

$$q_i^A = g_i^A(x^A, p^A), \tag{115}$$

where $x^A = p^A \cdot q^A$, while (115) has the same implication for all groups. Hence, if preferences in a life-cycle model are weakly separable over time periods, commodity demand functions conditional on $x$ and $p$ for each time period are guaranteed to exist. Similarly, if goods are separable from leisure, commodity demand functions of the usual type can be justified.

Tests of these forms of separability can be based on the restrictions on the substitution matrix implied by (115). If $i$ and $j$ are two goods in distinct groups, $i \in G$, $j \in H$, $G \neq H$, then the condition

$$s_{ij} = \mu_{GH} \frac{\partial q_i}{\partial x} \cdot \frac{\partial q_j}{\partial x}, \tag{116}$$

for some quantity $\mu_{GH}$ (independent of $i$ and $j$) is both necessary and sufficient for (114) to hold. If a general enough model of substitution can be estimated, (116) can be used to test separability, and Byron (1968), Jorgenson and Lau (1975) and Pudney (1981b), have used essentially this technique to find separability patterns between goods within a single period. Barnett (1979a) has tested the important separability restriction between goods and leisure using time series American data and decisively rejects it. If widely repeated, this result would suggest considerable misspecification in the traditional studies. It is also possible to use a single cross-section to test separability between goods and leisure. Consider the following cost function proposed by Muellbauer (1981b).

$$c(u, \omega, p) = d(p) + b(p)\omega + \{a(p)\}^{1-\delta}\omega^\delta u, \tag{117}$$

where $\omega$ is the wage $d(p)$, $b(p)$ and $a(p)$ are functions of $p$, homogenous of degrees, 1, 0 and 1 respectively. Shephard's Lemma gives immediately

$$\left.\begin{array}{l} q_i = \alpha_i + \beta_i\omega + \gamma_i\mu \\ \omega h = \alpha_0 + \beta_0\omega + \gamma_0\mu \end{array}\right), \tag{118}$$

for transfer income $\mu$, hours worked $h$ and parameters $\alpha$, $\beta$, $\gamma$ all constant in a single cross-section. It may be shown that (117) satisfies (114) for leisure vis-à-vis goods if and only if $b(p)$ is a constant, which for (118) implies that $\beta_i/\gamma_i$ be independent of $i$, $i = 1, \ldots, n$. This can be tested by first estimating (114) as a system by OLS equation by equation and then computing the Wald test for the

$(n-1)$ restrictions, $i = 1, \ldots, (n-1)$

$$\beta_i \gamma_n - \gamma_i \beta_n = 0. \tag{119}$$

This does not involve estimating the restricted nonlinear model. My own results on British data, Deaton (1981b), suggest relatively little conflict with separability, however, earlier work by Atkinson and Stern (1981) on the same data but using an ingenious adaptation of Becker's (1965) time allocation model, suggests the opposite. Blundell and Walker (1982), using a variant of (117) reject the hypothesis that wife's leisure is separable from goods. Separability between different time periods is much more difficult to test since it is virtually impossible to provide general unrestricted estimates of the substitution responses between individual commodities across different time periods.

Subgroup demand functions are only a part of what the applied econometrician needs from separability. Just as important is the question of whether it is possible to justify demand functions for commodity composites in terms of total expenditure and composite price indices. The Hicks (1936) composite commodity theorem allows this, but only at the price of assuming that there are no relative price changes within subgroups. Since there is no way of guaranteeing this, nor often even of checking it, more general conditions are clearly desirable. In fact, the separable structure (114) may be sufficient in many circumstances. Write $u_A$, $u_B$, etc. for the values of the felicity functions and $c_A(u_A, p^A)$ etc. for the subgroup cost functions corresponding to the $v_A(q^A)$ functions. Then the problem of choosing the group expenditure levels $x_A, x_B, \ldots$ can be written as

$$\max u = v(u_A, u_B, \ldots, u_N), \tag{120}$$
$$\text{s.t. } x = \sum_R c_R(u_R, p^R).$$

Write

$$c_R(u_R, p^R) = c_R(u_R, \bar{p}^R) \cdot \frac{c_R(u_R, p^R)}{c_R(u_R, \bar{p}^R)}, \tag{121}$$

for some fixed prices $\bar{p}^R$. For such a fixed vector, $c_R(u_R, \bar{p}^R)$ is a welfare indicator or quantity index, while the ratio $c_R(u_R, p^R)/c_R(u_R, \bar{p}^R)$ is a true (sub) cost-of-living price index comparing $p^R$ and $\bar{p}^R$ using $u_R$ as reference, see Pollak (1975). Finally, since $u_R = \psi_R(c_R(u_R, \bar{p}^R), \bar{p}^R)$, (120) may be written

$$\max u = v\{\psi_A(c_A(u_A, \bar{p}^A), \bar{p}^A), \psi_B( \ ), \qquad \qquad \}, \tag{122}$$
$$\text{s.t. } \sum c_R(u_R, \bar{p}_R) \cdot \frac{c_R(u_R, p^R)}{c_R(u_R, \bar{p}^R)} = x,$$

which is a standard utility maximization problem in which the constant price utility levels $c_R(u_R, \bar{p}^R)$ are the quantities and the indices $c_R(u_R, p^R)/c_R(u_R, \bar{p}^R)$ are the prices. Of course, neither of these quantities is directly observable and the foregoing analysis is useful only to the extent that $c_R(u_R, \bar{p}^R)$ is adequately approximated by the constant price composite $q^R \cdot \bar{p}^R$ and the price index by the implicit price deflator $p^R \cdot q^R / \bar{p}^R \cdot q^R$. The approximations will be exact under the conditions of the composite commodity theorem, but may be very good in many practical situations where prices are highly but not perfectly collinear. If so, the technique has the additional advantage of justifying the price and quantity indices typically available in the national accounts statistics. An ideal solution not relying on approximations requires quantity indices depending only on quantities and price indices depending only on prices. Given weak separability, this is only possible if *either* each subcost function is of the form $c_G(u_G, p^G) = \theta_G(u_G)b_G(p^G)$ so that the subgroup demands (11) display unit elasticity for all goods with respect to group outlay *or* each indirect felicity function takes the "*Gorman generalized polar form*"

$$u_G = F_G\left[x_G/b_G(p^G)\right] + a_G(p^G), \tag{123}$$

for suitable functions $F_G$, $b_G$ and $a_G$, the first monotone increasing, the latter two linearly homogeneous, *and* the utility function (114) or (120) must be additive in the individual felicity functions. Additivity is restrictive even between groups, and will be further discussed below, but (123) permits fairly general forms of Engel curves, e.g. the Working form, AIDS, PIGL and the translog (61) if $\Sigma_k \Sigma_j \beta_{kj} = 0$. See Blackorby, Boyce and Russell (1978) for an empirical application, and Anderson (1979) for an attempt to study the improvement over standard practice of actually computing the Gorman indices. In spite of this analysis, there seems to be a widespread belief in the profession that homothetic weak separability is *necessary* for the empirical implementation of two-stage budgeting (which is itself almost the only sensible way to deal with very large systems) – see the somewhat bizarre exchanges in the 1983 issue of the *Journal of Business and Economic Statistics*. In my view, homothetic separability is likely to be the *least* attractive of the alternatives given here; it is rarely sensible to *maintain* without testing that subgroup demands have unit group expenditure elasticities. In many cases, prices will be sufficiently collinear for the problem (122) to given an acceptably accurate representation. And if not, additivity between broad groups together with the very flexible Gorman generalized polar form should provide an excellent alternative. Even failing these possibilities, there are other types of separability with useful empirical properties, see Blackorby, Primont and Russell (1978) and Deaton and Muellbauer (1980, Chapter 5).

One final issue related to separability is worth noting. As pointed out by Blackorby, Primont and Russell (1977), flexible functional forms do not in general remain flexible under the global imposition of separability restrictions. Hence, a specific functional form which offers a local second-order approximation to an arbitrary utility function may not be able to similarly approximate, say, an arbitrary additive utility function once its parameters are restricted to render it globally additive. For example, Blackorby et al. show that weak separability of the translog implies either strong separability or homothetic separability so that the translog cannot model non-homothetic weak separability. The possibility of imposing and testing restrictions locally (say, at the sample mean) remains, but this is less attractive since it is difficult to discriminate between properties of the data generation process and the approximating properties of the functional form.

## 4.2. Strong separability and additivity

Strong separability restricts (114) to the case where the overall function is additive, i.e. for some monotone increasing $f$

$$u = f\left(\sum_R v_R(q^R)\right) \tag{124}$$

If each of the groups $q^R$ contains a single good, preferences are said to be *additive*, or that wants are independent. I deal with this case for simplicity since all the additional features over weak separability occur between groups rather than within them. The central feature of additivity is that *any* combination of goods forms a separable set from any other, so that (116) must hold without the $G$, $H$ labels on $\mu_{GH}$, i.e. for some $\mu$ and for all $i$, $j$ in different groups ($i \neq j$ under additivity)

$$s_{ij} = \mu \frac{\partial q_i}{\partial x} \frac{\partial q_j}{\partial x}. \tag{125}$$

The budget constraint (or homogeneity) can be used to complete this for all $i$ and $j$; in elasticity terms, the relationship is, Frisch (1959), Houthakker (1960)

$$e_{ij} = \phi \delta_{ij} e_i - e_i w_j (1 + \phi e_j), \tag{126}$$

for some scalar $\phi$, (uncompensated) cross-price elasticity $e_{ij}$, and total expenditure elasticity $e_i$. This formula shows immediately the strengths and weaknesses of additivity. Apart from the data $w_i$, knowledge of the $(n-1)$ independent $e_i$'s together with the quantity $\phi$ (obtainable from knowledge of one single price elasticity) is sufficient to determine the whole $(n \times n)$ array of price elasticities. Additivity can therefore be used to estimate price elasticities on data with little or no relative price variation, e.g. on cross-sections, on short-time series, or in centrally planned economies where relative prices are only infrequently altered. This was first realised by Pigou (1910) and the idea has a distinguished history in the subject, see Frisch (1932), (1959) and the enormous literature on the (additive) linear expenditure system [for Eastern European experience, see Szakolczai (1980) and Fedorenko and Rimashevskaya (1981)]. Conversely, however, there is very little reason to suppose that (126) is empirically valid. Note, in particular, that for $w_i$ small relative to $e_i$ (as is usually the case), $e_{ii} \simeq \phi e_i$ (as Pigou pointed out) and there seems no grounds for such a proportionality relationship to be generally valid. Indeed such tests as have been carried out, Barten (1969), Deaton (1974b) (1975a) (1975b), Theil (1975b), suggest that additivity is generally not true, even for broad categories of goods. Nevertheless, the assumption continues to be widely used, for example in the interesting cross-country work of Theil and Suhm (1982), no doubt because of its economy of parametrization ( = high level of restrictiveness). There is also a substantial industry in collecting estimates of the parameter $\phi$ under the (entirely baseless) supposition that it measures the inverse of the elasticity of the marginal utility of money.

Few of the practical objections to additivity apply to its use in an intertemporal context and it is standard practice to specify life-time preferences by (124) where the $R$'s refer to time periods, an example being Lluch's (1973) intertemporal linear expenditure system (ELES), although this is also additive within periods. On elegant way of exploiting additivity is again due to Gorman (1976) and utilizes the concept of a "consumer profit function". Define $\pi(p, r)$ by

$$\pi(p, r) = \max_q \{ - p \cdot q + r \cdot u; \ u = v(q) \}, \tag{127}$$

for concave $v(q)$, so that the consumer sells utility (to him or herself) at a price $r$ ( = the reciprocal of the marginal utility of money) using inputs $q$ at prices $p$. Now if $v(q)$ has the explicitly additive form $\Sigma v_R(q^R)$, so will $\pi(p, r)$, i.e.

$$\pi(p, r) = \sum_R \pi_R(r, p_R). \tag{128}$$

Now $\pi(p, r)$ also has the derivative property $q = - \nabla_p \pi(p, r)$ so that for $i$ belonging to group $R$,

$$q_i = - \frac{\partial \pi_R(r, p_R)}{\partial p_{Ri}}, \tag{129}$$

which depends only on within group prices and the single price of utility $r$ which is common to all groups and provides the link between them. In the intertemporal context, $r$ is the price of lifetime utility, which is constant under certainty or follows (approximately) a random walk under uncertainty, while $p_R$ is within period prices. Hence, as realized by MaCurdy and utilized in Heckman (1978), Heckman and MaCurdy (1980), and MaCurdy (1981), eq. (129) can be implemented on panel data by treating $r$ as a fixed effect so that only data on current magnitudes are required. Since these are typically the only data available, the technique is of considerable importance. See Browning, Deaton and Irish (1984) for further discussion of profit functions and additivity and for an application to British data (in which the simple life-cycle model of the simultaneous determination of consumption and labor supply has some difficulty in dealing with the evidence.)

Another important use of separability in general and of additivity in particular is as a vehicle for the structuring and interpretation of preference patterns. For example, in the "characteristics" model of consumer behaviour pioneered by Gorman (1956, 1980), Stone (1956) and Lancaster (1966), and recently estimated by Pudney (1981a), it is a transformation of the goods which generates utility, and it may be quite plausible to assume that preferences are separable or even additive in the transformed characteristics (food, shelter, mate, etc.) rather than in the market goods which have no direct role in satisfying wants. One possibility, extensively explored by Theil and his co-workers, e.g. Theil (1976) and Theil and Laitinen (1981) for a review, is that preferences are additive over characteristics given by a linear transform of the market goods. Theil and Laitinen use the Rotterdam model and, by a technique closely related to factor analysis, rotate the axes in goods space to obtain the "preference independence transform". Applied to the demand for beef, pork and chicken in the U.S., the model yields the transformed goods "inexpensive meat", "beef/pork contrast" and "antichicken", Theil (1976, p. 287). These characteristics may indeed reflect real aspects of preference structures in the U.S., but as is often the case with factor analytical techniques (see e.g. Armstrong (1967) for an amusing cautionary tale) there is room for some (largely unresolvable) scepticism about the validity and value of any specific interpretations.

## 5. Aggregation over consumers

Clearly, on micro or panel data, aggregation is not an issue, and as the use of such data increases, the aggregation problem will recede in importance. However, much demand analysis is carried out on macroeconomic aggregate or per capita data, and it is an open question as to whether this makes sense or not. The topic is a large one and I present only the briefest discussion here, see Deaton and Muellbauer (1980a, Chapter 6) for further discussion and references. At the most general level, average aggregate demand $\bar{q}_i$ is given by

$$\bar{q}_i = G_i(x^1, x^2, \ldots, x^h, \ldots, x^H, p), \tag{130}$$

for the $H$ outlays $x^h$ of household $h$. The function $G_i$ can be given virtually any properties whatever depending on the configuration of individual preferences. If, however, the outlay distribution were fixed in money terms, $x^h = k^h \bar{x}$ for constants $k^h$, (130) obviously gives

$$\bar{q}_i = G_i^*(\bar{x}, p), \tag{131}$$

although without restrictions on preferences, see e.g. Eisenberg (1961), Pearce (1964), Chipman (1974), and Jerison (1984), there is no reason to suppose that the $G_i^*$ functions possess any of the usual properties of Marshallian demands. Of course, if the *utility* (real outlay) distribution is fixed, Hicksian demands aggregate in the same way as (130) and (131) and there exist macro demand functions with all the usual properties. There is very little relevant empirical evidence on the movement over time of either the outlay or the utility distribution, but see Simmons (1980) for some conjectures for the U.K.

If the distribution of outlay is not to be restricted in any way, formulae such as (131) can only arise if mean preserving changes in the $x$-distribution have no effect on aggregate demand, i.e. if all individuals have identical marginal propensities to spend on each of the goods. This condition, of parallel linear Engel curves, dates back to Antonelli (1886), but is usually (justly) credited to Gorman (1953) (1961). As he showed, utility maximizing consumers have parallel linear Engel curves if and only if the individual cost functions have the form

$$c^h(u^h, p) = a^h(p) + b(p)u^h, \tag{132}$$

a specification known as the "Gorman polar form". Suitable choice of the $a^h(p)$ and $b(p)$ functions permits (132) to be a flexible functional form, Diewert (1980a), but the uniformity across households implied by the need for all Engel curves to be parallel seems implausible. However, it should be noted that a single cross-section is insufficient to disprove the condition since, in principle, and without the use of panel data, variation in the $a^h(p)$ functions due to non-outlay factors cannot be distinguished from the direct effects of variations in $x^h$. A somewhat weaker form of the aggregation condition, emphasized by Theil (1954) (1975 Chapter 4) is that the marginal propensities to consume be distributed independently of the $x^h$, see also Shapiro (1976) and Shapiro and Braithwait (1979). Note finally that if aggregation is to be possible for all possible income distributions, including those for which some people have zero income, then the parallel linear Engel curves must pass through the origin so that $a^h(p)$ in (132) is zero and preferences are identical and homothetic.

If, however, the casual evidence against any form of linear Engel curves is taken seriously exact aggregation requires the abandonment of (131), at least in principle. One set of possibilities has been pursued by Muellbauer (1975b) (1976a) (1976b) who examines conditions under which the aggregate budget share of each good can be expressed as a function of prices and a single indicator of $x$, not necessarily the mean. If, in addition, this indicator is made independent of prices, the cost functions must take the form

$$c^h(u^h, p) = k^h\{a(p)^\alpha(1-u^h) + b(p)^\alpha u^h\}^{1/\alpha}, \tag{133}$$

called by Muellbauer, "price-independent generalised linearity" (PIGL). With $\alpha = 1$, PIGL is essentially the Gorman polar form and the Engel curves are linear; otherwise, $\alpha$ controls the curvature of the Engel curves with, for example, the AIDS and Working-Leser forms as special cases when $\alpha = 0$. The macro relationships corresponding to (133) render $\bar{q}_i$ a function of both $x$ and of the mean of order $(1 - \alpha)$ of the outlay distribution. Hence, if $\alpha = -1$, the Engel curves are quadratic and the average aggregate demands depend upon the mean and variance of $x$. This opens up two new possibilities. On the one hand, the presumed (or estimated) curvature of the Engel curves can be used to formulate the appropriate index of dispersion for inclusion in the aggregate demands, see e.g. the papers by Berndt, Darrough and Diewert (1977) and by Simmons (1980) both of which use forms of (133). On the other hand, the income and hence outlay distribution changes very little over time, such models allow the dispersion terms to be absorbed into the function and justify the use of (131) interpreted as a conventional Marshallian demand function, see e.g. Deaton and Muellbauer (1980b). This position seems defensible in the light of the many studies which, using one technique or another, have failed to find any strong influence of the income distribution on consumer behaviour.

Recent theoretical work on aggregation has suggested that the generalized linearity and price independent generalised linearity forms of preference have a more fundamental role to play in aggregation than solving the problem posed by Muellbauer. Jerison (1984) has shown that the generalized linearity conditions are important for aggregation with fixed income distribution, while Freixas and Mas-Colell (1983) have proved the necessity of PIGL for the weak axiom of revealed preference to hold in aggregate if the income distribution is unrestricted. (Note that Hildenbrand's (1983) proof that WARP holds on aggregate data requires that the density of the income distribution be monotone declining and have support $(0, \infty)$, so that modal income is zero!).

In a more empirical vein, Lau (1982) has considered a more general form of aggregation than that required by (131). Lau considers individual demand functions of the form $g^h(x^h, p, a^h)$ for budget $x^h$, prices $p$ and attributes (e.g. demographics) $a^h$. His first requirement is that $\Sigma g^h(x^h, p, a^h)$ be symmetric in the $H$ $x^h$'s and $a^h$'s, i.e. be invariant to who has what $x$ and what $a$. This alone is sufficient to restrict demands to the form

$$g^h(x^h, p, a^h) = g(x^h, p, a^h) + k^h(p), \tag{134}$$

i.e. to be identical up to the addition of a function of prices alone. Lau then derives the conditions under which aggregate demands are a function of not the $H$ $x$'s and $a$'s, but of a smaller set of $m$ indices, $m < H$. Lau shows that

$$\Sigma g^h(x^h, p, a^h) = G\{p, f_1(x, a), f_2(x, a), \ldots, f_m(x, a)\}, \qquad (135)$$

with $f_i(x, a)$ non-constant symmetric functions of the $H$-vectors $x$ and $a$, implies that

$$g^h(x^h, p, a^h) = \sum_{k=1}^{m} h_k(p)\phi_k(x^h, a^h) + k^h(p), \qquad (136)$$

Gorman's (1981) theorem, see 3(a) above, tells us what form the $\phi_k$ functions can take, while Lau's theorem makes Gorman's results the more useful and important. Lau's theorem provides a useful compromise between conventional aggregation as represented by (131) on the one hand and complete agnosticism on the other. Distributional effects on demand are permitted, but in a limited way. Gorman's results tell us that to get these benefits, polynomial specifications are necessary which either link quantities to outlays or shares to the logarithms of outlays. The latter seem to work better in practice and are therefore recommended for use.

Finally, mention must be made of the important recent work of Stoker who, in a series of papers, particularly (1982) (1984), has forged new links between the statistical and economic theories of aggregation. This work goes well beyond demand analysis per se but has implications for the subject. Stoker (1982) shows that the estimated parameters from cross-section regressions will estimate the corresponding macro-effects not only under the Gorman perfect aggregation conditions, but also if the independent variables are jointly distributed within the exponential family of distributions. In the context of demand analysis, the marginal propensity to consume from a cross-section regression would con- sistently estimate the impact of a change in mean income on mean consumption either with linear Engel curves or with non-linear Engel curves and income distributed according to some exponential family distribution. Since one of the reasons we are interested in aggregation is to be able to move from micro to macro in this way, these results open up new possibilities. Stoker (1984) also carries out the process in reverse and derives completeness (or identification) conditions on the distribution of exogenous variables that allow recovery of micro behavior from macro relationships.

Much of the work reported in this section, by Muellbauer, Lau and Stoker, can be regarded as developing the appropriate techniques of allowing for the impacts of distribution on aggregate demand functions. That such effects could be potentially important has been known for a long time, see de Wolff (1941) for an early contribution. What still seems to be lacking so far is empirical evidence that such effects are actually important.

## 6. Behavior under quantity constraints

The existence and consequences of quantity constraints on purchases has recently been given much attention in the literature and the question of whether (or how) the labor market clears remains of central importance for much of economic analysis, see Ashenfelter (1980) for a good discussion in which rationing is taken seriously. If empirical studies of consumer behavior are to contribute to this discussion, they must be able to model the effects of quantity rationing on purchases in other markets and be able to test whether or not quantity constraints exist. Perhaps the most famous work on the theory of quantity constraints traces back to Samuelson's (1947) *Foundations* and the enunciation of the Le Chatelier principle by which substitution possibilities in all markets are reduced by the imposition of quantity restrictions in any. These effects were further studied in the later papers of Tobin and Houthakker (1951) and surveyed in Tobin (1952). All the results obtained are essentially local, given the effects on deviations or elasticities of imposition or changes in quantity restrictions. Applied work, however, requires theory which generates functional forms and, for this, global relationships between rationed and unrationed demands are required. In the presentation here, I follow the work of Neary and Roberts (1980) and Deaton (1981b).

The commodity vector $q$ is partitioned into $(q^0, q^1)$ where $q^0$ may or may not be constrained to take on values $z$. These may be outside impositions or they may essentially be "chosen" by the consumer. An example of the latter is when a consumer decides not to participate in the labor force; since hours cannot be negative, the commodity demand functions conditional on non-participation are those which arise from a quantity restriction of zero hours worked. The simplest case arises if $q^1$ forms a separable group, so that without quantity restrictions on $q^0$, it is possible to write

$$q_i^1 = g_i^1(x - p^0 \cdot q^0, p^1), \tag{137}$$

see eq. (115) above. Clearly, rationing makes no difference to (137) except that $z$ replaces $q^0$, so that testing for the existence of the quantity restrictions can be carried out by testing for the endogeneity of $q^0$ using a Wu (1973) or Hausman (1978) test with $p^0$ as the necessary vector of exogenous instruments not appearing in (137). Without separability matters are more complicated and, in addition to the variables in (137), the demand for $q^1$ depends on $z$ so that without quantity restrictions

$$q_i^1 = g_i^F(x, p^0, p^1), \tag{138}$$

while, under rationing,

$$q_i^1 = g_i^R(x - p^0 \cdot z, p^1, z). \tag{139}$$

Efficient estimation and testing requires that the relationship between $g^F$ and $g^R$ be fully understood. Once again, the cost function provides the answer. If $c(u, p^0, p^1)$ is the unrestricted cost function, i.e. that which generates (138), the *restricted* cost function $c*(u, p^0, p^1, z)$ is defined by

$$c*(u, p^0, p^1, z) = \min\{ p^0 \cdot q^0 + p^1 \cdot q^1; v(q^0, q^1) = u, q^0 = z \}$$
$$= p^0 \cdot z + \gamma(u, p^1, z), \tag{140}$$

where $\gamma$ does not depend upon $p^0$. Define the "virtual prices", $\tilde{p}^0$, Rothbarth (1941), as a function $\zeta^0(u, p^1, z)$ by the relation

$$\frac{\partial c\{ u, \zeta^0(u, p^1, z), p^1 \}}{\partial p_i^0} = z_i, \tag{141}$$

so that $\tilde{p}^0$ is the vector of prices which at $u$ and $p^1$ would cause $z$ to be freely chosen. At these prices, restricted and unrestricted costs must be identical, i.e.

$$c(u, \tilde{p}^0, p) = \tilde{p}^0 \cdot z + \gamma(u, p^1, z), \tag{142}$$

is an identity in $u$, $p^1$ and $z$ with $\tilde{p}^0 = \zeta^0(u, p^1, z)$. Hence, combining (140) and (142)

$$c*(u, p^0, p^1, z) = (p^0 - \tilde{p}^0) \cdot z + c(u, \tilde{p}^0, p). \tag{143}$$

With $\tilde{p}^0$ determined by (141), this equation is the bridge between restricted and unrestricted cost functions and, since (138) derives from differentiating $c(u, p^0, p)$ and (139) from differentiating $c*(u, p^0, p^1, z)$, it also gives full knowledge of the relationship between $g^F$ and $g^R$. This can be put to good theoretical use, to prove all the standard rationing results and a good deal more besides.

For empirical purposes, the ability to derive $g^R$ from $g^F$ allows the construction of a "matched pair" of demand functions, matched in the sense of deriving from the same preferences, and representing both free and constrained behavior. A first attempt, applied to housing expenditure in the U.K., and using the Muellbauer cost function (117) is given in Deaton (1981b). In that study I also found that allowing for quantity restrictions using a restricted cost function related to that for the AIDS, removed much of the conflict with homogeneity on post-war British data. Deaton and Muellbauer (1981) have also derived the matched functional form $g^F$ and $g^R$ for commodity demands for the case where there is quantity rationing in the labor market and where unrestricted labor supply equations take the linear functional forms frequently assumed in the labor supply literature.

## 7.  Other topics

In a review of even this length, only a minute fraction of demand analysis can be covered. However, rather than omit them altogether, I devote this last section to an acknowledgement of the existence of three areas closely linked to the preceeding analysis (and which many would argue are central), intertemporal demand analysis, the analysis of quality, and the use of demand analysis in welfare economics.

### 7.1.  Intertemporal demand analysis

Commodity choices over a lifetime can perhaps be modelled using the utility function

$$u = V\{q^1, q^2, \ldots, q^\tau, \ldots q^L, B/\pi^L\}, \qquad (144)$$

where the $q^\tau$ represent vectors of commodity demands for period $\tau$, $B$ is bequests at death which occurs with certainty at the end of period $L$, and $\pi^L$ is some appropriate price index to be applied to $B$. Utility is maximized subject to the appropriate constraint, i.e.

$$\sum_1^L \hat{p}^\tau \cdot q^\tau + \hat{\pi}^L(B/\pi^L) = W, \qquad (145)$$

where a  denotes discounting and $W$ is the discounted present value at 0 of present and future financial assets and either full income, if labor supply is included, or labor income, if labor supply is taken as fixed.

Clearly (144) (145) are together formally identical to the usual model so that the whole apparatus of cost functions, duality, functional forms and so on can be brought into play. However, the problem is nearly always given more structure by assuming (144) to be additively separable between periods so that demand analysis proper applies to the more disaggregated stage of two stage budgeting, while the allocation to broad groups (i.e. of expenditure between the periods) becomes the province of the consumption function, or more strictly, the life-cycle model. The apparatus of Section 4.2 can be brought into play to yield the new standard life-cycle results, see Browning, Deaton and Irish (1985), Hall (1981), Bewley (1977). Even a very short review of this consumption function literature would double the length of this chapter.

The presence of durable goods can also be allowed for by entering stocks at various dates into the intertemporal model (144). Under the assumption of perfect capital markets, constant proportional physical depreciation, and no divergence between buying and selling prices, these stocks can be priced at "user cost" defined by

$$p_t^* = \left[ p_t - p_{t+1}(1-\delta)/(1+r_{t+1}) \right], \tag{146}$$

when $p_t$ is the price of the good at time $t$, $\delta$ is the rate of physical depreciation and $r_t$ is the interest rate, see Diewert (1974b) or Deaton and Muellbauer (1980a Chapter 13) for full discussions of this model. If user cost pricing is followed, (although note the expectational element in $p_{t+1}$), durable goods can be treated like any other good with $p_t^* S_t$ (for stock $S_t$) as a dependent variable in a demand system, and $x_t$ (including $p_t^* S_t$ *not* the purchase of durables) and all prices and user costs as independent variables. The model is a very useful benchmark, but its assumptions are more than usually unrealistic and it is not surprising that it appears to be rejected in favour of alternative specifications, see Muellbauer (1981a). However, no fully satisfactory alternative formulation exists, and the literature contains a large number of quite distinct approaches. In many of these, commodity demands are modelled conditional on the stocks which, in turn, evolve with purchases, so that dynamic formulations are created in which long-run and short-run responses are distinct. The stock-adjustment models of Stone and Rowe (1957) (1958) and Chow (1957) (1960) are of this form, as is the very similar "state" adjustment model of Houthakker and Taylor (1966) who extend the formulation to all goods while extending the concept of stocks to include "stocks" of habits (since in these models, stocks are substituted out, it makes little difference what name is attached to them). There are also more sophisticated models in which utility functions are defined over instantaneous purchases and stocks, e.g. Phlips' (1972) "dynamic" linear expenditure system, and further refinements in which intertemporal functions are used to model the effects of current purchases on future welfare via their effects on future stocks, Phlips (1974, 1983 Part II). These models are extremely complicated to estimate and it is not clear that they capture any essential features not contained in the stock-adjustment model, on the one hand, and the user cost model on the other, see in particular the results of Spinnewyn (1979a) (1979b). It remains for future work to tackle the very considerable task of constructing models which can deal, in manageable form, with the problems posed by the existence of informational asymmetries [lemons, Akerlof (1970)], borrowing constraints, indivisibilities, technological diffusion, and so on.

## 7.2. *Choice of qualities*

The characteristics model of consumer behavior is a natural way of analysing choice of qualities and, indeed, Gorman's (1956, 1980) classic paper is concerned with quality differentials in the Iowa egg market. By specifying a technology linking quality with market goods, the model naturally leads to the characterization of shadow prices for qualities and these have played a central role in the "new household economics", see in particular, Becker (1976). A related but more direct method of dealing with quality was pioneered in the work of Fisher and Shell (1971), see also Muellbauer (1975a) and Gorman (1976) for reformulations and extensions. The model is formally identical to the Barten model of household composition discussed in Section 3 above with the $m$'s now interpreted as quality parameters "augmenting" the quantities in consumption. Under either formulation, competition between goods manufacturers will, under appropriate assumptions, induce a direct relationship between the price of each good (or variety) and an index of its quality attributes. These relationships are estimated by means of "hedonic" regressions in which (usually the logarithm of) price is regressed on physical attributes across different market goods, see e.g. Burstein (1961) and Dhrymes (1971) for studies of refrigerator prices, and Ohta and Griliches (1976), Cowling and Cubbin (1971) (1972), Cubbin (1975) and Deaton and Muellbauer (1980a p. 263–5) for results on car prices. These techniques date back to Griliches (1961) and ultimately to Court (1939). Choice among discrete varieties involves many closely related techniques, see Chapter 24 of this handbook.

Empirical studies of consumer demand for housing are a major area where quality differences are of great importance. However, until recently, much of the housing literature has consisted of two types of study, one regressing quantities of housing services against income and some index of housing prices, either individual or by locality, while the other follows the hedonic approach, regressing prices on the quantities of various attributes, e.g. number of rooms, size, presence of and type of heating, distance from transport, shops and so on. Serious attempts are currently being made to integrate these two approaches and this is a lively field with excellent data, immediate policy implications, and some first-rate work being done. Lack of space prevents my discussing it in detail; for a survey and further references see Mayo (1978).

## 7.3. Demand analysis and welfare economics

A large proportion of the results and formulae of welfare economics, from cost benefit analysis to optimal tax theory, depend for their implementation on the results of empirical demand analysis, particularly on estimates of substitution responses. Since the coherence of welfare theory depends on the validity of the standard model of behavior, the usefulness of applied demand work in this context depends crucially on the eventual solution of the problems with homogeneity (possible symmetry) and global regularity discussed in Section 2 above. But even without such difficulties, the relationship between the econometric estimates and their welfare application is not always clearly appreciated. In consequence, I review briefly here the estimation of three welfare measures, namely consumer surplus, cost-of-living indices, and equivalence scales.

I argued in Section 1 that it was convenient to regard the cost function as the centrepiece of applied demand analysis. It is even more convenient to do so in welfare analysis. Taking *consumer surplus* first, the compensating variation (CV) and equivalent variation (EV) are defined by, respectively,

$$CV = c(u^0, p^1) - c(u^0, p^0), \tag{147}$$

$$EV = c(u^1, p^1) - c(u^1, p^0), \tag{148}$$

so that both measure the money costs of a welfare affecting price change from $p^0$ to $p^1$, CV using $u^0$ as reference (compensation returns the consumer to the original welfare level) and EV using $u^1$ (it is equivalent to the change to $u^1$). Base and current reference true *cost-of-living index numbers* are defined analogously using ratios instead of differences, hence

$$P(p^1, p^0; u^0) = c(u^0, p^1)/c(u^0, p^0), \tag{149}$$

$$P(p^1, p^0; u^1) = c(u^1, p^1)/c(u^1, p^0), \tag{150}$$

are the base and current true indices. Note the CV, EV and the two price indices depend in no way on how utility is measured; they depend only on the indifference curve indexed by $u$, which could equally well be replaced by $\phi(u)$ for any monotone increasing $\phi$. Even so, the cost function is not observed directly and a procedure must be prescribed for constructing it from the (in principle) observable Marshallian demand functions. If the functional forms for these are known, and if homogeneity, symmetry and negativity are satisfied, the cost function can be obtained by solving the partial differential equations (12), often analytically, see e.g. Hausman (1981). Unobserved constants of integration affect only the measurability of $u$ so that complete knowledge of the Marshallian demands is equivalent to complete knowledge of consumer surplus and the index numbers. If analytical integration is impossible or difficult, numerical integration is straightforward (provided homogeneity and symmetry hold) and algorithms exist in the literature, see e.g. Samuelson (1948) and in much more detail, Vartia (1983). If the integrability conditions fail, consumer behavior is not according to the theory and it is not sensible to try to calculate the welfare indices in the first place, nor is it possible to do so. Geometrically, calculating CV or EV is simply a matter of integrating the area under a *Hicksian* demand curve; there is no valid theoretical or practical reason for ever integrating under a *Marshallian* demand curve. The very considerable literature discussing the practical difficulties of doing so (the path-dependence of the integral, for example) provides a remarkable example of the elaboration of secondary nonsense which can occur once a large primary category error has been accepted; the emperor with no clothes, although quite unaware of his total nakedness, is continuously distressed by his inability to tie his shoelaces. A much more real problem is the assumption that the functional forms of the Marshallian demands are known, so that working with a specific model inevitably understates the margin of ignorance about consumer surplus or index numbers. The tools of non-parametric demand analysis, as discussed in Section 2.7, can, however, be brought to bear to give bounding relationships on the cost function and hence on the welfare measures themselves, see Varian (1982b).

The construction of empirical scales is similar to the construction of price indices although there are a few special difficulties. For household characteristics $a^h$, the equivalence scale $M(a^h, a^0; u, p)$ is defined by

$$M(a^h, a^0, u, p) = c(u, p, a^h)/c(u, p, a^0),$$
(151)

for reference household characteristics $a^0$ and suitably chosen reference welfare level $u$ and price vector $p$. Models such as those discussed in Section 3.2 yield estimates of the parameters of $c(u, p, a)$ so that scales can be evaluated. However, the situation is not quite the same as for the price indices (149) and (150). For these, $c(u, p)$ only is required and this is identified by the functional forms for its tangents $h_i(u, p) = g_i\{c(u, p), p\}$. But for $c(u, p, a)$, we observe only the $p$-tangents together with their derivatives with respect to $a$, i.e. $\partial q_i/\partial a_j$, the demographic effects on demand, and this information is insufficient to identify the function. In particular, as emphasized by Pollak and Wales (1979), the cost functions $c(\phi(u, a), p, a)$ and $c(u, p, a)$ have identical behavioral consequences if $\partial\phi/\partial u > 0$ while giving quite different equivalence scales. Since $c(u, p, a)$ is formally identical to the restricted cost function discussed in Section 6 above, its derivatives with respect to $a$ can be interpreted as shadow prices [differentiate eq. (143)]. These could conceivably be measured from "economic" studies of fertility, in which case the equivalence scale would be fully identified just as are the price indices from $c(u, p)$. Failing such evidence, it is necessary to be very explicit about exactly what prior information is being used to identify the scales. In Deaton and Muellbauer (1981), the identification issue is discussed in detail and it is shown that the same empirical evidence yields systematically different scales for different models, e.g. those of Engel, Barten and Rothbarth discussed in 3.2. It is also argued that plausible identification assumptions can be made, so that demand analysis may, after all, have something to say about the economic costs of children.

# References

Afriat, S. N. (1967) "The Construction of Utility Functions From Expenditure Data", *International Economic Review*, 8, 67–77.

Afriat, S. N. (1973) "On a System of Inequalities in Demand Analysis: An Extension of the Classical Method", *International Economic Review*, 14, 460–472.

Afriat, S. N. (1976) *The Combinatorial Theory of Demand*. London: Input–output Co.

Afriat, S. N. (1977) *The Price Index*. Cambridge University Press.

Afriat, S. N. (1980) *Demand Functions and the Slutsky Matrix*. Princeton: Princeton University Press.

Afriat, S. N. (1981) "On the Constructability of Consistent Price Indices Between Several Periods Simultaneously", in: A. S. Deaton, ed., *Essays in the Theory and Measurement of Consumer Behaviour in Honour of Sir Richard Stone*. Cambridge: Cambridge University Press.

Aitchison, J. and J. A. C. Brown (1954–5) "A Synthesis of Engel Curve Theory", *Review of Economic Studies*, 22, 35–46.

Akerlof, G. (1970) "The Market for Lemons", *Quarterly Journal of Economics*, 84, 488–500.

Altfield, C. L. F. (1985) "Homogeneity and Endogeneity in Systems of Demand Equations", *Journal of Econometrics*, 27, 197–209.

Anderson, G. J. and R. W. Blundell (1982) "Estimation and Hypothesis Testing in Dynamic Singular Equation Systems", *Econometrica*, 50, 1559–1571.

Anderson, R. W. (1979) "Perfect Price Aggregation and Empirical Demand Analysis", *Econometrica*, 47, 1209–30.

Anderson, T. W. (1958) *An Introduction to Multivariate Statistical Analysis*. New York: John Wiley.

Aneuryn-Evans, G. B. and A. S. Deaton (1980) "Testing Linear versus Logarithmic Regressions", *Review of Economic Studies*, 47, 275–91.

Antonelli, G. B. (1886) *Sulla Teoria Matematica della Economia Politica*, Pisa: nella Tipografia del Folchetto. Republished as "On the Mathematical Theory of Political Economy", in: J. S. Chipman, L. Hurwicz, M. K. Richter and H. F. Sonnenschein, eds., *Preferences, Utility, and Demand*. New York: Harcourt Brace Jovanovich, 1971.

Armstrong, J. S. (1967) "Derivation of Theory by Means of Factor Analysis or Tom Swift and his Electric Factor Analysis Machine", *American Statistician*, 21(5), 17–21.

Ashenfelter, O. (1980) "Unemployment as Disequilibrium in a Model of Aggregate Labor Supply", *Econometrica*, 48, 547–564.

Atkinson, A. B. and N. Stern (1981) "On Labour Supply and Commodity Demands", in: A. S. Deaton, ed., *Essays in the Theory and Measurement of Consumer Behaviour*. New York: Cambridge University Press.

Barnett, W. A. (1979a) "The Joint Allocation of Leisure and Goods Expenditure", *Econometrica*, 47, 539–563.

Barnett, W. A. (1979b) "Theoretical Foundations for the Rotterdam Model", *Review of Economic Studies*, 46, 109–130.

Barnett, W. A. (1983a) "New Indices of Money Supply and the Flexible Laurent Demand System", *Journal of Economic and Business Statistics*, 1, 7–23.

Barnett, W. A. (1983b) "Definitions of 'Second Order Approximation' and 'Flexible Functional Form'", *Economics Letters*, 12, 31–35.

Barnett W. A. and A. Jonas (1983) "The Muntz–Szatz Demand System: An Application of a Globally Well-Behaved Series Expansion", *Economics Letters*, 11, 337–342.

Barnett W. A. and Y. W. Lee (1985) "The Regional Properties of the Miniflex Laurent, Generalized Leontief, and Translog Flexible Functional Forms". *Econometrica*, forthcoming.

Barten, A. P. (1964) "Family Composition, Prices and Expenditure Patterns", in: P. E. Hart, G. Mills and J. K. Whitaker, eds., *Economic Analysis for National Economic Planning*. London: Butterworth.

Barten, A. P. (1966) *Theorie en empirie van een volledig stelsel van vraagvergelijkingen*. Doctoral dissertation, Rotterdam.

Barten, A. P. (1967) "Evidence on the Slutsky Conditions for Demand Equations", *Review of Economics and Statistics*, 49, 77–84.

Barten, A. P. (1969) "Maximum Likelihood Estimation of a Complete System of Demand Equations", *European Economic Review*, 1, 7–73.

Barten, A. P. (1977) "The Systems of Consumer Demand Functions Approach: A Review", *Econometrica*, 45, 23–51.

Barten, A. P. and V. Böhm (1980) "Consumer Theory", in: K. J. Arrow and M. D. Intriligator, eds., *Handbook of Mathematical Economics*. Amsterdam: North-Holland.

Barten, A. P. and E. Geyskens (1975) "The Negativity Condition in Consumer Demand", *European Economic Review*, 6, 227–260.

Becker, G. S. (1965) "A Theory of the Allocation of Time", *Economic Journal*, 75, 493–517.

Becker, G. S. (1976) *The Economic Approach to Human Behaviour*. Chicago: University of Chicago Press.

Bera, A. K., R. P. Byron and C. M. Jarque (1981) "Further Evidence on Asymptotic Tests for Homogeneity and Symmetry in Large Demand Systems", *Economics Letters*, 8, 101–105.

Berndt, E. R., M. N. Darrough and W. E. Diewert (1977) "Flexible Functional Forms and Expenditure Distributions: An Application to Canadian Consumer Demand Functions", *International Economic Review*, 18, 651–675.

Berndt, E. R., B. H. Hall, R. E. Hall and J. A. Hausman (1974) "Estimation and Inference in Non-Linear Structural Models", *Annals of Economic and Social Measurement*, 3, 653–665.

Berndt, E. R. and M. S. Khaled (1979) "Parametric Productivity Measurement and the Choice Among Flexible Functional Forms", *Journal of Political Economy*, 84, 1220–1246.

Berndt, E. R. and N. E. Savin (1975) "Estimation and Hypothesis Testing in Singular Equation Systems With Autoregressive Disturbances", *Econometrica*, 43, 937–957.

Berndt, E. R. and N. E. Savin (1977) "Conflict Among Criteria For Testing Hypotheses in the Multivariate Linear Regression Model", *Econometrica*, 45, 1263–1277.

Bewley, T. (1977) "The Permanent Income Hypothesis: A Theoretical Formulation", *Journal of Economic Theory*, 16, 252–292.

Bhattacharrya, N. (1978) "Studies on Consumer Behaviour in India", in: *A Survey of Research in Economics, Vol. 7, Econometrics*, Indian Council of Social Science Research: New Delhi, Allied Publishers.

Blackorby, C., R. Boyce and R. R. Russell (1978) "Estimation of Demand Systems Generated by the Gorman Polar Form; A Generalization of the S-branch Utility Tree", *Econometrica*, 46, 345–363.

Blackorby, C., D. Primont and R. R. Russell (1977) "On Testing Separability Restrictions With Flexible Functional Forms", *Journal of Econometrics*, 5, 195–209.

Blackorby, C., D. Primont and R. R. Russell (1978) *Duality, Separability and Functional Structure*. New York: American Elsevier.

Blundell, R. W. and I. Walker (1982) "Modelling the Joint Determination of Household Labour Supplies and Commodity Demands", *Economic Journal*, 92, 351–364.

Breusch, T. S. and A. R. Pagan (1979) "A Simple Test for Heteroscedasticity and Random Coefficient Variation", *Econometrica*, 47, 1287–1294.

Brown, J. A. C. and A. S. Deaton (1972) "Models of Consumer Behaviour: A Survey", *Economic Journal*, 82, 1145–1236.

Browning, M. J., A. Deaton and M. Irish (1985) "A Profitable Approach to Labor Supply and Commodity Demands Over the Life-Cycle", *Econometrica*, forthcoming.

Burstein, M. L. (1961) "Measurement of the Quality Change in Consumer Durables", *Manchester School*, 29, 267–279.

Byron, R. P. (1968) "Methods for Estimating Demand Equations Using Prior Information: A Series of Experiments With Australia Data", *Australian Economic Papers*, 7, 227–248.

Byron, R. P. (1970a) "A Simple Method for Estimating Demand Systems Under Separable Utility Assumptions", *Review of Economic Studies*, 37, 261–274.

Byron, R. P. (1970b) "The Restricted Aitken Estimation of Sets of Demand Relations", *Econometrica*, 38, 816–830.

Byron, R. P. (1982) "A Note on the Estimation of Symmetric Systems", *Econometrica*, 50, 1573–1575.

Byron, R. P. and M. Rosalsky (1984) "Symmetry and Homogeneity Tests in Demand Analysis: A Size Correction Which Works". University of Florida at Gainsville, mimeo.

Carlevaro, F. (1976) "A Generalization of the Linear Expenditure System", in: L. Solari and J.-N. du Pasquier, eds., *Private and Enlarged Consumption*. North-Holland for ASEPELT, 73–92.

Cassell, C. M., C.-E. Sarndal and J. H. Wretman (1977) *Foundations of Inference in Survey Sampling*. New York: Wiley.

Caves, D. W. and L. R. Christensen (1980) "Global Properties of Flexible Functional Forms", *American Economic Review*, 70, 422–432.

Chipman, J. S. (1974) "Homothetic Preferences and Aggregation", *Journal of Economic Theory*, 8, 26–38.

Chow, G. (1957) *Demand for Automobiles in the U.S.: A Study in Consumer Durables*. Amsterdam: North-Holland.

Chow, G. (1960) "Statistical Demand Functions for Automobiles and Their Use for Forecasting", in: A. C. Harberger, ed., *The Demand for Durable Goods*. Chicago: University of Chicago Press.

Christensen, L. R., D. W. Jorgenson and L. J. Lau (1975) "Transcendental Logarithmic Utility Functions", *American Economic Review*, 65, 367–283.

Christensen, L. R. and M. E. Manser (1977) "Estimating U.S. Consumer Preferences for Meat With a Flexible Utility Function", *Journal of Econometrics*, 5, 37–53.

Conrad, K. and D. W. Jorgenson (1979) "Testing the Integrability of Consumer Demand Functions", *European Economic Review*, 12, 149–169.

Coondoo, D. (1975) "Effects of Household Composition on Consumption Pattern: A Note", *Arthaniti*, 17.

Court, A. T. (1939) "Hedonic Price Indexes with Automotive Examples", in: *The Dynamics of Automobile Demand*. New York: General Motors.

Cowling, K. and J. Cubbin (1971) "Price, Quality, and Advertising Competition", *Economica*, 82, 963–978.

Cowling, K. and J. Cubbin (1972) "Hedonic Price Indexes for U.K. Cars", *Economic Journal*, 82, 963–978.

Cramer, J. S. (1969) *Empirical Economics*. Amsterdam: North-Holland.

Cubbin, J. (1975) "Quality Change and Pricing Behaviour in the U.K. Car Industry 1956–1968", *Economica*, 42, 43–58.

Deaton, A. S. (1974a) "The Analysis of Consumer Demand in the United Kingdom, 1900–1970", *Econometrica*, 42, 341–367.

Deaton, A. S. (1974b) "A Reconsideration of the Empirical Implications of Additive Preferences", *Economic Journal*, 84, 338–348.

Deaton, A. S. (1975a) *Models and Projections of Demand in Post-War Britain*. London: Chapman & Hall.

Deaton, A. S. (1975b) "The Measurement of Income and Price Elasticities", *European Economic Review*, 6, 261–274.

Deaton, A. S. (1975c) The Structure of Demand 1920–1970, *The Fontana Economic History of Europe*. Collins: Fontana, 6(2).

Deaton, A. S. (1976) "A Simple Non-Additive Model of Demand", in: L. Solari and J.-N. du Pasquier, eds., *Private and Enlarged Consumption*. North-Holland for ASEPELT, 56–72.

Deaton, A. S. (1978) "Specification and Testing in Applied Demand Analysis", *Economic Journal*, 88, 524–536.

Deaton, A. S. (1979) "The Distance Function and Consumer Behaviour with Applications to Index Number and Optimal Taxation", *Review of Economic Studies*, 46, 391–405.

Deaton, A. S. (1981a) "Optimal Taxes and the Structure of Preferences", *Econometrica*, 49, 1245–1268.

Deaton, A. S. (1981b) "Theoretical and Empirical Approaches to Consumer Demand Under Rationing", in: A. S. Deaton, ed., *Essays in the Theory and Measurement of Consumer Behaviour*. New York: Cambridge University Press.

Deaton, A. S. (1981c) "Three Essays on a Sri Lankan Household Survey". Living Standards Measurement Study W.P. No. 11, Washington: The World Bank.

Deaton, A. S. (1982) "Model Selection Procedures, or Does the Consumption Function Exist?", in: G. Chow and P. Corsi, eds., *Evaluating the Reliability of Macroeconomic Models*. New York: Wiley.

Deaton, A. S. (1984) "Household Surveys as a Data Base for the Analysis of Optimality and Disequilibrium", *Sankhya: The Indian Journal of Statistics*, 46, Series B, forthcoming.

Deaton, A. S. and M. Irish (1984) "A Statistical Model for Zero Expenditures in Household Budgets", *Journal of Public Economics*, 23, 59–80.

Deaton, A. S. and J. Muellbauer (1980a) *Economics and Consumer Behavior*. New York: Cambridge University Press.

Deaton, A. S. and J. Muellbauer (1980b) "An Almost Ideal Demand System", *American Economic Review*, 70, 312–326.

Deaton, A. S. and J. Muellbauer (1981) "Functional Forms for Labour Supply and Commodity Demands with and without Quantity Constraints", *Econometrica*, 49, 1521–1532.

Deaton, A. S. and J. Muellbauer (1986) "Measuring Child Costs in Poor Countries", *Journal of Political Economy*, forthcoming.

Dhrymes, P. J. (1971) "Price and Quality Changes in Consumer Capital Goods: An Empirical Study", in: Z. Griliches, ed., *Price Indexes and Quality Change: Studies in New Methods of Measurement*. Cambridge: Harvard University Press.

Diewert, W. E. (1971) "An Application of the Shephard Duality Theorem: A Generalized Leontief Production Function", *Journal of Political Economy*, 79, 481–507.

Diewert, W. E. (1973a) "Afriat and Revealed Preference Theory", *Review of Economic Studies*, 40, 419–426.

Diewert, W. E. (1973b) "Functional Forms for Profit and Transformation Functions", *Journal of Economic Theory*, 6, 284–316.

Diewert, W. E. (1974a) "Applications of Duality Theory", Chapt. 3 in: M. D. Intriligator and D. A. Kendrick, eds., *Frontiers of Quantitive Economics*, American Elsevier: North-Holland, Vol. II.

Diewert, W. E. (1974b) "Intertemporal Consumer Theory and the Demand for Durables", *Econometrica*, 42, 497–516.

Diewert, W. E. (1980a) "Symmetry Conditions for Market Demand Functions", *Review of Economic Studies*, 47, 595–601.

Diewert, W. E. (1980b) "Duality Approaches to Microeconomic Theory", in: K. J. Arrow and M. J. Intriligator, eds., *Handbook of Mathematical Economics*. North-Holland.

Diewert, W. E. (1981) "The Economic Theory of Index Numbers: A Survey", in: A. S. Deaton, ed., *Essays in the Theory and Measurement of Consumer Behaviour in Honour of Sir Richard Stone*. Cambridge: Cambridge University Press.

Diewert, W. E. (1983) "The Theory of the Cost of Living Index and the Measurement of Welfare Change". University of British Columbia, mimeo.

Diewert, W. E. and C. Parkan (1978) "Tests for Consistency of Consumer Data and Nonparametric Index Numbers". University of British Columbia: Working Paper 78-27, mimeo.

DuMouchel, W. H. and G. J. Duncan (1983) "Using Sample Survey Weights in Multiple Regression Analyses of Statified Samples", *Journal of American Statistical Association*, 78, 535–543.

Eisenberg, E. (1961) "Aggregation of Utility Functions", *Management Science*, 7, 337–350.

Engel, E. (1895) "Die Lebenskosten Belgischer Arbeiterfamilien früher und jetzt", *International Statistical Institute Bulletin*, 9, 1–74.

Epstein, L. and A. Yatchew (1985). "Non-parametric Hypothesis Testing Procedures and Applications to Demand Analysis", University of Toronto, mimeo.

Evans, G. B. A. and N. E. Savin (1982) "Conflict Among the Criteria Revisited; the W, LR and LM Tests", *Econometrica*, 50, 737–748.

Federenko, N. P. and N. J. Rimashevskaya (1981) "The Analysis of Consumption and Demand in the USSR", in: A. S. Deaton, ed., *Essays in the Theory and Measurement of Consumer Behaviour*. New York: Cambridge University Press.

Fiebig, D. G. and H. Theil (1983) "The Two Perils of Symmetry Constrained Estimation of Demand Systems", *Economics Letters*, 13, 105–111.

Fisher, F. M. and K. Shell (1971) "Taste and Quality Change in the Pure Theory of the True Cost of Living Index", in: Z. Griliches, ed., *Price Indexes and Quality Changes: Studies in New Methods of Measurement*. Cambridge: Harvard University Press.

Forsyth, F. G. (1960) "The Relationship Between Family Size and Family Expenditure", *Journal of the Royal Statistical Society*, Series A, 123, 367–397.

Freixas, X. and A. Mas-Colell (1983) "Engel Curves Leading to the Weak Axiom in the Aggregate". Harvard University, mimeo.

Frisch, R. (1932) New Methods of Measuring Marginal Utility. Tübingen: J.C.B. Mohr.

Frisch, R. (1959) "A Complete Scheme for Computing All Direct and Cross Demand Elasticities in a Model with Many Sectors", *Econometrica*, 27, 367–397.

Gallant, R. A. (1975) "Seemingly Unrelated Non-Linear Regressions", *Journal of Econometrics*, 3, 35–50.

Gallant, R. A. (1981) "On the Bias in Flexible Functional Forms and an Essentially Unbiased Form: The Fourier Functional Form", *Journal of Econometrics*, 15, 211–245.

Gallant, R. A. and G. H. Golub (1983) "Imposing Curvature Restrictions on Flexible Functional Forms". North Carolina State University and Stanford University, mimeo.

Godambe, V. P. (1955) "A Unified Theory of Sampling From Finite Populations", *Journal of the Royal Statistical Society*, Series B, 17, 268–278.

Godambe, V. P. (1966) "A New Approach to Sampling from Finite Populations: Sufficiency and Linear Estimation", *Journal of the Royal Statistical Society*, Series B, 28, 310–319.

Goldberger, A. S. (1964) *Econometric Theory*. New York: Wiley.

Goldberger, A. S. (1967) "Functional Form and Utility: A Review of Consumer Demand Theory". Social Systems Research Institute, University of Wisconsin, mimeo.

Gorman, W. M. (1953) "Community Preference Fields", *Econometrica* 21, 63–80.

Gorman, W. M. (1956, 1980) "A Possible Procedure for Analysing Quality Differentials in the Egg Market", *Review of Economic Studies*, 47, 843–856.

Gorman, W. M. (1959) "Separable Utility and Aggregation", *Econometrica*, 27, 469–481.

Gorman, W. M. (1961) "On a Class of Preference Fields", *Metroeconomica*, 13, 53–56.

Gorman, W. M. (1968) "The Structure of Utility Functions", *Review of Economic Studies*, 5, 369–390.

Gorman, W. M. (1970) "Quasi Separable Preferences, Costs and Technologies". University of North Carolina, Chapel Hill, mimeo.

Gorman, W. M. (1976) "Tricks with Utility Functions", in: M. Artis and R. Nobay, eds., *Essays in Economic Analysis*. Cambridge: Cambridge University Press.

Gorman, W. M. (1981) "Some Engel Curves", in: A. S. Deaton, ed., *Essays in Theory and Measurement of Consumer Behaviour*. New York: Cambridge University Press.

Granger, C. W. J. and P. Newbold (1974) "Supurious Regressions in Econometrics", *Journal of Econometrics*, 2, 111–120.

Griffin, J. M. (1978) "Joint Production Technology: The Case of Petro-Chemicals", *Econometrica* 46, 379–396.

Griliches, Z. (1961) "Hedonic Price Indexes for Automobiles: An Econometric Analysis of Quality Change", in: Z. Griliches, ed., *Price Indexes and Quality Change: Studies in New Methods of Measurement*. Cambridge: Harvard University Press, 1971.

Guilkey, D. K. and C. A. Knox Lovell (1980) "On the Flexibility of the Translog Approximation", *International Economic Review*, 21, 137–147.

Guilkey, D. K. and P. Schmidt (1973) "Estimation of Seemingly Unrelated Regressions with Vector Auto-Regressive Errors", *Journal of the American Statistical Association*, 68, 642–647.

Haitovsky, Y. (1973) *Regression Estimation from Grouped Observations*. New York: Hafner.

Hall, R. E. (1978) "Stochastic Implications of the Life-Cycle Permanent Income Hypothesis: Theory and Evidence", *Journal of Political Economy*, 86, 971–987.

Hanoch, G. and M. R. Rothschild (1972) "Testing the Assumptions of Production Theory: A Non Parametric Approach", *Journal of Political Economy*, 80, 256–275.

Hausman, J. A. (1978) "Specification Tests in Econometrics", *Econometrica*, 46, 1251–1271.

Hausman, J. A. (1980) "The Effect of Wages, Taxes, and Fixed Costs on Women's Labor Force Participation", *Journal of Public Economics*, 14, 161–194.

Hausman, J. A. (1981) "Exact Consumer's Surplus and Deadweight Loss", *American Economic Review*, 71, 662–676.

Hausman, J. A. (1985) "The Econometrics of Non-Linear Budget Sets", *Econometrica*, forthcoming.

Hausman, J. A. and D. A. Wise (1980) "Discontinuous Budget Constraints and Estimation: The Demand for Housing", *Review of Economic Studies*, 47, 45–96.

Heckman, J. J. (1978) "A Partial Survey of Recent Research on the Labor Supply of Women", *American Economic Review*, pap & proc, 68, 200–207.

Heckman, J. J. (1979) "Sample Selection Bias as a Specification Error", *Econometrica*, 47, 153–161.

Heckman, J. J. and T. MaCurdy (1980) "A Life-Cycle Model of Female Labor Supply", *Review of Economic Studies*, 47, 47–74.

Henderson, A. M. (1949–1950a) "The Costs of Children", *Population Studies*, Parts I–III, 3, 130–150, 4, pp 267–298.

Henderson, A. M. (1949–1950b) "The Cost of a Family", *Review of Economic Studies*, 17, 127–148.

Hendry, D. F. (1980) "Econometrics: Alchemy or Science", *Economica*, 47, 387–406.

Hicks, J. R. (1936) *Value and Capital*. Oxford: Oxford Univeristy Press.

Hicks, J. R. (1956) *A Revision of Demand Theory*. Oxford: Oxford University Press.

Hildenbrand, W. (1983) "On the Law of Demand", *Econometrica*, 51, 997–1019.

Hoa, Tran van (1983) "The Integrability of Generalized Working Models", *Economics Letters*, 13, 101–104.

Hoa, Tran van, D. S. Ironmonger and I. Manning (1983) "Energy Consumption in Australia: Evidence from a Generalized Working Model", *Economics Letters*, 12, 383–389.

Houthakker, H. S. (1957) "An International Comparison of Household Expenditure Patterns Commemorating the Centenary of Engel's Law", *Econometrica*, 25, 532–551.

Houthakker, H. S. (1960) "Additive Preferences", *Econometrica*, 28, 224–256.

Houthakker, H. S. and L. D. Taylor (1966) *Consumer Demand in the United States, 1929–70, Analysis and Projections*. Cambridge: Harvard Univeristy Press, second edition 1970.

Howe, H. and P. Musgrove (1977) "An Analysis of ECIEL Household Budget Data for Bogota, Caracas, Guayaquil and Lima", in: C. Lluch, A. A. Powell and R. Williams, eds., *Patterns in Household Demand and Saving*. Oxford: Oxford University Press for the World Bank.

Howe, H., R. A. Pollak and T. J. Wales (1979) "Theory and Time Series Estimation of the Quadratic Expenditure System", *Econometrica*, 47, 1231–1247.

Hurwicz, L. and H. Uzawa (1971) "On the Integrability of Demand Functions", in: J. S. Chipman, L. Hurwicz, M. K. Richter and H. F. Sonnenschein, eds., *Preference, Utility and Demand*. New York: Harcourt, Brace, Jovanovich, 114–148.

Iyengar, N. S., L. R. Jain and T. N. Srinivasar (1968) "Economies of Scale in Household Consumption: A Case Study", *Indian Economic Journal*, Econometric Annual, 15, 465–477.

Jackson, C. (1968) "Revised Equivalence Scales for Estimating Equivalent Incomes for Budget Costs by Family Type", *BLS Bulletin*, U.S. Dept. of Labor, 1570–1572.

Jerison, M. (1984) "Aggregation and Pairwise Aggregation of Demand When the Distribution of Income is Fixed", *Journal of Economic Theory*, forthcoming.

Jorgenson, D. W. and L. J. Lau (1975) "The Structure of Consumer Preferences", *Annals of Economic and Social Measurement*, 4, 49–101.

Jorgenson, D. W. and L. J. Lau (1976) "Statistical Tests of the Theory of Consumer Behaviour", in: H. Albach, E. Helmstädter and R. Henn, eds., *Quantitative Wirtschaftsforschung*. Tübingen: J.C.B. Mohr.

Jorgenson, D. W., L. J. Lau and T. Stoker (1982) "The Transcendental Logarithmic Model of Aggregate Consumer Behavior", *Advances in Econometrics*, 1, JAI Press.

Kannai, Y. (1977) "Concavifiability and Constructions of Concave Utility Functions", *Journal of Mathematical Economics*, 4, 1–56.

Kay, J. A., M. J. Keen and C. N. Morris (1984) "Consumption, Income, and the Interpretation of Household Expenditure Data", *Journal of Public Economics*, 23, 169–181.

King, M. A. (1980) "An Econometric Model of Tenure Choice and Demand for Housing as a Joint Decision", *Journal of Public Economics*, 14, 137–159.

Klein, L. R. and H. Rubin (1947–48) "A Constant Utility Index of the Cost of Living", *Review of Economic Studies*, 15, 84–87.

Kuznets, S. (1962) "Quantitative Aspects of the Economic Growth of Nations: VII The Share and Structure of Consumption", *Economic Development and Cultural Change*, 10, 1–92.

Kuznets, S. (1966) *Modern Economic Growth*. New Haven: Yale University Press.

Laitinen, K. (1978) "Why is Demand Homogeneity so Often Rejected?", *Economics Letters*, 1, 187–191.

Lancaster, K. J. (1966) "A New Approach to Consumer Theory", *Journal of Political Economy*, 74, 132–157.

Lau, L. J. (1978) "Testing and Imposing Monotonicity, Convexity, and Quasi-Concavity", in: M. Fuss and D. McFadden, eds., *Production Economics: A Dual Approach to Theory and Applications*. Amsterdam: North-Holland.

Lau, L. J. (1982) "A Note on the Fundamental Theorem of Exact Aggregation", *Economics Letters*, 9, 119–126.

Lee, L. F. and M. M. Pitt (1983) "Specification and Estimation of Demand Systems with Limited Dependent Variables". University of Minnesota, mimeo.

Leser, C. E. V. (1963) "Forms of Engel Functions", *Econometrica*, 31, 694–703.

Lluch, C. (1971) "Consumer Demand Functions, Spain, 1958–64", *European Economic Review*, 2, 227–302.

Lluch, C. (1973) "The Extended Linear Expenditure System", *European Economic Review*, 4, 21–32.

Lluch, C., A. A. Powell and R. A. Williams (1977) *Patterns in Household Demand and Saving*. Oxford: Oxford University Press for the World Bank.

Lluch, C. and R. A. Williams (1974) "Consumer Demand Systems and Aggregate Consumption in the U.S.A.: An Application of the Extended Linear Expenditure System", *Canadian Journal of Economics*, 8, 49–66.

MaCurdy, T. E. (1981) "An Empirical Model of Labor Supply in a Life-Cycle Setting", *Journal of Political Economy*, 89, 1059–1085.

Malinvaud, E. (1970) *Statistical Methods of Econometrics*. Amsterdam: North-Holland.

Manser, M. E. and R. J. McDonald (1984) "An Analysis of the Substitution Bias in Measuring Inflation", Bureau of Labor Statistics, mimeo.

Marquardt, D. W. (1963) "An Algorithm for Least-Squares Estimation on Non-Linear Parameters", *Journal of the Society of Industrial and Applied Mathematics*, 11, 431–441.

Mayo, S. K. (1978) "Theory and Estimation in the Economics of Housing Demand", *Journal of Urban Economics*, 14, 137–159.

McClements, L. D. (1977) "Equivalence Scales for Children", *Journal of Public Economics*, 8, 191–210.

McFadden, D. (1978) "Costs, Revenue, and Profit Functions", in: M. Fuss and D. McFadden, eds., *Production Economics: A Dual Approach to Theory and Applications*. Amsterdam: North-Holland.

McGuire, T. W., J. W. Farley, R. E. Lucas and R. L. Winston (1968) "Estimation and Inference for Linear Models in which Subsets of the Dependent Variable are Constrained", *Journal of the American Statistical Association*, 63, 1201–1213.

Meisner, J. F. (1979) "The Sad Fate of the Asymptotic Slutsky Symmetry Test for Large Systems", *Economic Letters*, 2, 231–233.

Muellbauer, J. (1974) "Household Composition, Engel Curves and Welfare Comparisons Between Households: A Duality Approach", *European Economic Review*, 103–122.

Muellbauer, J. (1975a) "The Cost of Living and Taste and Quality Change", *Journal of Economic Theory*, 10, 269–283.

Muellbauer, J. (1975b) "Aggregation, Income Distribution and Consumer Demand", *Review of Economic Studies*, 62, 525–543.

Muellbauer, J. (1976a) "Community Preferences and the Representative Consumer", *Econometrica*, 44, 979–999.

Muellbauer, J. (1976b) "Economics and the Representative Consumer", in: L. Solari and J-N. du Pasquier, eds., *Private and Enlarged Consumption*. Amsterdam: North-Holland for ASEPELT, 29–53.

Muellbauer, J. (1976c) "Can We Base Welfare Comparisons Across Households on Behaviour?". London: Birkbeck College, mimeo.

Muellbauer, J. (1977) "Testing the Barten Model of Household Composition Effects and the Cost of Children", *Economic Journal*, 87, 460–487.

Muellbauer, J. (1980) "The Estimation of the Prais–Houthakker Model of Equivalence Scales", *Econometrica*, 48, 153–176.

Muellbauer, J. (1981a) "Testing Neoclassical Models of the Demand for Consumer Durables", in: A. S. Deaton, ed., *Essays in the Theory and Measurement of Consumer Behaviour*. New York: Cambridge University Press.

Muellbauer, J. (1981b) "Linear Aggregation in Neoclassical Labour Supply", *Review of Economic Studies*, 48, 21–36.

Musgrove, P. (1978) *Consumer Behavior in Latin America: Income and Spending of Families in Ten Andean Cities*. Washington: Brookings.

Neary, J. P. and K. W. S. Roberts (1980) "The Theory of Household Behaviour Under Rationing", *European Economic Review*, 13, 25–42.

Nicholson, J. L. (1949) "Variations in Working Class Family Expenditure", *Journal of the Royal Statistical Society*, Series A, 112, 359–411.

Ohta, M. and Z. Griliches (1976) "Automobile Prices Revisited: Extensions of the Hedonic Hypothesis", in: N. Terleckyj, ed., *Household Production and Consumption*. New York: National Bureau of Economic Research.

Orshansky, M. (1965) "Counting the Poor: Another Look at the Poverty Profile", *Social Security Bulletin*, 28, 3–29.

Parks, R. W. (1969) "Systems of Demand Equations: An Empirical Comparison of Alternative Functional Forms", *Econometrica*, 37, 629–650.

Pearce, I. F. (1964) *A Contribution to Demand Analysis*. Oxford University Press.

Phlips, L. (1972) "A Dynamic Version of the Linear Expenditure Model", *Review of Economics and Statistics*, 54, 450–458.

Phlips, L. (1974) *Applied Consumption Analysis*. Amsterdam and Oxford: North-Holland, second edition 1983.

Pigou, A. C. (1910) "A Method of Determining the Numerical Value of Elasticities of Demand", *Economic Journal*, 20, 636–640.

Pollak, R. A. (1975) "Subindexes in the Cost-of-Living Index", *International Economic Review*, 16, 135–150.

Pollak, R. A. and T. J. Wales (1978) "Estimation of Complete Demand Systems from Household Budget Data", *American Economic Review*, 68, 348–359.

Pollak, R. A. and T. J. Wales (1979) "Welfare Comparisons and Equivalence Scales", *American Economic Review*, pap & proc 69, 216–221.

Pollak, R. A. and T. J. Wales (1980) "Comparison of the Quadratic Expenditure System and Translog Demand Systems with Alternative Specifications of Demographic Effects", *Econometrica*, 48, 595–612.

Pollak, R. A. and T. J. Wales (1981) "Demographic Variables in Demand Analysis", *Econometrica*, 49, 1533–1551.

Powell, A. A. (1969) "Aitken Estimators as a Tool in Allocating Predetermined Aggregates", *Journal of the American Statistical Association*, 64, 913–922.

Prais, S. J. (1959) "A Comment", *Econometrica*, 27, 127–129.

Prais, S. J. and H. S. Houthakker (1955) *The Analysis of Family Budgets*. Cambridge: Cambridge University Press, second edition 1971.

Pudney, S. E. (1980) "Disaggregated Demand Analysis: The Estimation of a Class of Non-Linear Demand Systems", *Review of Economic Studies*, 47, 875–892.

Pudney, S. E. (1981a) "Instrumental Variable Estimation of a Characteristics Model of Demand", *Review of Economic Studies*, 48, 417–433.

Pudney, S. E. (1981b) "An Empirical Method of Approximating the Separable Structure of Consumer Preferences", *Review of Economic Studies*, 48, 561–577.

Quandt, R. E. (1983) "Computational Problems and Methods", *Handbook of Econometrics*. Chapter 12, Vol. 1.

Reece, W. S. and K. D. Zieschang (1985) "Consistent Estimation of the Impact of Tax Deductibility on the Level of Charitable Contributions", *Econometrica*, forthcoming.

Rothbarth, E. (1941) "The Measurement of Change in Real Income Under Conditions of Rationing", *Review of Economic Studies*, 8, 100–107.

Rothbarth, E. (1943) "Note on a Method of Determining Equivalent Income for Families of Different Composition", Appendix 4 in: C. Madge, ed., *War-Time Pattern of Saving and Spending*. Occasional paper No. 4., London: National Institute of Economic and Social Research.

Roy, R. (1942) *De l'Utilité, Contribution à la Théorie des Choix*. Paris: Hermann.

Russell, T. (1983) "On a Theorem of Gorman", *Economics Letters*, 11, 223–224.

Samuelson, P. A. (1938) "A Note on the Pure Theory of Consumer Behaviour", *Economica*, 5, 61–71.

Samuelson, P. A. (1947) *Foundations of Economic Analysis*. Cambridge: Harvard University Press.

Samuelson, P. A. (1947–48) "Some Implications of Linearity", *Review of Economic Studies*, 15, 88–90.

Samuelson, P. A. (1948) "Consumption Theory in Terms of Revealed Preference", *Economica*, 15, 243–253.

Samuelson, P. A. (1956) "Social Indifference Curves", *Quarterly Journal of Economics*, 70, 1–22.

Sargan, J. D. (1964) "Wages and Prices in the United Kingdom" in: P. E. Hart, C. Mills and J. K. Whitaker, eds., *Econometric Analysis for National Economic Planning*. London: Butterworths.

Sargan, J. D. (1971) "Production Functions", Part V in: P. R. G. Layard, J. D. Sargan, M. E. Ager and D. J. Jones, eds., *Qualified Manpower and Economic Performance*. London: Penguin Press.

Seneca, J. J. and M. K. Taussig (1971) "Family Equivalence Scales and Personal Income Tax Exemptions for Children", *Review of Economics and Statistics*, 53, 253–262.

Shapiro, P. (1977) "Aggregation and the Existence of a Social Utility Functions", *Review of Economic Studies*, 46, 653–665.

Shapiro, P. and S. Braithwait (1979) "Empirical Tests for the Existence of Group Utility Functions", *Review of Economic Studies*, 46, 653–665.

Shephard, R. (1953) *Cost and Production Functions*. Princeton: Princeton University Press.

Simmons, P. (1980) "Evidence on the Impact of Income Distribution on Consumer Demand in the U.K. 1955–68", *Review of Economic Studies*, 47, 893–906.

Singh, B. (1972) "On the Determination of Economies of Scale in Household Consumption", *International Economic Review*, 13, 257–270.

Singh, B. (1973) "The Effect of Household Composition on its Consumption Pattern", *Sankhya*, Series B, 35, 207–226.

Singh B. and A. L. Nagar (1973) "Determination of Consumer Unit Scales", *Econometrica*, 41, 347–355.

Spinnewyn, F. (1979a) "Rational Habit Formation", *European Economic Review*, 15, 91–109.

Spinnewyn, F. (1979b) "The Cost of Consumption and Wealth in a Model with Habit Formation", *Economics Letters*, 2, 145–148.

Srivastava, V. K. and T. D. Dwivedi (1979) "Estimation of Seemingly Unrelated Regression Equations: A Brief Survey", *Journal of Econometrics*, 10, 15–32.

Stoker, T. (1982) "The Use of Cross-Section Data to Characterize Macro Functions", *Journal of the American Statistical Association*, 77, 369–380.

Stoker, T. (1985) "Completeness, Distribution Restrictions and the Form of Aggregate Functions", *Econometrica*, forthcoming.

Stone, J. R. N. (1954) "Linear Expenditure Systems and Demand Analysis: An Application to the Pattern of British Demand", *Economic Journal*, 64, 511–527.

Stone, J. R. N. (1956) *Quantity and Price Indexes in National Accounts*. Paris: OEEC.

Stone, R. and D. A. Rowe (1957) "The Market Demand for Durable Goods", *Econometrica*, 25, 423–443.

Stone, R. and D. A. Rowe (1958) "Dynamic Demand Functions: Some Econometric Results", *Economic Journal*, 27, 256–70.

Summers, R. (1959) "A Note on Least Squares Bias in Household Expenditure Analysis", *Econometrica*, 27, 121–126.

Sydenstricker, E. and W. I. King (1921) "The Measurement of the Relative Economic Status of Families", *Quarterly Publication of the American Statistical Association*, 17, 842–857.

Szakolczai, G. (1980) "Limits to Redistribution: The Hungarian Experience", in: D. A. Collard, R. Lecomber and M. Slater, eds., *Income Distribution, the Limits to Redistribution*. Bristol: Scientechnica.

Theil, H. (1954) *Linear Aggregation of Economic Relations*. Amsterdam: North-Holland.

Theil, H. (1965) "The Information Approach to Demand Analysis", *Econometrica*, 33, 67–87.

Theil, H. (1971a) *Principles of Econometrics*. Amsterdam: North-Holland.

Theil, H. (1971b) "An Economic Theory of the Second Moments of Disturbances of Behavioural Equations", *American Economic Review*, 61, 190–194.

Theil, H. (1974) "A Theory of Rational Random Behavior", *Journal of the American Statistical Association*, 69, 310–314.

Theil, H. (1975a) "The Theory of Rational Random Behavior and its Application to Demand Analysis", *European Economic Review*, 6, 217–226.

Theil, H. (1975b) *Theory and Measurement of Consumer Demand*. North-Holland, Vol. I.

Theil, H. (1976) *Theory and Measurement of Consumer Demand*. North-Holland, Vol. II.

Theil, H. (1979) *The System-Wide Approach to Microeconomics*. Chicago: University of Chicago Press.

Theil, H. and K. Laitinen (1981) "The Independence Transformation: A Review and Some Further Explorations", in: A. S. Deaton, ed., *Essays in the Theory and Measurement of Consumer Behaviour*. New York: Cambridge University Press.

Theil, H. and M. Rosalsky (1984) "More on Symmetry-Constrained Estimation". University of Florida at Gainesville, mimeo.

Theil, H. and F. E. Suhm (1981) *International Consumption Comparisons: A System-Wide Approach*. Amsterdam: North-Holland.

Thursby, J. and C. A. Knox Lovell (1978) "An Investigation of the Kmenta Approximation to the CES Function", *International Economic Review*, 19, 363–377.

Tobin, J. (1952) "A Survey of the Theory of Rationing", *Econometrica*, 20, 512–553.

Tobin, J. (1958) "Estimation of Relationships for Limited Dependent Variables", *Econometrica*, 26, 24–36.

Tobin, J. and H. S. Houthakker (1951) "The Effects of Rationing on Demand Elasticities", *Review of Economic Studies*, 18, 140–153.

Tornqvist, L. (1941) "Review", *Ekonomisk Tidskrift*, 43, 216–225.

Varian, H. R. (1978) "A Note on Locally Constant Income Elasticities", *Economics Letters*, 1, 5–9.

Varian, H. R. (1982) "The Nonparametric Approach to Demand Analysis", *Econometrica*, 50, 945–973.

Varian, H. R. (1983) "Nonparametric Tests of Consumer Behavior", *Review of Economic Studies*, 50, 99–110.

Varian, H. R. (1984) "Nonparametric Analysis of Optimizing Behavior with Measurement Error". University of Michigan, mimeo.

Vartia, Y. O. (1983) "Efficient Methods of Measuring Welfare Change and Compensated Income in Terms of Market Demand Functions", *Econometrica*, 51, 79–98.

Wales, T. J. (1977) "On the Flexibility of Flexible Functional Forms: An Empirical Approach", *Journal of Econometrics*, 5, 183–193.

Wales, T. J. and A. D. Woodland (1983) "Estimation of Consumer Demand Systems with Binding Non-Negativity Constraints", *Journal of Econometrics*, 21, 263–285.

White, H. (1980) "A Heteroskedasticity-Consistent Covariance Matrix Estimator and a Direct Test for Heteroskedasticity", *Econometrica*, 48, 817–838.

Willig, R. (1976) "Integrability Implications for Locally Constant Demand Elasticities", *Journal of Economic Theory*, 12, 391–401.

de Wolff, P. (1941) "Income Elasticity of Demand, a Micro-Economic and a Macro-Economic Interpretation", *Economic Journal*, 51, 104–145.

Woodland, A. (1979) "Stochastic Specification and the Estimation of Share Equations", *Journal of Econometrics*, 10, 361–383.

Working, H. (1943) "Statistical Laws of Family Expenditure", *Journal of the American Statistical Association*, 38, 43–56.

Wu, D-M. (1973) "Alternative Tests of Independence Between Stochastic Regressors and Disturbances", *Econometrica*, 41, 733–750.

Yoshihara, K. (1969) "Demand Functions: An Application to the Japanese Expenditure Pattern", *Econometrica*, 37, 257–274.

Zellner, A. (1962) "An Efficient Method of Estimating Seemingly Unrelated Regressions and Tests for Aggregation Bias", *Journal of the American Statistical Association*, 57, 348–368.

*Chapter 31*

# ECONOMETRIC METHODS FOR MODELING PRODUCER BEHAVIOR

DALE W. JORGENSON

*Harvard University*

## Contents

*Handbook of Econometrics, Volume III, Edited by Z. Griliches and M.D. Intriligator*
© *Elsevier Science Publishers BV, 1986*

## 1. Introduction

The purpose of this chapter is to provide an exposition of econometric methods for modeling producer behavior. The objective of econometric modeling is to determine the nature of substitution among inputs, the character of differences in technology, and the role of economies of scale. The principal contribution of recent advances in methodology has been to exploit the potential of economic theory in achieving this objective.

Important innovations in specifying econometric models have arisen from the dual formulation of the theory of production. The chief advantage of this formulation is in generating demands and supplies as explicit functions of relative prices. By using duality in production theory, these functions can be specified without imposing arbitrary restrictions on patterns of production.

The econometric modeling of producer behavior requires parametric forms for demand and supply functions. Patterns of production can be represented in terms of unknown parameters that specify the responses of demands and supplies to changes in prices, technology, and scale. New measures of substitution, technical change, and economies of scale have provided greater flexibility in the empirical determination of production patterns.

Econometric models of producer behavior take the form of systems of demand and supply functions. All the dependent variables in these functions depend on the same set of independent variables. However, the variables and the parameters may enter the functions in a nonlinear manner. Efficient estimation of these parameters has necessitated the development of statistical methods for systems of nonlinear simultaneous equations.

The new methodology for modeling producer behavior has generated a rapidly expanding body of empirical work. We illustrate the application of this methodology by summarizing empirical studies of substitution, technical change, and economies of scale. In this introductory section we first review recent methodological developments and then provide a brief overview of the paper.

### 1.1. Production theory

The economic theory of production – as presented in such classic treatises as Hick's *Value and Capital* (1946) and Samuelson's *Foundations of Economic Analysis* (1983) – is based on the maximization of profit, subject to a production function. The objective of this theory is to characterize demand and supply functions, using only the restrictions on producer behavior that arise from optimization. The principal analytical tool employed for this purpose is the implicit function theorem.[1]

Unfortunately, the characterization of demands and supplies as implicit functions of relative prices is inconvenient for econometric applications. In specifying an econometric model of producer behavior the demands and supplies must be expressed as explicit functions. These functions can be parametrized by treating measures of substitution, technical change, and economies of scale as unknown parameters to be estimated on the basis of empirical data.

The traditional approach to modeling producer behavior begins with the assumption that the production function is additive and homogeneous. Under these restrictions demand and supply functions can be derived explicitly from the production function and the necessary conditions for producer equilibrium. However, this approach has the disadvantage of imposing constraints on patterns of production – thereby frustrating the objective of determining these patterns empirically.

The traditional approach was originated by Cobb and Douglas (1928) and was employed in empirical research by Douglas and his associates for almost two decades.[2] The limitations of this approach were made strikingly apparent by Arrow, Chenery, Minhas, and Solow (1961, henceforward ACMS), who pointed out that the Cobb–Douglas production function imposes a priori restrictions on patterns of substitution among inputs. In particular, elasticities of substitution among all inputs must be equal to unity.

The constant elasticity of substitution (CES) production function introduced by ACMS adds flexibility to the traditional approach by treating the elasticity of substitution as an unknown parameter.[3] However, the CES production function retains the assumptions of additivity and homogeneity and imposes very stringent limitations on patterns of substitution. McFadden (1963) and Uzawa (1962) have shown, essentially, that elasticities of substitution among all inputs must be the same.

The dual formulation of production theory has made it possible to overcome the limitations of the traditional approach to econometric modeling. This formulation was introduced by Hotelling (1932) and later revived and extended by Samuelson (1954, 1960)[4] and Shephard (1953, 1970).[5] The key features of the

---

[1] This approach to production theory is employed by Carlson (1939), Frisch (1965), and Schneider (1934). The English edition of Frisch's book is a translation from the ninth edition of his lectures, published in Norwegian in 1962; the first edition of these lectures dates back to 1926.

[2] These studies are summarized by Douglas (1948). See also: Douglas (1967, 1976). Early econometric studies of producer behavior, including those based on the Cobb–Douglas production function, have been surveyed by Heady and Dillon (1961) and Walters (1963). Samuelson (1979) discusses the impact of Douglas's research.

[3] Econometric studies based on the CES production function have been surveyed by Griliches (1967), Jorgenson (1974), Kennedy and Thirlwall (1972), Nadiri (1970), and Nerlove (1967).

dual formulation are, first, to characterize the production function by means of a dual representation such as a price or cost function and, second, to generate explicit demand and supply functions as derivatives of the price or cost function.[6]

The dual formulation of production theory embodies the same implications of optimizing behavior as the theory presented by Hicks (1946) and Samuelson (1983). However, the dual formulation has a crucial advantage in the development of econometric methodology: Demands and supplies can be generated as explicit functions of relative prices without imposing the arbitrary constraints on production patterns required in the traditional methodology. In addition, the implications of production theory can be incorporated more readily into an econometric model.

## 1.2. Parametric form

Patterns of producer behavior can be described most usefully in terms of the behavior of the derivatives of demand and supply functions.[7] For example, measures of substitution can be specified in terms of the response of demand patterns to changes in input prices. Similarly, measures of technical change can be specified in terms of the response of these patterns to changes in technology. The classic formulation of production theory at this level of specificity can be found in Hicks's *Theory of Wages* (1963).

Hicks (1963) introduced the elasticity of substitution as a measure of substitutability. The elasticity of substitution is the proportional change in the ratio of two inputs with respect to a proportional change in their relative price. Two inputs have a high degree of substitutability if this measure exceeds unity and a low degree of substitutability if the measure is less than unity. The unitary elasticity of substitution employed in the Cobb–Douglas production function is a borderline case between high and low degrees of substitutability.

Similarly, Hicks introduced the bias of technical change as a measure of the impact of changes in technology on patterns of demand for inputs. The bias of technical change is the response of the share of an input in the value of output to a change in the level of technology. If the bias is positive, changes in technology

[4] Hotelling (1932) and Samuelson (1954) develop the dual formulation of production theory on the basis of the Legendre transformation. This approach is employed by Jorgenson and Lau (1974a, 1974b) and Lau (1976, 1978a).

[5] Shephard utilizes distance functions to characterize the duality between cost and production functions. This approach is employed by Diewert (1974a, 1982), Hanoch (1978), McFadden (1978), and Uzawa (1964).

[6] Surveys of duality in the theory of production are presented by Diewert (1982) and Samuelson (1983).

[7] This approach to the selection of parametric forms is discussed by Diewert (1974a), Fuss, McFadden, and Mundlak (1978), and Lau (1974).

increase demand for the input and are said to use the input; if the bias is negative, changes in technology decrease demand for the input and are said to save input. If technical change neither uses nor saves an input, the change is neutral in the sense of Hicks.

By treating measures of substitution and technical change as fixed parameters the system of demand and supply functions can be generated by integration. Provided that the resulting functions are themselves integrable, the underlying price or cost function can be obtained by a second integration. As we have already pointed out, Hicks's elasticity of substitution is unsatisfactory for this purpose, since it leads to arbitrary restrictions on patterns of producer behavior.

The introduction of a new measure of substitution, the share elasticity, by Christensen, Jorgenson, and Lau (1971, 1973) and Samuelson (1973) has made it possible to overcome the limitations of parametric forms based on constant elasticities of substitution.[8] Share elasticities, like biases of technical change, can be defined in terms of shares of inputs in the value of output. The share elasticity of a given input is the response of the share of that input to a proportional change in the price of an input.

By taking share elasticities and biases of technical change as fixed parameters, demand functions for inputs with constant share elasticities and constant biases of technical change can be obtained by integration. The shares of each input in the value of output can be taken to be linear functions of the logarithms of input prices and of the level of technology. The share elasticities and biases of technical change can be estimated as unknown parameters of these functions.

The constant share elasticity (CSE) form of input demand functions can be integrated a second time to obtain the underlying price or cost function. For example, the logarithm of the price of output can be expressed as a quadratic function of the logarithms of the input prices and the level of technology. The price of output can be expressed as a transcendental or, more specifically, an exponential function of the logarithms of the input prices.[9] Accordingly, Christensen, Jorgenson, and Lau refer to this parametric form as the translog price function.[10]

## 1.3.  Statistical method

Econometric models of producer behavior take the form of systems of demand and supply functions. All the dependent variables in these functions depend on

---

[8]A more detailed discussion of this measure is presented in Section 2.2 below.

[9]An alternative approach, originated by Diewert (1971, 1973, 1974b), employs the square roots of the input prices rather than the logarithms and results in the "generalized Leontief" parametric form.

[10]Surveys of parametric forms employed in econometric modeling of producer behavior are presented by Fuss, McFadden, and Mundlak (1978) and Lau (1986).

the same set of independent variables – for example, relative prices and the level of technology. The variables may enter these functions in a nonlinear manner, as in the translog demand functions proposed by Christensen, Jorgenson, and Lau. The functions may also be nonlinear in the parameters. Finally, the parameters may be subject to nonlinear constraints arising from the theory of production.

The selection of a statistical method for estimation of systems of demand and supply functions depends on the character of the data set. For cross section data on individual producing units, the prices that determine demands and supplies can be treated as exogenous variables. The unknown parameters can be estimated by means of nonlinear multivariate regression techniques. Methods of estimation appropriate for this purpose were introduced by Jennrich (1969) and Malinvaud (1970, 1980).[11]

For time series data on aggregates such as industry groups, the prices that determine demands and supplies can be treated as endogenous variables. The unknown parameters of an econometric model of producer behavior can be estimated by techniques appropriate for systems of nonlinear simultaneous equations. One possible approach is to apply the method of full information maximum likelihood. However, this approach has proved to be impractical, since it requires the likelihood function for the full econometric model, not only for the model of producer behavior.

Jorgenson and Laffont (1974) have developed limited information methods for estimating the systems of nonlinear simultaneous equations that arise in modeling producer behavior. Amemiya (1974) proposed to estimate a single nonlinear structural equation by the method of nonlinear two stage least squares. The first step in this procedure is to linearize the equation and to apply the method of two stage least squares to the linearized equation. Using the resulting estimates of the coefficients of the structural equation, a second linearization can be obtained and the process can be repeated.

Jorgenson and Laffont extended Amemiya's approach to a system of nonlinear simultaneous equation by introducing the method of nonlinear three stage least squares. This method requires an estimate of the covariance matrix of the disturbances of the system of equations as well as an estimate of the coefficients of the equations. The procedure is initiated by linearizing the system and applying the method of three stage least squares to the linearized system. This process can be repeated, using a second linearization.[12]

It is essential to emphasize the role of constraints on the parameters of econometric models implied by the theory of production. These constraints may take the form of linear or nonlinear restrictions on the parameters of a single

---

[11] Methods for estimation of nonlinear multivariate regression models are summarized by Malinvaud (1980).

[12] Nonlinear two and three stage least squares methods are also discussed by Amemiya (1977), Gallant (1977), and Gallant and Jorgenson (1979).

equation or may involve restrictions on parameters that occur in several equations. An added complexity arises from the fact that the restrictions may take the form of equalities or inequalities. Estimation under inequality restrictions requires nonlinear programming techniques.[13]

The constraints that arise from the theory of production can be used to provide tests of the validity of the theory. Similarly, constraints that arise from simplification of the patterns of production can be tested statistically. Methods for statistical inference in multivariate nonlinear regression models were introduced by Jennrich (1969) and Malinvaud (1970, 1980). Methods for inference in systems of nonlinear simultaneous equations were developed by Gallant and Jorgenson (1979) and Gallant and Holly (1980).[14]

## 1.4. Overview of the paper

This paper begins with the simplest form of the econometric methodology for modeling producer behavior. This methodology is based on production under constant returns to scale. The dual representation of the production function is a price function, giving the price of output as a function of the prices of inputs and the level of technology. An econometric model of producer behavior is generated by differentiating the price function with respect to the prices and the level of technology.

We present the dual formulation of the theory of producer behavior under constant returns to scale in Section 2. We parameterize this model by taking measures of substitution and technical change to be constant parameters. We than derive the constraints on these parameters implied by the theory of production. In Section 3 we present statistical methods for estimating this model of producer behavior under linear and nonlinear restrictions. Finally, we illustrate the application of this model by studies of data on individual industries in Section 4.

In Section 5 we consider the extension of econometric modeling of producer behavior to nonconstant returns to scale. In regulated industries the price of output is set by regulatory authority. Given the demand for output as a function of the regulated price, the level of output can be taken as exogenous to the producing unit. Necessary conditions for producer equilibrium can be derived from cost minimization. The minimum value of total cost can be expressed as a function of the level of output and the prices of all inputs. This cost function provides a dual representation of the production function.

---

[13] Constrained estimation is discussed in more detail in Section 3.3 below.

[14] Surveys of methods for estimation of nonlinear multivariate regressions and systems of nonlinear simultaneous equations are given by Amemiya (1983) and Malinvaud (1980), especially Chs 9 and 20. Computational techniques are surveyed by Quandt (1983).

The dual formulation of the theory of producer behavior under nonconstant returns to scale parallels the theory under constant returns. However, the level of output replaces the level of technology as an exogenous determinant of production patterns. An econometric model can be parametrized by taking measures of substitution and economies of scale to be constant parameters. In Section 6 we illustrate this approach by means of studies of data on individual firms in regulated industries.

In Section 7 we conclude the paper by outlining frontiers for future research. Current empirical research has focused on the development of more elaborate and more detailed data sets. We consider, in particular, the modeling of consistent time series of inter-industry transactions tables and the application of the results to general equilibrium analysis of the impact of economic policy. We also discuss the analysis of panel data sets, that is, time series of cross sections of observations on individual producing units.

Current methodological research has focused on dynamic modeling of production. At least two promising approaches to this problem have been proposed; both employ optimal control models of producer behavior. The first is based on static expectations with all future prices taken to be equal to current prices. The second approach is based on stochastic optimization under rational expectations, utilizing information about expectations of future prices contained in current production patterns.

## 2.   Price functions

The purpose of this section is to present the simplest form of the econometric methodology for modeling producer behavior. We base this methodology on a production function with constant returns to scale. Producer equilibrium implies the existence of a price function, giving the price of output as a function of the prices of inputs and the level of technology. The price function is dual to the production function and provides an alternative and equivalent description of technology.

An econometric model of producer behavior takes the form of a system of simultaneous equations, determining the distributive shares of the inputs and the rate of technical change. Measures of substitution and technical change give the responses of the distributive shares and the rate of technical change to changes in prices and the level of technology. To generate an econometric model of producer behavior we treat these measures as unknown parameters to be estimated.

The economic theory of production implies restrictions on the parameters of an econometric model of producer behavior. These restrictions take the form of linear and nonlinear constraints on the parameters. Statistical methods employed in modeling producer behavior involve the estimation of systems of nonlinear simultaneous equations with parameters subject to constraints. These constraints give rise to tests of the theory of production and tests of restrictions on patterns of substitution and technical change.

## 2.1. Duality

In order to present the theory of production we first require some notation. We denote the quantity of output by $y$ and the quantities of $J$ inputs by $x_j (j = 1, 2 \ldots J)$. Similarly, we denote the price of output by $q$ and the prices of the $J$ inputs by $p_j (j = 1, 2 \ldots J)$. We find it convenient to employ vector notation for the input quantities and prices:

$x = (x_1, x_2 \ldots x_J) -$ vector of input quantities.

$p = (p_1, p_2 \ldots p_J) -$ vector of input prices.

We assume that the technology can be represented by a *production function*, say $F$, where:

$$y = F(x, t), \tag{2.1}$$

and $t$ is an index of the level of technology. In the analysis of time series data for a single producing unit the level of technology can be represented by time. In the analysis of cross section data for different producing units the level of technology can be represented by one-zero dummy variables corresponding to the different units.[15]

We can define the *shares* of inputs in the value of output by:

$$v_j = \frac{p_j x_j}{qy}, \qquad (j = 1, 2 \ldots J).$$

Under competitive markets for output and all inputs the necessary conditions for producer equilibrium are given by equalities between the share of each input in the value of output and the elasticity of output with respect to that input:

$$v = \frac{\partial \ln y}{\partial \ln x}(x, t), \tag{2.2}$$

where

$v = (v_1, v_2 \ldots v_J) -$ vector of value shares.

$\ln x = (\ln x_1, \ln x_2 \ldots \ln x_J) -$ vector of logarithms of input quantities.

---

[15] Time series and cross section differences in technology have been incorporated into a model of substitution and technical change in U.S. agriculture by Binswanger (1974a, 1974b, 1978c). Binswanger's study is summarized in Section 4.2 below.

Under constant returns to scale the elasticities and the value shares for all inputs sum to unity:

$$i'v = i'\frac{\partial \ln y}{\partial \ln x} = 1,$$

where $i$ is a vector of ones. The value of output is equal to the sum of the values of the inputs.

Finally, we can define the *rate of technical change*, say $v_t$, as the rate of growth of the quantity of output holding all inputs constant:

$$v_t = \frac{\partial \ln y}{\partial t}(x, t). \tag{2.3}$$

It is important to note that this definition does not impose any restriction on patterns of substitution among inputs.

Given the identity between the value of output and the value of all inputs and given equalities between the value share of each input and the elasticity of output with respect to that input, we can express the price of output as a function, say $Q$, of the prices of all inputs and the level of technology:

$$q = Q(p, t). \tag{2.4}$$

We refer to this as the *price function* for the producing unit.

The price function $Q$ is dual to the production function $F$ and provides an alternative and equivalent description of the technology of the producing unit.[16] We can formalize this description in terms of the following properties of the price function:

1. *Positivity.* The price function is positive for positive input prices.
2. *Homogeneity.* The price function is homogeneous of degree one in the input prices.
3. *Monotonicity.* The price function is increasing the input prices.
4. *Concavity.* The price function is concave in the input prices.

Given differentiability of the price function, we can express the value shares of all inputs as elasticities of the price function with respect to the input prices:

$$v = \frac{\partial \ln q}{\partial \ln p}(p, t), \tag{2.5}$$

---

[16] The dual formulation of production theory under constant returns to scale is due to Samuelson (1954).

where:

$\ln p = (\ln p_1, \ln p_2 \ldots \ln p_J)$ – vector of logarithms of input prices.

Further, we can express the negative of the rate of technical change as the rate of growth of the price of output, holding the prices of all inputs constant:

$$- v_t = \frac{\partial \ln q}{\partial t}(p, t). \tag{2.6}$$

Since the price function $Q$ is homogeneous of degree one in the input prices, the value shares and the rate of technical change are homogeneous of degree zero and the value shares sum to unity:

$$i'v = i'\frac{\partial \ln q}{\partial \ln p} = 1.$$

Since the price function is increasing in the input prices the value shares must be nonnegative,

$$v \geqq 0.$$

Since the value shares sum to unity, we can write:

$$v \geq 0,$$

where $v \geq 0$ implies $v \geqq 0$ and $v \neq 0$.

## 2.2.  Substitution and technical change

We have represented the value shares of all inputs and the rate of technical change as functions of the input prices and the level of technology. We can introduce measures of substitution and technical change to characterize these functions in detail. For this purpose we differentiate the logarithm of the price function twice with respect to the logarithms of input prices to obtain measures of substitution:

$$U_{pp} = \frac{\partial^2 \ln q}{\partial \ln p^2}(p, t) = \frac{\partial v}{\partial \ln p}(p, t). \tag{2.7}$$

We refer to the measures of substitution (2.7) as *share elasticities*, since they give the response of the value shares of all inputs to proportional changes in the input prices. If a share elasticity is positive, the corresponding value share increases with the input price. If a share elasticity is negative, the value share decreases with the input price. Finally, if a share elasticity is zero, the value share is independent of the price.[17]

Second, we can differentiate the logarithm of the price function twice with respect to the logarithms of input prices and the level of technology to obtain measures of technical change:

$$u_{pt} = \frac{\partial^2 \ln q}{\partial \ln p \, \partial t}(p,t) = \frac{\partial v}{\partial t} = -\frac{\partial v_t}{\partial \ln p}(p,t). \tag{2.8}$$

We refer to these measures as *biases of technical change*. If a bias of technical change is positive, the corresponding value share increases with a change in the level of technology and we say that technical change is *input-using*. If a bias of technical change is negative, the value share decreases with a change in technology and technical change is *input-saving*. Finally, if a bias is zero, the value share is independent of technology; in this case we say that technical change is *neutral*.[18]

Alternatively, the vector of biases of technical change $u_{pt}$ can be employed to derive the implications of changes in input prices for the rate of technical change. If a bias of technical change is positive, the rate of technical change decreases with the input price. If a bias is negative, the rate of technical change increases with the input price. Finally, if a bias is zero so that technical change is neutral, the rate of technical change is independent of the price.

To complete the description of technical change we can differentiate the logarithm of the price function twice with respect to the level of technology:

$$u_{tt} = \frac{\partial^2 \ln q}{\partial t^2}(p,t) = -\frac{\partial v_t}{\partial t}(p,t). \tag{2.9}$$

We refer to this measure as the *deceleration* of technical change, since it is the negative of rate of change of the rate of technical change. If the deceleration is positive, negative, or zero, the rate of technical change is decreasing, increasing, or independent of the level of technology.

The matrix of second-order logarithmic derivatives of the logarithm of the price function $Q$ must be symmetric. This matrix includes the matrix of share elasticities $U_{pp}$, the vector of biases of technical change $u_{pt}$, and the deceleration of technical change $u_{tt}$. Concavity of the price function in the input prices implies

---

[17] The share elasticity was introduced by Christensen, Jorgenson, and Lau (1971, 1973) and Samuelson (1973).
[18] This definition of the bias of technical change is due to Hicks (1963). Alternative definitions of biases of technical change are compared by Binswanger (1978b).

that matrix of second-order derivatives, say $H$, is nonpositive definite, so that the matrix $U_{pp} + vv' - V$ is nonpositive definite, where:

$$\frac{1}{q} N \cdot H \cdot N = U_{pp} + vv' - V;$$

the price of output $q$ is positive and the matrices $N$ and $V$ are diagonal:

$$N = \begin{bmatrix} p_1 & 0 & \cdots & 0 \\ 0 & p_2 & \cdots & 0 \\ \vdots & \vdots & & \vdots \\ 0 & 0 & \cdots & p_J \end{bmatrix}, \quad V = \begin{bmatrix} v_1 & 0 & \cdots & 0 \\ 0 & v_2 & \cdots & 0 \\ \vdots & \vdots & & \vdots \\ 0 & 0 & \cdots & v_J \end{bmatrix}.$$

We can define substitution and complementarity of inputs in terms of the matrix of share elasticities $U_{pp}$ and the vector of value shares $v$. We say that two inputs are *substitutes* if the corresponding element of the matrix $U_{pp} + vv' - V$ is negative. Similarly, we say that two inputs are *complements* if the corresponding element of this matrix is positive. If the element of this matrix corresponding to the two inputs is zero, we say that the inputs are *independent*. The definition of substitution and complementarity is symmetric in the two inputs, reflecting the symmetry of the matrix $U_{pp} + vv' - V$. If there are only two inputs, nonpositive definiteness of this matrix implies that the inputs cannot be complements.[19]

We next consider restrictions on patterns of substitution and technical change implied by separability of the price function $Q$. The most important applications of separability are associated with aggregation over inputs. Under separability the price of output can be represented as a function of the prices of a smaller number of inputs by introducing price indexes for input aggregates. By treating the price of each aggregate as a function of the prices of the inputs making up the aggregate, we can generate a second stage of the model.

We say that the price function $Q$ is *separable* in the $K$ input prices $\{p_1, p_2 \cdots p_K\}$ if and only if the price function can be represented in the form:

$$q = Q[P(p_1, p_2 \cdots p_K), p_{K+1} \cdots p_J, t], \tag{2.10}$$

where the function $P$ is independent of the $J - K$ input prices $\{p_{K+1}, p_{K+2} \cdots p_J\}$ and the level of technology $t$.[20] We say that the price function is *homothetically separable* if the function $P$ in (2.10) is homogeneous of degree one.[21] Separability of the price function implies homothetic separability.[22]

---

[19] Alternative definitions of substitution and complementarity are discussed by Samuelson (1974).
[20] The concept of separability is due to Leontief (1947a, 1947b) and Sono (1961).
[21] The concept of homothetic separability was introduced by Shephard (1953, 1970).
[22] A proof of this proposition is given by Lau (1969, 1978a).

The price function $Q$ is homothetically separable in the $K$ input prices $\{p_1, p_2 \cdots p_K\}$ if and only if the production function $F$ is homothetically separable in the $K$ input quantities $\{x_1, x_2 \ldots x_K\}$:

$$y = F[G(x_1, x_2 \ldots x_K), x_{K+1} \ldots x_J, t], \tag{2.11}$$

where the function $G$ is homogenous of degree one and independent of $J - K$ quantities $\{x_{K+1}, x_{K+2} \ldots x_J\}$ and the level of technology $t$.[23]

We can interpret the function $P$ in the definition of separability of the price function as a price index; similarly, we can interpret the function $G$ as a quantity index. The price index is dual to the quantity index and has properties analogous to those of the price function:

1.  *Positivity.*   The price index is positive for positive input prices.
2.  *Homogeneity.*   The price index is homogeneous of degree one in the input prices.
3.  *Monotonicity.*   The price index is increasing in the input prices.
4.  *Concavity.*   The price index is concave in the input prices.

The total cost of the $K$ inputs included in the price index $P$, say $c$, is the sum of expenditures on all $K$ inputs:

$$c = \sum_{k=1}^{K} p_k x_k.$$

We can define the quantity index $G$ for this aggregate as the ratio of total cost to the price index $P$:

$$G = \frac{c}{P}. \tag{2.12}$$

The product of the price and quantity indexes for the aggregate is equal to the cost of the $K$ inputs.[24]

We can analyze the implications of homothetic separability by introducing price and quantity indexes of aggregate input and defining the value share of aggregate input in terms of these indexes. An aggregate input can be treated in precisely the same way as any other input, so that price and quantity indexes can be used to reduce the dimensionality of the space of input prices and quantities. The price index generates a second stage of the model, by treating the price of each aggregate as a function of the prices of the inputs making up the aggregate.[25]

---

[23] A proof of this proposition is given by Lau (1978a).

[24] This characterization of price and quantity indexes was originated by Shephard (1953, 1970).

[25] Gorman (1959) has analyzed the relationship between aggregation over commodities and two stage allocation. A presentation of the theory of two stage allocation and references to the literature are given by Blackorby, Primont, and Russell (1978).

## 2.3. Parametrization

In the theory of producer behavior the dependent variables are value shares of all inputs and the rate of technical change. The independent variables are prices of inputs and the level of technology. The purpose of an econometric model of producer behavior is to characterize the value shares and the rate of technical change as functions of the input prices and the level of technology.

To generate an econometric model of producer behavior a natural approach is to treat the measures of substitution and technical change as unknown parameters to be estimated. For this purpose we introduce the parameters:

$$B_{pp} = U_{pp}, \qquad \beta_{pt} = u_{pt}, \qquad \beta_{tt} = u_{tt}, \tag{2.13}$$

where $B_{pp}$ is a matrix of constant share elasticities, $\beta_{pt}$ is a vector of constant biases of technical change, and $\beta_{tt}$ is a constant deceleration of technical change.[26]

We can regard the matrix of share elasticities, the vector of biases of technical change, and the deceleration of technical change as a system of second-order partial differential equations. We can integrate this system to obtain a system of first-order partial differential equations:

$$v = \alpha_p + B_{pp}\ln p + \beta_{pt} \cdot t,$$
$$- v_t = \alpha_t + \beta'_{pt}\ln p + \beta_{tt} \cdot t, \tag{2.14}$$

where the parameters $- \alpha_p, \alpha_t$ – are constants of integration.

To provide an interpretation of the parameters $- \alpha_p, \alpha_t$ – we first normalize the input prices. We can set the prices equal to unity where the level of technology $t$ is equal to zero. This represents a choice of origin for measuring the level of technology and a choice of scale for measuring the quantities and prices of inputs. The vector of parameters $\alpha_p$ is the vector of value shares and the parameter $\alpha_t$ is the negative of the rate of technical charge where the level of technology $t$ is zero.

Similarly, we can integrate the system of first-order partial differential eqs. (2.14) to obtain the price function:

$$\ln p = \alpha'_0 + a'_p\ln p + \alpha_t \cdot t + \tfrac{1}{2}\ln p' B_{pp}\ln p + \ln p'\beta_{pt} \cdot t + \tfrac{1}{2}\beta_{tt} \cdot t^2, \tag{2.15}$$

where the parameter $\alpha_0$ is a constant of integration. Normalizing the price of

---

[26] Share elasticities were introduced as constant parameters of an econometric model of producer behavior by Christensen, Jorgenson, and Lau (1971, 1973). Constant share elasticities, biases, and deceleration of technical change are employed by Jorgenson and Fraumeni (1981) and Jorgenson (1983, 1984b). Binswanger (1974a, 1974b, 1978c) uses a different definition of biases of technical change in parametrizing an econometric model with constant share elasticities.

output so that it is equal to unity where $t$ is zero, we can set this parameter equal to zero. This represents a choice of scale for measuring the quantity and price of output.

For the price function (2.15) the price of output is a transcendental or, more specifically, an exponential function of the logarithms of the input prices. We refer to this form as the *transcendental logarithmic* price function or, more simply, the translog price function, indicating the role of the variables. We can also characterize this price function as the *constant share elasticity* or CSE price function, indicating the role of the fixed parameters. In this representation the scalars – $\alpha_t$, $\beta_t$ – the vectors – $\alpha_p$, $\beta_{pt}$ – and the matrix $B_{pp}$ are constant parameters that reflect the underlying technology. Differences in levels of technology among time periods for a given producing unit or among producing units at a given point of time are represented by differences in the level of technology $t$.

For the translog price function the negative of the average rates of technical change at any two levels of technology, say $t$ and $t-1$, can be expressed as the difference between successive logarithms of the price of output, less a weighted average of the differences between successive logarithms of the input prices with weights given by the average value shares:

$$ - \bar{v}_t = \ln q(t) - \ln q(t-1) - \bar{v}'\left[\ln p(t) - \ln p(t-1)\right]. \tag{2.16} $$

In the expression (2.16) $\bar{v}_t$ is the average rate of technical change,

$$ \bar{v}_t = \tfrac{1}{2}\left[v_t(t) + v_t(t-1)\right], $$

and the vector of average value shares $\bar{v}$ is given by:

$$ \bar{v} = \tfrac{1}{2}\left[v(t) + v(t-1)\right]. $$

We refer to the expression (2.16), introduced by Christensen and Jorgenson (1970), as the *translog rate of technical change*.

We have derived the translog price function as an exact representation of a model of producer behavior with constant share elasticities and constant biases and deceleration of technical change.[27] An alternative approach to the translog price function, based on a Taylor's series approximation to an arbitrary price function, was originated by Christensen, Jorgenson, and Lau (1971, 1973). Diewert (1976, 1980) has shown that the translog rate of technical change (2.16) is exact for the translog price function and the converse.

Diewert (1971, 1973, 1974b) introduced the Taylor's series approach for parametrizing models of producer behavior based on the dual formulation of the

---

[27]Arrow, Chenery, Minhas, and Solow (1961) have derived the CES production function as an exact representation of a model of producer behavior with a constant elasticity of substitution.

theory of production. He utilized this approach to generate the "generalized Leontief" parametric form, based on square root rather than logarithmic transformations of prices. Earlier, Heady and Dillon (1961) had employed Taylor's series approximations to generate parametric forms for the production function, using both square root and logarithmic transformations of the quantities of inputs.

The limitations of Taylor's series approximations have been emphasized by Gallant (1981) and Elbadawi, Gallant, and Souza (1983). Taylor's series provide only a local approximation to an arbitrary price or production function. The behavior of the error of approximation must be specified in formulating an econometric model of producer behavior. To remedy these deficiencies Gallant (1981) has introduced global approximations based on Fourier series.[28]

## 2.4. Integrability

The next stop in generating our econometric model of producer behavior is to incorporate the implications of the econometric theory of production. These implications take the form of restrictions on the system of eqs. (2.14), consisting of value shares of all inputs $v$ and the rate of technical change $v_t$. These restrictions are required to obtain a price function $Q$ with the properties we have listed above. Under these restrictions we say that the system of equations is *integrable*. A complete set of conditions for integrability is the following:

### 2.4.1. Homogeneity

The value shares and the rate of technical change are homogeneous of degree zero in the input prices.

We first represent the value shares and the rate of technical change as a system of eqs. (2.14). Homogeneity of the price function implies that the parameters $- B_{pp}, \beta_{pt} -$ in this system must satisfy the restrictions:

$$B_{pp}i = 0,$$

$$\beta'_{pt}i = 0, \tag{2.17}$$

where $i$ is a vector of ones. For $J$ inputs there are $J+1$ restrictions implied by homogeneity.

---

[28]An alternative approach to the generation of the translog parametric form for the production function by means of the Taylor's series was originated by Kmenta (1967). Kmenta employs a Taylor's series expansion in terms of the parameters of the CES production function. This approach imposes the same restrictions on patterns of production as those implied by the constancy of the elasticity of substitution. The Kmenta approximation is employed by Griliches and Ringstad (1971) and Sargan (1971), among others, in estimating the elasticity of substitution.

### 2.4.2. *Product exhaustion*

The sum of the value shares is equal to unity.

Product exhaustion implies that the value of the $J$ inputs is equal to the value of the product. Product exhaustion implies that the parameters $- \alpha_p$, $B_{pp}$, $\beta_{pt}$ – must satisfy the restrictions:

$$\alpha'_p i = 1,$$

$$B'_{pp} i = 0,$$

$$\beta'_{pt} i = 0. \qquad (2.18)$$

For $J$ inputs there are $J + 2$ restrictions implied by product exhaustion.

### 2.4.3. *Symmetry*

The matrix of share elasticities, biases of technical change, and the deceleration of technical change must be symmetric.

A necessary and sufficient condition for symmetry is that the matrix of parameters must satisfy the restrictions:

$$\begin{bmatrix} B_{pp} & \beta_{pt} \\ \beta'_{pt} & \beta_{tt} \end{bmatrix} = \begin{bmatrix} B_{pp} & \beta_{pt} \\ \beta'_{pt} & \beta_{tt} \end{bmatrix}'. \qquad (2.19)$$

For $J$ inputs the total number of symmetry restrictions is $\frac{1}{2}J(J+1)$.

### 2.4.4. *Nonnegativity*

The value shares must be nonnegative. Nonnegativity is implied by monotonicity of the price function:

$$\frac{\partial \ln q}{\partial \ln p} \geqq 0.$$

For the translog price function the conditions for monotonicity take the form:

$$\frac{\partial \ln q}{\partial \ln p} = \alpha_p + B_{pp} \ln p + \beta_{pt} \cdot t \geqq 0. \qquad (2.20)$$

Since the translog price function is quadratic in the logarithms of the input prices, we can always choose prices so that the monotonicity of the price function is violated. Accordingly, we cannot impose restrictions on the parameters that would imply nonnegativity of the value shares for all prices and levels of technology. Instead, we consider restrictions that imply monotonicity of the value shares wherever they are nonnegative.

### 2.4.5. Monotonicity

The matrix of share elasticities must be nonpositive definite.

Concavity of the price function implies that the matrix $B_{pp} + vv' - V$ is nonpositive definite. Without violating the product exhaustion and nonnegativity restrictions we can set the matrix $vv' - V$ equal to zero. For example, we can choose one of the value shares equal to unity and all the others equal to zero. A necessary condition for the matrix $B_{pp} + vv' - V$ to be nonpositive definite is that the matrix of constant share elasticities $B_{pp}$ must be nonpositive definite. This condition is also sufficient, since the matrix $vv' - V$ is nonpositive definite and the sum of two nonpositive definite matrixes is nonpositive definite.[29]

We can impose concavity on the translog price functions by representing the matrix of constant share elasticities $B_{pp}$ in terms of its Cholesky factorization:

$$B_{pp} = TDT',$$

where $T$ is a unit lower triangular matrix and $D$ is a diagonal matrix. For $J$ inputs we can write the matrix $B_{pp}$ in terms of its Cholesky factorization as follows:

$$B_{pp} = \begin{bmatrix} \delta_1 & \lambda_{21}\delta_1 & \cdots & \lambda_{J1}\delta_1 \\ \lambda_{21}\delta_1 & \lambda_{21}\lambda_{21}\delta_1 + \delta_2 & \cdots & \lambda_{J1}\lambda_{21}\delta_1 + \lambda_{J2}\delta_2 \\ \vdots & \vdots & & \vdots \\ \lambda_{J1}\delta_1 & \lambda_{J1}\lambda_{21}\delta_1 + \lambda_{J2}\delta_2 & \cdots & \lambda_{J1}\lambda_{J1}\delta_1 + \lambda_{J2}\lambda_{J2}\delta_2 + \cdots + \delta_J \end{bmatrix}$$

where:

$$T = \begin{bmatrix} 1 & 0 & \cdots & 0 \\ \lambda_{21} & 1 & \cdots & 0 \\ \vdots & \vdots & & \vdots \\ \lambda_{J1} & \lambda_{J2} & \cdots & 1 \end{bmatrix}, \quad D = \begin{bmatrix} \delta_1 & 0 & \cdots & 0 \\ 0 & \delta_2 & \cdots & 0 \\ \vdots & \vdots & & \vdots \\ 0 & 0 & \cdots & \delta_J \end{bmatrix}$$

The matrix of constant share elasticities $B_{pp}$ must satisfy restrictions implied by symmetry and product exhaustion. These restrictions imply that the parameters of

---

[29] This approach to global concavity was originated by Jorgenson and Fraumeni (1981). Caves and Christensen (1980) have compared regions where concavity obtains for alternative parametric forms.

the Cholesky factorization must satisfy the following conditions:

$$1 + \lambda_{21} + \lambda_{31} + \cdots + \lambda_{J1} = 0,$$
$$1 + \lambda_{32} + \lambda_{42} + \cdots + \lambda_{J2} = 0,$$
$$\dots\dots\dots\dots\dots\dots\dots\dots$$
$$\delta_J \qquad\qquad\qquad\qquad = 0.$$

Under these conditions there is a one-to-one transformation between the elements of the matrix of share elasticities $B_{pp}$ and the parameters of the Cholesky factorization – $T$, $D$. The matrix of share elasticities is nonpositive definite if and only if the diagonal elements $\{\delta_1, \delta_2 \ldots \delta_{J-1}\}$ of the matrix $D$ are nonpositive.[30]

## 3.  Statistical methods

Our model of producer behavior is generated from a translog price function for each producing unit. To formulate an econometric model of production and technical change we add a stochastic component to the equations for the value shares and the rate of technical change. We associate this component with unobservable random disturbances at the level of the producing unit. The producer maximizes profits for given input prices, but the value shares of inputs are subject to a random disturbance.

The random disturbances in an econometric model of producer behavior may result from errors in implementation of production plans, random elements in the technology not reflected in the model of producer behavior, or errors of measurement in the value shares. We assume that each of the equations for the value shares and the rate of technical change has two additive components. The first is a nonrandom function of the input prices and the level of technology; the second is an unobservable random disturbance that is functionally independent of these variables.[31]

### 3.1.  Stochastic specification

To represent an econometric model of production and technical change we require some additional notation. We consider observations on the relative distribution of the value of output among all inputs and the rate of technical

[30] The Cholesky factorization was first proposed for imposing local concavity restrictions by Lau (1978b).

[31] Different stochastic specifications are compared by Appelbaum (1978), Burgess (1975), and Geary and McDonnell (1980). The implications of alternative stochastic specifications are discussed in detail by Fuss, McFadden, and Mundlak (1978).

change. We index the observations by levels of technology $(t = 1, 2 \ldots T)$. We employ a level of technology indexed by time as an illustration throughout the following discussion. The vector of value shares in the $t$th time period is denoted $v^t(t = 1, 2 \ldots T)$. Similarly, the rate of technical change in the $t$th time period is denoted $v_t^t$. The vector of input prices in the $t$th time period is denoted $p_t(t = 1, 2 \ldots T)$. Similarly, the vector of logarithms of input prices is denoted $\ln p_t(t = 1, 2 \ldots T)$.

We obtain an econometric model of production and technical change corresponding to the translog price function by adding random disturbances to the equations for the value shares and the rate of technical change:

$$
\begin{aligned}
v^t &= \alpha_p + B_{pp} \ln p_t + \beta_{pt} \cdot t + \varepsilon^t, \\
v_t^t &= \alpha_t + \beta_{pt}' \ln p_t + \beta_{tt} \cdot t + \varepsilon_t^t, \qquad (t = 1, 2 \ldots T),
\end{aligned}
\tag{3.1}
$$

where $\varepsilon^t$ is the vector of unobservable random disturbances for the value shares of the $t$th time period and $\varepsilon_t^t$ is the corresponding disturbance for the rate of technical change. Since the value shares for all inputs sum to unity in each time period, the random disturbances corresponding to the $J$ value shares sum to zero in each time period:

$$
i' \varepsilon^t = 0, \qquad (t = 1, 2 \ldots T),
\tag{3.2}
$$

so that these disturbances are not distributed independently.

We assume that the unobservable random disturbances for all $J + 1$ equations have expected value equal to zero for all observations:

$$
E \begin{bmatrix} \varepsilon^t \\ \varepsilon_t^t \end{bmatrix} = 0, \qquad (t = 1, 2 \ldots T).
\tag{3.3}
$$

We also assume that the disturbances have a covariance matrix that is the same for all observations; since the random disturbances corresponding to the $J$ value shares sum to zero, this matrix is nonnegative definite with rank at most equal to $J$. We assume that the covariance matrix of the random disturbances corresponding to the value shares and the rate of technical change, say $\Sigma$, has rank $J$, where:

$$
V = \begin{bmatrix} \varepsilon^t \\ \varepsilon_t^t \end{bmatrix} = \Sigma, \qquad (t = 1, 2 \ldots T).
$$

Finally, we assume that the random disturbances corresponding to distinct observations in the same or distinct equations are uncorrelated. Under this assumption the covariance matrix of random disturbances for all observations has the Kronecker product form:

$$V \begin{bmatrix} \varepsilon_1^1 \\ \varepsilon_1^2 \\ \vdots \\ \varepsilon_1^T \\ \varepsilon_2^1 \\ \vdots \\ \varepsilon_t^T \end{bmatrix} = \Sigma \otimes I. \tag{3.4}$$

### 3.2. Autocorrelation

The rate of technical change $v_t^t$ is not directly observable; we assume that the equation for the translog price index of the rate of technical change can be written:

$$- \bar{v}_t^t = \alpha_t + \beta_{pt}' \overline{\ln p_t} + \beta_{tt} \cdot \bar{t} + \bar{\varepsilon}_t^t, \qquad (t = 1, 2 \ldots T), \tag{3.5}$$

where $\bar{\varepsilon}_t^t$ is the average disturbance in the two periods:

$$\bar{\varepsilon}_t^t = \tfrac{1}{2} \left[ \varepsilon_t^t + \varepsilon_t^{t-1} \right], \qquad (t = 1, 2 \ldots T).$$

Similarly, $\overline{\ln p_t}$ is a vector of averages of the logarithms of the input prices and $\bar{t}$ is the average of time as an index of technology in the two periods.

Using our new notation, the equations for the value shares of all inputs can be written:

$$\bar{v}^t = \alpha_p + B_{pp} \overline{\ln p_t} + \beta_{pt} \cdot \bar{t} + \bar{\varepsilon}^t, \qquad (t = 1, 2 \ldots T), \tag{3.6}$$

where $\bar{\varepsilon}^t$ is a vector of averages of the disturbances in the two periods. As before, the average value shares sum to unity, so that the average disturbances for the equations corresponding to value shares sum to zero:

$$i' \bar{\varepsilon}^t = 0, \qquad (t = 1, 2 \ldots T). \tag{3.7}$$

The covariance matrix of the average disturbances corresponding to the equation for the rate of technical change for all observations is proportional to a

Laurent matrix:

$$
V \begin{bmatrix} \bar{\varepsilon}_t^2 \\ \bar{\varepsilon}_t^3 \\ \vdots \\ \bar{\varepsilon}_t^T \end{bmatrix} \sim \Omega,
\tag{3.8}
$$

where:

$$
\Omega = \begin{bmatrix}
\frac{1}{2} & \frac{1}{4} & 0 & \cdots & 0 \\
\frac{1}{4} & \frac{1}{2} & \frac{1}{4} & \cdots & 0 \\
0 & \frac{1}{4} & \frac{1}{2} & \cdots & 0 \\
\vdots & \vdots & \vdots & & \vdots \\
0 & 0 & 0 & \cdots & \frac{1}{2}
\end{bmatrix}.
$$

The covariance matrix of the average disturbance corresponding to the equation for each value share is proportional to the same Laurent matrix. The covariance matrix of the average disturbances for all observations has the Kronecker product form:

$$
V \begin{bmatrix} \bar{\varepsilon}_1^2 \\ \bar{\varepsilon}_1^3 \\ \vdots \\ \bar{\varepsilon}_1^T \\ \bar{\varepsilon}_2^2 \\ \vdots \\ \bar{\varepsilon}_t^T \end{bmatrix} = \Sigma \otimes \Omega.
\tag{3.9}
$$

Since the matrix $\Omega$ in (3.9) is known, the equations for the average rate of technical change and the average value shares can be transformed to eliminate autocorrelation. The matrix $\Omega$ is positive definite, so that there is a matrix $P$ such that:

$$
P\Omega P' = I,
$$
$$
P'P = \Omega^{-1}.
$$

To construct the matrix $P$ we first invert the matrix $\Omega$ to obtain the inverse matrix $\Omega^{-1}$, a positive definite matrix. We then calculate the Cholesky factorization of the inverse matrix $\Omega^{-1}$,

$$\Omega^{-1} = TDT'.$$

where $T$ is a unit lower triangular matrix and $D$ is a diagonal matrix with positive elements along the main diagonal. Finally, we can write the matrix $P$ in the form:

$$P = D^{1/2}T'.$$

where $D^{1/2}$ is a diagonal matrix with elements along the main diagonal equal to the square roots of the corresponding elements of $D$.

We can transform equations for the average rates of technical change by the matrix $P = D^{1/2}T'$ to obtain equations with uncorrelated random disturbances:

$$D^{1/2}T'\begin{bmatrix} \bar{v}_t^2 \\ \bar{v}_t^3 \\ \vdots \\ \bar{v}_t^T \end{bmatrix} = D^{1/2}T'\begin{bmatrix} 1 & \ln p_{12} & \cdots & 2-\frac{1}{2} \\ 1 & \ln p_{13} & \cdots & 3-\frac{1}{2} \\ \vdots & \vdots & & \vdots \\ 1 & \ln p_{1T} & \cdots & T-\frac{1}{2} \end{bmatrix}\begin{bmatrix} \alpha_t \\ \beta_{1t} \\ \vdots \\ \beta_{tt} \end{bmatrix} + D^{1/2}T'\begin{bmatrix} \bar{\varepsilon}_t^2 \\ \bar{\varepsilon}_t^3 \\ \vdots \\ \bar{\varepsilon}_t^T \end{bmatrix},$$

$$(3.10)$$

since:

$$P\Omega P' = D^{1/2}T'\Omega(D^{1/2}T')' = I.$$

The transformation $P = D^{1/2}T'$ is applied to data on the average rates of technical change $\bar{v}_t$ and data on the average values of the variables that appear on the right hand side of the corresponding equation.

We can apply the transformation $P = D^{1/2}T'$ to the equations for average value shares to obtain equations with uncorrelated disturbances. As before, the transformation is also applied to data on the average values of variables that appear on the right hand side of the corresponding equations. The covariance matrix of the transformed disturbances from the equations for the average value shares and the equation for the average rates of technical change has the Kronecker product form:

$$(I \otimes D^{1/2}T')(\Sigma \otimes \Omega)(I \otimes D^{1/2}T')' = \Sigma \otimes I.$$

$$(3.11)$$

To estimate the unknown parameters of the translog price function we combine the first $J-1$ equations for the average value shares with the equation for the average rate of technical change to obtain a complete econometric model of production and technical change. We can estimate the parameters of the equation for the remaining average value share, using the product exhaustion restrictions on these parameters. The complete model involves $\frac{1}{2}J(J+3)$ unknown parameters. A total of $\frac{1}{2}(J^2+4J+5)$ additional parameters can be estimated as functions of these parameters, using the homogeneity, product exhaustion, and symmetry restrictions.[32]

## 3.3. Identification and estimation

We next discuss the estimation of the econometric model of production and technical change given in (3.5) and (3.6). The assumption that the input prices and the level of technology are exogenous variables implies that the model becomes a nonlinear multivariate regression model with additive errors, so that nonlinear regression techniques can be employed. This specification is appropriate for cross section data and individual producing units. For aggregate time series data the existence of supply functions for all inputs makes it essential to treat the prices as endogenous. Under this assumption the model becomes a system of nonlinear simultaneous equations.

To estimate the complete model of production and technical change by the method of full information maximum likelihood it would be necessary to specify the full econometric model, not merely the model of producer behavior. Accordingly, to estimate the model of production in (3.5) and (3.6) we consider limited information techniques. For nonlinear multivariate regression models we can employ the method of maximum likelihood proposed by Malinvaud (1980).[33] For systems of nonlinear simultaneous equations we outline the estimation of the model by the nonlinear three stage least squares (NL3SLS) method originated by Jorgenson and Laffont (1974). Wherever the right hand side variables can be treated as exogenous, this method reduces to limited information maximum likelihood for nonlinear multivariate regression models.

Application of NL3SLS to our model of production and technical change would be straightforward, except for the fact that the covariance matrix of the disturbances is singular. We obtain NL3SLS estimators of the complete system by dropping one equation and estimating the resulting system of $J$ equations by NL3SLS. The parameter estimates are invariant to the choice of the equation omitted in the model.

The NL3SLS estimator can be employed to estimate all parameters of the model of production and technical change, provided that these parameters are

---

[32] This approach to estimation is presented by Jorgenson and Fraumeni (1981).

[33] Maximum likelihood estimation by means of the "seemingly unrelated regressions" model analyzed by Zellner (1962) would not be appropriate here, since the symmetry constraints we have described in Section 2.4 cannot be written in the bilinear form considered by Zellner.

identified. The necessary order condition for identification is that:

$$\tfrac{1}{2}(J+3) < (J-1)\min(V, T-1), \tag{3.12}$$

where $V$ is the number of instruments. A necessary and sufficient rank condition is given below; this amounts to the nonlinear analogue of the absence of multicollinearity.

Our objective is to estimate the unknown parameters – $\alpha_p$, $B_{pp}$, $\beta_{pt}$ – subject to the restrictions implied by homogeneity, product exhaustion, symmetry, and monotonicity. By dropping the equation for one of the value shares, we can eliminate the restrictions implied by summability. These restrictions can be used in estimating the parameters that occur in the equation that has been dropped. We impose the restrictions implied by homogeneity and symmetry as equalities. The restrictions implied by monotonicity take the form of inequalities.

We can write the model of production and technical change in (3.5) and (3.6) in the form:

$$\begin{aligned}
v_1 &= f_1(\gamma) + \varepsilon_1, \\
v_2 &= f_2(\gamma) + \varepsilon_2, \\
&\;\cdots\cdots\cdots \\
v_J &= f_J(\gamma) + \varepsilon_J,
\end{aligned} \tag{3.13}$$

where $v_j(j = 1, 2 \ldots J-1)$ is the vector of observations on the distributive share of the $j$th input for all time periods, transformed to eliminate autocorrelation, $v_J$ is the corresponding vector of observations on the rates of technical change; the vector $\gamma$ includes the parameters – $\alpha_p, \alpha_t, B_{pp}, \beta_{pt}, \beta_{tt}$; $f_j(j = 1, \ldots, 2 \ldots J)$ is a vector of nonlinear functions of these parameters; finally, $\varepsilon_j(j = 1, 2 \ldots J)$ is the vector of disturbances in the $j$th equation, transformed to eliminate autocorrelation.

We can stack the equations in (3.13), obtaining:

$$v = f(\gamma) + \varepsilon, \tag{3.14}$$

where:

$$v = \begin{bmatrix} v_1 \\ v_2 \\ \vdots \\ v_J \end{bmatrix}, \qquad f = \begin{bmatrix} f_1 \\ f_2 \\ \vdots \\ f_J \end{bmatrix}, \qquad \varepsilon = \begin{bmatrix} \varepsilon_1 \\ \varepsilon_2 \\ \vdots \\ \varepsilon_J \end{bmatrix}.$$

By the assumptions in Section 3.1 above the random vector $\varepsilon$ has mean zero and covariance matrix $\Sigma_\varepsilon \otimes I$ where $\Sigma_\varepsilon$ is obtained from the covariance $\Sigma$ in (3.11) by striking the row and column corresponding to the omitted equation.

The nonlinear three stage least squares (NL3SLS) estimator for the model of production and technical change is obtained by minimizing the weighted sum of squared residuals:

$$S(\gamma) = [v - f(\gamma)]' [\hat{\Sigma}_\varepsilon^{-1} \otimes Z(Z'Z)^{-1}Z'][v - f(\gamma)], \tag{3.15}$$

with respect to the vector of unknown parameters $\gamma$, where $Z$ is the matrix of $T-1$ observations on the $V$ instrumental variables. Provided that the parameters are identified, we can apply the Gauss–Newton method to minimize (3.15). First, we linearize the model (3.14), obtaining:

$$v = f(\gamma_0) + \frac{\partial f}{\partial \gamma}(\gamma_0)\Delta\gamma + u, \tag{3.16}$$

where $\gamma_0$ is the initial value of the vector of unknown parameters $\gamma$ and

$$\Delta\gamma = \gamma_1 - \gamma_0,$$

where $\gamma_1$ is the revised value of this vector. The fitted residuals $u$ depend on the initial and revised values.

To revise the initial values we apply Zellner and Theil's (1962) three stage least squares method to the linearized model, obtaining:

$$\Delta\gamma = \left\{ \frac{\partial f}{\partial \gamma}(\gamma_0)' \left( \hat{\Sigma}_\varepsilon^{-1} \otimes Z(Z'Z)^{-1}Z' \right) \frac{\partial f}{\partial \gamma}(\gamma_0) \right\}^{-1}$$
$$\cdot \frac{\partial f}{\partial \phi}(\gamma_0)' \left\{ \hat{\Sigma}_\varepsilon^{-1} \otimes Z(Z'Z)^{-1}Z' \right\}[v - f(\gamma_0)]. \tag{3.17}$$

If $S(\gamma_0) > S(\gamma_1)$, a further iteration is performed by replacing $\gamma_0$ by $\gamma_1$ in (3.16) and (3.17); resulting in a further revised value, say $\gamma_2$, and so on. If this condition is not satisfied, we divide the revision $\Delta\gamma$ by two and evaluate the criteria $S(\gamma)$ again; we continue reducing the revision $\Delta\gamma$ until the criterion improves or the convergence criterion $\max_j \Delta\gamma_j/\gamma_j$ is less than some prespecified limit. If the criterion improves, we continue with further iterations. If not, we stop the iterative process and employ the current value of the vector of unknown parameters as our NL3SLS estimator.[34]

---

[34] Computational techniques for constrained and unconstrained estimation of nonlinear multivariate regression models are discussed by Malinvaud (1980). Techniques for computation of unconstrained estimators for systems of nonlinear simultaneous equations are discussed by Berndt, Hall, Hall, and Hausman (1974) and Belsley (1974, 1979).

The final step in estimation of the model of production and technical change is to minimize the criterion function (3.15) subject to the restrictions implied by monotonicity of the distributive shares. We have eliminated the restrictions that take the form of equalities. Monotonicity of the distributive shares implies inequality restrictions on the parameters of the Cholesky factorization of the matrix of constant share elasticities $B_{pp}$. The diagonal elements of the matrix $D$ in this factorization must be nonpositive.

We can represent the inequality constrains on the matrix of share elasticities $B_{pp}$ in the form:

$$\phi_j(\gamma) \geqq 0, \qquad (j=1,2...J-1), \tag{3.18}$$

where $J-1$ is the number of restrictions. We obtain the inequality constrained nonlinear three stage least squares estimator for the model by minimizing the criterion function subject to the constraints (3.18). This estimator corresponds to the saddlepoint of the Lagrangian function:

$$L = S(\gamma) + \lambda'\phi, \tag{3.19}$$

where $\lambda$ is a vector of $J-1$ Lagrange multipliers and $\phi$ is a vector of $J-1$ constraints.

The Kuhn–Tucker (1951) conditions for a saddlepoint of the Lagrangian (3.19) are the first-order conditions:

$$\frac{\partial L}{\partial \gamma} = \frac{\partial S(\gamma)}{\partial \gamma} + \lambda'\frac{\partial \phi}{\partial \gamma} = 0, \tag{3.20}$$

and the complementary slackness condition:

$$\lambda'\phi = 0, \qquad \lambda \geq 0. \tag{3.21}$$

To find a saddlepoint of the Lagrangian (3.19) we begin by linearizing the model of production and technical change (3.14) as in (3.16). Second, we linearize the constraints as:

$$\phi(\gamma) = \frac{\partial \phi}{\partial \gamma}\Delta\gamma + \phi(\gamma_0), \tag{3.22}$$

where $\gamma_0$ is a vector of initial values of the unknown parameters. We apply Liew's (1976) inequality constrained three stage least squares method to the linearized model, obtaining

$$\Delta\gamma^* = \Delta\gamma + \left\{ \frac{\partial f}{\partial \gamma}(\gamma_0)'\left(\hat{\Sigma}_\epsilon^{-1} \otimes Z(Z'Z)^{-1}Z'\right)\frac{\partial f}{\partial \gamma}(\gamma_0)\right\}^{-1}\frac{\partial \phi}{\partial \gamma}(\gamma_0)'\lambda^*, \tag{3.23}$$

where $\Delta\delta$ is the change in the values of the parameters (3.17) and $\lambda^*$ is the solution of the linear complementarity problem:

$$\frac{\partial\phi}{\partial\gamma}(\gamma_0)\left\{\frac{\partial f}{\partial\gamma}(\gamma_0)'\left(\hat{\Sigma}_\varepsilon^{-1}\otimes Z(Z'Z)^{-1}Z'\right)\frac{\partial f}{\partial\gamma}(\gamma_0)\right\}^{-1}\frac{\partial\phi}{\partial\gamma}(\gamma_0)\lambda$$

$$+\frac{\partial\phi}{\partial\gamma}(\gamma_0)\Delta\gamma-\phi(\gamma_0)\geq 0,$$

where:

$$\left[\frac{\partial\phi}{\partial\gamma}(\gamma_0)\left\{\frac{\partial f}{\partial\gamma}(\gamma_0),\left(\hat{\Sigma}_\varepsilon^{-1}\otimes Z(Z'Z)^{-1}Z'\right)\frac{\partial f}{\partial\gamma}(\gamma_0)\right\}^{-1}\frac{\partial\phi}{\partial\gamma}(\gamma_0)\right.$$

$$\left.+\frac{\partial\phi}{\partial\gamma}(\gamma_0)\Delta\gamma-\phi(\gamma_0)\right]'\lambda=0, \qquad \lambda\geq 0.$$

Given an initial value of the unknown parameters $\gamma_0$ that satisfies the $J-1$ constraints (3.18), if $S(\gamma_1)<S(\gamma_0)$ and $\delta_1$ satisfies the constraints, the iterative process continues by linearizing the model (3.14) as in (3.16) and the constraints (3.18) as in (3.22) at the revised value of the vector of unknown parameters $\gamma_1=\gamma_0+\Delta\gamma$. If not, we shrink $\Delta\gamma$ as before, continuing until an improvement is found subject to the constraints or $\max_j \Delta\gamma_j/\gamma_j$ is less than a convergence criterion.

The nonlinear three stage least squares estimator obtained by minimizing the criterion function (3.15) is a consistent estimator of the vector of unknown parameters $\gamma$. A consistent estimator of the covariance matrix $\Sigma_\varepsilon$ with typical element is $\sigma_{jk}$ is given by

$$\hat{\sigma}_{jk}=\frac{1}{T}\left[v_j-f_j(\hat{\gamma})\right]'\left[v_k-f_k(\hat{\gamma})\right], \qquad (j,k=1,2\ldots J). \tag{3.24}$$

Under suitable regularity conditions the estimator $\hat{\gamma}$ is asymptotically normal with covariance matrix:

$$V(\hat{\gamma})=\left\{\frac{\partial f}{\partial\gamma}(\gamma)'\left(\Sigma_\varepsilon^{-1}\otimes Z(Z'Z)^{-1}Z'\right)\frac{\partial f}{\partial\gamma}(\gamma)\right\}^{-1}. \tag{3.25}$$

We obtain a consistent estimator of this matrix by inserting the consistent estimators $\hat{\gamma}$ and $\hat{\Sigma}_\varepsilon$ in place of the parameters $\gamma$ and $\Sigma_\varepsilon$. The nonlinear three stage least squares estimator is efficient in the class of instrumental variables estimators using $Z$ as the matrix of instrumental variables.[35]

---

[35] The method of nonlinear three stage least squares introduced by Jorgenson and Laffont (1974) was extended to nonlinear inequality constrained estimation by Jorgenson, Lau, and Stoker (1982), esp. pp. 196–204.

The rank condition necessary and sufficient for identifiability of the vector of unknown parameters $\gamma$ is the nonsingularity of the following matrix in the neighborhood of the true parameter vector:

$$\frac{\partial f}{\partial \gamma}(\gamma)' \left( \Sigma_\varepsilon^{-1} \otimes Z(Z'Z)^{-1}Z' \right) \frac{\partial f}{\partial \gamma}(\gamma). \tag{3.26}$$

The order condition (3.12) given above is necessary for the nonsingularity of this matrix.

Finally, we can consider the problem of testing equality restrictions on the vector of unknown parameters $\gamma$. For example, suppose that the maintained hypothesis is that there are $r = \frac{1}{2}J(J+3)$ elements in this vector after solving out the homogeneity, product exhaustion, and symmetry restrictions. Additional equality restrictions can be expressed in the form:

$$\gamma = g(\delta), \tag{3.27}$$

where $\delta$ is a vector of unknown parameters with $S$ elements, $s < r$. We can test the hypothesis:

$$H: \gamma = g(\delta),$$

against the alternative:

$$A: \gamma \neq g(\delta).$$

Test statistics appropriate for this purpose have been analyzed by Gallant and Jorgenson (1979) and Gallant and Holly (1980).[36]

A statistic for testing equality restrictions in the form (3.27) can be constructed by analogy with the likelihood ratio principle. First, we can evaluate the criterion function (3.15) at the minimizing value $\hat{\gamma}$, obtaining:

$$S(\hat{\gamma}) = [v - f(\hat{\gamma})]' \left[ \hat{\Sigma}_\varepsilon^{-1} \otimes Z(Z'Z)^{-1}Z' \right] [v - f(\hat{\gamma})].$$

Second, we can replace the vector of unknown parameters $\gamma$ by the function $g(\delta)$ in (3.27):

$$S(\delta) = \{ v - f[g(\delta)] \}' \left[ \hat{\Sigma}_\varepsilon^{-1} \otimes Z(Z'Z)^{-1}Z' \right] \{ v - f[g(\delta)] \};$$

---

[36]A nonstatistical approach to testing the theory of production has been presented by Afriat (1972), Diewert and Parkan (1983), Hanoch and Rothschild (1972), and Varian (1984).

minimizing the criterion function with respect to $\delta$, we obtain the minimizing value $\hat{\delta}$, the constrained estimator of $\gamma$, $g(\hat{\delta})$, and the constrained value of the criterion itself $S(\hat{\delta})$.

The appropriate test statistic, say $T(\hat{\gamma}, \hat{\delta})$, is equal to the difference between the constrained and unconstrained values of the criterion function:

$$T(\hat{\gamma}, \hat{\delta}) = S(\hat{\delta}) - S(\hat{\gamma}). \tag{3.28}$$

Gallant and Jorgenson (1979) show that this statistic is distributed asymptotically as chi-squared with $r - s$ degrees of freedom. Wherever the right hand side variables can be treated as exogenous, this statistic reduces to the likelihood ratio statistic for nonlinear multivariate regression models proposed by Malinvaud (1980). The resulting statistic is distributed asymptotically as chi-squared.[37]

## 4. Applications of price functions

We first illustrate the econometric modeling of substitution among inputs in Section 4.1 by presenting an econometric model for nine industrial sectors of the U.S. economy implemented by Berndt and Jorgenson (1973). The Berndt–Jorgenson model is based on a price function for each sector, giving the price of output as a function of the prices of capital and labor inputs and the prices of inputs of energy and materials. Technical change is assumed to be neutral, so that all biases of technical change are set equal to zero.

In Section 4.2 we illustrate the econometric modeling of both substitution and technical change. We present an econometric model of producer behavior that has been implemented for thirty-five industrial sectors of the U.S. economy by Jorgenson and Fraumeni (1981). In this model the rate of technical change and the distributive shares of productive inputs are determined simultaneously as functions of relative prices. Although the rate of technical change is endogenous, this model must be carefully distinguished from models of induced technical change.

Aggregation over inputs has proved to be an extremely important technique for simplifying the description of technology for empirical implementation. The corresponding restrictions can be used to generate a two stage model of producer behavior. Each stage can be parametrized separately; alternatively, the validity of alternative simplifications can be assessed by testing the restrictions. In Section 4.3 we conclude with illustrations of aggregation over inputs in studies by Berndt and Jorgenson (1973) and Berndt and Wood (1975).

---

[37]Statistics for testing linear inequality restrictions in linear multivariate regression models have been developed by Gourieroux, Holly, and Montfort (1982); statistics for testing nonlinear inequality restrictions in nonlinear multivariate regression models are given by Gourieroux, Holly, and Monfort (1980).

## 4.1.  Substitution

In the Berndt–Jorgenson (1973) model, production is divided among nine sectors of the U.S. economy:

1. Agriculture, nonfuel mining, and construction.
2. Manufacturing, excluding petroleum refining.
3. Transportation.
4. Communications, trade, and services.
5. Coal mining.
6. Crude petroleum and natural gas.
7. Petroleum refining.
8. Electric utilities.
9. Gas utilities.

The nine producing sectors of the U.S. economy included in the Berndt–Jorgenson model can be divided among five sectors that produce energy commodities – coal, crude petroleum and natural gas, refined petroleum, electricity, and natural gas as a product of gas utilities – and four sectors that produce nonenergy commodities – agriculture, manufacturing, transportation, and communications. For each sector output is defined as the total domestic supply of the corresponding commodity group, so that the input into the sector includes competitive imports of the commodity, inputs of energy, and inputs of nonenergy commodities.

The Berndt–Jorgenson model of producer behavior includes a system of equations for each of the nine producing sectors giving the shares of capital, labor, energy and materials inputs in the value of output as functions of the prices of the four inputs. To formulate an econometric model stochastic components are added to this system of equations. The rate of technical change is taken to be exogenous, so that the adjustment for autocorrelation described in Section 3.2 is not required. However, all prices are treated as endogenous variables; estimates of the unknown parameters of the econometric model are based on the nonlinear three stage least squares estimator presented in Section 3.3.

The endogenous variables in the Berndt–Jorgenson model of producer behavior include value shares of capital, labor, energy, and materials inputs for each sector. Three equations can be estimated for each sector, corresponding to three of the value shares, as in (2.14). The unknown parameters include three elements of the vector $\{\alpha_p\}$ and six share elasticities in the matrix $\{B_{pp}\}$, which is constrained to be symmetric, so that there is a total of nine unknown parameters. Berndt and Jorgenson estimate these parameters from time series data for the period 1947–1971 for each industry; the estimates are presented by Hudson and Jorgenson (1974).

As a further illustration of modeling of substitution among inputs, we consider an econometric model of the total manufacturing sector of the U.S. economy implemented by Berndt and Wood (1975). This sector combines the manufacturing and petroleum refining sectors of the Berndt–Jorgenson model. Berndt and Wood generate this model by expressing the price of aggregate input as a function of the prices of capital, labor, energy, and materials inputs into total manufacturing. They find that capital and energy inputs are complements, while all other pairs of inputs are substitutes.

By comparison with the results of Berndt and Wood, Hudson and Jorgenson (1978) have classified patterns of substitution and complementarity among inputs for the four nonenergy sectors of the Berndt–Jorgenson model. For agriculture, nonfuel mining and construction, capital and energy are complements and all other pairs of inputs are substitutes. For manufacturing, excluding petroleum refining, energy is complementary with capital and materials, while other pairs of inputs are substitutes. For transportation energy is complementary with capital and labor while other pairs of inputs are substitutes. Finally, for communications, trade and services, energy and materials are complements and all other pairs of inputs are substitutes.

Berndt and Wood have considered further simplification of the Berndt–Jorgenson model of producer behavior by imposing separability restrictions on patterns of substitution among capital, labor, energy, and materials inputs.[38] This would reduce the number of input prices at the first stage of the model through the introduction of additional input aggregates. For this purpose additional stages in the allocation of the value of sectoral output among inputs would be required. Berndt and Wood consider all possible pairs of capital, labor, energy, and materials inputs, but find that only the input aggregate consisting of capital and energy is consistent with the empirical evidence.[39]

Berndt and Morrison (1979) have disaggregated the Berndt–Wood data on labor input between blue collar and white collar labor and have studied the substitution among the two types of labor and capital, energy, and materials inputs for U.S. total manufacturing, using a translog price function. Anderson (1981) has reanalyzed the Berndt–Wood data set, testing alternative specifications of the model of substitution among inputs. Gallant (1981) has fitted an alternative model of substitution among inputs to these data, based on the Fourier functional form for the price function. Elbadawi, Gallant, and Souza (1983) have employed this approach in estimating price elasticities of demand for inputs, using the Berndt–Wood data as a basis for Monte Carlo simulations of the performance of alternative functional forms.

---

[38] Restrictions on patterns of substitution implied by homothetic separability have been discussed by Berndt and Christensen (1973a), Jorgenson and Lau (1975), Russell (1975), and Blackorby and Russell (1976).

[39] The methodology for testing separability restrictions was originated by Jorgenson and Lau (1975). This methodology has been discussed by Blackorby, Primont and Russell (1977) and by Denny and Fuss (1977). An alternative approach has been developed by Woodland (1978).

Cameron and Schwartz (1979), Denny, May, and Pinto (1978), Fuss (1977a), and McRae (1981) have constructed econometric models of substitution among capital, labor, energy, and materials inputs based on translog functional forms for total manufacturing in Canada. Technical change is assumed to be neutral, as in the study of U.S. total manufacturing by Berndt and Wood (1975), but nonconstant returns to scale are permitted. McRae and Webster (1982) have compared models of substitution among inputs in Canadian manufacturing, estimated from data for different time periods.

Friede (1979) has analyzed substitution among capital, labor, energy, and materials inputs for total manufacturing in the Federal Republic of Germany. He assumes that technical change is neutral and utilizes a translog price function. He has disaggregated the results to the level of fourteen industrial groups, covering the whole of the West German economy. He has separated materials inputs into two groups – manufacturing and transportation services as one group and other nonenergy inputs as a second group. Ozatalay, Grubaugh, and Long (1979) have modeled substitution among capital, labor, energy and materials inputs, on the basis of a translog price function. They use time series data for total manufacturing for the period 1963–74 in seven countries – Canada, Japan, the Netherlands, Norway, Sweden, the U.S., and West Germany.

Longva and Olsen (1983) have analyzed substitution among capital, labor, energy, and materials inputs for total manufacturing in Norway. They assume that technical change is neutral and utilize a generalized Leontief price function. They have disaggregated the results to the level of nineteen industry groups. These groups do not include the whole of the Norwegian economy; eight additional industries are included in a complete multi-sectoral model of production for Norway. Dargay (1983) has constructed econometric models of substitution among capital, labor, energy, and materials inputs based on translog functional forms for total manufacturing in Sweden. She assumes that technical change is neutral, but permits nonconstant returns to scale. She has disaggregated the results to the level of twelve industry groups within Swedish manufacturing.

Although the breakdown of inputs among capital, labor, energy, and materials has come to predominate in econometric models of production at the industry level, Humphrey and Wolkowitz (1976) have grouped energy and materials inputs into a single aggregate input in a study of substitution among inputs in several U.S. manufacturing industries that utilizes translog price functions. Friedlaender and Spady (1980) have disaggregated transportation services between trucking and rail service and have grouped other inputs into capital, labor and materials inputs. Their study is based on cross section data for ninety-six three-digit industries in the United States for 1972 and employs a translog functional form with fixed inputs.

Parks (1971) has employed a breakdown of intermediate inputs among agricultural materials, imported materials and commercial services, and transportation services in a study of Swedish manufacturing based on the generalized Leontief functional form. Denny and May (1978) have disaggregated labor input between while collar and blue collar labor, capital input between equipment and structures, and have grouped all other inputs into a single aggregate input for Canadian total manufacturing, using a translog functional form. Frenger (1978) has analyzed substitution among capital, labor, and materials inputs for three industries in Norway, breaking down intermediate inputs in a different way for each industry, and utilizing a generalized Leontief functional form.

Griffin (1977a, 1977b, 1977c, 1978) has estimated econometric models of substitution among inputs for individual industries based on translog functional forms. For this purpose he has employed data generated by process models of the U.S. electric power generation, petroleum refining, and petrochemical industries constructed by Thompson, et al. (1977). Griffin (1979) and Kopp and Smith (1980a, 1980b, 1981a, 1981b) have analyzed the effects of alternative aggregations of intermediate inputs on measures of substitution among inputs in the steel industry. For this purpose they have utilized data generated from a process analysis model of the U.S. steel industry constructed by Russell and Vaughan (1976).[40]

Although we have concentrated attention on substitution among capital, labor, energy, and materials inputs, there exists a sizable literature on substitution among capital, labor, and energy inputs alone. In this literature the price function is assumed to be homothetically separable in the prices of these inputs. This requires that all possible pairs of the inputs – capital and labor, capital and energy, and labor and energy – are separable from materials inputs. As we have observed above, only capital-energy separability is consistent with the results of Berndt and Wood (1975) for U.S. total manufacturing.

Appelbaum (1979b) has analyzed substitution among capital, labor, and energy inputs in the petroleum and natural gas industry of the United States, based on the data of Berndt and Jorgenson. Field and Grebenstein (1980) have analyzed substitution among physical capital, working capital, labor, and energy for ten two-digit U.S. manufacturing industries on the basis of translog price functions, using cross section data for individual states for 1971.

Griffin and Gregory (1976) have modeled substitution among capital, labor, and energy inputs for total manufacturing in nine major industrialized countries – Belgium, Denmark, France, Italy, the Netherlands, Norway, the U.K., the U.S., and West Germany – using a translog price function. They pool four cross sections for these countries for the years 1955, 1960, 1965, and 1969, allowing for differences in technology among countries by means of one-zero

---

[40] The advantages and disadvantages of summarizing data from process analysis models by means of econometric models have been discussed by Maddala and Roberts (1980, 1981) and Griffin (1980, 1981c).

dummy variables. Their results differ substantially from those of Berndt and Jorgenson and Berndt and Wood. These differences have led to an extensive discussion among Berndt and Wood (1979, 1981), Griffin (1981a, 1981b), and Kang and Brown (1981), attempting to reconcile the alternative approaches.

Substitution among capital, labor, and energy inputs requires a price function that is homothetically separable in the prices of these inputs. An alternative specification is that the price function is homothetically separable in the prices of capital, labor, and natural resource inputs. This specification has been utilized by Humphrey and Moroney (1975), Moroney and Toeves (1977, 1979) and Moroney and Trapani (1981a, 1981b) in studies of substitution among these inputs for individual manufacturing industries in the U.S. based on translog price functions.

A third alternative specification is that the price function is separable in the prices of capital and labor inputs. Berndt and Christensen (1973b, 1974) have used translog price functions employing this specification in studies of substitution among individual types of capital and labor inputs for U.S. total manufacturing. Berndt and Christensen (1973b) have divided capital input between structures and equipment inputs and have tested the separability of the two types of capital input from labor input. Berndt and Christensen (1974) have divided labor input between blue collar and white collar inputs and have tested the separability of the two types of labor input from capital input. Hamermesh and Grant (1979) have surveyed the literature on econometric modeling of substitution among different types of labor input.

Woodland (1975) has analyzed substitution among structures, equipment and labor inputs for Canadian manufacturing, using generalized Leontief price functions. Woodland (1978) has presented an alternative approach to testing separability and has applied it in modeling substitution among two types of capital input and two types of labor input for U.S. total manufacturing, using the translog parametric form. Field and Berndt (1981) and Berndt and Wood (1979, 1981) have surveyed econometric models of substitution among inputs. They focus on substitution among capital, labor, energy and materials inputs at the level of individual industries.

## 4.2. Technical change

The Jorgenson–Fraumeni (1981) model is based on a production function characterized by constant returns to scale for each of thirty-five industrial sectors of the U.S. economy. Output is a function of inputs of primary factors of production – capital and labor services – inputs of energy and materials, and time as an index of the level of technology. While the rate of technical change is endogenous in this econometric model, the model must be carefully distinguished from models of induced technical change, such as those analyzed by Hicks (1963), Kennedy (1964), Samuelson (1965), von Weizsacker (1962), and many others. In those models the biases of technical change are endogenous and depend on relative prices. As Samuelson (1965) has pointed out, models of induced technical change require intertemporal optimization since technical change at any point of time affects future production possibilities.[41]

In the Jorgenson–Fraumeni model of producer behavior myopic decision rules can be derived by treating the price of capital input as a rental price of capital services.[42] The rate of technical change at any point of time is a function of relative prices, but does not affect future production possibilities. This greatly simplifies the modeling of producer behavior and facilitates the implementation of the econometric model. Given myopic decision rules for producers in each industrial sector, all of the implications of the economic theory of production can be described in terms of the properties of the sectoral price functions given in Section 2.1.[43]

The Jorgenson–Fraumeni model of producer behavior consists of a system of equations giving the shares of capital, labor, energy, and materials inputs in the value of output and the rate of technical change as functions of relative prices and time. To formulate an econometric model a stochastic component is added to these equations. Since the rate of technical change is not directly observable, we consider a form of the model with autocorrelated disturbances; the data are transformed to eliminate the autocorrelation. The prices are treated as endogenous variables and the unknown parameters are estimated by the method of nonlinear three stage least squares presented in Section 3.3.

The endogenous variables in the Jorgenson–Fraumeni model include value shares of sectoral inputs for four commodity groups and the sectoral rate of technical change. Four equations can be estimated for each industry, corresponding to three of the value shares and the rate of technical change. As unknown parameters there are three elements of the vector $\{\alpha_p\}$, the scalar $\{\alpha_t\}$, six share elasticities in the matrix $\{B_{pp}\}$, which is constrained to be symmetric, three biases of technical change in the vector $\{\beta_{pt}\}$, and the scalar $\{\beta_{tt}\}$, so that there is a total of fourteen unknown parameters for each industry. Jorgenson and Fraumeni estimate these parameters from time series data for the period 1958–1974 for each industry, subject to the inequality restrictions implied by monotonicity of the sectoral input value shares.[44]

The estimated share elasticities with respect to price $\{B_{pp}\}$ describe the implications of patterns of substitution for the distribution of the value of output among capital, labor, energy, and materials inputs. Positive share elasticities

---

[41]A review of the literature on induced technical change is given by Binswanger (1978a).

[42]The model of capital as a factor of production was originated by Walras (1954). This model has been discussed by Diewert (1980) and by Jorgenson (1973a, 1980).

[43]Myopic decision rules are derived by Jorgenson (1973b).

[44]Data on energy and materials are based on annual interindustry transactions tables for the United States compiled by Jack Faucett Associates (1977). Data on labor and capital are based on estimates by Fraumeni and Jorgenson (1980).

imply that the corresponding value shares increase with an increase in price; negative share elasticities imply that the value shares decrease with price; zero share elasticities correspond to value shares that are independent of price. The concavity constraints on the sectoral price functions contribute substantially to the precision of the estimates, but require that the share of each input be nonincreasing in the price of the input itself.

The empirical findings on patterns of substitution reveal some striking similarities among industries.[45] The elasticities of the shares of capital with respect to the price of labor are nonnegative for thirty-three of the thirty-five industries, so that the shares of capital are nondecreasing in the price of labor for these thirty-three sectors. Similarly, elasticities of the share of capital with respect to the price of energy are nonnegative for thirty-four industries and elasticities with respect to the price of materials are nonnegative for all thirty-five industries. The share elasticities of labor with respect to the prices of energy and materials are nonnegative for nineteen and for all thirty-five industries, respectively. Finally, the share elasticities of energy with respect to the price of materials are nonnegative for thirty of the thirty-five industries.

We continue the interpretation of the empirical results with estimated biases of technical change with respect to price $\{\beta_{pt}\}$. These parameters can be interpreted as changes in the sectoral value shares (2.14) with respect to time, holding prices constant. This component of change in the value shares can be attributed to changes in technology rather than to substitution among inputs. For example, if the bias of technical change with respect to the price of capital input is positive, we say that technical change is capital-using; if the bias is negative, we say that technical change is capital-saving.

Considering the rate of technical change (2.14), the biases of technical change $\{\beta_{pt}\}$ can be interpreted in an alternative and equivalent way. These parameters are changes in the negative of the rate of technical change with respect to changes in prices. As substitution among inputs takes place in response to price changes, the rate of technical change is altered. For example, if the bias of technical change with respect to capital input is positive, an increase in the price of capital input decreases the rate of technical change; if the bias is negative, an increase in the price of capital input increases the rate of technical change.

A classification of industries by patterns of the biases of technical change is given in Table 1. The pattern that occurs with greatest frequency is capital-using, labor-using, energy-using, and materials-saving technical change. This pattern occurs for nineteen of the thirty-five industries for which biases are fitted. Technical change is capital-using for twenty-five of the thirty-five industries, labor-using for thirty-one industries, energy-using for twenty-nine industries, and materials-using for only two industries.

[45] Parameter estimates are given by Jorgenson and Fraumeni (1983), pp. 255–264.

Table 1
Classification of industries by biases of technical change.

| Pattern of biases | Industries |
|---|---|
| Capital using<br>Labor using<br>Energy using<br>Material saving | Agriculture, metal mining, crude petroleum and natural gas, nonmetallic mining, textiles, apparel, lumber, furniture, printing, leather, fabricated metals, electrical machinery, motor vehicles, instruments, miscellaneous manufacturing, transportation, trade, finance, insurance and real estate, services |
| Capital using<br>Labor using<br>Energy saving<br>Material saving | Coal mining, tobacco manufacturers, communications, government enterprises |
| Capital using<br>Labor saving<br>Energy using<br>Material saving | Petroleum refining |
| Capital using<br>Labor saving<br>Energy saving<br>Material using | Construction |
| Capital saving<br>Labor saving<br>Energy using<br>Material saving | Electric utilities |
| Capital saving<br>Labor using<br>Energy saving<br>Material saving | Primary metals |
| Capital saving<br>Labor using<br>Energy using<br>Material saving | Paper, chemicals, rubber, stone, clay and glass, machinery except electrical, transportation equipment and ordnance, gas utilities |
| Capital saving<br>Labor saving<br>Energy using<br>Material using | Food |

[a]*Source*: Jorgenson and Fraumeni (1983), p. 264.

The patterns of biases of technical change given in Table 1 have important implications for the relationship between relative prices and the rate of economic growth. An increase in the price of materials increases the rate of technical change in thirty-three of the thirty-five industries. By contrast, increases in the prices of capital, labor, and energy reduced the rates of technical change in twenty-five, thirty-one, and twenty-nine industries, respectively. The substantial increases in energy prices since 1973 have had the effect of reducing sectoral rates of technical change, slowing the aggregate rate of technical change, and diminishing the rate of growth for the U.S. economy as a whole.[46]

While the empirical results suggest a considerable degree of similarity across industries, it is necessary to emphasize that the Jorgenson–Fraumeni model of producer behavior requires important simplifying assumptions. First, conditions for producer equilibrium under perfect competition are employed for all industries. Second, constant returns to scale at the industry level are assumed. Finally, a description of technology that leads to myopic decision rules is employed. These assumptions must be justified primarily by their usefulness in implementing production models that are uniform for all thirty-five industrial sectors of the U.S. economy.

Binswanger (1974a, 1974b, 1978c) has analyzed substitution and technical change for U.S. agriculture, using cross sections of data for individual states for 1949, 1954, 1959, and 1964. Binswanger was the first to estimate biases of technical change based on the translog price function. He permits technology to differ among time periods and among groups of states within the United States. He divides capital inputs between land and machinery and divides intermediate inputs between fertilizer and other purchased inputs. He considers substitution among these four inputs and labor input.

Binswanger employs time series data on U.S. agriculture as a whole for the period 1912–1964 to estimate biases of technical change on an annual basis. Brown and Christensen (1981) have analyzed time series data on U.S. agriculture for the period 1947–1974. They divide labor services between hired labor and self-employed labor and capital input between land and all other – machinery, structures, and inventories. Other purchased inputs are treated as a single aggregate. They model substitution and technical change with fixed inputs, using a translog functional form.

Berndt and Khaled (1979) have augmented the Berndt–Wood data set for U.S. manufacturing to include data on output. They estimate biases of technical change and permit nonconstant returns to scale. They employ a Box–Cox transformation of data on input prices, generating a functional form that includes the translog, generalized Leontief, and quadratic as special cases. The Box–Cox transformation is also employed by Appelbaum (1979a) and by Caves, Christensen, and Trethaway (1980). Denny (1974) has proposed a closely related approach to parametrization based on mean value functions.

Kopp and Diewert (1982) have employed a translog parametric form to study technical and allocative efficiency. For this purpose they have analyzed data on U.S. total manufacturing for the period 1947–71 compiled by Berndt and Wood

---

[46] The implications of patterns of biases of technical change are discussed in more detail by Jorgenson (1981).

(1975) and augmented by Berndt and Khaled (1979). Technical change is not required to be neutral and nonconstant returns to scale are permitted. They have interpreted the resulting model of producer behavior as a representation of average practice. They have then re-scaled the parameters to obtain a "frontier" representing best practice and have employed the results to obtain measures of technical and allocative efficiency for each year in the sample.[47]

Wills (1979) has modeled substitution and technical change for the U.S. steel industry, using a translog price function. Norsworthy and Harper (1981) have extended and augmented the Berndt–Wood data set for total manufacturing and have modeled substitution and technical change, using a translog price function. Woodward (1983) has reanalyzed these data and has derived estimates of rates of factor augmentation for capital, labor, energy, and materials inputs, using a translog price function.

Jorgenson (1984b) has modeled substitution and technical change for thirty-five industries of the United States for the period 1958–1979, dividing energy inputs between electricity and nonelectrical energy inputs. He employs translog price functions with capital, labor, two kinds of energy, and materials inputs and finds that technical change is electricity-using and nonelectrical energy-using for most U.S. industries. Nakamura (1984) has developed a similar model for twelve sectors covering the whole of the economy for the Federal Republic of Germany for the period 1960–1974. He has disaggregated intermediate inputs among energy, materials, and services.

We have already discussed the work of Kopp and Smith on substitution among inputs, based on data generated by process models of the U.S. steel industry. Kopp and Smith (1981c, 1982) have also analyzed the performance of different measures of technical change, also using data generated by these models. They show that measures of biased technical change based on the methodology developed by Binswanger can be explained by the proportion of investment in specific technologies.

Econometric models of substitution among inputs at the level of individual industries have incorporated intermediate inputs – broken down between energy and materials inputs – along with capital and labor inputs. However, models of substitution and technical change have also been constructed at the level of the economy as a whole. Output can be divided between consumption and investment goods, as in the original study of the translog price function by Christensen, Jorgenson, and Lau (1971, 1973), and input can be divided between capital and labor services.

Hall (1973) has considered nonjointness of production of investment and consumption goods outputs for the United States. Kohli (1981, 1983) has also

---

[47]A survey of the literature on frontier representations of technology is given by Forsund, Lovell, and Schmidt (1980).

studied nonjointness in production for the United States. Burgess (1974) has added imports as an input to inputs of capital and labor services. Denny and Pinto (1978) developed a model with this same breakdown of inputs for Canada. Conrad and Jorgenson (1977, 1978) have considered nonjointness of production and alternative models of technical change for the Federal Republic of Germany.

### 4.3.  Two stage allocation

Aggregation over inputs has proved to be a very important means for simplifying the description of technology in modeling producer behavior. The price of output can be represented as a function of a smaller number of input prices by introducing price indexes for input aggregates. These price indexes can be used to generate a second stage of the model by treating the price of each aggregate as a function of the prices of the inputs making up the aggregate. We can parametrize each stage of the model separately.

The Berndt–Jorgenson (1973) model of producer behavior is based on two stage allocation of the value of output of each sector. In the first stage the value of sectoral output is allocated among capital, labor, energy, and materials inputs, where materials include inputs of nonenergy commodities and competitive imports. In the second stage the value of energy expenditure is allocated among expenditures on individual types of energy and the value of materials expenditure is allocated among expenditures on competitive imports and nonenergy commodities.

The first stage of the econometric model is generated from a price function for each sector. The price of sectoral output is a function of the prices of capital and labor inputs and the prices of inputs of energy and materials. The second stage of the model is generated from price indexes for energy and materials inputs. The price of energy is a function of the prices of five types of energy inputs, while the price of materials is a function of the prices of four types of nonenergy inputs and the price of competitive imports.

The Berndt–Jorgenson model of producer behavior consists of three systems of equations. The first system gives the shares of capital, labor, energy and materials inputs in the value of output, the second system gives the shares of energy inputs in the value of energy input, and the third system gives the shares of nonenergy inputs and competitive imports in the value of materials inputs. To formulate an econometric model stochastic components are added to these systems of equations. The rate of technical change is taken to be exogenous; all prices – including the prices of energy and materials inputs for each sector – are treated as endogenous variables. Estimates of the unknown parameters of all three systems of equations are based on the nonlinear three stage least squares estimator.

The Berndt–Jorgenson model illustrates the use of two stage allocation to simplify the description of producer behavior. By imposing the assumption that the price of aggregate input is separable in the prices of individual energy and materials inputs, the price function that generates the first stage of the model can be expressed in terms of four input prices rather than twelve. However, simplifications of the first stage of the model requires the introduction of a second stage, consisting of price functions for energy and materials inputs. Each of these price functions can be expressed in terms of five prices of individual inputs.

Fuss (1977a) has constructed a two stage model of Canadian total manufacturing using translog functional forms. He treats substitution among coal, liquid petroleum gas, fuel oil, natural gas, electricity, and gasoline as a second stage of the model. Friede (1979) has developed two stage models based on translog price functions for fourteen industries of the Federal Republic of Germany. In these models the second stage consists of three separate models – one for substitution among individual types of energy and two for substitution among individual types of nonenergy inputs. Dargay (1983) has constructed a two stage model of twelve Swedish manufacturing industries utilizing a translog functional form. She has analyzed substitution among electricity, oil, and solid fuels inputs at the second stage of the model.

Nakamura (1984) has constructed three stage models for twelve industries of the Federal Republic of Germany, using translog price functions. The first stage encompasses substitution and technical change among capital, labor, energy, materials, and services inputs. The second stage consists of three models – a model for substitution among individual types of energy, a model for substitution among individual types of materials, and a model for substitution among individual types of services. The third stage consists of models for substitution between domestically produced input and the corresponding imported input of each type.

Pindyck (1979a, 1979b) has constructed a two stage model of total manufacturing for ten industrialized countries – Canada, France, Italy, Japan, the Netherlands, Norway, Sweden, the U.K., the U.S., and West Germany – using a translog price function. He employs annual data for the period 1959–1973 in estimating a model for substitution among four energy inputs – coal, oil, natural gas, and electricity. He uses annual data for the period 1963–73 in estimating a model for substitution among capital, labor, and energy inputs. Magnus (1979) and Magnus and Woodland (1984) have constructed a two stage model for total manufacturing in the Netherlands along the same lines. Similarly, Ehud and Melnik (1981) have developed a two stage model for the Israeli economy.

Halvorsen (1977) and Halvorsen and Ford (1979) have constructed a two stage model for substitution among capital, labor, and energy inputs for nineteen two-digit U.S. manufacturing industries on the basis of translog price functions. For this purpose they employ cross section data for individual states in 1971. The second stage of the model provides a disaggregation of energy input among inputs of coal, oil, natural gas, and electricity. Halvorsen (1978) has analyzed substitution among different types of energy on the basis of cross section data for 1958, 1962, and 1971.

## 5. Cost functions

In Section 2 we have considered producer behavior under constant returns to scale. The production function (2.1) is homogeneous of degree one, so that a proportional change in all inputs results in a change in output in the same proportion. Necessary conditions for producer equilibrium (2.2) are that the value share of each input is equal to the elasticity of output with respect to that input. Under constant returns to scale the value shares and the elasticities sum to unity.

In this Section we consider producer behavior under increasing returns to scale. Under increasing returns and competitive markets for output and all inputs, producer equilibrium is not defined by profit maximization, since no maximum of profit exists. However, in regulated industries the price of output is set by regulatory authority. Given demand for output as a function of the regulated price, the level of output is exogenous to the producing unit.

With output fixed from the point of view of the producer, necessary conditions for equilibrium can be derived from cost minimization. Where total cost is defined as the sum of expenditures on all inputs, the minimum value of cost can be expressed as as function of the level of output and the prices of all inputs. We refer to this function as the cost function. We have described the theory of production under constant returns to scale in terms of properties of the price function (2.4); similarly, we can describe the theory under increasing returns in terms of properties of the cost function.

### 5.1. Duality

Utilizing the notation of Section 2, we can define total cost, say $c$, as the sum of expenditures on all inputs:

$$c = \sum_{j=1}^{J} p_j x_j.$$

We next define the shares of inputs in total cost by:

$$v_j = \frac{p_j x_j}{c}, \qquad (j=1,2\ldots J).$$

With output fixed from the point of view of the producing unit and competitive markets for all inputs, the necessary conditions for producer equilibrium are given by equalities between the shares of each input in total cost and the ratio of the elasticity of output with respect to that input and the sum of all such elasticities:

$$v = \frac{\dfrac{\partial \ln y}{\partial \ln x}}{i' \dfrac{\partial \ln y}{\partial \ln x}},$$

(5.1)

where $i$ is a vector of ones and:

$$v = (v_1, v_2 \cdots v_J) - \text{vector of cost shares.}$$

Given the definition of total cost and the necessary conditions for producer equilibrium, we can express total cost, say $c$, as a function of the prices of all inputs and the level of output:

$$c = C(p, y).$$

(5.2)

We refer to this as the *cost function*. The cost function $C$ is dual to the production function $F$ and provides an alternative and equivalent description of the technology of the producing unit.[48]

We can formalize the theory of production in terms of the following properties of the cost function:

*1. Positivity.* The cost function is positive for positive input prices and a positive level of output.

*2. Homogeneity.* The cost function is homogeneous of degree one in the input prices.

*3. Monotonicity.* The cost function is increasing in the input prices and in the level of output.

*4. Concavity.* The cost function is concave in the input prices.

Given differentiability of the cost function, we can express the cost shares of all inputs as elasticities of the cost function with respect to the input prices:

$$v = \frac{\partial \ln c}{\partial \ln p}(p, y).$$

(5.3)

Further, we can define an index of returns to scale as the elasticity of the cost function with respect to the level of output:

$$v_y = \frac{\partial \ln c}{\partial \ln y}(p, y).$$

(5.4)

---

[48] Duality between cost and production functions is due to Shephard (1953, 1970).

Following Frisch (1965), we can refer to this elasticity as the *cost flexibility*.

The cost flexibility $v_y$ is the reciprocal of the *degree of returns to scale*, defined as the elasticity of output with respect to a proportional increase in all inputs:

$$v_y = \frac{1}{i' \dfrac{\partial \ln y}{\partial \ln x}}.$$  (5.5)

If output increases more than in proportion to the increase in inputs, cost increases less than in proportion to the increase in output.

Since the cost function $C$ is homogeneous of degree one in the input prices, the cost shares and the cost flexibility are homogeneous of degree zero and the cost shares sum to unity:

$$i'v = i' \frac{\partial \ln c}{\partial \ln p} = 1.$$

Since the cost function is increasing in the input prices, the cost shares must be nonnegative and not all zero:

$$v \geq 0.$$

The cost function is also increasing in the level of output, so that the cost flexibility is positive:

$$v_y > 0.$$

### 5.2.   Substitution and economies of scale

We have represented the cost shares of all inputs and the cost flexibility as functions of the input prices and the level of output. We can characterize these functions in terms of measures of substitution and economies of scale. We obtain *share elasticities* by differentiating the logarithm of the cost function twice with respect to the logarithms of input prices:

$$U_{pp} = \frac{\partial^2 \ln c}{\partial \ln p^2}(p, y) = \frac{\partial v}{\partial \ln p}(p, y).$$  (5.6)

These measures of substitution give the response of the cost shares of all inputs to proportional changes in the input prices.

Second, we can differentiate the logarithm of the cost function twice with respect to the logarithms of the input prices and the level of output to obtain measures of economies of scale:

$$u_{py} = \frac{\partial^2 \ln c}{\partial \ln p \, \partial \ln y}(p, y) = \frac{\partial v}{\partial \ln y}(p, y) = \frac{\partial v_y}{\partial \ln p}(p, y). \qquad (5.7)$$

We refer to these measures as *biases of scale*. The vector of biases of scale $u_{py}$ can be employed to derive the implications of economies of scale for the relative distribution of total cost among inputs. If a scale bias is positive, the cost share of the corresponding input increases with a change in the level of output. If a scale bias is negative, the cost share decreases with a change in output. Finally, if a scale bias is zero, the cost share is independent of output.

Alternatively, the vector of biases of scale $u_{py}$ can be employed to derive the implications of changes in input prices for the cost flexibility. If the scale bias is positive, the cost flexibility increases with the input price. If the scale bias is negative, the cost flexibility decreases with the input price. Finally, if the bias is zero, the cost flexibility is independent of the input price.

To complete the description of economies of scale we can differentiate the logarithm of the cost function twice with respect to the level of output:

$$u_{yy} = \frac{\partial^2 \ln c}{\partial \ln y^2}(p, y) = \frac{\partial v_y}{\partial \ln y}(p, y). \qquad (5.8)$$

If this measure is positive, zero, or negative, the cost flexibility is increasing, decreasing, or independent of the level of output.

The matrix of second-order logarithmic derivatives of the logarithms of the cost function $C$ must be symmetric. This matrix includes the matrix of share elasticities $U_{pp}$, the vector of biases of scale $u_{py}$, and the derivative of the cost flexibility with respect to the logarithm of output $u_{yy}$. Concavity of the cost function in the input prices implies that the matrix of second-order derivatives, say $H$, is nonpositive definite, so that the matrix $U_{pp} + vv' - V$ is nonpositive definite, where:

$$\frac{1}{c} N \cdot H \cdot N = U_{pp} + vv' - V;$$

Total cost $c$ is positive and the diagonal matrices $N$ and $V$ are defined in terms of the input prices $p$ and the cost shares $v$, as in Section 2. Two inputs are *substitutes* if the corresponding element of the matrix $U_{pp} + vv' - V$ is negative, *complements* if the element is negative, and *independent* if the element is zero.

In Section 2.2 above we have introduced price and quantity indexes of aggregate input implied by homothetic separability of the price function. We can analyze the implications of homothetic separability of the cost function by introducing price and quantity indexes of aggregate input and defining the cost share of aggregate input in terms of these indexes. An aggregate input can be treated in precisely the same way as any other input, so that price and quantity indexes can be used to reduce the dimensionality of the space of input prices and quantities.

We say that the cost function $C$ is *homothetic* if and only if the cost function is separable in the prices of all $J$ inputs $\{p_1, p_2 \cdots p_J\}$, so that:

$$c = C[P(p_1, p_2 \cdots p_J), y],\qquad(5.9)$$

where the function $P$ is homogeneous of degree one and independent of the level of output $y$. The cost function is homothetic if and only if the production function is *homothetic*, where

$$y = F[G(x_1, x_2 \cdots x_J)],\qquad(5.10)$$

where the function $G$ is homogeneous of degree one.[49]

Since the cost function is homogeneous of degree one in the input prices, it is homogeneous of degree one in the function $P$, which can be interpreted as the price index for a single aggregate input; the function $G$ is the corresponding quantity index. Furthermore, the cost function can be represented as the product of the price index of aggregate input $P$ and a function, say $H$, of the level of output:

$$c = P(p_1, p_2 \cdots p_J) \cdot H(y).\qquad(5.11)$$

Under homotheticity, the cost flexibility $v_y$ is independent of the input prices:

$$v_y = \frac{\partial \ln H}{\partial \ln y}(y).\qquad(5.12)$$

If the cost flexibility is also independent of the level of output, the cost function is homogeneous in the level of output and the production function is homogeneous in the quantity index of aggregate input $G$. The degree of homogeneity of the production function is the degree of returns to scale and is equal to the reciprocal of the cost flexibility. Under constant returns to scale the degree of returns to scale and the cost flexibility are equal to unity.

---

[49] The concept of homotheticity was introduced by Shephard (1953). Shephard shows that homotheticity of the cost function is equivalent to homotheticity of the production function.

## 5.3. *Parametrization and integrability*

In Section 2.3 we have generated an econometric model of producer behavior by treating the measures of substitution and technical change as unknown parameters to be estimated. In this Section we generate an econometric model of cost and production by introducing the parameters:

$$B_{pp} = U_{pp}, \qquad \beta_{py} = u_{py}, \qquad \beta_{yy} = u_{yy}, \tag{5.13}$$

where $B_{pp}$ is a matrix of constant share elasticities, $\beta_{py}$ is a vector of constant biases of scale, and $\beta_{yy}$ is a constant derivative of the cost flexibility with respect to the logarithm of output.

We can regard the matrix of share elasticities, the vector of biases of scale, and the derivative of the cost flexibility with respect to the logarithm of output as a system of second-order partial differential equations. We can integrate this system to obtain a system of first-order partial differential equations:

$$v = \alpha_p + B_{pp}\ln p + \beta_{py}\ln y,$$
$$v_y = \alpha_y + \beta'_{py}\ln p + \beta_{yy}\ln y, \tag{5.14}$$

where the parameters – $\alpha_p$, $\alpha_y$ – are constants of integration. Choosing scales for measuring the quantities and prices of output and the inputs, we can consider values of input prices and level of output equal to unity. At these values the vector of parameters $\alpha_p$ is equal to the vector of cost shares and the parameters $\alpha_y$ is equal to the cost flexibility.

We can integrate the system of first-order partial differential eqs. (5.14) to obtain the cost function:

$$\ln c = \alpha_0 + \alpha_p\ln p + \alpha_y\ln y + \tfrac{1}{2}\ln p'B_{pp}\ln p$$
$$+ \ln p'\beta_{py}\ln y + \tfrac{1}{2}\beta_{yy}(\ln y)^2, \tag{5.15}$$

where the parameter $\alpha_0$ is a constant of integration. This parameter is equal to the logarithm of total cost where the input prices and level of output are equal to unity. We can refer to this form as the translog cost function, indicating the role of the variables, or the constant share elasticity (CSE) cost function, indicating the role of the parameters.

To incorporate the implications of the economic theory of production we consider restrictions on the system of eqs. (5.14) required to obtain a cost function with properties listed above. A complete set of conditions for integrability is the following:

### 5.3.1. Homogeneity

The cost shares and the cost flexibility are homogeneous of degree zero in the input prices.

Homogeneity of degree zero of the cost shares and the cost flexibility implies that the parameters – $B_{pp}$ and $\beta_{py}$ – must satisfy the restrictions:

$$B_{pp}i = 0$$
$$\beta'_{py}i = 0. \tag{5.16}$$

where $i$ is a vector of ones. For $J$ inputs there are $J+1$ restrictions implied by homogeneity.

### 5.3.2. Cost exhaustion

The sum of the cost shares is equal to unity.

Cost exhaustion implies that the value of the $J$ inputs is equal to total cost. Cost exhaustion implies that the parameters – $\alpha_p, B_{pp}, \beta_{py}$ – must satisfy the restrictions:

$$\alpha'_p i = 1,$$
$$B'_{pp}i = 0, \tag{5.17}$$
$$\beta'_{py}i = 0.$$

For $J$ inputs there are $J+2$ restrictions implied by cost exhaustion.

### 5.3.3. Symmetry

The matrix of share elasticities, biases of scale, and the derivative of the cost flexibility with respect to the logarithm output must be symmetric.

A necessary and sufficient condition for symmetry is that the matrix of parameters must satisfy the restrictions:

$$\begin{bmatrix} B_{pp} & \beta_{py} \\ \beta'_{py} & \beta_{yy} \end{bmatrix} = \begin{bmatrix} B_{pp} & \beta_{py} \\ \beta'_{py} & \beta_{yy} \end{bmatrix}'. \tag{5.18}$$

For $J$ inputs the total number of symmetry restrictions is $\frac{1}{2}J(J+1)$.

## 5.3.4. Nonnegativity

The cost shares and the cost flexibility must be nonnegative.

Since the translog cost function is quadratic in the logarithms of the input prices and the level of output, we cannot impose restrictions on the parameters that imply nonnegativity of the cost shares and the cost flexibility. Instead, we consider restrictions on the parameters that imply monotonicity of the cost shares wherever they are nonnegative.

## 5.3.5. Monotonicity

The matrix of share elasticities $B_{pp} + vv' - V$ is nonpositive definite.

The conditions on the parameters assuring concavity of the cost function wherever the cost shares are nonnegative are precisely analogous to the conditions given in Section 2.4 for concavity of the price function wherever the value shares are nonnegative. These conditions can be expressed in terms of the Cholesky factorization of the matrix of constant share elasticities $B_{pp}$.

## 5.4. Stochastic specification

To formulate an econometric model of cost and production we add a stochastic component to the equations for the cost shares and the cost function itself. To represent the econometric model we require some additional notation. Where there are $K$ producing units we index the observations by producing unit ($k = 1, 2 \ldots K$). The vector of cost shares for the $k$th unit is denoted $v_k$ and total cost of the unit is $c_k$ ($k = 1, 2 \ldots K$). The vector of input prices faced by the $k$th unit is denoted $p_k$ and the vector of logarithms of input prices is $\ln p_k$($k = 1, 2 \ldots K$). Finally, the level of output of the $i$th unit is denoted $y_k$($k = 1, 2 \ldots K$).

We obtain an econometric model of cost and production corresponding to the translog cost function by adding random disturbances to the equations for the cost shares and the cost function:

$$v_k = \alpha_p + B_{pp}\ln p_k + \beta_{py}\ln y + \varepsilon_k, \tag{5.19}$$

$$\ln c_k = \alpha_0 + \alpha_p \ln p_k + \alpha_y \ln y_k + \tfrac{1}{2}\ln p_k B_{pp}\ln p_k$$

$$+ \ln p_k \beta_{py}\ln y_k + \tfrac{1}{2}(\ln y_k)^2 + \varepsilon_k^c, \qquad (k = 1, 2 \ldots K),$$

where $\varepsilon_k$ is the vector of unobservable random disturbances for the cost shares of the $k$th producing unity and $\varepsilon_k^c$ is the corresponding disturbance for the cost function ($k = 1, 2 \ldots K$). Since the cost shares for all inputs sum to unity for each producing unit, the random disturbances corresponding to the $J$ cost shares sum to zero for each unit:

$$i'\varepsilon_k = 0 \qquad (k = 1, 2 \ldots K), \tag{5.20}$$

so that these disturbances are not distributed independently.

We assume that the unobservable random disturbances for all $J + 1$ equations have expected value equal to zero for all observations:

$$E\begin{bmatrix} \varepsilon_k \\ \varepsilon_k^c \end{bmatrix} = 0, \qquad (k = 1, 2 \ldots K). \tag{5.21}$$

We also assume that the disturbances have a covariance matrix that is the same for all producing units and has rank $J$, where:

$$V\begin{bmatrix} \varepsilon_k \\ \varepsilon_k^c \end{bmatrix} = \Sigma, \qquad (k = 1, 2 \ldots K).$$

Finally, we assume that random disturbances corresponding to distinct observations are uncorrelated, so that the covariance matrix of random disturbances for all observations has the Kronecker product form:

$$V\begin{bmatrix} \varepsilon_{11} \\ \varepsilon_{12} \\ \vdots \\ \varepsilon_{1K} \\ \varepsilon_{21} \\ \vdots \\ \varepsilon_K^c \end{bmatrix} = \Sigma \otimes I. \tag{5.22}$$

We can test the validity of restrictions on economies of scale by expressing them in terms of the parameters of an econometric model of cost and production. Under homotheticity the cost flexibility is independent of the input prices. A necessary and sufficient condition for homotheticity is given by:

$$\beta_{py} = 0; \tag{5.23}$$

the vector of biases of scale is equal to zero. Under homogeneity the cost flexibility is independent of output, so that:

$$\beta_{yy} = 0;$$

the derivative of the flexibility with respect to the logarithm of output is zero. Finally, under constant returns to scale, the cost flexibility is equal to unity; given the restrictions implied by homogeneity, constant returns requires:

$$\alpha_y = 1. \tag{5.24}$$

## 6. Applications of cost functions

To illustrate the econometric modeling of economies of scale in Section 6.1, we present an econometric model that has been implemented for the electric power industry in the United States by Christensen and Greene (1976). This model is based on cost functions for cross sections of individual electric utilities in 1955 and 1970. Total cost of steam generation is a function of the level of output and the prices of capital, labor, and fuel inputs. Steam generation accounts for more than ninety percent of total power generation for each of the firms in the Christensen–Greene sample.

A key feature of the electric power industry in the United States is that individual firms are subject to price regulation. The regulatory authority sets the price for electric power. Electric utilities are required to supply the electric power that is demanded at the regulated price. This model must be carefully distinguished from the model of a regulated firm proposed by Averch and Johnson (1962).[50] In the Averch–Johnson model firms are subject to an upper limit on the rate of return rather than price regulation. Firms minimize costs under rate of return regulation only if the regulatory constraint is not binding.

The literature on econometric modeling of scale economies in U.S. transportation and communications industries parallels the literature on the U.S. electric power industry. Transportation and communications firms, like electric utilities, are subject to price regulation and are required to supply all the services that are demanded at the regulated price. However, the modeling of transportation and communications services is complicated by joint production of several outputs. We review econometric models with multiple outputs in Section 6.2.

### 6.1. Economies of scale

The Christensen–Greene model of the electric power industry consists of a system of equations giving the shares of all inputs in total cost and total cost itself as

---

[50]A model of a regulated firm based on cost minimization was introduced by Nerlove (1963). Surveys of the literature on the Averch–Johnson model have been given by Bailey (1973) and Baumol and Klevorick (1970).

functions of relative prices and the level of output. To formulate an econometric model Christensen and Greene add a stochastic component to these equations. They treat the prices and levels of output as exogenous variables and estimate the unknown parameters by the method of maximum likelihood for nonlinear multivariate regression models.

The endogenous variables in the Christensen–Greene model are the cost shares of capital, labor, and fuel inputs and total cost. Christensen and Greene estimate three equations for each cross section, corresponding to two of the cost shares and the cost function. As unknown parameters they estimate two elements of the vector $\alpha_p$, the two scalars – $\alpha_0$ and $\alpha_y$ – three elements of the matrix of share elasticities $B_{pp}$, two biases of scale in the vector $\beta_{py}$, and the scalar $\beta_{yy}$. They estimate a total of ten unknown parameters for each of two cross sections of electric utilities for the years 1955 and 1970.[51] Estimates of the remaining parameters of the model are calculated by using the cost exhaustion, homogeneity, and symmetry restrictions. They report that the monotonicity and concavity restrictions are met at every observation in both cross section data sets.

The hypothesis of constant returns to scale can be tested by first considering the hypothesis that the cost function is homothetic; under this hypothesis the cost flexibility is independent of the input prices. Given homotheticity the additional hypothesis that the cost function is homogeneous can be tested; under this hypothesis the cost flexibility is independent of output as well as prices. These hypotheses can be nested, so that the test of homogeneity is conditional on the test of homotheticity. Likelihood ratio statistics for these hypotheses are distributed, asymptotically, as chi-squared.

We present the results of Christensen and Greene for 1955 and 1970 in Table 2. Test statistics for the hypotheses of homotheticity and homogeneity for both cross section data sets and critical values for chi-squared are also presented in Table 2. Homotheticity can be rejected, so that both homotheticity and homogeneity are inconsistent with the evidence; homogeneity, given homotheticity, is also rejected. If all other parameters involving the level of output were set equal to zero, the parameter $\alpha_y$ would be the reciprocal of the degree of returns to scale. For both 1955 and 1970 data sets this parameter is significantly different from unity.

Christensen and Greene employ the fitted cost functions presented in Table 2 to characterize scale economies for individual firms in each of the two cross sections. For both years the cost functions are U-shaped with a minimum point occurring at very large levels of output. In 1955 118 of the 124 firms have

---

[51] Christensen and Greene have assembled data on cross sections of individual firms for 1955 and 1970. The quantity of output is measured in billions of kilowatt hours (kwh). The quantity of fuel input is measured by British thermal units (Btu). Fuel prices per million Btu are averaged by weighting the price of each fuel by the corresponding share in total consumption. The price of labor input is measured as the ratio of total salaries and wages and employee pensions and benefits to the number of full-time employees plus half the number of part-time employees. The price of capital input is estimated as the sum of interest and depreciation.

Table 2
Cost function for U.S. electric power industry (parameter estimates, 1955 and 1970;
t-ratios in parentheses).[a]

| Parameter | 1955 | 1970 |
|---|---|---|
| $\alpha_0$ | 8.412 | 7.14 |
| | (31.52) | (32.45) |
| $\alpha_y$ | 0.386 | 0.587 |
| | (6.22) | (20.87) |
| $\alpha_K$ | 0.094 | 0.208 |
| | (0.94) | (2.95) |
| $\alpha_L$ | 0.348 | −0.151 |
| | (4.21) | (−1.85) |
| $\alpha_E$ | 0.558 | 0.943 |
| | (8.57) | (14.64) |
| $\beta_{yy}$ | 0.059 | 0.049 |
| | (5.76) | (12.94) |
| $\beta_{Ky}$ | −0.008 | 0.003 |
| | (−1.79) | (−1.23) |
| $\beta_{Ly}$ | −0.016 | −0.018 |
| | (−10.10) | (−8.25) |
| $\beta_{Ey}$ | 0.024 | 0.021 |
| | (5.14) | (6.64) |
| $\beta_{KK}$ | 0.175 | 0.118 |
| | (5.51) | (6.17) |
| $\beta_{LL}$ | 0.038 | 0.081 |
| | (2.03) | (5.00) |
| $\beta_{EE}$ | 0.176 | 0.178 |
| | (6.83) | (10.79) |
| $\beta_{KL}$ | −0.018 | −0.011 |
| | (−1.01) | (−0.749) |
| $\beta_{KE}$ | −0.159 | −0.107 |
| | (−6.05) | (−7.48) |
| $\beta_{LE}$ | −0.020 | −0.070 |
| | (−2.08) | (−6.30) |

Test statistics for restrictions on economies of scale[b]

| Statistic | Homotheticity | Homogeneity |
|---|---|---|
| 1955 | 78.22 | 102.27 |
| 1970 | 57.91 | 157.46 |
| Critical Value (1%) | 9.21 | 11.35 |

[a]Source: Christensen and Greene (1976, Table 4, p. 666).
[b]Source: Christensen and Greene (1976, Table 5, p. 666).

significant economies of scale; only six firms have no significant economies or diseconomies of scale, but these firms produce 25.9 percent of the output of the sample. In 1970 ninety-seven of the 114 firms have significant economies of scale, sixteen have none, and one has significant scale diseconomies.

Econometric modeling of economies of scale in the U.S. electric power industry has generated a very extensive literature. The results through 1978 have been surveyed by Cowing and Smith (1978). More recently, the Christensen–Greene data base has been extended by Greene (1983) to incorporate cross sections of individual electric utilities for 1955, 1960, 1965, 1970, and 1975. By including both the logarithm of output and time as an index of technology in the translog total cost function (5.15), Greene is able to characterize economies of scale and technical change simultaneously.

Stevenson (1980) has employed a translog total cost function incorporating output and time to analyze cross sections of electric utilities for 1964 and 1972. Gollop and Roberts (1981) have used a similar approach to study annual data on eleven electric utilities in the United States for the period 1958–1975. They use the results to decompose the growth of total cost among economies of scale, technical change, and growth in input prices. Griffin (1977b) has modeled substitution among different types of fuel in steam electricity generation using four cross sections of twenty OECD countries. Halvorsen (1978) has analyzed substitution among different fuel types, using cross section data for the United States in 1972.

Cowing, Reifschneider, and Stevenson (1983) have employed a translog total cost function similar to that of Christensen and Greene to analyze data for eighty-one electric utilities for the period 1964–1975. For this purpose they have grouped the data into four cross sections, each consisting of three-year totals for all firms. If disturbances in the equations for the cost shares (5.19) are associated with errors in optimization, costs must increase relative to the minimum level given by the cost function (5.15). Accordingly, Cowing, Reifschneider and Stevenson employ a disturbance for the cost function that is constrained to be positive.[52]

An alternative to the Christensen–Greene model for electric utilities has been developed by Fuss (1977b, 1978). In Fuss's model the cost function is permitted to differ ex ante, before a plant is constructed, and ex post, after the plant is in place.[53] Fuss employs a generalized Leontief cost function with four input prices – structures, equipment, fuel, and labor. He models substitution among inputs and economies of scale for seventy-nine steam generation plants for the period 1948–61.

We have observed that a model of the behavior of a regulated firm based on cost minimization must be carefully distinguished from the model originated by Averch and Johnson (1962). In addition to allowing a given rate of return, regulatory authorities may permit electric utilities to adjust the regulated price of output for changes in the cost of specific inputs. In the electric power industry a

---

[52] Statistical methods for models of production with disturbances constrained to be positive or negative are discussed by Aigner, Amemiya and Poirier (1976) and Greene (1980).

[53] A model of production with differences between ex ante and ex post substitution possibilities was introduced by Houthakker (1956). This model has been further developed by Johansen (1972) and Sato (1975) and has been discussed by Hildenbrand (1981) and Koopmans (1977). Recent applications are given by Forsund and Hjalmarsson (1979, 1983), and Forsund and Jansen (1983).

common form of adjustment is to permit utilities to change prices with changes in fuel costs.

Peterson (1975) has employed a translog cost function for the electric utility industry to test the Averch–Johnson hypothesis. For this purpose he introduces three measures of the effectiveness of regulation into the cost function: a one-zero dummy variable distinguishing between states with and without a regulatory commission, a similar variable differentiating between alternative methods for evaluation of public utility property for rate making purposes, and a variable representing differences between the rate of return allowed by the regulatory authority and the cost of capital. He analyzes annual observations on fifty-six steam generating plans for the period 1966 to 1968.

Cowing (1978) has employed a quadratic parametric form to test the Averch–Johnson hypothesis for regulated firms. He introduces both the cost of capital and the rate of return allowed by the regulatory authority as determinants of input demands. Cowing analyzes data on 114 steam generation plants constructed during each of three time periods – 1947–50, 1955–59, and 1960–65. Gollop and Karlson (1978) have employed a translog cost function that incorporates a measure of the effectiveness of regulatory adjustments for changes in fuel costs. This measure is the ratio of costs that may be recovered under the fuel cost adjustment mechanism to all fuel costs. Gollop and Karlson analyze data for cross sections of individual electric utilities for the years 1970, 1971, and 1972.

Atkinson and Halvorsen (1980) have employed a translog parametric form to test the effects of both rate of return regulation and fuel cost adjustment mechanisms. For this purpose they have analyzed cross section data for electric utilities in 1973. Gollop and Roberts (1983) have studied the effectiveness of regulations on sulfur dioxide emissions in the electric utility industry. They employ a translog cost function that depends on a measure of regulatory effectiveness. This measure is based on the legally mandated reduction in emissions and on the enforcement of emission standards. Gollop and Roberts analyze cross sections of fifty-six electric utilities for each of the years 1973–1979 and employ the results to study the impact of environmental regulation on productivity growth.

## 6.2.  Multiple outputs

Brown, Caves, and Christensen (1979) have introduced a model for joint production of freight and transportation services in the railroad industry based on the translog cost function (5.15).[54] A cost flexibility (5.4) can be defined for each output. Scale biases and derivatives of the cost flexibilities with respect to each

---

[54]A review of the literature on regulation with joint production is given by Bailey and Friedlaender (1982).

output can be taken to be constant parameters. The resulting cost function depends on logarithms of input prices and logarithms of the quantities of each output. Caves, Christensen, and Trethaway (1980) have extended this approach by introducing Box–Cox transformations of the quantities of the outputs in place of logarithmic transformations. This generalized translog cost function permits complete specialization in the production of a single output.

The generalized translog cost function has been applied to cross sections of Class I railroads in the United States for 1955, 1963, and 1974 by Caves, Christensen, and Swanson (1980). They consider five categories of inputs: labor, way and structures, equipment, fuel, and materials. For freight transportation services they take ton-miles and average length of freight haul as measures of output. Passenger services are measured by passenger-miles and average length of passenger trip. They employ the results to measure productivity growth in the U.S. railroad industry for the period 1951–74. Caves, Christensen, and Swanson (1981) have employed data for cross sections of Class I railroads in the United States to fit a variable cost function, treating way and structures as a fixed input and combining equipment and materials into a single variable input. They have employed the results in measuring productivity growth for the period 1951–74.

Friedlaender and Spady (1981) and Harmatuck (1979) have utilized a translog total cost function to analyze cross section data on Class I railroads in the United States. Jara–Diaz and Winston (1981) have employed a quadratic cost function to analyze data on Class III railroads, with measures of output disaggregated to the level of individual point-to-point shipments. Brautigan, Daugherty, and Turnquist (1982) have used a translog variable cost function to analyze monthly data for nine years for a single railroad. Speed of shipment and quality of service are included in the cost function as measures of the characteristics of output.

The U.S. trucking industry, like the U.S. railroad industry, is subject to price regulation. Spady and Friedlaender (1978) have employed a translog cost function to analyze data on a cross section of 168 trucking firms in 1972. They have disaggregated inputs into four categories – labor, fuel, capital, and purchased transportation. Freight transportation services are measured in ton-miles. To take into account the heterogeneity of freight transportation services, five additional characteristics of output are included in the cost function – average shipment size, average length of haul, percentage of less than truckload traffic, insurance costs, and average load per truck.

Friedlaender, Spady, and Chiang (1981) have employed the approach of Spady and Friedlaender (1978) to analyze cross sections of 154, 161, and 47 trucking firms in 1972. Inputs are disaggregated in the same four categories, while an additional characteristic of output is included, namely, terminal density, defined as ton-miles per terminal. Separate models are estimated for each of the three samples. Friedlaender and Spady (1981) have employed the results in analyzing the impact of changes in regulatory policy. Harmatuck (1981) has employed a translog cost function to analyze a cross section of 100 trucking firms in 1977. He has included data on the number and size of truck load and less-than-truckload shipments and average length of haul as measures of output. He disaggregates input among five activities – line haul, pickup and delivery, billing and collecting, platform handling, and all other.

Finally, Chiang and Friedlaender (1985) have disaggregated the output of trucking firms into four categories – less than truckload hauls of under 250 miles, between 250–500 miles, and over 500 miles, and truck load traffic – all measured in ton miles. Inputs are disaggregated among five categories – labor, fuel, revenue equipment, "other" capital, and purchased transportation. Characteristics of output similar to those included in earlier studies by Chiang, Friedlaender, and Spady are incorporated into the cost function, together with measures of the network configuration of each firm. They have employed this model to analyze a cross section of 105 trucking firms for 1976.

The U.S. air transportation industry, like the U.S. railroad and trucking industries, is subject to price regulation. Caves, Christensen, and Trethaway (1983) have employed a translog cost function to analyze a panel data set for all U.S. truck and local service airlines for the period 1970–81. Winston (1985) has provided a survey of econometric models of producer behavior in the transportation industries, including railroads, trucking, and airlines.

In the United States the communications industries, like the transportation industries, are largely privately owned but subject to price regulation. Nadiri and Schankerman (1981) have employed a translog cost function to analyze time series data for 1947–76 on the U.S. Bell System. They include the operating telephone companies and Long Lines, but exclude the manufacturing activities of Western Electric and the research and development activities of Bell Laboratories. Output is an aggregate of four service categories; inputs of capital, labor, and materials are distinguished. A time trend is included in the cost function as an index of technology; the stock of research and development is included as a separate measure of the level of technology.

Christensen, Cummings, and Schoech (1983) have employed alternative specifications of the translog cost functions to analyze time series data for the U.S. Bell System for 1947–1977. They employ a distributed lag of research and development expenditures by the Bell System to represent the level of technology. As alternative representations they consider the proportion of telephones with access to direct distance dialing, the percentage of telephones connected to central offices with modern switching facilities, and a more comprehensive measure of research and development. They also consider specifications with capital input held fixed and with experienced labor and management held fixed. Evans and Heckman (1983, 1984) have provided an alternative analysis of the same data set. They have studied economies of scope in the joint production of telecommunications services.

Bell Canada is the largest telecommunications firm in Canada. Fuss, and Waverman (1981) have employed a translog cost function to analyze time series data on Bell Canada for the period 1952–1975. Three outputs are distinguished: message toll service, other total service, and local and miscellaneous service. Capital, labor, and materials are treated as separate categories of input. The level of technology is represented by a time trend. Denny, Fuss, Everson, and Waverman (1981) have analyzed time series data for the period 1952–1976. The percentage of telephones with access to direct dialing and the percentage of telephones connected to central offices with modern switching facilities are incorporated into the cost function as measures of the level of technology. Kiss, Karabadjian, and Lefebvre (1983) have compared alternative specifications of output and the level of technology. Fuss (1983) has provided a survey of econometric modeling of telecommunications services.

## 7.   Conclusion

The purpose of this concluding section is to suggest possible directions for future research on econometric modeling of producer behavior. We first discuss the application of econometric models of production in general equilibrium analysis. The primary focus of empirical research has been on the characterization of technology for individual producing units. Application of the results typically involves models for both demand and supply for each commodity. The ultimate objective of econometric modeling of production is to construct general equilibrium models encompassing demand and supplies for a wide range of products and factors of production.

A second direction for future research on producer behavior is to exploit statistical techniques appropriate for panel data. Panel data sets consist of observations on several producing units at many points of time. Empirical research on patterns of substitution and technical change has been based on time series observations on a single producing unit or on cross section observations on different units at a given point of time. Research on economics of scale has been based primarily on cross section observations.

Our exposition of econometric methods has emphasized areas of research where the methodology has crystallized. An important area for future research is the implementation of dynamic models of technology. These models are based on substitution possibilities among outputs and inputs at different points of time. A number of promising avenues for investigation have been suggested in the literature on the theory of production. We conclude the paper with a brief review of possible approaches to the dynamic modeling of producer behavior.

## 7.1. General equilibrium modeling

At the outset of our discussion it is essential to recognize that the predominant tradition in general equilibrium modeling does not employ econometric methods. This tradition originated with the seminal work of Leontief (1951), beginning with the implementation of the static input–output model. Leontief (1953) gave a further impetus to the development of general equilibrium modeling by introducing a dynamic input–output model. Empirical work associated with input–output analysis is based on estimating the unknown parameters of a general equilibrium model from a single interindustry transactions table.

The usefulness of the "fixed coefficients" assumption that underlies input–output analysis is hardly subject to dispute. By linearizing technology it is possible to solve at one stroke the two fundamental problems that arise in the practical implementation of general equilibrium models. First, the resulting general equilibrium model can be solved as a system of linear equations with constant coefficients. Second, the unknown parameters describing technology can be estimated from a single data point.

The first successful implementation of a general equilibrium model without the fixed coefficients assumption of input–output analysis is due to Johansen (1974). Johansen retained the fixed coefficients assumption in modeling demands for intermediate goods, but employed linear logarithmic or Cobb–Douglas production functions in modeling the substitution between capital and labor services and technical change. Linear logarithmic production functions imply that relative shares of inputs in the value of output are fixed, so that the unknown parameters characterizing substitution between capital and labor inputs can be estimated from a single data point.

In modeling producer behavior Johansen employed econometric methods only in estimating constant rates of technical change. The essential features of Johansen's approach have been preserved in the general equilibrium models surveyed by Fullerton, Henderson, and Shoven (1984). The unknown parameters describing technology in these models are determined by "calibration" to a single data point. Data from a single interindustry transactions table are supplemented by a small number of parameters estimated econometrically. The obvious disadvantage of this approach is that arbitrary constraints on patterns of production are required in order to make calibration possible.

An alternative approach to modeling producer behavior for general equilibrium models is through complete systems of demand functions for inputs in each industrial sector. Each system gives quantities demanded as functions of prices and output. This approach to general equilibrium of modeling producer behavior was originated by Berndt and Jorgenson (1973). As in the descriptions of technology by Leontief and Johansen, production is characterized by constant returns to scale in each sector. As a consequence, commodity prices can be expressed as functions of factor prices, using the nonsubstitution theorem of Samuelson (1951). This greatly facilitates the solution of the econometric general equilibrium model constructed by Hudson and Jorgenson (1974) by permitting a substantial reduction in dimensionality of the space of prices to be determined by the model.

The implementation of econometric models of producer behavior for general equilibrium analysis is very demanding in terms of data requirements. These models require the construction of a consistent time series of interindustry transactions tables. By comparison, the noneconometric approaches of Leontief and Johansen require only a single interindustry transactions table. Second, the implementation of systems of input demand functions requires methods for the estimation of parameters in systems of nonlinear simultaneous equations. Finally, the restrictions implied by the economic theory of producer behavior require estimation under both equality and inequality constraints.

Jorgenson and Fraumeni (1981) have constructed an econometric model of producer behavior for thirty-five industrial sectors of the U.S. economy. The next research objective is to disaggregate the demands for energy and materials by constructing a hierarchy of models for allocation within the energy and materials aggregates. A second research objective is to incorporate the production models for all thirty-five industrial sectors into an econometric general equilibrium model of production for the U.S. economy along the lines suggested by Jorgenson (1983, 1984a). A general equilibrium model will make it possible to analyze the implications of sectoral patterns of substitution and technical change for the behavior of the U.S. economy as a whole.

## 7.2. Panel data

The approach to modeling economies of scale originated by Christensen and Greene (1976) is based on the underlying assumption that individual producing units at the same point of time have the same technology. Separate models of production are fitted for each time period, implying that the same producing unit has a different technology at different points of time. A more symmetrical treatment of observations at different points of time is suggested by the model of substitution and technical change in U.S. agriculture developed by Binswanger (1974a, 1974b, 1978c). In this model technology is permitted to differ among time periods and among producing units.

Caves, Christensen, and Trethaway (1984) have employed a translog cost function to analyze a panel data set for all U.S. trunk and local service airlines for the period 1970–81. Individual airlines are observed in some or all years during the period. Differences in technology among years and among producing units are incorporated through one–zero dummy variables that enter the cost function. One set of dummy variables corresponds to the individual producing units. A second set of dummy variables corresponds to the time periods.

Although airlines provide both freight and passenger service, the revenues for passenger service greatly predominate in the total, so that output is defined as an aggregate of five categories of transportation services. Inputs are broken down into three categories – labor, fuel, and capital and materials. The number of points served by an airline is included in the cost functions as a measure of the size of the network. Average stage length and average load factor are included as additional characteristics of output specific to the airline.

Caves, Christensen, and Trethaway introduce a distinction between economies of scale and economies of density. Economies of scale are defined in terms of the sum of the elasticities of total cost with respect to output and points served, holding input prices and other characteristics of output constant. Economies of density are defined in terms of the elasticity of total cost with respect to output, holding points served, input prices, and other characteristics of output constant. Caves, Christensen, and Trethaway find constant returns to scale and increasing returns to density in airline service.

The model of panel data employed by Caves, Christensen, and Trethaway in analyzing air transportation service is based on "fixed effects". The characteristics of output specific to a producing unit can be estimated by employing one-zero dummy variables for each producing unit. An alternative approach based on "random effects" of output characteristics is utilized by Caves, Christensen, Trethaway, and Windle (1984) in modeling rail transportation service. They consider a panel data set for forty-three Class I railroads in the United States for the period 1951–1975.

Caves, Christensen, Trethaway, and Windle employ a generalized translog cost function in modeling the joint production of freight and passenger transportation services by rail. They treat the effects of characteristics of output specific to each railroad as a random variable. They estimate the resulting model by panel data techniques originated by Mundlak (1963, 1978). The number of route miles served by a railroad is included in the cost function as a measure of the size of the network. Length of haul for freight and length of trip for passengers are included as additional characteristics of output.

Economies of density in the production of rail transportation services are defined in terms of the elasticity of total cost with respect to output, holding route miles, input prices, firm-specific effects, and other characteristics of output fixed. Economies of scale are defined holding only input prices and other characteristics of output fixed. The impact of changes in outputs, route miles, and firm specific effects can be estimated by panel data techniques. Economies of density and scale can be estimated from a single cross section by omitting firm-specific dummy variables.

Panel data techniques require the construction of a consistent time series of observation on individual producing units. By comparison, the cross section methods developed by Christensen and Greene require only a cross section of observations for a single time period. The next research objective in characterizing economies of scale and economies of density is to develop panel data sets for regulated industries – electricity generation, transportation, and communications – and to apply panel data techniques in the analysis of economies of scale and economies of density.

## 7.3. *Dynamic models of production*

The simplest intertemporal model of production is based on capital as a factor of production. A less restrictive model generates costs of adjustment from changes in the level of capital input through investment. As the level of investment increases, the amount of marketable output that can be produced from given levels of all inputs is reduced. Marketable output and investment can be treated as outputs that are jointly produced from capital and other inputs. Models of production based on costs of adjustment have been analyzed, for example, by Lucas (1967) and Uzawa (1969).

Optimal production planning with costs of adjustment requires the use of optimal control techniques. The optimal production plan at each point of time depends on the initial level of capital input, so that capital is a "quasi-fixed" input. Obviously, labor and other inputs can also be treated as quasi-fixed in models of production based on costs of adjustment. The optimal production plan at each point of time depends on the initial levels of all quasi-fixed inputs.

The optimal production plan with costs of adjustment depends on all future prices of outputs and inputs of the production process. Unlike the prices of outputs and inputs at each point of time employed in the production studies we have reviewed, future prices cannot be observed on the basis of market transactions. To simplify the incorporation of future prices into econometric models of production, a possible approach is to treat these prices as if they were known with certainty. A further simplification is to take all future prices to be equal to current prices, so that expectations are "static".

Dynamic models of production based on static expectations have been employed by Denny, Fuss, and Waverman (1981), Epstein and Denny (1983), and Morrison and Berndt (1980). Denny, Fuss, and Waverman have constructed models of substitution among capital, labor, energy, and materials inputs for two-digit industries in Canada and the United States. Epstein and Denny have analyzed substitution among these same inputs for total manufacturing in the United States. Morrison and Berndt have utilized a similar data set with labor input divided between blue collar and white collar labor. Berndt, Morrison, and Watkins (1981) have surveyed dynamic models of production.

The obvious objection to dynamic models of production based on static expectations is that current prices change from period to period, but expectations are based on unchanging future prices. An alternative approach is to base the dynamic optimization on forecasts of future prices. Since these forecasts are subject to random errors, it is natural to require that the optimization process take into account the uncertainty that accompanies forecasts of future prices. Two alternative approaches to optimization under uncertainty have been proposed.

We first consider the approach to optimization under uncertainty based on certainty equivalence. Provided that the objective function for producers is quadratic and constraints are linear, optimization under uncertainty can be replaced by a corresponding optimization problem under certainty. This gives rise to linear demand functions for inputs with prices replaced by their certainty equivalents. This approach has been developed in considerable detail by Hansen and Sargent (1980, 1981) and has been employed in modeling producer behavior by Epstein and Yatchew (1985), Meese (1980), and Sargent (1978).

An alternative approach to optimization under uncertainty is to employ the information about expectations of future prices contained in current input levels. This approach has the advantage that it is not limited to quadratic objective functions and linear constraints. Pindyck and Rotemberg (1983a) have utilized this approach in analyzing the Berndt–Wood (1975) data set for U.S. manufacturing, treating capital and labor input as quasi-fixed. They employ a translog variable cost function to represent technology, adding costs of adjustment that are quadratic in the current and lagged values of the quasi-fixed inputs. Pindyck and Rotemberg (1983b) have employed a similar approach to the analysis of production with two kinds of capital input and two types of labor input.

## References

Afriat, S. (1972) "Efficiency Estimates of Production Functions", *International Economic Review*, October, 13(3), 568–598.

Aigner, D. J., T. Amemiya and D. J. Poirier (1976) "On the Estimation of Production Frontiers: Maximum Likelihood Estimation of the Parameters of a Discontinuous Density Function", *International Economic Review*, June, 17(2), 377–396.

Amemiya, T. (1974) "The Nonlinear Two-Stage Least Squares Estimator", *Journal of Econometrics*, July, 2(2), 105–110.

Amemiya, T. (1977) "The Maximum Likelihood Estimator and the Nonlinear Three-Stage Least Squares Estimator in the General Nonlinear Simultaneous Equation Model", *Econometrica*, May, 45(4), 955–968.

Amemiya, T. (1983) "Nonlinear Regression Models", this *Handbook*, 1, 333–389.

Anderson, R. G. (1981) "On The Specification of Conditional Factor Demand Functions in Recent Studies of U.S. Manufacturing", in: E. R. Berndt and B. C. Field, eds., 119–144.

Applebaum, E. (1978) "Testing Neoclassical Production Theory," *Journal of Econometrics*, February, 7(1), 87–102.

Applebaum, E. (1979a) "On the Choice of Functional Forms", *International Economic Review*, June, 20(2), 449–458.

Applebaum, E. (1979b) "Testing Price Taking Behavior", *Journal of Econometrics*, February, 9(3), 283–294.

Arrow, K. J., H. B. Chenery, B. S. Minhas and R. M. Solow (1961) "Capital-Labor Substitution and Economic Efficiency", *Review of Economics and Statistics*, August, 63(3), 225–247.

Atkinson, S. E. and R. Halvorsen (1980) "A Test of Relative and Absolute Price Efficiency in Regulated Utilities", *Review of Economics and Statistics*, February, 62(1), 81–88.

Averch, H. and L. L. Johnson (1962) "Behavior of the Firm Under Regulatory Constraint", *American Economic Review*, December, 52(5), 1052–1069.

Bailey, E. E. (1973) *Economic Theory of Regulatory Constraint*. Lexington: Lexington Books.

Bailey, E. E. and A. F. Friedlaender (1982) "Market Structure and Multiproduct Industries", *Journal of Economic Literature*, September, 20(3), 1024–1048.

Baumol, W. J. and A. K. Klevorick (1970) "Input Choices and Rate-of-Return Regulation: An Overview of the Discussion", *Bell Journal of Economics and Management Science*, Autumn, 1(2), 162–190.

Belsley, D. A. (1974) "Estimation of Systems of Simultaneous Equations and Computational Applications of GREMLIN", *Annals of Social and Economic Measurement*, October, 3(4), 551–614.

Belsley, D. A. (1979) "On The Computational Competitiveness of Full-Information Maximum-Likelihood and Three-Stage Least-Squares in the Estimation of Nonlinear, Simultaneous-Equations Models", *Journal of Econometrics*, February, 9(3), 315–342.

Berndt, E. R. and L. R. Christensen (1973a) "The Internal Structure of Functional Relationships: Separability, Substitution, and Aggregation", *Review of Economic Studies*, July, 40(3), 123, 403–410.

Berndt, E. R. and L. R. Christensen (1973b) "The Translog Function and the Substitution of Equipment, Structures, and Labor in U.S. Manufacturing, 1929–1968", *Journal of Econometrics*, March, 1(1), 81–114.

Berndt, E. R. and L. R. Christensen (1974) "Testing for the Existence of a Consistent Aggregate Index of Labor Inputs", *American Economic Review*, June, 64(3), 391–404.

Berndt, E. R. and B. C. Field, eds. (1981) *Modeling and Measuring Natural Resource Substitution*. Cambridge: M.I.T. Press.

Berndt, E. R., B. H. Hall, R. E. Hall and J. A. Hausman (1974) "Estimation and Inference in Nonlinear Structural Models", *Annals of Social and Economic Measurement*, October, 3(4), 653–665.

Berndt, E. R. and D. W. Jorgenson (1973) "Production Structure", in: D. W. Jorgenson and H. S. Houthakker, eds., *U.S. Energy Resources and Economic Growth*. Washington: Energy Policy Project.

Berndt, E. R. and M. Khaled (1979) "Parametric Productivity Measurement and Choice Among Flexible Functional Forms", *Journal of Political Economy*, December, 87(6), 1220–1245.

Berndt, E. R. and C. J. Morrison (1979) "Income Redistribution and Employment Effects of Rising Energy Prices", *Resources and Energy*, October, 2(2), 131–150.

Berndt, E. R., C. J. Morrison and G. C. Watkins (1981) "Dynamic Models of Energy Demand: An Assessment and Comparison", in: E. R. Berndt and B. C. Field, eds., 259–289.

Berndt, E. R. and D. O. Wood (1975) "Technology, Prices, and the Derived Demand for Energy", *Review of Economics and Statistics*, August, 57(3), 376–384.

Berndt, E. R. and D. O. Wood (1979) "Engineering and Econometric Interpretations of Energy-Capital Complementarity", *American Economic Review*, June, 69(3), 342–354.

Berndt, E. R. and D. O. Wood (1981) "Engineering and Econometric Interpretations of Energy-Capital Complementarity: Reply and Further Results", *American Economic Review*, December, 71(5), 1105–1110.

Binswanger, H. P. (1974a) "A Cost-Function Approach to the Measurement of Elasticities of Factor Demand and Elasticities of Substitution", *American Journal of Agricultural Economics*, May, 56(2), 377–386.

Binswanger, H. P. (1974b) "The Measurement of Technical Change Biases with Many Factors of Production," *American Economic Review*, December, 64(5), 964–976.

Binswanger, H. P. (1978a) "Induced Technical Change: Evolution of Thought", in: H. P. Binswanger and V. W. Ruttan, eds., 13–43.

Binswanger, H. P. (1978b) "Issues in Modeling Induced Technical Change", in: H. P. Binswanger and V. W. Ruttan, eds., 128–163.

Binswanger, H. P. (1978c) "Measured Biases of Technical Change: The United States", in: H. P. Binswanger and V. W. Ruttan, eds., 215–242.

Binswanger, H. P. and V. W. Ruttan, eds. (1978) *Induced Innovation*. Baltimore: Johns Hopkins University Press.

Blackorby, C., D. Primont and R. R. Russell (1977) "On Testing Separability Restrictions with Flexible Functional Forms", *Journal of Econometrics*, March, 5(2), 195–209.

Blackorby, C., D. Primont and R. R. Russell (1978) *Duality, Separability, and Functional Structure*. Amsterdam: North-Holland.

Blackorby, C. and R. R. Russell (1976) "Functional Structure and the Allen Partial Elasticities of Substitution: An Application of Duality Theory", *Review of Economic Studies*, 43(2), 134, 285–292.

Braeutigan, R. R., A. F. Daughety and M. A. Turnquist (1982) "The Estimation of a Hybrid Cost Function for a Railroad Firm", *Review of Economics and Statistics*, August, 64(3), 394–404.

Brown, M., ed. (1967) *The Theory and Empirical Analysis of Production*. New York: Columbia University Press.

Brown, R. S., D. W. Caves and L. R. Christensen (1979) "Modeling the Structure of Cost and Production for Multiproduct Firms", *Southern Economic Journal*, July, 46(3), 256–273.

Brown, R. S. and L. R. Christensen (1981) "Estimating Elasticities of Substitution in a Model of Partial Static Equilibrium: An Application to U.S. Agriculture, 1947 to 1974", in: E. R. Berndt and B. C. Field, eds., 209–229.

Burgess, D. F. (1974) "A Cost Minimization Approach to Import Demand Equations", *Review of Economics and Statistics*, May, 56(2), 224–234.

Burgess, D. F. (1975) "Duality Theory and Pitfalls in the Specification of Technology", *Journal of Econometrics*, May, 3(2), 105–121.

Cameron, T. A. and S. L. Schwartz (1979) "Sectoral Energy Demand in Canadian Manufacturing Industries", *Energy Economics*, April, 1(2), 112–118.

Carlson, S. (1939) *A Study on the Pure Theory of Production*. London: King.

Caves, D. W. and L. R. Christensen (1980) "Global Properties of Flexible Functional Forms", *American Economic Review*, June, 70(3), 422–432.

Caves, D. W., L. R. Christensen and J. A. Swanson (1980) "Productivity in U.S. Railroads, 1951–1974", *Bell Journal of Economics*, Spring 1980, 11(1), 166–181.

Caves, D. W., L. R. Christensen and J. A. Swanson (1981) "Productivity Growth, Scale Economies and Capacity Utilization in U.S. Railroads, 1955–1974", *American Economic Review*, December, 71(5), 994–1002.

Caves, D. W., L. R. Christensen and M. W. Trethaway (1980) "Flexible Cost Functions for Multiproduct Firms", *Review of Economics and Statistics*, August, 62(3), 477–481.

Caves, D. W., L. R. Christensen and M. W. Trethaway (1984) "Economics of Density Versus Economics of Scale: Why Trunk and Local Airline Costs Differ", *Rand Journal of Economics*, Winter, 15(4), 471–489.

Caves, D. W., L. R. Christensen, M. W. Trethaway and R. Windle (1984) "Network Effects and the Measurement of Returns to Scale and Density for U.S. Railroads", in: A. F. Daughety, ed., *Analytical Studies in Transport Economics*, forthcoming.

Chiang, S. J. W. and A. F. Friedlaender (1985) "Trucking Technology and Marked Structure", *Review of Economics and Statistics*, May, 67(2), 250–258.

Christ, C., et. al. (1963) *Measurement in Economics*. Stanford: Stanford University Press.

Christensen, L. R., D. Cummings and P. E. Schoech (1983) "Econometric Estimation of Scale Economies in Telecommunications", in: L. Courville, A. de Fontenay and R. Dobell, eds., 27–53.

Christensen, L. R. and W. H. Greene (1976) "Economies of Scale in U.S. Electric Power Generation", *Journal of Political Economy*, August, 84(4), 655–676.

Christensen, L. R. and D. W. Jorgenson (1970) "U.S. Real Product and Real Factor Input, 1929–1967", *Review of Income and Wealth*, March, 16(1), 19–50.

Christensen, L. R., D. W. Jorgenson and L. J. Lau (1971) "Conjugate Duality and the Transcendental Logarithmic Production Function", *Econometrica*, July, 39(3), 255–256.

Christensen, L. R., D. W. Jorgenson and L. J. Lau (1973) "Transcendental Logarithmic Production Frontiers", *Review of Economics and Statistics*, February, 55(1), 28–45.

Cobb, C. W. and P. H. Douglas (1928) "A Theory of Production", *American Economic Review*, March, 18(2), 139–165.

Conrad, K. and D. W. Jorgenson (1977) "Tests of a Model of Production for the Federal Republic of Germany, 1950–1973", *European Economic Review*, October, 10(1), 51–75.

Conrad, K. and D. W. Jorgenson (1978) "The Structure of Technology: Nonjointness and Commodity Augmentation, Federal Republic of Germany, 1950–1973", *Empirical Economics*, 3(2), 91–113.

Courville, L., A. de Fontenay and R. Dobell, eds. (1983) *Economic Analysis of Telecommunications.* Amsterdam: North-Holland.

Cowing, T. G. (1978) "The Effectiveness of Rate-of-Return Regulation: An Empirical Test Using Profit Functions", in: M. Fuss and D. McFadden, eds., 2, 215–246.

Cowing, T. G. and V. K. Smith (1978) "The Estimation of a Production Technology: A Survey of Econometric Analyses of Steam Electric Generation", *Land Economics*, May, 54(2), 158–168.

Cowing, T. G. and R. E. Stevenson, eds. (1981) *Productivity Measurement in Regulated Industries.* New York: Academic Press.

Cowing, T. G., D. Reifschneider and R. E. Stevenson, "A Comparison of Alternative Frontier Cost Function Specifications", in: A. Dogramaci, ed., 63–92.

Dargay, J. (1983) "The Demand for Energy in Swedish Manufacturing," in B.-C. Ysander, ed., *Energy in Swedish Manufacturing.* Stockholm: Industrial Institute for Economic and Social Research, 57–128.

Denny, M. (1974) "The Relationship Between Functional Forms for the Production System", *Canadian Journal of Economics*, February, 7(1), 21–31.

Denny, M. and M. Fuss (1977) "The Use of Approximation Analysis to Test for Separability and the Existence of Consistent Aggregates", *American Economic Review*, June, 67(3), 404–418.

Denny, M., M. Fuss, C. Everson and L. Waverman (1981) "Estimating the Effects of Technological Innovation in Telecommunications: The Production Structure of Bell Canada", *Canadian Journal of Economics*, February, 14(1), 24–43.

Denny, M., M. Fuss and L. Waverman (1981) "The Substitution Possibilities for Energy: Evidence from U.S. and Canadian Manufacturing Industries", in: E. R. Berndt and B. C. Field, eds., 230–258.

Denny, M. and J. D. May (1978) "Homotheticity and Real Value-Added in Canadian Manufacturing", in: M. Fuss and D. McFadden, eds., 2, 53–70.

Denny, M., J. D. May and C. Pinto (1978) "The Demand for Energy in Canadian Manufacturing: Prologue to an Energy Policy", *Canadian Journal of Economics*, May, 11(2), 300–313.

Denny, M. and C. Pinto, "An Aggregate Model with Multi-Product Technologies", in: M. Fuss and D. McFadden, eds., 2, 249–268.

Diewert, W. E. (1971) "An Application of the Shephard Duality Theorem, A Generalized Leontief Production Function", *Journal of Political Economy*, May/June, 79(3), 481–507.

Diewert, W. E. (1973) "Functional Forms for Profit and Transformation Functions", *Journal of Economic Theory*, June, 6(3), 284–316.

Diewert, W. E. (1974a) "Applications of Duality Theory", in: M. D. Intrilligator and D. A. Kendrick, eds., 106–171.

Diewert, W. E. (1974b) "Functional Forms for Revenue and Factor Requirement Functions", *International Economic Review*, February, 15(1), 119–130.

Diewert, W. E. (1976) "Exact and Superlative Index Numbers", *Journal of Econometrics*, May, 4(2), 115–145.

Diewert, W. E. (1980) "Aggregation Problems in the Measurement of Capital", in: D. Usher, ed., *The Measurement of Capital*. Chicago: University of Chicago Press, 433–528.

Diewert, W. E. (1982) "Duality Approaches to Microeconomic Theory", in: K. J. Arrow and M. D. Intrilligator, eds., *Handbook of Mathematical Economics*, 2, 535–591.

Diewert, W. E. and C. Parkan (1983) "Linear Programming Tests of Regularity Conditions for Production Functions", in: W. Eichhorn, R. Henn, K. Neumann and R. W. Shephard, eds., 131–158.

Dogramaci, A., ed. (1983) *Developments in Econometric Analyses of Productivity.* Boston: Kluwer–Nijhoff.

Douglas, P. W. (1948) "Are There Laws of Production?", *American Economic Review*, March, 38(1), 1–41.

Douglas, P. W. (1967) "Comments on the Cobb–Douglas Production Function", in: M. Brown, ed., 15–22.

Douglas, P. W. (1976) "The Cobb–Douglas Production Function Once Again: Its History, Its Testing, and Some Empirical Values," October, 84(5), 903–916.

Ehud, R. I. and A. Melnik (1981) "The Substitution of Capital, Labor and Energy in the Israeli Economy", *Resources and Energy*, November, 3(3), 247–258.

Eichhorn, W., R. Henn, K. Neumann and R. W. Shephard, eds. (1983) *Quantitative Studies on Production and Prices*. Wurzburg: Physica–Verlag.

Elbadawi, I., A. R. Gallant and G. Souza (1983) "An Elasticity Can Be Estimated Consistently Without a Priori Knowledge of Functional Form", *Econometrica*, November, 51(6), 1731–1752.

Epstein, L. G. and A. Yatchew (1985) "The Empirical Determination of Technology and Expectations: A Simplified Procedure", *Journal of Econometrics*, February, 27(2), 235–258.

Evans, D. S. and J. J. Heckman (1983) "Multi-Product Cost Function Estimates and Natural Monopoly Tests for the Bell System", in: D. S. Evans, ed., *Breaking up Bell*. Amsterdam: North-Holland, 253–282.

Evans, D. S. and J. J. Heckman (1984) "A Test for Subadditivity of the Cost Function with an Application to the Bell System", *American Economic Review*, September, 74(4), 615–623.

Faucett, Jack and Associates (1977) *Development of 35-Order Input–Output Tables, 1958–1974*. Washington: Federal Emergency Management Agency.

Field, B. C. and E. R. Berndt (1981) "An Introductiory Review of Research on the Economics of Natural Resource Substitution", in: E. R. Berndt and B. C. Field, eds., 1–14.

Field, B. C. and C. Grebenstein (1980) "Substituting for Energy in U.S. Manufacturing", *Review of Economics and Statistics*, May, 62(2), 207–212.

Forsund, F. R. and L. Hjalmarsson (1979) "Frontier Production Functions and Technical Progress: A Study of General Milk Processing Swedish Dairy Plants", *Econometrica*, July, 47(4), 883–901.

Forsund, F. R. and L. Hjalmarsson (1983) "Technical Progress and Structural Change in the Swedish Cement Industry 1955–1979", *Econometrica*, September, 51(5), 1449–1467.

Forsund, F. R. and E. S. Jansen (1983) "Technical Progress and Structural Change in the Norwegian Primary Aluminum Industry", *Scandinavian Journal of Economics*, 85(2), 113–126.

Forsund, F. R., C. A. K. Lovell and P. Schmidt (1980) "A Survey of Frontier Production Functions and of Their Relationship to Efficiency Measurement", *Journal of Econometrics*, May, 13(1), 5–25.

Fraumeni, B. M. and D. W. Jorgenson (1980) "The Role of Capital in U.S. Economic Growth, 1948–1976", in: G. von Furstenberg, ed., 9–250.

Frenger, P. (1978) "Factor Substitution in the Interindustry Model and the Use of Inconsistent Aggregation", in: M. Fuss and D. McFadden, eds., 2, 269–310.

Friede, G. (1979) *Investigation of Producer Behavior in the Federal Republic of Germany Using the Translog Price Function*. Cambridge: Oelgeschlager, Gunn and Hain.

Friedlaender, A. F. and R. H. Spady (1980) "A Derived Demand Function for Freight Transportation", *Review of Economics and Statistics*, August, 62(3), 432–441.

Friedlaender, A. F. and R. H. Spady (1981) *Freight Transport Regulation*. Cambridge: M.I.T. Press.

Friedlaender, A. F., R. H. Spady and S. J. W. Chiang (1981) "Regulation and the Structure of Technology in the Trucking Industry", in: T. G. Cowing and R. E. Stevenson, eds., 77–106.

Frisch, R. (1965) *Theory of Production*. Chicago: Rand McNally.

Fullerton, D., Y. K. Henderson and J. B. Shoven, "A Comparison of Methodologies in Empirical General Equilibrium Models of Taxation", in: H. E. Scarf and J. B. Shoven, eds., 367–410.

Fuss, M. (1977a) "The Demand for Energy in Canadian Manufacturing: An Example of the Estimation of Production Structures with Many Inputs", *Journal of Econometrics*, January, 5(1), 89–116.

Fuss, M. (1977b) "The Structure of Technology Over Time: A Model for Testing the Putty–Clay Hypothesis", *Econometrica*, November, 45(8), 1797–1821.

Fuss, M. (1978) "Factor Substitution in Electricity Generation: A Test of the Putty–Clay Hypothesis", in: M. Fuss and D. McFadden, eds., 2, 187–214.

Fuss, M. (1983) "A Survey of Recent Results in the Analysis of Production Conditions in Telecommunications", in: L. Courville, A. de Fontenay and R. Dobell, eds., 3–26.

Fuss, M. and D. McFadden, eds. (1978) *Production Economics*. Amsterdam, North-Holland, 2 Vols.

Fuss, M., D. McFadden and Y. Mundlak (1978) "A Survey of Functional Forms in the Economic Analysis of Production", in: M. Fuss and D. McFadden, eds., 1, 219–268.

Fuss, M. and L. Waverman (1981) "Regulation and the Multiproduct Firm: The Case of Telecommunications in Canada", in: G. Fromm, ed., *Studies in Public Regulation*. Cambridge: M.I.T. Press, 277–313.

Gallant, A. R. (1977) "Three-Stage Least Squares Estimation for a System of Simultaneous, Nonlinear, Implicit Equations", *Journal of Econometrics*, January, 5(1), 71–88.

Gallant, A. R. (1981) "On the Bias in Flexible Functional Forms and an Essentially Unbiased Form", *Journal of Econometrics*, February, 15(2), 211–246.

Gallant, A. R. and A. Holly (1980) "Statistical Inference in an Implicit, Nonlinear, Simultaneous Equations Model in the Context of Maximum Likelihood Estimation", *Econometrica*, April, 48(3), 697–720.

Gallant, A. R. and D. W. Jorgenson (1979) "Statistical Inference for a System of Simultaneous, Nonlinear, Implicit Equations in the Context of Instrumental Variable Estimation", *Journal of Econometrics*, October/December, 11(2/3), 275–302.

Geary, P. T. and E. J. McDonnell (1980) "Implications of the Specification of Technologies: Further Evidence", *Journal of Econometrics*, October, 14(2), 247–255.

Gollop, F. M. and S. M. Karlson (1978) "The Impact of the Fuel Adjustment Mechanism on Economic Efficiency", *Review of Economics and Statistics*, November, 60(4), 574–584.

Gollop, F. M. and M. J. Roberts (1981) "The Sources of Economic Growth in the U.S. Electric Power Industry", in: T. G. Cowing and R. E. Stevenson, eds., 107–145.

Gollop, F. M. and M. J. Roberts (1983) "Environmental Regulations and Productivity Growth: The Case of Fossil-Fueled Electric Power Generation", *Journal of Political Economy*, August, 91(4), 654–674.

Gorman, W. M. (1959) "Separable Utility and Aggregation", *Econometrica*, July, 27(3), 469–481.

Gourieroux, C., A. Holly and A. Monfort (1980) "Kuhn–Tucker, Likelihood Ratio and Wald Tests for Nonlinear Models with Constraints on the Parameters". Harvard University, Harvard Institute for Economic Research, Discussion Paper No. 770, June.

Gourieroux, C., A. Holly and A. Monfort (1982) "Likelihood Ratio Test, Wald Test, and Kuhn–Tucker Test in Linear Models with Inequality Constraints on the Regression Parameters", *Econometrica*, January, 50(1), 63–80.

Greene, W. H. (1980) "Maximum Likelihood Estimation of Econometric Frontier Functions", *Journal of Econometrics*, May, 13(1), 27–56.

Greene, W. H. (1983) "Simultaneous Estimation of Factor Substitution, Economies of Scale, Productivity, and Non-Neutral Technical Change", in: A. Dogramaci, ed., 121–144.

Griffin, J. M. (1977a) "The Econometrics of Joint Production: Another Approach", *Review of Economics and Statistics*, November, 59(4), 389–397.

Griffin, J. M. (1977b) "Interfuel Substitution Possibilities: A Translog Application to Pooled Data", *International Economic Review*, October, 18(3), 755–770.

Griffin, J. M. (1977c) "Long-Run Production Modeling with Pseudo Data: Electric Power Generation", *Bell Journal of Economics*, Spring 1977, 8(1), 112–127.

Griffin, J. M. (1978) "Joint Production Technology: The Case of Petrochemicals", *Econometrica*, March, 46(1), 379–396.

Griffin, J. M. (1979) "Statistical Cost Analysis Revisited", *Quarterly Journal of Economics*, February, 93(1), 107–129.

Griffin, J. M. (1980) "Alternative Functional Forms and Errors of Pseudo Data Estimation: A Reply", *Review of Economics and Statistics*, May, 62(2), 327–328.

Griffin, J. M. (1981a) "The Energy-Capital Complementarity Controversy: A Progress Report on Reconciliation Attempts", in: E. R. Berndt and B. C. Field, eds., 70–80.

Griffin, J. M. (1981b) "Engineering and Econometric Interpretations of Energy-Capital Complementarity: Comment", *American Economic Review*, December, 71(5), 1100–1104.

Griffin, J. M. (1981c) "Statistical Cost Analysis Revisited: Reply", *Quarterly Journal of Economics*, February, 96(1), 183–187.

Griffin, J. M. and P. R. Gregory (1976) "An Intercountry Translog Model of Energy Substitution Responses", *American Economic Review*, December, 66(5), 845–857.

Griliches, Z. (1967) "Production Functions in Manufacturing: Some Empirical Results", in: M. Brown, ed., 275–322.

Griliches, Z. and V. Ringstad (1971) *Economies of Scale and the Form of the Production Function*. Amsterdam: North-Holland.

Hall, R. E. (1973) "The Specification of Technology with Several Kinds of Output", *Journal of Political Economy*, July/August, 81(4), 878–892.

Halvorsen, R. (1977) "Energy Substitution in U.S. Manufacturing", *Review of Economics and Statistics*, November, 59(4), 381–388.

Halvorsen, R. (1978) *Econometric Studies of U.S. Energy Demand*. Lexington: Lexington Books.

Halvorsen, R. and J. Ford, "Substitution Among Energy, Capital and Labor Inputs in U.S. Manufacturing", in: R. S. Pindyck, ed., *Advances in the Economics of Energy and Resources*. Greenwich: JAI Press, 1, 51–75.

Hamermesh, D. S. and J. Grant (1979) "Econometric Studies of Labor–Labor Substitution and Their Implications for Policy", *Journal of Human Resources*, Fall, 14(4), 518–542.

Hanoch, G. (1978) "Symmetric Duality and Polar Production Functions", in: M. Fuss and D. McFadden, eds., 1, 111–132.

Hanoch, G. and M. Rothschild (1972) "Testing the Assumptions of Production Theory: A Nonparametric Approach", *Journal of Political Economy*, March/April, 80(2), 256–275.

Hansen, L. P. and T. J. Sargent (1980) "Formulating and Estimating Dynamic Linear Rational Expectations Models", *Journal of Economic Dynamics and Control*, February, 2(1), 7–46.

Hansen, L. P. and T. J. Sargent (1981) "Linear Rational Expectations Models for Dynamically Interrelated Variables", in: R. E. Lucas and T. J. Sargent, eds., *Rational Expectations and Econometric Practice*. Minneapolis: University of Minnesota Press, 1, 127–156.

Harmatuck, Donald J. (1979) "A Policy-Sensitive Railway Cost Function", *Logistics and Transportation Review*, April, 15(2), 277–315.

Harmatuck, Donald J. (1981) "A Multiproduct Cost Function for the Trucking Industry", *Journal of Transportation Economics and Policy*, May, 15(2), 135–153.

Heady, E. O. and J. L. Dillon (1961) *Agricultural Production Functions*. Ames: Iowa State University Press.

Hicks, J. R. (1946) *Value and Capital*. 2nd ed. (1st ed. 1939), Oxford: Oxford University Press.

Hicks, J. R. (1963) *The Theory of Wages*. 2nd ed. (1st ed. 1932), London: Macmillan.

Hildenbrand, W. (1981) "Short-Run Production Functions Based on Microdata", *Econometrica*, September, 49(5), 1095–1125.

Hotelling, H. S. (1932) "Edgeworth's Taxation Paradox and the Nature of Demand and Supply Functions", *Journal of Political Economy*, October, 40(5), 577–616.

Houthakker, H. S. (1955–1956) "The Pareto Distribution and the Cobb–Douglas Production Function in Activity Analysis", *Review of Economic Studies*, 23(1), 60, 27–31.

Hudson, E. A. and D. W. Jorgenson (1974) "U.S. Energy Policy and Economic Growth, 1975–2000", *Bell Journal of Economics and Management Science*, Autumn, 5(2), 461–514.

Hudson, E. A. and D. W. Jorgenson (1978) "The Economic Impact of Policies to Reduce U.S. Energy Growth," *Resources and Energy*, November, 1(3), 205–230.

Humphrey, D. B. and J. R. Moroney (1975) "Substitution Among Capital, Labor, and Natural Resource Products in American Manufacturing", *Journal of Political Economy*, February, 83(1), 57–82.

Humphrey, D. B. and B. Wolkowitz (1976) "Substituting Intermediates for Capital and Labor with Alternative Functional Forms: An Aggregate Study", *Applied Economics*, March, 8(1), 59–68.

Intriligator, M. D. and D. A. Kendrick, eds. (1974) *Frontiers in Quantitative Economics*. Amsterdam: North-Holland, Vol. 2.

Jara–Diaz, S. and C. Winston (1981) "Multiproduct Transportation Cost Functions: Scale and Scope in Railway Operations", in: N. Blattner, ed., *Eighth European Association for Research in Industrial Economics*, Basle: University of Basle, 1, 437–469.

Jennrich, R. I. (1969) "Asymptotic Properties of Nonlinear Least Squares Estimations", *Annals of Mathematical Statistics*, April, 40(2), 633–643.

Johansen, L. (1972) *Production Functions*. Amsterdam: North-Holland.

Johansen, L. (1974) *A Multi-Sectoral Study of Economic Growth*. 2nd ed. (1st ed. 1960), Amsterdam, North-Holland.

Jorgenson, D. W. (1973a) "The Economic Theory of Replacement and Depreciation", in: W. Sellekaerts, ed., *Econometrics and Economic Theory*. New York: Macmillan, 189–221.

Jorgenson, D. W. (1973b) "Technology and Decision Rules in the Theory of Investment Behavior", *Quarterly Journal of Economics*, November 1973, 87(4), 523–543.

Jorgenson, D. W. (1974) "Investment and Production: A Review", in: M. D. Intriligator and D. A. Kendrick, eds., 341–366.

Jorgenson, D. W. (1980) "Accounting for Capital", in: G. von Furstenberg, ed., 251–319.

Jorgenson, D. W. (1981) "Energy Prices and Productivity Growth", *Scandinavian Journal of Econom-*

*ics*, 83(2), 165–179.

Jorgenson, D. W. (1983) "Modeling Production for General Equilibrium Analysis", *Scandinavian Journal of Economics*, 85(2), 101–112.

Jorgenson, D. W. (1984a) "Econometric Methods for Applied General Equilibrium Analysis", in: H. E. Scarf and J. B. Shoven, eds., 139–203.

Jorgenson, D. W. (1984b) "The Role of Energy in Productivity Growth", in: J. W. Kendrick, ed., *International Comparisons of Productivity and Causes of the Slowdown*. Cambridge: Ballinger, 279–323.

Jorgenson, D. W. and B. M. Fraumeni (1981) "Relative Prices and Technical Change", in: E. R. Berndt and B. C. Field, eds., 17–47; revised and reprinted in: W. Eichhorn, R. Henn, K. Neumann and R. W. Shephard, eds., 241–269.

Jorgenson, D. W. and J.-J. Laffont (1974) "Efficient Estimation of Non-Linear Simultaneous Equations with Additive Disturbances", *Annals of Social and Economic Measurement*, October, 3(4), 615–640.

Jorgenson, D. W. and L. J. Lau (1974a) "Duality and Differentiability in Production", *Journal of Economic Theory*, September, 9(1), 23–42.

Jorgenson, D. W. and L. J. Lau (1974b) "The Duality of Technology and Economic Behavior", *Review of Economic Studies*, April, 41(2), 126, 181–200.

Jorgenson, D. W. and L. J. Lau (1975) "The Structure of Consumer Preferences", *Annals of Social and Economic Measurement*, January, 4(1), 49–101.

Jorgenson, D. W., L. J. Lau and T. M. Stoker (1982) "The Transcendental Logarithmic Model of Aggregate Consumer Behavior", in: R. L. Basmann and G. Rhodes, eds., *Advances in Econometrics*. Greenwich: JAI Press, 1, 97–238.

Kang, H. and G. M. Brown (1981) "Partial and Full Elasticities of Substitution and the Energy-Capital Complementarity Controversy", in: E. R. Berndt and B. C. Field, eds., 81–90.

Kennedy, C. (1964) "Induced Bias in Innovation and the Theory of Distribution", *Economic Journal*, September, 74(298), 541–547.

Kennedy, C. and A. P. Thirlwall (1972) "Technical Progress: A Survey", *Economic Journal*, March, 82(325), 11–72.

Kiss, F., S. Karabadjian and B. J. Lefebvre (1983) "Economies of Scale and Scope in Bell Canada", in: L. Courville, A. de Fontenay and R. Dobell, eds., 55–82.

Kmenta, J. (1967) "On Estimation of the CES Production Function", *International Economic Review*, June, 8(2), 180–189.

Kohli, U. R. (1981) "Nonjointness and Factor Intensity in U.S. Production", *International Economic Review*, February, 22(1), 3–18.

Kohli, U. R. (1983) "Non-joint Technologies", *Review of Economic Studies*, January, 50(1), 160, 209–219.

Kopp, R. J. and W. E. Diewert (1982) "The Decomposition of Frontier Cost Function Deviations into Measures of Technical and Allocative Efficiency", *Journal of Econometrics*, August, 19(2/3), 319–332.

Kopp, R. J. and V. K. Smith (1980a) "Input Substitution, Aggregation, and Engineering Descriptions of Production Activities", *Economics Letters*, 5(4), 289–296.

Kopp, R. J. and V. K. Smith (1980b) "Measuring Factor Substitution with Neoclassical Models: An Experimental Evaluation", *Bell Journal of Economics*, Autumn, 11(2), 631–655.

Kopp, R. J. and V. K. Smith (1981a) "Measuring the Prospects of Resource Substitution Under Input and Technology Aggregation", in: E. R. Berndt and B. C. Field, eds., 145–174.

Kopp, R. J. and V. K. Smith (1981b) "Productivity Measurement and Environmental Regulation: An Engineering-Econometric Analysis", in: T. G. Cowing and R. E. Stevenson, eds., 249–283.

Kopp, R. J. and V. K. Smith (1981c) "Neoclassical Modeling of Nonneutral Technological Change: An Experimental Appraisal", *Scandinavian Journal of Economics*, 85(2), 127–146.

Kopp, R. J. and V. K. Smith (1982) "Neoclassical Measurement of Ex Ante Resource Substitution: An Experimental Evaluation", in: J. R. Moroney, ed., *Advances in the Economics of Energy and Resources*. Greenwich: JAI Press, 4, 183–198.

Koopmans, T. C. (1977) "Examples of Production Relations Based on Microdata", in: G. C. Harcourt, ed., *The Microeconomic Foundations of Macroeconomics*. London: Macmillan, 144–171.

Kuhn, H. W. and A. W. Tucker (1951) "Nonlinear Programming", in: J. Neyman, ed., *Proceedings of the Second Berkeley Symposium on Mathematical Statistics and Probability*. Berkeley: University of California Press, 481–492.

Lau, L. J. (1969) "Duality and the Structure of Utility Functions", *Journal of Economic Theory*, December, 1(4), 374–396.

Lau, L. J. (1974) "Applications of Duality Theory: Comments", in: M. D. Intriligator and D. A. Kendrick, eds., 176–199.

Lau, L. J. (1976) "A Characterization of the Normalized Restricted Profit Function", *Journal of Economic Theory*, February, 12(1), 131–163.

Lau, L. J. (1978a) "Applications of Profit Functions", in: M. Fuss and D. McFadden, eds., 1, 133–216.

Lau, L. J. (1978b) "Testing and Imposing Monotonicity, Convexity and Quasi-Convexity Constraints", in: M. Fuss and D. McFadden, eds., 1, 409–453.

Lau, L. J. (1986) "Functional Forms in Econometric Model Building", this *Handbook*, Vol. 3.

Leontief, W. W. (1947a) "Introduction to a Theory of the Internal Structure of Functional Relationships", *Econometrica*, October, 15(4), 361–373.

Leontief, W. W. (1947b) "A Note on the Interrelation of Subsets of Independent Variables of a Continuous Function with Continuous First Derivatives", *Bulletin of the American Mathematical Society*, April, 53(4), 343–350.

Leontief, W. W. (1951) *The Structure of the American Economy, 1919–1939*. 2nd ed. (1st ed. 1941), New York: Oxford University Press.

Leontief, W. W., ed. (1953) *Studies in the Structure of the American Economy*. New York: Oxford University Press.

Liew, C. K. (1976) "A Two-Stage Least-Squares Estimator with Inequality Restrictions on the Parameters", *Review of Economics and Statistics*, May, 58(2), 234–238.

Longva, S. and O. Olsen (1983) "Producer Behaviour in the MSG Model", in: O. Bjerkholt, S. Longva, O. Olsen and S. Strom, eds., *Analysis of Supply and Demand of Electricity in the Norwegian Economy*. Oslo: Central Statistical Bureau, 52–83.

Lucas, R. E. (1967) "Adjustment Costs and the Theory of Supply", *Journal of Political Economy*, August, Pt. 1, 75(4), 321–334.

Maddala, G. S. and R. B. Roberts (1980) "Alternative Functional Forms and Errors of Pseudo Data Estimation", *Review of Economics and Statistics*, May, 62(2), 323–326.

Maddala, G. S. and R. B. Roberts (1981) "Statistical Cost Analysis Revisited: Comment", *Quarterly Journal of Economics*, February, 96(1), 177–182.

Magnus, J. R. (1979) "Substitution Between Energy and Non-Energy Inputs in the Netherlands, 1950–1976", *International Economic Review*, June, 20(2), 465–484.

Magnus, J. R. and A. D. Woodland (1980) "Interfuel Substitution and Separability in Dutch Manufacturing: A Multivariate Error Components Approach", London School of Economics, November.

Malinvaud, E. (1970) "The Consistency of Non-Linear Regressions", *Annals of Mathematical Statistics*, June, 41(3), 456–469.

Malinvaud, E. (1980) *Statistical Methods of Econometrics*. 3rd ed. (1st ed. 1966), trans. A. Silvey, Amsterdam: North-Holland.

McFadden, D. (1963) "Further Results on CES Production Functions", *Review of Economic Studies*, June, 30(2), 83, 73–83.

McFadden, D. (1978) "Cost, Revenue, and Profit Functions", in: M. Fuss and D. McFadden, eds., 1, 1–110.

McRae, R. N. (1981) "Regional Demand for Energy by Canadian Manufacturing Industries", *International Journal of Energy Systems*, January, 1(1), 38–48.

McRae, R. N. and A. R. Webster (1982) "The Robustness of a Translog Model to Describe Regional Energy Demand by Canadian Manufacturing Industries", *Resources and Energy*, March, 4(1), 1–25.

Meese, R. (1980) "Dynamic Factor Demand Schedules for Labor and Capital Under Rational Expectations", *Journal of Econometrics*, September, 14(1), 141–158.

Moroney, J. R. and A. Toevs (1977) "Factor Costs and Factor Use: An Analysis of Labor, Capital, and Natural Resources", *Southern Economic Journal*, October, 44(2), 222–239.

Moroney, J. R. and A. Toevs (1979) "Input Prices, Substitution, and Product Inflation", in: R. S. Pindyck, ed., *Advances in the Economics of Energy and Resources*. Greenwich: JAI Press, 1, 27–50.

Moroney, J. R. and J. M. Trapani (1981a) "Alternative Models of Substitution and Technical Change in Natural Resource Intensive Industries", in: E. R. Berndt and B. C. Field, eds., 48–69.

Moroney, J. R. and J. M. Trapani (1981b) "Factor Demand and Substitution in Mineral-Intensive Industries", *Bell Journal of Economics*, Spring, 12(1), 272–285.

Morrison, C. J. and E. R. Berndt (1981) "Short-run Labor Productivity in a Dynamic Model", *Journal of Econometrics*, August, 16(3), 339–366.

Mundlak, Y. (1963) "Estimation of Production and Behavioral Functions from a Combination of Cross-Section and Time Series Data", in: C. Christ, et al., 138–166.

Mundlak, Y. (1978) "On the Pooling of Time Series and Cross Section Data", *Econometrica*, January, 46(1), 69–86.

Nadiri, M. I. (1970) "Some Approaches to the Theory and Measurement of Total Factor Productivity: A Survey", *Journal of Economic Literature*, December, 8(4), 1137–1178.

Nadiri, M. I. and M. Schankerman (1981) "The Structure of Production, Technological Change, and the Rate of Growth of Total Factor Productivity in the U.S. Bell System", in: T. G. Cowing and R. E. Stevenson, eds., 219–248.

Nakamura, S. (1984) *An Inter-Industry Translog Model of Prices and Technical Change for the West German Economy*. Berlin: Springer-Verlag.

Nerlove, M. (1963) "Returns to Scale in Electricity Supply", in: C. Christ, et al., 167–200.

Nerlove, M. (1967) "Recent Empirical Studies of the CES and Related Production Functions", in: M. Brown, ed., 55–122.

Norsworthy, J. R. and M. J. Harper (1981) "Dynamic Models of Energy Substitution in U.S. Manufacturing", in: E. R. Berndt and B. C. Field, eds., 177–208.

Ozatalay, S., S. S. Grumbaugh and T. V. Long III, "Energy Substitution and National Energy Policy", *American Economic Review*, May, 69(2), 369–371.

Parks, R. W. (1971) "Responsiveness of Factor Utilization in Swedish Manufacturing, 1870–1950", *Review of Economics and Statistics*, May, 53(2), 129–139.

Peterson, H. C. (1975) "An Empirical Test of Regulatory Effects", *Bell Journal of Economics*, Spring, 6(1), 111–126.

Pindyck, R. S. (1979a) "Interfuel Substitution and Industrial Demand for Energy", *Review of Economics and Statistics*, May, 61(2), 169–179.

Pindyck, R. S. (1979b) *The Structure of World Energy Demand*. Cambridge: M.I.T. Press.

Pindyck, R. S. and J. J. Rotemberg (1983a) "Dynamic Factor Demands and the Effects of Energy Price Shocks", *American Economic Review*, December, 73(5), 1066–1079.

Pindyck, R. S. and J. J. Rotemberg (1983b) "Dynamic Factor Demands Under Rational Expectations", *Scandinavian Journal of Economics*, 85(2), 223–239.

Quandt, R. E. (1983) "Computational Problems and Methods", this *Handbook*, 1, 701–764.

Russell, C. S. and W. J. Vaughan (1976) *Steel Production*. Baltimore: Johns Hopkins University Press.

Russell, R. R. (1975) "Functional Separability and Partial Elasticities of Substitution", *Review of Economic Studies*, January, 42(1), 129, 79–86.

Samuelson, P. A. (1951) "Abstract of a Theorem Concerning Substitutability in Open Leontief Models", in: T. C. Koopmans, ed., *Activity Analysis of Production and Allocation*. Wiley: New York, 142–146.

Samuelson, P. A. (1953–1954) "Prices of Factors and Goods in General Equilibrium", *Review of Economic Studies*, 21(1), 54, 1–20.

Samuelson, P. A. (1960) "Structure of a Minimum Equilibrium System", in: R. W. Pfouts, ed., *Essays in Economics and Econometrics*. Chapel Hill: University of North Carolina Press, 1–33.

Samuelson, P. A. (1965) "A Theory of Induced Innovation Along Kennedy–Weizsacker Lines", *Review of Economics and Statistics*, November, 47(4), 343–356.

Samuelson, P. A. (1973) "Relative Shares and Elasticities Simplified: Comment", *American Economic Review*, September, 63(4), 770–771.

Samuelson, P. A. (1974) "Complementarity–An Essay on the 40th Anniversary of the Hicks–Allen Revolution in Demand Theory", *Journal of Economic Literature*, December, 12(4), 1255–1289.

Samuelson, P. A. (1979) "Paul Douglas's Measurement of Production Functions and Marginal Productivities", *Journal of Political Economy*, October, Part 1, 87(5), 923–939.

Samuelson, P. A. (1983) *Foundations of Economic Analysis*. 2nd ed. (1st ed. 1947), Cambridge: Harvard University Press.

Sargan, J. D. (1971) "Production Functions", in: R. Layard, ed., *Qualified Manpower and Economic*

*Performance*. London: Allan Lane, 145–204.

Sargent, T. J. (1978) "Estimation of Dynamic Labor Demand Schedules Under Rational Expectations", *Journal of Political Economy*, December, 86(6), 1009–1045.

Sato, K. (1975) *Production Functions and Aggregation*. Amsterdam: North-Holland.

Scarf, H. E. and J. B. Shoven, eds. (1984) *Applied General Equilibrium Analysis*. Cambridge: Cambridge University Press.

Schneider, E. (1934) *Theorie der Produktion*. Wien: Springer.

Shephard, R. W. (1953) *Cost and Production Functions*. Princeton: Princeton University Press.

Shephard, R. W. (1970) *Theory of Cost and Production Functions*. Princeton: Princeton University Press.

Sono, M. (1961) "The Effect of Price Changes on the Demand and Supply of Separable Goods", *International Economic Review*, September, 2(3), 239–271.

Spady, R. H. and A. F. Friedlaender (1978) "Hedonic Cost Functions for the Regulated Trucking Industry", *Bell Journal of Economics*, Spring, 9(1), 159–179.

Stevenson, R. E. (1980) "Measuring Technological Bias", *American Economic Review*, March, 70(1), 162–173.

Thompson, R. G., et al. (1977) *Environment and Energy in Petroleum Refining, Electric Power, and Chemical Industries*. Houston: Gulf Publishing.

Uzawa, H. (1962) "Production Functions with Constant Elasticity of Substitution", *Review of Economic Studies*, October, 29(4), 81, 291–299.

Uzawa, H. (1964) "Duality Principles in the Theory of Cost and Production", *International Economic Review*, May, 5(2), 216–220.

Uzawa, H. (1969) "Time Preference and the Penrose Effect in a Two-Class Model of Economic Growth", *Journal of Political Economy*, July/August, Pt. 2, 77(4), 628–652.

Varian, H. (1984) "The Nonparametric Approach to Production Analysis", *Econometrica*, May, 52(2), 579–598.

von Furstenberg, G., ed. (1980) *Capital, Efficiency, and Growth*. Cambridge: Ballinger.

von Weizsacker, C. C. (1962) "A New Technical Progress Function". Massachusetts Institute of Technology, Department of Economics.

Walras, L. (1954) *Elements of Pure Economics*. trans. W. Jaffe, Homewood: Irwin.

Walters, A. A. (1963) "Production and Cost Functions: An Econometric Survey", *Econometrica*, January–April, 31(1), 1–66.

Wills, J. (1979) "Technical Change in the U.S. Primary Metals Industry", *Journal of Econometrics*, April, 10(1), 85–98.

Winston, C. (1985) "Conceptual Developments in the Economics of Transportation: An Interpretive Survey", *Journal of Economic Literature*, March, 23(1), 57–94.

Woodland, A. D. (1975) "Substitution of Structures, Equipment, and Labor in Canadian Production", *International Economic Review*, February, 16(1), 171–187.

Woodland, A. D. (1978) "On Testing Weak Separability", *Journal of Econometrics*, December, 8(3), 383–398.

Woodward, G. T. (1983) "A Factor Augmenting Approach for Studying Capital Measurement, Obsolescence, and the Recent Productivity Slowdown", in: A. Dogramaci, ed., 93–120.

Zellner, A. (1962) "An Efficient Method of Estimating Seemingly Unrelated Regressions and Tests for Aggregation Bias", *Journal of the American Statistical Association*, June, 58(2), 348–368.

Zellner, A. and H. Theil (1962) "Three-Stage Least Squares: Simultaneous Estimation of Simultaneous Equations", *Econometrica*, January, 30(1), 54–78.

*Chapter 32*

# LABOR ECONOMETRICS*

JAMES J. HECKMAN

*University of Chicago*

THOMAS E. MACURDY

*Stanford University*

## Contents

*Heckman's research on this project was supported by National Science Foundation Grant No. SES-8107963 and NIH Grants R01-HD16846 and R01-HD19226. MaCurdy's research on this project was supported by National Science Foundation Grant No. SES-8308664 and a grant from the Alfred P. Sloan Foundation. This paper has benefited greatly from comments generously given by Ricardo Barros, Mark Gritz, Joe Hotz, and Frank Howland.

*Handbook of Econometrics, Volume III, Edited by Z. Griliches and M.D. Intriligator*

## 0.  Introduction

In the past twenty years, the field of labor economics has been enriched by two developments: (a) the evolution of formal neoclassical models of the labor market and (b) the infusion of a variety of sources of microdata. This essay outlines the econometric framework developed by labor economists who have built theoretically motivated models to explain the new data.

The study of female labor supply stimulated early research in labor econometrics. In any microdata study of female labor supply, two facts are readily apparent: that many women do not work, and that wages are often not available for nonworking women. To account for the first fact in a theoretically coherent framework, it is necessary to model corner solutions (choices at the extensive margin) along with conventional interior solutions (choices at the intensive margin) and to develop an econometrics sufficiently rich to account for both types of choices by agents. Although there were precedents for the required type of econometric model in work in consumer theory by Tobin (1958) and his students [e.g. Rosett (1959)], it is fair to say that labor economists have substantially improved the original Tobin framework and have extended it in various important ways to accommodate a variety of models and types of data. To account for the second fact that wages are missing in a nonrandom fashion for nonworking women, it is necessary to develop models for censored random variables. The research on censored regression models developed in labor economics had no precedent in econometrics and was largely neglected by statisticians (See the essay by Griliches in this volume).

The econometric framework developed for the analysis of female labor supply underlies more recent models of job search [Yoon (1981), Kiefer and Neumann (1979), Flinn and Heckman (1982)], occupational choice [Roy (1951), Tinbergen (1951), Siow (1984), Willis and Rosen (1979), Heckman and Sedlacek (1984)], job turnover [Mincer and Jovanovic (1981), Borjas and Rosen (1981), Flinn (1984)], migration [Robinson and Tomes (1982)], unionism [Lee (1978), Strauss and Schmidt (1976), Robinson and Tomes (1984)] and training evaluation [Heckman and Robb (1985)].

All of the recent models presented in labor econometrics are special cases of an index function model. The origins of this model can be traced to Karl Pearson's (1901) work on the mathematical theory of evolution. See D. J. Kevles (1985, p. 31) for one discussion of Pearson's work. In Pearson's framework, discrete and censored random variables are the manifestations of underlying continuous random variables subject to various sampling schemes. Discrete random variables are indicators of whether or not certain latent continuous variables lie above or below given thresholds. Censored random variables are direct observations on the underlying random variables given that certain selection criteria are met. Assuming that the underlying continuous random variables are normally distributed leads to the theory of biserial and tetrachoric correlation. [See Kendall and Stuart (1967, Vol. II), for a review of this theory.] Later work in mathematical psychology by Thurstone (1927) and Bock and Jones (1968) utilized the index function framework to produce mathematical models of choice among discrete alternatives and stimulated a considerable body of ongoing research in economics [See McFadden's paper in Volume II for a survey of this work and Lord and Novick (1968) for an excellent discussion of index function models used in psychometrics].

The index function model cast in terms of underlying continuous latent variables provides the empirical counterpart of many theoretical models in labor economics. For example, it is both natural and analytically convenient to formulate labor supply or job search models in terms of unobserved reservation wages which can often be plausibly modeled as continuous random variables. When reservation wages exceed market wages, people do not work. If the opposite occurs, people work and wages are observed. A variety of models that are special cases of the reservation wage framework will be presented below in Section 3.

The great virtue of research in labor econometrics is that the problems and the solutions in the field are the outgrowth of research on well-posed economic problems. In this area, the economic problems lead and the proposed statistical solutions follow in response to specific theoretical and empirical challenges. This imparts a vitality and originality to the field that is not found in many other branches of econometrics.

One format for presenting recent developments in labor econometrics is to chart the history of the subject, starting with the earliest models, and leading up to more recent developments. This is the strategy we have pursued in previous joint work [Heckman and MaCurdy (1981); Heckman, Killingsworth and MaCurdy (1981)]. The disadvantage of such a format is that basic statistical ideas become intertwined with specific economic models, and general econometric points are sometimes difficult to extract.

This paper follows another format. We first state the basic statistical and econometric principles. We then apply them in a series of worked examples. This format has obvious pedagogical advantages. At the same time, it artificially separates economic problems from econometric theory and does not convey the flow of research problems that stimulated the econometric models.

This paper is in three parts: Part 1 presents a general introduction to the index function framework; Part 2 presents methods for estimating index function models; and Part 3 makes the discussion concrete by presenting a series of models in labor economics that are special cases of the index function framework.

# 1.   The index function model

## 1.1.   Introduction

The critical assumption at the heart of index function models is that unobserved or partially observed continuous random variables generate observed discrete, censored, and truncated random variables. The goal of econometric analysis conducted for these models is to recover the parameters of the distributions of the underlying continuous random variables.

The notion that continuous latent variables generate observed discrete, censored and truncated random variables is natural in many contexts. For example, in the discrete choice literature surveyed by McFadden (1985), the difference between the utility of one option and the utility of another is often naturally interpreted as a continuous random variable, especially if, as is sometimes plausible, utility depends on continuously distributed characteristics. When the difference of utilities exceeds a threshold (zero in this example), the first option is selected. The underlying utilities of choices are never directly observed.

As another example, many models in labor economics are characterized by a "reservation wage" property. Unemployed persons continue to search until their reservation wage – a latent variable – is less their the offered wage. The difference between reservation wages and offered wages is a continuous random variable if some of the characteristics generating reservation wages are continuous random variables. The decision to stop searching is characterized by a continuous latent variable falling below a threshold (zero). Observed wages are censored random variables with the censoring rule characterized by a continuous random variable (the difference between reservation wages and market wages) crossing a threshold. Further examples of index functions generated by economic models are presented in Section 3.

From the vantage point of context-free statistics, using continuous latent variables to generate discrete, censored or truncated random variables introduces unnecessary complications into the statistical analysis. Despite its ancient heritage, the index function approach is no longer widely used or advocated in the modern statistics literature. [See, e.g. Bishop, Fienberg and Holland (1975) or Haberman (1978), Volumes I and II.][1] Given their disinterest in behavioral models, many statisticians prefer direct parameterizations of discrete data and censored data models that typically possess no behavioral interpretation. Some statisticians have argued that econometric models that incorporate behavioral

---

[1] Such models are still widely used in the psychometric literature. See Lord and Novick (1968) or Bock and Jones (1968).

theory are needlessly complicated. For this reason labor economics has been the locus of recent research activity on index function models.

## 1.2. Some definitions and basic ideas

Index functions are defined as continuously distributed random variables. It is helpful to distinguish two types of index functions: those corresponding to continuous random variables that are not directly observed in a given context ($Z$) and those corresponding to continuous random variables that are partially observed ($Y$) in a sense to be made precise below. In the subsequent discussion, the set $\Omega$ represents the support (or the domain of definition) of $(Y, Z)$; the set $\Theta$ denotes the support of $Z$, and $\Psi$ is the support of $Y$; $\Omega$ is the Cartesian product of $\Psi$ and $\Theta$.[2]

### 1.2.1. Quantal response models

We begin with the most elementary index function model. This model ignores the existence of $Y$ and focuses on discrete variables whose outcomes register the occurrence of various states of the world. Let $\Theta_i$ be a nontrivial subset of $\Theta$. Although we do not directly observe $Z$, we know if

$$Z \in \Theta_i.$$

If this event occurs, we denote it by setting an indicator function $\delta_i$ equal to one. More formally,

$$\delta_i = \begin{cases} 1 & \text{if } Z \in \Theta_i, \\ 0 & \text{otherwise.} \end{cases} \tag{1.2.1}$$

When $\delta_i = 1$, state $i$ occurs. The distribution of $Z$ induces a distribution on the $\delta_i$ because

$$\Pr(\delta_i = 1) = \Pr(Z \in \Theta_i). \tag{1.2.2}$$

The discrete choice models surveyed by McFadden (1985) can be cast in this framework. Let $Z$ be a $J \times 1$ vector of utilities, $Z = (V(1), \ldots, V(J))'$. The event that option $i$ is selected is the event that $V(i)$ is maximal in the set $\{V(j)\}_{j=1}^{J}$. In the space of the distribution of utilities, the event that $V(i)$ is maximal corre-

[2] $\Omega$, $\Theta$ and $\Psi$ and all partitions of these sets considered in this paper are assumed to be Borel sets.

sponds to the subspace of $Z$ defined by the inequalities

$$V(j) - V(i) \le 0, \qquad j = 1, \ldots, J.$$

Then in this notation

$$\Theta_i = \{ Z \mid V(j) - V(i) \le 0, \ j = 1, \ldots, J \},$$

and

$$\Pr(\delta_i = 1) = \Pr(Z \in \Theta_i).$$

Introducing exogenous variables ($X$) into this model raises only minor conceptual issues.[3] The distribution of $Z$ can be defined conditional on $X$, and the regions of definition of $\delta_i$ can also be allowed to depend on $X$ (so $\Theta_i = \Theta_i(X)$). The conditional probability that $\delta_i = 1$ given $X$ is

$$\Pr(\delta_i = 1 \mid X) = \Pr(Z \in \Theta_i \mid X). \tag{1.2.3}$$

### 1.2.2.  Models involving endogenous discrete and continuous random variables

We now consider a selection mechanism which records observations on $Y$ only if $(Y, Z)$ lies in some subspace of $\Omega$. More formally, we define the observed value of $Y$ as $Y^*$ with

$$Y^* = Y \qquad \text{if } (Y, Z) \in \Omega_i, \tag{1.2.4}$$

where $\Omega_i$ is a subspace of $\Omega$. We establish the convention that

$$Y^* = 0 \qquad \text{if } (Y, Z) \notin \Omega_i. \tag{1.2.5}$$

This convention is innocuous because the probability that $Y = 0$ is zero as a consequence of the assumption that $Y$ is an absolutely continuous random variable.

A special case of this selection mechanism produces truncated random variables. $Y^*$ is a truncated random variable if the event $(Y, Z) \in \Omega_i$ implies that $Y$ must lie in a strict subset of its support $\Psi$. Thus $Y^*$ is observed only in certain ranges of values of $Y$. For example, negative income tax experiments sample only low income persons. Letting $Y$ be income, $Y^*$ is only observed in data from such experiments if $Y$ is below the cut off point for inclusion of observations into the experiments.

---

[3] Exogenous random variables are always observed and have a marginal density that shares no parameters in common with the conditional distribution of the endogenous variables given the exogenous variables.

Observed values of $Y$ produced by the general selection mechanism (1.2.4) without restrictions on the range of $Y^*$ are censored random variables. As an example of a censored random variable, consider the analysis of Cain and Watts (1973). Let $Y$ be hours of work, $Z_1$ be wage rates, and $Z_2$ denote unearned income, where the $Z_i$ are assumed to be unobserved in this context. Negative income tax experiments observe $Y$ only for low income people (i.e. people for whom $Z_1Y + Z_2$ is sufficiently low). While sampled hours of work – $Y^*$ – may take on all values assumed by $Y$, the density of $Y^*$ may differ greatly from the density of $Y$.

A useful extension of the selection mechanism presented in eq. (1.2.4) is a multi-state model which defines observed values of $Y$ for various states of the world indexed by $i$, $i = 1, \ldots, I$. For state $i$ we define the observed value of $Y$ as

$$Y_i^* = Y \qquad \text{if } (Y, Z) \in \Omega_i, \qquad i = 1, \ldots, I_1, \tag{1.2.6}$$

where the $\Omega_i$'s are subsets of $\Omega$, and $I_1 \ (\leq I)$ is the number of states in which $Y$ is observed. In the remaining states ($I - I_1$ in number), $Y$ is not observed. We define an indicator variable $\delta_i$ by

$$\delta_i = \begin{cases} 1 & \text{if } (Y, Z) \in \Omega_i \\ 0 & \text{if } (Y, Z) \notin \Omega_i \end{cases}, \qquad i = 1, \ldots, I.$$

To avoid uninteresting complications, it is assumed that $\cup_{i=1}^{I} \Omega_i = \Omega$, and that the sets $\Omega_i$ and $\Omega_j$ are disjoint for $i \neq j$. Without any loss of generality, we may set

$$Y_i^* = 0 \quad \text{if } \delta_i = 0. \tag{1.2.7}$$

The variable $Y^* \equiv \sum_{i=1}^{I_1} Y_i^*$ equals $Y$ if it is observed (i.e. if $\delta_i = 1$ for some $i = 1, \ldots, I_1$), and $Y^* = 0$ if any of the states $i = I_1 + 1, \ldots, I$ occur. In other words, $Y$ is observed when $\sum_{i=1}^{I_1} \delta_i = 1$.

To obtain specifications of various density functions that are useful in the econometrics of labor supply, rationing and state contingent demand theory, let $f(y, z)$ be the joint density of $(Y, Z)$. Denote the conditional support of $Z$ when $(Y, Z) \in \Omega_i$ as $\Theta_{i|y}$ which is defined so that, for any fixed $Y = y \in \Psi_i$, the set of admissible $Y$ values in $\Omega_i$, the event $Z \in \Theta_{i|y}$ necessarily implies $\delta_i = 1$; the set $\Theta_{i|y}$ in general depends on $Y = y$. In this notation, the density of $Y_i^*$ conditional on $\delta_i = 1$ is

$$g_i(y_i^*) = \begin{cases} \dfrac{\int_{\Theta_{i|y_i^*}} f(y_i^*, z)\,\mathrm{d}z}{\Pr(\delta_i = 1)} & \text{for } y_i^* \in \Psi_i \\ 0 & \text{for } y_i^* \notin \Psi_i \end{cases} \qquad i = 1, \ldots, I_1, \tag{1.2.8}$$

with

$$\Pr(\delta_i = 1) = \Pr((Y, Z) \in \Omega_i) = \int_{\Omega_i} f(y, z) \, \mathrm{d}y \, \mathrm{d}z,$$

where the notation $\int_{\Theta_{i|y}}$ and $\int_{\Omega}$ denotes integration over the sets $\Theta$ given $y$ and $\Omega$, respectively – i.e.

$$\int_{\Theta_{i|y}} f(y, z) \, \mathrm{d}z = \int_{\{z | z \in \Theta_{i|y}\}} f(y, z) \, \mathrm{d}z,$$

and

$$\int_{\Omega_i} f(y, z) \, \mathrm{d}y \, \mathrm{d}z = \int_{\{(y, z) | (y, z) \in \Omega_i\}} f(y, z) \, \mathrm{d}y \, \mathrm{d}z.$$

The function $g_i(\cdot)$ is the conditional density of $Y$ given that selection rule (1.2.6) is satisfied. As a consequence of convention (1.2.7), the distribution of $Y_i^*$ when $\delta_i = 0$ has point mass at $Y_i^* = 0$ (i.e. $\Pr(Y_i^* = 0 | \delta_i = 0) = 1$).
    The joint density of $Y_i^*$ and $\delta_i$ is

$$g_i(y_i^*, \delta_i) = \left[ g_i(y_i^*) \Pr(\delta_i = 1) \right]^{\delta_i} \left[ J(y_i^*) \Pr(\delta_i = 0) \right]^{1 - \delta_i}, \qquad i = 1, \dots, I_1,$$
$$(1.2.9)$$

where $J(y_i^*) = 1$ if $y_i^* = 0$ and $J(y_i^*) = 0$ otherwise, where $\Pr(\delta_i = 0) = 1 - \Pr(\delta_i = 1)$, and where we adopt the convention that zero raised to the zero-th power equals one (i.e. when $\delta_i = 0$ and $y_i^* \notin \Psi_i$ so $g_i(y_i^*) = 0$, then $[g_i(y_i^*) \Pr(\delta_i = 1)]^0 = 1$).[4]
    From (1.2.9) the conditional density of $Y^*$ given that state $i = 1, \dots, I_1$ occurs is

$$h(y^* | \delta_i = 1) = g_i(y^*).$$

$Y^*$ is defined to be degenerate at zero if one of the other states $i = I_1 + 1, \dots, I$ occurs. A compact expression for the conditional density of $Y^*$ is

$$h(y^* | \delta_1, \dots, \delta_I) = \prod_{i=1}^{I_1} \left[ g_i(y^*) \right]^{\delta_i}, \quad \text{if } \sum_{i=1}^{I_1} \delta_i = 1 \qquad (1.2.10)$$

---

[4] We use the term "density" in the sense of the product measure $\mathrm{d}[y_1 + K_0(y_1)] \times \mathrm{d}[K_0(z) + K_1(z)]$ on $R^1 \times [0, 1]$ where $\mathrm{d}y$ is Lebesgue measure on $R^1$ and $K_\alpha(z)$ is the probability distribution that assigns the point $\alpha$ in $R^1$ unit mass.

with $Y^* = 0$ with probability one when $\delta_i = 1$ for some value of $i = I_1 + 1, \ldots, I$. The joint density of $Y^*$ and $\delta_1, \ldots, \delta_I$ is the product of the conditional density of $Y^*$ (1.2.10) and the joint probability of $\delta_1, \ldots, \delta_I$; i.e.

$$h(y^*, \delta_1, \ldots, \delta_I) = \prod_{i=1}^{I_1} \left[ g_i(y^*) \Pr(\delta_i = 1) \right]^{\delta_i} \prod_{i=I_1+1}^{I} \left[ J(y_i^*) \Pr(\delta_i = 1) \right]^{\delta_i}.$$

$$(1.2.11)$$

In some problems the particular state of the world in which an observation occurs is unknown (i.e. the $\delta_i$'s are not separately observed); it is only known that one of a subset of states has occurred. Given information on $Y^*$, one can determine whether or not one of the first $I_1$ states has occurred – since $Y^* \neq 0$ indicates $\delta_i = 1$ for some $i \leq I_1$ and $Y^* = 0$ indicates $\delta_i = 1$ for some $i > I_1$ – but it may not be possible to determine the particular $i$ for which $\delta_i = 1$.

For example, suppose that when $Y^* \neq 0$, one only knows that either $\bar{\delta}_1 \equiv \sum_{i=1}^{I_a} \delta_i = 1$ or $\bar{\delta}_2 \equiv \sum_{i=I_a+1}^{I_1} \delta_i = 1$. Suppose further that when $Y^* = 0$, it is only known that $\bar{\delta}_3 \equiv \sum_{i=I_1+1}^{I} \delta_i = 1$. The densities (1.2.10) and (1.2.11) cannot directly be used as a basis for inference in this situation. (Unless, of course, $I_a = 1$, $I_1 = 2$, and $I = 3$.)

The densities appropriate for analyzing data on $Y^*$ and the $\bar{\delta}_j$'s are obtained by conditioning on the available knowledge about states. The desired densities are derived by computing the expected value of (1.2.10) to eliminate the individual $\delta_i$'s that are not observed. In particular, the marginal density of $y^*$ given $\bar{\delta}_1 = 1$ is given by the law of iterated expectations as

$$k(y^* | \bar{\delta}_1 = 1) = E\left( h(y^* | \delta_1, \ldots, \delta_I) | \bar{\delta}_1 = 1 \right)$$

$$= \sum_{i=1}^{I_a} h(y^* | \delta_i = 1) \Pr(\delta_i = 1 | \bar{\delta}_1 = 1)$$

$$= \sum_{i=1}^{I_a} g_i(y^*) \Pr(\delta_i = 1) / \Pr(\bar{\delta}_1 = 1).$$

Analogously, the density of $Y^*$ given $\bar{\delta}_2 = 1$ is

$$k(y^* | \bar{\delta}_2 = 1) = \sum_{i=I_a+1}^{I_1} g_i(y^*) \Pr(\delta_i = 1) / \Pr(\bar{\delta}_2 = 1).[5]$$

When $\bar{\delta}_3 = 1$, $Y^*$ is degenerate at zero. Thus the density of $Y^*$ conditional on the

---

[5] These derivations use the fact that the sets $\Omega_i$ are mutually exclusive so $\Pr(\bar{\delta}_1 = 1) = \sum_{i=1}^{I_a} \Pr(\delta_i = 1)$ and $E(\delta_i | \bar{\delta}_1 = 1) = \Pr(\delta_i = 1 | \bar{\delta}_1 = 1) = \Pr(\delta_i = 1) / \Pr(\bar{\delta}_1 = 1)$, with completely analogous results holding for $\bar{\delta}_2$ and $\bar{\delta}_3$.

$\bar{\delta}_j$'s is given by

$$h(y^*|\bar{\delta}_1,\bar{\delta}_2,\bar{\delta}_3) = \left[\sum_{i=1}^{I_a} g_i(y^*)\Pr(\delta_i=1)/\Pr(\bar{\delta}_1=1)\right]^{\bar{\delta}_1}$$

$$\cdot\left[\sum_{i=I_a+1}^{I_1} g_i(y^*)\Pr(\delta_i=1)/\Pr(\bar{\delta}_2=1)\right]^{\bar{\delta}_2}\left[\prod_{i=1}^{I_1} J(y_i^*)\right]^{\bar{\delta}_3},$$

$$(1.2.12)$$

where $Y^*$ has point mass at zero when $\bar{\delta}_3=1$ (i.e. $\Pr(Y^*=0|\bar{\delta}_3=1)=1$). Multiplying the conditional density (1.2.12) by the probability of the events $\bar{\delta}_1$, $\bar{\delta}_2$, $\bar{\delta}_3$ generates the joint density for $Y^*$ and the $\bar{\delta}_j$'s:

$$h(y^*,\bar{\delta}_1,\bar{\delta}_2,\bar{\delta}_3) = \left[\sum_{i=1}^{I_a} g_i(y^*)\Pr(\delta_i=1)\right]^{\bar{\delta}_1}$$

$$\cdot\left[\sum_{i=I_a+1}^{I_1} g_i(y^*)\Pr(\delta_i=1)\right]^{\bar{\delta}_2}\left[\prod_{i=l}^{I_1} J(y_i^*)\Pr(\bar{\delta}_3=1)\right]^{\bar{\delta}_3}.$$

$$(1.2.13)$$

Densities of the form (1.2.8)–(1.2.13) appear repeatedly in the models for the analysis of labor supply presented in Section 3.3.

All the densities in the preceding analysis can be modified to depend on exogenous variables $X$, as can the support of the selection region (i.e. $\Omega_i = \Omega_i(X)$). Writing $f(y, z|X)$ to denote the appropriate conditional density, only obvious notational modifications are required to introduce such variables.

## 1.3. Sampling plans

A variety of different sampling plans are used to collect the data available to labor economists. The econometric implications of data collected from such sampling plans have received a great deal of attention in the discrete choice and labor econometrics literatures. In this subsection we define the concepts of simple random samples, truncated random samples, censored random samples, stratified random samples, and choice based samples. To this end we let $h(X)$ denote the population density of the exogenous variables $X$, so that the joint density of $(Y, \delta, X)$ is

$$f(Y, \delta, X) = f(Y, \delta|X)h(X),$$

$$(1.3.1)$$

with c.d.f.

$$F(Y, \delta, X). \tag{1.3.2}$$

From the definition of exogeneity, the marginal density of $X$ contains no parameters in common with the conditional density of $(Y, \delta)$ given $X$.

In the cases considered here, the underlying population is assumed to be infinite and generated by probability density (1.3.1) and c.d.f. (1.3.2). If the sampling is such that it produces a *simple random sample*, successive observations must (a) be independent and (b) each observation must be a realization from a common density (1.3.1). In this textbook case, the sample likelihood is the product of terms of the form (1.3.1) with realized values of $(Y, \delta, X)$ substituted in place of the random variables.

Next suppose that from a simple random sample, observations on $(Y, \delta, X)$ are retained only if these random variables lie in some open subset of the support of $(Y, \delta, X)$. More precisely suppose that observations on $(Y, \delta, X)$ are retained only if

$$(Y, \delta, X) \in \Lambda_1 \subset \Lambda, \tag{1.3.3}$$

where $\Lambda$ is the support of random variables $(Y, \delta, X)$.

In the classical statistical literature [See, e.g. Kendall and Stuart, Vol. II, (1967)] no regressors are assumed to appear in the model. In this case, a sample is defined to be *censored* if the *number* of observations not in $\Lambda_1$ is recorded (so $\delta$ is known for all observations). If this information is not retained, the sample is *truncated*. When regressors are present, there are several ways to extend these definitions allowing either $\delta$ or $X$ to be recorded when $(Y, \delta, X) \notin \Lambda_1$. In this paper we adopt the following conventions. If information on $(\delta, X)$ for all $(Y, \delta, X) \notin \Lambda_1$ is retained (but $Y$ is not known), we call the sample censored. If information on $(\delta, X)$ is not retained for $(Y, \delta, X) \notin \Lambda_1$, the sample is truncated. Note that in these definitions $\Lambda_1$ can consist of disconnected sets of $\Lambda$.

One operational difference between censored and truncated samples is that for censored samples it is possible to consistently estimate the population probability that $(Y, \delta, X) \in \Lambda_1$, whereas for truncated samples these probabilities cannot be consistently estimated as sample sizes become large. In neither sample is it possible to directly estimate the conditional distribution of $(Y, \delta, X)$ given $(Y, \delta, X) \notin \Lambda_1$ using an empirical c.d.f. for this subsample.[6]

---

[6]It is possible to estimate this conditional distribution using the subsample generated by the requirement that $(Y, \delta, X) \in \Lambda_1$ for certain specific functional form assumptions for $F$. Such forms for $F$ are termed "recoverable" in the literature. See Heckman and Singer (1986) for further discussion of this issue of recoverability.

In the special case in which the subset $\Lambda_1$ only restricts the support of $X$, (exogenous truncated and censored samples), the econometric analysis can proceed conditional on $X$. In light of the assumed exogeneity of $X$, the only possible econometric problem is a loss in efficiency of proposed estimators.

Truncated and censored samples are special cases of the more general notion of a *stratified sample*. In place of the special sampling rule (1.3.3), in a general stratified sample, the rule for selecting independent observations is such that even in an infinite sample the probability that $(Y, \delta, X) \in \Lambda_i \subset \Lambda$ does not equal the population probability that $(Y, \delta, X) \in \Lambda_i$ where $\cup_{i=1}^{I} \Lambda_i = \Lambda$, and $\Lambda_i$ and $\Lambda_j$ are disjoint for all $i \neq j$. It is helpful to further distinguish between *exogenously stratified* and *endogenously stratified* samples.

In an exogenously stratified sample, selection occurs solely on the $X$ in the sense that the sample distribution of $X$ does not converge to the population distribution of $X$ even as the sample size is increased. This may occur because data are systematically missing for $X$ in certain regions of the support, or more generally because some subsets of the support of $X$ are oversampled. However, conditional on $X$, the sample distribution of $(Y, \delta \mid X)$ converges to the population distribution. By virtue of the assumed exogeneity of $X$, such a sampling scheme creates no special econometric problems.

In an endogenously stratified sample, selection occurs on $(Y, \delta)$ (and also possibly on the $X$), and the sampling rule is such that the sample distribution of $(Y, \delta)$ does not converge to the population distribution $F(Y, \delta)$ (conditional or unconditional on $X$). This can occur because data are missing for certain values of $Y$ or $\delta$ (or both), or because some subsets of the support of these random variables are oversampled. The special case of an endogenously stratified sample in which, conditional on $(Y, \delta)$, the population density of $X$ characterizes the data, i.e.

$$h(X \mid Y, \delta) = \frac{f(Y, \delta, X)}{f(Y, \delta)}, \tag{1.3.4}$$

is termed *choice based sampling* in the literature. [See McFadden in Volume II or the excellent survey article by Manski and McFadden (1981).[7]] In a general endogenously stratified sample, (1.3.4) need not characterize the density of the data produced by an infinite repetition of the sampling rule. Moreover, in both choice based and more general endogenously stratified samples, the sample distribution of $X$ depends on the parameters of the conditional distribution of $(Y, \delta)$ given $X$ so, as a consequence of the sampling rules, $X$ is no longer

---

[7]Strictly speaking, the choice based sampling literature focuses on a model in which $Y$ is integrated out of the model so that $\delta$ and $X$ are the relevant random variables.

exogenous in such samples, and its distribution is informative on the structural parameters of the model.

Truncated and censored samples are special cases of a general stratified sample. A truncated sample is produced from a general stratified sample for which the sampling weight for the event $(Y, \delta, X) \notin \Lambda_1$ is identically zero. In a censored sample, the sampling weight for the event $(Y, \delta, X) \notin \Lambda_1$ is the same as the population probability of the event.

Note that in a truncated sample, observed $Y$ may or may not be a truncated random variable. For example, if $\Lambda_1$ only restricts $\delta$, and $\delta$ does not restrict the support of $Y$, observed $Y$ is a censored random variable. On the other hand, if $\Lambda_1$ restricts the support of $Y$, observed $Y$ is a truncated random variable. Similarly in a censored sample, $Y$ may or may not be censored. For example, if $\Lambda_1$ is defined only by a restriction on values that $\delta$ can assume, and $\delta$ does not restrict the support of $Y$, observed $Y$ is censored. If $\Lambda_1$ is defined by a restriction on the support of $Y$, observed $Y$ is truncated even though the sample is censored. An unfortunate and sometimes confusing nomenclature thus appears in the literature. The concepts of censored and truncated random variables are to be carefully distinguished from the concepts of censored and truncated random samples.

Truncated and censored sample selection rule (1.3.3) is essentially identical to the selection rule (1.2.6) (augmented to include $X$ in the manner suggested at the end of subsection 1.2). Thus the econometric analysis of models generated by rules such as (1.2.6) can be applied without modification to the analysis of models estimated on truncated and censored samples. The same can be said of the econometric analysis of models fit on all stratified samples for which the sampling rule can be expressed as some restriction on the support of $(Y, Z, \delta, X)$. In the recent research in labor econometrics, all of the sample selection rules considered can be written in this form, and an analysis based on samples generated by (augmented) versions of (1.2.6) captures the essence of the recent literature.[8]

## 2. Estimation

The conventional approach to estimating the parameters of index function models postulates specific functional forms for $f(y, z)$ or $f(y, z|X)$ and estimates the parameters of these densities by the method of maximum likelihood or by the method of moments. Pearson (1901) invoked a normality assumption in his original work on index function models and this assumption is still often used in

---

[8] We note, however, that it is possible to construct examples of stratified sample selection rules that cannot be cast in this format. For example, selection rules that weight various strata in different (nonzero) proportions than the population proportions cannot be cast in the form of selection rule (1.2.6).

recent work in labor econometrics. The normality assumption has come under attack in the recent literature because when implications of it have been subject to empirical test they have often been rejected.

It is essential to separate conceptual ideas that are valid for any index function model from results special to the normal model. Most of the conceptual framework underlying the normal index model is valid in a general nonnormal setting. In this section we focus on general ideas and refer the reader to specific papers in the literature where relevant details of normal models are presented.

For two reasons we do not discuss estimation of index function models by the method of maximum likelihood. First, once the appropriate densities are derived, there is little to say about the method beyond what already appears in the literature. [See Amemiya (1985).] We devote attention to the derivation of the appropriate densities in Section 3. Second, it is our experience that the conditions required to secure identification of an index function model are more easily understood when stated in a regression or method of moments framework. Discussions of identifiability that appeal to the nonsingularity of an information matrix have no intuitive appeal and often degenerate into empty tautologies. For these reasons we focus attention on regression and method of moments procedures.

## 2.1. Regression function characterizations

We begin by presenting a regression function characterization of the econometric problems encountered in the analysis of data collected from truncated, censored and stratified samples and models with truncated and censored random variables. We start with a simple two equation linear regression specification for the underlying index functions and derive the conditional expectations of the observed counterparts of the index variables. More elaborate models are then developed. We next present several procedures for estimating the parameters of the regression specifications.

### 2.1.1. A prototypical regression specification

A special case of the index function framework set out in Section 1 writes $Y$ and $Z$ as scalar random variables which are assumed to be linear functions of a common set of exogenous variables $X$ and unobservables $U$ and $V$ respectively.[9]

---

[9] By exogenous variables we mean that $X$ is observed and is distributed independently of $(U, V)$ and that the parameters of the distribution of $X$ are not functions of the parameters $(\beta, \gamma)$ or the parameters of the distribution of $(U, V)$.

We write

$$Y = X\beta + U, \tag{2.1.1}$$
$$Z = X\gamma + V, \tag{2.1.2}$$

where $(\beta, \gamma)$ is a pair of suitably dimensioned parameter vectors, and $Y$ is observed only if $Z \in \Theta_1$, a proper subset of the support of $Z$. For expositional convenience we initially assume that the sample selection rule depends only on the value of $Z$ and not directly on $Y$. In terms of the notation of Section 1, we begin by considering a case in which $Y$ is observed if $(Y, Z) \in \Omega_1$ where $\Omega_1$ is a subset of the support of $(Y, Z)$ defined by $\Omega_1 = \{(Y, Z)| -\infty \le Y \le \infty, \ Z \in \Theta_1\}$. For the moment, we also restrict attention to a two-state model. State 1 occurs if $Z \in \Theta_1$ and state 0 is observed if $Z \notin \Theta_1$. We later generalize the analysis to consider inclusion rules that depend explicitly on $Y$ and we also consider multi-state models.

The joint density of $(U, V)$, denoted by $f(u, v)$, depends on parameters $\psi$ and may depend on the exogenous variables $X$. Since elements of $\beta$, $\gamma$, and $\psi$ may be zero, there is no loss of generality in assuming that a common $X$ vector enters (2.1.1), (2.1.2) and the density of $(U, V)$.

As in Section 1, we define the indicator function

$$\delta = \begin{cases} 1 & \text{if } Z \in \Theta_1; \\ 0 & \text{otherwise.} \end{cases}$$

In a censored regression model in which $Y$ is observed only if $\delta = 1$, we define $Y^* = Y$ if $\delta = 1$ and use the convention that $Y^* = 0$ if $\delta = 0$. In shorthand notation

$$Y^* = \delta X\beta + \delta U \equiv X^*\beta + U^*.$$

The conditional expectation of $Y$ given $\delta = 1$ and $X$ is

$$E(Y|\delta = 1, X) = X\beta + M, \tag{2.1.3}$$

where

$$M = M(X\gamma, \psi) \equiv E(U|\delta = 1, X),$$

is the conditional expectation of $U$ given that $X$ and $Z \in \Theta_1$. If the disturbance $U$ is independent of $V$, $M = 0$. If the disturbances are not independent, $M$ is in general a nontrivial function of $X$ and the parameters of the model $(\gamma, \psi)$.

Note that since $Y^* = \delta Y$, by the law of iterated expectations

$$E(Y^*|X) = E(Y^*|\delta = 0, X)\Pr(\delta = 0|X) + E(Y^*|\delta = 1, X)\Pr(\delta = 1|X)$$
$$= (X\beta + M)\Pr(\delta = 1|X). \tag{2.1.4}$$

Applying the analysis of Section 1, the conditional distribution of $U$ given $X$ and $Z \in \Theta_1$ is

$$f(u|Z \in \Theta_1, X) = \frac{\int_{\Theta_1} f(u, z - X\gamma) \, \mathrm{d}z}{P_1}, \tag{2.1.5}$$

where $P_1 \equiv \Pr(Z \in \Theta_1 | X)$ is the probability that $\delta = 1$ given $X$. $P_1$ is defined as

$$P_1 \equiv P_1(X\gamma, \psi) = \int_{-\infty}^{\infty} \int_{\Theta_1} f(u, z - X\gamma) \, \mathrm{d}z \, \mathrm{d}u$$

$$= \int_{\Theta_1} f_v(z - X\gamma) \, \mathrm{d}z, \tag{2.1.6}$$

where $f_v(\cdot)$ denotes the marginal density of $V$. Hence,

$$M = \frac{\int_{-\infty}^{\infty} \int_{\Theta_1} u f(u, z - X\gamma) \, \mathrm{d}z \, \mathrm{d}u}{P_1}. \tag{2.1.7}$$

A regression of $Y$ on $X$ using a sample of observations restricted to have $\delta = 1$ omits the term $M$ from the regression function (2.1.3), and familiar specification bias error arguments apply.

For example, consider a variable $X_j$ that appears in both equations (so the $j$th coefficients of $\beta$ and $\gamma$ are nonzero). A regression of $Y$ on $X$ fit on samples restricted to satisfy $\delta = 1$ that does not include $M$ as a regressor produces coefficients that do not converge to $\beta$. Letting "ˆ" denote the OLS coefficient,

$$\text{plim } \hat{\beta}_j = \beta_j + L_{MX_j},$$

where $L_{MX_j}$ is the probability limit of the coefficient of $X_j$ in a projection of $M$ on $X$.[10] Note that if a variable $X_k$ that does not appear in (2.1.1) is introduced into a least squares equation that omits $M$, the least squares coefficient converges to

$$\text{plim } \hat{\beta}_k = L_{MX_k},$$

so $X_k$ may proxy $M$.

The essential feature of both examples is that in samples selected so that $\delta = 1$, $X$ is no longer exogenous with respect to the disturbance term $U^*$ ( $= \delta U$ )

---

[10] It is not the case that $L_{MX_j} = (\partial M / \partial X_j)$, although the approximation may be very close. See Byron and Bera (1983).

although it is defined to be exogenous with respect to $U$. The distribution of $U^*$ depends on $X$ (see the expression for $M$ below (2.1.3)). As $X$ is varied, the mean of the distribution of $U^*$ is changed. Estimated regression coefficients combine the desired ceteris paribus effect of $X$ on $Y$ (holding $U^*$ fixed) with the effect of changes in $X$ on the mean of $U^*$.

Characterizing a sample as a subsample from a larger random sample generated by having $Z \in \Theta_1$ encompasses two distinct ideas that are sometimes confused in the literature. The first idea is that of self-selection. For example, in a simple model of labor supply an individual chooses either to work or not to work. An index function $Z$ representing the difference between the utility of working and of not working can be used to characterize this decision. From an initial random sample, a sample of workers is not random since $Z \geq 0$ for each worker. The second idea is a more general concept – that of sample selection – which includes the first idea as a special case. From a simple random sample, some rule is used to generate the sample used in an empirical analysis. These rules may or may not be the consequences of choices made by the individuals being studied.

Econometric solutions to the general sample selection bias problem and the self-selection bias problem are identical. Both the early work on female labor supply and the later analysis of "experimental data" generated from stratified samples sought to eliminate the effects of sample selection bias on estimated structural labor supply and earnings functions.

It has been our experience that many statisticians and some econometricians find these ideas quite alien. From the context-free view of mathematical statistics, it seems odd to define a sample of workers as a selected sample if the object of the empirical analysis is to estimate hours of work equations "After all," the argument is sometimes made, "nonworkers give us no information about the determinants of working hours."

This view ignores the fact that meaningful behavioral theories postulate a common decision process used by all agents (e.g. utility maximization). In neoclassical labor supply theory all agents are assumed to possess preference orderings over goods and leisure. Some agents choose not to work, but nonworkers still possess well-defined preference functions. Equations like (2.1.1) are defined for all agents in the population and it is the estimation of the parameters of the *population distribution* of preferences that is the goal of *structural* econometric analysis. Estimating functions on samples selected on the basis of choices biases the estimates of the parameters of the distribution of population preferences unless explicit account is taken of the sample selection rule in the estimation procedure.[11]

---

[11] Many statisticians implicitly adopt the extreme view that nonworkers come from a different population than workers and that there is no commonality of decision processes and/or parameter values in the two populations. In some contexts (e.g. in a single cross section) these two views are empirically indistinguishable. See the discussion of recoverability in Heckman and Singer (1986).

### 2.1.2. *Specification for selection corrections*

In order to make the preceding theory empirically operational it is necessary to know $M$ (up to a vector of estimable parameters). One way to acquire this information is to postulate a specific functional form for it directly. Doing so makes clear that conventional regression corrections for sample selection bias depend critically on assumptions about the correct functional form of the underlying regression eq. (2.1.1) and the functional form of $M$.

The second and more commonly utilized approach used to generate $M$ postulates specific functional forms for the density of $(U, V)$ and derives the conditional expectation of $U$ given $\delta$ and $X$. Since in practice this density is usually unknown, it is not obvious that this route for selecting $M$ is any less ad hoc than the first.

One commonly utilized assumption postulates a linear regression relationship for the conditional expectation of $U$ given $V$:

$$E(U|V, X) = \tau V,   \tag{2.1.8}$$

where $\tau$ is a regression coefficient. For example, (2.1.8) is generated if $U$ and $V$ are bivariate normal random variables and $X$ is exogenous with respect to $U$ and $V$. Many other joint densities for $(U, V)$ also yield linear representation (2.1.8). [See Kagan, Linnik and Rao (1973)].

Equation (2.1.8) implies that the selection term $M$ can be written as

$$M \equiv E(U|\delta = 1, X) = \tau E(V|\delta = 1, X).   \tag{2.1.9}$$

Knowledge of the marginal distribution of $V$ determines the functional form of the selection bias term.

Letting $f_v(v)$ denote the marginal density of $V$, it follows from the analysis of Section 1 that

$$E(V|\delta = 1, X) = \frac{\int_{\Gamma_1} v f_v(v)\, dv}{P_1},   \tag{2.1.10}$$

where the set $\Gamma_1 \equiv \{V: V + X\gamma \in \Theta_1\}$, and

$$P_1 \equiv \mathrm{Prob}(Z \in \Theta_1|X) = \mathrm{Prob}(V \in \Gamma_1|X) = \int_{\Gamma_1} f_v(v)\, dv.   \tag{2.1.11}$$

One commonly used specification of $\Theta_1$ writes $\Theta_1 = \{Z: Z \geq 0\}$, so $\Gamma_1 = \{V:$

$V \geq - X\gamma\}$. In this case (2.1.10) and (2.1.11) become

$$E(V|\delta = 1, X) = E(V|V \geq - X\gamma, X) = \frac{\int_{-X\gamma}^{\infty} v f_v(v)\, dv}{P_1},$$  (2.1.12)

and

$$P_1 = \text{Prob}(\delta = 1|X) = \int_{-X\gamma}^{\infty} f_v(v)\, dv = 1 - F_v(- X\gamma),$$  (2.1.13)

respectively, where $F_v(\cdot)$ is the cumulative distribution function of $V$. Since $Z$ is not directly observed, it is permissible to arbitrarily normalize the variance of the disturbance of the selection rule equation because division by a positive constant does not change the probability content of the inequality that defines $\Gamma_1$. Thus, $E(U|\delta = 1, X)$ is the same if one replaces $f_v(v)$ with $f_v(\sigma v)/\sigma$ and reinterprets $\Gamma_1$ as $\{\sigma V: V + (X\gamma^*)/\sigma \in \Theta_1\}$ using any $\sigma > 0$ where $\gamma^* \equiv \sigma\gamma$. The normalization for $E(V^2)$ that we adopt depends on the particular distribution under consideration.

Numerous choices for $f_v(v)$ have been advanced in the literature yielding a wide variety of functional forms for (2.1.12). Table 1 presents various specifications of $f_v(v)$ and the implied specifications for $E(V|\delta = 1, X) = E(V|V \geq - X\gamma, X)$ proposed in work by Heckman (1976b, 1979), Goldberger (1983), Lee (1982), and Olson (1980). Substituting the formulae for the truncated means presented in the third column of the table into relation (2.1.4) produces an array of useful expressions for the sample selection term $M$. All of the functions appearing in these formulae – including the gamma, the incomplete gamma, and the distribution functions – are available on most computers.

Inserting any of these expressions for $M$ into eqs. (2.1.3) or (2.1.4) yields an explicit specification for the regression relation associated with $Y$ (or $Y^*$) given the selection rule generating the data. In order to generate (2.1.9) one requires a formula for the probability that $\delta = 1$ given $X$ to complete the specification for $E(Y^*)$. Formula (2.1.13) gives the specification of this probability in terms of the cumulative distribution function of $V$.

In place of the linear conditional expectation (2.1.8), Lee (1982) suggests a more general nonlinear conditional expectation of $U$ given $V$. Drawing on well-known results in the statistics literature, Lee suggests application of Edgeworth-type expansions. For the bivariate Gram–Charlier series expansion, the conditional expectation of $U$ given $V$ and exogenous $X$ is

$$E(U|V, X) = \rho V + \frac{B(V)}{A(V)},$$  (2.1.14)

Table 1
Truncated mean formulae for selected zero-mean distributions.

| Distribution (Mean, Variance, Sign of Skewness)[e] | Density $f_v(v)$ | Truncated means[a] $E(v \mid v \geq -X\gamma)$ |
|---|---|---|
| Normal (0,1,0) | $(2\pi)^{-1/2}e^{-v^2/2}$ | $f_v(X\gamma)/F_v(X\gamma)$ |
| Student's $t$[b] $\left(0, \dfrac{n}{n-2}, 0\right)$ | $(n\pi)^{-1/2}\Gamma\left(\dfrac{n+1}{2}\right)\left[1+\dfrac{v^2}{n}\right]^{-(n+1)/2}\Big/\Gamma\left(\dfrac{n}{2}\right)$ | $\dfrac{n+(X\gamma)^2}{n-1}f_v(X\gamma)/F_v(X\gamma)$ |
| Chi-square[c] $(0,2n,+)$ | $2^{-n/2}(v+n)^{(n/2)-1}e^{-(v+n)/2}/\Gamma\left(\dfrac{n}{2}\right)$ for $-n \leq v < \infty$ | $\left[2G\left(\dfrac{n}{2}+1, \dfrac{n-X\gamma}{2}\right) - nG\left(\dfrac{n}{2}, \dfrac{n-X\gamma}{2}\right)\right]\Big/\left[\Gamma\left(\dfrac{n}{2}\right)(1-F_v(-X\gamma))\right]$ for $n \geq X_\gamma$ |
| Logistic $\left(0, \dfrac{\pi^2}{3}, 0\right)$ | $\dfrac{e^v}{(1+e^v)^2}$ | $-[\ln F_v(X\gamma)-X\gamma F_v(-X\gamma)]/F_v(X\gamma)$ |
| Laplace $(0,2,0)$ | $\tfrac{1}{2}e^{-|v|}$ | $1-X\gamma$ for $X\gamma \leq 0$; $(1+X\gamma)/(2e^{X\gamma}-1)$ for $X\gamma > 0$ |
| Uniform $(0,1,0)$ | $1/\sqrt{12}$ for $|v| \leq \sqrt{3}$ | $(\sqrt{3} - X\gamma)/2$ for $|X\gamma| \leq \sqrt{3}$ |
| Log-normal[d] $(0, e^2 - e, +)$ | $(2\pi)^{-1/2}(e^{(1/2)}+v)^{-1}e^{-[\ln(e^{1/2}+v)]^2/2}$ for $v \geq -e^{1/2}$ | $e^{1/2}\left[\dfrac{\Phi\big(1-\ln(e^{1/2}-X\gamma)\big)}{\Phi\big(-\ln(e^{1/2}-X\gamma)\big)}-1\right]$ for $X\gamma \leq e^{1/2}$ |

[a] The function $F_v(a) \equiv \int_{-\infty}^{a} f_v(v)\,dv$ in these formulae is the cumulative distribution function.

[b] The parameter $n$ denotes degrees of freedom. For Student's $t$, it is assumed that $n > 2$. The function $\Gamma(a) \equiv \int_0^\infty y^{a-1}e^{-y}\,dy$ is the gamma function.

[c] The function $G(a,b) \equiv \int_b^\infty y^{a-1}e^{-y}\,dy$ is the incomplete gamma function.

[d] The function $\Phi(\cdot)$ represents the standardized normal cumulative distribution function.

[e] Skewness is defined as mean minus the median.

with

$$A(V) \equiv 1 + [\mu_{03}\Lambda_3(V)/6] + [\mu_{04} - 3](\Lambda_4(V)/24),$$
$$B(V) \equiv [\mu_{12} - \rho\mu_{03}](\Lambda_2(V)/2) + [\mu_{13} - \rho\mu_{04}](\Lambda_3(V)/6),$$

where $\rho$ is the correlation coefficient of $U$ and $V$, $\mu_{ij} \equiv E(U^iV^j)$ are cross moments of $U$ and $V$, and the functions $\Lambda_2(V) \equiv V^2 - 1$, $\Lambda_3(V) \equiv V^3 - 3V$, and $\Lambda_4(V) \equiv V^4 - 6V^2 + 3$ are Hermite polynomials. Assuming the event $V \geq - X\gamma$ determines whether or not $Y$ is observed, the selection term is

$$M = E(U|V \geq - X\gamma, X) = \frac{\int_{-X\gamma}^{\infty}\left(\rho v + \dfrac{B(v)}{A(v)}\right)f_v(v)\,\mathrm{d}v}{1 - F_v(- X\gamma)}. \tag{2.1.15}$$

This expression does not have a simple analytical solution except in very special cases. Lee (1982) invokes the assumption that $V$ is a standard normal random variable, in which case $A(V) = 1$ (since $\mu_{03} = \mu_{04} - 3 = 0$) and the conditional mean is

$$E(U|V) = V\rho + (V^2 - 1)(\mu_{12}/2) + (V^3 - 3V)(\mu_{13} - 3\rho)/6. \tag{2.1.16}$$

For this specification, (2.1.15) reduces to

$$M = \frac{\phi(X\gamma)}{\Phi(X\gamma)}\tau_1 + \frac{- X\gamma\phi(X\gamma)}{\Phi(X\gamma)}\tau_2 + \left[(X\gamma)^2 - 1\right]\frac{\phi(X\gamma)}{\Phi(X\gamma)}\tau_3, \tag{2.1.17}$$

where $\phi(\cdot)$ and $\Phi(\cdot)$ are, respectively, the density function and the cumulative distribution functions associated with a standard normal distribution, and $\tau_1$, $\tau_2$, and $\tau_3$ are parameters.[12]

---

[12] The requirement that $V$ is normally distributed is not as restrictive as it may first appear. In particular, suppose that the distribution of $V$, $F_v(\cdot)$ is not normal. Defining $J(\cdot)$ as the transformation $\Phi^{-1} \circ F_v$, the random variable $J(V)$ is normally distributed with mean zero and a variance equal to one. Define a new unobserved dependent variable $Z_J$ by the equation

$$Z_J = - J(- X\gamma) + J(V). \tag{$*$}$$

Since $J(\cdot)$ is monotonic, the events $Z_J \geq 0$ and $Z \geq 0$ are equivalent. All the analysis in the text continues to apply if eq. $(*)$ is substituted in place of eq. (2.1.2) and the quantities $X\gamma$ and $V$ are replaced everywhere by $- J(- X\gamma)$ and $J(V)$, respectively. Notice that expression (2.1.17) for $M$ obtained by replacing $X\gamma$ by $- J(- X\gamma)$ does not arise by making a change of variables from $V$ to $J(V)$ in performing the integration appearing in (2.1.15). Thus, (2.1.17) does not arise from a Gram–Charlier expansion of the bivariate density for $U$ and nonnormal $V$; instead, it is derived from a Gram–Charlier expansion applied to the bivariate density of $U$ and normal $J(V)$.

An obvious generalization of (2.1.8) or (2.1.16) assumes that

$$E(U|V, X) = \sum_{k=1}^{K} \tau_k g_k(V),$$

(2.1.18)

where the $g_k(\cdot)$'s are known functions. The functional form implied for the selection term is

$$M = E(U|V \geq -X\gamma, X) = \sum_{k=1}^{K} \tau_k E(g_k|V \geq -X\gamma, X)$$

$$\equiv \sum_{k=1}^{K} \tau_k m_k(X).$$

(2.1.19)

Specifying a particular functional form for the $g_k$'s and the marginal distribution for $V$ produces an entire class of sample selection corrections that includes Lee's procedure as a special case.

Cosslett (1984) presents a more robust procedure that can be cast in the format of eq. (2.1.19). With his methods it is possible to consistently estimate the distribution of $V$, the functions $m_k$, the parameters $\tau_k$, and $K$ the number of terms in the expansion. In independent work Gallant and Nychka (1984) present a more robust procedure for correcting models for sample selection bias assuming that the joint density of $(U, V)$ is twice continuously differentiable. Their analysis does not require specifications like (2.1.8), (2.1.14) or (2.1.18) or prior specification of the distribution of $V$.

### 2.1.3.   Multi-state generalizations

Among many possible generalizations of the preceding analysis, one of the most empirically fruitful considers the situation in which the dependent variable $Y$ is generated by a different linear equation for each state of the world. This model includes the "switching regression" model of Quandt (1958, 1972). The occurrence of a particular state of the world results from $Z$ falling into one of the mutually exclusive and exhaustive subsets of $\Theta$, $\Theta_i$, $i = 0, \ldots, I$. The event $Z \in \Theta_i$ signals the occurrence of the $i$th state of the world. We also suppose that $Y$ is observed in states $i = 1, \ldots, I$ and is not observed in state $i = 0$. In state $i > 0$, the equation for $Y$ is

$$Y = X\beta_i + U_i,$$

(2.1.20)

where the $U_i$'s are error terms with $E(U_i) = 0$. Define $U \equiv (U_1,\ldots,U_I)$, and let $f_{uv}(U,V)$ be the joint density of $U$ and the disturbance $V$ of the equation determining $Z$. The value of the discrete dependent variable

$$\delta_i = \begin{cases} 1 & \text{if } Z \in \Theta_i, \\ 0 & \text{otherwise,} \end{cases} \tag{2.1.21}$$

records whether or not state $i$ occurs. In this notation the equation determining the censored version of $Y$ may be written as

$$Y^* = \sum_{i=1}^{I} \delta_i(X\beta_i + U_i), \tag{2.1.22}$$

where we continue to adopt the convention that $Y^* = 0$ when $Y$ is not observed (i.e. when $Z \in \Theta_0$).

It is useful to distinguish two cases of this model. In the first case all states of the world are observed by the analyst, so that the values of the $\delta_i$'s are known by the econometrician for all $i$. In the second case not all of the $\delta_i$'s are known by the econometrician. The analysis of the first case closely parallels the analysis presented for the simple two-state model.

For the first case, the regression function for observed $Y$ given $\delta_i = 1$, $X$, and $i \neq 0$, is

$$E(Y|\delta_i = 1, X) = X\beta_i + M_i, \tag{2.1.23}$$

with

$$M_i \equiv E(U_i|Z \in \Theta_i, X) = \frac{\int_{-\infty}^{\infty} \int_{\Theta_i} u_i f_{u_i v}(u_i, z - X\gamma) \, dz \, du_i}{P_i}, \tag{2.1.24}$$

where $f_{u_i v}(\cdot, \cdot)$ denotes the joint density of $U_i$ and $V$, and $P_i = \text{Prob}(Z \in \Theta_i | X)$ is the probability that state $i$ occurs.

Paralleling the analysis of Section 2.1.2, one can develop explicit specifications for each selection bias correction term $M_i$ by using formulae such as (2.1.8), (2.1.14) or (2.1.18). With the convention that $Y^* = 0$ when $\delta_0 = 1$, the regression functions (2.1.23) can be combined into a single relation

$$E(Y^*|\delta_0, \delta_1,\ldots,\delta_I, X) = \sum_{i=1}^{I} \delta_i(X\beta_i + M_i). \tag{2.1.25}$$

In the second case considered here not all states of the world are observed by the econometrician. It often happens that it is known if $Y$ is observed, and the value of $Y$ is known if it is observed, but it is not known which of a number of possible states has occurred. In such a case, one might observe whether $\delta_0 = 1$ or $\delta_0 = 0$ (i.e. whether $\sum_{i=1}^{I}\delta_i = 0$ or $\sum_{i=1}^{I}\delta_i = 1$), but not individual values of the $\delta_i$'s for $i = 1,\ldots, I$. Examples of such situations are given in our discussion of labor supply presented in Section 3.3.

To determine the appropriate regression equation for $Y$ in this second case, it is necessary to compute the expected value of $Y$ given by (2.1.22) conditional on $\delta_0 = 0$ and $X$. This expectation is

$$E(Y|\delta_0 = 0, X) = \sum_{i=1}^{I} (X\beta_i + M_i)P_i/(1 - P_0), \tag{2.1.26}$$

where $P_i = \text{Prob}(Z \in \Theta_i|X)$.[13] Relation (2.1.26) is the regression of $Y$ on $X$ for the case in which $Y$ is observed but the particular state occupied by an observation is not observed.

Using (2.1.22), and recalling that $Y^* = Y(1 - \delta_0)$ is a censored random variable, the regression of $Y^*$ on $X$ is

$$E(Y^*|X) = \sum_{i=1}^{I} (X\beta_i + M_i)P_i. \tag{2.1.27}$$

If $Y$ is observed for all states of the world, then $Y^* = Y$, $\delta_0 \equiv 0$, and (2.1.26) and (2.1.27) are identical because the set $\Theta_0$ is the null set so that $P_0 = 0$ and $\sum_{i=1}^{I}P_i = 1$.i

### 2.1.4. Generalization of the regression framework

Extensions of the basic framework presented above provide a rich structure for analyzing a wide variety of problems in labor econometrics. We briefly consider three useful generalizations.

The first relaxes the linearity assumption maintained in the specification of the equations determining the dependent variables $Y$ and $Z$. In eqs. (2.1.1) and (2.1.2) substitute $h_Y(X, \beta)$ for $X\beta$ and $h_Z(X, \gamma)$ for $X\gamma$ where $h_Y(\cdot, \cdot)$ and

[13] In order to obtain (2.1.26) we use the fact that the $\Theta_i$'s are nonintersecting sets so that

$$E\left(\delta_i \middle| \sum_{j=1}^{I}\delta_j = 1, X\right) = \text{Prob}\left(\delta_i = 1 \middle| \sum_{j=1}^{I}\delta_j = 1, X\right)$$

$$= \text{Prob}\left(Z \in \Theta_i \middle| Z \in \bigcup_{j=1}^{I}\Theta_j, X\right) = P_i \middle/ \sum_{j=1}^{I}P_j = P_i/(1 - P_0).$$

$h_Z(\cdot, \cdot)$ are known nonlinear functions of exogenous variables and parameters. Modifying the preceding analysis and formulae to accommodate this change in specification only requires replacing the quantities $X\beta$ and $X\gamma$ everywhere by the functions $h_Y$ and $h_Z$. A completely analogous modification of the multi-state model introduces nonlinear specifications for the conditional expectation of $Y$ in the various states.

A second generalization extends the preceding framework of Sections 2.1.1–2.1.3 by interpreting $Y$, $Z$ and the errors $U$ and $V$ as vectors. This extension enables the analyst to consider a multiplicity of behavioral functions as well as a broad range of sampling rules. No conceptual problems are raised by this generalization but severe computational problems must be faced. Now the sets $\Theta_i$ are multidimensional. Tallis (1961) derives the conditional means relevant for the linear multivariate normal model, but it remains a challenge to find other multivariate specifications that yield tractable analytical results. Moreover, work on estimating the multivariate normal model has just begun [e.g. see Catsiapsis and Robinson (1982)]. A current area of research is the development of computationally tractable specifications for the means of the disturbance vector $U$ conditional on the occurrence of alternative states of the world.

A third generalization allows the sample selection rule to depend directly on realized values of $Y$. For this case, the sets $\Theta_i$ are replaced by the sets $\Omega_i$ where $(Y, Z) \in \Omega_i$ designates the occupation of state $i$. The integrals in the preceding formulae are now defined over the $\Omega_i$. In place of the expression for the selection term $M$ in (2.1.7), use the more general formula

$$E(U|(Y, Z) \in \Omega_1, X) = \frac{\int_{\Omega_1} (y - X\beta) f_{uv}(y - X\beta, z - X\gamma) \, dz \, dy}{P_1},$$

where

$$P_1 = \int_{\Omega_1} f_{uv}(y - X\beta, z - X\gamma) \, dz \, dy,$$

is the probability that $\delta_1 = 1$ given $X$. This formula specializes to the expression (2.1.7) for $M$ when $\Omega_1 = \{(Y, Z): -\infty \leq Y \leq \infty \text{ and } Z \in \Theta_1\}$, i.e. when $Z$ alone determines whether state 1 occurs.

### 2.1.5. Methods for estimating the regression specifications

We next consider estimating the regression specifications associated with the elementary two-state model (2.1.1) and (2.1.2). This simple specification is by far the most widely used model encountered in the literature. Estimation procedures available for this two-state model can be directly generalized to more complicated models.

For the two-state model, expression (2.1.3) implies that the regression equation for $Y$ conditional on $X$ and $\delta = 1$ is given by

$$Y = X\beta + M + e,$$

where $e \equiv U - E(U|\delta = 1, X)$ is a disturbance with $E(e|\delta = 1, X) = 0$. Choosing specification (2.1.9), (2.1.17) or one based on (2.1.19) for the selection term $M$ leads to

$$M = m\tau \quad \text{with} \quad m = m(X\gamma, \psi), \tag{2.1.28}$$

where the $\psi$ are unknown parameters of the density function for $V$. If, for example, specification (2.1.9) is chosen, $m(X\gamma, \psi) = E(V|V \geq -X\gamma)$ which can be any one of the truncated mean formulae presented in Table 1. If, on the other hand, specification (2.1.19) is chosen, $\tau$ and $m$ are to be interpreted as vectors with $\tau' = (\tau_1, \dots, \tau_K)$ and $m = (m_1, \dots, m_K)$. The regression equation for $Y$ is

$$Y = X\beta + m\tau + e. \tag{2.1.29}$$

The implied regression equation for the censored dependent variable $Y^* = \delta Y$ is

$$Y^* = (X\beta + m\tau)(1 - F_v(-X\gamma; \psi)) + \varepsilon, \tag{2.1.30}$$

where $\varepsilon$ is a disturbance with $E(\varepsilon|X) = 0$ and we now make explicit the dependence of $F_v$ on $\psi$.

The appropriate procedure for estimating the parameters of regression eqs. (2.1.29) and (2.1.30) depends on the sampling plan that generates the available data. It is important to distinguish between two types of samples discussed in Section 1: truncated samples which include data on $Y$ and $X$ only for observations for which the value of the dependent variable $Y$ is actually known (i.e. where $Z \geq 0$ for the model under consideration here), and censored samples which include data on $Y^*$ and $X$ from a simple random sample of $\delta$, $X$ and $Y^*$.

For a truncated sample, nonlinear least squares applied to regression eq. (2.1.29) can be used to estimate the coefficients of $\beta$ and $\tau$ and the parameters $\gamma$ and $\psi$ which enter this equation through the function $m$. More specifically, defining the function $g$ and the parameter vector $\theta$ as $g(X, \theta) \equiv X\beta + m(X\gamma, \psi)\tau$ and $\theta' = (\beta', \tau', \gamma', \psi')$, eq. (2.1.29) can be written as

$$Y = g(X, \theta) + e. \tag{2.1.31}$$

Since the disturbance $e$ has a zero mean conditional on $X$ and $\delta = 1$ and is distributed independently across the observations in the truncated sample, under standard conditions [see Amemiya (1985)] nonlinear least squares estimators of the parameters of this equation are both consistent and asymptotically normally distributed.

In general, the disturbance $e$ is heteroscedastic, and the functional form of the heteroscedasticity is unknown unless the joint density $f_{uv}$ is specified. As a consequence, when calculating the large-sample covariance matrix of $\hat{\theta}$, it is necessary to use methods proposed by Eicker (1963, 1967) and White (1981) to consistently estimate this covariance matrix in the presence of arbitrary heteroscedasticity. The literature demonstrates that the estimator $\hat{\theta}$ is approximately normally distributed in large samples with the true value $\theta$ as its mean and a variance–covariance matrix given by $H^{-1}RH^{-1}$ with

$$
H = \sum_{n=1}^{N} \frac{\partial g_n}{\partial \theta}\bigg|_{\hat{\theta}} \frac{\partial g_n}{\partial \theta'}\bigg|_{\hat{\theta}} \quad \text{and} \quad R = \sum_{n=1}^{N} \frac{\partial g_n}{\partial \theta}\bigg|_{\hat{\theta}} \frac{\partial g_n}{\partial \theta'}\bigg|_{\hat{\theta}} \hat{e}_n^2, \tag{2.1.32}
$$

where $N$ is the size of the truncated sample, $\partial g_n / \partial \theta|_{\hat{\theta}}$ denotes the gradient vector of $g$ for the $n$th observation evaluated at $\hat{\theta}$, and $\hat{e}_n$ symbolizes the least square residual for observation $n$. Thus

$$
\hat{\theta} \sim N(\theta, H^{-1}RH^{-1}). \tag{2.1.33}
$$

For censored samples, two regression methods are available for estimating the parameters $\beta$, $\tau$, $\gamma$, and $\psi$. First, one can apply the nonlinear least squares procedure just described to estimate regression eq. (2.1.30). In particular, reinterpreting the function $g$ as $g(X, \theta) = [X\beta + m(X\gamma, \psi)\tau](1 - F_v(-X\gamma; \psi))$, it is straightforward to write eq. (2.1.30) in the form of an equation analogous to (2.1.31) with $Y^*$ and $\varepsilon$ replacing $Y$ and $e$. Since the disturbance $\varepsilon$ has a zero mean conditional on $X$ and is distributed independently across the observations making up the censored sample, under standard regularity conditions nonlinear least squares applied to this equation yields a consistent estimator $\hat{\theta}$ with a large-sample normal distribution. To account for potential heteroscedasticity compute the asymptotic variance–covariance matrix of $\hat{\theta}$ using the formula in (2.1.33) with the matrices $H$ and $R$ calculated by summing over the $N^*$ observations of the censored sample.

A second type of regression procedure can be implemented on censored samples. A two-step procedure can be applied to estimate the equation for $Y$ given by (2.1.29). In the first step, obtain consistent estimates of the parameters $\gamma$ and $\psi$ from a discrete choice analysis which estimates the parameters of $P_1$. From these estimates it is possible to consistently estimate $m$ (or the variables in the vector $m$). More specifically, define $\theta_2' \equiv (\gamma', \psi')$ as a parameter vector which uniquely determines $m$ as a function of $X$. The log likelihood function for the independently distributed discrete variables $\delta_n$ given $X_n$, $n = 1, \ldots, N^*$ is

$$
\sum_{n=1}^{N^*} \left[ \delta_n \ln(1 - F_v(-X_n\gamma; \psi)) + (1 - \delta_n)\ln(F_v(-X_n\gamma; \psi)) \right]. \tag{2.1.34}
$$

Under general conditions [See Amemiya (1985) for one statement of these conditions], maximum likelihood estimators of $\gamma$ and $\psi$ are consistent, and with maximum likelihood estimates $\tilde{\theta}_2$ one can construct $\tilde{m}_n = m(X_n \tilde{\gamma}, \tilde{\psi})$ for each observation. In step two of the proposed estimation procedure, replace the unobserved variable $m$ in regression eq. (2.1.29) by its constructed counterpart $\tilde{m}$ and apply linear least-squares to the resulting equation using only data from the subsample in which $Y$ and $X$ are observed. Provided that the model is identified, the second step produces estimators for the parameters $\theta_1' = (\beta', \tau')$ that are both consistent and asymptotically normally distributed.

When calculating the appropriate large-sample covariance matrix for least squares estimator $\tilde{\theta}_1$, one must account for the fact that in general the disturbances of the regression equation are heteroscedastic and that the variables $\tilde{m}$ are estimated quantities. A consistent estimator for the covariance matrix which accounts for both of these features is given by

$$C = Q_1^{-1} Q_2 Q_1^{-1} + Q_1^{-1} Q_3 Q_4 Q_3' Q_1^{-1}, \tag{2.1.35}$$

where $Q_4$ is the covariance matrix for $\tilde{\theta}_2$ estimated by maximum likelihood [minus the inverse of the Hessian matrix of (2.1.34)], and the matrices $Q_1$, $Q_2$, and $Q_3$ are defined by

$$Q_1 = \sum_{n=1}^{N} w_n w_n', \qquad Q_2 = \sum_{n=1}^{N} w_n w_n' \tilde{e}_n^2, \qquad \text{and} \qquad Q_3 = \sum_{n=1}^{N} w_n \frac{\partial e_n}{\partial \theta_2'} \bigg|_{\tilde{\theta}}, \tag{2.1.36}$$

where the row vector $w_n \equiv (X_n, \tilde{m}_n)'$ denotes the regressors for the $n$th observation, the variable $\tilde{e}_n$ symbolizes the least-squares residual, and the row vector $\partial e_n / \partial \theta_2' |_{\tilde{\theta}}$ is the gradient of the function $e_n = Y_n - X_n \beta - m_n \alpha$ with respect to $\gamma$ and $\psi$ evaluated at the maximum likelihood estimates $\tilde{\gamma}$ and $\tilde{\psi}$ and at the least squares estimates $\tilde{\beta}$ and $\tilde{\tau}$ – i.e.

$$\frac{\partial e_n}{\partial \theta_2'} \bigg|_{\tilde{\theta}} = -\frac{\partial m_n \tau}{\partial \theta_2'} \bigg|_{\tilde{\theta}} = -\tilde{\tau}' \frac{\partial m_n'}{\partial \theta_2'} \bigg|_{\tilde{\theta}_2}.^{14}$$

--------

[14] To derive the expression for the matrix $C$ given by (2.1.35), we use the following result. Let $L_n = L(\delta_n, X_n)$ denote the $n$th observation on the gradient of the likelihood function (2.1.34) with respect to $\theta_2$, with this gradient viewed as a function of the data and the true value of $\theta_2$; and let $w_{on}$ and $e_{on}$ be $w_n$ and $e_n$ evaluated at the true parameter values. Then $E(w_{on} e_{on} L_n' | \delta_n = 1, X_n) = w_{on} E(e_{on} | \delta_n = 1, X_n) L_n'(\delta_n = 1, X_n) = 0$.

The large-sample distribution for the two-step estimator is thus

$$\tilde{\theta}_1 \sim N(\theta_1, C).$$

$$(2.1.38)$$

## 2.2. Dummy endogenous variable models

One specialization of the general model presented in Section 2.1 is of special importance in labor economics. The multi-state equation system $(2.1.20)–(2.1.22)$ is at the heart of a variety of models of the impact of unions, training, occupational choice, schooling, the choice of region of residence and the choice of industry on wages. These models have attracted considerable attention in the recent literature.

This section considers certain aspects of model formulation for this class of models. Simple consistent estimators are presented for an empirically interesting subclass of these models. These estimators require fewer assumptions than are required for distribution dependent maximum likelihood methods or for the sample selection bias corrections ($M$ functions) discussed in Section 2.1.

In order to focus on essential ideas, we consider a two-equation, two-state model with a single scalar dummy right-hand side variable that can assume two values. $Y$ is assumed to be observed in both states so that we also abstract from censoring. Generalization of this model to the vector case is performed in Heckman (1976a, 1978, Appendix), Schmidt (1981), and Lee (1981).

### 2.2.1. Specification of a two-equation system

Two versions of the dummy endogenous variable model are commonly confused in the literature: fixed coefficient models and random coefficient models. These specifications should be carefully distinguished because different assumptions are required to consistently estimate the parameters of these two distinct models. The fixed coefficient model requires fewer assumptions.

In the fixed coefficient model

$$Y = X\beta + \delta\alpha + U,$$

$$(2.2.1)$$

$$Z = X\gamma + V,$$

$$(2.2.2)$$

where

$$\delta = \begin{cases} 1 & \text{if } Z \geq 0, \\ 0 & \text{otherwise,} \end{cases}$$

$U$ and $V$ are mean zero random disturbances, and $X$ is exogenous with respect to $U$. Simultaneous equation bias is present in (2.2.1) when $U$ is correlated with $\delta$.

In the random coefficient model the effect of $\delta$ on $Y$ (holding $U$ fixed) varies in the population. In place of (2.2.1) we write

$$Y = X\beta + \delta(\alpha + \varepsilon) + U, \tag{2.2.3}$$

where $\varepsilon$ is a mean zero error term.[15] Equation (2.2.2) is unchanged except now $V$ may be correlated with $\varepsilon$ as well as $U$. The response to $\delta = 1$ differs in the population, with successively sampled observations assumed to be random draws from a common distribution for $(U, \varepsilon, V)$. In this model $X$ is assumed to be exogenous with respect to $(U, \varepsilon)$. Regrouping terms, specification (2.2.3) may be rewritten as

$$Y = X\beta + \delta\alpha + (U + \delta\varepsilon). \tag{2.2.4}$$

Unless $\delta$ is uncorrelated with $\varepsilon$ (which occurs in some interesting economic models – see Section 3.2), the expectation of the composite error term $U + \varepsilon\delta$ in (2.2.4) is nonzero because $E(\delta\varepsilon) \neq 0$. This aspect of the random coefficient model makes its econometric analysis fundamentally different from the econometric analysis of the fixed coefficient model. Simultaneous equations bias is present in the random coefficient model if the composite error term in (2.2.4) is correlated with $\delta$.

Both the random coefficient model and the fixed coefficient model are special cases of the multi-state "switching" model presented in Section 2.1.3. Rewriting random coefficient specification (2.2.3) as

$$Y = \delta(\alpha + X\beta + U + \varepsilon) + (1 - \delta)(X\beta + U), \tag{2.2.5}$$

this equation is of the form of multi-state eq. (2.1.22). The equivalence of (2.2.5) and (2.1.22) follows directly from specializing the multi-state framework so that: (i) $\delta_0 \equiv 0$ (so that there is no censoring and $Y = Y^*$); (ii) $I = 2$ (which along with (i) implies that there are two states); (iii) $\delta = 1$ indicates the occurrence of state 1 and the events $\delta_1 = 1$ and $\delta_2 = 0$ (with $1 - \delta = 1$ indicating the realization of state 2); and (iv) $X\beta_2 = X\beta$, $U_1 = U$, $X\beta_1 = X\beta + \alpha$, and $U_2 = U + \varepsilon$. In this notation eq. (2.2.3) may be written as

$$Y = X\beta_2 + \delta X(\beta_1 - \beta_2) + U_1 + (U_1 - U_2)\delta. \tag{2.2.6}$$

One empirically fruitful generalization of this model relaxes (iv) by letting both slope and intercept coefficients differ in the two regimes. Equation (2.2.6) with

---

[15] Individuals may or may not know their own value of $\varepsilon$. "Randomness" as used here refers to the econometrician's ignorance of $\varepsilon$.

condition (iv) modified so that $\beta_1$ and $\beta_2$ are freely specified can also be used to represent this generalization.

Fixed coefficient specification (2.2.1) specializes the random coefficient model further by setting $\varepsilon = 0$ so $U_1 - U_2 = 0$ in (2.2.6). In the fixed coefficient model, $U_1 \equiv U_2$ so that the unobservables in the state specific eqs. (2.1.20) are identical in each state. Examples of economic models which produce this specification are given below in Section 3.2.

The random coefficient and the fixed coefficient models are sometimes confused in the literature. For example, recent research on the union effects on wage rates has been unclear about the distinction [e.g. see Freeman (1984)]. Many of the cross section estimates of the union impact on wage rates have been produced from the random coefficient model [e.g. see Lee (1978)] whereas most of the recent longitudinal estimates are based on a fixed coefficient model, or a model that can be transformed into that format [e.g. see Chamberlain (1982)]. Estimates from these two data sources are not directly comparable because they are based on different model specifications.[16]

Before we consider methods for estimating both models, we mention one aspect of model formulation that has led to considerable confusion in the recent literature. Consider an extension of equation system (2.2.1)–(2.2.2) in which dummy variables appear on the right-hand side of each equation

$$Y = X\beta + \alpha_1\delta_2 + U, \qquad (2.2.7a)$$

$$Z = X\gamma + \alpha_2\delta_1 + V, \qquad (2.2.7b)$$

where

$$\delta_1 = \begin{cases} 1 & \text{if } Y \geq 0, \\ 0 & \text{otherwise,} \end{cases}$$

and

$$\delta_2 = \begin{cases} 1 & \text{if } Z \geq 0, \\ 0 & \text{otherwise.} \end{cases}$$

Without imposing further restrictions on the support of the random variables $(U, V)$, this model makes no statistical sense unless

$$\alpha_1\alpha_2 = 0. \qquad (2.2.8)$$

[See Heckman (1978) or Schmidt (1981)]. This assumption – termed the "principal

---

[16] For further discussion of this point, see Heckman and Robb (1985).

assumption" in the literature – rules out contradictions such as the possibility that $Y \geq 0$ but $\delta_1 = 0$, or other such contradictions between the signs of the elements of $(Y, Z)$ and the values assumed by the elements of $(\delta_1, \delta_2)$.

The principal assumption is a logical requirement that any well-formulated behavioral model must satisfy. An apparent source of confusion on this point arises from interpreting (2.2.7) as well-specified *behavioral* relationships. In the absence of a precise specification determining the behavioral content of (2.2.7), it is incomplete. The principal assumption forces the analyst to estimate a well-specified behavioral and statistical model. This point is developed in the context of a closely related model in an appendix to this paper.

### 2.2.2. Estimation of the fixed coefficient model

In this subsection we consider methods for consistently estimating the fixed coefficient dummy endogenous variable model and examine the identifiability assumptions that must be invoked in order to recover the parameters of this model. We do not discuss estimation of discrete choice eq. (2.2.2) and we focus solely on estimating (2.2.1). An attractive feature of some of the estimators discussed below is that the parameters of (2.2.1) can be identified even when no regressor appears in (2.2.2) or when the conditions required to define (2.2.2) as a conventional discrete choice model are not satisfied. It is sometimes possible to decouple the estimation of these two equations.

### 2.2.2.1. Instrumental variable estimation.

Equation (2.2.1) is a standard linear simultaneous equation with $\delta$ as an endogenous variable. A simple method for estimating the parameters of this equation is a conventional instrumental variable procedure. Since $E(U|X) = 0$, $X$ and functions of $X$ are valid instrumental variables. If there is at least one variable in $X$ with a nonzero $\gamma$ coefficient in (2.2.2) such that the variable (or some known transformation of it) is linearly independent of $X$ included in (2.2.1), then this variable (or its transformation) can be used as an instrumental variable for $\delta$ in the estimation of (2.2.1).

These conditions for identification are very weak. The functional forms of the distributions of $U$ or $V$ need not be specified. The variables $X$ (or more precisely $X\gamma$) need not be distributed independently of $V$ so that (2.2.2) is not required to be a well-specified discrete choice model.

If (2.2.2) is a well-specified discrete choice model, then the elements of $X$ and a consistent estimator of $E(\delta|X) = P(\delta = 1|X)$ constitute an optimal choice for the instrumental variables according to well-known results in the analysis of nonlinear two-stage least squares [e.g. see Amemiya (1985, Chapter 8)]. Choosing $X$ and simple polynomials in $X$ as instruments can often achieve comparable asymptotic efficiency. Conventional formulae for the sampling error of instrumental variable estimators fully apply in this context.

*2.2.2.2. Conditioning on X.* The $M$ function regression estimators presented in Section 2.1 are based on the conditional expectation of $Y$ given $X$ and $\delta$. It is often possible to consistently estimate the parameters of (2.2.1) using the conditional expectation of $Y$ given only $X$.

From the specification of (2.2.1), we have

$$E(Y|X) = X\beta + \alpha E(\delta|X). \qquad (2.2.9)$$

Notice that (if $X$ is distributed independently of $V$)

$$E(\delta|X) = 1 - F_v(-X\gamma). \qquad (2.2.10)$$

Given knowledge of the functional form of $F_v$, one can estimate (2.2.9) by nonlinear least squares. The standard errors for this procedure are given by (2.1.32) and (2.1.33) where $g_n$ in these formulae is defined as $g_n = Y_n - X_n\beta - \alpha(1 - F_v(-X_n\gamma))$.

One benefit of this direct estimation procedure is that the estimator is consistent even if $\delta$ is measured with error because measurements on $\delta$ are never directly used in the estimation procedure. Notice that the procedure requires specification of the distribution of $V$ (or at least its estimation). Specification of the distribution of $U$ or the joint distribution of $U$ and $V$ is not required.

*2.2.2.3. Invoking a distributional assumption about U.* The coefficients of (2.2.1) can be identified if some assumptions are made about the distribution of $U$. No assumption need be made about the distribution of $V$ or its stochastic dependence with $U$. It is not required to precisely specify discrete choice eq. (2.2.2) or to use nonlinearities or exclusion restrictions involving exogenous variables which are utilized in the two estimation strategies just presented. No exogenous variables need appear in either equation.

If $U$ is normal, $\alpha$ and $\beta$ are identified given standard rank conditions even if no regressor appears in the index function equation determining the dummy variable (2.2.2). Heckman and Robb (1985) establish that if $E(U^3) = E(U^5) = 0$, which is implied by, but weaker than, assuming symmetry or normality of $U$, $\alpha$ and $\beta$ are identified even if no regressor appears in the index function (2.2.2). It is thus possible to estimate (2.2.1) without a regressor in the index function equation determining $\delta$ or without making any assumption about the marginal distribution of $V$ provided that stronger assumptions are maintained about the marginal distribution of $U$.

In order to see how identification is secured in this case, consider a simplified version of (2.2.1) with only an intercept and dummy variable $\delta$

$$Y = \beta_0 + \delta\alpha + U. \qquad (2.2.11)$$

Assume $E(U^3) = 0 = E(U^5)$. With observations indexed by $n$, the method of moments estimator solves for $\hat{\alpha}$ from the pair of moment equations that equate sample moments to their population values:

$$\frac{1}{N} \sum_{n=1}^{N} [(Y_n - \bar{Y}) - \hat{\alpha}(\delta_n - \bar{\delta})]^3 = 0, \qquad (2.2.12a)$$

and

$$\frac{1}{N} \sum_{n=1}^{N} [(Y_n - \bar{Y}) - \hat{\alpha}(\delta_n - \bar{\delta})]^5 = 0, \qquad (2.2.12b)$$

where $\bar{Y}$ and $\bar{\delta}$ are sample means of $Y$ and $\delta$ respectively. There is only one consistent root that satisfies both equations. The inconsistent roots of (2.2.12a) do not converge to the inconsistent roots of (2.2.12b). Choosing a value of $\alpha$ to minimize a suitably weighted sum of squared discrepancies from (2.2.12a) and (2.2.12b) (or choosing any other metric) solves the small sample problem that for any finite $N$ (2.2.12a) and (2.2.12b) cannot be simultaneously satisfied. For proof of these assertions and discussion of alternative moment conditions on $U$ to secure identification of the fixed coefficient model, see Heckman and Robb (1985).

### 2.2.3.  *Estimation of the random coefficient model*

Many of the robust consistent estimators for the fixed coefficient model are inconsistent when applied to estimate $\alpha$ in the random coefficient model.[17] The reason this is so is that in general the composite error term of (2.2.4) does not possess a zero conditional (on $X$) or unconditional mean. More precisely, $E(\delta\varepsilon|X) \neq 0$ and $E(\delta\varepsilon) \neq 0$ even though $E(U|X) = 0$ and $E(U) = 0$.[18] The instrumental variable estimator of Section 2.2.2.1 is inconsistent because $E(U + \delta\varepsilon|X) \neq 0$ and so $X$ and functions of $X$ are not valid instruments. The nonlinear least squares estimator of Section 2.2.2.2 that conditions on $X$ is also in general inconsistent. Instead of (2.2.9), the conditional expectation of $Y$ given $X$ for eq. (2.2.4) is

$$E(Y|X) = X\beta + \alpha E(\delta|X) + E(\delta\varepsilon|X). \qquad (2.2.13)$$

[17] In certain problems the coefficient of interest is $\alpha + E(\varepsilon|\delta = 1)$. Reparameterizing (2.2.4) to make this rather than $\alpha$ as the parameter of econometric interest effectively converts the random coefficient model back into a fixed coefficient model when no regressors appear in index function (2.2.2).

[18] However, some of the models presented in Section 3.2 have a zero unconditional mean for $\delta\varepsilon$. This can occur when $\varepsilon$ is unknown at the time an agent makes decisions about $\delta$.

Inconsistency of the nonlinear least squares estimator arises because the unobserved omitted term $E(\delta \varepsilon | X)$ is correlated with the regressors in eq. (2.2.9).

*2.2.3.1. Selectivity corrected regression estimators.* The analysis of Section 2.1.5 provides two regression methods for estimating the parameters of a random coefficient model. From eq. (2.2.6), a general specification of this model is

$$Y = \delta(X\beta_1 + U_1) + (1 - \delta)(X\beta_2 + U_2). \tag{2.2.14}$$

Relation (2.1.25) for the multi-state model of Section 2.1 implies that the regression equation for $Y$ on $\delta$ and $X$ is

$$Y = \delta(X\beta_1 + M_1) + (1 - \delta)(X\beta_2 + M_2) + e, \tag{2.2.15}$$

where $M_1 \equiv E(U_1 | \delta = 1, X)$, $M_2 \equiv E(U_2 | \delta = 0, X)$, and $e = \delta(U_1 - M_1) + (1 - \delta)(U_2 - M_2)$. Using selection specification (2.1.28),

$$M_i = m_i \tau_i \qquad \text{where} \qquad m_i \equiv m_i(X\gamma, \psi), \qquad i = 1, 2, \tag{2.2.16}$$

where the functional forms of the elements of the row vectors $m_1$ and $m_2$ depend on the particular specification chosen from Section 2.1.2.[19] Substituting (2.2.16) into (2.2.15), the regression equation for $Y$ becomes

$$Y = X_1^* \beta_1 + X_2^* \beta_2 + m_1^* \tau_1 + m_2^* \tau_2 + e, \tag{2.2.17}$$

where

$$X_1^* \equiv \delta X, \quad X_2^* \equiv (1 - \delta)X, \quad m_1^* \equiv \delta m_1 \qquad \text{and} \qquad m_2^* \equiv (1 - \delta)m_2.$$

Given familiar regularity conditions, the nonlinear least-squares estimator of (2.2.17) is consistent and approximately distributed according to the large-sample normal distribution given by (2.1.33), where the matrices $H$ and $R$ are defined by (2.1.32) with $g_n$ in these formulae given by

$$g_n = X_{1n}^* \beta_1 + X_{2n}^* \beta_2 + \delta m_1(X_n \gamma, \psi)\tau_1 + (1 - \delta)m_2(X_n \gamma, \psi)\tau_2.$$

A second approach adapts the two-step estimation scheme outlined in Section 2.1.5. Using maximum likelihood estimates $\tilde{\theta}_2$ of the parameter vector $\theta_2 \equiv (\gamma', \psi')'$, construct estimates of $\tilde{m}_{in} = m_i(X_n \tilde{\gamma}, \tilde{\psi})$, $i = 1, 2$, for each observation.

---

[19] Inspection of eq. (2.2.2) and the process generating $\delta$ reveals that the events $\delta = 1$ and $\delta = 0$ correspond to the conditions $V \geq -X\gamma$ and $-V > X\gamma$; and, consequently, the functions $M_1$ and $M_2$ have forms completely analogous to the selection correction $M$ whose specification is the topic of Section 2.1.2.

Replacing unobserved $m_1$ and $m_2$ in (2.2.17) by their observed counterparts $\tilde{m}_1$ and $\tilde{m}_2$, the application of linear least-squares to the resulting equation yields an estimate $\tilde{\theta}_1$ of the parameter vector $\theta_1 \equiv (\beta_1', \beta_2', \tau_1', \tau_2')'$. Given standard assumptions, the estimator $\tilde{\theta}_1$ is consistent and approximately normally distributed in large samples. The covariance matrix $C$ in (2.1.38) in this case is given by (2.1.35), and the matrices $Q_1$, $Q_2$ and $Q_3$ are as defined by (2.1.36) with $w_n = (X_{1n}^*, X_{2n}^*, \tilde{m}_{1n}^*, \tilde{m}_{2n}^*)'$ and where

$$\left.\frac{\partial e_n}{\partial \theta_2'}\right|_{\tilde{\theta}} = -\delta\tilde{\tau}_1' \left.\frac{\partial m_{1n}'}{\partial \theta_2'}\right|_{\tilde{\theta}} - (1-\delta)\tilde{\tau}_2' \left.\frac{\partial m_{2n}'}{\partial \theta_2'}\right|_{\tilde{\theta}_2}. \tag{2.2.18}$$

## 3. Applications of the index function model

This section applies the index function framework to specific problems in labor economics. These applications give economic content to the statistical framework presented above and demonstrate that a wide range of behavioral models can be represented as index function models.

Three prototypical models are considered. We first present models with a "reservation wage" property. In a variety of models for the analysis of unemployment, job turnover and labor force participation, an agent's decision process can be characterized by the rule "stay in the current state until an offered wage exceeds a reservation wage." The second prototype we consider is a dummy endogenous variable model that has been used to estimate the impact of schooling, training, occupational choice, migration, unionism and job turnover on wages. The third model we discuss is one for labor force participation and hours of work in the presence of taxes and fixed costs of work.

### 3.1. Models with the reservation wage property

Many models possess a reservation wage property, including models for the analysis of unemployment spells [e.g. Kiefer and Neumann, (1979), Yoon (1981, 1984), Flinn and Heckman (1982)], for labor force participation episodes [e.g. Heckman and Willis (1977); Heckman (1981), Heckman and MaCurdy (1980), Killingsworth (1983)], for job histories [e.g. Johnson (1978), Jovanovic (1979), Miller (1984), Flinn (1984)] and for fertility and labor supply [Moffit (1984), Hotz and Miller (1984)]. Agents continue in a state until an opportunity arises (e.g. an offered wage) that exceeds the reservation wage for leaving the state currently occupied. The index function framework has been used to formulate and estimate such models.

### 3.1.1. A model of labor force participation

Agents at age $t$ are assumed to possess a quasiconcave twice differentiable one period utility function defined over goods ($C(t)$) and leisure ($L(t)$). Denote this utility function by $U(C(t), L(t))$. We define leisure hours so that $0 \leq L(t) \leq 1$. An agent is assumed to be able to freely choose his hours of work at a parametric wage $W(t)$. There are no fixed costs of work or taxes. At each age agents receive unearned income, $R(t)$, assumed to be nonnegative. Furthermore, to simplify the exposition, we assume that there is no saving or borrowing, and decisions are taken in an environment of certainty. Labor force participation models without lending and borrowing constraints have been estimated by Heckman and MaCurdy (1980) and Moffitt (1984).

In the simple model considered here, an agent does not work if his or her reservation wage or value of time at home (the marginal rate of substitution between goods and leisure evaluated at the no work position) exceeds the market wage $W(t)$. The reservation wage in the absence of savings is

$$W_R(t) = U_2(R(t), 1) / U_1(R(t), 1),$$

where $U_1(\cdot)$ and $U_2(\cdot)$ denote partial derivatives. The market wage $W(t)$ is assumed to be known to the agent but it is observed by the econometrician only if the agent works.

In terms of the index function apparatus presented in Section 1,

$$Z(t) = W(t) - W_R(t). \tag{3.1.1}$$

If $Z(t) \geq 0$ the agent works, $\delta(t) = 1$, and the wage rate $W(t)$ is observed. Thus the observed wage is a censored random variable

$$Y^*(t) = W(t)\delta(t). \tag{3.1.2}$$

The analysis of Sections 1 and 2 can be directly applied to formulate likelihood functions for this model and to estimate its parameters.

For comparison with other economic models possessing the reservation wage property, it is useful to consider the implications of this simple labor force participation model for the duration of nonemployment. A nonworking spell begins at $t_1$ and ends at $t_2$ provided that $Z(t_1 - 1) > 0$, $Z(t_1 + j) \leq 0$, for $j = 0, \ldots, t_2 - t_1$ and $Z(t_2 + 1) > 0$. Reversing their direction, these inequalities also characterize an employment spell that begins at $t_1$ and ends at $t_2$. Assuming that unobservables in the model are distributed independently of each other in different time periods, the (conditional) probability that a spell that begins at $t_1$ lasts $t_2 - t_1 + 1$ periods is

$$\prod_{t=t_1}^{t_2} \Pr(Z(t) \le 0) \cdot \Pr(Z(t_2 + 1) > 0). \tag{3.1.3}$$

Precisely the same sort of specification arises in econometric models of search unemployment.

As a specific example of a deterministic model of labor force participation, assume that

$$U(C(t), L(t)) = C^\alpha(t) + A(t) L^\gamma(t).$$

Setting $A(t) = \exp\{X(t)\beta_1 + e(t)\}$ where $e(t)$ is a mean zero disturbance, the reservation wage is

$$\ln W_R(t) = X(t)\beta_1 + \ln(\gamma/\alpha) + (1 - \alpha)\ln R(t) + e(t).$$

The equation for log wage rates can be written as

$$\ln W(t) = X(t)\beta_2 + U(t).$$

Define an index function for this example as $Z(t) = \ln W(t) - \ln W_R(t)$, so that

$$Z(t) = X(t)(\beta_2 - \beta_1) - \ln(\gamma/\alpha) - (1 - \alpha)\ln R(t) + V(t),$$

where $V(t) = U(t) - e(t)$. Define another index function $Y$ as

$$Y(t) = \ln W(t) = X(t)\beta_2 + U(t),$$

and a censored random variable $Y^*(t)$ by

$$Y^*(t) \equiv Y(t)\delta(t) = \delta(t)X(t)\beta_2 + \delta(t)U(t).$$

Assuming that $(X(t), R(t))$ is distributed independently of $V(t)$, and letting $\sigma_v^2 = \mathrm{Var}(V(t))$, the conditional probability that $\delta(t) = 1$ given $X(t)$ and $R(t)$ is

$$\Pr(\delta(t) = 1 | X(t), R(t)) = 1 - G_v\left(\frac{X(t)(\beta_1 - \beta_2) + \ln \gamma/\alpha + (1 - \alpha)\ln R(t)}{\sigma_v}\right),$$

where $G_v$ is the c.d.f. of $V(t)/\sigma_v$. If $V(t)$ is distributed independently across all $t$, the probability that a spell of employment begins at $t = t_1$ and ends at $t = t_2$ conditional on $t_1$ is

$$\left[ \prod_{t=t_1}^{t=t_2} \Pr(\delta(t) = 1 | X(t), R(t)) \right] \Pr(\delta(t_2 + 1) = 0 | X(t_2), R(t_2)).$$

Assuming a functional form for $G_v$, under standard conditions it is possible to use discrete choice methods to consistently estimate $(\beta_1 - \beta_2)/\sigma_v$ (except for the intercept) and $(1 - \alpha)/\sigma_v$. Using the $M$ function regression estimators discussed in Section 2.1, under standard conditions it is possible to estimate $\beta_2$ consistently. Provided that there is one regressor in $X$ with a nonzero $\beta_2$ coefficient and with a zero coefficient in $\beta_1$, it is possible to estimate $\sigma_v$ and $\alpha$ from the discrete choice analysis. Hence it is possible to consistently estimate $\beta_1$. These exclusion restrictions provide one method for identifying the parameters of the model. In the context of a one period model of labor supply, such exclusion restrictions are plausible.

In dynamic models of labor supply with savings such exclusion restrictions are implausible. This is so because the equilibrium reservation wage function determining labor force participation in any period depends on the wages received in all periods in which agents work. Variables that determine wage rates in working periods determine reservation wages in all periods. Conventional simultaneous equation exclusion restrictions cannot be used to secure identification in this model. Identifiability can be achieved by exploiting the (nonlinear) restrictions produced by economic theory as embodied in particular functional forms. Precisely the same problem arises in econometric models of search unemployment, a topic to which we turn next.

### 3.1.2. A model of search unemployment

The index function model provides the framework required to give econometric content to the conventional model of search unemployment. As in the labor force participation example just presented, agents continue on in a state of search unemployment until they receive an offered wage that exceeds their reservation wage. Accepted wages are thus censored random variables. The only novelty in the application of the index function to the unemployment problem is that a different economic theory is used to produce the reservation wage.

In the most elementary version of the search model, agents are income maximizers. An unemployed agent's decision problem is very simple. If cost $c$ is incurred in a period, the agent receives a job offer but the wage that comes with the offer is unknown before the offer arrives. This uncertainty is fundamental to the problem. Successive wage offers are assumed to independent realizations from a known absolutely continuous wage distribution $F(w)$ with $E|W| < \infty$. Assuming a positive real interest rate $r$, no search on the job, and jobs that last forever (so there is no quitting from jobs), Lippman and McCall (1976) show that the value of search at time $t$, $V(t)$, is implicitly determined by the functional equation

$$V(t) = \max\left\{0; -c + \frac{1}{1+r}E\max\left[\frac{W}{r}; V(t+1)\right]\right\}, \tag{3.1.4}$$

where the expectation is computed with respect to the distribution of $W$.

The decision process is quite simple. A searching agent spends $c$ in period $t$ and faces two options in period $t+1$: to accept a job which offers a per period wage of $W$ with present value $W/r$, or to continue searching, which option has value $V(t+1)$. In period $t$, $W$ is uncertain. Assuming that the nonmarket alternative has a fixed nonstochastic value of 0, if $V$ falls below 0, the agent ceases to search. Lippman and McCall (1976) call the nonsearching state "out of the labor force".

Under very general conditions (see Robbins (1970) for one statement of these conditions), the solution to the agent's decision making problem has a reservation wage characterization: search until the value of the option currently in hand ($W/r$) exceeds the value of continuing on in the state, $V(t+1)$. For a time homogenous (stationary) environment, the solution to the search problem has a reservation wage characterization.[20]

Focusing on the time homogeneous case to simplify the exposition, note that $V(t) = V(t+1)$ and that eq. (3.1.4) implies

$$rV + (1+r)c = \frac{1}{r}\int_{rV}^{\infty}(w - rV)\,dF(w) \qquad \text{for} \quad rV \geq 0. \tag{3.1.5}$$

The reservation wage is $W_R = rV$. This function clearly depends on $c$, $r$ and the parameters of the wage offer distribution. Conventional exclusion restrictions of the sort invoked in the labor force participation example presented in the previous section cannot be invoked for this model.

Solving (3.1.5) for $W_R = rV$ and inserting the function so obtained into eqs. (3.1.1) and (3.1.2) produces a statistical model that is identical to the deterministic labor force participation model.

Except for special cases for $F$, closed form expressions for $W_R$ are not available.[21] Consequently, structural estimation of these models requires numerical evaluation of implicit functions (like $V(t)$ in (3.1.4)) as input to evaluation of sample likelihoods. To date, these computational problems have inhibited wide

---

[20] The reservation wage property characterizes other models as well. See Lippman and McCall (1976).

[21] See Yoon (1981) for an approximate closed form expression of $W_R$.

scale use of structural models derived from dynamic optimizing theory and have caused many analysts to adopt simplifying approximations.[22]

The density of accepted wages is

$$g(w^*) = \frac{f(w)}{1 - F(W_R)}, \qquad w^* \geq W_R, \tag{3.1.6}$$

which is truncated. Assuming that no serially correlated unobservables generate the wage offer distribution, the probability that an unemployment spell lasts $j - 1$ periods and terminates in period $j$ is

$$[F(W_R)]^{j-1}[1 - F(W_R)]. \tag{3.1.7}$$

The joint density of durations and accepted wages is the product of (3.1.6) and (3.1.7), or

$$h(w^*, j) = [F(W_R)]^{j-1} f(w^*), \tag{3.1.8a}$$

where

$$w^* \geq W_R. \tag{3.1.8b}$$

In general the distribution of wages, $F(w)$, cannot be identified. While the truncated distribution $G(w^*)$ is identified, $F(w)$ cannot be recovered without invoking some untestable assumption about $F$. If offered wages are normally distributed, $F$ is recoverable. If, on the other hand, offered wages are Pareto random variables, $F$ is not identified. Conditions under which $F$ can be recovered from $G$ are presented in Heckman and Singer (1985).

Even if $F$ is recoverable, not all of the parameters of the simple search model can be identified. From eq. (3.1.5) it should be clear that even if $rV$ and $F$ were known exactly, an infinity of nonnegative values of $r$ and $c$ solve that equation. From data on accepted wages and durations it is not possible to estimate both $r$ and $c$ without further restrictions.[23] One normalization sets $r$ at a known value.[24]

---

[22] Coleman (1984) presents indirect reduced form estimation procedures which offer a low cost alternative to costly direct maximum likelihood procedures. Flinn and Heckman (1982), Miller (1985), Wolpin (1984), and Rust (1984) discuss explicit solutions to such dynamic problems. Kiefer and Neumann (1979), Yoon (1981, 1984), and Hotz and Miller (1984) present approximate solutions.

[23] A potential source of such restrictions makes $r$ and $c$ known functions of exogenous variables.

[24] Kiefer and Neumann (1979) achieve identification in this manner.

Even if $r$ is fixed, the parameter $c$ can only be identified by exploiting inequality (3.1.8b).[25]

If a temporally persistent heterogeneity component $\eta$ is introduced into the model (say due to unobserved components of $c$ or $r$), the analysis becomes somewhat more difficult. To show this write $W_R$ as an explicit function of $\eta$, $W_R = W_R(\eta)$. In place of (3.1.8b) there is an implied restriction on the support of $\eta$

$$\{\eta \mid 0 \leq W_R(\eta) \leq w^*\}, \tag{3.1.9}$$

i.e. $\eta$ is now restricted to produce a nonnegative reservation wage that is less than (or equal to) the offered accepted wage. Modifying density (3.1.8a) to reflect this dependence and letting $\psi(\eta)$ be the density of $\eta$ leads to

$$h(w^*, j) = \int_{\{\eta \mid 0 \leq W_R(\eta) \leq w^*\}} \left[ \prod_{t=1}^{j-1} F(W_R(\eta)) \right] f(w^*) \psi(\eta) \, d\eta. \tag{3.1.10}$$

Unless restriction (3.1.9) is utilized, the model is not identified.[26]

### 3.1.3. Models of job turnover

The index function model can also be used to provide a precise econometric framework for models of on-the-job learning and job turnover developed by Johnson (1978), Jovanovic (1979), Flinn (1984) and Miller (1985). In this class of models, agents learn about their true productivity on a job by working at the job. We consider the most elementary version of these models and assume that workers are paid their realized marginal product, but that this product is due, in part, to random factors beyond the control of the agent. Agents learn about their true productivity by a standard Bayesian learning process. They have beliefs about the value of their alternatives elsewhere. Ex ante all jobs look alike in the simplest model and have value $V_A$.

The value of a job which currently pays wage $W(t)$ in the $t$th period on the job is $V(W(t))$. An agent's decision at the end of period $t$ given $W(t)$ is to decide whether to stay on the job the next period or to go on to pursue an alternative opportunity. In this formulation, assuming no cost of mobility and a positive real interest rate $r$,

$$V(W(t)) = W(t) + \frac{1}{1+r} \max\{ E_t V(W(t+1)); V_A \}, \tag{3.1.11}$$

---

[25] See Flinn and Heckman (1982) for further discussion of this point.
[26] For further discussion of identification in this model, see Flinn and Heckman (1982).

where the expectation is taken with respect to the distribution induced by the information available in period $t$ which may include the entire history of wage payments on the job. If $V_A > E_t V(W(t+1))$, the agent changes jobs. Otherwise, he continues on the job for one more period.

This setup can be represented by an index function model. Wages are observed at a job in period $t+1$ if $E_t(V(W(t+1))) > V_A$.

$$Z(t) = E_t(V(W(t+1))) - V_A,$$

is the index function characterizing job turnover behavior. If $Z(t) \geq 0$, $\delta(t) = 1$ and the agent stays on the current job. Otherwise, the agent leaves. Wages observed at job duration $t$ are censored random variables $Y^*(t) = W(t)\delta(t)$. As in the model of search unemployment computation of sample likelihoods requires numerical evaluation of functional equations like (3.1.11).[27]

## 3.2. Prototypical dummy endogenous variable models

In this subsection we consider some examples of well posed economic models that can be cast in terms of the dummy endogenous variable framework presented in Section 2.2. We consider fixed and random coefficient versions of these models for both certain and uncertain environments. We focus only on the simplest models in order to convey essential ideas.

### 3.2.1. The impact of training on earnings

Consider a model of the impact of training on earnings in which a trainee's decision to enroll is based on a comparison of the present value of earnings with and without training in an environment of perfect foresight. Our analysis of this model serves as a prototype for the analysis of the closely related problems of assessing the impact of schooling, unions, and occupational choice on earnings.

Let the annual earnings of an individual in year $t$ be

$$W(t) = \begin{cases} X(t)\beta + \delta\alpha + U(t), & t > k \\ X(t)\beta + U(t), & t \leq k. \end{cases} \tag{3.2.1}$$

In writing this equation, we suppose that all individuals have access to training at only one period in their life (period $k$) and that anyone can participate in

---

[27] Miller (1984) provides a discussion and an example of estimation of this class of models.

training if he or she chooses to do so. However, once the opportunity to train has passed, it never reoccurs. Training takes one period to complete.[28]

Income maximizing agents are assumed to discount all earnings streams by a common discount factor $1/(1+r)$. From (3.2.1) training raises earnings by an amount $\alpha$ per period. While taking training, the individual receives subsidy $S$ which may be negative, (e.g. tuition payments). Income in period $k$ is foregone for trainees. To simplify the algebra we assume that people live forever.

As of period $k$, the present value of earnings for an individual who does not receive training is

$$PV(0) = \sum_{j=0}^{\infty} \left(\frac{1}{1+r}\right)^j W(k+j).$$

The present value of earnings for a trainee is

$$PV(1) = S + \sum_{j=1}^{\infty} \left(\frac{1}{1+r}\right)^j W(k+j) + \sum_{j=1}^{\infty} \frac{\alpha}{(1+r)^j}.$$

The present value maximizing enrollment rule has a person enroll in the program if $PV(1) > PV(0)$. Letting $Z$ be the index function for enrollment,

$$Z = PV(1) - PV(0) = S - W(k) + \frac{\alpha}{r}, \tag{3.2.2}$$

and

$$\delta = \begin{cases} 1 & \text{if } S - W(k) + \dfrac{\alpha}{r} > 0 \\ 0 & \text{otherwise} \end{cases}. \tag{3.2.3}$$

Because $W(k)$ is not observed for trainees, it is convenient to substitute for $W(k)$ in (3.2.2) using (3.2.1). In addition some components of subsidy $S$ may not be observed by the econometrician. Suppose

$$S = Q\psi + \eta, \tag{3.2.4}$$

where $Q$ is observed by the econometrician and $\eta$ is not. Collecting terms, we

---

[28] The assumption that enrollment decisions are made solely on the basis of an individual's choice process is clearly an abstraction. More plausibly, the training decision is the joint outcome of decisions taken by the prospective trainee, the training agency and other agents. See Heckman and Robb (1985) for a discussion of more general models.

have

$$\delta = \begin{cases} 1 & \text{if } Q\psi + \dfrac{\alpha}{r} - X(k)\beta + \eta - U(k) \geq 0 \\ 0 & \text{otherwise} \end{cases}. \tag{3.2.5}$$

In terms of the dummy endogenous variable framework presented in Section 2.2, (3.2.1) corresponds to eq. (2.2.1), and (3.2.5) corresponds to (2.2.2).

This framework can be modified to represent a variety of different choice processes. For example, $\alpha$ may represent the union–nonunion wage differential. The variable $S$ in this case may represent a membership bribe or enrollment fee. In applying this model to the unionism problem an alternative selection mechanism might be introduced since it is unlikely that income is foregone in any period or that a person has only one opportunity in his or her lifetime to join a union. In addition, it is implausible that membership is determined solely by the prospective trainee's decision if rents accrue to union membership.[29]

As another example, this model can be applied to schooling choices. In this application, $\alpha$ is the effect of schooling on earnings and it is likely that schooling takes more than one period. Moreover, a vector $\delta$ is more appropriate since agents can choose among a variety of schooling levels.

This framework can also be applied to analyze binary migration decisions, occupational choice or industrial choice. In such applications, $\alpha$ is the per period return that accrues to migration, choice of occupation or choice of industry respectively. As in the schooling application noted above, it is often plausible that $\delta$ is a vector. Furthermore, the content of the latent variable $Z$ changes from context to context; $S$ should be altered to represent a cost of migration, or a cost of movement among occupations or industries, and income may or may not be foregone in a period of transition among states. In each of these applications, the income maximizing framework can be replaced by a utility maximizing model.

### 3.2.2. *A random coefficient specification*

In place of eq. (3.2.1), a random coefficient earnings function is

$$\begin{aligned} W(t) &= X(t)\beta + \delta(\alpha + \varepsilon) + U(t) \\ &= X(t)\beta + \delta\alpha + U(t) + \varepsilon\delta. \end{aligned} \tag{3.2.6}$$

using the notation of eq. (2.2.3). This model captures the notion of a variable effect of training (or unionism or migration or occupational choice, etc.) on earnings.

---

[29]See Abowd and Farber (1982) for a discussion of this problem.

If agents know $\varepsilon$ when they make their decisions about $\delta$, the following modification to (3.2.5) characterizes the decision process:

$$\delta = \begin{cases} 1 & \text{if } Q\psi + \dfrac{\alpha}{r} - X(k)\beta + \eta - U(k) + \varepsilon/r > 0 \\ 0 & \text{otherwise} \end{cases}. \qquad (3.2.7)$$

The fact that $\varepsilon$ appears in the disturbance terms in (3.2.6) and (3.2.7) creates another source of covariance between $\delta$ and the error term in the earnings equation that is not present in the fixed coefficient dummy endogenous variable model.

The random coefficient model captures the key idea underlying the model of self selection introduced by Roy (1951) that has been revived and extended in recent work by Lee (1978) and Willis and Rosen (1979). In Roy's model, it is solely population variation in $X(k)$, $\varepsilon$, and $U(k)$ that determines $\delta$ (so $\eta = Q = 0$ in (3.2.7)).[30]

As noted in Section 2, the fixed coefficient and random coefficient dummy endogenous variable models are frequently confused in the literature. In the context of studies of the union impact on wages, Robinson and Tomes (1984), find that a sample selection bias correction (or $M$-function) estimator of $\alpha$ and an instrumental variable estimator produce virtually the same estimate of the coefficient. As noted in Section 2.2.3, the instrumental variable estimator is inconsistent for the random coefficient model while the sample selection bias estimator is not. Both are consistent for $\alpha$ in the fixed coefficient model. The fact that the same estimate is obtained from the two different procedures indicates that a fixed coefficient model of unionism describes their data. (It is straightforward to develop a statistical test that discriminates between these two models that is based on this principle.)

*3.2.2.1.  Introducing uncertainty.*   In many applications of the dummy endogenous variable model it is unlikely that prospective trainees (union members, migrants, etc.) know all components of future earnings and the costs and benefits of their contemplated action at the time they decide whether or not to take the action. More likely, decisions are made in an environment of uncertainty.

Ignoring risk aversion, the natural generalization of decision rules (3.2.3) and (3.2.5) assumes that prospective trainees (union members, migrants, etc.) compare the expectation of $PV(0)$ evaluated at the end of period $k-1$ with the expectation of $PV(1)$ evaluated at the same date. This leads to the formulation

$$\delta = \begin{cases} 1 & \text{if } E_{k-1}\left[S - W(k) + \dfrac{\alpha + \varepsilon}{r}\right] > 0 \\ 0 & \text{otherwise} \end{cases}, \qquad (3.2.8)$$

[30] For further discussion of this model and its applications see Heckman and Sedlacek (1985).

where $E_{k-1}$ denotes the expectation of the argument in brackets conditional on the information available in period $k-1$. "$\varepsilon$" is a degenerate constant in the fixed coefficient dummy endogenous variable model, but is not degenerate in the general random coefficient specification.

Introducing uncertainty can sometimes simplify the econometrics of a problem. (See Zellner, et al. (1966)). In the random coefficient model suppose that agents do not know the value of $\varepsilon$ they will obtain when $\delta = 1$. For example, suppose $E_{k-1}(\varepsilon) = 0$. In this case trainees, union members, etc. do not know their idiosyncratic gain to training, union membership, etc., before participating in the activity. The random variable $\varepsilon$ does not appear in selection eq. (3.2.8) and is *not* a source of covariation between $\delta$ and the composite disturbance term in (3.2.6). In this case earnings eq. (3.2.6) becomes a more conventional random coefficient model in which the random coefficient is not correlated with its associated variable. (See Heckman and Robb (1985).)

If an agent's best guess of $\varepsilon$ is the population mean in eq. (3.2.8), then $E(\varepsilon|\delta = 1) = 0$ so $E(\varepsilon\delta) = 0$ and the error component $\varepsilon\delta$ creates no new econometric problem not already present in the fixed coefficient framework. Consistent estimators for the fixed coefficient model also consistently estimate $\alpha$ and $\beta$ in this version of the random coefficients model. In many contexts it is implausible that $\varepsilon$ is known at the time decisions are taken, so that the more robust fixed coefficient estimators may be applicable to random coefficient models.[31]

## 3.3. Hours of work and labor supply

The index function framework has found wide application in the recent empirical literature on labor supply. Because this work is surveyed elsewhere [Heckman and MaCurdy (1981) and Moffitt and Kehrer (1981)], our discussion of this topic is not comprehensive. We briefly review how recent models of labor supply dealing with labor force participation, fixed costs of work, and taxes can be fit within the general index function framework.

### 3.3.1. An elementary model of labor supply

We initially consider a simple model of hours of work and labor force participation that ignores fixed costs and taxes. Let $W$ be the wage rate facing a consumer, $C$ is a Hick's composite commodity of goods and $L$ is a Hicks' composite commodity of nonmarket time. The consumer's strictly quasi-concave preference

---

[31] In the more general case in which future earnings are not known, the optimal forecasting rule for $W(k)$ depends on the time series process generating $U(t)$. For an extensive discussion of more general decision processes under uncertainty see Heckman and Robb (1985). An uncertainty model provides yet another rationalization for the results reported in Robinson and Tomes (1984).

function is $U(C, L, \nu)$, where $\nu$ is a "taste shifter." For a population of consumers, the density of $W$ and $\nu$ is written as $k(w, \nu)$. The maximum amount of leisure is $T$. Income in the absence of work is $R$, and is assumed to be exogenous with respect to $\nu$ and any unobservables generating $W$.

A consumer works only if the best work alternative is better than the best nonwork alternative (i.e. full leisure). In the simple model, this comparison can be reduced to a local comparison between the marginal value of leisure at the no work position (the slope of the consumer's highest attainable indifference curve at zero hours of work) and the wage rate.

The marginal rate of substitution (MRS) along an equilibrium interior solution hours of work path is obtained by solving the implicit equation

$$\text{MRS} = \frac{U_2(R + \text{MRS} \cdot H, T - H, \nu)}{U_1(R + \text{MRS} \cdot H, T - H, \nu)}. \tag{3.3.1}$$

for MRS, where $H$ is hours of work and $C = R + \text{MRS} \cdot H$. In equilibrium the wage equals MRS. The reservation wage is $\text{MRS}(R, 0, \nu)$. The consumer works if

$$\text{MRS}(R, 0, \nu) < W; \tag{3.3.2}$$

otherwise, he does not. If condition (3.3.2) is satisfied, the labor supply function is determined by solving the equation $\text{MRS}(R, H, \nu) = W$ for $H$ to obtain

$$H = H(R, W, \nu). \tag{3.3.3}$$

Consider a population of consumers who all face wage $W$ and receive unearned income $R$ but who have different $\nu$'s. The density $k(\nu|W)$ is the conditional density of "tastes for work" over the population with a given value of $W$. Letting $\Gamma_1$ denote the subset of the support of $\nu$ which satisfies $\text{MRS}(R, 0, \nu) < W$ for a given $W$, the fraction of the population that works is

$$P(W, R) = \int_{\Gamma_1} k(\nu|W) \, d\nu = \Pr[\text{MRS}(R, 0, \nu) < W | W, R]. \tag{3.3.4}$$

The mean hours worked for those employed is

$$E[H|\text{MRS}(R, 0, \nu) < W, W, R] = \frac{\int_{\Gamma_1} H(R, W, \nu) k(\nu|W, R) \, d\nu}{P(W, R)}. \tag{3.3.5}$$

The mean hours worked in the entire population is

$$E(H) = \int_{\Gamma_1} H(R, W, \nu) k(\nu|W, R) \, d\nu, \tag{3.3.6}$$

[remember $H(R, W, \nu) = 0$ for $\nu \notin \Gamma_1$].

The model of Heckman (1974) offers an example of this framework. Write the marginal rate of substitution function given by (3.3.1) in semilog form as

$$\ln \text{MRS}(R, H, \nu) = \alpha_0 + \alpha_1 R + \alpha_2 X_2 + \gamma H + \nu, \qquad (3.3.7)$$

where $\nu$ is a mean zero, normally distributed error term. Market wage rates are written as

$$\ln W = \beta_0 + \beta_1 X_1 + \eta, \qquad (3.3.8)$$

where $\eta$ is a normally distributed error term with zero mean. Equating (3.3.7) and (3.3.8) for equilibrium hours of work for those observations satisfying $\ln W > \text{MRS}(R, 0, \nu)$, one obtains

$$H = \frac{1}{\gamma} [\ln W - \ln \text{MRS}(R, 0, \nu)]$$

$$= \frac{1}{\gamma} (\beta_0 + \beta_1 X_1 + \eta - \alpha_0 - \alpha_1 R - \alpha_2 X_2 - \nu)$$

$$= \frac{1}{\gamma} (\beta_0 - \alpha_0 + \beta_1 X_1 - \alpha_1 R - \alpha_2 X_2) + \frac{1}{\gamma} (\eta - \nu). \qquad (3.3.9)$$

In terms of the conceptual apparatus of Sections 1 and 2, one can interpret this labor supply model as a two-state model. State 0 corresponds to the state in which the consumer does not work which we signify by setting the indicator variable $\delta = 0$. When $\delta = 1$ a consumer works and state 1 occurs. Two index functions characterize the model where $Y' = (Y_1, Y_2)$ is a two element vector with

$$Y_1 = H \qquad \text{and} \qquad Y_2 = \ln W.$$

The consumer works ($\delta = 1$) when $(Y_1, Y_2) \in \Omega_1$ where $\Omega_1 = \{(Y_1, Y_2) | Y_1 > 0, -\infty \leq Y_2 \leq \infty\}$ is a subset of the support of $(Y_1, Y_2)$. Note that the exogenous variables $X$ include $X_1$, $X_2$ and $R$. The joint distribution of the errors $\nu$ and $\eta$ induces a joint distribution $f(y_1, y_2 | X)$ for $Y$ via eqs. (3.3.8) and (3.3.9). Letting $Y^* = \delta Y$ denote the observed value of $Y$, $Y_1^* = H^*$ represents a consumer's actual hours of work and $Y_2^*$ equals $\ln W$ when the consumer works and equals zero otherwise.

By analogy with eq. (1.2.8), the joint density of hours and wages conditional on $X$ and working is given by

$$g(y^*|\delta=1, X) = \frac{f(y_1^*, y_2^*|X)}{\int_{\Omega_1} f(y_1, y_2|X)\,\mathrm{d}y_1\mathrm{d}y_2}$$

$$= \frac{f(y_1^*, y_2^*|X)}{\int_{-\infty}^{\infty}\int_0^{\infty} f(y_1, y_2|X)\,\mathrm{d}y_1\mathrm{d}y_2}. \tag{3.3.10}$$

From eq. (1.2.9), the distribution of $Y^*$ given $X$ is

$$g(y^*, \delta|X) = \left[f(y_1^*, y_2^*|X)\right]^{\delta}\left[(1-\Pr(\delta=1|X))J(y_1^*, y_2^*)\right]^{1-\delta}, \tag{3.3.11}$$

where $\Pr(\delta=1|X)$ denotes the probability that the consumer works given $X$, i.e.

$$\Pr(\delta=1|X) = \int_{\Omega_1} f(y_1, y_2|X)\,\mathrm{d}y_1\mathrm{d}y_2$$

$$= \int_{-\infty}^{\infty}\int_0^{\infty} f(y_1, y_2|X)\,\mathrm{d}y_1\mathrm{d}y_2. \tag{3.3.12}$$

and where $J(y_1^*, y_2^*)=1$ if $y_1^*=0=y_2^*$ and $=0$ otherwise. When $f(\cdot)$ is a bivariate normal density, the density $g(y^*, \delta|X)$ is sometimes called a bivariate Tobit model. Provided that one variable in $X$ appears in (3.3.8) that does not appear in (3.3.7), $\gamma$ can be consistently estimated by maximum likelihood using the bivariate Tobit model.

### 3.3.2. A general model of labor supply with fixed costs and taxes

In this section we extend the simple model presented above to incorporate fixed costs of work (such as commuting costs) and regressive taxes. We present a general methodology to analyze cases in which marginal comparisons do not fully characterize labor supply behavior. We synthesize the suggestions of Burtless and Hausman (1978), Hausman (1980), Wales and Woodland (1979), and Cogan (1981).

Fixed costs of work or regressive taxes produce a nonconvex budget constraint. Figure 1 depicts the case considered here.[32] This figure represents a situation in which a consumer must pay a fixed money cost equal to $F$ in order to work. $R_1$ is his nonlabor income if he does not work. Marginal tax rate of $t_A$

---

[32] Generalization to more than two branches involves no new principle. Constraint sets like $R_2 SN$ are alleged to be common in negative income tax experiments and in certain social programs.

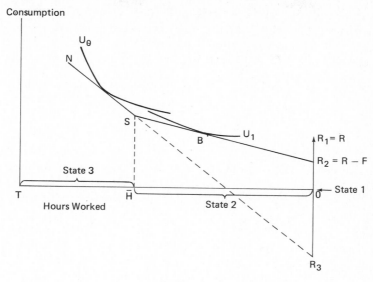

Figure 1

applies to the branch $R_2 S$ defined up to $\bar{H}$ hours, and a lower marginal rate $t_B$ applies to branch $SN$.

Assuming that no one would ever choose to work $T$ or more hours, a consumer facing this budget set may choose to be in one of three possible states of the world: the no work position at kink point $R_1$ (which we define as state 1), or an interior equilibrium on either segment $R_2 S$ or segment $SN$ (defined as states 2 and 3, respectively).[33] A consumer in state 1 receives initial after-tax income $R_1$. In state 2, a consumer receives unearned income $R_2$ and works at an after-tax wage rate equal to $W_2 \equiv W(1 - t_A)$ where $W$ is the gross wage. A consumer in state 3 earns after-tax wage rate $W_3 \equiv W(1 - t_B)$ and can be viewed as receiving the equivalent of $R_3$ as unearned income. Initially we assume that $W$ is exogenous and known for each consumer.

In the analysis of kinked-nonconvex budget constraints, a local comparison between the reservation wage and the market wage does not adequately characterize the work–no work decision as it did in the model of Section 3.3.1. Due to the nonconvexity of the constraint set, existence of an interior solution on a branch does *not* imply that equilibrium will occur on the branch. Thus in Figure 1, point $B$ associated with indifference curve $U_1$ is a possible interior equilibrium on branch $R_2 S$ that is clearly not the global optimum.

---

[33] The kink at $S$ is not treated as a state of the world because preferences are assumed to be twice differentiable and quasiconcave.

A general approach for determining the portion of the budget constraint on which a consumer locates is the following. Write the direct preference function as $U(C, L, \nu)$ where $\nu$ represents taste shifters. Form the indirect preference function $V(R, W, \nu)$. Using Roy's identity for interior solutions, the labor supply function may be written as

$$H = \frac{V_W}{V_R} = H(R, W, \nu).$$

While the arguments of the functions $U(\cdot)$, $V(\cdot)$, and $H(\cdot)$ may differ across consumers, the functional forms are assumed to be the same for each consumer.

If a consumer is at an interior equilibrium on either segment $R_2 S$ or $SN$, then the equilibrium is defined by a tangency of an indifference curve and the budget constraint. Since this tangency indicates a point of maximum attainable utility, the indifference curve at this point represents a level of utility given by $V(R_i, W_i, \nu)$ where $R_i$ and $W_i$ are, respectively, the after-tax unearned income and wage rate associated with segment $i$. Thus, hours of work for an interior equilibrium are given by $V_W/V_R$ evaluated at $R_i$ and $W_i$. For this candidate equilibrium to be admissible, the implied hours of work must lie between the two endpoints of the interval (i.e. equilibrium must occur on the budget segment). A consumer does not work if utility at kink $R_1$, $U(R_1, T, \nu)$, is greater than both $V(R_2, W_2, \nu)$ and $V(R_3, W_3, \nu)$, provided that these latter utility values represent admissible solutions located on the budget constraint.

More specifically, define the labor supply functions $H_{(1)}$, $H_{(2)}$ and $H_{(3)}$ as $H_{(1)} = 0$ and

$$H_{(i)} = \frac{V_W(R_i, W_i, \nu)}{V_R(R_i, W_i, \nu)} = H(R_i, W_i, \nu), \qquad i = 2,3; \tag{3.3.13}$$

and define the admissible utility levels $V_{(1)}$, $V_{(2)}$, and $V_{(3)}$ as $V_{(1)} = U(R_1, T, \nu)$, assumed to be greater than zero, and

$$V_{(2)} = \begin{cases} V(R_2, W_2, \nu) & \text{if } 0 < H_{(2)} \le \overline{H} \\ 0 & \text{otherwise} \end{cases}, \tag{3.3.14}$$

and

$$V_{(3)} = \begin{cases} V(R_3, W_3, \nu) & \text{if } \overline{H} < H_{(3)} \le T \\ 0 & \text{otherwise} \end{cases}. \tag{3.3.15}$$

We assume the $U(\cdot)$ is chosen so that $U(\cdot) > 0$ for all $C$, $L$, and $\nu$. A consumer whose $\nu$ lies in the set

$$\Gamma_1 = \{ \nu \mid V_{(1)} \geq V_{(2)} \quad \text{and} \quad V_{(1)} \geq V_{(3)} \}, \tag{3.3.16}$$

will not work and occupies state 1. If $\nu$ lies in the set

$$\Gamma_2 = \{ \nu \mid V_{(2)} > V_{(1)} \quad \text{and} \quad V_{(2)} \geq V_{(3)} \}, \tag{3.3.17}$$

a consumer is at an interior solution on segment $R_2 S$ and occupies state 2. Finally, a consumer is at equilibrium in state 3 on segment $SN$ if $\nu$ is an element of the set

$$\Gamma_3 = \{ \nu \mid V_{(3)} > V_{(1)} \quad \text{and} \quad V_{(3)} > V_{(2)} \}. \tag{3.3.18}$$

The sets $\Gamma_1$, $\Gamma_2$, and $\Gamma_3$ do not intersect, and their union is the relevant subspace of the support of $\nu$. These sets are thus mutually exclusive.[34] The functions $H_{(i)}$ determine the hours of work for individuals for whom $\nu \in \Gamma_i$.

Choosing a specification for the preference function and a distribution for "tastes" in the population, $\phi(\nu)$, produces a complete statistical characterization of labor supply behavior. The probability that a consumer is in state $i$ is

$$\Pr(\nu \in \Gamma_i) = \int_{\Gamma_i} \phi(\nu) \, d\nu. \tag{3.3.19}$$

The expected hours of work of a consumer who is known to be in state $i$ is

$$E(H \mid \nu \in \Gamma_i) = E\left(H_{(i)} \mid \nu \in \Gamma_i\right)$$
$$= \frac{\int_{\Gamma_i} H_{(i)} \phi(\nu) \, d\nu}{\Pr(\nu \in \Gamma_i)}. \tag{3.3.20}$$

The expected hours of work for a randomly chosen individual is

$$E(H) = \sum_{i=1}^{3} E\left(H_{(i)} \mid \nu \in \Gamma_i\right) \Pr(\nu \in \Gamma_i). \tag{3.3.21}$$

We have thus far assumed: (i) that data on potential wage rates are available for all individuals including nonworkers, and (ii) that wage rates are exogenous

---

[34] Certain values for $\nu$ may be excluded if they imply such phenomena as negative values of $U$ or $V$ or nonconvex preferences. In this case we use the conditional density of $\nu$ excluding those values.

variables. Relaxing these assumptions does not raise any major conceptual problems and makes the analysis relevant to a wider array of empirical situations.

Suppose that market wage rates are described by the function

$$W = W(X, \eta), \qquad (3.3.22)$$

where $X$ includes a consumer's measured characteristics, and $\eta$ is an error term representing unmeasured characteristics. Substituting $W(X, \eta)$ for $W$ in the preceding discussion, the extended partitions

$$\Gamma_i = \{ (\nu, \eta) \mid V_{(i)} \ge V_{(j)} \qquad \text{for all } j \}, \qquad (3.3.23)$$

(recall that equality holds on a set of measure zero) replace the characterization of the sets $\Gamma_i$ for known wages given by (3.3.16)–(3.3.18). A consumer for whom $(\nu, \eta) \in \Gamma_i$ occupies state $i$. The probability of such an event is

$$\Pr((\nu, \eta) \in \Gamma_i) = \int_{\Gamma_i} \phi(\nu, \eta) \, d\nu \, d\eta, \qquad (3.3.24)$$

where $\phi(\nu, \eta)$ is the joint density of $\nu$ and $\eta$.

The labor supply functions for each state are changed by substituting $W(X, \eta)$ for $W$ in constructing the arguments of the functions for states 2 and 3 given by (3.3.13).[35] In place of (3.3.21), the expression for expected hours of work becomes

$$E(H) = \sum_{i=1}^{3} E(H_{(i)} \mid (\nu, \eta) \in \Gamma_i) \Pr((\nu, \eta) \in \Gamma_i), \qquad (3.3.25)$$

where

$$E(H_{(i)} \mid (\nu, \eta) \in \Gamma_i) = \frac{\int_{\Gamma_i} H_{(i)} \phi(\nu, \eta) \, d\nu \, d\eta}{\Pr((\nu, \eta) \in \Gamma_i)}. \qquad (3.3.26)$$

Using the expression for $E(H)$ given by (3.3.25) in a regression analysis permits wages to be endogenous and does not require that wage offer data be available for all observations. The parameters of (3.3.25) or (3.3.26) can be estimated using the nonlinear least-squares procedure described in Section 2.1. To identify all the parameters of the model, the wage equation must also be estimated using data on workers appropriately adjusting for sample selection bias. An alternative strategy is to jointly estimate hours and wage equations.

---

[35] Note that the arguments $W_2$, $W_3$, $R_2$ and $R_3$ each depend on $W$.

Thus far we have assumed that hours of work and wages are not measured with error. The needed modifications required in the preceding analysis to accommodate measurement error are presented in Heckman and MaCurdy (1981).

To illustrate the required modifications when measurement error is present, suppose that we express the model in terms of $\nu$ and $\eta$ and that errors in the variables plague the available data on hours of work. When $H > 0$, suppose that measured hours, which we denote by $H^+$, are related to true hours by the equation $H^+ = H + e$ where $e$ is a measurement error distributed independently of the explanatory variables $X$. When such errors in variables are present, data on hours of work (i.e. $H^+$ when $H > 0$ and $H$ when $H = 0$) do not allocate working individuals to the correct branch of the budget constraint. Consequently, the states of the world a consumer occupies can no longer be directly observed.

This model translates into a three index function model of the sort described in Section 1.2. Two index functions, $Y' = (Y_1, Y_2) \equiv (H^+, W)$ are observed in some states, and one index function, $Z = \nu$, is never directly observed. Given an assumption about the joint distribution of the random errors $\nu$, $\eta$, and $e$, a transformation from these errors to the variables $\nu$, $W$, and $H^+$ using eq. (3.3.13) and the relation $H^+ = H(R, W, \nu) + e$ produces a joint density function $f(Y, Z)$. There are three states of the world in this model (so $I = 3$ in the notation of Section 1.2). The $i$th state occurs when $\delta_i = 1$ which arises if $(Y, Z) \in \Omega_i$ where

$$\Omega_1 = \{(Y, Z) | V_{(1)} \geq V_{(2)} \quad \text{and} \quad V_{(1)} \geq V_{(3)}\},$$
$$\Omega_2 = \{(Y, Z) | V_{(2)} > V_{(1)} \quad \text{and} \quad V_{(2)} \geq V_{(3)}\},$$

and

$$\Omega_3 = \{(Y, Z) | V_{(3)} > V_{(1)} \quad \text{and} \quad V_{(3)} > V_{(2)}\}.$$

$Y$ is observed in the work states 2 and 3, but not when $\delta_1 = 1$. Thus, adopting the convention of Section 1, the observed version of $Y$ is given by $Y^* = (\delta_2 + \delta_3)Y$. In this notation, the appropriate density functions for this model are given by formulae (1.2.12) and (1.2.13), with $\bar{\delta}_1 = \delta_2 + \delta_3$, $\bar{\delta}_2 \equiv 0$, and $\bar{\delta}_3 = \delta_1$.

## 4. Summary

This paper presents and extends the index function model of Karl Pearson (1901) that underlies all recent models in labor econometrics. In this framework, censored, truncated and discrete random variables are interpreted as the manifestation of various sampling schemes for underlying index function models. A unified derivation of the densities and regression representations for index function models is presented. Methods of estimation are discussed with an emphasis on regression and instrumental variable procedures.

We demonstrate how a variety of substantive models in labor economics can be given an econometric representation within the index function framework. Models for the analysis of unemployment, labor force participation, job turnover, the impact of interventions on earnings (and other outcomes) and hours of work are formulated as special cases of the general index function model. By casting these diverse models in a common mold we demonstrate the essential commonalities in the econometric approach required for their formulation and estimation.

## Appendix: The principal assumption

This appendix discusses the principal assumption in the context of a more conventional discrete choice model. We write

$$Z_1 = X\beta_1 + \delta_2\alpha_1 + V_1, \tag{A.1a}$$

$$Z_2 = X\beta_2 + \delta_1\alpha_2 + V_2, \tag{A.1b}$$

$$E(V_1) = E(V_2) = 0, \qquad \text{Var}(V_1) = \text{Var}(V_2) = 1.$$

$$Z_1 \geq 0 \qquad \text{iff } \delta_1 = 1,$$

$$Z_1 < 0 \qquad \text{iff } \delta_1 = 0,$$

$$Z_2 \geq 0 \qquad \text{iff } \delta_2 = 1,$$

$$Z_2 < 0 \qquad \text{iff } \delta_2 = 0.$$

In this model $Z_1$ and $Z_2$ are not observed. Unless

$$\alpha_1\alpha_2 = 0, \tag{A.2}$$

it is possible that $Z_1 \geq 0$ but $\delta_1 = 0$ or that $Z_2 \geq 0$ but $\delta_2 = 0$.

An argument that is often made against this model is that condition (A.2) rules out "true simultaneity" among outcomes. By analogy with the conventional simultaneous equations literature, replacing $\delta_2$ with $Z_2$ and $\delta_1$ with $Z_1$ in eq. (A.1) generates a statistically meaningful model without need to invoke condition (A.2). Appealing to this literature, the principal assumption looks artificial.

To examine this issue more closely, we present a well-specified model of consumer choice in which condition (A.2) naturally emerges. Let $X = 1$ (so there are no exogenous variables in the model) and write the utility ordering over outcomes as

$$U(\delta_1, \delta_2) = \eta_1\delta_1 + \eta_2\delta_2 + \eta_3\delta_1\delta_2 + \varepsilon_1\delta_1 + \varepsilon_2\delta_2 + \varepsilon_3\delta_1\delta_2, \qquad (A.3)$$

where $(\eta_1, \eta_2, \eta_3)$ is a vector of parameters and $(\varepsilon_1, \varepsilon_2, \varepsilon_3)$ is a vector of mean zero continuous unobserved random variables.

The outcome $\delta_1 = 1$ of the choice process arises if either $U(1,1)$ or $U(1,0)$ is maximal in the choice set (i.e. $\max(U(1,1), U(1,0)) \geq \max(U(0,1), U(0,0))$). For a separable model with no interactions ($\eta_3 = 0$ and $\varepsilon_3 \equiv 0$), this condition can be stated as

$$\delta_1 = 1 \qquad \text{iff } \eta_1 + \varepsilon_1 \geq 0.$$

Setting $\eta_1 = \beta_1$, $\alpha_1 = 0$ and $\varepsilon_1 = V_1$ produces eq. (A.1a). Condition (A.2) is satisfied. By a parallel argument for $\delta_2$, (A.1b) is produced. Condition (A.2) is satisfied because both $\alpha_1 = 0$ and $\alpha_2 = 0$.

For a general nonseparable choice problem ($\eta_3 \neq 0$ or $\varepsilon_3 \neq 0$ or both) equation system (A.1) still represents the choice process but once more $\alpha_1 = \alpha_2 = 0$. For example, suppose that $\varepsilon_3 \equiv 0$. In this case

$$\delta_1 = 1 \qquad \text{iff } \max(U(1,1), U(1,0)) \geq \max(U(0,1), U(0,0)).$$

For the case $\eta_3 > 0$,

$$\delta_1 = 1 \qquad \text{iff } \eta_1 + \varepsilon_1|_{\varepsilon_2 < -(\eta_2 + \eta_3)} > 0, \qquad (A.4)$$

or

$$\eta_1 + \eta_2 + \eta_3 + \varepsilon_1 + \varepsilon_2|_{-(\eta_2 + \eta_3) < \varepsilon_2 < -\eta_2} > 0,$$

or

$$\eta_1 + \eta_3 + \varepsilon_1|_{\varepsilon_2 > -\eta_2} > 0,$$

where $X|_y$ denotes the conditional random variable $X$ given $Y = y$. The probability that $\delta_1 = 1$ can be represented by eq. (A.1) with $\alpha_1 = 0$. In this model the distribution of $(V_1, V_2)$ is of a different functional form than is the distribution of $(\varepsilon_1, \varepsilon_2)$.

In this example there is genuine interaction in the utility of outcomes and eqs. (A.1) still characterize the choice process. The model satisfies condition (A.2). Even if $\alpha_1 = \alpha_2 = 0$, there is genuine simultaneity in choice.

Unconditional representation (A.1) (with $\alpha_1 \neq 0$ or $\alpha_2 \neq 0$) sometimes characterizes a choice process of interest and sometimes does not. Often partitions of the support of $(V_1, V_2)$ required to define $\delta_1$ and $\delta_2$ are not rectangular and so the unconditional representation of the choice process with $\alpha_1 \neq 0$ or $\alpha_2 \neq 0$ is not appropriate, but any well-posed simultaneous choice process can be represented by equation system (A.1).

An apparent source of confusion arises from interpreting (A.1) as a well-specified behavioral relationship. Thus it might be assumed that the utility of agent 1 depends on the actions of agent 2, and vice versa. In the absence of any behavioral mechanism for determining the precise nature of the interaction between two actors (such as (A.3)), the model is incomplete. Assuming that player 1 is dominant (so $\alpha_1 = 0$) is one way to supply the missing behavioral relationship. (Dominance here means that player 1 temporally has the first move.) Another way to complete the model is to postulate a dynamic sequence so that current utilities depend on previous outcomes (so $\alpha_1 = \alpha_2 = 0$, see Heckman (1981)). Bjorn and Vuong (1984) complete the model by suggesting a game theoretic relationship between the players. In all of these completions of the model, (A.2) is satisfied.

## References

Abowd, J. and H. Farber (1982) "Jobs Queues and the Union Status of Workers", *Industrial and Labor Relations Review*, 35, 354–367.

Amemiya, T. (1985) *Advanced Econometrics*. Harvard University Press, forthcoming.

Bishop, Y., S. Fienberg and P. Holland (1975) *Discrete Multivariate Analysis*. Cambridge: MIT Press.

Bock and Jones (1968) *The Measurement and Prediction of Judgment and Choice*. San Francisco: Holden-Day.

Borjas, G. and S. Rosen (1981) "Income Prospects and Job Mobility of Younger Men", in: R. Ehrenburg, ed., *Research in Labor Economics*. London: JAI Press, 3.

Burtless, G. and J. Hausman (1978) "The Effect of Taxation on Labor Supply: Evaluating the Gary Negative Income Tax Experiment", *Journal of Political Economy*, 86(6), 1103–1131.

Byron, R. and A. K. Bera (1983) "Least Squares Approximations to Unknown Regression Functions, A Comment", *International Economic Review*, 24(1), 255–260.

Cain, G. and H. Watts, eds. (1973) *Income Maintenance and Labor Supply*. Chicago: Markham.

Catsiapsis, B. and C. Robinson (1982) "Sample Selection Bias with Multiple Selection Rules: An Application to Student Aid Grants", *Journal of Econometrics*, 18, 351–368.

Chamberlain, G. (1982) "Multivariate Regression Models for Panel Data", *Journal of Econometrics*, 18, 5–46.

Cogan, J. (1981) "Fixed Costs and Labor Supply", *Econometrica*, 49(4), 945–963.

Coleman, T. (1981) "Dynamic Models of Labor Supply". University of Chicago, unpublished manuscript.

Coleman, T. (1984) "Two Essays on the Labor Market". University of California, unpublished Ph.D. dissertation.

Cosslett, S. (1984) "Distribution-Free Estimator of Regression Model with Sample Selectivity". University of Florida, unpublished manuscript.

Eicker, F. (1963) "Asymptotic Normality and Consistency of the Least Squares Estimators for Families of Linear Regressions", *Annals of Mathematical Statistics*, 34, 446–456.

Eicker, F. (1967) "Limit Theorems for Regressions with Unequal and Dependent Errors", in: *Proceedings of the Fifth Berkeley Symposium on Mathematical Statistics and Probability*. Berkeley: University of California Press, 1, 59–82.

Flinn, C. (1984) "Behavioral Models of Wage Growth and Job Change Over the Life Cycle". University of Chicago, unpublished Ph.D. dissertation.

Flinn, C. and J. Heckman (1982) "New Methods for Analyzing Structural Models of Labor Force Dynamics", *Journal of Econometrics*, 18, 115–168.

Freeman, R. (1984) "Longitudinal Analysis of the Effects of Trade Unions", *Journal of Labor Economics*, 2, 1–26.

Gallant, R. and D. Nychka (1984) "Consistent Estimation of the Censored Regression Model", unpublished manuscript, North Carolina State University.

Goldberger, A. (1983) "Abnormal Selection Bias", in: S. Karlin, T. Amemiya and L. Goodman, eds., *Studies in Econometrics, Time Series and Multivariate Statistics*. New York: Academic Press, 67–84.

Griliches, Z. (1986) "Economic Data Issues", in this volume.

Haberman, S. (1978) *Analysis of Qualitative Data*, New York: Academic Press, I and II.

Hausman, J. (1980) "The Effects of Wages, Taxes, and Fixed Costs on Women's Labor Force Participation", *Journal of Public Economics*, 14, 161–194.

Heckman, J. (1974) "Shadow Prices, Market Wages and Labor Supply", *Econometrica*, 42(4), 679–694.

Heckman, J. (1976a) "Simultaneous Equations Models with Continuous and Discrete Endogenous Variables and Structural Shifts", in: S. Goldfeld and R. Quandt, eds., *Studies in Nonlinear Estimation*. Cambridge: Ballinger.

Heckman, J. (1976b) "The Common Structure of Statistical Models of Truncation, Sample Selection and Limited Dependent Variables and a Simple Estimator for Such Models", *Annals of Economic and Social Measurement*, Fall, 5(4), 475–492.

Heckman, J. (1978) "Dummy Endogenous Variables in a Simultaneous Equations System", *Econometrica*, 46, 931–961.

Heckman, J. (1979) "Sample Selection Bias as a Specification Error", *Econometrica*, 47, 153–162.

Heckman, J. (1981) "Statistical Models for Discrete Panel Data", in: C. Manski and D. McFadden, eds., *Structural Analysis of Discrete Data with Economic Applications*. Cambridge: MIT Press.

Heckman, J., M. Killingsworth and T. MaCurdy (1981) "Empirical Evidence on Static Labour Supply Models: A Survey of Recent Developments", in: Z. Hornstein, J. Grice and A. Webb, eds., *The Economics of the Labour Market*. London: Her Majesty's Stationery Office, 75–122.

Heckman, J. and T. MaCurdy (1980) "A Life Cycle Model of Female Labor Supply", *Review of Economic Studies*, 47, 47–74.

Heckman, J. and T. MaCurdy (1981) "New Methods for Estimating Labor Supply Functions: A Survey", in: R. Ehrenberg, ed., *Research in Labor Economics*. London: JAI Press, 4.

Heckman, J. and R. Robb (1985) "Alternative Methods for Evaluating the Impact of Training on Earnings", in: J. Heckman and B. Singer, eds., *Longitudinal Analysis of Labor Market Data*. Cambridge: Cambridge University Press.

Heckman, J. and G. Sedlacek (1985) "Heterogeneity, Aggregation and Market Wage Functions: An Empirical Model of Self Selection in the Labor Market", *Journal of Political Economy*, 93, December.

Heckman, J. and B. Singer (1986) "Econometric Analysis of Longitudinal Data", in this volume.

Heckman, J. and R. Willis (1977) "A Beta Logistic Model for Analysis of Sequential Labor Force Participation by Married Women", *Journal of Political Economy*, 85, 27–58.

Hotz, J. and R. Miller (1984) "A Dynamic Model of Fertility and Labor Supply". Carnegie–Mellon University, unpublished manuscript.

Johnson, W. (1978) "A Theory of Job Shopping", *Quarterly Journal of Economics*.

Jovanovic, B. (1979) "Firm Specific Capital and Turnover", *Journal of Political Economy*, December, 87(6), 1246–1260.

Kagan, A., T. Linnik and C. R. Rao (1973) *Some Characterization Theorems in Mathematical Statistics*. New York: Wiley.

Kendall, M. and A. Stuart (1967) *The Advanced Theory of Statistics*. London: Griffen, II.

Kevles, D. J. (1985) *In the Name of Eugenics*, New York: Knopf.

Kiefer, N. and G. Neumann (1979) "An Empirical Job Search Model with a Test of the Constant Reservation Wage Hypothesis", *Journal of Political Economy*, February, 87(1), 89–108.

Killingsworth, M. (1983) *Labour Supply*. Cambridge: Cambridge University Press.

Lee, L. F. (1978) "Unionism and Wage Rates: A Simultaneous Equations Model with Qualitative and Limited Dependent Variables", *International Economic Review*, 19, 415–433.

Lee, L. F. (1981) "Simultaneous Equation Models with Discrete and Censored Variables", in: C. Manski and D. McFadden, eds., *Structural Analysis of Discrete Data with Economic Applications*. Cambridge: MIT Press.

Lee, L. F. (1982) "Some Approaches to the Correction of Selectivity Bias", *Review of Economic Studies*, 49, 355–372.

Lippman, S. and J. McCall (1976) "The Economics of Job Search: A Survey, Part I", *Economic Inquiry*, 14, 155–189.

Lord, F. and M. Novick (1968) *Statistical Theories of Mental Test Scores*. Reading: Addison-Wesley Publishing Company.

Manski, C. and D. McFadden (1981) "Alternative Estimates and Sample Designs for Discrete Choice Analysis", in: C. Manski and D. McFadden, eds., *Structural Analysis of Discrete Data with Econometric Applications*. Cambridge: MIT Press.

McFadden, D. (1985) "Econometric Analysis of Qualitative Response Models", in: Z. Griliches and J. Intriligator, eds., *Handbook of Econometrics*. North-Holland, II.

Miller, R. (1984) "An Estimate of a Job Matching Model", *Journal of Political Economy*, Vol. 92, December.

Mincer, J. and B. Jovanovic (1981) "Labor Mobility and Wages", in: S. Rosen, ed., *Studies in Labor Markets*. Chicago: University of Chicago Press.

Moffitt, R. (1984) "Profiles of Fertility, Labor Supply and Wages of Married Women: A Complete Life-Cycle Model", *Review of Economic Studies*, 51, 263–278.

Moffitt, R. and K. Kehrer (1981) "The Effect of Tax and Transfer Programs on Labor Supply: The Evidence from the Income Maintenance Experiments", in: R. Ehrenberg, ed., *Research in Labor Economics*. London: JAI Press, 4.

Olson, R. (1980) "A Least Squares Correction for Selectivity Bias", *Econometrica*, 48, 1815–1820.

Pearson, K. (1901) "Mathematical Contributions to the Theory of Evolution", *Philosophical Transactions*, 195, 1–47.

Quandt, R. (1958) "The Estimation of the Parameters of a Linear Regression System Obeying Two Separate Regimes", *Journal of the American Statistical Association*, 53, 873–880.

Quandt, R. (1972) "A New Approach to Estimating Switching Regressions", *Journal of the American Statistical Association*, 67, 306–310.

Robbins, H. (1970) "Optimal Stopping", *American Mathematical Monthly*, 77, 333–343.

Robinson, C. and N. Tomes (1982) "Self Selection and Interprovincial Migration in Canada", *Canadian Journal of Economics*, 15(3), 474–502.

Robinson, C. and N. Tomes (1984) "Union Wage Differentials in the Public and Private Sectors: A Simultaneous Equations Specification", *Journal of Labor Economics*, 2(1), 106–127.

Rossett, R. (1959) "A Statistical Model of Friction in Economics", *Econometrica*, 27(2), 263–267.

Roy, A. (1951) "Some Thoughts on the Distribution of Earnings", *Oxford Economic Papers*, 3, 135–146.

Rust, J. (1984) "Maximum Likelihood Estimation of Controlled Discrete Choice Processes". SSRI No. 8407, University of Wisconsin, May 1984.

Schmidt, P. (1981) "Constraints on the Parameters in Simultaneous Tobit and Probit Models", in: C. Manski, and D. McFadden, eds., *Structural Analysis of Discrete Data with Econometric Applications*. Cambridge: MIT Press.

Siow, A. (1984) "Occupational Choice Under Uncertainty", *Econometrica*, 52(3), 631–646.

Strauss, R. and P. Schmidt (1976) "The Effects of Unions on Earnings and Earnings on Unions: A Mixed Logit Approach", *International Economic Review*, 17(1), 204–212.

Tallis, G. M. (1961) "The Moment Generating Function of the Truncated Multivariate Distribution", *Journal of the Royal Statistical Society*, Series B, 23, 233–239.

Thurstone, L. (1927) "A Law of Comparative Judgment", *Psychological Review*, 37, 273–286.

Tinbergen, J. (1951) "Some Remarks on the Distribution of Labour Incomes", *International Economic Papers*, 195–207.

Tobin, J. (1958) "Estimation of Relationships for Limited Dependent Variables", *Econometrica*, 26, 24–36.

Wales, T. J. and A. D. Woodland (1979) "Labour Supply and Progressive Taxes", *Review of Economic Studies*, 46, 83–95.

White, H. (1981) "Consequences and Detection of Misspecified Nonlinear Regression Models", *Journal of the American Statistical Association*, 76, 419–433.

Willis, R. and S. Rosen (1979) "Education and Self Selection", *Journal of Political Economy*, 87, S7–S36.

Wolpin, K. (1984) "An Estimable Dynamic Stochastic Model of Fertility and Child Mortality", *Journal of Political Economy*, Vol. 92, August.

Yoon, B. (1981) "A Model of Unemployment Duration with Variable Search Intensity", *Review of Economics and Statistics*, November, 63(4), 599–609.

Yoon, B. (1984) "A Nonstationary Hazard Model of Unemployment Duration". New York: SUNY, Department of Economics, unpublished manuscript.

Zellner, A., J. Kmenta and J. Dreze (1966) "Specification and Estimation of Cobb Douglas Production Function Models", *Econometrica*, 34, 784–795.

*Chapter 33*

# EVALUATING THE PREDICTIVE ACCURACY OF MODELS

RAY C. FAIR

## Contents

*Handbook of Econometrics, Volume III, Edited by Z. Griliches and M.D. Intriligator*
© *Elsevier Science Publishers BV, 1986*

## 1.  Introduction

Methods for evaluating the predictive accuracy of econometric models are discussed in this chapter. Since most models used in practice are nonlinear, the nonlinear case will be considered from the beginning. The model is written as:

$$f_i(y_t, x_t, \alpha_i) = u_{it}, \qquad (i = 1, \ldots, n), \qquad (t = 1, \ldots, T), \tag{1}$$

where $y_t$ is an $n$-dimensional vector of endogenous variables, $x_t$ is a vector of predetermined variables (including lagged endogenous variables), $\alpha_i$ is a vector of unknown coefficients, and $u_{it}$ is the error term for equation $i$ for period $t$. The first $m$ equations are assumed to be stochastic, with the remaining $u_{it}(i = m + 1, \ldots, n)$ identically zero for all $t$.

The emphasis in this chapter is on methods rather than results. No attempt is made to review the results of comparing alternative models. This review would be an enormous undertaking and is beyond the scope of this *Handbook*. Also, as will be argued, most of the methods that have been used in the past to compare models are flawed, and so it is not clear that an extensive review of results based on these methods is worth anyone's effort. The numerical solution of nonlinear models is reviewed in Section 2, including stochastic simulation procedures. This is background material for the rest of the chapter. The standard methods that have been used to evaluate ex ante and ex post predictive accuracy are discussed in Sections 3 and 4, respectively. The main problems with these methods, as will be discussed, are that they (1) do not account for exogenous variable uncertainty, (2) do not account for the fact that forecast-error variances vary across time, and (3) do not treat the possible existence of misspecification in a systematic way. Section 5 discusses a method that I have recently developed that attempts to handle these problems, a method based on successive reestimation and stochastic simulation of the model. Section 6 contains a brief conclusion.

It is important to note that this chapter is not a chapter on forecasting techniques. It is concerned only with methods for *evaluating* and *comparing* econometric models with respect to their predictive accuracy. The use of these methods should allow one (in the long run) to decide which model best approximates the true structure of the economy and how much confidence to place on the predictions from a given model. The hope is that one will end up with a model that for a wide range of loss functions produces better forecasts than do other techniques. At some point along the way one will have to evaluate and compare other methods of forecasting, but it is probably too early to do this. At any rate, this issue is beyond the scope of this chapter.[1]

---

[1] For a good recent text on forecasting techniques for time series, see Granger and Newbold (1977).

## 2. Numerical solution of nonlinear models

The Gauss–Seidel technique is generally used to solve nonlinear models. [See Chapter 14 (Quandt) for a discussion of this technique.] Given a set of estimates of the coefficients, given values for the predetermined variables, and given values for the error terms, the technique can be used to solve for the endogenous variables. Although in general there is no guarantee that the technique will converge, in practice it has worked quite well.

A "static" simulation is one in which the actual values of the predetermined variables are used for the solution each period. A "dynamic" simulation is one in which the predicted values of the endogenous variables from the solutions for previous periods are used for the values of the lagged endogenous variables for the solution for the current period. An "ex post" simulation or forecast is one in which the actual values of the exogenous variables are used. An "ex ante" simulation or forecast is one in which guessed values of the exogenous variables are used. A simulation is "outside-sample" if the simulation period is not included within the estimation period; otherwise the simulation is "within-sample." In forecasting situations in which the future is truly unknown, the simulations must be ex ante, outside-sample, and (if the simulation is for more than one period) dynamic.

If one set of values of the error terms is used, the simulation is said to be "deterministic." The expected values of most error terms in most models are zero, and so in most cases the errors terms are set to zero for the solution. Although it is well known [see Howrey and Kelejian (1971)] that for nonlinear models the solution values of the endogenous variables from deterministic simulations are not equal to the expected values of the variables, in practice most simulations are deterministic. It is possible, however, to solve for the expected values of the endogenous variables by means of "stochastic" simulation, and this procedure will now be described. As will be seen later in this chapter, stochastic simulation is useful for purposes other than merely solving for the expected values.

Stochastic simulation requires that an assumption be made about the distributions of the error terms and the coefficient estimates. In practice these distributions are almost always assumed to be normal, although in principle other assumptions can be made. For purposes of the present discussion the normality assumption will be made. In particular, it is assumed that $u_t = (u_{1t}, \ldots, u_{mt})'$ is independently and identically distributed as multivariate $N(0, \Sigma)$. Given the estimation technique, the coefficient estimates, and the data, one can estimate the covariance matrix of the error terms and the covariance matrix of the coefficient estimates. Denote these two matrices as $\hat{\Sigma}$ and $\hat{V}$, respectively. The dimension of $\hat{\Sigma}$ is $m \times m$, and the dimension of $\hat{V}$ is $K \times K$, where $K$ is the total number of coefficients in the model. $\hat{\Sigma}$ can be computed as $(1/T)\hat{U}\hat{U}'$, where $\hat{U}$ is the $m \times T$ matrix of values of the estimated error terms. The computation of $\hat{V}$ depends on the estimation technique used. Given $\hat{V}$ and given the normality assumption, an estimate of the distribution of the coefficient estimates is $N(\hat{\alpha}, \hat{V})$, where $\hat{\alpha}$ is the $K \times 1$ vector of the coefficient estimates.

Let $u_t^*$ denote a particular draw of the $m$ error terms for period $t$ from the $N(0, \hat{\Sigma})$ distribution, and let $\alpha^*$ denote a particular draw of the $K$ coefficients from the $N(\hat{\alpha}, \hat{V})$ distribution. Given $u_t^*$ for each period $t$ of the simulation and given $\alpha^*$, one can solve the model. This is merely a deterministic simulation for the given values of the error terms and coefficients. Call this simulation a "trial". Another trial can be made by drawing a new set of values of $u_t^*$ for each period $t$ and a new set of values of $\alpha^*$. This can be done as many times as desired. From each trial one obtains a prediction of each endogenous variable for each period. Let $\tilde{y}_{itk}^j$ denote the value on the $j$th trial of the $k$-period-ahead prediction of variable $i$ from a simulation beginning in period $t$.[2] For $J$ trials, the estimate of the expected value of the variable, denoted $\tilde{\tilde{y}}_{itk}$, is:

$$\tilde{\tilde{y}}_{itk} = \frac{1}{J} \sum_{j=1}^{J} \tilde{y}_{itk}^j. \tag{2}$$

In a number of studies stochastic simulation with respect to the error terms only has been performed, which means drawing only from the distribution of the error terms for a given trial. These studies include Nagar (1969); Evans, Klein, and Saito (1972); Fromm, Klein, and Schink (1972); Green, Liebenberg, and Hirsch (1972); Sowey (1973); Cooper and Fischer (1972); Cooper (1974); Garbade (1975); Bianchi, Calzolari, and Corsi (1976); and Calzolari and Corsi (1977). Studies in which stochastic simulation with respect to both the error terms and coefficient estimates has been performed include Cooper and Fischer (1974); Schink (1971), (1974); Haitovsky and Wallace (1972); Muench, Rolnick, Wallace, and Weiler (1974); and Fair (1980).

One important empirical conclusion that can be drawn from stochastic simulation studies to date is that the values computed from deterministic simulations are quite close to the mean predicted values computed from stochastic simulations. In other words, the bias that results from using deterministic simulation to solve nonlinear models appears to be small. This conclusion has been reached by Nagar (1969), Sowey (1973), Cooper (1974), Bianchi, Calzolani, and Corsi (1976), and Calzolani and Corsi (1977) for stochastic simulation with respect to the error terms only and by Fair (1980) for stochastic simulation with respect to both error terms and coefficients.

A standard way of drawing values of $\alpha^*$ from the $N(\hat{\alpha}, \hat{V})$ distribution is to (1) factor numerically (using a subroutine package) $\hat{V}$ into $PP'$, (2) draw (again using

---

[2] Note that $t$ denotes the first period of the simulation, so that $\tilde{y}_{itk}^j$ is the prediction for period $t + k - 1$.

a subroutine package) $K$ values of a standard normal random variable with mean 0 and variance 1, and (3) compute $\alpha^*$ as $\hat{\alpha} + Pe$, where $e$ is the $K \times 1$ vector of the standard normal draws. Since $Eee' = I$, then $E(\alpha^* - \hat{\alpha})(\alpha^* - \hat{\alpha})' = EPee'P' = \hat{V}$, which is as desired for the distribution of $\alpha^*$. A similar procedure can be used to draw values of $u_t^*$ from the $N(0, \hat{\Sigma})$ distribution: $\hat{\Sigma}$ is factored into $PP'$, and $u_t^*$ is computed as $Pe$, where $e$ is a $m \times 1$ vector of standard normal draws.

An alternative procedure for drawing values of the error terms, due to McCarthy (1972), has also been used in practice. For this procedure one begins with the $m \times T$ matrix of estimated error terms, $\hat{U}$. $T$ standard normal random variables are then drawn, and $u_t^*$ is computed as $T^{-1/2}\hat{U}e$, where $e$ is a $T \times 1$ vector of the standard normal draws. It is easy to show that the covariance matrix of $u_t^*$ is $\hat{\Sigma}$, where, as above, $\hat{\Sigma}$ is $(1/T)\hat{U}\hat{U}'$.

An alternative procedure is also available for drawing values of the coefficients. Given the estimation period (say, 1 through $T$) and given $\hat{\Sigma}$, one can draw $T$ values of $u_t^*(t = 1, \ldots, T)$. One can then add these errors to the model and solve the model over the estimation period (static simulation, using the original values of the coefficient estimates). The predicted values of the endogenous variables from this solution can be taken to be a new data base, from which a new set of coefficients can be estimated. This set can then be taken to be one draw of the coefficients. This procedure is more expensive than drawing from the $N(\hat{\alpha}, \hat{V})$ distribution, since reestimation is required for each draw, but it has the advantage of not being based on a fixed estimate of the distribution of the coefficient estimates. It is, of course, based on a fixed value of $\hat{\Sigma}$ and a fixed set of original coefficient estimates.

It should finally be noted with respect to the solution of models that in actual forecasting situations most models are subjectively adjusted before the forecasts are computed. The adjustments take the form of either using values other than zero for the future error terms or using values other than the estimated values for the coefficients. Different values of the same coefficient are sometimes used for different periods. Adjusting the values of constant terms is equivalent to adjusting values of the error terms, given that a different value of the constant term can be used each period.[3] Adjustments of this type are sometimes called "add factors". With enough add factors it is possible, of course, to have the forecasts from a model be whatever the user wants, subject to the restriction that the identities must be satisfied. Most add factors are subjective in that the procedure by which they were chosen cannot be replicated by others. A few add factors are objective. For example, the procedure of setting the future values of the error terms equal to the average of the past two estimated values is an objective one. This procedure,

---

[3]Although much of the discussion in the literature is couched in terms of constant-term adjustments, Intriligator (1978, p. 516) prefers to interpret the adjustments as the user's estimates of the future values of the error terms.

along with another type of mechanical adjustment procedure, is used for some of the results in Haitovsky, Treyz, and Su (1974). See also Green, Liebenberg, and Hirsch (1972) for other examples.

## 3.   Evaluation of ex ante forecasts

The three most common measures of predictive accuracy are root mean squared error (RMSE), mean absolute error (MAE), and Theil's inequality coefficient[4] ($U$). Let $\hat{y}_{it}$ be the forecast of variable $i$ for period $t$, and let $y_{it}$ be the actual value. Assume that observations on $\hat{y}_{it}$ and $y_{it}$ are available for $t = 1, \ldots, T$. Then the measures for this variable are:

$$\text{RMSE} = \sqrt{\frac{1}{T} \sum_{t=1}^{T} (y_{it} - \hat{y}_{it})^2}, \tag{3}$$

$$\text{MAE} = \frac{1}{T} \sum_{t=1}^{T} |y_{it} - \hat{y}_{it}|, \tag{4}$$

$$U = \frac{\sqrt{\dfrac{1}{T} \sum_{t=1}^{T} (\Delta y_{it} - \Delta \hat{y}_{it})^2}}{\sqrt{\dfrac{1}{T} \sum_{t=1}^{T} (\Delta y_{it})^2}}, \tag{5}$$

where $\Delta$ in (5) denotes either absolute or percentage change. All three measures are zero if the forecasts are perfect. The MAE measure penalizes large errors less than does the RMSE measure. The value of $U$ is one for a no-change forecast ($\Delta \hat{y}_{it} = 0$). A value of $U$ greater than one means that the forecast is less accurate than the simple forecast of no change.

An important practical problem that arises in evaluating ex ante forecasting accuracy is the problem of data revisions. Given that the data for many variables are revised a number of times before becoming "final", it is not clear whether the forecast values should be compared to the first-released values, to the final values, or to some set in between. There is no obvious answer to this problem. If the revision for a particular variable is a benchmark revision, where the level of the variable is revised beginning at least a few periods before the start of the prediction period, then a common procedure is to adjust the forecast value by

[4]See Theil (1966, p. 28).

adding the forecasted change ($\Delta \hat{y}_{it}$), which is based on the old data, to the new lagged value ($y_{it-1}$) and then comparing the adjusted forecast value to the new data. If, say, the revision took the form of adding a constant amount $\bar{y}_i$ to each of the old values of $y_{it}$, then this procedure merely adds the same $\bar{y}_i$ to each of the forecasted values of $y_{it}$. This procedure is often followed even if the revisions are not all benchmark revisions, on the implicit assumption that they are more like benchmark revisions than other kinds. Following this procedure also means that if forecast changes are being evaluated, as in the $U$ measure, then no adjustments are needed.

There are a number of studies that have examined ex ante forecasting accuracy using one or more of the above measures. Some of the more recent studies are McNees (1973, 1974, 1975, 1976) and Zarnowitz (1979). It is usually the case that forecasts from both model builders and nonmodel builders are examined and compared. A common "base" set of forecasts to use for comparison purposes is the set from the ASA/NBER Business Outlook Survey. A general conclusion from these studies is that there is no obvious "winner" among the various forecasters [see, for example, Zarnowitz (1979, pp. 23, 30)]. The relative performance of the forecasters varies considerably across variables and length ahead of the forecast, and the differences among the forecasters for a given variable and length ahead are generally small. This means that there is yet little evidence that the forecasts from model builders are more accurate than, say, the forecasts from the ASA/NBER Survey.

Ex ante forecasting comparisons are unfortunately of little interest from the point of view of examining the predictive accuracy of models. There are two reasons for this. The first is that the ex ante forecasts are based on guessed rather than actual values of the exogenous variables. Given only the actual and forecast values of the endogenous variables, there is no way of separating a given error into that part due to bad guesses and that part due to other factors. A model should not necessarily be penalized for bad exogenous-variable guesses from its users. More will be said about this in Section 5. The second, and more important, reason is that almost all the forecasts examined in these studies are generated from subjectively adjusted models, (i.e. subjective add factors are used). It is thus the accuracy of the forecasting performance of the model builders rather than of the models that is being examined.

Before concluding this section it is of interest to consider two further points regarding the subjective adjustment of models. First, there is some indirect evidence that the use of add factors is quite important in practice. The studies of Evans, Haitovsky, and Treyz (1972) and Haitovsky and Treyz (1972) analyzing the Wharton and OBE models found that the ex ante forecasts from the model builders were more accurate than the ex post forecasts from the models, even when the same add factors that were used for the ex ante forecasts were used for the ex post forecasts. In other words, the use of actual rather than guessed values of the exogenous variables decreased the accuracy of the forecasts. This general conclusion can also be drawn from the results for the BEA model in Table 3 in Hirsch, Grimm, and Narasimham (1974). This conclusion is consistent with the view that the add factors are (in a loose sense) more important than the model in determining the ex ante forecasts: what one would otherwise consider to be an improvement for the model, namely the use of more accurate exogenous-variable values, worsens the forecasting accuracy.

Second, there is some evidence that the accuracy of non-subjectively adjusted ex ante forecasts is improved by the use of actual rather than guessed values of the exogenous variables. During the period 1970III–1973II, I made ex ante forecasts using a short-run forecasting model [Fair (1971)]. No add factors were used for these forecasts. The accuracy of these forecasts is examined in Fair (1974), and the results indicate that the accuracy of the forecasts is generally improved when actual rather than guessed values of the exogenous variables are used.

It is finally of interest to note, although nothing really follows from this, that the (non-subjectively adjusted) ex ante forecasts from my forecasting model were on average less accurate than the subjectively adjusted forecasts [McNees (1973)], whereas the ex post forecasts, (i.e. the forecasts based on the actual values of the exogenous variables) were on average about the same degree of accuracy as the subjectively adjusted forecasts [Fair (1974)].

## 4. Evaluation of ex post forecasts

The measures in (3)–(5) have also been widely used to evaluate the accuracy of ex post forecasts. One of the more well known comparisons of ex post forecasting accuracy is described in Fromm and Klein (1976), where eleven models are analyzed. The standard procedure for ex post comparisons is to compute ex post forecasts over a common simulation period, calculate for each model and variable an error measure, and compare the values of the error measure across models. If the forecasts are outside-sample, there is usually some attempt to have the ends of the estimation periods for the models be approximately the same. It is generally the case that forecasting accuracy deteriorates the further away the forecast period is from the estimation period, and this is the reason for wanting to make the estimation periods as similar as possible for different models.

The use of the RMSE measure, or one of the other measures, to evaluate ex post forecasts is straightforward, and there is little more to be said about this. Sometimes the accuracy of a given model is compared to the accuracy of a "naive" model, where the naive model can range from the simple assumption of no change in each variable to an autoregressive moving average (ARIMA) process for each variable. (The comparison with the no-change model is, of course, already implicit in the $U$ measure.) It is sometimes the case that turning-point observations are examined separately, where by "turning point" is meant a point at which the change in a variable switches sign. There is nothing inherent in the statistical specification of models that would lead one to examine turning points separately, but there is a strand of the literature in which turning-point accuracy has been emphasized.

Although the use of the RMSE or similar measure is widespread, there are two serious problems associated with the general procedure. The first concerns the exogenous variables. Models differ both in the number and types of variables that are taken to be exogenous and in the sensitivity of the predicted values of the endogenous variables to the exogenous-variable values. The procedure does not take these differences into account. If one model is less "endogenous" than another (say that prices are taken to be exogenous in one model but not in another), then it has an unfair advantage in the calculation of the error measures. The other problem concerns the fact that forecast error variances vary across time. Forecast error variances vary across time both because of nonlinearities in the model and because of variation in the exogenous variables. Although RMSEs are in some loose sense estimates of the averages of the variances across time, no rigorous statistical interpretation can be placed on them: they are not estimates of any parameters of the model.

There is another problem associated with within-sample calculations of the error measures, which is the possible existence of data mining. If in the process of constructing a model one has, by running many regressions, searched diligently for the best fitting equation for each variable, there is a danger that the equations chosen, while providing good fits within the estimation period, are poor approximations to the true structure. Within-sample error calculations are not likely to discover this, and so they may give a very misleading impression of the true accuracy of the model. Outside-sample error calculations should, of course, pick this up, and this is the reason that more weight is generally placed on outside-sample results.

Nelson (1972) used an alternative procedure in addition to the RMSE procedure in his ex post evaluation of the FRB-MIT-PENN (FMP) model. For each of a number of endogenous variables he obtained a series of static predictions using both the FMP model and an ARIMA model. He then regressed the actual value of each variable on the two predicted values over the period for which the predictions were made. Ignoring the fact that the FMP model is nonlinear, the predictions from the model are conditional expectations based on a given information set. If the FMP model makes efficient use of this information, then no further information should be contained in the ARIMA predictions. The ARIMA model for each variable uses only a subset of the information, namely, that contained in the past history of the variable. Therefore, if the FMP model has made efficient use of the information, the coefficient for the ARIMA predicted values should be zero. Nelson found that in general the estimates of this coefficient were significantly different from zero. This test, while interesting, cannot be used to compare models that differ in the number and types of variables that are taken to be exogenous. In order to test the hypothesis of efficient information use, the information set used by one model must be contained in the set used by the other model, and this is in general not true for models that differ in their exogenous variables.

## 5.  An alternative method for evaluating predictive accuracy

The method discussed in this section takes account of exogenous-variable uncertainty and of the fact that forecast error variances vary across time. It also deals in a systematic way with the question of the possible misspecification of the model. It accounts for the four main sources of uncertainty of a forecast: uncertainty due to (1) the error terms, (2) the coefficient estimates, (3) the exogenous-variable forecasts, and (4) the possible misspecification of the model. The method is discussed in detail in Fair (1980). The following is an outline of its main features.

Estimating the uncertainty from the error terms and coefficients can be done by means of stochastic simulation. Let $\sigma^2_{itk}$ denote the variance of the forecast error for a $k$-period-ahead forecast of variable $i$ from a simulation beginning in period $t$. Given the $J$ trials discussed in Section 2, a stochastic-simulation estimate of $\sigma^2_{itk}$ (denoted $\tilde{\sigma}^2_{itk}$) is:

$$\tilde{\sigma}^2_{itk} = \frac{1}{J} \sum_{j=1}^{J} \left( \tilde{y}^j_{itk} - \tilde{\bar{y}}_{itk} \right)^2, \tag{6}$$

where $\tilde{\bar{y}}_{itk}$ is determined by (2). If an estimate of the uncertainty from the error terms only is desired, then the trials consist only of draws from the distribution of the error terms.[5]

There are two polar assumptions that can be made about the uncertainty of the exogenous variables. One is, of course, that there is no exogenous-variable uncertainty. The other is that the exogenous-variable forecasts are in some way as uncertain as the endogenous-variable forecasts. Under this second assumption one could, for example, estimate an autoregressive equation for each exogenous variable and add these equations to the model. This expanded model, which would have no exogenous variables, could then be used for the stochastic-simula-

---

[5] Note that it is implicitly assumed here that the variances of the forecast errors exist. For some estimation techniques this is not always the case. If in a given application the variances do not exist, then one should estimate other measures of dispersion of the distribution, such as the interquartile range or mean absolute deviation.

tion estimates of the variances. While the first assumption is clearly likely to underestimate exogenous-variable uncertainty in most applications, the second assumption is likely to overestimate it. This is particularly true for fiscal-policy variables in macroeconomic models, where government-budget data are usually quite useful for purposes of forecasting up to at least about eight quarters ahead. The best approximation is thus likely to lie somewhere in between these two assumptions.

The assumption that was made for the results in Fair (1980) was in between the two polar assumptions. The procedure that was followed was to estimate an eighth-order autoregressive equation for each exogenous variable (including a constant and time in the equation) and then to take the estimated standard error from this regression as the estimate of the degree of uncertainty attached to forecasting the change in this variable for each period. This procedure ignores the uncertainty of the coefficient estimates in the autoregressive equations, which is one of the reasons it is not as extreme as the second polar assumption. In an earlier stochastic-simulation study of Haitovsky and Wallace (1972), third-order autoregressive equations were estimated for the exogenous variables, and these equations were then added to the model. This procedure is consistent with the second polar assumption above *except* that for purposes of the stochastic simulations Haitovsky and Wallace took the variances of the error terms to be one-half of the estimated variances. They defend this procedure (pp. 267–268) on the grounds that the uncertainty from the exogenous-variable forecasts is likely to be less than is reflected in the autoregressive equations.

Another possible procedure that could be used for the exogenous variables would be to gather from various forecasting services data on their ex ante forecasting errors of the exogenous variables (exogenous to you, not necessarily to the forecasting service). From these errors for various periods one could estimate a standard error for each exogenous variable and then use these errors for the stochastic-simulation draws.

For purposes of describing the present method, all that needs to be assumed is that *some* procedure is available for estimating exogenous-variable uncertainty. If equations for the exogenous variables are not added to the model, but instead some in between procedure is followed, then each stochastic-simulation trial consists of draws of error terms, coefficients, and exogenous-variable errors. If equations are added, then each trial consists of draws of error terms and coefficients from both the structural equations and the exogenous-variable equations. In either case, let $\tilde{\tilde{\sigma}}_{itk}^2$ denote the stochastic-simulation estimate of the variance of the forecast error that takes into account exogenous-variable uncertainty. $\tilde{\tilde{\sigma}}_{itk}^2$ differs from $\tilde{\sigma}_{itk}^2$ in (6) in that the trials for $\tilde{\tilde{\sigma}}_{itk}^2$ include draws of exogenous-variable errors.

Estimating the uncertainty from the possible misspecification of the model is the most difficult and costly part of the method. It requires successive reestimation and stochastic simulation of the model. It is based on a comparison of estimated variances computed by means of stochastic simulation with estimated variances computed from outside-sample forecast errors.

Consider for now stochastic simulation with respect to the structural error terms and coefficients only (no exogenous-variable uncertainty). Assume that the forecast period begins one period after the end of the estimation period, and call this period $t$. As noted above, from this stochastic simulation one obtains an estimate of the variance of the forecast error, $\tilde{\sigma}^2_{itk}$. One also obtains from this simulation an estimate of the *expected value* of the $k$-period-ahead forecast of variable $i$: $\tilde{\tilde{y}}_{itk}$ in equation (2). The difference between this estimate and the actual value, $y_{it+k-1}$, is the mean forecast error:

$$\hat{\varepsilon}_{itk} = y_{it+k-1} - \tilde{\tilde{y}}_{itk}. \tag{7}$$

If it is assumed that $\tilde{\tilde{y}}_{itk}$ exactly equals the true expected value, $\bar{y}_{itk}$, then $\hat{\varepsilon}_{itk}$ in (7) is a sample draw from a distribution with a known mean of zero and variance $\sigma^2_{itk}$. The square of this error, $\hat{\varepsilon}^2_{itk}$, is thus under this assumption an unbiased estimate of $\sigma^2_{itk}$. One thus has two estimates of $\sigma^2_{itk}$, one computed from the mean forecast error and one computed by stochastic simulation. Let $d_{itk}$ denote the difference between these two estimates:

$$d_{itk} = \hat{\varepsilon}^2_{itk} - \tilde{\sigma}^2_{itk}. \tag{8}$$

If it is further assumed that $\tilde{\sigma}^2_{itk}$ exactly equals the true value, then $d_{itk}$ is the difference between the estimated variance based on the mean forecast error and the true variance. Therefore, under the two assumptions of no error in the stochastic-simulation estimates, the expected value of $d_{itk}$ is zero.

The assumption of no stochastic-simulation error, i.e. $\tilde{\tilde{y}}_{itk} = \bar{y}_{itk}$ and $\tilde{\sigma}^2_{itk} = \sigma^2_{itk}$, is obviously only approximately correct at best. Even with an infinite number of draws the assumption would not be correct because the draws are from estimated rather than known distributions. It does seem, however, that the error introduced by this assumption is likely to be small relative to the error introduced by the fact that some assumption must be made about the mean of the distribution of $d_{itk}$. Because of this, nothing more will be said about stochastic-simulation error. The emphasis instead is on the possible assumptions about the mean of the distribution of $d_{itk}$, given the assumptions of no stochastic-simulation error.

The procedure just described uses a given estimation period and a given forecast period. Assume for sake of an example that one has data from period 1 through 100. The model can then be estimated through, say, period 70, with the forecast period beginning with period 71. Stochastic simulation for the forecast period will yield for each $i$ and $k$ a value of $d_{i71k}$ in (8). The model can then be reestimated through period 71, with the forecast period now beginning with period 72. Stochastic simulation for this forecast period will yield for each $i$ and $k$ a value of $d_{i72k}$ in (8). This process can be repeated through the estimation period ending with period 99. For the one-period-ahead forecast ($k=1$) the procedure will yield for each variable $i$ 30 values of $d_{it1}$ ($t=71,\ldots,100$); for the two-period-ahead forecast ($k=2$) it will yield 29 values of $d_{it2}$ ($t=72,\ldots,100$); and so on. If the assumption of no simulation error holds for all $t$, then the expected value of $d_{itk}$ is zero for all $t$.

The discussion so far is based on the assumption that the model is correctly specified. Misspecification has two effects on $d_{itk}$ in (8). First, if the model is misspecified, the estimated covariance matrices that are used for the stochastic simulation will not in general be unbiased estimates of the true covariance matrices. The estimated variances computed by means of stochastic simulation will thus in general be biased. Second, the estimated variances computed from the forecast errors will in general be biased estimates of the true variances. Since misspecification affects both estimates, the effect on $d_{itk}$ is ambiguous. It is possible for misspecification to affect the two estimates in the same way and thus leave the expected value of the difference between them equal to zero. In general, however, this does not seem likely, and so in general one would not expect the expected value of $d_{itk}$ to be zero for a misspecified model. The expected value may be negative rather than positive for a misspecified model, although in general it seems more likely that it will be positive. Because of the possibility of data mining, misspecification seems more likely to have a larger positive effect on the outside sample forecast errors than on the (within-sample) estimated covariance matrices.

An examination of how the $d_{itk}$ values change over time (for a given $i$ and $k$) may reveal information about the strengths and weaknesses of the model that one would otherwise not have. This information may then be useful in future work on the model. The individual values may thus be of interest in their own right aside from their possible use in estimating total predictive uncertainty.

For the total uncertainty estimates some assumption has to be made about how misspecification affects the expected value of $d_{itk}$. For the results in Fair (1980a) it was assumed that the expected value of $d_{itk}$ is constant across time: for a given $i$ and $k$, misspecification was assumed to affect the mean of the distribution of $d_{itk}$ in the same way for all $t$. Other possible assumptions are, of course, possible. One could, for example, assume that the mean of the distribution is a function of other variables. (A simple assumption in this respect is that the mean follows a linear time trend.) Given this assumption, the mean can be then estimated from a regression of $d_{itk}$ on the variables. For the assumption of a constant mean, this regression is merely a regression on a constant (i.e. the estimated constant term is merely the mean of the $d_{itk}$ values).[6] The predicted value from this regression for period $t$, denoted $\hat{d}_{itk}$, is the estimated mean for period $t$.

---

[6] For the results in Fair (1980) a slightly different assumption than that of a constant mean was made for variables with trends. For these variables it was assumed that the mean of $d_{itk}$ is proportional to $\tilde{\bar{y}}_{itk}^2$, i.e. that the mean of $d_{itk}/\tilde{\bar{y}}_{itk}^2$ is constant across time.

An estimate of the total variance of the forecast error, denoted $\hat{\sigma}^2_{itk}$, is the sum of $\tilde{\tilde{\sigma}}^2_{itk}$ – the stochastic-simulation estimate of the variance due to the error terms, coefficient estimates, and exogenous variables – and $\hat{d}_{itk}$:

$$\hat{\sigma}^2_{itk} = \tilde{\tilde{\sigma}}^2_{itk} + \hat{d}_{itk}. \tag{9}$$

Since the procedure in arriving at $\hat{\sigma}^2_{itk}$ takes into account the four main sources of uncertainty of a forecast, the values of $\hat{\sigma}^2_{itk}$ can be compared across models for a given $i$, $k$, and $t$. If, for example, one model has consistently smaller values of $\hat{\sigma}^2_{itk}$ then another, this would be fairly strong evidence for concluding that it is a more accurate model, i.e. a better approximation to the true structure.

This completes the outline of the method. It may be useful to review the main steps involved in computing $\hat{\sigma}^2_{itk}$ in (9). Assume that data are available for periods 1 through $T$ and that one is interested in estimating the uncertainty of an eight-period-ahead forecast that began in period $T+1$, (i.e. in computing $\hat{\sigma}^2_{itk}$ for $t = T+1$ and $k = 1, \ldots, 8$). Given a base set of values for the exogenous variables for periods $T+1$ through $T+8$, one can compute $\tilde{\tilde{\sigma}}^2_{itk}$ for $t = T+1$ and $k = 1, \ldots, 8$ by means of stochastic simulation. Each trial consists of one eight-period dynamic simulation and requires draws of the error terms, coefficients, and exogenous-variable errors. These draws are based on the estimate of the model through period $T$. This is the relative inexpensive part of the method. The expensive part consists of the successive reestimation and stochastic simulation of the model that are needed in computing the $d_{itk}$ values. In the above example, the model would be estimated 30 times and stochastically simulated 30 times in computing the $d_{itk}$ values. After these values are computed for, say, periods $T - r$ through $T$, then $\hat{d}_{itk}$ can be computed for $t = T+1$ and $k = 1, \ldots, 8$ using whatever assumption has been made about the distribution of $d_{itk}$. This allows $\hat{\sigma}^2_{itk}$ in (9) to be computed for $t = T+1$ and $k = 1, \ldots, 8$.

In the successive reestimation of the model, the first period of the estimation period may or may not be increased by one each time. The criterion that one should use in deciding this is to pick the procedure that seems likely to correspond to the chosen assumption about the distribution of $d_{itk}$ being the best approximation to the truth. It is also possible to take the distance between the last period of the estimation period and the first period of the forecast period to be other than one, as was done above.

It is important to note that the above estimate of the mean of the $d_{itk}$ distribution is not in general efficient because the error term in the $d_{itk}$ regression is in general heteroscedastic. Even under the null hypothesis of no misspecification, the variance of the $d_{itk}$ distribution is not constant across time. It is true, however, that $\hat{\varepsilon}_{itk}/(\tilde{\sigma}^2_{itk} + \hat{d}_{itk})^{1/2}$ has unit variance under the null hypothesis, and so it may not be a bad approximation to assume that $\hat{\varepsilon}^2_{itk}/(\tilde{\sigma}^2_{itk} + \hat{d}_{itk})$ has a constant variance across time. This then suggests the following iterative procedure. 1) For each $i$ and $k$, calculate $\hat{d}_{itk}$ from the $d_{itk}$ regression, as discussed above; 2) divide each observation in the $d_{itk}$ regression by $\tilde{\sigma}_{itk}^2 + \hat{d}_{itk}$, run another regression, and calculate $\hat{d}_{itk}$ from this regression; 3) repeat step 2) until the successive estimates of $\hat{d}_{itk}$ are within some prescribed tolerance level. Litterman (1980) has carried out this procedure for a number of models for the case in which the only explanatory variable in the $d_{itk}$ regression is the constant term (i.e. for the case in which the null hypothesis is that the mean of the $d_{itk}$ distribution is constant across time).

If one is willing to assume that $\hat{\varepsilon}_{itk}$ is normally distributed, which is at best only an approximation, then Litterman (1979) has shown that the above iterative procedure produces maximum likelihood estimates. He has used this assumption in Litterman (1980) to test the hypothesis (using a likelihood ratio test) that the mean of the $d_{itk}$ distribution is the same in the first and second halves of the sample period. The hypothesis was rejected at the 5 percent level in only 3 of 24 tests. These results thus suggest that the assumption of a constant mean of the $d_{itk}$ distribution may not be a bad approximation in many cases. This conclusion was also reached for the results in Fair (1982), where plots of $d_{itk}$ values were examined across time (for a given $i$ and $k$). There was little evidence from these plots that the mean was changing over time.

The mean of the $d_{itk}$ distribution can be interpreted as a measure of the average unexplained forecast error variance, (i.e. that part not explained by $\tilde{\sigma}_{itk}^2$) rather than as a measure of misspecification. Using this interpretation, Litterman (1980) has examined whether the use of the estimated means of the $d_{itk}$ distributions lead to more accurate estimates of the forecast error variances. The results of his tests, which are based on the normality assumption, show that substantially more accurate estimates are obtained using the estimated means. Litterman's overall results are thus quite encouraging regarding the potential usefulness of the method discussed in this section.

Aside from Litterman's use of the method to compare various versions of Sims' (1980) model, I have used the method to compare my model [Fair (1976)], Sargent's (1976) model, Sims' model, and an eighth-order autoregressive model. The results of this comparison are presented in Fair (1979).

## 6. Conclusion

It should be clear from this chapter that the comparison of the predictive accuracy of alternative models is not a straightforward exercise. The difficulty of evaluating alternative models is undoubtedly one of the main reasons there is currently so little agreement about which model best approximates the true structure of the economy. If it were easy to decide whether one model is more accurate than another, there would probably be by now a generally agreed upon model of, for example, the U.S. economy. With further work on methods like the one described in Section 5, however, it may be possible in the not-too-distant future to begin a more systematic comparison of models. Perhaps in ten or twenty years time the use of these methods will have considerably narrowed the current range of disagreements.

## References

Bianchi, C., G. Calzolari and P. Corsi (1976) "Divergences in the Results of Stochastic and Deterministic Simulation of an Italian Non Linear Econometric Model", in: L. Dekker, ed., _Simulation of Systems_. Amsterdam: North-Holland Publishing Co.

Calzolari, G. and P. Corsi (1977) "Stochastic Simulation as a Validation Tool for Econometric Models". Paper presented at IIASA Seminar, Laxenburg, Vienna, September 13–15.

Cooper, J. P. (1974) _Development of the Monetary Sector, Prediction and Policy Analysis in the FRB-MIT-Penn Model_. Lexington: D. C. Heath & Co.

Cooper, J. P. and S. Fischer (1972) "Stochastic Simulation of Monetary Rules in Two Macroeconometric Models", _Journal of the American Statistical Association_, 67, 750–760.

Cooper, J. P. and S. Fischer (1974) "Monetary and Fiscal Policy in the Fully Stochastic St. Louis Econometric Model", _Journal of Money, Credit and Banking_, 6, 1–22.

Evans, Michael K., Yoel Haitovsky and George I. Treyz, assisted by Vincent Su (1972) "An Analysis of the Forecasting Properties of U.S. Econometric Models", in: B. G. Hickman, ed., _Econometric Models of Cyclical Behavior_. New York: Columbia University Press, 949–1139.

Evans, M. K., L. R. Klein and M. Saito (1972) "Short-Run Prediction and Long-Run Simulation of the Wharton Model", in: B. G. Hickman, ed., _Econometric Models of Cyclical Behavior_. New York: Columbia University Press, 139–185.

Fair, Ray C. (1971) _A Short-Run Forecasting Model of the United States Economy_. Lexington: D. C. Heath & Co.

Fair, Ray C. (1974) "An Evaluation of a Short-Run Forecasting Model", _International Economic Review_, 15, 285–303.

Fair, Ray C. (1976) _A Model of Macroeconomic Activity. Volume II: The Empirical Model_. Cambridge: Ballinger Publishing Co.

Fair, Ray C. (1979) "An Analysis of the Accuracy of Four Macroeconometric Models", _Journal of Political Economy_, 87, 701–718.

Fair, Ray C. (1980) "Estimating the Expected Predictive Accuracy of Econometric Models," _International Economic Review_, 21, 355–378.

Fair, Ray C. (1982) "The Effects of Misspecification on Predictive Accuracy," in: G. C. Chow and P. Corsi, eds., _Evaluating the Reliability of Macro-economic Models_. New York: John Wiley & Sons, 193–213.

Fromm, Gary and Lawrence R. Klein (1976) "The NBER/NSF Model Comparison Seminar: An Analysis of Results", _Annals of Economic and Social Measurement_, Winter, 5, 1–28.

Fromm, Gary, L. R. Klein and G. R. Schink (1972) "Short- and Long-Term Simulations with the Brookings Model", in: B. G. Hickman, ed., _Econometric Models of Cyclical Behavior_. New York: Columbia University Press, 201–292.

Garbade, K. D. (1975) _Discretionary Control of Aggregate Economic Activity_. Lexington: D. C. Heath & Co.

Granger, C. W. J. and Paul Newbold (1977) _Forecasting Economic Time Series_. New York: Academic Press.

Green, G. R., M. Liebenberg and A. A. Hirsch (1972) "Short- and Long-Term Simulations with the OBE Econometric Model", in: B. G. Hickman, ed., _Econometric Models of Cyclical Behavior_. New York: Columbia University Press, 25–123.

Haitovsky, Yoel and George Treyz (1972) "Forecasts with Quarterly Macroeconometric Models: Equation Adjustments, and Benchmark Predictions: The U.S. Experience", _The Review of Economics_ *and Statistics*, 54, 317–325.

Haitovsky, Yoel, G. Treyz and V. Su (1974) *Forecasts with Quarterly Macroeconometric Models*. New York: National Bureau of Economic Research, Columbia University Press.

Haitovsky, Y. and N. Wallace (1972) "A Study of Discretionary and Non-discretionary Monetary and Fiscal Policies in the Context of Stochastic Macroeconometric Models", in: V. Zarnowitz, ed., *The Business Cycle Today*. New York: Columbia University Press.

Hirsch, Albert A., Bruce T. Grimm and Gorti V. L. Narasimham (1974) "Some Multiplier and Error Characteristics of the BEA Quarterly Model", *International Economic Review*, 15, 616–631.

Howrey, E. P. and H. H. Kelejian (1971) "Simulation versus Analytical Solutions: The Case of Econometric Models", in: T. H. Naylor, ed., *Computer Simulation Experiments with Models of Economic Systems*. New York: Wiley.

Intriligator, Michael D. (1978) *Econometric Models, Techniques, and Applications*. Amsterdam: North-Holland Publishing Co.

Litterman, Robert B. (1979) "Techniques of Forecasting Using Vector Autoregression". Working Paper No. 115, Federal Reserve Bank of Minneapolis, November.

Litterman, Robert B. (1980) "Improving the Measurement of Predictive Accuracy", mimeo.

McCarthy, Michael D. (1972) "Some Notes on the Generation of Pseudo-Structural Errors for Use in Stochastic Simulation Studies", in: B. G. Hickman, ed., *Econometric Models of Cyclical Behavior*. New York: Columbia University Press, 185–191.

McNees, Stephen K. (1973) "The Predictive Accuracy of Econometric Forecasts", *New England Economic Review*, September/October, 3–22.

McNees, Stephen K. (1974) "How Accurate Are Economic Forecasts?", *New England Economic Review*, November/December, 2–19.

McNees, Stephen K. (1975) "An Evaluation of Economic Forecasts", *New England Economic Review*, November/December, 3–39.

McNees, Stephen K. (1976) "An Evaluation of Economic Forecasts: Extension and Update", *New England Economic Review*. September/October, 30–44.

Muench, T., A. Rolnick, N. Wallace and W. Weiler (1974) "Tests for Structural Change and Prediction Intervals for the Reduced Forms of the Two Structural Models of the U.S.: The FRB-MIT and Michigan Quarterly Models", *Annals of Economic and Social Measurement*, 3, 491–519.

Nagar, A. L. (1969) "Stochastic Simulation of the Brookings Econometric Model", in: J. S. Duesenberry, G. Fromm, L. R. Klein, and E. Kuh, eds., *The Brookings Model: Some Further Results*. Chicago: Rand McNally & Co.

Nelson, Charles R. (1972) "The Prediction Performance of the FRB-MIT-PENN Model of the U.S. Economy", *The American Economic Review*, 62, 902–917.

Sargent, Thomas J. (1976) "A Classical Macroeconometric Model for the United States", *Journal of Political Economy*, 84, 207–237.

Schink, G. R. (1971) "Small Sample Estimates of the Variance–Covariance Matrix Forecast Error for Large Econometric Models: The Stochastic Simulation Technique", Ph.D. Dissertation, University of Pennsylvania.

Schink, G. R. (1974) "Estimation of Small Sample Forecast Error for Nonlinear Dynamic Models: A Stochastic Simulation Approach", mimeo.

Sims, Christopher A. (1980) "Macroeconomics and Reality", *Econometrica*, 48, 1–48.

Sowey, E. R. (1973) "Stochastic Simulation for Macroeconomic Models: Methodology and Interpretation", in: A. A. Powell and R. W. Williams, eds., *Econometric Studies of Macro and Monetary Relations*. Amsterdam: North-Holland Publishing Co.

Theil, Henri (1966) *Applied Economic Forecasting*. Amsterdam: North-Holland Publishing Co.

Zarnowitz, Victor (1979) "An Analysis of Annual and Multiperiod Quarterly Forecasts of Aggregate Income, Output, and the Price Level", *Journal of Business*, 52, 1–33.

*Chapter 34*

# NEW ECONOMETRIC APPROACHES TO STABILIZATION POLICY IN STOCHASTIC MODELS OF MACROECONOMIC FLUCTUATIONS

JOHN B. TAYLOR*

*Stanford University*

## Contents

*Grants from the National Science Foundation and the Guggenheim Foundation are gratefully acknowledged. I am also grateful to Olivier Blanchard, Gregory Chow, Avinash Dixit, George Evans, Zvi Griliches, Sandy Grossman, Ben McCallum, David Papell, Larry Reed, Philip Reny, and Ken West for helpful discussions and comments on an earlier draft.

*Handbook of Econometrics, Volume III, Edited by Z. Griliches and M.D. Intriligator*
© *Elsevier Science Publishers BV, 1986*

## 1.  Introduction

During the last 15 years econometric techniques for evaluating macroeconomic policy using dynamic stochastic models in which expectations are consistent, or rational, have been developed extensively. Designed to solve, control, estimate, or test such models, these techniques have become essential for theoretical and applied research in macroeconomics. Many recent macro policy debates have taken place in the setting of dynamic rational expectations models. At their best they provide a realistic framework for evaluating policy and empirically testing assumptions and theories. At their worst, they serve as a benchmark from which the effect of alternative assumptions can be examined. Both "new Keynesian" theories with sticky prices and rational expectations, as well as "new Classical" theories with perfectly flexible prices and rational expectations fall within the domain of such models. Although the models entail very specific assumptions about expectation formation and about the stochastic processes generating the macroeconomic time series, they may serve as an approximation in other circumstances where the assumptions do not literally hold.

The aim of this chapter is to describe and explain these recently developed policy evaluation techniques. The focus is on discrete time stochastic models, though some effort is made to relate the methods to the geometric approach (i.e. phase diagrams and saddlepoint manifolds) commonly used in theoretical continuous time models. The exposition centers around a number of specific prototype rational expectations models. These models are useful for motivating the solution methods and are of some practical interest *per se*. Moreover, the techniques for analyzing these prototype models can be adapted fairly easily to more general models. Rational expectations techniques are much like techniques to solve differential equations: once some of the basic ideas, skills, and tricks are learned, applying them to more general or higher order models is straightforward and, as in many differential equations texts, might be left as exercises.

Solution methods for several prototype models are discussed in Section 2. The effects of anticipated, unanticipated, temporary, or permanent changes in the policy variables are calculated. The stochastic steady state solution is derived, and the possibility of non-uniqueness is discussed. Evaluation of policy rules and estimation techniques oriented toward the prototype models are discussed in Sections 3 and 4. Techniques for general linear and nonlinear models are discussed in Sections 5 and 6.

## 2.  Solution concepts and techniques

The *sine qua non* of a rational expectations model is the appearance of forecasts of events based on information available before the events take place. Many different techniques have been developed to solve such models. Some of these techniques are designed for large models with very general structures. Others are designed to be used in full information estimation where a premium is placed on computing reduced form parameters in terms of structural parameters as quickly and efficiently as possible. Others are short-cut methods designed to exploit special features of a particular model. Still others are designed for exposition where a premium is placed on analytic tractability and intuitive appeal. Graphical methods fall in this last category.

In this section, I examine the basic solution concept and explain how to obtain the solutions of some typical linear rational expectations models. For expositional purposes I feel the method of undetermined coefficients is most useful. This method is used in time series analysis to convert stochastic difference equations into deterministic difference equations in the coefficients of the infinite moving average representation. [See Anderson (1971, p. 236) or Harvey (1981, p. 38)]. The difference equations in the coefficients have exactly the same form as a deterministic version of the original model, so that the method can make use of techniques available to solve deterministic difference equations. This method was used by Muth (1961) in his original exposition of the rational expectations assumption. It provides a general unified treatment of most stochastic rational expectations models without requiring knowledge of any advanced techniques, and it clearly reveals the nature of the assumptions necessary for existence and uniqueness of solutions. It also allows for different viewpoint dates for expectations, and provides an easy way to distinguish between the effects of anticipated versus unanticipated policy shifts. The method gives the solution in terms of an infinite moving average representation which is also convenient for comparing a model's properties with the data as represented in estimated infinite moving average representations. An example of such a comparison appears in Taylor (1980b). An infinite moving average representation, however, is not useful for maximum likelihood estimation for which a finite ARMA model is needed. Although it is usually easy to convert an infinite moving average model into a finite ARMA model, there are computationally more advantageous ways to compute the ARMA model directly as we will describe below.

## 2.1. Scalar models

Let $y_t$ be a random variable satisfying the relationship

$$y_t = \alpha \mathop{\mathrm{E}}_t y_{t+1} + \delta u_t, \tag{2.1}$$

where $\alpha$ and $\delta$ are parameters and $\mathrm{E}_t$ is the conditional expectation based on all information through period $t$. The variable $u_t$ is an exogenous shift variable or "shock" to the equation. It is assumed to follow a general linear process with the representation

$$u_t = \sum_{i=0}^{\infty} \theta_i \varepsilon_{t-i},$$ (2.2)

where $\theta_i = 0, 1, 2, \ldots$ is a sequence of parameters, and where $\varepsilon_t$ is a serially uncorrelated random variable with zero mean. The shift variable could represent a policy variable or a stochastic error term as in an econometric equation. In the latter case, $\delta$ would normally be set to 1.

The information upon which the expectation in (2.1) is conditioned includes past and current observations on $\varepsilon_t$ as well as the values of $\alpha$, $\delta$, and $\theta_i$. The presence of the expected value of a *future* endogenous variable $E_t y_{t+1}$ is emphasized in this prototype model because the dynamic properties that this variable gives to the model persist in more complicated models and raise many important conceptual issues. Solving the model means finding a stochastic process for the random variable $y_t$ that satisfies eq. (2.1). The forecasts generated by this process will then be equal to the expectations that appear in the model. In this sense, expectations are consistent with the model, or equivalently, expectations are rational.

*A macroeconomic example.*   An important illustration of eq. (2.1) is a classical full-employment macro model with flexible prices. In such a model the real rate of interest and real output are unaffected by monetary policy and thus they can be considered fixed constants. The demand for real money balances – normally a function of the nominal interest rate and total output – is therefore a function only of the expected inflation rate. If $p_t$ is the log of the price level and $m_t$ is the log of the money supply, then the demand for real money can be represented as

$$m_t - p_t = -\beta \left( \mathop{E}_t p_{t+1} - p_t \right),$$ (2.3)

with $\beta > 0$. In other words, the demand for real money balances depends negatively on the expected rate of inflation, as approximated by the expected first difference of the log of the price level. Eq. (2.3) can be written in the form of eq. (2.1) by setting $\alpha = \beta/(1+\beta)$ and $\delta = 1/(1+\beta)$, and by letting $y_t = p_t$ and $u_t = m_t$. In this example the variable $u_t$ represents shifts in the supply of money, as generated by the process (2.2). Alternatively, we could add an error term $v_t$ to the right hand side of eq. (2.3), to represent shifts in the demand for money. Eq. (2.3) was originally introduced in the seminal work by Cagan (1956), but with adaptive, rather than rational expectations. The more recent rational expectations version has been used by many researchers including Sargent and Wallace (1973).

## 2.1.1.  Some economic policy interpretations of the shocks

The stochastic process for the shock variable $u_t$ is assumed in eq. (2.2) to have a general form. This form includes any stationary ARMA process [see Harvey (1981), p. 27, for example]. For empirical applications this generality is necessary because both policy variables and shocks to equations frequently have complicated time series properties. In many policy applications (where $u_t$ in (2.2) is a policy variable), one is interested in "thought experiments" in which the policy variable is shifted in a special way and the response of the endogenous variables is examined. In standard econometric model methodology, such thought experiments require one to calculate policy multipliers [see Chow (1983), p. 147, for example]. In forward-looking rational expectations models, the multipliers depend not only on whether the shift in the policy variable is temporary or permanent, but also on whether it is anticipated or unanticipated. Eq. (2.2) can be given a special form to characterize these different thought experiments, as the following examples indicate.

*Temporary versus permanent shocks.*   The shock $u_t$ is purely temporary when $\theta_0 = 1$ and $\theta_i = 0$ for $i > 0$. Then any shock $u_t$ is expected to disappear in the period immediately after it has occurred; that is $E_t u_{t+i} = 0$ for $i > 0$ at every realization of $u_t$. At the other extreme the shock $u_t$ is permanent when $\theta_i = 1$ for $i > 0$. Then any shock $u_t$ is expected to remain forever; that is $E_t u_{t+i} = u_t$ for $i > 0$ at every realization of $u_t$. In this permanent case the $u_t$ process can be written as $u_t = u_{t-1} + \varepsilon_t$. (Although $u_t$ is not a stationary process in this case, the solution can still be used for thought experiments, or transformed into a stationary series by first-differencing.)

By setting $\theta_i = \rho^i$, a range of intermediate persistence assumptions can be modeled as $\rho$ varies from 0 to 1. For $0 < \rho < 1$ the shock $u_t$ is assumed to phase out geometrically. In this case the $u_t$ process is simply $u_t = \rho u_{t-1} + \varepsilon_t$, a first order autoregressive model. When $\rho = 0$, the disturbances are purely temporary. When $\rho = 1$, they are permanent.

*Anticipated versus unanticipated shocks.*   In policy applications it is also important to distinguish between anticipated and unanticipated shocks. Time delays between the realization of the shock and its incorporation in the current information set can be introduced for this purpose by setting $\theta_i = 0$ for values of $i$ up to the length of time of anticipation. For example, in the case of a purely temporary shock, we can set $\theta_0 = 0$, $\theta_1 = 1$, $\theta_i = 0$ for $i > 1$ so that $u_t = \varepsilon_{t-1}$. This would characterize a temporary shock which is anticipated one period in advance. In other words the expectation of $u_{t+1}$ at time $t$ is equal to $u_{t+1}$ because $\varepsilon_t = u_{t+1}$ is in the information set at time $t$. More generally a temporary shock anticipated $k$ periods in advance would be represented by $u_t = \varepsilon_{t-k}$.

A permanent shock which is anticipated $k$ periods in advance would be modeled by setting $\theta_i = 0$ for $i = 1, \ldots, k - 1$ and $\theta_i = 1$ for $i = k, \; k + 1, \ldots$ .

Table 1
Summary of alternative policies and their effects.

Model: $\qquad y_t = \alpha \, \mathrm{E}_t \, y_{t+1} + \delta u_t, \ |\alpha| < 1.$

Policy: $\qquad u_t = \sum_{i=0}^{\infty} \theta_i \varepsilon_{t-i} \Rightarrow \theta_i = \dfrac{\mathrm{d} u_{t+i}}{\mathrm{d}\varepsilon_t}, \ i = 0, 1, \ldots$.

Solution Form: $\qquad y_t = \sum_{i=0}^{\infty} \gamma_i \varepsilon_{t-i} \Rightarrow \gamma_i = \dfrac{\mathrm{d} y_{t+i}}{\mathrm{d}\varepsilon_t}, \ i = 0, 1, \ldots$.

Stochastics: $\varepsilon_t$ is serially uncorrelated with zero mean.
Thought Experiment: One time unit impulse to $\varepsilon_t$.
Theorem: For every integer $k \geq 0$.
  if

$$\theta_i = \begin{cases} 0 & \text{for } i < k, \\ \rho^{i-k} & \text{for } i \geq k, \end{cases}$$

  then

$$\gamma_i = \begin{cases} \dfrac{\delta \alpha^{-(i-k)}}{1 - \alpha\rho} & \text{for } i < k, \\[3mm] \dfrac{\delta \rho^{i-k}}{1 - \alpha\rho} & \text{for } i \geq k. \end{cases}$$

Interpretation:
  Policy is *anticipated* $k$ periods in advance,
    $k = 0$ means *unanticipated*.
  Policy is phased-out at geometric rate $\rho$, $0 \leq \rho \leq 1$,
    $\rho = 0$ means purely *temporary* (N.B. $\rho^0 = 1$ when $\rho = 0$),
    $\rho = 1$ means *permanent*.

Similarly, a shock which is anticipated $k$ periods in advance and which is then expected to phase out gradually would be modeled by setting $\theta_i = 0$ for $i = 1, \ldots, k-1$ and $\theta_i = \rho^{i-k}$ for $i = k, k+1, \ldots$, with $0 < \rho < 1$. In this case (2.2) can be written alternatively as $u_t = \rho u_{t-1} + \varepsilon_{t-k}$, a first-order autoregressive model with a time delay.

The various categories of shocks and their mathematical representations are summarized in Table 1. Although in practice, we interpret $\varepsilon_t$ in eq. (2.2) as a continually perturbed random variable, for these thought experiments we examine the effect of a one-time unit impulse to $\varepsilon_t$. The solution for $y_t$ derived below can be used to calculate the effects on $y_t$ of such single realizations of $\varepsilon_t$.

### 2.1.2. Finding the solution

In order to find a solution for $y_t$ (that is, a stochastic process for $y_t$ which satisfies the model (2.1) and (2.2)), we begin by representing $y_t$ in the unrestricted infinite moving average form

$$y_t = \sum_{i=0}^{\infty} \gamma_i \varepsilon_{t-i}. \qquad (2.4)$$

Finding a solution for $y_t$ then requires determining values for the undetermined coefficients $\gamma_i$ such that eq. (2.1) and (2.2) are satisfied. Current and past $\varepsilon_t$ represent the entire history of the perturbations to the model. Eq. (2.4) simply states that $y_t$ is a general function of all possible events that may potentially influence $y_t$. The linear form is used in (2.4) because the model (2.2) is linear. Note that the solution for $y_t$ in eq. (2.4) can easily be used to calculate the effect of a one time unit shock to $\varepsilon_t$. The dynamic impact of such a shock is simply $\mathrm{d}y_{t+s}/\mathrm{d}\varepsilon_t = \gamma_s$.

To find the unknown coefficients, the most direct procedure is to substitute for $y_t$ and $\mathrm{E}_t y_{t+1}$ in (2.1) using (2.4), and solve for the $\gamma_i$ in terms of $\alpha$, $\delta$ and $\theta_i$. The conditional expectation $\mathrm{E}_t y_{t+1}$ is obtained by leading (2.4) by one period and taking expectations, making use of the equalities $\mathrm{E}_t \varepsilon_{t+i} = 0$ for $i > 0$. The first equality follows from the assumption that $\varepsilon_t$ has a zero unconditional mean and is uncorrelated; the second follows from the fact that $\varepsilon_{t+i}$ for $i < 0$ is in the conditioning set at time $t$. The conditional expectation is

$$\mathrm{E}_t y_{t+1} = \sum_{i=1}^{\infty} \gamma_i \varepsilon_{t-i+1}. \tag{2.5}$$

Substituting (2.2), (2.4) and (2.5) into (2.1) results in

$$\sum_{i=0}^{\infty} \gamma_i \varepsilon_{t-1} = \alpha \sum_{i=1}^{\infty} \gamma_i \varepsilon_{t-i+1} + \delta \sum_{i=0}^{\infty} \theta_i \varepsilon_{t-i}. \tag{2.6}$$

Equating the coefficients of $\varepsilon_t, \varepsilon_{t-1}, \varepsilon_{t-2}, \ldots$ on both sides of the equality (2.6) results in the set of equations

$$\gamma_i = \alpha \gamma_{i+1} + \delta \theta_i \qquad i = 0, 1, 2, \ldots . \tag{2.7}$$

The first equation in (2.7) for $i = 0$ equates the coefficients of $\varepsilon_t$ on both sides of (2.6); the second equation similarly equates the coefficient for $\varepsilon_{t-1}$ and so on.

Note that (2.7) is a deterministic difference equation in the $\gamma_i$ coefficients with $\theta_i$ as a forcing variable. This deterministic difference equation has the same structure as the stochastic difference eq. (2.1). It can be thought of as a deterministic perfect foresight model of the "variable" $\gamma_i$. Hence, the problem of solving a *stochastic* difference equation with conditional expectations of future variables has been converted into a problem of solving a *deterministic* difference equation.

### 2.1.3. The solution in the case of unanticipated shocks

Consider first the most elementary case where $u_t = \varepsilon_t$. That is, $\theta_i = 0$ for $i \geq 1$. This is the case of unanticipated shocks which are temporary. Then eq. (2.7) can be written

$$\gamma_0 = \alpha\gamma_1 + \delta. \tag{2.8}$$

$$\gamma_{i+1} = \frac{1}{\alpha}\gamma_i \qquad i = 1, 2, \ldots . \tag{2.9}$$

From eq. (2.9) all the $\gamma_i$ for $i > 1$ can be obtained once we have $\gamma_1$. However, eq. (2.8) gives only one equation in the two unknowns $\gamma_0$ and $\gamma_1$. Hence without further information we cannot determine the $\gamma_i$ coefficients uniquely. The number of unknowns is one greater than the number of equations. This indeterminacy is what leads to non-uniqueness in rational expectations models and has been studied by many researchers including Blanchard (1979), Flood and Garber (1980), McCallum (1983), Gourieroux, Laffont, and Monfort (1982), Taylor (1977), and Whiteman (1983).

If $|\alpha| \le 1$ then the requirement that $y_t$ is a stationary process will be sufficient to yield a unique solution. (The case where $|\alpha| > 1$ is considered below in Section 2.1.4.). To see this suppose that $\gamma_1 \ne 0$. Since eq. (2.9) is an unstable difference equation, the $\gamma_i$ coefficients will explode as $i$ gets large. But then $y_t$ would not be a stationary stochastic process. The only value for $\gamma_1$ that will prevent the $\gamma_i$ from exploding is $\gamma_1 = 0$. From (2.9) this in turn implies that $\gamma_i = 0$ for all $i > 1$. From eq. (2.8) we then have that $\gamma_0 = \delta$. Hence, the unique stationary solution is simply $y_t = \delta\varepsilon_t$. In this case, the impact of a unit shock $\mathrm{d}y_{t+s}/\mathrm{d}\varepsilon_t$ is equal to $\delta$ for $s = 0$ and is equal to 0 for $s \ge 1$. This simple impact effect is illustrated in Figure 1a. (The more interesting charts in Figures 1b, 1c, and 1d will be described below).

### Example

In the case of the Cagan money demand equation this means that the price $p_t = (1+\beta)^{-1}m_t$. Because $\beta > 0$, a temporary unanticipated increase in the money supply increases the price level by less than the increase in money. This is due to the fact that the price level is expected to decrease to its normal value (zero) next period, thereby generating an expected deflation. The expected deflation increases the demand for money so that real balances must increase. Hence, the price $p_t$ rises by less than $m_t$. This is illustrated in Figure 2a.

For the more general case of unanticipated shifts in $u_t$ that are expected to phase-out gradually we set $\theta_i = \rho^i$, where $\rho < 1$. Eq. (2.7) then becomes

$$\gamma_{i+1} = \frac{1}{\alpha}\gamma_i - \frac{\delta\rho^i}{\alpha} \qquad i = 0, 1, 2, 3, \ldots . \tag{2.10}$$

Again, this is a standard deterministic difference equation. In this more general case, we can obtain the solution $\gamma_i$ by deriving the solution to the homogeneous part $\gamma_i^{(H)}$ and the particular solution to the non-homogeneous part $\gamma_i^{(P)}$.

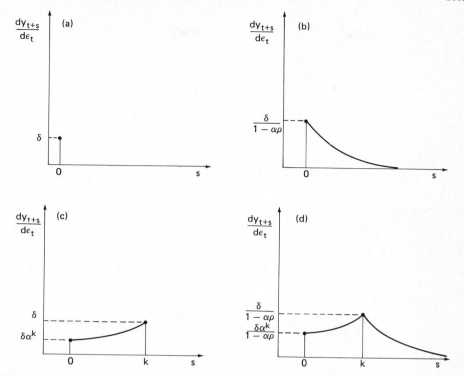

Figure 1(a). Effect on $y_t$ of an unanticipated unit shift in $u_t$ which is temporary ($u_t = \varepsilon_t$). (b). Effect on $y_t$ of an unanticipated unit shift in $u_t$ which is phased-out gradually ($u_t = \rho u_{t-1} + \varepsilon_t$). (c). Effect on $y_t$ of an anticipated unit shift in $u_t$ which is temporary (anticipated at time 0 and to occur at time $k$) ($u_t = \varepsilon_{t-k}$). (d). Effect on $y_t$ of an anticipated shift in $u_t$ which is phased-out gradually (anticipated at time 0 and to occur at time $k$) ($u_t = \rho u_{t-1} + \varepsilon_{t-k}$).

The solution to (2.10) is the sum of the homogeneous solution and the particular solution $\gamma_i = \gamma_i^{(H)} + \gamma_i^{(P)}$. [See Baumol (1970) for example, for a description of this solution technique for deterministic difference equations]. The homogeneous part is

$$\gamma_{i+1}^{(H)} = \frac{1}{\alpha}\gamma_i^{(H)} \qquad i = 0,1,2,\ldots, \tag{2.11}$$

with solution $\gamma_{i+1}^{(H)} = (1/\alpha)^{i+1}\gamma_0^{(H)}$. As in the earlier discussion if $|\alpha| < 1$ then for stationarity we require that $\gamma_0^{(H)} = 0$. For any other value of $\gamma_0^{(H)}$ the homogeneous solution will explode. Stationarity therefore implies that $\gamma_i^{(H)} = 0$ for $i = 0,1,2,\ldots$ .

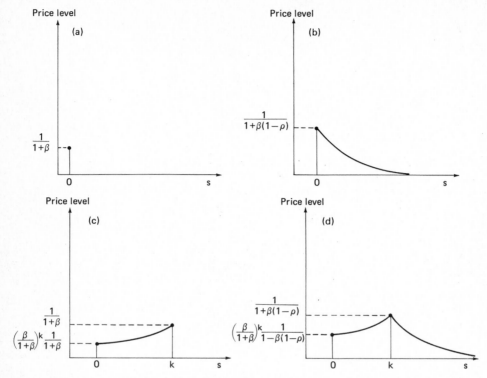

Figure 2(a). Price level effect of an unanticipated unit increase in $m_t$ which lasts for one period. (b). Price level effect of an unanticipated increase in $m_t$ which is phased-out gradually. (c). Price level effect of an anticipated unit increase in $m_{t+k}$ which lasts for one period. The increase is anticipated $k$ periods in advance. (d). Price level of an anticipated unit increase in $m_{t+k}$ which is phased-out gradually. The increase is anticipated $k$ periods in advance.

To find the particular solution we substitute $\gamma_i^{(P)} = hb^i$ into (2.10) and solve for the unknown coefficients $h$ and $b$. This gives:

$$b = \rho, \tag{2.12}$$

$$h = \delta(1 - \alpha\rho)^{-1}.$$

Because the homogeneous solution is identically equal to zero, the sum of the homogeneous and the particular solutions is simply

$$\gamma_i = \frac{\delta\rho^i}{1 - \alpha\rho}, \qquad i = 0, 1, 2, \dots. \tag{2.13}$$

In terms of the representation for $y_t$ this means that

$$y_t = \frac{\delta}{1 - \alpha\rho} \sum_{i=0}^{\infty} \rho^i \varepsilon_{t-i}$$

$$= \frac{\delta}{1 - \alpha\rho} u_t. \tag{2.14}$$

The variable $y_t$ is proportional to the shock $u_t$ at all $t$. The effect of a unit shock $\varepsilon_t$ is shown in Figure 1b. Note that $y_t$ follows the same type of first order stochastic process that $u_t$ does; that is,

$$y_t = \rho y_{t-1} + \frac{\delta \varepsilon_t}{1 - \alpha\rho}. \tag{2.15}$$

*Example*

For the money demand example, eq. (2.14) implies that

$$p_t = \frac{1}{1 - \left(\dfrac{\beta}{1+\beta}\right)\rho} \left(\frac{1}{1+\beta}\right) m_t$$

$$= \left(\frac{1}{1 + \beta(1-\rho)}\right) m_t. \tag{2.16}$$

As long as $\rho < 1$ the increase in the price level will be less than the increase in the money supply. The dynamic impact on $p_t$ of a unit shock to the money supply is shown in Figure 2b. The price level increases by less than the increase in the money supply because of the expected deflation that occurs as the price level gradually returns to its equilibrium value of 0. The expected deflation causes an increase in the demand for real money balances which is satisfied by having the price level rise less than the money supply. For the special case that $\rho = 1$, a permanent increase in the money supply, the price level moves proportionately to money as in the simple quantity theory. In that case there is no change in the expected rate of inflation since the price level remains at its new level.

### 2.1.4. A digression on the possibility of non-uniqueness

If $|\alpha| > 1$, then simply requiring that $y_t$ is a stationary process will not yield a unique solution. In this case eq. (2.9) is stable, and any value of $\gamma_1$ will give a stationary time series. There is a continuum of solutions and it is necessary to place additional restrictions on the model if one wants to obtain a unique solution for the $\gamma_i$. There does not seem to be any completely satisfactory approach to take in this case.

One possibility raised by Taylor (1977) is to require that the process for $y_t$ have a minimum variance. Consider the case where $u_t$ is uncorrelated. The variance of $y_t$ is given by

$$\operatorname{Var} y_t = \gamma_0^2 + (\gamma_0 - \delta)^2 (\alpha^2 - 1)^{-1}. \tag{2.17}$$

where the variance of $\varepsilon_t$ is supposed to be 1. The minimum occurs at $\gamma_0 = \delta_\alpha^{-2}$ from which the remaining $\gamma_i$ can be calculated. Although the minimum variance condition is a natural extension of the stationarity (finite variance) condition, it is difficult to give it an economic rationale.

An alternative rule for selecting a solution was proposed by McCallum (1983), and is called the "minimum state variable technique". In this case it chooses a representation for $y_t$ which involves the smallest number of $\varepsilon_t$ terms; hence, it would give $y_t = \delta \varepsilon_t$. McCallum (1983) examines this selection rule in several different applications.

Chow (1983, p. 361) has proposed that the uniqueness issue be resolved empirically by representing the model in a more general form. To see this substitute eq. (2.8) with $\delta = 1$ and eq. (2.9) into eq. (2.4) for an arbitrary $\gamma_1$. That is, from eq. (2.4) we write

$$y_t = \sum_{i=0}^{\infty} \gamma_i \varepsilon_{t-1}$$
$$= (\alpha \gamma_1 + 1) \varepsilon_t + \gamma_1 \varepsilon_{t-1} + (\gamma_1/\alpha) \varepsilon_{t-2} + (\gamma_1/\alpha^2) \varepsilon_{t-3} + \cdots. \tag{2.18}$$

Lagging (2.18) by one time period, multiplying by $\alpha^{-1}$ and subtracting from (2.18) gives

$$y_t = \frac{1}{\alpha} y_{t-1} + (\alpha \gamma_1 + 1) \varepsilon_t - \frac{1}{\alpha} \varepsilon_{t-1}, \tag{2.19}$$

which is ARMA (1,1) model with a free parameter $\gamma_1$. Clearly if $\gamma_1 = 0$ then this more general solution reduces to the solution discussed above. But, rather than imposing this condition, Chow (1983) has suggested that the parameter $\gamma_1$ be estimated, and has developed an appropriate econometric technique. Evans and Honkapohja (1984) use a similar procedure for representing ARMA models in terms of a free parameter.

Are there any economic examples where $|\alpha| > 1$? In the case of the Cagan money demand equation, $\alpha = \beta/(1 + \beta)$ which is always less than 1 since $\beta$ is a positive parameter. One economic example where $\alpha > 1$ is a flexible-price macroeconomic model with money in the production function. To see this consider the following equations:

$$m_t - p_t = az_t - \beta i_t. \tag{2.20}$$

$$z_t = -c\left(i_t - \left(\mathop{E}_t p_{t+1} - p_t\right)\right), \tag{2.21}$$

$$z_t = d(m_t - p_t). \tag{2.22}$$

where $z_t$ is real output, $i_t$ is the nominal interest rate, and the other variables are as defined in the earlier discussion of the Cagan model. The first equation is the money demand equation. The second equation indicates that real output is negatively related to the real rate of interest (an "*IS*" equation). In the third equation $z_t$ is positively related to real money balances. The difference between this model and the Cagan model (in eq. (2.3)) is that output is a positive function of real money balances. The model can be written in the form of eq. (2.1) with

$$\alpha = \frac{\beta}{1 + \beta - d(a + \beta c^{-1})}. \tag{2.23}$$

Eq. (2.23) is equal to the value of $\alpha$ in the Cagan model when $d = 0$. In the more general case where $d > 0$ and money is a factor in the production function, the parameter $\alpha$ can be greater than one. This example was explored in Taylor (1977). Another economic example which arises in an overlapping generation model of money was investigated by Blanchard (1979).

Although there are examples of non-uniqueness such as these in the literature, most theoretical and empirical applications in economics have the property that there is a unique stationary solution. However, some researchers, such as Gourieroux, Laffont, and Monfort (1982), have even questioned the appeal to stationarity. Sargent and Wallace (1973) have suggested that the stability requirement effectively rules out speculative bubbles. But there are examples in history where speculative bubbles have occurred and some analysts feel they are quite common. There have been attempts to model speculative bubbles as movements of $y_t$ along a self-fulfilling nonstationary (explosive) path. Blanchard and Watson (1982) have developed a model of speculative bubbles in which there is a positive probability that the bubble will burst. Flood and Garber (1980) have examined whether the periods toward the end of the eastern European hyperinflations in the 1920s could be described as self-fulfilling speculative bubbles. To date, however, the vast majority of rational expectations research has assumed that there is a unique stationary solution. For the rest of this paper we assume that $|\alpha| < 1$, or the equivalent in higher order models, and we assume that the solution is stationary.

### 2.1.5. Finding the solution in the case of anticipated shocks

Consider now the case where the shock is anticipated $k$ periods in advance and is purely temporary. That is, $u_t = \varepsilon_{t-k}$ so that $\theta_k = 1$ and $\theta_i = 0$ for $i \neq k$. The difference equations in the unknown parameters can be written as:

$$\gamma_i = \alpha \gamma_{i+1} \qquad i = 0, 1, 2, \ldots k - 1. \tag{2.24}$$

$$\gamma_{k+1} = \frac{1}{\alpha} \gamma_k - \frac{\delta}{\alpha}. \tag{2.25}$$

$$\gamma_{i+1} = \frac{1}{\alpha} \gamma_i \qquad i = k+1, \, k+2, \ldots . \tag{2.26}$$

The set of equations in (2.26) is identical in form to what we considered earlier except that the initial condition is at $k + 1$. For stationarity we therefore require that $\gamma_{k+1} = 0$. This implies from eq. (2.25) that $\gamma_k = \delta$. The remaining coefficients are obtained by working back using (2.24) starting with $\gamma_k = \delta$. This gives $\gamma_i = \delta \alpha^{k-i}$, $i = 0, 1, 2, \ldots k - 1$.

The pattern of the $\gamma_i$ coefficients is shown in Figure 1c. These coefficients give the impact of $\varepsilon_t$ on $y_{t+s}$, for $s > 0$, or equivalently the *impact of the news* that the shock $u_t$ will occur $k$ periods later. The size of $\gamma_0$ depends on how far in the future the shock is anticipated. The farther in advance the shock is known (that is, the larger is $k$), the smaller will be the current impact of the news.

*Example*

For the demand for money example we have

$$p_t = \delta \left[ \alpha^k \varepsilon_t + \alpha^{k-1} \varepsilon_{t-1} + \cdots + \alpha \varepsilon_{t-(k-1)} + \varepsilon_{t-k} \right]. \tag{2.27}$$

Substituting $\alpha = \beta/(1+\beta)$, $\delta = 1/(1+\beta)$, and $\varepsilon_t = u_{t+k} = m_{t+k}$ into (2.27) we get

$$p_t = \left( \frac{1}{1+\beta} \right) \sum_{i=0}^{k} \left( \frac{\beta}{1+\beta} \right)^{k-i} m_{t+k-i}. \tag{2.28}$$

Note how this reduces to $p_t = (1+\beta)^{-1} m_t$ in the case of unanticipated shocks ($k = 0$), as we calculated earlier. When the temporary increase in the money supply is anticipated in advance, the price level "jumps" at the date of announcement and then gradually increases until the money supply *does* increase. This is illustrated in Figure 2c.

Finally, we consider the case where the shock is anticipated in advance, but is expected to be permanent or to phase-out gradually. Then, suppose that $\theta_i = 0$ for

$i = 1, \ldots, k - 1$ and $\theta_i = \rho^{i-k}$ for $i \geq k$. Eq. (2.7) becomes

$$\gamma_i = \alpha \gamma_{i+1} \qquad i = 0, 1, 2, \ldots, k - 1, \tag{2.29}$$

$$\gamma_{i+1} = \frac{1}{\alpha} \gamma_i - \frac{\delta \rho^{i-k}}{\alpha} \qquad i = k, k + 1, \ldots. \tag{2.30}$$

Note that eq. (2.30) is identical to eq. (2.10) except that the initial condition starts at $k$ rather than 0. The homogeneous part of (2.30) is

$$\gamma_{i+1}^{(H)} = \frac{1}{\alpha} \gamma_i^{(H)} \qquad i = k, k + 1, \ldots. \tag{2.31}$$

In order to prevent the $\gamma_i^{(H)}$ from exploding as $i$ increases it is necessary that $\gamma_k^{(H)} = 0$. Therefore $\gamma_i^{(H)} = 0$ for $i = k, k + 1, \ldots$. The unknown coefficients $h$ and $b$ of the particular solution $\gamma_i^{(P)} = h b^{i-k}$ are

$$h = \delta (1 - \alpha p)^{-1},$$
$$b = \rho. \tag{2.32}$$

Since the homogeneous part is zero we have that

$$\gamma_i = \frac{\delta \rho^{i-k}}{1 - \alpha \rho} \qquad i = k, k + 1, \ldots. \tag{2.33}$$

The remaining coefficients can be obtained by using (2.29) backwards starting with $\gamma_k = \delta (1 - \alpha \rho)^{-1}$. The solution for $y_t$ is

$$y_t = \frac{\delta}{1 - \alpha \rho} \left( \alpha^k \varepsilon_t + \alpha^{k-1} \varepsilon_{t-1} + \cdots + \alpha \varepsilon_{t-k+1} + \varepsilon_{t-k} \right.$$
$$\left. + \rho \varepsilon_{t-k-1} + \rho^2 \varepsilon_{t-k-2} + \cdots \right). \tag{2.34}$$

After the immediate impact of the announcement, $y_t$ will grow smoothly until it equals $\delta (1 - \alpha \rho)^{-1}$ at the time that $u_t$ increases. The effect then phases out geometrically. This pattern is illustrated in Figure 1d.

*Example*

For the money demand model, the effect on the price level $p_t$ is shown in Figure 2d. As before the anticipation of an increase in the money supply causes the price level to jump. The price level then increases gradually until the increase in money actually occurs. During the period before the actual increase in money, the level of real balances is below equilibrium because of the expected inflation. The initial increase becomes larger as the phase-out parameter $\rho$ gets larger. For the permanent case where $\rho = 1$ the price level eventually increases by the same amount that the money supply increases.

### 2.1.6. General ARMA processes for the shocks

The above solution procedure can be generalized to handle the case where (2.2) is an autoregressive moving average (ARMA) model. We consider only unanticipated shocks where there is no time delay. Suppose the error process is

$$u_t = \rho_1 u_{t-1} + \cdots + \rho_p u_{t-p} + \varepsilon_t + \psi_1 \varepsilon_{t-1} + \cdots + \psi_q \varepsilon_{t-q}, \tag{2.35}$$

an ARMA $(p, q)$ model. The coefficients in the linear process for $u_t$ in the form of (2.2) can be derived from:

$$\theta_j = \psi_j + \sum_{i=1}^{\min(j,p)} \rho_i \theta_{j-1} \qquad j = 0, 1, 2, \ldots, q,$$

$$\theta_j = \sum_{i=1}^{\min(j,p)} \rho_i \theta_{j-i} \qquad j > q. \tag{2.36}$$

where $\psi_0 = 1$. See Harvey (1981, p. 38), for example.

Starting with $j = M \equiv \max(p, q+1)$ the $\theta_j$ coefficients in (2.36) are determined by a $p$th order difference equation. The $p$ initial conditions $(\theta_{M-1}, \ldots, \theta_{M-p})$ for this difference equation are given by the $p$ equations that preceed the $\theta_M$ equation in (2.36).

To obtain the $\gamma_i$ coefficients, (2.36) can be substituted into eq. (2.7). As before, the solution to the homogeneous part is $\gamma_i^{(H)} = 0$ for all $i$. The particular solution to the non-homogeneous part will have the same form as (2.36) for $j \geq M$. That is,

$$\gamma_j = \sum_{i=1}^{p} \rho_i \gamma_{j-1} \qquad j = M, M+1, \ldots. \tag{2.37}$$

The initial conditions $(\gamma_{M-1}, \ldots, \gamma_{M-p})$ for (2.37), as well as the remaining $\gamma$ values $(\gamma_{M-p-1}, \ldots, \gamma_0)$ can then be obtained by substitution of $\theta_i$ for $i = 0, \ldots, M-1$ into (2.37). That is,

$$\gamma_{i+1} = \frac{1}{\alpha} \gamma_i - \frac{\delta}{\alpha} \theta_i \qquad i = 0, 1, \ldots, M-1. \tag{2.38}$$

Comparing the form of (2.37) and (2.38) with (2.36) indicates that the $\gamma_i$ coefficients can be interpreted as the infinite moving average representation of an ARMA $(p, M-1)$ model. That is, the solution for $y_t$ is an ARMA $(p, M-1)$ model with an autoregressive part equal to the autoregressive part of the $u_t$ process defined in eq. (2.35). This result is found in Gourieroux, Laffont, and Monfort (1982). The methods of Hansen and Sargent (1980) and Taylor (1980a) can also be used to compute the ARMA representations directly as summarized in Section 2.4 below.

*Example :   $p = 3, q = 1$*

In this case $M = 3$ and eq. (2.36) becomes

$$\theta_0 = 1,$$
$$\theta_1 = \psi_1 + \rho_1\theta_0,$$
$$\theta_2 = \rho_1\theta_1 + \rho_2\theta_0,$$
$$\theta_i = \rho_1\theta_{i-1} + \rho_2\theta_{i-2} + \rho_3\theta_{i-3} \qquad i = 3, 4, \ldots. \tag{2.39}$$

The $\gamma$ coefficients are then given by

$$\gamma_i = \rho_1\gamma_{i-1} + \rho_2\gamma_{i-2} + \rho_3\gamma_{i-3} \qquad i = 3, 4, \ldots. \tag{2.40}$$

and the initial conditions $\gamma_0$, $\gamma_1$ and $\gamma_2$ are given by solving the three linear equations

$$\gamma_1 = \frac{1}{\alpha}\gamma_0 - \frac{\delta}{\alpha},$$
$$\gamma_2 = \frac{1}{\alpha}\gamma_1 - \frac{\delta}{\alpha}(\psi_1 + \rho_1), \tag{2.41}$$
$$\gamma_2 = \left(\rho_1 - \alpha^{-1}\right)^{-1}\left(\rho_2\gamma_1 + \rho_3\gamma_0 - \frac{\delta}{2}\left(\rho_1^2 + \rho_1\psi_1 + \rho_2\right)\right).$$

Eqs. (2.40) and (2.41) imply that $y_t$ is an ARMA $(3,2)$ model.

### 2.1.7.   Different viewpoint dates

In some applications of rational expectation models the forecast of future variables might be made at different points in time. For example, a generalization of (2.1) is

$$y_t = \alpha_1 \mathop{E}_{t} y_{t+1} + \alpha_2 \mathop{E}_{t-1} y_{t+1} + \alpha_3 \mathop{E}_{t-1} y_t + u_t. \tag{2.42}$$

Substituting for $y_t$ and expected $y_t$ from (2.4) into (2.42) results in a set of equations for the $\gamma$ coefficients much like the equations that we studied above. Suppose $u_t = \rho u_{t-1} + \varepsilon_t$. Then, the equations for $\gamma$ are

$$\gamma_0 = \alpha_1\gamma_1 + \delta,$$
$$\gamma_{i+1} = \left(\frac{1 - \alpha_3}{\alpha_1 + \alpha_2}\right)\gamma_i - \frac{\delta\rho^i}{\alpha_1 + \alpha_2} \qquad i = 1, 2, \ldots. \tag{2.43}$$

Hence, we can use the same procedures for solving this set of difference equations. The solution is

$$\gamma_0 = \alpha_1 b\rho + \delta,$$
$$\gamma_i = b\rho^i \qquad i = 1,2,\dots.$$

where $b = \delta/(1 - \alpha_3 - \rho\alpha_2 - \rho\alpha_1)$. Note that this reduces to (2.13) when $\alpha_2 = \alpha_3 = 0$.

### 2.1.8. *Geometric interpretation*

The solution of the difference eq. (2.7) that underlies this technique has an intuitive graphical interpretation which corresponds to the phase diagram method used to solve continuous time models with rational expectations. [See Calvo (1980) or Dixit (1980) for example]. Eq. (2.7) can be written

$$\gamma_{i+1} - \gamma_i = \left(\frac{1}{\alpha} - 1\right)\gamma_i - \frac{\delta}{\alpha}\theta \qquad i = 0,1,\dots. \tag{2.44}$$

The set of values for which $\gamma_i$ is not changing are given by setting the right-hand side of (2.44) to zero. These values of $(\gamma_i, \theta_i)$ are plotted in Figure 3. In the case where $\theta_i = \rho^i$, for $0 < \rho < 1$ there is a difference equation representation for $\theta_i$ of the form

$$\theta_{i+1} - \theta_i = (\rho - 1)\theta_i, \tag{2.45}$$

where $\theta_0 = 1$. The set of points where $\theta$ is not changing is a vertical line at $\theta_i = 0$ in Figure 3. The forces which move $\gamma$ and $\theta$ in different directions are also shown in Figure 3. Points above (below) the upward sloping line cause $\gamma_i$ to increase (decrease). Points to the right (left) of the vertical line cause $\theta_i$ to decrease (increase). In order to prevent the $\gamma_i$ from exploding we found in Section 2.1.3

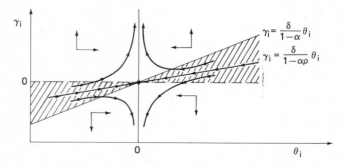

Figure 3.    Illustration of the rational expectations solution and the saddle path. Along the saddle path the motion is towards the origin at geometric rate $\rho$. That is, $\theta_i = \rho\theta_{i-1}$.

that it was necessary for $\gamma_i = (\delta/1 - \alpha\rho)\theta_i$. This linear equation is shown as the straight line with the arrows in Figure 3. This line balances off the unstable vertical forces and uses the stable horizontal forces to bring $\gamma_i$ back to the values $\gamma_i = 0$ and $\theta_i = 0$ and $i \to \infty$. For this reason it is called a saddle point and corresponds to the notion of a saddle path in differential equation models [see Birkhoff and Rota (1962), for example].

Figure 3 is special in the sense that one of the zero-change lines is perfectly vertical. This is due to the fact that the shock variable $u_t$ is exogenous to $y_t$. If we interpret (2.1) and (2.2) as a two variable system with variables $y_t$ and $u_t$ as the two variables, then the system is recursive in that $u_t$ affects $y_t$ in the current period and there are no effects of past $y_t$ on $u_t$. In Section 2.2 we consider a more general two variable system in which $u_t$ is endogenous.

In using Figure 3 for thought experiments about the effect of one time shocks, recall that $\gamma_i$ is $\mathrm{d}y_{t+i}/\mathrm{d}\varepsilon_i$ and $\theta_i$ is $\mathrm{d}\dot{u}_{t+i}/\mathrm{d}\varepsilon_i$. The vertical axis thereby gives the paths of the endogenous variable $y_t$ corresponding to a shock $\varepsilon_t$ to the policy eq. (2.2). The horizontal axis gives the path of the policy variable. The points in Figure 3 can be therefore viewed as displacements of $y_t$ and $u_t$ from their steady state values in response to a one-time unit shock.

The arrows in Figure 3 show that the saddle path line must have a slope greater than zero and a slope less than the zero-change line for $\gamma$. That is, the saddle path line must lie in the shaded region of Figure 3. Only in this region is the direction of motion toward the origin. The geometric technique to determine whether the saddle path is upward or downward sloping is frequently used in practice to obtain the sign of an impact effect of policy. [See Calvo (1980), for example].

In Figure 4 the same diagram is used to determine the qualitative movement of $y_t$ in response to a shock to $u_t$ which is anticipated $k$ periods in advance and which is expected to then phase out geometrically. This is the case considered

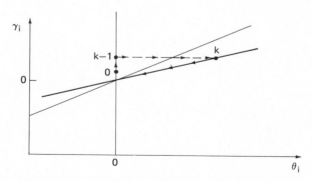

Figure 4.  Illustration of the effect of an anticipated shock to $u_t$ which is then expected to be phased out gradually at geometric rate $\rho$. The shock is anticipated $k$ periods in advance. This thought experiment corresponds to the chart in Figure 1(d).

above in Section 2.1.5. The endogenous variable $y$ initially jumps at time 0 when the future increase in $u$ becomes known; it then moves along an explosive path through period $k$ when $u$ increases by 1 unit. From time $k$ on the motion is along the saddle path as $y$ and $u$ approach their steady state values of zero.

### 2.1.9.  *Nonstationary forcing variables*

In many economic applications the forcing variables are nonstationary. For example the money supply is a highly nonstationary series. One typically wants to estimate the effects of changes in the *growth rate* of the money supply. What happens when the growth rate is reduced gradually? What if the reduction in growth is anticipated? Letting $u_t$ be the log of the money supply $m_t$, these alternatives can be analyzed by writing the *growth rate* of money as $g_t = m_t - m_{t-1}$ and assuming that

$$g_t - g_{t-1} = \rho(g_{t-1} - g_{t-2}) + \varepsilon_{t-k}.$$

Thus, the *change* in the growth rate is anticipated $k$ periods in advance. The new growth rate is phased in at a geometric rate $\rho$. By solving the model for the particular solution corresponding to this equation, one can solve for the price level and the inflation rate. In this case, the inflation rate is nonstationary, but the change in the inflation rate is stationary.

### 2.2.  *Bivariate models*

Let $y_{1t}$ and $y_{2t}$ be given by

$$
\begin{aligned}
y_{1t} &= \alpha_1 \underset{t}{E} y_{1t+1} + \beta_{10} y_{2t} + \beta_{11} y_{2t-1} + \delta_1 u_t, \\
y_{2t} &= \alpha_2 \underset{t}{E} y_{1t+1} + \beta_{20} y_{1t} + \beta_{21} y_{2t-1} + \delta_2 u_t,
\end{aligned}
\tag{2.46}
$$

where $u_t$ is a shock variable of the form (2.2). Model (2.46) is a special bivariate model in that there are no lagged values of $y_{1t}$ and no lead values of $y_{2t}$. This asymmetry is meant to convey the continuous time idea that one variable $y_{1t}$ is a "jump" variable, unaffected by its past while $y_{2t}$ is a more slowly adjusting variable that is influenced by its past values. Of course in discrete time all variables tend to jump from one period to the next so that the terminology is not exact. Nevertheless, the distinction is important in practice. Most commonly, $y_{1t}$ would be a price and $y_{2t}$ a stock which cannot change without large costs in the short run.

We assume in (2.46) that there is only one shock $u_t$. This is for notational convenience. The generalization to a bivariate shock $(u_{1t}, u_{2t})$ where $u_{1t}$ appears in the first equation and $u_{2t}$ in the second equation is straightforward, as should be clear below.

Because (2.46) has this special form it can be reduced to a first order 2-dimensional vector process:

$$
\begin{pmatrix} 1 & -\beta_{11} \\ -\beta_{20} & -\beta_{21} \end{pmatrix} \begin{pmatrix} y_{1t} \\ y_{2t-1} \end{pmatrix} = \begin{pmatrix} \alpha_1 & \beta_{10} \\ \alpha_2 & -1 \end{pmatrix} \begin{pmatrix} \underset{t}{E} y_{1t+1} \\ y_{2t} \end{pmatrix} + \begin{pmatrix} \delta_1 \\ \delta_2 \end{pmatrix} u_t. \tag{2.47}
$$

This particular way to construct a first order process follows that of Blanchard and Kahn (1980). A generalization to the case of viewpoint dates earlier than time $t$ is fairly straightforward. If $y_{1t-1}$ or $E_t y_{2t+1}$ also appeared in (2.46) then a first-order model would have to be more than 2 dimensional.

### 2.2.1. Some examples

There are many interesting examples of this simple bivariate model. Five of these are summarized below.

*Example 1: Exchange rate overshooting*

Dornbusch (1976) considered the following type of model of a small open economy [see also Wilson (1979) and Buiter and Miller (1983)]:

$$
m_t - p_t = -\alpha \left( \underset{t}{E} e_{t+1} - e_t \right),
$$

$$
p_t - p_{t-1} = \beta(e_t - p_t),
$$

where $e_t$ is the log of the exchange rate, and $p_t$ and $m_t$ are as defined in the Cagan model. The first equation is simply the demand for money as a function of the nominal interest rate. In a small open economy with perfect capital mobility the nominal interest rate is equal to the world interest rate (assumed fixed) plus the expected rate of depreciation $E_t e_{t+1} - e_t$. The second equation describes the slow adjustment of prices in response to the excess demand for goods. Excess demand is assumed to be a negative function of the relative price of home goods. Here prices adjust slowly and the exchange rate is a jump variable. This model is of the form (2.47) with $y_{1t} = e_t$, $y_{2t} = p_t$, $\alpha_1 = 1$, $\beta_{10} = -1/\alpha$, $\beta_{11} = 0$, $\delta_1 = 1/\alpha$, $\alpha_2 = 0$, $\beta_{20} = \beta/(1+\beta)$, $\beta_{21} = 1/(1+\beta)$, $\delta_2 = 0$.

*Example 2: Open economy portfolio balance model*

Kouri (1976), Rodriquez (1980), and Papell (1984) have considered the following type of rational expectations model which is based on a portfolio demand for foreign assets rather than on perfect capital mobility:

$$e_t + f_t = \alpha\left( \underset{t}{E} e_{t+1} - e_t \right) + u_t,$$

$$f_t - f_{t-1} = \beta e_t.$$

The first equation represents the demand for foreign assets $f_t$ (in logs) evaluated in domestic currency, as a function of the expected rate of depreciation. Here $u_t$ is a shock. The second equation is the "current account" (the proportional change in the stock of foreign assets) as a function of the exchange rate. Prices are assumed to be fixed and out of the picture. This model reduces to (2.47) with $y_{1t} = e_t$, $y_{2t} = f_t$, $\alpha_1 = \alpha(1 + \alpha)$, $\beta_{10} = 1/(1 + \alpha)$, $\beta_{11} = 0$, $\delta_1 = 1/1 + \alpha$, $\alpha_2 = 0$, $\beta_{20} = \beta$, $\beta_{21} = -1$, $\delta_2 = 0$.

*Example 3:   Money and capital*

Fischer (1979) developed the following type of model of money and capital.

$$y_t = \gamma k_{t-1},$$

$$r_t = -(1 - \gamma) k_{t-1},$$

$$m_t - p_t = -a_1 \underset{t}{E} r_{t+1} - a_2\left( \underset{t}{E} p_{t+1} - p_t \right) + y_t,$$

$$k_t = b_1 \underset{t}{E} r_{t+1} + b_2\left( \underset{t}{E} p_{t+1} - p_t \right) + y_t.$$

The first two equations describe output $y_t$ and the marginal efficiency of capital $r_t$ as a function of the stock of capital at the end of period $t-1$. The third and fourth equations are a pair of portfolio demand equations for capital and real money balances as a function of the rates of return on these two assets. Lucas (1976) considered a very similar model. Substituting the first two equations into the third and fourth we get model (2.47) with

$$y_{1t} = p_t, \qquad y_{2t} = k_t, \qquad \alpha_1 = \frac{a_2}{1 + a_2}, \qquad \beta_{10} = \frac{-a_1(1 - \gamma)}{1 + a_2},$$

$$\beta_{11} = 0, \qquad \delta_1 = \frac{1}{1 + a_2}, \qquad \alpha_2 = \frac{b_2}{(1 + b_1(1 - \gamma))},$$

$$\beta_{20} = \frac{-b_2}{(1 + b_1(1 - \gamma))}, \qquad \beta_{21} = \frac{\gamma}{(1 + b_1(1 - \gamma))}.$$

*Example 4:    Staggered contracts model*

The model $y_t = a_1 E_t y_{t+1} + a_2 y_{t-1} + \delta u_t$ of a contract wage $y_t$ can occur in a staggered wage setting model as in Taylor (1980a). The future wage appears because workers and firms forecast the wage set by other workers and firms. The lagged wage appears because contracts last two periods. This model can be put in the form of (2.47) by stacking the $y$'s into a vector:

$$\begin{pmatrix} 1 & -a_2 \\ -1 & 0 \end{pmatrix} \begin{pmatrix} y_t \\ y_{t-1} \end{pmatrix} = \begin{pmatrix} a_1 & 0 \\ 0 & -1 \end{pmatrix} \begin{pmatrix} E_t y_{t+1} \\ y_t \end{pmatrix} + \begin{pmatrix} \delta \\ 0 \end{pmatrix} u_t.$$

*Example 5:    Optimal control problem*

Hansen and Sargent (1980) consider the following optimal control problem. A firm chooses a contingency plan for a single factor of production (labor) $n_t$ to maximize expected profits.

$$E_t \sum_{j=0}^{\infty} \beta^j \left[ p_{t+j} y_{t+j} - \frac{\delta}{2} (n_{t+j} - n_{t+j-1})^2 - w_{t+j} n_{t+j} \right],$$

subject to the linear production function $y_t = \gamma n_t$. The random variables $p_t$ and $w_t$ are the price of output and the wage, respectively. The first order conditions of this maximization problem are:

$$\beta E_t n_{t+1} - (1 + \beta) n_t + n_{t-1} = \frac{\beta}{\delta} (w_t - \gamma p_t).$$

This model is essentially the same as that in Example (4) where $u_t = w_t - \gamma p_t$.

### 2.2.2.    Finding the solution

Equation (2.47) is a vector version of the univariate eq. (2.1). The technique for finding a solution to (2.47) is directly analogous with the univariate case.

The solution can be represented as

$$y_{1t} = \sum_{i=0}^{\infty} \gamma_{1i} \varepsilon_{t-i},$$

$$y_{2t} = \sum_{i=0}^{\infty} \gamma_{2i} \varepsilon_{t-i}. \tag{2.48}$$

These representations for the endogenous variables are an obvious generalization of eqs. (2.4).

Utilizing matrix notation we rewrite (2.47) as

$$Bz_t = C\mathop{E}_t z_{t+1} + \delta u_t, \tag{2.49}$$

$$\mathop{E}_t z_{t+1} = Az_t + du_t, \tag{2.50}$$

where the definitions of the matrices $B$ and $C$, and the vectors $z_t$ and $\delta$ in (2.49) should be clear, and where $A = C^{-1}B$ and $d = -C^{-1}\delta$. Let $\gamma_i = (\gamma_{1i}, \gamma_{2i-1})'$, $i = 0, 1, 2, \dots$ and set $\gamma_{2,-1} = 0$. Substitution of (2.2) and (2.48) into (2.50) gives

$$\gamma_{i+1} = A\gamma_i + d\theta_i \qquad i = 0, 1, 2, \dots . \tag{2.51}$$

Eq. (2.51) is analogous to eq. (2.7). For $i = 0$ we have three unknown elements of the unknown vectors $\gamma_0 = (\gamma_{10}, 0)'$ and $\gamma_1 = (\gamma_{11}, \gamma_{20})'$. The 3 unknowns are $\gamma_{10}$, $\gamma_{11}$ and $\gamma_{20}$. However, there are only two equations (at $i = 0$) in (2.51) that can be used to solve for these three parameters. Much as in the scalar case considering $i = 1$ gives two more equations, but it also gives two more unknowns ($\gamma_{12}, \gamma_{21}$); the same is true for $i = 2$ and so on. To determine the solution for the $\gamma_i$ process we therefore need another equation. As in the scalar case this third equation comes by imposing stationarity on the process for $y_{1t}$ and $y_{2t}$ or equivalently in this context by preventing either element of $\gamma_i$ from exploding. For uniqueness we will require that one root of $A$ be greater than one in modulus, and one root be less than one in modulus. The additional equation thus comes from choosing $\gamma_1 = (\gamma_{11}, \gamma_{20})'$ so that $\gamma_i$ does not explode as $i \to \infty$. This condition implies a unique linear relationship between $\gamma_{11}$ and $\gamma_{20}$. This relationship is the extra equation. It is the analogue of setting the scalar $\gamma_1 = 0$ in model (2.1).

To see this, we decompose the matrix $A$ into $H^{-1}\Lambda H$ where $\Lambda$ is a diagonal matrix with $\lambda_1$ and $\lambda_2$ on the diagonal. $H$ is the matrix whose rows are the characteristic vectors of $A$. Assume that the roots are distinct and that $|\lambda_1| > 1$ and $|\lambda_2| < 1$. Let $\mu_i \equiv (\mu_{1i}, \mu_{2i})' = H\gamma_i$. Then the homogeneous part of (2.51) is

$$\gamma_{i+1} = H^{-1}\Lambda H \gamma_i, \qquad i = 1, 2, \dots , \tag{2.52}$$

so that

$$\mu_{i+1} = \Lambda \mu_i \qquad i = 1, 2, \dots ,$$

or

$$\begin{aligned}
\mu_{1i+1} &= \lambda_1 \mu_{1i} & i &= 1, 2, \dots , \\
\mu_{2i+1} &= \lambda_2 \mu_{2i} & i &= 1, 2, \dots .
\end{aligned} \tag{2.53}$$

For stability of $\mu_{1i}$ as $i \to \infty$ we therefore require that $\mu_{11} = 0$ which in turn implies that $\mu_{1i} = 0$ for all $i > 1$. In other words we want

$$\mu_{11} = h_{11}\gamma_{11} + h_{12}\gamma_{20} = 0, \tag{2.54}$$

where $(h_{11}, h_{12})$ is the first row of $H$ and is the characteristic vector of $A$ corresponding to the unstable root $\lambda_1$. Eq. (2.54) is the extra equation. When combined with (2.51) at $i = 0$ we have 3 linear equations that can be solved for $\gamma_{10}$, $\gamma_{11}$ and $\gamma_{20}$. From these we can use (2.51) or equivalently (2.53) to obtain the remaining $\gamma_i$ for $i > 1$. In particular $\mu_{1i} = 0$ implies that

$$\gamma_{1i} = -\frac{h_{12}}{h_{11}}\gamma_{2i-1} \qquad i = 1, 2, \ldots, \cdot \tag{2.55}$$

From the second equation in (2.53) we have that

$$h_{21}\gamma_{1i+1} + h_{22}\gamma_{2i} = \lambda_2(h_{21}\gamma_{1i} + h_{22}\gamma_{2i-1}).$$

Substituting for $\gamma_{1i+1}$ and $\gamma_{1i}$ from (2.55) this gives

$$\gamma_{2i+1} = \lambda_2\gamma_{2i} \qquad i = 0, 1, 2, \ldots. \tag{2.56}$$

Given the initial values $\gamma_{21}$ we compute the remaining coefficients from (2.55) and (2.56).

### 2.2.3. The solution in the case of unanticipated shocks

When the shock $u_t$ is unanticipated and purely temporary, $\theta_0 = 1$ and $\theta_i = 0$ for all $i > 0$. In this case eq. (2.51) for $i = 0$ is

$$\gamma_{11} = a_{11}\gamma_{10} + d_1,$$
$$\gamma_{20} = a_{21}\gamma_{10} + d_2, \tag{2.57}$$

and the difference equation described by (2.51) for $i > 0$ is homogeneous. Hence the solution given by (2.55), (2.56), and (2.57) is the complete solution.

For the more general case where $\theta_i = \rho^i$, eq. (2.57) still holds but the difference equation in (2.51) for $i \geq 1$ has a nonhomogeneous part. The particular solution to the nonhomogeneous part is of the form $\gamma_i^{(P)} = gb^i$ where $g$ is a $2 \times 1$ vector. Substituting this form into (2.51) for $i \geq 1$ and equating coefficients we obtain the particular solution

$$\gamma_i^{(P)} = (\rho I - A)^{-1} d\rho^i, \qquad i = 1, 2, \ldots. \tag{2.58}$$

Since eq. (2.55) is the requirement for stability of the homogeneous solution, the complete solution can be obtained by substituting $\gamma_{11}^{(H)} = \gamma_{11} - \gamma_{11}^{(P)}$ and $\gamma_{20}^{(H)} = \gamma_{20} - \gamma_{20}^{(P)}$ into (2.54) to obtain

$$\gamma_{11} - \gamma_{11}^{(P)} = -\frac{h_{12}}{h_{11}}\left(\gamma_{20} - \gamma_{20}^{(P)}\right). \tag{2.59}$$

Eq. (2.59) can be combined with (2.57) to obtain $\gamma_{10}$, $\gamma_{11}$, and $\gamma_{20}$. The remaining coefficients are obtained by adding the appropriate elements of particular solutions (2.58) to the homogeneous solutions of (2.56) and (2.57).

### 2.2.4.   The solution in the case of anticipated shocks

For the case where the shock is anticipated $k$ periods in advance, but is purely temporary ($\theta_0 = 0$ for $i = 1, \ldots, k-1$, $\theta_i = 0$ for $i = k+1, \ldots$), we break up the difference eq. (2.51) as:

$$\gamma_{i+1} = A\gamma_i \qquad i = 0, 1, \ldots, k-1. \tag{2.60}$$

$$\gamma_{k+1} = A\gamma_k + d. \tag{2.61}$$

$$\gamma_{i+1} = A\gamma_i \qquad i = k+1, k+2, \ldots. \tag{2.62}$$

Looking at the equations in (2.62) it is clear that for stationarity, $\gamma_{k+1} = (\gamma_{1k+1}, \gamma_{2k})'$ must satisfy the same relationship that the vector $\gamma_1$ satisfied in eq. (2.55). That is,

$$\gamma_{1k+1} = -\frac{h_{12}}{h_{11}}\gamma_{2k}. \tag{2.63}$$

Once $\gamma_{2k}$ and $\gamma_{1k+1}$ have been determined the $\gamma$ values for $i > k$ can be computed as above in eqs. (2.55) and (2.56). That is,

$$\gamma_{1i+1} = -\frac{h_{12}}{h_{11}}\gamma_{2i} \qquad i = k, \ldots, \tag{2.64}$$

$$\gamma_{2i+1} = \lambda_2\gamma_{2i} \qquad i = k, \ldots. \tag{2.65}$$

To determine $\gamma_{2k}$ and $\gamma_{1k+1}$ we solve eq. (2.63) jointly with the $2(k+1)$ equations in (2.60) and (2.61) for the $2(k+1)+1$ unknowns $\gamma_{11}, \ldots, \gamma_{1k+1}$ and $\gamma_{20}, \ldots, \gamma_{2k}$. (Note how this reduces to the result obtained for the unanticipated case above when $k = 0$). A convenient way to solve these equations is to first solve the three equations consisting of the two equations from:

$$\gamma_{k+1} = A^{k+1}\gamma_0 + d, \tag{2.66}$$

(obtained by "forecasting" $\gamma_i$ out $k$ periods) and eq. (2.61) for $\gamma_{2k}$, $\gamma_{1k+1}$ and $\gamma_{10}$. Then the remaining coefficients can be obtained from the difference equations in (2.60) starting with the calculated value for $\gamma_{10}$.

The case where $\theta_i = 0$ for $i = 1, \ldots, k-1$ and $\theta_k = \rho^{k-i}$ for $i = k$, $k-1$ can be solved by adding the particular solution to the nonhomogeneous equation

$$\gamma_{i+1} = A\gamma_i + d\rho^{(i-k)} \qquad i = k, k+1, k+2, \ldots, \tag{2.67}$$

in place of (2.62) and solving for the remaining coefficients using eqs. (2.60) and (2.61) as above. The particular solution of (2.67) is

$$\gamma_i^{(P)} = (\rho I - A)^{-1} d\rho^{i-k} \qquad i = k, k+1, k+2, \ldots \,. \tag{2.68}$$

### 2.2.5. *The exchange rate overshooting example*

The preceding calculations can be usefully illustrated with Example 1 of Section 2.2.1.: the two variable "overshooting" model in which the exchange rate ($y_{1t} = e_t$) is the jump variable and the price level ($y_{2t} = p_t$) is the slowly moving variable. For this model eq. (2.50) is

$$\begin{pmatrix} \mathrm{E}e_{t+1} \\ t \\ p_t \end{pmatrix} = A \begin{pmatrix} e_t \\ p_{t-1} \end{pmatrix} + dm_t, \tag{2.69}$$

where the matrix

$$A = \frac{1}{1+\beta} \begin{pmatrix} 1 + \beta\left(1 + \dfrac{1}{\alpha}\right) & \dfrac{1}{\alpha} \\ \beta & 1 \end{pmatrix}, \tag{2.70}$$

and the vector $d = (-1/\alpha, 0)'$. Suppose that $\alpha = 1$ and $\beta = 1$. Then the characteristic roots of $A$ are

$$\lambda = 1 \pm -0.707. \tag{2.71}$$

The characteristic vector associated with the unstable root is obtained from

$$(h_{11}, h_{12}) A = \lambda_1 (h_{11}, h_{12}), \tag{2.72}$$

this gives $-h_{12}/h_{11} = -0.414$ so that according to eq. (2.56) the coefficients of the (homogeneous) solution must satisfy

$$\gamma_{1i+1} = -0.414\gamma_{2i} \qquad i = 0, 1, \ldots . \tag{2.73}$$

Using the stable root we have

$$\gamma_{2i+1} = 0.293\gamma_{2i} \qquad i = 0, 1, \ldots . \tag{2.74}$$

The particular solution is given by the vector $(\rho I - A)^{-1} \, d\rho^{i-k}$ as in eq. (2.68). That is

$$\gamma_{1i}^{(P)} = \frac{(0.5 - \rho)\rho^{i-k}}{(1.5 - \rho)(0.5 - \rho) - 0.25} \qquad i = k, k+1, k+2, \ldots, \tag{2.75}$$

$$\gamma_{2i-1}^{(P)} = \frac{-0.5\rho^{i-k}}{(1.5 - \rho)(0.5 - \rho) - 0.25} \qquad i = k, k+1, k+2, \ldots, \tag{2.76}$$

where $k$ is the number of periods in advance that the shock to the money supply is anticipated ($k = 0$ for unanticipated shocks).

In Tables 2, 3, and 4 and in Figures 5, 6, and 7, respectively, the effects of temporary unanticipated money shocks ($k = 0$, $\rho = 0$), permanent unanticipated money shocks ($k = 0$, $\rho = 1$), and permanent money shocks anticipated 3 periods

Table 2

Effect of an unanticipated temporary increase in money on the exchange rate and the price level ($k = 0$, $\rho = 0$).

| Period after shock: | $i$ | 0 | 1 | 2 | 3 | 4 |
|---|---|---|---|---|---|---|
| Effect on exchange rate: | $\gamma_{1i}$ | 0.59 | $-0.12$ | $-0.04$ | $-0.01$ | $-0.00$ |
| Effect on price level: | $\gamma_{2i}$ | 0.29 | 0.09 | 0.03 | 0.01 | 0.00 |

Table 3

Effect of unanticipated permanent increase in money on the exchange rate and the price level ($k = 0$, $\rho = 1$).

| Period after shock: | $i$ | 0 | 1 | 2 | 3 | 4 |
|---|---|---|---|---|---|---|
| Effect on exchange rate: | $\gamma_{1i}$ | 1.41 | 1.12 | 1.04 | 1.01 | 1.00 |
| particular solution: | $\gamma_{1i}^{(P)}$ | – | 1 | 1 | 1 | 1 |
| homogeneous solution: | $\gamma_{1i}^{(H)}$ | – | 0.12 | 0.04 | 0.01 | 0.00 |
| Effect on price level: | $\gamma_{2i}$ | 0.71 | 0.91 | 0.97 | 0.99 | 1.00 |
| particular solution: | $\gamma_{2i}^{(P)}$ | 1 | 1 | 1 | 1 | 1 |
| homogeneous solution: | $\gamma_{2i}^{(H)}$ | $-0.29$ | $-0.09$ | $-0.03$ | $-0.01$ | $-0.00$ |

Table 4
Effect of a permanent increase in money anticipated 3 periods in advance on the exchange rate and the price level ($k = 3$, $\rho = 1$).

| Period after the shock: | $i$ | 0 | 1 | 2 | 3 | 4 | 5 | 6 |
|---|---|---|---|---|---|---|---|---|
| Effect on the exchange rate: | $\gamma_{1i}$ | 0.28 | 0.43 | 0.71 | 1.21 | 1.06 | 1.02 | 1.00 |
| particular solution: | $\gamma_{1i}^{(P)}$ | – | – | – | – | 1.00 | 1.00 | 1.00 |
| homogeneous solution: | $\gamma_{1i}^{(H)}$ | – | – | – | – | 0.06 | 0.02 | 0.01 |
| Effect on the price level: | $\gamma_{2i}$ | 0.14 | 0.28 | 0.50 | 0.85 | 0.96 | 0.99 | 1.00 |
| particular solution: | $\gamma_{2i}^{(P)}$ | – | – | – | 1.00 | 1.00 | 1.00 | 1.00 |
| homogeneous solution: | $\gamma_{2i}^{(H)}$ | – | – | – | −0.15 | −0.04 | −0.01 | −0.00 |

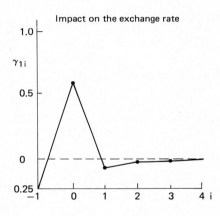

Impact on the exchange rate

$\gamma_{1i}$

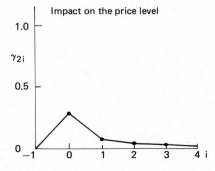

Impact on the price level

$\gamma_{2i}$

Figure 5.  Temporary unanticipated increase in money.

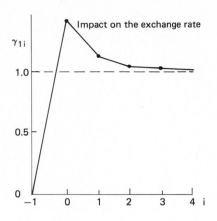

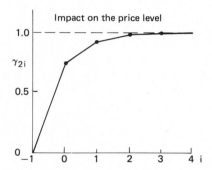

Figure 6.   Permanent unanticipated increase in money.

in advance ($k = 3$, $\rho = 1$) are shown. In each case the increase in money is by 1 percent.

A temporary unanticipated increase in money causes the exchange rate to depreciate ($e$ rises) and the price level to increase in the first period. Subsequently, the price level converges monotonically back to equilibrium. In the second period, $e$ falls below its equilibrium value and then gradually rises again back to zero (Table 2 and Figure 5).

A permanent unanticipated increase in money of 1 percent eventually causes the exchange rate to depreciate by 1 percent and the price level to rise by 1 percent. But in the short run $e$ rises above the long-run equilibrium and then gradually falls back to zero. This is the best illustration of overshooting (Table 3 and Figure 6).

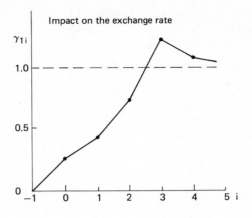

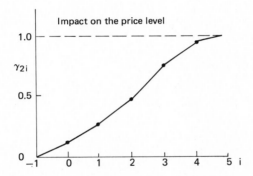

Figure 7.   Permanent increase in money, anticipated 3 periods in advance.

If the increase in the money supply is anticipated in advance, then the price level rises and the exchange rate depreciates at the announcement date. Subsequently, the price level and $e$ continue to rise. The exchange rate reaches its lowest value ($e$ reaches its highest value) on the announcement date, and then appreciates back to its new long-run value of 1 (Table 4 and Figure 7). Note that $p$ and $e$ are on explosive paths from period 0 until period 3.

### 2.2.6.   Geometric interpretation

The solution of the bivariate model has a helpful geometric interpretation. Writing out eq. (2.51) with $\theta_i = 0$ in scalar form as two different equations and subtracting $\gamma_{1i}$ and $\gamma_{2i-1}$ from the first and second equation respectively results in

$$\Delta\gamma_{1i+1} \equiv \gamma_{1i+1} - \gamma_{1i} = (a_{11} - 1)\gamma_{1i} + a_{12}\gamma_{2i-1},$$
$$\Delta\gamma_{2i} \equiv \gamma_{2i} - \gamma_{2i-1} = a_{21}\gamma_{1i} + (a_{22} - 1)\gamma_{2i-1}. \tag{2.77}$$

According to (2.77) there are two linear relationships between $\gamma_{1i}$ and $\gamma_{2i-1}$ consistent with no change in the coefficients: $\Delta\gamma_{1i-1} = 0$ and $\Delta\gamma_{2i} = 0$. For example, in the exchange rate model in eq. (2.69), the equations in (2.77) become

$$\Delta\gamma_{1i+1} = \frac{\beta}{\alpha(1+\beta)}\gamma_{1i} + \frac{1}{\alpha(1+\beta)}\gamma_{2i-1},$$
$$\Delta\gamma_{2i} = \frac{\beta}{1+\beta}\gamma_{1i} - \frac{\beta}{1+\beta}\gamma_{2i-1}. \tag{2.78}$$

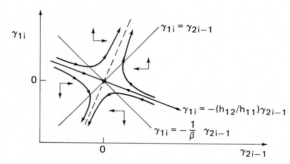

Figure 8. Geometric interpretation of the solution in the bivariate model. The darker line is the saddle point path along which the impact coefficients converge to the equilibrium value of $(0,0)$.

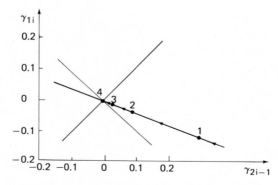

Figure 9. Solution values for the case of temporary-unanticipated shocks. ($k = 0$, $\rho = 0$). The numbered points are the values of $i$. See also Table 2 and Figure 5.

The two no-change lines are

$$\gamma_{1i} = -\frac{1}{\beta}\gamma_{2i-1},$$

$$\gamma_{1i} = \gamma_{2i-1},$$

(2.79)

and are plotted in Figure 8. The arrows in Figure 8 show the directions of motion according to eq. (2.78) when the no-change relationships in (2.79) are not satisfied. It is clear from these arrows that if the $\gamma$ coefficients are to converge to

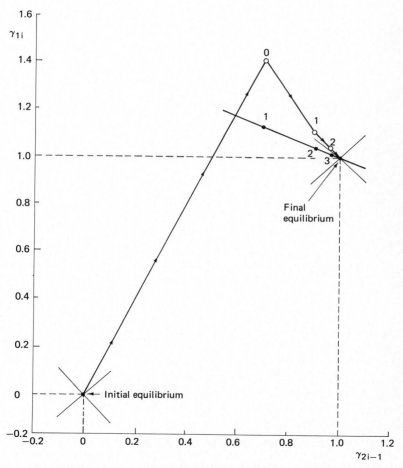

Figure 10.   Solution values for a permanent unanticipated increase in the money supply. The open circles give the $(\gamma_{1i}, \gamma_{2i})$ pairs starting with $i = 0$.

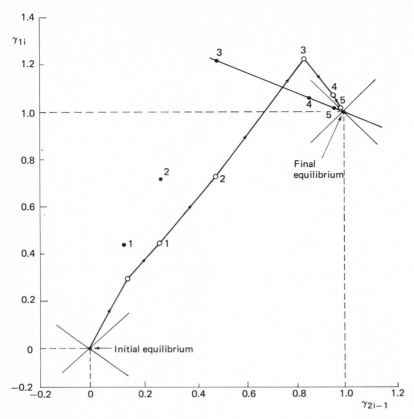

Figure 11.   Solution values for an anticipated permanent increase in the money supply. The open
circles give the $\gamma_{1i}, \gamma_{2i}$ pairs starting with $i = 0$.

their equilibrium value $(0,0)$ they must move along the "saddle point" path
shown by the darker line in Figure 8. Points off this line will lead to ever-increas-
ing values of the $\gamma$ coefficients. The linear combination of $\gamma_{1i}$ and $\gamma_{2i-1}$ along this
saddle point path is given by the characteristic vector associated with the unstable
root $\lambda_1$ as given in general by eq. (2.55) and for this example in eq. (2.73). Note
how Figure 8 immediately shows that the saddle point path is downward sloping.
In Figure 9 the solution values for the impacts on the exchange rate and the price
level are shown for the case of a temporary shock as considered in Table 2 and
Figure 5. In Figures 10 and 11, the solution values are shown for the case where
the increase in money is permanent. The permanent increase shifts the reference
point from $(0,0)$ to $(1,1)$. The point $(1,1)$ is simply the value of the particular

solution in this case. Figure 10 is the case where the permanent increase is unanticipated; Figure 11 is the anticipated case.

Note that these diagrams do not give the impact on the exchange rate and the price level in the *same* period; they are one period out of synchronization. Hence, the points do not correspond to a scatter diagram of the effects of a change in money on the exchange rate and on the price level. It is a relatively simple matter to deduce a scatter diagram as shown by the open circles in Figures 10 and 11.

## 2.3. The use of operators, generating functions, and z-transforms

As the previous Sections have shown, the problem of solving rational expectations models is equivalent to solving nonhomogeneous deterministic difference equations. The homogeneous solution is obtained simply by requiring that the stochastic process for the endogenous variables be stationary. Once this is accomplished, most of the work comes in obtaining the particular solution to the nonhomogeneous part. Lag or lead operators, operator polynomials, and the power series associated with these polynomials (i.e. generating functions or z-transformations) have frequently been found useful in solving the nonhomogeneous part of difference equations [see Baumol (1970), for economic examples]. These methods have also been useful in rational expectations analysis. Futia (1981) and Whiteman (1983) have exploited the algebra of z-transforms in solving a wide range of linear rational expectations models.

To illustrate the use of operators, let $F^s x_t = x_{t+s}$ be the forward lead operator. Then the scalar equation in the impact coefficients that we considered in eq. (2.7), can be written

$$(1 - \alpha F)\gamma_i = \delta\theta_i \qquad i = 0, 1, 2, \ldots. \tag{2.80}$$

Consider the case where $\theta_i = \rho^i$ and solve for $\gamma_i$ by operating on both sides by the inverse of the polynomial $(1 - \alpha F)$. We then have

$$\gamma_i = \frac{\delta\rho^i}{1 - \alpha F}$$

$$= \frac{\delta\rho^i}{1 - \alpha\rho} \qquad i = 0, 1, 2 \ldots. \tag{2.81}$$

the last equality follows from the algebra of operator polynomials [see for example Baumol (1970)]. The result is identical to what we found in Section 2.1 using the method of undetermined coefficients to obtain the particular solution. The procedure easily generalizes to the bivariate case and yields the particular solution shown in eq. (2.58). It also generalizes to handle other time series specifications of $\theta_i$.

The operator notation used in (2.80) is standard in difference equation analysis. In some applications of rational expectations models, a non-standard operator has been used directly on the basic model (2.1). To see this redefine the operator $F$ as $F\mathrm{E}_t y_t = \mathrm{E}_t y_{t+1}$. That is, $F$ moves the date on the variable but the viewpoint date in the expectation is held constant. Then eq. (2.1) can be written (note that $\mathrm{E}_t y_t = y_t$):

$$(1 - \alpha F)\,\mathrm{E}_t y_t = \delta_t. \tag{2.82}$$

Formally, we can apply the inverse of $(1 - \alpha F)$ to (2.82) to obtain

$$
\begin{aligned}
\mathrm{E}_t y_t &= \delta (1 - \alpha F)^{-1} u_t \\
&= \delta \left( 1 + \alpha F + (\alpha F)^2 + \cdots \right) u_t \\
&= \delta \left( u_t + \alpha \mathrm{E}_t u_{t+1} + \alpha^2 \mathrm{E} u_{t+2} + \cdots \right) \\
&= \delta \left( u_t + \alpha \rho u_t + (\alpha \rho)^2 u_t + \cdots \right) \\
&= \frac{\delta}{1 - \alpha \rho} u_t,
\end{aligned}
\tag{2.83}
$$

and where we again assume that $u_t = \rho u_{t-1} + \varepsilon_t$. Eq. (2.83) gives the same answer that the previous methods did (again note that $\mathrm{E}_t y_t = y_t$). As Sargent (1979, p. 337) has discussed, the use of this type of operator on conditional expectations can lead to confusion or mistakes, if it is interpreted as a typical lag operator that shifts all time indexes, including the viewpoint dates. The use of operators on conventional difference operations like (2.6) is much more straightforward, and perhaps it is best to think of the algebra in (2.82) and (2.83) in terms of (2.80) and (2.81).

Whiteman's (1983) use of the generating functions associated with the operator polynomials can be illustrated by writing the power series corresponding to eqs. (2.2) and (2.4):

$$\gamma(z) = \sum_{i=0}^{\infty} \gamma_i z^i,$$

$$\theta(z) = \sum_{i=0}^{\infty} \theta_i z^i.$$

These are the $z$-transforms [see Dhrymes (1971) for a short introduction to $z$-transforms and their use in econometrics]. Equating the coefficients of $\varepsilon_{t-i}$ in eq. (2.6) is thus the same as equating the coefficients of powers of $z$. That is, (2.6) means that

$$\gamma(z) = \alpha z^{-1}(\gamma(z) - \gamma_0) + \delta\theta(z). \tag{2.84}$$

Solving (2.84) for $\gamma(z)$ we have

$$\gamma(z) = (1 - \alpha^{-1}z)^{-1}(\gamma_0 - \delta\alpha^{-1}z\theta(z)). \tag{2.85}$$

As in Section 2.1, eq. (2.85) has a free parameter $\gamma_0$ which must be determined before $\gamma(z)$ can be evaluated. For $y_t$ to be a stationary process, it is necessary that $\gamma(z)$ be a convergent power series (or equivalently an analytic function) for $|z| < 1$. The term $(1 - \alpha^{-1}z)^{-1}$ on the right-hand side of (2.85) is divergent if $\alpha^{-1} > 1$. Hence, the second term in parentheses must have a factor to "cancel out" this divergent series. For the case of serially uncorrelated shocks, $\theta(z)$ is a constant $\theta_0 = 1$ so that it is obvious that $\gamma_0 = \delta$ will cancel out the divergent series. We then have $\gamma(z) = \delta$ which corresponds with the results in Section 2.1. Whiteman (1983) shows that in general $\gamma(z)$ will be convergent when $|\alpha| < 1$ if $\gamma_0 = \delta\theta(\alpha)$. For the unanticipated autoregressive shocks this implies that $\gamma(z) = \delta(1 - \rho\alpha)^{-1}(1 - \rho z)$ which is the $z$-transform of the solution we obtained earlier. When $|\alpha| > 1$ there is no natural way to determine $\gamma_0$, so we are left with non-uniqueness as in Section 2.1.

## 2.4. Higher order representations and factorization techniques

We noted in Section 2.2 that a first-order bivariate model with one lead variable could be interpreted as a second-order scalar model with a lead and a lag. That is,

$$y_t = \alpha_1 \underset{t}{E} y_{t+1} + \alpha_2 y_{t-1} + \delta u_t, \tag{2.86}$$

can be written as a bivariate model and solved using the saddle point stability method. An alternative approach followed by Sargent (1979), Hansen and Sargent (1980) and Taylor (1980a) is to work with (2.86) directly. That the two approaches give the same result can be shown formally.

Substitute for $y_t$, $y_{t-1}$, and $E_t y_{t+1}$ in eq. (2.86) using (2.4) to obtain the equations

$$\gamma_1 = \frac{1}{\alpha_1}(\gamma_0 - \delta\theta_0), \tag{2.87}$$

$$\gamma_{i+1} = \frac{1}{\alpha_1}\gamma_i - \frac{\alpha_2}{\alpha_1}\gamma_{i-1} - \frac{\delta}{\alpha_1}\theta_i, \qquad i = 1, 2, \dots . \tag{2.88}$$

As above, we need one more equation to solve for all the $\gamma$ coefficients. Consider first the homogeneous part of (2.88). Its characteristic polynomial is

$$z^2 - \frac{1}{\alpha_1} z + \frac{\alpha_2}{\alpha_1}, \tag{2.89}$$

which can be *factored* into

$$(\lambda_1 - z)(\lambda_2 - z), \tag{2.90}$$

where $\lambda_1$ and $\lambda_2$ are the roots of (2.89). The solution to the homogeneous part is $\gamma_i^{(H)} = k_1 \lambda_1^i + k_2 \lambda_2^i$. As we discussed above, in many economic applications one root, say $\lambda_1$, will be larger than 1 in modulus and the other will be smaller than 1 in modulus. Thus, the desired solution to the homogeneous part is achieved by setting $k_1 = 0$ so that $\gamma_i^{(H)} = k_2 \lambda_2^i$ where $k_2$ equals the initial condition $\gamma_0^{(H)}$. Equivalently we can interpret the setting of $k_1 = 0$ as reducing the characteristic polynomial (2.89) to $(z - \lambda_2)$. Thus, the $\gamma$ coefficients satisfy

$$\gamma_i = \lambda_2 \gamma_{i-1} \qquad i = 1, 2, \ldots . \tag{2.91}$$

Equivalently, we have "factored out" $(z - \lambda_1)$ from the characteristic polynomial.

For the case where $u_t$ is uncorrelated so that $\theta_i = 0$ for $i > 0$, difference equation in (2.88) is homogeneous. We can solve for $\gamma_0$ by using $\gamma_1 = \lambda_2 \gamma_0$ along with eq. (2.87). This gives $\gamma_0 = \delta(1 - \alpha_1 \lambda_2)^{-1} \lambda_2^i \quad i = 0, 1, \ldots$

To see how this result compares with the saddle-point approach, write (2.88) as

$$\begin{pmatrix} \gamma_{i+1} \\ \gamma_i \end{pmatrix} = \begin{pmatrix} \dfrac{1}{\alpha_1} & -\dfrac{\alpha_2}{\alpha_1} \\ 1 & 0 \end{pmatrix} \begin{pmatrix} \gamma_i \\ \gamma_{i-1} \end{pmatrix} - \begin{pmatrix} \dfrac{\delta}{\alpha_1} \\ 0 \end{pmatrix} \theta_i \qquad i = 1, 2, \ldots . \tag{2.92}$$

The characteristic equation of the matrix $A$ is $\lambda^2 - (1/\alpha_1)\lambda - \alpha_2/\alpha_1 = 0$. Hence, the roots of $A$ are identical to the roots of the characteristic polynomial associated with the second-order difference eq. (2.88). [This is a well-known result shown for the general $p$th order difference equation in Anderson (1971)].

The characteristic vector of the matrix $A$ associated with the unstable root $\lambda_1$ is found from the equation $(h_{11}, h_{12})A = \lambda_1(h_{11}, h_{12})$. Thus, the saddle point path is given by

$$\gamma_i = -\frac{h_{12}}{h_{11}} \gamma_{i-1} = \left( \frac{1}{\alpha_1} - \lambda_1 \right) \gamma_{i-1}. \tag{2.93}$$

For the two methods to be equivalent, we need to show that (2.91) and (2.93) are equivalent, or that $\lambda_2 = 1/\alpha_1 - \lambda_1$. This follows immediately from the fact that the sum of the roots $(\lambda_1 + \lambda_2)$ of a second-order polynomial equals the coefficients of the linear term in the polynomial: $\lambda_1 + \lambda_2 = 1/\alpha_1$.

For the case where $\theta_i = \rho^i$, we need to compare the particular solutions as well. For the second-order scalar model we guess the form $\gamma_i^{(P)} = ab^i$. Substituting this into (2.88) we find that $b = \rho$ and $a = \delta(1 - \alpha_1\rho - \alpha_2\rho^{-1})^{-1}$. To see that this gives the same value for the particular solution that emerges from the matrix formulation in eq. (2.58), note that

$$
(\rho I - A)^{-1} \mathrm{d}\rho^i = \begin{pmatrix} \rho - \dfrac{1}{\alpha_1} & -\dfrac{\alpha_2^{-1}}{\alpha_1} \\ -1 & \rho \end{pmatrix} \begin{pmatrix} -\dfrac{\delta}{\alpha_1} \\ 0 \end{pmatrix} \rho^i
$$

$$
= \frac{1}{\rho^2 - \dfrac{1}{\alpha_1}\rho + \dfrac{\alpha_2}{\alpha_1}} \begin{pmatrix} -\rho\dfrac{\delta}{\alpha_1} \\ -\dfrac{\delta}{\alpha_1} \end{pmatrix} \rho^i. \tag{2.94}
$$

Eq. (2.94) gives the particular solution for the vector $(\gamma_i^{(P)}, \gamma_{i-1}^{(P)})$, which corresponds to the vector $\gamma_i^{(P)}$ in eq. (2.58). Hence

$$
\gamma_i^{(P)} = \frac{-\rho\alpha_1^{-1}\delta\rho^i}{\rho^2 - \rho\alpha_1^{-1} + \alpha_2\alpha_1^{-1}}
$$

$$
= \frac{\delta\rho^i}{1 - \alpha_1\rho - \alpha_2\rho^{-1}},
$$

which is the particular solution obtained from the second-order scalar representation.

Rather than obtaining the solution of the homogeneous system by factoring the characteristic equation, one can equivalently factor the polynomial in the time shift operators. Because the operator polynomials also provide a convenient way to obtain the nonhomogeneous solution (as was illustrated in Section 2.3), this approach essentially combines the homogeneous solution and the nonhomogeneous solution in a notationally and computationally convenient way.

Write (2.88) as

$$
\left( -L^{-1} + \frac{1}{\alpha_2} - \frac{\alpha_2}{\alpha_1}L \right) \gamma_i = \frac{\delta}{\alpha_1}\theta_i. \tag{2.95}
$$

Let $H(L) = L^{-1} - 1/\alpha_1 + (\alpha_2/\alpha_1)L$ be the polynomial on the left-hand side of

(2.95) and let $P(z) = z^2 - 1/(\alpha_1)z + \alpha_2/\alpha_1$ be the characteristic polynomial in (2.89). The polynomial $H(L)$ can be factored into

$$\mu(1 - \phi L^{-1})(1 - \psi L), \tag{2.96}$$

where $\phi = -\mu^{-1}$, $\psi = -\mu^{-1}\alpha_2\alpha_1^{-1}$, and where $\mu$ is one of the solutions of $P(\mu) = 0$; that is one of the roots of $P(\cdot)$. This can be seen by equating the coefficient of $H(L)$ and the polynomial in (2.96). Continuing to assume that only one of the roots of $P(\cdot)$ is greater than one in modulus (say $\lambda_1$) we set $\phi = \lambda_1^{-1} < 1$. Since the product of the roots of $P(\cdot)$ equals $\alpha_2\alpha_1^{-1}$ we immediately have that $\psi = \lambda_2$. Thus, there is a unique factorization of the polynomial with $\phi$ and $\psi$ both less than one in modulus.

Because $\psi = \lambda_2$, the stable solution (2.97) to the homogeneous difference equation can be written

$$(1 - \psi L)\gamma_i^{(H)} = 0. \tag{2.97}$$

The particular solution also can be written using the operator notation:

$$\gamma_i^{(P)} = \frac{\delta\alpha_1^{-1}\rho^i}{\mu(1 - \phi L^{-1})(1 - \psi L)}. \tag{2.98}$$

The complete solution is given by $\gamma_i = \gamma_i^{(H)} + \gamma_i^{(P)}$ which implies that

$$(1 - \lambda_2 L)\gamma_i = (1 - \lambda_2 L)\gamma_i^{(H)} + (1 - \lambda_2 L^{-1})\gamma_i^{(P)}. \tag{2.99}$$

The first term on the right-hand side of (2.99) equals zero. Therefore the complete solution is given by

$$\gamma_i = \lambda_2\gamma_{i-1} + \frac{\delta\alpha_1^{-1}\rho^i}{\lambda_1(1 - \lambda_1^{-1}L^{-1})}$$

$$= \lambda_2\gamma_{i-1} + \frac{\delta\alpha_1^{-1}\rho^i}{\lambda_1(1 - \rho\lambda_1^{-1})}. \tag{2.100}$$

This solution is equivalent to that derived by adding the particular solution in (2.95) to the solution of the homogeneous solution of (2.91).

Note that this procedure or solving (2.95) can be stated quite simply in two steps: (1) factor the lag polynomial into two stable polynomials, one involving positive powers of $L$ (lags) and the other involving negative powers of $L$ (leads), and (2) operate on both sides of (2.95) by the inverse of the polynomial involving negative powers of $L$.

It is clear from (2.94) that the $\gamma_i$ weights are such that the solution for $y_t$ can be represented as a first-order autoregressive process with a serially correlated error:

$$y_t = \lambda_2 y_{t-1} + \delta \alpha_1^{-1} (\lambda_1 - \rho)^{-1} u_t, \tag{2.101}$$

where

$$u_t = \rho u_{t-1} + \varepsilon_t.$$

In the papers by Sargent (1979), Taylor (1980a) and Hansen and Sargent (1980), the difference equation in (2.95) was written $\gamma_i = E_t y_{t+i}$ and $\theta_i = E_t u_{t+i}$, a form which can be obtained by taking conditional expectations in eq. (2.86). In other words rather than working with the moving average coefficients they worked directly with the conditional expectations. As discussed in Section 2.3 this requires the use of a non-standard lag operator.

## 2.5. Rational expectations solutions as boundary value problems

It is useful to note that the problem of solving rational expectations models can be thought of as a boundary value problem where final conditions as well as initial conditions are given. To see this consider the homogeneous equation

$$\gamma_{i+1} = \frac{1}{\alpha} \gamma_i \qquad i = 0, 1, \ldots. \tag{2.102}$$

The stationarity conditions place a restriction on the "final" value $\lim_{j \to \infty} \gamma_j = 0$ rather than on the "initial" value $\gamma_0$. As an approximation we want $\gamma_j = 0$ for large $j$. A traditional method to solve boundary value problems is "shooting": One guesses a value for $\gamma_0$ and then uses (2.102) to project (shoot) a value of $\gamma_j$ for some large $j$. If the resulting $\gamma_j \neq 0$ (or if $\gamma_j$ is further from 0 than some tolerance range) then a new value (chosen in some systematic fashion) of $\gamma_0$ is tried until one gets $\gamma_j$ sufficiently close to zero. It is obvious in this case that $\gamma_0 = 0$ so it would be impractical to use such a method. But in nonlinear models the approach can be quite useful as we discuss in Section 6.

This approach obviously generalizes to higher order systems; for example the homogeneous part of (2.88) is

$$\gamma_{i+1} = \frac{1}{\alpha_1}\gamma_i - \frac{\alpha_2}{\alpha_1}\gamma_{i-1} \qquad i = 0,1,2,\ldots. \tag{2.103}$$

with $\gamma_{-1} = 0$ as one initial condition and $\gamma_j = 0$ for some large $j$ as the one "final" condition. This is a two point boundary problem which can be solved in the same way as (2.102).

## 3. Econometric evaluation of policy rules

Perhaps the main motivation behind the development of rational expectations models was the desire to improve policy evaluation procedures. Lucas (1976) argued that the parameters of the models conventionally used for policy evaluation – either through model simulation or formal optimal control – would shift when policy changed. The main reason for this shift is that expectations mechanisms are adaptive, or backward looking, in conventional models and thereby unresponsive to those changes in policy that would be expected to change expectations of future events. Hence, the policy evaluation results using conventional models would be misleading.

The Lucas criticism of conventional policy evaluation has typically been taken as destructive. Yet, implicit in the Lucas' criticism is a constructive way to improve on conventional evaluation techniques by modeling economic phenomena in terms of "structural" parameters; by "structural" one simply means invariant with respect to policy intervention. Whether a parameter is invariant or not is partly a matter of researcher's judgment, of course, so that any attempt to take the Lucas critique seriously by building structural models is subject to a similar critique that the researcher's assumption about which parameters are structural is wrong. If taken to this extreme that no feasible structural modeling is possible, the Lucas critique does indeed become purely destructive and perhaps even stifling.

Hansen and Sargent (1980), Kydland and Prescott (1982), Taylor (1982), and Christiano (1983) have examined policy problems where only the parameters of utility functions or production functions can be considered invariant or structural. Taylor (1979, 1980b) has considered models where the parameters of the wage and price setting functions are invariant or structural.

The thought experiments described in Section 2 whereby multiplier responses are examined should be part of any policy evaluation technique. But it is unrealistic to think of policy as consisting of such one-shot changes in the policy instrument settings. They never occur. Rather, one wants to consider changes in the way the policymakers respond to events – that is, changes in their policy rules. For this we can make use of stochastic equilibrium solutions examined in Section 2. We illustrate this below.

## 3.1. Policy evaluation for a univariate model

Consider the following policy problem which is based on model (2.1). Suppose that an econometric policy advisor knows that the demand for money is given by

$$m_t - p_t = -\beta(E_t p_{t+1} - p_t) + u_t. \tag{3.1}$$

Here there are two shocks to the system, the supply of money $m_t$ and the demand for money $u_t$. Suppose that $u_t = \rho u_{t-1} + \varepsilon_t$, and that in the past the money supply was fixed: $m_t = 0$; suppose that under this fixed money policy, prices were thought to be too volatile. The policy advisor is asked by the Central Bank for advice on how $m_t$ can be used in the future to reduce the fluctuations in the price level. Note that the policy advisor is not asked just what to do today or tomorrow, but what to do for the indefinite future. Advice thus should be given as a contingency rule rather than as a fixed path for the money supply.

Using the solution technique of Section 2, the behavior of $p_t$ during the past is

$$p_t = \rho p_{t-1} - \frac{\varepsilon_t}{1 + \beta(1 - \rho)}. \tag{3.2}$$

Conventional policy evaluation might proceed as follows: first, the econometrician would have estimated $\rho$ in the reduced form relation (3.2) over the sample period. The estimated equation would then serve as a model of expectations to be substituted into (3.1); that is, $E_t p_{t+1} = \rho p_t$ would be substituted into

$$m_t - p_t = -\beta(\rho p_t - p_t) + u_t. \tag{3.3}$$

The conventional econometricians model of the price level would then be

$$p_t = \frac{m_t - u_t}{1 + \beta(1 - \rho)}. \tag{3.4}$$

Considering a feedback policy rule of the form $m_t = g u_{t-1}$ eq. (3.4) implies

$$\text{var } p_t = \frac{1}{[1 + \beta(1 - \rho)]^2 (1 - \rho^2)} \sigma_\varepsilon^2 [g^2 + 1 - 2g\rho]. \tag{3.5}$$

If there were no cost to varying the money supply, then eq. (3.5) indicates that the best choice for $g$ to minimize fluctuation in $p_t$ is $g = \rho$.

But we know that (3.5) is incorrect if $g \neq 0$. The error was to assume that $E_t p_{t+1} = \rho p_t$ regardless of the choice of policy. This is the expectations error that rational expectations was designed to avoid. The correct approach would have been to substitute $m_t = g u_{t-1}$ directly into (3.1) and calculate the stochastic equilibrium for $p_t$. This results in

$$p_t = \frac{-1 - \beta(1 - g)}{(1 + \beta)(1 + \beta(1 - \rho))} u_t + \frac{g}{1 + \beta} u_{t-1}. \tag{3.6}$$

Note how the parameters of (3.6) depend on the parameters of the policy rule. The variance of $p_t$ is

$$\operatorname{Var} p_t = \frac{1}{(1 + \beta)^2 (1 - \rho^2)} \left[ \frac{(1 + \beta(1 - g))^2}{(1 + \beta(1 - \rho))^2} - \frac{2g(1 + \beta(1 - g))\rho}{1 + \beta(1 - \rho)} + g^2 \right] \sigma_\varepsilon^2. \tag{3.7}$$

The optimal policy is found by minimizing $\operatorname{Var} p_t$ with respect to $g$.

This simple policy problem suggests the following approach to macro policy evaluation: (1) Derive a stochastic equilibrium solution which shows how the endogenous variables behave as a function of the parameters of the policy rule; (2) Specify a welfare function in terms of the moments of the stochastic equilibrium, and (3) Maximize the welfare function across the parameters of the policy rule. In this example the welfare function is simply $\operatorname{Var} p$. In more general models there will be several target variables. For example, in Taylor (1979) an optimal policy rule to minimize a weighted average of the variance of real output and the variance of inflation was calculated.

Although eq. (3.1) was not derived explicitly from an individual optimization problem, the same procedure could be used when the model is directly linked to parameters of a utility function. For instance, the model of Example (5) in Section 2.2 in which the parameters depend on a firm's utility function could be handled in the same way as the model in (3.1).

### 3.2.   *The Lucas critique and the Cowles Commission critique*

The Lucas critique can be usefully thought of as a dynamic extension of the critique developed by the Cowles Commission researchers in the late 1940s and early 1950s and which gave rise to the enormous literature on simultaneous equations. At that time it was recognized that reduced forms could not be used for many policy evaluation questions. Rather one should model structural relationships. The parameters of the reduced form are, of course, functions of the structural parameters in the standard Cowles Commission setup. The discussion by Marschak (1953), for example, is remarkably similar to the more recent rational expectations critiques; Marschak did not consider expectations variables, and in this sense the rational expectations critique is a new extension. But earlier analyses like Marschak's are an effort to explain why structural modeling is necessary, and thus has much in common with more recent research.

## 3.3. Game-theoretic approaches

In the policy evaluation procedure discussed above, the government acts like a dominant player with respect to the private sector. The government sets $g$ and the private sector takes $g$ as given. The government then maximizes its social welfare function across different values of $g$. One can imagine alternatively a game theoretic setup in which the government and the private sector each are maximizing utility. Chow (1983), Kydland (1975), Lucas and Sargent (1981), and Epple, Hansen, and Roberds (1983) have considered this alternative approach. It is possible to specify the game theoretic model as a choice of parameters of decision rules in the steady state or as a formal non-steady state dynamic optimization problem with initial conditions partly determining the outcome. Alternative solution concepts including Nash equilibria have been examined.

The game-theoretic approach naturally leads to the important time inconsistency problem raised by Kydland and Prescott (1977) and Calvo (1979). Once the government announces its policy, it will be optimal to change it in the future. The consistent solution in which everyone expects the government to change is generally suboptimal. Focussing on rules as in Section 3.1 effectively eliminates the time inconsistency issue. But even then, there can be temptation to change the rule.

## 4. Statistical inference

The statistical inference issues that arise in rational expectations models can be illustrated in a model like that of Section 2.

### 4.1. Full information estimation

Consider the problem of estimating the parameters of the structural model

$$y_t = \alpha \mathop{\mathrm{E}}_t y_{t+1} + \delta x_t + v_t, \tag{4.1}$$

where $v_t$ is a serially uncorrelated random variable. Assume (for example) that $x_t$ has a finite moving average representation:

$$x_t = \varepsilon_t + \theta_1 \varepsilon_{t-1} + \cdots + \theta_q \varepsilon_{t-q}, \tag{4.2}$$

where $\varepsilon_t$ is serially uncorrelated and assume that $\text{Cov}(v_t, \varepsilon_s) = 0$ for all $t$ and $s$.

To obtain the full information maximum likelihood estimate of the structural system (4.1) and (4.2) we need to reduce (4.1) to a form which does not involve expectations variables. This can be done by solving the model using one of the techniques described in Section 2. Using the method of undetermined coefficients, for example, the solution for $y_t$ is

$$y_t = \gamma_0 \varepsilon_t + \cdots + \gamma_q \varepsilon_{t-q} + v_t, \tag{4.3}$$

where the $\gamma$ parameters are given by

$$\begin{pmatrix} \gamma_0 \\ \gamma_1 \\ \vdots \\ \gamma_q \end{pmatrix} = \delta \begin{pmatrix} 1 & \alpha & \alpha^2 & \cdots & \alpha^{q-1} & \alpha^q \\ 0 & 1 & \alpha & \cdots & \alpha^{q-2} & \alpha^{q-1} \\ \vdots & & & & & \vdots \\ 0 & & & & 1 & \alpha \\ 0 & 0 & 0 & \cdots & 0 & 1 \end{pmatrix} \begin{pmatrix} 1 \\ \theta_1 \\ \vdots \\ \theta_q \end{pmatrix}. \tag{4.4}$$

Eqs. (4.2) and (4.3) together form a two dimensional vector model.

$$\begin{pmatrix} y_t \\ x_t \end{pmatrix} = \begin{pmatrix} \gamma_0 & 1 \\ 1 & 0 \end{pmatrix} \begin{pmatrix} \varepsilon_t \\ v_t \end{pmatrix} + \begin{pmatrix} \gamma_1 & 0 \\ \theta_1 & 0 \end{pmatrix} \begin{pmatrix} \varepsilon_{t-1} \\ v_{t-1} \end{pmatrix}$$

$$+ \cdots + \begin{pmatrix} \gamma_q & 0 \\ \theta_q & 0 \end{pmatrix} \begin{pmatrix} \varepsilon_{t-q} \\ v_{t-q} \end{pmatrix}. \tag{4.5}$$

Eq. (4.5) is an estimatable reduced form system corresponding to the structural form in (4.1) and (4.2).

If we assume that $(v_t, \varepsilon_t)$ is distributed normally and independently, then the full-information maximum likelihood estimate of $(\theta_1, \ldots, \theta_q, \alpha, \delta)$ can be obtained using existing methods to estimate multivariate ARMA models. See Chow (1983, Section 6.7 and 11.6). Note that the coefficients of the ARMA model (4.5) are constrained. There are *cross-equation* restrictions in that the $\theta$ and $\gamma$ parameters are related to each other by (4.4). In addition, relative to a fully unconstrained ARMA model, the off-diagonal elements of the autoregression are equal to zero.

Full information estimation maximum likelihood methods for linear rational expectations models have been examined by Chow (1983), Muth (1981), Wallis (1980), Hansen and Sargent (1980, 1981), Dagli and Taylor (1985), Mishkin

(1983), Taylor (1979, 1980a), and Wickens (1982). As in this example, the basic approach is to find a constrained reduced form and maximize the likelihood function subject to the constraints. Hansen and Sargent (1980, 1981) have emphasized these cross-equation constraints in their expositions of rational expectations estimation methods. In Muth (1981), Wickens (1982) and Taylor (1979) multivariate models were examined in which expectations are dated at $t-1$ rather than $t$ and $E_{t-1}y_t$ appears in (4.1) rather than $E_t y_{t+1}$. More general multivariate models with leads and lags are examined in the other papers.

For full information estimation, it is also important that the relationship between the structural parameters and the reduced form parameters can be easily evaluated. In this example the mapping from the structural parameters to the reduced form parameters is easy to evaluate. In more complex models the mapping does not have a closed form; usually because the roots of high-order polynomials must be evaluated.

## 4.2. Identification

There has been relatively little formal work on identification in rational expectations models. As in conventional econometric models, identification involves the properties of the mapping from the structural parameters to the reduced form parameters. The model is identified if the structural parameters can be uniquely obtained from the reduced form parameters. Over-identification and under-identification are similarly defined as in conventional econometric models. In rational expectations models the mapping from reduced form to structural parameters is much more complicated than in conventional models and hence it has been difficult to derive a simple set of conditions which have much generality. The conditions can usually be derived in particular applications as we can illustrate using the previous example.

When $q=0$, there is one reduced form parameter $\gamma_0$, which can be estimated from (4.2) and (4.3), recalling that Cov $(v_t, \varepsilon_t) = 0$, and two structural parameters $\delta$ and $\alpha$ in eq. (4.4). Hence, the model is not identified. In this case, $\delta = \gamma_0$ is identified from the regression of $y_t$ on the exogenous $x_t$, but $\alpha$ is not identified. When $q=1$, there are three reduced form parameters $\gamma_0$, $\gamma_1$ and $\theta_1$ which can be estimated from (4.2) and (4.3), and three structural parameters $\delta$, $\alpha$, and $\theta_1$. ($\theta_1$ is both a structural and reduced form parameter since $x_t$ is exogenous). Hence, the model is exactly identified according to a simple order condition. More generally, there are $q+2$ structural parameters $(\delta, \alpha, \theta_1, \ldots, \theta_q)$ and $2q+1$ reduced form parameters $(\gamma_0, \gamma_1, \ldots, \gamma_q, \theta_1, \ldots, \theta_q)$ in this model. According to the order conditions, therefore, the model is overidentified if $q>1$.

Treatments of identification in more general models focus on the properties of the cross-equation restrictions in more complex versions of eq. (4.4). Wallis (1980) gives conditions for identification for a class of rational expectations models; the conditions may be checked in particular applications. Blanchard (1982) has derived a simple set of identification restrictions for the case where $x_t$ in (4.2) is autoregressive and has generalized this to higher order multivariate versions of (4.1) and (4.2).

### 4.3.  Hypothesis testing

Tests of the rational expectations assumption have generally been constructed as a test of the cross-equation constraints. These constraints arise because of the rational expectations assumption. In the previous example, the null hypothesis that the cross-equation constraints in (4.5) hold can be tested against the alternative that (4.5) is a fully unconstrained moving average model by using a likelihood ratio test. Note, however, that this is a joint test of rational expectations and the specification of the model. Testing rational expectations against a specific alternative like adaptive expectations usually leads to non-nested hypotheses.

In more general linear models, the same types of cross-equation restrictions arise, and tests of the model can be performed analogously. However, for large systems the fully unconstrained ARMA model may be difficult to estimate because of the large number of parameters.

### 4.4.  Limited information estimation methods

Three different types of "limited information" estimates have been used for rational expectations models. These can be described using the model in (4.1) and (4.2). One method investigated by Wallis estimates (4.2) separately in order to obtain the parameters $\theta_1, \ldots, \theta_q$. These estimates then are taken as given (as known parameters) in estimating (4.3). Clearly this estimator is less efficient than the full information estimator, but in more complex problems the procedure saves considerable time and effort. This method has been suggested by Wallis (1980) and has been used by Papell (1984) and others in applied work.

A second method proposed by Chow (1983) and investigated by Chow and Reny (1983) was mentioned earlier in our discussion of nonuniqueness. This method does not impose the saddle point stability constraints on the model. It leads to an easier computation problem than does imposing the saddle point constraints. If the investigator does not have any reason to impose this constraint, then this could prove quite practical.

A third procedure is to estimate eq. (4.1) as a single equation using instrumental variables. Much work has been done in this area in recent years, and because of computational costs of full information methods it has been used frequently in applied research. Consider again the problem of estimating eq. (4.1). Let $e_t = E_t y_{t+1} - y_{t+1}$ be the forecast error for the prediction of $y_t$. Substitute $E_t y_{t+1}$ into

(4.1) to get

$$y_t = \alpha y_{t+1} + \delta x_t + v_t - \alpha e_{t+1}. \tag{4.6}$$

By finding instruments of variables for $y_{t+1}$ that are uncorrelated with $v_t$ and $e_{t+1}$ one can estimate (4.6) using the method of instrumental variables. In fact this estimate would simply be the two stage least squares estimate with $y_{t+1}$ treated as if it were a right-hand side endogenous variable in a conventional simultaneous equation model. Lagged values of $x_t$ could serve as instruments here. This estimate was first proposed by McCallum (1976).

Several extensions of McCallum's method have been proposed to deal with serial correlation problems including Cumby, Huizinga and Obstfeld (1983), McCallum (1979), Hayashi and Sims (1983), Hansen (1982), and Hansen and Singleton (1982). A useful comparison of the efficiency of these estimators is found in Cumby, Huizinga and Obstfeld (1983).

## 5. General linear models

A general linear rational expectations model can be written as

$$B_0 y_t + B_1 y_{t-1} + \cdots + B_p y_{t-p} + A_1 \mathop{E}_{t} y_{t+1} + \cdots + A_q \mathop{E}_{t} y_{t+q} = C u_t, \tag{5.1}$$

where $y_t$ is a vector of endogenous variables, $u_t$ is a vector of exogenous variables or shocks, and $A_i$, $B_i$ and $C$ are matrices containing parameters.

Two alternative approaches have been taken to solve this type of model. Once it is solved, the policy evaluation and estimation methods discussed above can be applied. One approach is to write the model as a large first-order vector system directly analogous to the 2-dimensional vector model in eq. (2.50). The other approach is to solve (5.1) directly by generalizing the approach taken to the second-order scalar model in eq. (2.86). The first approach is the most straightforward. The disadvantage is that it can easily lead to very large (although sparse) matrices with high-order polynomials to solve to obtain the characteristic roots. This type of generalization is used by Blanchard and Kahn (1980) and Anderson and Moore (1984) to solve deterministic rational expectations models.

### 5.1. A general first-order vector model

Equation (5.1) can be written as

$$\mathop{E}_{t} z_{t+1} = A z_t + D u_t, \tag{5.2}$$

by stacking $y_t, y_{t-1}, \ldots, y_{t-p}$ into the vector $z_t$ much as in eq. (2.50). (It is necessary that $A_q$ be nonsingular to write (5.1) as (5.2)). Anderson and Moore (1984) have developed an algorithm that reduces equations with a singular $A_q$ into an equivalent form with a nonsingular matrix coefficient of $y_{t+q}$ and have applied it to an econometric model of the U.S. money market. (Alternatively, Preston and Pagan (1982, pp. 297–304) have suggested that a "shuffle" algorithm described by Luenberger (1977) be used for this purpose). In eq. (5.2) let $z_t$ be an $n$-dimensional vector and let $u_t$ be an $m$ dimensional vector of stochastic disturbances. The matrix $A$ is $n \times n$ and the matrix $D$ is $n \overset{\ast}{\times} m$.

We describe the solution for the case of unanticipated temporary shocks: $u_t = \varepsilon_t$ where $\varepsilon_t$ is a serially uncorrelated vector with a zero mean. Alternative assumptions about $u_t$ can be handled by the methods discussed in Section 2.2. The solution for $z_t$ can be written in the general form:

$$z_t = \sum_{i=0}^{\infty} \Gamma_i \varepsilon_{t-i}, \tag{5.3}$$

where the $\Gamma_i$ are $n \times m$ matrices of unknown coefficients. Substituting (5.3) into (5.2) we get

$$\Gamma_1 = A\Gamma_0 + D,$$
$$\Gamma_{i+1} = A\Gamma_i \qquad i = 1, 2, \ldots . \tag{5.4}$$

Note that these matrix difference equations hold for each column of $\Gamma_i$ separately; that is

$$\gamma_1 = A\gamma_0 + d,$$
$$\gamma_{i+1} = A\gamma_i \qquad i = 1, 2, \ldots, \tag{5.5}$$

where $\gamma_i$ is any one of the $n \times 1$ column vectors in $\Gamma_i$ and where $d$ is the corresponding column of $D$. Eq. (5.5) is a deterministic first-order vector difference equation analogous to the stochastic difference equation in (5.2). The solution for the $\Gamma_i$ is obtained by solving for each of the columns of $\Gamma_i$ separately using (5.5).

The analogy from the 2-dimensional case is now clear. There are $n$ equations in (5.5). In a given application we will know some of the elements of $\gamma_0$, but not all of them. Hence, there will generally be more than $n$ unknowns in (5.5). The number of unknowns is $2n - k$ where $k$ is the number of values of $\gamma_0$ which we know. For example, in the simple bivariate case of Section 2 where $n = 2$, we know that the second element of $\gamma_0$ equals 0. Thus, $k = 1$ and there are 3 unknowns and 2 equations.

To get a unique solution in the general case, we therefore need $(2n - k) - n = n - k$ additional equations. These additional equations can be obtained by requiring that the solution for $y_t$ be stationary or equivalently in this context that the $\gamma_i$ do not explode. If there are exactly $n - k$ distinct roots of $A$ which are greater than one in modulus, then the saddle point manifold will give exactly the number of additional equations necessary for a solution. The solution will be unique. If there are less than $n - k$ roots then we have the same nonuniqueness problem discussed in Section 2.

Suppose this root condition for uniqueness is satisfied. Let the $n - k$ roots of $A$ that are greater than one in modulus be $\lambda_1, \ldots, \lambda_{n-k}$. Diagonalize $A$ as $H^{-1}\Lambda H = A$. Then

$$H\gamma_{i+1} = \Lambda H\gamma_i \qquad i = 1, 2, \ldots . \tag{5.6}$$

$$\begin{pmatrix} H_{11} & H_{12} \\ H_{21} & H_{22} \end{pmatrix} \begin{pmatrix} \gamma_{i+1}^{(1)} \\ \gamma_{i+1}^{(2)} \end{pmatrix} = \begin{pmatrix} \Lambda_1 & 0 \\ 0 & \Lambda_2 \end{pmatrix} \begin{pmatrix} H_{11} & H_{12} \\ H_{21} & H_{22} \end{pmatrix} \begin{pmatrix} \gamma_i^{(1)} \\ \gamma_i^{(2)} \end{pmatrix} \qquad i = 1, 2, \ldots, \tag{5.7}$$

where $\Lambda_1$ is a diagonal matrix with all the unstable roots on the diagonal. The $\gamma$ vectors are partitioned accordingly and the rows $(H_{11}, H_{12})$ of $H$ are the characteristic vectors associated with the unstable roots. Thus, for stability we require

$$H_{11}\gamma_1^{(1)} + H_{12}\gamma_1^{(2)} = 0. \tag{5.8}$$

These $n - k$ equations define the saddle point manifold and are the additional $n - k$ equations needed for a solution. Having solved for $\gamma_1$ and the unknown elements of $\gamma_0$ we then obtain the remaining $\gamma_i$ coefficients from

$$\gamma_i^{(1)} = -H_{11}^{-1}H_{12}\gamma_i^{(2)} \qquad i = 2, \ldots, \tag{5.9}$$

$$\gamma_{i+1}^{(2)} = \Lambda_2\gamma_i^{(2)} \qquad i = 1, 2, \ldots . \tag{5.10}$$

## 5.2. Higher order vector models

Alternatively the solution of (5.1) can be obtained directly without forming a large first order system. This method is essentially a generalization of the scalar method used in Section 2.4. Very briefly, by substituting the general solution of $y_t$ into (5.1) and examining the equation in the $\Gamma_i$ coefficients the solution can be obtained by factoring the characteristic polynomial associated with these equations.

This approach has been used by Hansen and Sargent (1981) in an optimal control example where $p = q$ and $B_i = hA_i'$. In that case, the factorization can be shown to be unique by an appeal to the factorization theorems for spectral density matrices. A similar result was used in Taylor (1980a) in the case of a factoring spectral density functions.

In general econometric applications, these special properties on the $A_i$ and $B_i$ matrices do not hold. Whiteman (1983) has a proof that a unique factorization exists under conditions analogous to those placed on the roots of the model in Section 5.1. Dagli and Taylor (1983) have investigated an iterative method to factor the polynomials in the lag operator in order to obtain a solution. This factorization method was used by Rehm (1982) to estimate a 7-equation rational expectations model of the U.S. using full information maximum likelihood.

## 6. Techniques for nonlinear models

As yet there has been relatively little research with nonlinear rational expectations models. The research that does exist has been concerned more with solution and policy evaluation rather than with estimation. Fair and Taylor (1983) have investigated a full-information estimation method for a non-linear model based on a solution procedure described below. However, this method is extremely expensive to use given current computer technology. Hansen and Singleton (1982) have developed and applied a limited-information estimator for nonlinear models.

There are a number of alternative solution procedures for nonlinear models that have been investigated in the literature. They generally focus on deterministic models, but can be used for stochastic analysis by stochastic simulation techniques.

Three methods are reviewed here: (1) a "multiple shooting" method, adopted for rational expectations models from two-point boundary problems in the differential equation literature by Lipton, Poterba, Sachs, and Summers (1982), (2) an "extended path" method based on an iterative Gauss–Seidel algorithm examined by Fair and Taylor (1983), and (3) a nonlinear stable manifold method examined by Bona and Grossman (1983). This is an area where there is likely to be much research in the future.

A general nonlinear rational expectation model can be written

$$f_i\left(y_t, y_{t-1}, \ldots, y_{t-p}, \mathop{E}_t y_{t+1}, \ldots, \mathop{E}_t y_{t+q}, \alpha_i, x_t\right) = u_{it}, \tag{6.1}$$

for $i = 1, \ldots, n$, where $y_t$ is an $n$ dimensional vector of endogenous variables at time $t$, $x_t$ is a vector of exogenous variables, $\alpha_i$ is a vector of parameters, and $u_{it}$ is a vector of disturbances. In some write-ups, (e.g. Fair–Taylor) the viewpoint date on the expectations in (6.1) is based on information through period $t - 1$

rather than through period $t$. For continuity with the rest of this paper, we continue to assume that the information is through period $t$, but the methods can easily be adjusted for different viewpoint dates. We also distinguish between exogenous variables and disturbances, because some of the nonlinear algorithms can be based on known future values of $x_t$ rather than on forecasts of these from a model like (2.2).

### 6.1. Multiple shooting method

We described the shooting method to solve linear rational expectations models in Section 2.5. This approach is quite useful in nonlinear models. The initial conditions are the values for the lagged dependent variables and the final conditions are given by the long-run equilibrium of the system. In this case, a system of nonlinear equations must be solved using an iterative scheme such as Newton's method. One difficulty with this technique is that (6.1) is explosive when solved forward so that very small deviations of the endogenous variables from the solution can lead to very large final values. If this is a problem then the shooting method can be broken up in the series of shootings (multiple shooting) over intervals smaller than $(0, j)$. For example three intervals would be $(0, j_1)$, $(j_1, j_2)$ and $(j_2, j)$ for $0 < j_1 < j_2 < j$. In effect the relationship between the final values and the initial values is broken up into a relationship between intermediate values of these variables. The intervals can be made arbitrarily small. This approach has been used by Summers (1981) and others to solve rational expectations models of investment and in a number of other applications. It seems to work very well.

### 6.2. Extended path method

This approach has been examined by Fair and Taylor (1983) and used to solve large-scale nonlinear models. Briefly it works as follows. Guess values for the $E_t y_{t+j}$ in eq. (6.1) for $j = 1, \ldots, J$. Use these values to solve the model to obtain a new path for $y_{t+j}$. Replace the initial guess with the new solution and repeat the process until the path $y_{t+j}$, $j = 1, \ldots, J$ converges, or changes by less than some tolerance range. Finally, extend the path from $J$ to $J+1$ and repeat the previous sequence of iterations. If the values of $y_{t+j}$ on this extended path are within the tolerance range for the values of $J+1$, then stop; otherwise extend the path one more period to $J+2$ and so on. Since the model is nonlinear, the Gauss–Seidel method is used to solve (6.1) for each iteration given a guess for $y_{t+j}$. There are no general proofs available to show that this method works for an arbitrary nonlinear model. When applied to the linear model in Section (2.1) with $|\alpha| < 1$ the method is shown to converge in Fair and Taylor (1983). When $|\alpha| > 1$, the iterations diverge. A convergence proof for the general linear model is not yet available, but many experiments have indicated that convergence is achieved under the usual saddle path assumptions. This method is expensive but is fairly easy to use. An empirical application of the method to a modified version of the Fair model is found in Fair and Taylor (1983) and to a system with time varying parameters in Taylor (1983). Carlozzi and Taylor (1984) have used the method to calculate stochastic equilibria. This method also appears to work well.

## 6.3.  Nonlinear saddle path manifold method

In Section (2.4) we noted that the solution of the second-order linear difference eq. (2.88) is achieved by placing the solution on the stable path associated with the saddle point line. For nonlinear models one can use the same approach after linearizing the system. The saddle point manifold is then linear. Such a linearization, however, can only yield a local approximation.

Bona and Grossman (1983) have experimented with a method that computes a nonlinear saddle-point path. Consider a deterministic univariate second-order version of (6.1):

$$f(y_{t+1}, y_t, y_{t-1}) = 0, \qquad i = 1, 2, \ldots . \tag{6.2}$$

A solution will be of the form

$$y_t = g(y_{t-1}), \tag{6.3}$$

where we have one initial condition $y_0$. Note that eq. (6.2) is a nonlinear version of the homogeneous part of eq. (2.88) and eq. (6.3) is a nonlinear version of the saddle path dynamics (2.91).

Bona and Grossman (1983) compute $g(\cdot)$ by a series of successive approximations. If eq. (6.3) is to hold for all values of the argument of $g$ then

$$f(g(g(x)), g(x), x) = 0, \tag{6.4}$$

must hold for every value of $x$ (at least within the range of interest). In the application considered by Bona and Grossman (1983) there is a natural way to write (6.4) as

$$g(x) = h(g(g(x)), g(x), x), \tag{6.5}$$

for some function $h(\cdot)$. For a given $x$ eq. (6.5) may be solved using successive approximations:

$$g_{n+1}(x) = h\big(g_n(g_n(x)), g_n(x), x\big), \qquad n = 0, 1, 2, \ldots. \tag{6.6}$$

The initial function $g_0(x)$ can be chosen to equal the linear stable manifold associated with the linear approximation of $f(\cdot)$ at $x$.

Since this sequence of successive approximations must be made at every $x$, there are two alternative ways to proceed. One can make the calculations recursively for each point $y_t$ of interest; that is, obtain a function $g$ for $x = y_0$, a new function for $x = y_1$ and so on. Alternatively, one could evaluate $g$ over a grid of the entire range of possible values of $x$, and form a "meta function" $g$ which is piecewise linear and formed by linear interpolation for the value of $x$ between the grid points. Bona and Grossman (1983) use the first procedure to numerically solve a macroeconomic model of the form (6.2).

It is helpful to note that when applied to linear models the method reduces to a type of undetermined coefficients method used by Lucas (1975) and McCallum (1983) to solve rational expectations models (a different method of undetermined coefficients than that applied to linear process (2.4) in Section 2 above). To see this, substitute a linear function $y_t = g y_{t-1}$ into

$$y_{t+1} = \frac{\alpha_2}{\alpha_1} y_t - \frac{1}{\alpha_1} y_{t-1}, \tag{6.7}$$

the deterministic difference equation already considered in eq. (2.88). The resulting equation is

$$\left( g^2 - \frac{2}{\alpha_1} g + \frac{1}{\alpha_1} \right) y_{t-1} = 0. \tag{6.8}$$

Setting the term in parenthesis equal to zero, yields the characteristic polynomial of (6.7) which appears in eq. (2.89). Under the usual assumption that one root is inside and one root is outside the unit circle a unique stable value of $g$ is found and is equal to stable root $\lambda_2$ of (2.89).

## 7. Concluding remarks

As its title suggests, the aim of this chapter has been to review and tie together in an expository way the extensive volume of recent research on econometric techniques for macroeconomic policy evaluation. The table of contents gives a good summary of the subjects that I have chosen to review. In conclusion it is perhaps useful to point out in what ways the title is either overly inclusive or not inclusive enough relative to the subjects actually reviewed.

All of the methods reviewed – estimation, solution, testing, optimization – involve the rational expectations assumption. In fact the title would somewhat more accurately identify the methods reviewed if the work "new" were replaced by "rational expectations". Some other new econometric techniques not reviewed here that have macroeconomic policy applications include the multivariate time series methods (vector auto-regressions, causality, exogeneity) reviewed by Geweke (1983) in Volume 1 of the *Handbook of Econometrics*, the control theory methods reviewed by Kendrick (1981) in Volume 1 of the *Handbook of Mathematical Economics*, and the prediction methods reviewed by Fair (1986) in this volume. On the other hand some of the estimation and testing techniques reviewed here were designed for other applications even though they have proven useful for policy.

Some of the topics included were touched on only briefly. In particular the short treatment of limited information estimation techniques, time inconsistency, and stochastic general equilibrium models with optimizing agents does not give justice to the large volume of research in these areas.

Most of the research reviewed here is currently very active and the techniques are still being developed. (About $\frac{2}{3}$ of the papers in the bibliography were published between the time I agreed to write the review in 1979 and the period in 1984 when I wrote it.) The development of computationally tractable ways to deal with large and in particular non-linear models is an important area that needs more work. But in my view the most useful direction for future research in this area will be in the applications of the techniques that have already been developed to practical policy problems.

## References

Anderson, Gary and George Moore (1984) "An Efficient Procedure for Solving Linear Perfect Foresight Models". Board of Governors of the Federal Reserve Board, unpublished manuscript.

Anderson, T. W. (1971) *The Statistical Analysis of Time Series*. New York: Wiley.

Baumol, W. J. (1970) *Economic Dynamics: An Introduction*, 3d ed. New York: Macmillan.

Birkhoff, Garret and G. C. Rota (1962) *Ordinary Differential Equations*. Waltham: Blaisdell, 2nd Edition.

Blanchard, Olivier J. (1979) "Backward and Forward Solutions for Economies with Rational Expectations", *American Economic Review*, 69, 114–118.

Blanchard, Olivier J. (1982) "Identification in Dynamic Linear Models with Rational Expectations". Technical Paper No. 24, National Bureau of Economic Research.

Blanchard, Oliver and Charles Kahn (1980) "The Solution of Linear Difference Models under Rational Expectations", *Econometrica*, 48, 1305–1311.

Blanchard, Olivier and Mark Watson (1982) "Rational Expectations, Bubbles and Financial Markets", in: P. Wachtel, ed., *Crises in The Economic and Financial Structure*. Lexington: Lexington Books.

Bona, Jerry and Sanford Grossman (1983) "Price and Interest Rate Dynamics in a Transactions Based Model of Money Demand". University of Chicago, unpublished paper.

Buiter, Willem H. and Marcus Miller (1983) "Real Exchange Rate Overshooting and the Output Cost of Bringing Down Inflation: Some Further Results", in: J. A. Frenkel, ed., *Exchange Rates and International Macroeconomics*. Chicago: University of Chicago Press for National Bureau of Economic Research.

Cagan, Phillip (1956) "The Monetary Dynamics of Hyperinflation", in: M. Friedman, ed., *Studies in the Quantity Theory of Money*. Chicago: University of Chicago Press.

Calvo, Guillermo (1978) "On The Time Consistency of Optimal Policy in a Monetary Economy", *Econometrica*, 46, 1411–1428.

Calvo, Guillermo (1980) "Tax-Financed Government Spending in a Neo-Classical Model with Sticky Wages and Rational Expectations", *Journal of Economic Dynamics and Control*, 2, 61–78.

Carlozzi, Nicholas and John B. Taylor (1984) "International Capital Mobility and the Coordination of Monetary Rules", in: J. Bandhari, ed., *Exchange Rate Management under Uncertainty*. MIT Press, forthcoming.

Chow, G. C. (1983) *Econometrics*. New York: McGraw Hill.

Chow, Gregory and Philip J. Reny (1984) "On Two Methods for Solving and Estimating Linear Simultaneous Equations with Rational Expectations". Princeton University, unpublished paper.

Christiano, Lawrence J. (1984) "Can Automatic Stabilizers be Destablizing: An Old Question Revisited", *Carnegie–Rochester Conference Series on Public Policy*, 20, 147–206.

Cumby, Robert E., John Huizinga and Maurice Obstfeld (1983) "Two-Step Two-Stage Least Squares Estimation in Models with Rational Expectations", *Journal of Econometrics*, 21, 333–355.

Dagli, C. Ates and John B. Taylor (1985) "Estimation and Solution of Linear Rational Expectations Models Using a Polynomial Matrix Factorization", *Journal of Economic Dynamics and Control*, forthcoming.

Dhrymes, Pheobus J. (1971) *Distributed Lags: Problems of Estimation and Formulation*. San Francisco: Holden–Day.

Dixit, Avinash (1980) "A Solution Technique for Rational Expectations Models with Applications to Exchange Rate and Interest Rate Determination". Princeton University, unpublished paper.

Dornbusch, Rudiger (1976) "Expectations and Exchange Rate Dynamics", *Journal of Political Economy*, 84, 1161–1176.

Epple, Dennis, Lars P. Hansen and William Roberds (1983) "Linear Quadratic Games of Resource Depletion", in: Thomas J. Sargent, ed., *Energy, Foresight, and Strategy*. Washington: Resources for the Future.

Evans, George and Seppo Honkapohja (1984) "A Complete Characterization of ARMA Solutions to Linear Rational Expectations Models". Technical Report No. 439, Institute for Mathematical Studies in the Social Sciences, Stanford University.

Fair, Ray (1986) "Evaluating the Predictive Accuracy of Models", in: Z. Griliches and M. Intriligator, eds., *Handbook of Econometrics*. Amsterdam: North-Holland, Vol. III.

Fair, Ray and John B. Taylor (1983) "Solution and Maximum Likelihood Estimation of Dynamic Nonlinear Rational Expectations Models", *Econometrica*, 51, 1169–1185.

Fischer, Stanley (1979) "Anticipations and the Nonneutrality of Money", *Journal of Political Economy*, 87, 225–252.

Flood, R. P. and P. M. Garber (1980) "Market Fundamentals versus Price-Level Bubbles: The First Tests", *Journal of Political Economy*, 88, 745–770.

Futia, Carl A. (1981) "Rational Expectations in Stationary Linear Models", *Econometrica*, 49, 171–192.

Geweke, John (1984) "Inference and Causality in Economic Time Series Models", in: Z. Griliches and M. Intriligator, eds., *Handbook of Econometrics*. Amsterdam: North-Holland, Vol. II.

Gourieroux, C., J. J. Laffont and A. Monfort (1982) "Rational Expectations in Linear Models: Analysis of Solutions", *Econometrica*, 50, 409–425.

Hansen, Lars P. (1982) "Large Sample Properties of Generalized Method of Moments Estimators", *Econometrica*, 50, 1029–1054.

Hansen, Lars P. and Thomas J. Sargent (1980) "Formulating and Estimating Dynamic Linear Rational Expectations Models", *Journal of Economic Dynamics and Control*, 2, 7–46.

Hansen, Lars P. and Thomas J. Sargent (1981) "Linear Rational Expectations Models for Dynamically Interrelated Variables", in: R. E. Lucas and T. J. Sargent, eds., *Rational Expectations and Econometric Practice*. Minneapolis: University of Minnesota Press.

Hansen, L. P. and K. Singleton (1982) "Generalized Instrumental Variables Estimation of Nonlinear Rational Expectations Models", *Econometrica*, 50, 1269–1286.

Harvey, Andrew C. (1981) *Time Series Models*. New York: Halsted Press.

Hayashi, Fumio and Christopher Sims (1983) "Nearly Efficient Estimation of Time Series Models with Predetermined, but not Exogenous, Instruments", *Econometrica*, 51, 783–798.

Kendrick, David (1981) "Control Theory with Applications to Economics", in: K. Arrow and M. Intriligator, eds., Amsterdam: North-Holland, Vol. I.

Kouri, Pentti J. K. (1976) "The Exchange Rate and the Balance of Payments in the Short Run and in the Long Run: A Monetary Approach", *Scandinavian Journal of Economics*, 78, 280–304.

Kydland, Finn E. (1975) "Noncooperative and Dominant Player Solutions in Discrete Dynamic Games", *International Economic Review*, 16, 321–335.

Kydland, Finn and Edward C. Prescott (1977) "Rules Rather Than Discretion: The Inconsistency of Optimal Plans", *Journal of Political Economy*, 85, 473–491.

Kydland, Finn and Edward C. Prescott (1982) "Time to Build and Aggregate Fluctuations", *Econometrica*, 50, 1345–1370.

Lipton, David, James Poterba, Jeffrey Sachs and Lawrence Summers (1982) "Multiple Shooting in Rational Expectations Models", *Econometrica*, 50, 1329–1333.

Lucas, Robert E. Jr. (1975) "An Equilibrium Model of the Business Cycle", *Journal of Political Economy*, 83, 1113–1144.

Lucas, Robert E. Jr. (1976) "Econometric Policy Evaluation: A Critique", in: K. Brunner and A. H. Meltzer, eds., *Carnegie Rochester Conference Series on Public Policy*. Amsterdam: North-Holland, 19–46.

Lucas, Robert E. Jr. and Thomas J. Sargent (1981) "Introduction", to their *Rational Expectations and Econometric Practice*. University of Minneapolis.

Luenberger, David G. (1977) "Dynamic Equations in Descriptor Form", *IEEE Transactions on Automatic Control*, AC-22, 312–321.

Marschak, Jacob (1953) "Economic Measurements for Policy and Prediction", in: W. C. Hood and T. C. Koopmans, eds., *Studies in Econometric Method*. Cowles Foundation Memograph 14, New Haven: Yale University Press.

McCallum, Bennett T. (1976) "Rational Expectations and the Natural Rate Hypothesis: Some Consistent Estimates", *Econometrica*, 44, 43–52.

McCallum. Bennett T. (1979) "Topics Concerning the Formulation, Estimation, and Use of Macro-econometric Models with Rational Expectation", *American Statistical Association*, *Proceedings of the Business and Economics Section*, 65–72.

McCallum, Bennett T. (1983) "On Non-Uniqueness in Rational Expectations: An Attempt at Perspective", *Journal of Monetary Economics*, 11, 139–168.

Mishkin, Frederic S. (1983) *A Rational Expectations Approach to Macroeconometrics: Testing Policy Ineffectiveness and Efficient-Markets Models*. Chicago: University of Chicago Press.

Muth, John F. (1961) "Rational Expectations and The Theory of Price Movements", *Econometrica*, 29, 315–335.

Muth, John F. (1981) "Estimation of Economic Relationships Containing Latent Expectations Variables", reprinted in: R. E. Lucas and T. J. Sargent, eds., *Rational Expectations and Econometric Practice*. Minneapolis: University of Minnesota Press.

Papell, David (1984) "Anticipated and Unanticipated Disturbances: The Dynamics of The Exchange Rate and The Current Account", *Journal of International Money and Finance*, forthcoming.

Preston, A. J. and A. R. Pagan (1982) *The Theory of Economic Policy*. Cambridge: Cambridge University Press.

Rehm, Dawn (1982) *Staggered Contracts, Capital Flows, and Macroeconomic Stability in The Open Economy*. Ph.D. Dissertation, Columbia University.

Rodriquez, Carlos A. (1980) "The Role of Trade Flows in Exchange Rate Determination: A Rational Expectations Approach", *Journal of Political Economy*, 88, 1148–1158.

Sargent, Thomas J. (1979) *Macroeconomic Theory*. New York: Academic Press.

Sargent, Thomas J. and Neil Wallace (1973) "Rational Expectations and The Dynamics of Hyperinfla-tion", *International Economic Review*, 14, 328–350.

Sargent, Thomas J. and Neil Wallace (1975) "'Rational' Expectations, The Optimal Monetary Instrument, and The Optimal Money Supply Rule", *Journal of Political Economy*, 83, 241–254.

Summers, Lawrence H. (1981) "Taxation and Corporate Investment: A *q*-Theory Approach", *Brookings Papers on Economic Activity*, 1, 67–127.

Taylor, John B. (1977) "Conditions for Unique Solutions in Stochastic Macroeconomic Models with Rational Expectations", *Econometrica*, 45, 1377–1385.

Taylor, John B. (1979) "Estimation and Control of a Macroeconomic Model with Rational Expectations", *Econometrica*, 47, 1267–1286.

Taylor, John B. (1980a) "Aggregate Dynamics and Staggered Contracts", *Journal of Political Economy*, 88, 1–23.

Taylor, John B. (1980b) "Output and Price Stability: An International Comparison", *Journal of Economic Dynamics and Control*, 2, 109–132.

Taylor, John B. (1982) "The Swedish Investment Fund System as a Stabilization Policy Rule", *Brookings Papers on Economic Activity*, 1, 57–99.

Taylor, John B. (1983) "Union Wage Settlements During a Disinflation", *American Economic Review*, 73, 981–993.

Wallis, Kenneth F. (1980) "Econometric Implications of The Rational Expectations Hypothesis", *Econometrica*, 48, 49–73.

Whiteman, Charles H. (1983) *Linear Rational Expectations Models: A User's Guide*. Minneapolis: University of Minnesota.

Wickens, M. (1982) "The Efficient Estimation of Econometric Models with Rational Expectations", *Review of Economic Studies*, 49, 55–68.

Wilson, Charles (1979) "Anticipated Shocks and Exchange Rate Dynamics", *Journal of Political Economy*, 87, 639–647.

*Chapter 35*

# ECONOMIC POLICY FORMATION: THEORY AND IMPLEMENTATION (APPLIED ECONOMETRICS IN THE PUBLIC SECTOR)

LAWRENCE R. KLEIN

*University of Pennsylvania*

## Contents

*Handbook of Econometrics, Volume III, Edited by Z. Griliches and M.D. Intriligator*
© *Elsevier Science Publishers BV, 1986*

## 1. Some contemporary policy issues

Mainstream economic policy, known basically as demand management, and its econometric implementation are jointly under debate now. The main criticism comes from monetarists, who focus on versions of the quantity theory of money, from advocates of the theory of rational expectations, and more recently from supply side economists. All these criticisms will be considered in this paper, as well as the criticisms of public policy makers, who are always looking for precision in their choice procedures, even when the subject matter is inherently stochastic and relatively "noisy".

Demand management is usually identified as Keynesian economic policy, i.e. as the type that is inspired by the aggregative Keynesian model of effective demand. Also, the mainstream econometric models are called Keynesian type models; so the present state of world wide stagflation is frequently attributed to the use of Keynesian econometric models for the implementation of Keynesian policies.

These are popular and not scientific views. In this presentation the objective will be to put policy measures in a more general perspective, only some of which are purely demand management and aggregative. Also, the evolution of econometric models for policy application to many supply-side characteristics will be stressed. To a certain extent, the orientation will be towards experience derived from the application of U.S. models to U.S. economic policy, but the issue and methods to be discussed will be more general.

For purposes of exposition, two types of policy will be examined, (1) overall *macro* policies, and (2) specific *structural* policies. Macro policies refer to traditional monetary and fiscal policies, principally of central governments, but the model applications to local government policies are also relevant. As the world economy becomes more interdependent, more economies are recognizing their openness; therefore, trade/payments policies are also part of the complement known as macro policy.

By structural policy I mean policies that are aimed at specific segments of the economy, specific groups of people, specific production sectors, distributions of aggregative magnitudes or markets. Economists like to focus on macro policies because they have overall impacts and leave the distributive market process unaffected, able to do its seemingly efficient work. Most economists look upon the free competitive market as an ideal and do not want to make specific policies that interfere with its smooth working. They may, however, want to intervene with structural policy in order to preserve or guarantee the working of the idealized market process.

Macro policies are quite familiar. Monetary policy is carried out by the central bank and sometimes with a treasury ministry. Also, the legislative branch of democratic governments influence or shape monetary policy. Central executive offices of government also participate in the formation of monetary policy. It is a many sided policy activity. The principal policy instruments are bank reserves and discount rates. Reserves may be controlled through open market operations or the setting of reserve requirements. Policies directed at the instrument levels have as objectives specified time paths of monetary aggregates or interest rates.

At the present time, there is a great deal of interest in controlling monetary aggregates through control of reserves, but some countries continue to emphasize interest rate control through the discount window. On the whole, monetary authorities tend to emphasize one approach or the other; i.e. they try to control monetary aggregates along *monetarist* doctrinal lines or they try to control interest rates through discount policy, but in the spirit of a generalized approach to economic policy there is no reason why central monetary authorities cannot have multiple targets through the medium of multiple instruments. This approach along the lines of modern control theory will be exemplified below.

Monetary policy is of particular importance because it can be changed on short notice, with little or no legislative delay. It may be favored as a flexible policy but is often constrained, in an open economy, by the balance of payments position and the consequent stability of the exchange value of a country's currency. Therefore, we might add a third kind of financial target, namely, an exchange value target. Flexibility is thus restricted in an open economy. Monetary policies that may seem appropriate for a given domestic situation may be constrained by a prevalent international situation.

There are many monetary aggregates extending all the way from the monetary base, to checking accounts, to savings accounts, to liquid money market instruments, to more general credit instruments. The credit instruments may also be distinguished between private and public sectors of issuance. The plethora of monetary aggregates has posed problems, both for the implementation of policy and for the structure of econometric models used in that connection. The various aggregates all behave differently with respect to reserves and the monetary base. The authorities may be able to control these latter concepts quite well, but the targets of interest all react differently. Furthermore, monetary aggregates are not necessarily being targeted because of their inherent interest but because they are thought to be related to nominal income aggregates and the general price level. The more relevant the monetary aggregate for influencing income and the price level, the more difficult is it to control it through the instruments that the authorities can effect.

Benjamin Friedman has found, for the United States, that the most relevant aggregate in the sense of having a stable velocity coefficient is total credit, but this is the least controllable.[1] The most controllable aggregate, currency plus checking

---

[1] Benjamin Friedman, "The Relative Stability of Money and Credit 'Velocities' in the United States: Evidence and Some Speculations", National Bureau of Economic Research, working paper No. 645, March, 1981.

accounts, has the most variable velocity. Between these extremes it appears that the further is the aggregate from control, the less variable is its associated velocity. This is more a problem for the implementation of monetary policy that for the construction of models.

But a problem for both policy formation and modeling is the recent introduction of new monetary instruments and technical changes in the operation of credit markets. Electronic banking, the use of credit cards, the issuance of more sophisticated securities to the average citizen are all innovations that befuddle the monetary authorities and the econometrician. Authorities find that new instruments are practically outside their control for protracted periods of time, especially when they are first introduced. They upset traditional patterns of seasonal variation and generally enlarge the bands of uncertainty that are associated with policy measures. They are problematic for econometricians because they establish new modes of behavior and have little observational experience on which to base sample estimates.

Side by side with monetary policy goes the conduct of fiscal policy. For many years – during and after the Great Depression – fiscal policy was central as far as macro policy was concerned. It was only when interest rates got significantly above depression floor levels that monetary policy was actively used and shown to be fairly powerful.

Fiscal policy is usually, but not necessarily, less flexible than monetary policy because both the legislative and executive branches of government must approve major changes in public revenues and expenditures. In a parliamentary system, a government cannot survive unless its fiscal policy is approved by parliament, but this very process frequently delays effective policy implementation. In a legislative system of the American type, a lack of agreement may not bring down a government, but it may seriously delay the implementation of policy. On the other hand, central banking authorities can intervene in the functioning of financial markets on a moment's notice.

On the side of fiscal policy, there are two major kinds of instruments, public spending and taxing. Although taxing is less flexible than monetary management, it is considerably more flexible than are many kinds of expenditure policy. In connection with expenditures, it is useful to distinguish between purchases of goods or services and transfer payments. The latter are often as flexible as many kinds of taxation instruments.

It is generally safer to focus on tax instruments and pay somewhat less attention to expenditure policy. Tax changes have the flexibility of being made retroactive when desirable. This can be done with some expenditures, but not all. Tax changes can be made effective right after enactment. Expenditure changes, for goods or services, especially if they are increases, can be long in the complete making. Appropriate projects must be designed, approved, and executed. Often it is difficult to find or construct appropriate large projects.

Tax policy can be spread among several alternatives such as personal direct taxes, (either income or expenditure) business income taxes, or indirect taxes. At present, much interest attaches to indirect taxes because of their ease of collection, if increases are being contemplated, or because of their immediate effect on price indexes, if decreases are in order. Those taxes that are levied by local, as opposed to national governments, are difficult to include in national economic analysis because of their diversity of form, status, and amount.

Some tax policies are general, affecting most people or most sectors of the economy all at once. But, *specific*, in contrast to *general*, taxes are important for the implementation of structural policies. An expenditure tax focuses on stimulating personal savings. Special depreciation allowances or investment tax credits aim at stimulating private fixed capital formation. Special allowances for R&D, scientific research, or capital gains are advocated as important for helping the process of entrepreneurial innovation in high technology or venture capital lines. These structural policies are frequently cited in present discussions of industrial policy.

A favorite proposal for strictly anti-inflationary policy is the linkage of tax changes, either as rewards (cuts) or penalties (increases), to compliance by businesses and households with prescribed wage/price guidelines. Few have ever been successfully applied on a broad continuing scale, but this approach, known as incomes policies, social contracts, or TIPS (tax based incomes policies), is widely discussed in the scholarly literature.

These monetary and fiscal policies are the conventional macro instruments of overall policies. They are important and powerful; they must be included in any government's policy spectrum, but are they adequate to deal with the challenge of contemporary problems? Do they deal effectively with such problems as:

–severe unemployment among certain designated demographic groups;
–delivery of energy;
–conservation of energy;
–protection of the environment;
–public health and safety;
–provision of adequate agricultural supply;
–maintenance of healthy trade balance?

Structural policies, as distinct from macro policies, seem to be called for in order to deal effectively with these specific issues.

If these are the kinds of problems that economic policy makers face, it is worthwhile considering the kinds of policy decisions with instruments that have to be used in order to address these issues appropriately, and consider the kind of economic model that would be useful in this connection.

For dealing with youth unemployment and related structural problems in labor markets, the relevant policies are minimum wage legislation, skill training grants, and provision of vocational education. These are typical things that ought to be done to reduce youth unemployment. These policy actions require legislative support with either executive or legislative initiative.

In the case of energy policy, the requisite actions are concerned with pricing of fuels, rules for fuel allocation, controls on imports, protection of the terrain against excessive exploitation. These are specific structural issues and will be scarcely touched by macro policies. These energy issues also effect the environment, but there are additional considerations that arise from non-energy sources. Tax and other punitive measures must be implemented in order to protect the environment, but, at the same time, monitor the economic costs involved. The same is true for policies to protect public health and safety. These structural policies need to be implemented but not without due regard to costs that have serious inflationary consequences. The whole area of public regulation of enterprise is under scrutiny at the present time, not only for the advantages that might be rendered, but also for the fostering of competition, raising incentives, and containing cost elements. It is not a standard procedure to consider the associated inflationary content of regulatory policy.

Ever since the large harvest failures of the first half of the 1970s (1972 and 1975, especially) economists have become aware of the fact that special attention must be paid to agriculture in order to insure a basic flow of supplies and moderation in world price movements. Appropriate policies involve acreage limitations (or expansions), crop subsidies, export licenses, import quotas, and similar specific measures. They all have bearing on general inflation problems through the medium of food prices, as components of consumer price indexes, and of imports on trade balances.

Overall trade policy is mainly guided by the high minded principle of fostering of conditions for the achievement of multilateral free trade. This is a macro concept, on average, and has had recent manifestation in the implementation of the "Tokyo Round" of tariff reductions, together with pleas for moderation of non-tariff barriers to trade. Nevertheless, there are many specific breaches of the principle, and specific protectionist policies are again a matter of concern. Trade policy, whether it is liberal or protectionist, will actually be implemented through a set of structural measures. It might mean aggressive marketing in search of export sales, provision of credit facilities, improved port/storage facilities, and a whole group of related policy actions that will, in the eyes of each country by itself, help to preserve or improve its net export position.

We see then that economic policy properly understood in the context of economic problems of the day goes far beyond the macro setting of tax rates, overall expenditure levels, or establishing growth rates for some monetary aggregates. It is a complex network of specific measures, decrees, regulations (or their absence), and recommendations coming from all branches of the public sector. In many cases they require government coordination. Bureaus, offices, departments, ministries, head of state, and an untold number of public bodies participate in this process. It does not look at all like the simple target-instrument approach of macroeconomics, yet macroeconometric modeling, if pursued at the appropriate level of detail, does have much to contribute. That will be the subject of sections of this paper that follow.

## 2. Formal political economy

The preceding section has just described the issues and actors in a very summary outline. Let us now examine some of the underlying doctrine. The translation of economic theory into policy is as old as our subject, but the modern formalism is conveniently dated from the *Keynesian Revolution*. Clear distinction should be made between Keynesian theory and Keynesian policy, but as far as macro policy is concerned, it derives from Keynesian theory.

The principal thrust of Keynesian theory was that savings-investment balance at full employment would be achieved through adjustment of the aggregative activity level of the economy. It was interpreted, at an early stage, in a framework of interest-inelastic investment and interest-elastic demand for cash. This particular view and setting gave a secondary role to monetary policy. Direct effects on the spending or activity stream were most readily achieved through fiscal policy, either adding or subtracting directly from the flow of activity through public spending or affecting it indirectly through changes in taxation. Thinking therefore centered around the achievement of balance in the economy, at full employment, by the appropriate choice of fiscal measures. In a formal sense, let us consider the simple *model*

$$C = f(Y - T) \qquad \text{consumption function}$$
$$T = tY \qquad \text{tax function}$$
$$I = g(\Delta Y) \qquad \text{investment function}$$
$$Y = C + I + G \qquad \text{output definition}$$

where

$G$ = public expenditures
$\Delta$ = time difference operator
$Y$ = total output (or income, or activity level).

Fiscal policy means the choice of an appropriate value of $t$ (tax rate), or level $G$ (expenditure), or mixture of both in order to achieve a target level of $Y$. This could also be a dynamic policy, by searching for achievement of a target path of $Y$ through time. To complement dynamic policy it is important to work with a richer dynamic specification of the economic model. Lag distributions of $Y-T$ or $\Delta Y$ in the $C$ and $I$ function would be appropriate. This kind of thinking inspired the approach to fiscal policy that began in the 1930's and still prevails today. It inspired thoughts about "fine tuning" or "steering" an economic system. It is obviously terribly simplified. It surely contains grains of truth, but what are the deficiencies?

In the first place, there is no explicit treatment of the price level or inflation rate in this system. Arguments against Keynesian policy pointed out the inflationary dangers from the outset. These dangers were minimal during the 1930's and did not become apparent on a widespread basis for about 30 years – after much successful application of fiscal policy, based on some monetary policy as time wore on. There is no doubt, however, that explicit analysis of price formation and great attention to the inflation problem must be guiding principles for policy formation from this time forward.

Another argument against literal acceptance of this version of crude Keynesianism is that it deals with unrealistic, simplistic concepts. Fiscal action is not directed towards "$t$" or "$G$". Fiscal action deals with complicated allowances, exemptions, bracket rates, capital gains taxation, value added taxation, expenditures for military hardware, agricultural subsidies, food stamps, aid to dependent children, and unemployment insurance benefits. These specific policy instruments have implications for the broad, general concepts represented by "$t$" and "$G$", but results can be quite misleading in making a translation from realistic to such macro theoretical concepts. The system used here for illustration is so simplified that there is no distinction between direct and indirect taxes or between personal and business taxes.

The Keynesian model of income determination can be extended to cover the pricing mechanism, labor input, labor supply, unemployment, wages, and monetary phenomena. There is a difference, however, between monetary analysis and monetarism. Just as the simple Keynesian model serves as the background for doctrinaire Keynesian fiscal policy, there is another polar position, namely, the monetarist model which goes beyond the thought that money matters, to the extreme that says that *only* money matters. The monetarist model has its simplest and crudest exposition in the following equation of exchange

$$Mv = Y.$$

For a steady, parametric, value of $v$ (velocity), there is a linear proportional correspondence between $M$ (nominal money supply) and $Y$ (nominal value of aggregate production or income). For every different $M$-concept, say $M_i$, we would have[2]

$$M_i v_i = Y.$$

[2] See the various concepts in the contribution by Benjamin Friedman, op. cit.

A search for a desired subscript $i$ may attach great importance to the corresponding stability of $v_i$. It is my experience, for example, that in the United States, $v_2$ is more stable than $v_1$.

More sophisticated concepts would be

$$Mv = \sum_{i=0}^{n} w_i Y_{-i},$$

or

$$Mv = \left( \sum_{i=0}^{n} w_i Y_{-i} \right)^{\alpha},$$

or

$$Mv = \left( \sum_{i=0}^{n} w_i X_{-i} \right)^{\alpha} \left( \sum_{i=0}^{m} q_i P_{-i} \right)^{\beta}.$$

The first says that $M$ is proportional to long run $Y$ or a distributed lag in $Y$. The second says that $M$ is proportional to a power of long run $Y$ or merely that a stable relationship exists between long run $Y$ and $M$. Finally, the third says that $M$ is a function of long run price as well as long run real income ($X$). In these relationships no attention is paid to subscripts for $M$, because the theory would be similar (not identical) for any $M$, and proponents of monetarist policy simply argue that a stable relationship should be found for the authorities for some $M_i$ concept, and that they should stick to it.

The distributed lag relationships in $P_{-i}$ and $X_{-i}$ are evidently significant generalizations of the crude quantity theory, but in a more general view, the principal thing that monetarists need for policy implementation of their theory is a *stable* demand function for money. If this stable function depends also on interest rates (in lag distributions), the theory can be only *partial*, and analysis then falls back on the kind of mainstream general macroeconometric model used in applications that are widely criticized by strict monetarists.[3]

The policy implications of the strict monetarist approach are clear and are, indeed, put forward as arguments for minimal policy intervention. The proponents are generally against activist fiscal policy except possibly for purposes of indexing when price movements get out of hand. According to the basic monetarist

---

[3] The lack of applicability of the monetarist type relationship, even generalized dynamically, to the United Kingdom is forcefully demonstrated by D. F. Hendry and N. R. Ericsson, "Assertion without Empirical Basis: An Econometric Appraisal of Friedman and Schwartz' 'Monetary Trends in...the United Kingdom,'" *Monetary Trends in the United Kingdom*, Bank of England Panel of Academic Consultants, Panel Paper No. 22 (October 1983), 45–101.

relationship, a rule should be established for the growth rate of $M$ according to the growth rate of $Y$, preferably the long run concept of $Y$. A steady growth of $M$, according to this rule, obviates the need for frequent intervention and leaves the economy to follow natural economic forces. This is a macro rule, in the extreme, and the monetarists would generally look for the competitive market economy to make all the necessary micro adjustments without personal intervention.

The theory for the steady growth of $M$ and $Y$ also serves as a theory for inflation policy, for if the competitive economy maintains long run real income $(\Sigma w_i X_{-i})$ at its full capacity level – not in every period, but on average over the cycle – then steady growth of $M$ implies a steady level for long run price $(\Sigma q_i P_{-i})$. The monetarist rule is actually intended as a policy rule for inflation control.

There are several lines of argument against this seemingly attractive policy for minimal intervention except at the most aggregative level, letting the free play of competitive forces do the main work of guiding the economy in detail. In the first place there is a real problem in defining $M_i$, as discussed already in the previous section. Banking and credit technology is rapidly changing. The various $M_i$ concepts are presently quite fluid, and there is no clear indication as to which $M_i$ to attempt to control. To choose the most tractable concept is not necessarily going to lead to the best economic policy.

Not only are the $M_i$ concepts under debate, but the measurement of any one of them is quite uncertain. Coverage of reporting banks, the sudden resort to new sources of funds (Euro-currency markets, e.g.), the attempts to live with inflation, and other disturbing factors have lead to very significant measurement errors, indicated in part at least by wide swings in data revision of various $M_i$ series. If the monetary authorities do not know $M_i$ with any great precision, how can they hit target values with the precision that is assumed by monetarists? It was previously remarked that policy makers do not actually choose values for "$t$" and "$G$". Similarly, they do not choose values for "$M_i$". They engage in open market buying and selling of government securities; they fix reserve requirements for specific deposits or specific classes of banks; they fix the discount rate and they make a variety of micro decisions about banking practices. In a fractional reserve system, there is a money multiplier connecting the reserve base that is controlled by monetary authorities to $M_i$, but the multiplier concept is undergoing great structural change at the present time, and authorities do not seem to be able to hit $M_i$ targets well.

A fundamental problem with either the Keynesian or the monetarist view of formal political economy is that they are based on simple models – models that are useful for expository analysis but inadequate to meet the tasks of economic policy. These simple models do not give a faithful representation of the economy; they do not explicitly involve the appropriate levels of action; they do not take account of enough processes in the economy. Imagine running the economy according to a strict monetarist rule or fine tuning applications of tax policy in the face of world shortages in energy markets and failing to take appropriate action simply because there are no energy parameters or energy processes in the expository system. This, in fact, is what people from the polar camps have said at various times in the past few years.

What is the appropriate model, if neither the Keynesian nor the monetarist models are appropriate? An eclectic view is at the base of this presentation. Some would argue against eclecticism on a priori grounds as being too diffuse, but it may be that an eclectic view is necessary in order to get an adequate model approximation to the complicated modern economy. Energy, agriculture, foreign trade, exchange rates, the spectrum of prices, the spectrum of interest rates, demography, and many other things must be taken into account simultaneously. This cannot be done except through the medium of large scale models. These systems are far different in scope and method from either of the polar cases. They have fiscal and monetary sectors, but they have many other sectors and many other policy options too.

As a general principle, I am arguing against the formulation of economic policy through the medium of small models – anything fewer than 25 simultaneous equations. Small models are inherently unable to deal with the demands for economic policy formation. An appropriate large-scale model can, in my opinion, be used in the policy process. An adequate system is not likely to be in the neighborhood of 25 equations, however. It is likely to have more than 100 equations, and many in use today have more than 500–1000 equations. The actual size will depend on the country, its openness, its data system, its variability, and other factors. The largest systems in regular use have about 5000 equations, and there is an upper limit set by manageability.[4]

It is difficult to present such a large system in compact display, but it is revealing to lay out its sectors:

Consumer demand
Fixed capital formation
Inventory accumulation
Foreign trade
Public spending on goods and services
Production of goods and services
Labor requirements
Price formation
Wage determination

[4] The Wharton Quarterly Model, regularly used for short run business cycle analysis had 1000 equations in 1980, and the medium term Wharton Annual Model had 1595 equations, exclusive of input–output relationships. The world system of Project LINK has more than 15,000 equations at the present time, and is still growing.

Labor supply and demography
Income formation
Money supply and credit
Interest rate determination
Tax receipts
Transfer payments
Inter industry production flows

In each of these sectors, there are several subsectors, some by type of product, some by type of end use, some by age-sex-race, some by country of origin or destination, some by credit market instrument, and some by level of government. The production sector may have a complete input–output system embedded in the model. Systems like these should not be classified as either Keynesian or monetarist. They are truly eclectic and are better viewed as approximations to the true but unknown Walrasian structure of the economy. These approximations are not unique. The whole process of model building is in a state of flux because at any time when one generational system is being used, another, better approximation to reality is being prepared. The outline of the equation structure for a system combining input–output relations with a macro model of income determination and final demand, is given in the appendix.

The next section will deal with the concrete policy making process through the medium of large scale models actually in use. They do not govern the policy process on an automatic basis, but they play a definite role. This is what this presentation is attempting to show.

There is, however, a new school of thought, arguing that economic policy will not get far in actual application because the smart population will counter public officials' policies, thus nullifying their effects. On occasion, this school of thought, called the *rational expectations* school, indicate that they think that the use of macroeconometric models to guide policy is vacuous, but on closer examination their argument is seen to be directed at any activist policy, whether through the model medium or not.

The argument, briefly put, of the rational expectations school is that economic agents (household, firms, and institutions) have the same information about economic performance as the public authorities and any action by the latter, on the basis of their information has already been anticipated and will simply lead to re-action by economic agents that will nullify the policy initiatives of the authorities. On occasion, it has been assumed that the hypothetical parameters of economic models are functions of policy variables and will change in a particular way when policy variables are changed.[5]

---

[5] R. Lucas, "Econometric Policy Evaluation: A Critique," *The Phillips Curve and Labor Markets*, eds., K. Brunner and A. K. Meltzer. (Amsterdam: North-Holland, 1976), 19–46.

Referring to a linear expression of the consumption function in the simple Keynesian model, they would assume

$$C = \alpha + \beta(Y - T)$$

$$\beta = \beta(T, G)$$

This argument seems to me to be highly contrived. It is true that a generalization of the typical model from fixed to variable parameters appears to be very promising, but there is little evidence that the generalization should make the coefficients depend in such a special way on exogeneous instrument variables.

The thought that economic models should be written in terms of the agent's perceptions of variables on the basis of their interpretation of history is sound. The earliest model building attempts proceeded from this premise and introduced lag distributions and various proxies to relate strategic parameter values, to information at the disposal of both economic agents and public authorities, but they did not make the blind intellectual jump to the conclusion that perceptions of the public at large and authorities are the same. It is well known that the public, at any time, holds widely dispersed views about anticipations for the economy. Many do not have sophisticated perceptions and do not share the perceptions of public authorities. Many do not have the qualifications or facilities to make detailed analysis of latest information or history of the economy.

Econometric models are based on theories and estimates of the way people *do* behave, not on the way they *ought* to behave under the conditions of some hypothesized decision making rules. In this respect, many models currently in use, contain data and variables on expressed expectations, i.e. those expected values that can be ascertained from sample surveys. In an interesting paper dealing with business price expectations, de Leeuw and McKelvey find that statistical evidence on expected prices contradict the hypothesis of rationality, as one might expect.[6]

The rise of the rational expectations school is associated with an assertion that the mainstream model, probably meaning the Keynesian model, has failed during the 1970s. It principally failed because of its inability to cope with a situation in which there are rising rates of inflation and rising rates of unemployment. In standard analysis the two ought to be inversely related, but recently they have been positively related. Charging that macroeconomic models have failed in this situation, Lucas and Sargent, exponents of the school of rational expectations, seek an equilibrium business cycle model consisting of optimizing behavior by

[6] F. de Leeuw and M. McKelvey, "Price Expectations by Business Firms," Brookings Papers on Economic Activity, 1981) 299–314. The findings in this article have been extended, and they now report that there is evidence in support of long run lack of bias in price expectations, a necessary but not sufficient condition for *rationality* of price expectations. See "Price Expectations of Business Firms: Bias in the Short and Long Run," *American Economic Review*, 74 (March 1984), 99–110.

economic agents and the clearing of markets.[7] Many, if not most, macroeconometric models are constructed piece-by-piece along these lines and have been for the past 30 or more years. Rather than reject a whole body of analysis or demand wholly new modelling approaches, it may be more fruitful to look more carefully at the eclectic model that has, in fact, been in use for some time. If such models have appropriate allowance for supply side disturbances, they can do quite well in interpreting the events of the 1970s and even anticipated them in many instances.[8]

## 3.  Some policy projections

Rather than move in the direction of the school of rational expectations, I suggest that we turn from the oversimplified model and the highly aggregative policy instruments to the eclectic system that has large supply side content, together with conventional demand side analysis and examine structural as well as macro policies.

In the 1960s, aggregative policies of Keynesian demand management worked very well. The 1964 tax cut in the United States was a textbook example and refutes the claim of the rational expectations school that parametric shifts will nullify policy action. It also refutes the idea that we know so little about the response pattern of the economy that we should refrain from activist policies.

Both the Wharton and Brookings Models were used for simulations of the 1964 tax cut.[9] A typical policy simulation with the Wharton Model is shown in the accompanying table.

This is a typical policy simulation with an econometric model, solving the system dynamically, with and without a policy implementation. The results in the above table estimate that the policy added about $10 billion (1958 $) to real GNP and sacrificed about $7 billion in tax revenues. Actually, by 1965, the expansion of the (income) tax base brought revenues back to their pre-tax cut position.

The Full Employment Act of 1946 in the United States was the legislation giving rise to the establishment of the Council of Economic Advisers. Similar commitments of other governments in the era following World War II and reconstruction led to the formulation of aggregative policies of demand manage-

---

[7]Robert S. Lucas and Thomas J. Sargent, "After Keynesian Macroeconomics", *After the Phillips Curve: Persistence of High Inflation and High Unemployment.* (Boston: Federal Reserve Bank of Boston, 1978), 49–72.

[8]L. R. Klein, "The Longevity of Economic Theory", *Quantitative Wirtschaftsforschung*, ed. by H. Albach et al. (Tübingen: J. C. B. Mohr (Paul Siebeck), 1977), 411–19; "Supply Side Constraints in Demand Oriented Systems: An Interpretation of the Oil Crisis", *Zeitschrift für Nationalökonomie*, 34 (1974), 45–56; "Five-year Experience of Linking National Econometric Models and of Forecasting International Trade", *Quantitative Studies of International Economic Relations.* H. Glejser, ed. (Amsterdam: North-Holland, 1976), 1–24.

[9]L. R. Klein, "Econometric Analysis of the Tax Cut of 1964," *The Brookings Model: Some Further Results*, ed. by J. Duesenberry et al. (Amsterdam: North-Holland, 1969).

Table 1
Comparative simulations of the tax cut of 1964 (The Wharton Model).

| | Real GNP (bill 1958 $) | | | Personal tax and nontax payments (bill of curr. $) | | |
|---|---|---|---|---|---|---|
| | Actual | Tax cut simulation | No tax cut simulation | Actual | Tax cut simulation | No tax cut simulation |
| 1964.1 | 569.7 | 567.0 | 563.1 | 60.7 | 61.3 | 64.0 |
| 1964.2 | 578.1 | 575.8 | 565.4 | 56.9 | 57.9 | 64.5 |
| 1964.3 | 585.0 | 581.0 | 569.6 | 59.1 | 59.0 | 65.6 |
| 1964.4 | 587.2 | 585.0 | 574.7 | 60.9 | 59.9 | 66.7 |

ment on a broad international scale. New legislation in the United States, under the name of the Humphrey–Hawkins Bill, established ambitious targets for unemployment and inflation during the early part of the 1980s. The bill, however, states frankly that aggregative policy alone will not be able to accomplish the objectives. Structural policies will be needed, and to formulate those, with meaning, it will be necessary to draw upon the theory of a more extensive model, manely, the Keynes–Leontief model.

The Wharton Annual Model is of the Keynes–Leontief type. It combines a model of income generation and final demand determination with a complete input–output system of 65 sectors and a great deal of demographic detail. It is described in general terms in the preceding section and laid out in equation form in the appendix. To show how some structural policies for medium term analysis work out in this system, I have prepared a table with a baseline projection for the 1980s, together with an alternative simulation in which the investment tax credit has been increased (doubled to 1982 and raised by one-third thereafter), in order to stimulate capital formation, general personal income taxes have been reduced by about 6% and a tax has been placed on gasoline (50¢ per gallon).[10] To offset the gasoline tax on consumers, sales taxes have been cut back, with some grants in aid to state and local governments increased to offset the revenue loss of the sales taxes.

These policies mix aggregative fiscal measures with some structural measures to get at the Nation's energy problem. Also, tax changes have been directed specifically at investment in order to improve the growth of productivity and hold down inflation for the medium term. It is an interesting policy scenario because it simultaneously includes both stimulative and restrictive measures. Also, it aims to steer the economy in a particular direction towards energy conservation and inducement of productivity.

As the figures in Table 2 show, the policy simulation produces results that induce more real output, at a lower price level. Lower unemployment accompa-

---

[10] The investment tax credit provides tax relief to business, figured as a percentage of an equipment purchase, if capital formation is undertaken. The percentage has varied, but is now about 10 percent.

Table 2
Estimated policy projections of the Wharton Annual Model 1980–89
(Deviation of policy simulation from baseline)
Selected economic indicators

|  | 1980 | 1981 | 1982 | 1983 | 1984 | 1985 | 1986 | 1987 | 1988 | 1989 |
|---|---|---|---|---|---|---|---|---|---|---|
| GNP (bill $ 1972) | −1 | 14 | 35 | 44 | 50 | 51 | 131 | 48 | 48 | 46 |
| GNP deflator (index points) | −0.4 | −0.7 | −1.4 | −1.7 | −2.1 | −2.4 | −3.0 | −3.6 | −4.6 | −5.7 |
| Unemployment Rate (percentage points) | 0.0 | −0.5 | −1.2 | −1.6 | −1.8 | −1.9 | −1.7 | −1.5 | −1.3 | −1.1 |
| Productivity change (percentage points) | −0.1 | 0.6 | 0.5 | 0.0 | 0.0 | −0.1 | 0.0 | 0.1 | 0.1 | 0.0 |
| Net Exports (bill $) | 0.8 | 6.8 | 4.7 | 10.5 | 6.2 | 2.2 | 0.8 | 0.9 | −0.5 | −1.6 |
| Federal surplus (bill $) | −2.7 | 1.1 | −0.2 | −1.0 | 4.5 | 0.1 | −2.5 | −0.6 | −9.2 | −2.7 |
| Energy ratio (thou BTU/Real GNP) | −0.9 | −0.8 | −0.6 | −0.5 | −0.3 | −0.3 | −0.2 | −0.3 | −0.2 | −0.2 |
| Nonresidential Investment (bill $ 1972) | 0.9 | 4.1 | 8.4 | 11.3 | 13.8 | 14.8 | 16.0 | 16.7 | 17.2 | 17.2 |

nies the higher output, and the improvement in productivity contributes to the lower price index. The lowering of indirect taxes offsets the inflationary impact of higher gasoline taxes.

A cutback in energy use, as a result of the higher gasoline tax, results in a lower BTU/GNP ratio. This holds back energy imports and makes the trade balance slightly better in the policy alternative case.

A contributing factor to the productivity increase is the higher rate of capital formation in the policy alternative. There are no surprises in this example. The results come out as one would guess on the basis of a priori analysis, but the main contribution of the econometric approach is to try to *quantify* the outcome and provide a basis for *net* assessment of both the positive and negative sides of the policy. Also, the differences from the base-line case are not very large. Econometric models generally project moderate gains. To some extent, they *underestimate* change in a systematic way, but they also suggest that the present inflationary situation is deep seated and will not be markedly cured all at once by the range of policies that is being considered.

## 4. The theory of economic policy

The framework introduced by Tinbergen is the most fruitful starting point.[11] He proposed the designation of two kinds of variables, *targets* and *instruments*. A

---

[11] J. Tinbergen, *On The Theory of Economic Policy*. (Amsterdam: North Holland, 1952).

target is an endogenous (dependent) variable in a multivariate–multiequation representation of the economy. An instrument is an exogenous (independent) variable that is controlled or influenced by policy making authorities in order to lead the economy to targets. Not all endogenous variables are targets; not all exogenous variables are instruments.

In the large eclectic model, with more than 500 endogenous variables, policy makers cannot possible comprehend the fine movements in all such magnitudes. Some systems in use have thousands of endogenous variables. At the national economy level, top policy makers may want to focus on the following: GDP growth rate, overall inflation rate, trade balance, exchange rate, unemployment rate, interest rate. There may be intermediate or intervening targets, too, as in our energy policy today – to reduce the volume of oil imports. This is partly a goal on its own, but partly a means of improving the exchange value of the dollar, the trade balance, and the inflation rate. There may be layers of targets in recursive fashion, and in this way policy makers can extend the scope of variables considered as targets, but it is not practical to extend the scope much beyond 10 targets or so. This refers to policy makers at the top. Elsewhere in the economy, different ministers or executives are looking at a number of more specialized targets – traffic safety, agricultural yield, size of welfare rolls, number of housing completions, etc.

The large scale eclectic model has many hundreds or thousands of equations with an equal number of endogenous variables, but there will also be many exogenous variables. A crude rule of thumb might be that there are about as many exogenous as endogenous variables in an econometric model.[12] Perhaps we are too lax in theory building and resign ourselves to accept too many variables in the exogenous category because we have not undertaken the task of explaining them. All government spending variables and all demographic variables, for example, are not exogenous, yet they are often not explicitly modeled, but are left to be explained by the political scientist and sociologist. This practice is rapidly changing. Many variables that were formerly accepted as exogenous are now being given explicit and careful endogenous explanation in carefully designed additional equations; nevertheless, there remains a large number of exogenous variables in the eclectic, large scale model. There are, at least, hundreds.

Only a few of the many exogenous variables are suitable for consideration as instruments. In the first place, public authorities cannot effectively control very many at once. Just as coordinated thought processes can comprehend only a few targets at a time, so can they comprehend only a few instruments at a time. Moreover, some exogenous variables cannot, in principle, be controlled effec-

---

[12] The Wharton Quarterly Model (1980) has 432 stochastic equations, 568 identities, and 401 exogenous variables. The Wharton Annual Model (1980) had 647 stochastic equations, 948 identities and 626 exogenous variables. Exclusive of identities, (and input–output relations) these each have approximate balance between endogenous and exogenous variables.

tively. The many dimensions of weather and climate that are so important for determining agricultural output are the clearest examples of non-controllable exogenous variables – with or without cloud seeding.

The econometric model within which these concepts are being considered will be written as

$$F\left(y', y'_{-1}\ldots y'_{-p}, x'x'_{-1}\ldots x'_{-q}, w'w'_{-1}\ldots w'_{-r}, z', z'_{-1}, z'_{-s}, \Theta'\right) = e \qquad (1)$$

$F$ = column vector of functions:

$$f_1, f_2, \ldots, f_n.$$

$y$ = column vector of target (endogenous) variables:

$$y_1, y_2, \ldots, y_{n_1}.$$

$x$ = column vector of non-target (endogenous) variables:

$$x_1 x_2, \ldots, x_{n_2}$$
$$n_1 + n_2 = n$$

$w$ = column vector of instrument (exogenous) variables:

$$w_1, w_2, \ldots, w_{m_1}$$

$z$ = column vector of non-instrument (exogenous) variables:

$$z_1, z_2, \ldots, z_{m_2}$$
$$m_1 + m_2 = m$$

$\Theta$ = column vector of parameters
$e$ = column vector of errors:

$$e_1, e_2, \ldots, e_n$$

In this system, there are $n$ stochastic equations, with unknown coefficients, in $n$ endogenous variables and $m$ exogenous variables. A subset of the endogenous variables will be targets ($n_1 \leq n$), and a subset of the exogenous variables will be instruments ($m_1 \leq m$).

The parameters are unknown, but estimated by the statistician from observable data or a priori information. The estimated values will be denoted by $\hat{\Theta}$. Also, for any application situation, values must be assigned to the random variables $e$. Either the assumed mean ($E(e) = 0$) will be assigned, or values of $e$ will be generated by some random drawings, or fixed at some a priori non-zero values. But, given values for $e$ and $\hat{\Theta}$, together with initial conditions, econometricians can generally "solve" this equation system. Such solutions or integrals will be used in the policy formation process in a key way.

First, let us consider Tinbergen's special case of equality between the number of instruments and targets, $n_1 = m_1$. Look first at the simplest possible case with one instrument, one target, and one estimated parameter. If the $f$-function expresses a single-valued relationship between $y$ and $w$, we can invert it to give

$$w = g(y, \hat{\Theta}).$$

For a particular target value of $y(y^*)$, we can find the appropriate instrument value $w = w^*$ from the solution of

$$w^* = g(y^*, \hat{\Theta}).$$

If the $f$-function were simple proportional, we can write the answer in closed form as

$$y = \hat{\Theta} w$$

$$w^* = \frac{1}{\hat{\Theta}} y^*.$$

For any desired value of $y$ we can thus find the appropriate action that the authorities must take by making $w = w^*$. This will enable us to hit the target exactly. The only exception to this remark would be that a legitimate target $y^*$ required an unattainable or inadmissable $w^*$. Apart from such inadmissible solutions, we say that for this case the straightforward rule is to interchange the roles of exogenous and endogenous variable and resolve the system, that is to say, treat the $n_1 = m_1$ instruments as though they were unknown endogenous variables and the $n_1 = m_1$ targets as though they were known exogenous variables. Then solve the system for all the endogenous as functions of the exogenous variables so classified.

It is obvious and easy to interchange the roles of endogenous and exogenous variables by inverting the single equation and solving for the latter, given the target value of the former. In a large complicated system linear or not, it is easy to indicate how this may be done or even to write closed form linear expressions for doing it in linear systems, but it is not easy to implement in most large scale models.

For the linear static case, $n_1 = m_1$, we can write

$$\begin{pmatrix} A_{11} & A_{12} \\ A_{21} & A_{22} \end{pmatrix} \begin{pmatrix} y \\ x \end{pmatrix} + \begin{pmatrix} B_{11} & B_{12} \\ B_{21} & B_{22} \end{pmatrix} \begin{pmatrix} w \\ z \end{pmatrix} = e$$

$A_{11}$ is $n_1 \times n_1$; $A_{12}$ is $n_1 \times n_2$; $A_{21}$ is $n_2 \times n_1$; $A_{22}$ is $n_2 \times n_2$

$B_{11}$ is $n_1 \times m_1$; $B_{12}$ is $n_1 \times m_2$; $B_{21}$ is $n_2 \times m_1$; $B_{22}$ is $n_2 \times m_2$

The solution for the desired instruments $w^*$ in terms of the targets $y^*$ and of $z$ is

$$\begin{pmatrix} B_{11} & A_{12} \\ B_{21} & A_{22} \end{pmatrix} \begin{pmatrix} w^* \\ x \end{pmatrix} + \begin{pmatrix} A_{11} & B_{12} \\ A_{21} & B_{22} \end{pmatrix} \begin{pmatrix} y^* \\ z \end{pmatrix} = e$$

$$\begin{pmatrix} w^* \\ x \end{pmatrix} = - \begin{pmatrix} B_{11} & A_{12} \\ B_{21} & A_{22} \end{pmatrix}^{-1} \begin{pmatrix} A_{11} & B_{12} \\ A_{21} & B_{22} \end{pmatrix} \begin{pmatrix} y^* \\ z \end{pmatrix} + \begin{pmatrix} B_{11} & A_{12} \\ B_{12} & A_{22} \end{pmatrix}^{-1} e$$

The relevant values come from the first $n_1$ rows of this solution.

This solution is not always easy to evaluate in practice. Whether the system is linear or nonlinear, the usual technique employed in most econometric centers is to solve the equations by iterative steps in what is known as the Gauss–Seidel algorithm. An efficient working of this algorithm in large dynamic systems designed for standard calculations of simulation, forecasting, multiplier analysis and similar operations requires definite rules of ordering, normalizing, and choosing step sizes.[13] It is awkward and tedious to re-do that whole procedure for a transformed system in which some variables have been interchanged, unless they are standardized.

It is simpler and more direct to solve the problem by searching (systematically) for instruments that bring the $n_1$ values of $y$ as "close" as possible to their targets $y^*$. There are many ways of doing this, but one would be to find the minimum value of

$$L = \sum_{i=1}^{n_1} u_i (y_i - y_i^*)^2$$

subject to    $\hat{F} = \hat{e}$

where    $\hat{F}$ = estimated value of $F$ for $\Theta = \hat{\Theta}$

$\hat{e}$ = assigned values to error vector

In the theory of *optimal* economic policy, $L$ is called a loss function and is arbitrarily made a quadratic in this example. Other loss functions could equally well be chosen. The $u_i$ are weights in the loss function and should be positive.

If there is an admissible solution and if $n_1 = m_1$, the optimal value of the loss function should become zero.

[13] L. R. Klein, *A Textbook of Econometrics*, (New York: Prentice-Hall, 1974), p. 239.

A more interesting optimization problem arises if $n_1 \geq m_1$; i.e. if there are more targets than instruments. In this case, the optimization procedure will not, in general, bring one all the way to target values, but only to a "minimum distance" from the target. If $m_1 > n_1$, it would be possible, in principle, to assign arbitrary values to $m_1 - n_1$ (superfluous) instruments and solve for the remaining $n_1$ instruments as functions of the $n_1$ target values of $y$. Thus, the problem of excess instruments can be reduced to the special problem of equal numbers of instruments and targets.

It should be noted that the structural model is a *dynamic* system, and it is unlikely that a static loss function would be appropriate. In general, economic policy makers have targeted paths for $y$. A whole stream of $y$-values are generally to be targeted over a policy planning horizon. In addition, the loss function could be generalized in other dimensions, too. There will usually be a loss associated with instrumentation. Policy makers find it painful to make activist decisions about running the economy, especially in the industrial democracies; therefore, $L$ should be made to depend on $w - w^*$ as well as on $y - y^*$. In the quadratic case, covariation between $y_i - y_i^*$ might also be considered, but this may well be beyond the comprehension of the typical policy maker.

A better statement of the optimal policy problem will then be

$$L = \sum_{t=1}^{h} \left\{ \sum_{i=1}^{n_1} u_i \left( y_{it} - y_{it}^* \right)^2 + \sum_{i=1}^{m_1} v_i \left( w_{it} - w_{it}^* \right)^2 \right\} = \min.$$

w.r.t. $w_{it}$ subject to $\hat{f} = \hat{e}_t$,

$$t = 1, 2, \ldots, h.$$

The $v_i$ are weights associated with instrumentation losses. If future values are to be discounted it may be desirable to vary $u_i$ and $v_i$ with $t$. A simple way would be to write

$$u_i / (1+\rho)^t; \ v_i / (1+\rho)^t,$$

where $\rho$ is the rate of discount.

A particular problem in the application of the dynamic formulation is known as the end-point problem. Decisions made at time point $h$ (end of the horizon) may imply awkward paths for the system beyond $h$ because it is a dynamic system whose near term movements ($h+1$, $h+2$, ...) will depend on the (initial) conditions of the system up to time $h$. It may be advisable to carry the optimization exercise beyond $h$, even though policy focuses on the behaviour of the system only through period $h$.

Many examples have been worked out for application of this approach to policy making – few in prospect (as genuine extrapolations into the future) but many in retrospect, assessing what policy should have been.[14] A noteworthy series of experimental policies dealt with attempts to have alleviated the stagflation of the late 1960s and the 1970s in the United States; in other words, could a combination of fiscal and monetary policies have been chosen that would have led to full (or fuller) employment without (so much) inflation over the period 1967–75?

The answers, from optimal control theory applications among many models, suggest that better levels of employment and production could have been achieved with very little additional inflationary pressures but that it would not have been feasible to bring down inflation significantly at the same time. Some degree of stagflation appears to have been inevitable, given the prevailing exogenous framework.

Such retrospective applications are interesting and useful, but they leave one a great distance from the application of such sophisticated measures to the positive formulation of economic policy. There are differences between the actual and optimal paths, but if tolerance intervals of error for econometric forecasts were properly evaluated, it is not likely that the two solutions would be significantly apart for the whole simulation path. If the two solutions are actually far apart, it is often required to use extremely wide ranges of policy choice, wider and more frequently changing than would be politically acceptable.

Two types of errors must be considered for evaluation of tolerance intervals,

$$\text{var}(\hat{\Theta})$$

$$\text{var}(e).$$

The correct parameter values are not known, they must be estimated from small statistical samples and have fairly sizable errors. Also, there is behavioral error, arising from the fact that models cannot completely describe the economy. Appropriate valuation of such errors does not invalidate the use of models for some kinds of applications, but the errors do preclude "fine tuning".

A more serious problem is that the optimum problem is evaluated for a fixed system of constraints; i.e. subject to

$$\hat{F} = \hat{e}$$

[14]A. Hirsch, S. Hymans, and H. Shapiro, "Econometric Review of Alternative Fiscal and Monetary Policy, 1971–75," *Review of Economics and Statistics*, LX (August, 1978), 334–45.
L. R. Klein and V. Su, "Recent Economic Fluctuations and Stabilization Policies: An Optimal Control Approach," *Quantitative Economics and Development*, (New York: Academic Press, 1980) eds, L. R. Klein, M. Nerlove, and S. C. Tsiang.
M. B. Zarrop, S. Holly, B. Rutem, J. H. Westcott, and M. O'Connell, "Control of the LBS Econometric Model Via a Control Model," *Optimal Control for Econometric Models*, ed by S. Holly, et al. (London: Macmillan, 1979), 23–64.

The problem of optimal policy may, in fact, be one of varying constraints, respecifying *F*.

It has been found that the problem of coping with stagflation is intractable in the sense that macro policies cannot bring both unemployment and inflation close to desired targets simultaneously. On the other hand, there may exist policies that do so if the constraint system is modified. By introducing a special TIPS policy that ties both wage rates and profit rates to productivity

$$X/hL = \text{real output per worker-hour}$$

it has been found that highly favorable simulations can be constructed that simultaneously come close to full employment and low inflation targets. These simulation solutions were found with the same (Wharton) model that resisted full target approach using the methods of optimal control. The wage and profits (price) equations of the model had to be re-specified to admit

$$\Delta \ln w = \Delta \ln( X/hL )$$

$$\Delta \ln( PR/K ) = \Delta \ln( X/hL )$$

$$PR = \text{corporate profits}$$

$$K = \text{stock of corporate capital}$$

Equations for wages and prices, estimated over the sample period had to be removed, in favor of the insertion of these.[15]

A creative policy search with simulation exercises was able to get the economy to performance points that could not be reached with feasible applications of optimal control methods. This will not always be the case, but will frequently be so. Most contemporary problems cannot be fully solved by simple manipulation of a few macro instruments, and the formalism of optimal control theory has very limited use in practice. Simulation search for "good" policies, realistically formulated in terms of parameter values that policy makers actually influence is likely to remain as the dominant way that econometric models are used in the policy process.

That is not to say that optimal control theory is useless. It shows a great deal about model structure and instrument efficiency. By varying weights in the loss function and then minimizing, this method can show how sensitive the uses of policy instruments are. Also, some general propositions can be developed. The more uncertainty is attached to model specification and estimation, the less should be the amplitude of variation of instrument settings. Thus, William Brainard has shown, in purely theoretical analysis of the optimum problem, that

[15]L. R. Klein and V. Duggal, "Guidelines in Economic Stabilization: A New Consideration," *Wharton Quarterly*, VI (Summer, 1971), 20–24.

Table 3
Growth assumptions and budget deficit fiscal policy planning, USA February 1984[a]

|  | 1984 | 1985 | 1986 | 1987 | 1988 | 1989 |
|---|---|---|---|---|---|---|
| Real GNP estimates or assumptions (%) | | | | | | |
|   administration | 5.3 | 4.1 | 4.0 | 4.0 | 4.0 | 4.0 |
| Congressional Budget Office | | | | | | |
|   Baseline | 5.4 | 4.1 | 3.5 | 3.5 | 3.5 | 3.5 |
|   Low alternative | 4.9 | 3.6 | −0.9 | 2.1 | 3.8 | 3.1 |
| Estimated deficit ($ billion) | | | Fiscal Years | | | |
|   administration | 186 | 192 | 211 | 233 | 241 | 248 |
| Congressional Budget Office | | | | | | |
|   Baseline | 189 | 197 | 217 | 245 | 272 | 308 |
|   Low alternative | 196 | 209 | 267 | 329 | 357 | 390 |

[a]*Source*: Baseline Budget Projections for Fiscal Years 1985–1989 Congressional Budget Office, Washington, D. C. February 1984 Testimony of Rudolph G. Penner, Committee on Appropriations, U.S. Senate, February 22, 1984.

policy makers ought to hold instruments cautiously to a narrow range (intervene less) if there is great uncertainty.[16] This is a valuable advice developed from the analysis of optimal policy.

A particular case of uncertainty concerns the business cycle. The baseline solution for $y_t$ should reflect whatever cyclical variation is present in the actual economy if predictions of $y_t$ are at all accurate. For example, the existence of a cycle in the United States has been well documented by the National Bureau of Economic Research and has been shown to be evident in the solutions of macro econometric models.[17]

Although the baseline solution of a macro economy extending over 5 to 10 years should reflect a normal cyclical pattern unless some specific inputs are included that wipe out the cycle, that is not the usual practice in public policy planning. Policy makers are reluctant to forecast a downturn in their own planning horizon. The accompanying table illustrates this point in connection with U.S. budget planning in early 1984. The official baseline path assumes steady growth of the economy, contrary to historical evidence about the existence and persistence of a 4-year American cycle. An argument in support of this practice has been that the exact timing of the cyclical turning points is in doubt. If they

[16]W. Brainard, "Uncertainty and the Effectiveness of Policy," *American Economic Review* LVIII May 1967), 411–25. See also L. Johansen, "Targets and Instruments Under Uncertainty," Institute of Economics, Oslo, 1972. Brainard's results do not, in all theoretical cases, lead to the conclusion that instrument variability be reduced as uncertainty is increased, but that is the result for the usual case.

[17]See I. and F. Adelman, "The Dynamic Properties of the Klein–Goldberger Model," *Econometrica* 27 (October 1959). 596–625. See also, *Econometric Models of Cyclical Behavior*, ed. B. G. Hickman (New York: Columbia University Press, 1972).

are not known with great precision, it is argued that it is better not to introduce them at all. An appropriate standard error of estimate is probably no larger than $\pm 1.0$ year; therefore, they ought to be introduced with an estimated degree of certainty.

The Congressional Budget Office in the United States has a fairly steady expansion path for its baseline case, but introduces a cycle downturn for 1986, in a low growth alternative case, between 4 and 5 years after the last downturn. It would seem more appropriate to consider this as a baseline case, with the steady growth projection an upper limit for a more favorable budget projection.

A series of randomly disturbed simulations of an estimated model

$$F = e_t^{(i)} \qquad t = 1, 2, \ldots H \qquad i = 1, 2, \ldots R,$$

with $R$ replications of random error disturbances, generates solutions of the estimated equation system $F$. Each replication produces

$$\begin{pmatrix} y_1^{(i)} \\ y_2^{(i)} \\ \vdots \\ y_H^{(i)} \end{pmatrix} \quad \text{given} \quad \begin{pmatrix} z_1 \\ z_2 \\ \vdots \\ z_H \end{pmatrix} \quad \text{and initial conditions.}$$

The $R$ stochastic projections will, on average, have cycles with random timing and amplitude. They will produce $R$ budget deficit estimates. The mean and variance of these estimates can be used to construct an interval that includes a given fraction of cases, which can be used to generate a high, low, and average case for budget deficit values. The stochastic replications need not allow only for drawings of $e_t^{(i)}$; they can also be used to estimate distributions of parameter estimates for $F$.[18] This is an expensive and time consuming way to generate policy intervals, but it is a sound way to proceed in the face of uncertainty for momentous macro problems.

It is evident from the table that provision for a business cycle, no matter how uncertain its timing may be, is quite important. The higher and steadier growth assumptions of the American administration produces, by far, the lowest fiscal deficits in budgetary planning. A slight lowering of the steady path (by only 0.5 percentage points, 1986–89) produces much larger deficits, and if a business cycle correction is built into the calculations, the rise in the deficit is very big. In the cyclical case, we have practically a doubling of the deficit in five years, while in

---

[18] The technique employed in G. Schink, *Estimation of Forecast Error in a Dynamic and/or Non-Linear Econometric Model* (Ph.D. dissertation, University of Pennsylvania (1971)) can be used for joint variation of parameters and disturbances.

the cycle-free case the rise is no more than about 50 percent in the same time period.

Also, optimal control theory can be used to good advantage in the choice of exogenous inputs for long range simulations. Suppose that values for

$$w_t, z_t$$

are needed for $t = T+1, T+2, T+3, \ldots T+30$ where $T+30$ is 30 years from now (in the 21st century). We have little concrete basis for choice of

$$w_{T+30}, z_{T+30}.$$

By optimizing about a *balanced growth path* for the endogenous variables, with respect to choice of key exogenous variables, we may be able to indicate sensible choices of these latter variables for a *baseline* path, about which to examine alternatives. These and other analytical uses will draw heavily on optimal control theory, but it is unlikely that such theory will figure importantly in the positive setting of economic policy.

The role of the baseline (balanced growth) solution for policy making in the medium or long term is to establish a reference point about which policy induced deviations can be estimated. The baseline solution is not, strictly speaking, a forecast, but it is a policy reference set of points. Many policy problems are long term. Energy availability, other natural resource supplies, social insurance reform, and international debt settlement are typical long term problems that use econometric policy analysis at the present time.

At the present time, the theory of economic policy serves as a background for development of policy but not for its actual implementation. There is too much uncertainty about the choice of loss function and about the constraint system to rely on this approach to policy formation in any mechanistic way.[19] Instead, economic policy is likely to be formulated, in part at least, through comparison of alternative simulations of econometric models.

In the typical formulation of policy, the following steps are taken:

(i) definition of a problem, usually to determine the effects of external events and of policy actions;
(ii) carry out model simulations in the form of historical and future projections that take account of the problem through changes in exogenous variables, parameter values, or system specification;
(iii) estimation of quantitative effects of policies as differences between simulations with and without the indicated changes;

[19] See, in this respect, the conclusions of the Royal Commission (headed by R. J. Ball) Committee on Policy Optimisation, *Report*, (London: HMSO, 1978).

(iv) presentation of results to policy decision makers for consideration in competition with estimates from many different sources.

Policy is rarely based on econometric information alone, but it is nearly always based on perusal of relevant econometric estimates together with other assessments of quantitative policy effects. Among econometric models, several will often be used as checking devices for confirmation or questioning of policy decisions.

It is important in policy formulation to have a baseline projection. For the short run, this will be a forecast of up to 3 years' horizon. For the longer run, it will be a model projection that is based on plausible assumptions about inputs of exogenous variables and policy related parameters. For the longer run projections, the inputs will usually be smooth, but for short run forecasts the inputs will usually move with perceptions of monthly, quarterly, or annual information sources in a more irregular or cyclical pattern.

The model forecast or baseline projection serves not only as a reference point from which to judge policy effects. It also serves as a standard of credibility. That is to say, past performance of forecast accuracy is important in establishing the credibility of any model.

Judgmental information, quantitative reduced form extrapolations (without benefit of a formal model) and estimated models will all be put together for joint information and discussion. Models are significant parts of this information source but by no means the whole. In many respects, model results will be used for confirmation or substantiation of decisions based on more general sources of information.

Models are most useful when they present alternative simulations of familiar types of changes that have been considered on repetitive occasions in the past, so that there is an historical data base on which to build simulation analyses. A new tax, a new expenditure program, the use of a new monetary instrument, or, in general, the implementation of a new policy that calls on uses of models that have not been examined in the past are the most questionable. There may be no historical data base in such situations from which to judge model performance.

In new situations, external a priori information for parameter values or for respecification with new (numerical) parameter values is needed. These new estimates of parameters should be supplied by engineers or scientists for technical relations, by legal experts for new tax relationships, or by whatever expertise can be found for other relationships. The resulting simulations with non sample based parameter estimates are simply explorations of alternatives and not forecasts or projections.

Much attention has been paid, in the United States, recently to changes in laws for taxing capital gains. There is no suitable sample that is readily available with many observations at different levels of capital gains taxation. Instead, one would be well advised to look at other countries' experience in order to estimate marginal consequences of changing the tax laws for treatment of capital gains. In addition, one could investigate state-to-state cross section estimates to see how capital gains taxes might influence spending behavior. Similar analyses across countries may also be of some help. Finally, we might try to insert questions into a field survey on people's attitudes towards the use of capital gains. These are all basic approaches and should be investigated simultaneously. There is nothing straightforward to do in a new situation, but some usable pieces of econometric information may be obtained and it might help in policy formation. Recently, claims were made about the great benefits to be derived from the liberalization of capital gains rates in the United States, but these claims were not backed by econometric research that could be professionally defended. For the ingenious econometric researcher, there is much to gain on a tentative basis, but care and patience are necessary.

In all this analysis, the pure forecast and forecasting ability of models play key roles. Forecasts are worthwhile in their own right, but they are especially valuable when examined from the viewpoint of accuracy because users of model results are going to look at forecast accuracy as means of validating models. It is extremely important to gain the confidence of model users, and this is most likely to be done through the establishment of credibility. This comes about through relative accuracy of the forecast. Can forecasts from models be made at least as accurately as by other methods and are the forecasts superior at critical points, such as business cycle turning points?

These questions are partly answered by the accuracy researches of Stephen McNees and others.[20] The answer is that models do no worse than other methods and tend to do better at cyclical turning points and over larger stretches of time horizon. The acceptability of model results by those who pay for them in the commercial market lends greater support to their usefulness and credibility. This supports their use in the policy process through the familiar technique of alternative/comparative simulation.

International policy uses provide a new dimension for applications of econometric models. Comprehensive models of the world economy are relatively new; so it is meaningful to examine their use in the policy process. The world model that is implemented through Project LINK has been used in a number of international policy studies, and an interpretation of some leading cases may be helpful.

---

[20] Stephen McNees, "The Forecasting Record for the 1970s," *New England Economic Review*, (September/October, 1979), 33–53.
Vincent Su, "An Error Analysis of Econometric and Noneconometric Forecasts," *American Economic Review*, 68, (May, 1978), 360–72.

Some of the problems for which the LINK model has been used are: exchange rate policy, agricultural policy associated with grain failures, oil pricing policy, coordinated fiscal policies, coordinated monetary policies.

When the LINK system was first constructed, the Bretton Woods system of fixed exchange rates was still in force. It was appropriate to make exchange rates exogenous in such an environment. At the present time exchange rate equations have been added in order to estimate currency rates endogenously. An interesting application of optimal control theory can be used for exchange rate estimation and especially for developing the concept of *equilibrium* exchange rates. Such equilibrium rates give meaning to the concept of the degree of *over- or under-evaluation* of rates, which may be significant for the determining of fiscal intervention in the foreign exchange market.

In a system of multiple models, for given exchange rates there is a solution, model by model, for

$$(PX) * X_i - (PM)_i * M_i = \text{trade balance for } i\text{th country}$$

$$(PX)_i = \text{export price}$$

$$X_i = \text{export volume (goods/services)}$$

$$(PM)_i = \text{import price}$$

$$M_i = \text{import volume (goods/services)}$$

These are all endogenous variables in a multi-model world system. The equilibrium exchange rate problem is to set targets for each trade balance at levels that countries could tolerate at either positive or negative values for protracted periods of time – or zero balance could also be imposed. The problem is then transformed according to Tinbergen's approach, and assumed values are given to the trade balance, as though they are exogenous, while solutions are obtained for

$$(EXR)_i = \text{exchange rate of the } i\text{th country}$$

The exchange rates are usually denominated in terms of local currency units per U.S. dollar. For the United States, the trade balance is determined as a residual by virtue of the accounting restraints.

$$\sum_i (PX)_i * X_i = \sum_i (PM)_i * M_i$$

and the exchange rate in terms of U.S. dollars is, by definition, 1.0.

As noted earlier, this problem, although straightforward from a conceptual point of view, is difficult to carry out in practice, especially for a system as large and complicated as LINK; therefore, it has to be solved empirically from the criterion

$$\sum_i \left\{ \left[ (PX)_i * X_i - (PM)_i * M_i \right] - \left[ (PX)_i * X_i - (PM)_i * M_i \right]^* \right\}^2$$

$$= \min = 0,$$

with the entire LINK system functioning as a set of constraints. The minimization is done with respect to the values of the exchange rates (instruments). With modern computer technology, hardware, and software, this is a feasible problem. Its importance for policy is to give some operational content to the concept of equilibrium exchange rate values.

Optimal control algorithms built for project LINK to handle the multi-model optimization problem have been successfully implemented to calculate Ronald McKinnon's proposals for exchange rate stabilization through monetary policy.[21]

As a result of attempts by major countries to stop inflation, stringent monetary measures were introduced during October, 1979, and again during March, 1980. American interest rates ascended rapidly reaching a rate of some 20% for short term money. One country after another quickly followed suit, primarily to protect foreign capital holdings and to prevent capital from flowing out in search of high yields. An internationally coordinated policy to reduce rates was considered in LINK simulations. Such international coordination would diminish the possibility of the existence of destabilizing capital flows across borders. Policy variables (or near substitutes) were introduced in each of the major country models. The resulting simulations were compared with a baseline case. Some world results are shown, in the aggregate, in Table 4.

The results in Table 4 are purely aggregative. There is no implication that all participants in a coordinated policy program benefit. The net beneficial results are obtained by summing gains and losses. Some countries might not gain, individually, in a coordinated framework, but on balance they would probably gain if coordination were frequently used for a variety of policies and if the whole world economy were stabilized as a result of coordinated implementation of policy.

Coordinated policy changes of easier credit conditions helps growth in the industrial countries. It helps inflation in the short run by lowering interest cost, directly. Higher inflation rates caused by enhanced levels of activity are restrained

---

[21] Ronald I. McKinnon, *An International Standard for Monetary Stabilization*, (Washington, D. C.: Institute for International Economics), March, 1984.
Peter Pauly and Christian E. Petersen, "An Empirical Evaluation of the McKinnon Proposal" *Issues in International Monetary Policy*, *Project LINK Conference Proceedings*, (San Francisco: Federal Reserve Bank), 1985.

Table 4
Effects of coordinated monetary policy, LINK system world aggregates
(Deviation of policy simulation from baseline)

|  | 1979 | 1980 | 1981 | 1982 | 1983 | 1984 |
|---|---|---|---|---|---|---|
| Value of world trade | | | | | | |
| (bill $) | 15 | 53 | 85 | 106 | 125 | 149 |
| Volume of world trade | | | | | | |
| (bill $, 1970) | 4.7 | 14.4 | 20.2 | 22.8 | 24.7 | 26.9 |
| OECD (13 LINK countries) | | | | | | |
| GDP growth rate (%) | 1.9 | 1.9 | 1.0 | −0.2 | −0.5 | −0.4 |
| Consumer price inflation rate (%) | −0.2 | −0.5 | −0.4 | 0.1 | 0.3 | 0.3 |

by the overall improvement in productivity. This latter development comes about because easier credit terms stimulate capital formation. This, in turn, helps productivity growth measured as changes in output per worker. A pro-inflationary influence enters through the attainment of higher levels of capacity utilization, but it is the function of the models to balance out the pro and counter inflationary effects.

Policy is not determined at the international level, yet national forums consult simulations such as this coordinated lowering of interest rates and the frequent repetition of such econometric calculations can ultimately stimulate policy thinking along these lines in several major countries. A number of fiscal and exchange rate simulations along coordinated international lines have been made over the past few years.[22,23]

## 5. Prospects

Economic policy guidance through the use of econometric models is clearly practiced on a large scale, over a wide range of countries. Fine tuning through the use of overall macro policies having to do with fiscal, monetary, and trade matters

[22] L. R. Klein, P. Beaumont, and V. Su, "Coordination of International Fiscal Policies and Exchange Rate Revaluations," *Modelling the International Transmission Mechanism*. ed. J. Sawyer (Amsterdam: North-Holland, 1979), 143–59.
H. Georgiadis, L. R. Klein, and V. Su, "International Coordination of Economic Policies," *Greek Economic Review* I (August, 1979), 27–47.
L. R. Klein, R. Simes, and P. Voisin, "Coordinated Monetary Policy and the World Economy," *Prévision et Analyse économique*, 2 (October 1981), 75–104.
[23] A new and promising approach is to make international policy coordination a dynamic game. See Gilles Oudiz and Jeffrey Sachs, "Macroeconomic Policy Coordination among the Industrial Countries" *Brookings Papers on Economic Activity* (1, 1984) 1–64.

has been carried quite far, possibly as far as it can in terms of methodological development. There will always be new cases to consider, but the techniques are not likely to be significantly improved upon. To some extent, formal methods of optimal control can be further developed towards applicability. But significant new directions can be taken through the development of more supply side content in models to deal with the plethora of structural policy issues that now confront economies of the world. This situation is likely to develop·further along supply side lines. The bringing into play of joint Leontief–Keynes models with fully articulated input–output systems, demographic detail, resource constraints and environmental conditions are likely to be important for the development of more specific policy decisions requiring the use of more micro details from models. This is likely to be the next wave of policy applications, focusing on energy policy, environmental policy, food policy, and other specific issues. It is clear that econometric methods are going to play a major role in this phase of development.

### Appendix: An outline of a combined (Keynes–Leontief) input–output/macro model

The first five sectors listed on p. 2067 are the components of *final demand* as they are laid out in the simple versions of the Keynesian macro model, extending the cases cited earlier by the explicit introduction of inventory investment and foreign trade. When the Keynesian system is extended to cover price and wage formation, then the production function, labor requirements, labor supply and income determination must also be included. These, together, make up the main components of national income. Interest income and monetary relationships to generate interest rates must also be included. This outlines, in brief form, the standard macro components of the mainstream econometric model. The interindustry relationships making up the input–output system round out the total model.

The flow of goods, in a numeraire unit, from sector $i$ to sector $j$ is denoted as

$$X_{ij}.$$

Correspondingly, the total gross output of $j$ is $X_j$. The technical coefficients of input–output analysis are defined as

$$a_{ij} = X_{ij}/X_j$$

and the basic identity of input–output analysis becomes

$$X_i = \sum_{j=1}^{n} X_{ij} + F_i = \sum_{j=1}^{n} a_{ij} X_j + F_i,$$

where $F_i$ is final demand, and the total number of sectors is $n$. In matrix notation this becomes

$$(I - A)X = F.$$

$X$ is a column vector of gross outputs, and $F$ is a column vector of final demand. $F$ can be decomposed into

$$F_c + F_I + F_G + F_E - F_M = F$$

where $F_c$ is total consumer demand, $F_I$ is total investment demand (including inventory investment), $F_G$ is public spending, $F_E$ is export demand, and $F_M$ is import demand. The detail of decomposition of $F$ used here is only illustrative. Many subcategories are used in a large system in applied econometrics.

The elements of $F$ sum to GNP. If we denote each row of $F$ as

$$F_i = F_{iC} + F_{iI} + F_{iG} + F_{iE} - F_{iM}$$

and divide each component by its column total, we get

$$a_{iC} = \frac{F_{iC}}{F_C} \, ; \, a_{iI} = \frac{F_{iI}}{F_I} \, ; \, a_{iG} = \frac{F_{iG}}{F_G} \, ; \, a_{iE} = \frac{F_{iE}}{F_E} \, ; \, a_{iM} = \frac{F_{iM}}{F_M} \, .$$

The array of elements of these final demand coefficients make up a rectangular matrix, called $C$. If we denote the column

$$\mathscr{G} = \begin{pmatrix} F_C \\ F_I \\ F_G \\ F_E \\ - F_M \end{pmatrix}$$

by $\mathscr{G}$ (standing for GNP), we can write

$$F = C\mathscr{G}$$

or

$$(I - A)X = C\mathscr{G}$$
$$X = (I - A)^{-1}C\mathscr{G}$$

This gives a (row) transformation expressing each sector's gross output as a weighted sum of the components of GNP. It shows how a model of $\mathcal{G}$ (GNP) values can be transformed into individual sector outputs if we make use of the matrix of input–output and final demand coefficients.

This transformation is extended from gross output values to *values-added* by sector. The transformation is

$$X = BY$$

where

$$B = \begin{pmatrix} \dfrac{1}{1 - \sum\limits_{i=1}^{n} a_{i1}} & & 0 \\ & \ddots & \\ 0 & & \dfrac{1}{1 - \sum\limits_{i=1}^{n} a_{in}} \end{pmatrix}$$

We observe also that the sum of $Y_i$ gives the GNP total, too.

$$Y = B^{-1}(I - A)^{-1} C\mathcal{G}.$$

This gives the (row) transformation between elements of $\mathcal{G}$ and elements of $Y$, where both column vectors are different decompositions of the GNP, one on the side of spending and the other on the side of production.

If we construct synthetic price deflators for values added $P_y$ and for final demand $P_g$, we find the relationship

$$P_y' Y = P_g' G.$$

This can be transformed into

$$P_y' B^{-1}(I - A)^{-1} C\mathcal{G} = P_g'\mathcal{G}.$$

By equating corresponding terms in the elements of $\mathcal{G}$ we have the (column) transformations

$$P_{gi} = \sum_{j=1}^{n} h_{ji} P_{yj} \qquad i = C, I, G, X, M$$

A typical element of $B^{-1}(I - A)^{-1} C$ is denoted as $h_{ji}$.

Industry outputs are weighted sums of final expenditures, and expenditure deflators are weighted sums of sector output prices. Prices, in this model, are determined by mark-up relations, over costs, at the industry sector level and transformed into expenditure deflators. Demand is determined at the final expenditure level and transformed into output levels.

The relation

$$X = BY$$

provides a set of simple transformations to convert from sector gross output values to sector values added. There is a corresponding price transformation

$$(I - A')P_x = B^{-1}P_y$$

$$P_y = B(I - A')P_x$$

This is derived as follows:

$$P_{xj}X_j = P_{yj}Y_j + \sum_{i=1}^{n} P_{xi}X_{ij}$$

$$P_{xj}X_j = P_{yj}Y_j + \sum_{i=1}^{n} P_{xi}a_{ij}X_j$$

$$P_{xj} = P_{yj}\frac{Y_j}{X_j} + \sum_{i=1}^{n} P_{xi}a_{ij}$$

The ratio $Y_j/X_j$ (value added to gross output of sector $j$) can be written as

$$\frac{Y_j}{X_j} = 1 - \sum_{i=1}^{n} a_{ij};$$

they are the reciprocals of the diagonal elements of $B$. In matrix notation we have

$$P_x = B^{-1}P_y + A'P_x$$

or, more compactly

$$P_y = B(I - A')P_x.$$

This system of equations provides transformation from gross output prices to value added prices, or vice-versa. In the model's behavioral equations, there is first determination of $P_x$; then the above transformation derives $P_y$, and from these, we estimate $P_g$.

The integration of input–output analysis with macro models of final demand, income generation, and monetary relationships is seemingly straightforward and non-stochastic, to a large extent. This is, however, deceptive because the $a_{ij}$ from the inter-industry flow matrix, and the expenditure shares in the final demand coefficient matrix are not time invariant parameters; they are ratios of variables. This model has been generalized so that production functions are written as

$$X_j = F_j(X_{1j}, \ldots, X_{nj}, L_j, K_j, t)$$

and the input–output coefficients

$$\frac{X_{ij}}{X_j}$$

must be generated from a set of relationships describing behavior of the producing units of the economy. Similarly, the final demand ratios should be generated by behavioral relationships in consuming and market trading sectors of the economy. The main point is that these coefficients should all depend on *relative* prices.

The Wharton Model has been estimated for the case in which the functions $F_j$ are generalized CES functions for the intermediate output flows, while the original factors $L_j$ and $K_j$ are related to value added production in a Cobb–Douglas relationship.[24]

$$X_j = \left( \sum_{i=1}^{n} \delta_{ij} X_{ij}^{-\rho_j} \right)^{-\frac{1}{\rho_j}} + A_j L_j^{\alpha_j} K_j^{\beta_j} e^{\gamma_j}$$

$$\sigma_j = \frac{1}{1 + \rho_j}$$

$\sigma_j$ = elasticity of substitution.

The associated optimization equations for producer behavior are

$$\frac{X_{ij}}{X_{kj}} = \left( \frac{\delta_{ij}}{\delta_{kj}} \right)^{+\sigma_j} \left( \frac{p_j}{p_k} \right)^{-\sigma_j}$$

$$\frac{L_j}{K_j} = \frac{\alpha_j}{\beta_j} \left( \frac{w_j}{r_j} \right)^{-1}.$$

[24] R. S. Preston, "The Wharton Long Term Model: Input–Output Within the Context of a Macro Forecasting Model," *Econometric Model Performance*, ed. by L. R. Klein and E. Burmeister, (Philadelphia, University of Pennsylvania Press, 1976), 271–87. In a new generation of this system, the sector production functions are nested CES functions, with separate treatment for energy and non-energy components of $X_{ij}$.

This system has an implicit restriction that the elasticity of substitution between pairs of intermediate inputs is invariant for each sector, across input pairs. This assumption is presently being generalized as indicated in the preceding footnote. In other models, besides the Wharton Model, different production function specifications are being used for this kind of work, e.g. translog specifications.

The demand side coefficients of final expenditure have not yet been estimated in terms of complete systems, but they could be determined as specifications of complete expenditure systems.[25]

$$P_{xi}F_{ic} = \varepsilon_i P_{xi} + \eta_i \left( F_c - \sum_{j=1}^{n} P_{xj}\varepsilon_j \right)$$

All these equations are stochastic and dynamic, often with adaptive adjustment relations.

[25] See Theodore Gamaletsos, *Forecasting Sectoral Final Demand by a Dynamic Expenditure System*, (Athens: Center of Planning and Economic Research, 1980), for a generalization of this expenditure system.

# LIST OF THEOREMS

# INDEX